Fluid Flow in Fractured Porous Media

Fluid Flow in Fractured Porous Media

Special Issue Editors

Richeng Liu
Yujing Jiang

MDPI • Basel • Beijing • Wuhan • Barcelona • Belgrade

Special Issue Editors
Richeng Liu
China University of Mining and Technology
China

Yujing Jiang
Nagasaki University
Japan

Editorial Office
MDPI
St. Alban-Anlage 66
4052 Basel, Switzerland

This is a reprint of articles from the Special Issue published online in the open access journal *Processes* (ISSN 2227-9717) from 2018 to 2019 (available at: https://www.mdpi.com/journal/processes/special_issues/porous_media).

For citation purposes, cite each article independently as indicated on the article page online and as indicated below:

LastName, A.A.; LastName, B.B.; LastName, C.C. Article Title. *Journal Name* **Year**, *Article Number*, Page Range.

Volume 1
ISBN 978-3-03921-423-5 (Pbk)
ISBN 978-3-03921-424-2 (PDF)

Volume 1-2
ISBN 978-3-03921-475-4 (Pbk)
ISBN 978-3-03921-476-1 (PDF)

Contents

About the Special Issue Editors

Richeng Liu is Research Associate of Rock Mechanics and Rock Engineering at China University of Mining and Technology, China. His research is focused on fractal and nonlinear flow properties of complex rock fracture networks which are strongly connected with projects covering underground assessments, such as CO2 sequestration, enhanced oil recovery, and geothermal energy development. He has authored over 60 refereed journal publications. He has received numerous awards for his outstanding contributions to Engineering, including the Best Research Paper and Best Doctoral Thesis awards by the Japanese Society for Rock Mechanics (JSRM). He is also a JSPS Postdoctoral Research Fellow in Japan, supported by JSPS (Japan Society for the Promotion of Science) and has been awarded with the Young Elite Scientist Sponsorship Program by CAST (China Association for Science and Technology). He is the leading Guest Editor of *Processes*, *Water*, and *Computer Modeling in Engineering and Sciences*.

Yujing Jiang is Professor of Civil Engineering at Nagasaki University, Japan. His main research activities involve the experimental characterization and numerical modeling of fluid flow through fractured porous media during shearing. He has co-authored around 200 refereed journal papers. He is a member of the Engineering Academy of Japan, Japan.

processes
MDPI

Editorial

Recent Advances in Fluid Flow in Fractured Porous Media

Richeng Liu [1,2], Sha Lou [3,*] and Yujing Jiang [2]

[1] State Key Laboratory for Geomechanics and Deep Underground Engineering, China University of Mining and Technology, Xuzhou 221116, China; liuricheng@cumt.edu.cn
[2] School of Engineering, Nagasaki University, 1-14 Bunkyo-machi, Nagasaki 8528521, Japan; jiang@nagasaki-u.ac.jp
[3] Department of Hydraulic Engineering, Tongji University, Shanghai 200092, China
* Correspondence: lousha@tongji.edu.cn; Tel.: +86-21-65989220

Received: 28 April 2019; Accepted: 28 April 2019; Published: 2 May 2019

The fluid flows through both the fractures and pores in deep-seated rock masses, which is of special importance for the performance of underground facilities such as groundwater use and nuclear waste disposal [1–5]. Many researchers have contributed to modeling fluid flow through pores, fractures, and dual-porosity media, using the developed laboratory experimental systems, theories and numerical methods [6–9]. Although the fluid flow is affected by stresses that always exist deep underground [10–13], we should first understand the hydraulic characteristics of fractured porous media [14,15]. How to predict the permeability of a porous media and/or a fracture network, as well as how the interface of pores and fractures influences fluid flow, is still unknown [16,17]. This is our motivation to handle this Special Issue (SI) of Processes, focusing on reporting the recent advances in fluid flow in fractured porous media.

The special issue "Fluid Flow in Fractured Porous Media" has been closed on 31 December 2018, in which a total of 58 papers are collected from different countries and different affiliations. The accepted papers include a wide variety of topics such as granite permeability investigation, grouting, coal mining, roadway, concrete, and so on. In the previous review work [18], we have introduced the contributions of first published 27 papers [19–45] and we will continue to introduce the remaining 31 papers in this study [46–76]. The details are summarized as follows:

(1) Jiang, D. et al., Experimental Investigation on the Law of Grout Diffusion in Fractured Porous Rock Mass and Its Application [46].

This paper investigated the evolutions of grout diffusion in fractured porous rock masses using a developed visualization platform and a three-dimensional grouting experimental system. The results show that a parabolic shape exists when the grout flows from the hole to the bottom of the model. The grouting pressure has a larger influence on the grouting diffusion radius than the water-cement ratio.

(2) Qiu, Z. et al., Rheological Behavior and Modeling of a Crushed Sandstone—Mudstone Particle Mixture [47].

This paper presented the rheological deformation of a sandstone-mudstone particle mixture through confined uniaxial compressive tests using 24 samples. The results show that with increasing the uniaxial compressive, the rheological process experiences four stages: Linear stage, attenuation rheological stage, secondary attenuation rheological stage, and stable stage. The authors proposed a segmented rheological model for a coarse-grained soil.

(3) Yang, X. et al., Effect of Dry-Wet Cycling on the Mechanical Properties of Rocks: A Laboratory-Scale Experimental Study [48].

This paper experimentally depicted the effect of dry-wet cycling on mechanical properties of rocks in a lab-scale. The results show that the uniaxial compressive strength decreases by a rate of 34.21% and the elastic modulus decreases by a rate of 44.63% when the wet-cycle number increases from 0–15. The dry-wet cycling has a significant influence on the distributions of particle size and pore structure.

(4) Zhang, Z. et al., Changes in Particle Size Composition under Seepage Conditions of Reclaimed Soil in Xinjiang, China [49].

This paper used the geotechnical experimental method to investigate the changes in particle size composition of reclaimed soil in Xinjiang, China. The results show that the granulometric composition of the reclaimed soil changes dramatically with varying depths from 10 m–70 m. For a depth of 10 m, the granulometric composition is similar with that of undisturbed reclaimed soil; while when the depth is 70 m, the fine particle content is greatly different with that of undisturbed reclaimed soil.

(5) Irfan, M. et al., Geotechnical Properties of Effluent-Contaminated Cohesive Soils and Their Stabilization Using Industrial By-Products [50].

This paper reported the chemical impact of dyeing and tannery on high plastic clay and low plastic clay, and examined the time-dependent variations in the properties at different effluent percentages. The results show that with increasing time upon dyeing contamination, the unconfined compressive strength of both soils increases.

(6) Zhang, M. et al., High Mixing Efficiency by Modulating Inlet Frequency of Viscoelastic Fluid in Simplified Pore Structure [51].

This paper proposed an effective mixing approach to simulate the mixing of viscoelastic fluids flowing through a simplified pore T-junction model. The results show that the mixing degree is relatively low, i.e., 0.15 for Newtonian fluid and 0.25 for viscoelastic fluid, under a constant driving pressure. When the phase difference is 180°, the mixing degree has a maximum value of 0.82 for the viscoelastic fluid.

(7) Liu, S. et al., Experimental Development Process of a New Fluid–Solid Coupling Similar-Material Based on the Orthogonal Test [52].

This paper developed a new fluid-solid coupling similar material, which is mixed using river sand, calcium carbonate, talc powder, white cement, Vaseline, and antiwear hydraulic oil. The results indicate that the selected raw materials and proportioning method are suitable for usage in a coal mine floor water inrush.

(8) Liu, L. et al., The Impact of Oriented Perforations on Fracture Propagation and Complexity in Hydraulic Fracturing [53].

This paper developed a fully coupled FEM-based hydro-mechanical model to understand the interaction between hydraulic fracture and oriented perforation. Here, FEM represents the finite element method. The results show that when the perforation azimuth from the direction of the maximum principal stress is larger, the curvature of the fracture during hydraulic fracture reorientation is larger. For the liquid-based hydraulic fracturing, an increase in liquid viscosity from 10^{-3} Pa·s–1 Pa·s increases the critical perforation angle from 70°–80°.

(9) Wang, F. et al., Experimental Study on Feasibility of Enhanced Gas Recovery through CO_2 Flooding in Tight Sandstone Gas Reservoirs [54].

This paper studied the feasibility of enhanced gas recovery through CO_2 flooding in tight sandstone gas reservoirs. The results show that the CO_2 flooding increases the gas recovery by a rate of 18.36% with respect to the depletion development method. A significant influence is observed from the presence of a dip angle in tight sandstone gas reservoirs.

(10) Zhuo, Q. et al., The Effect of Collision Angle on the Collision and Adhesion Behavior of Coal Particles and Bubbles [55].

This paper exhibited the effect of collision angle on collision and adhesion behaviors of coal particles and bubbles. The results show that with the increment of collision angle, the distribution ranges of initial settlement position and particle central distribution interval expands outward. The resistance layer influences the velocity of particles and the collision angle affects the adhesion efficiency significantly.

(11) Wang, H. et al., Feasibility of Using Gangue and Fly Ash as Filling Slurry Materials [56].

In this paper, gangue and fly ash samples are used as filling materials through tests on the optimal ratio of the slurry components using an Intelligent Torque Rheometer. The results show that the optimal slurry composition of coal gangue, fly ash and gelling agent should have a weight ratio of 8:3:1. The paste-like slurry can effectively reduce the surface subsidence.

(12) Huang, S. et al., Effects of Cyclic Wetting-Drying Conditions on Elastic Modulus and Compressive Strength of Sandstone and Mudstone [57].

This paper investigated the effect of wet-dry cycling on the evolutions of elastic modulus and compressive strengths of sandstone and mudstone. The results show that the larger the number of wet-dry cycling, the smaller the values of both elastic modulus and uniaxial compressive strength of sandstone and mudstone. The mechanical properties have a linear relationship with the porosity.

(13) Zhang, Y. et al., Experimental Study on Compression Deformation and Permeability Characteristics of Grading Broken Gangue under Stress [58].

This paper experimentally studied the compression deformations and permeability of grading broken gangue under stresses. The results show that the stress of broken gangue has an exponential relationship with the strain, and the compression modulus is positively correlated with the compression rate. It is more difficult to compress for the samples with discontinuous grading than the continuous grading samples.

(14) Sun, X. et al., Investigation of Deep Mine Shaft Stability in Alternating Hard and Soft Rock Strata Using Three-Dimensional Numerical Modeling [59].

This paper used the geological strength index (GSI) method to characterize the deformation and stress of surrounding rocks and the shaft lining under soft and hard strata in Anju coal mine, Shandong Province, China. The results show that when the shaft is excavated without supports, the shear failure and tensile failure occur in shallow surrounding rock shafts. The maximum shear stress and plastic zone appear around the interface between the soft and hard rocks.

(15) Huang, M. et al., 2D Plane Strain Consolidation Process of Unsaturated Soil with Vertical Impeded Drainage Boundaries [60].

This paper estimated the 2D plane strain consolidation processes of unsaturated soils under vertical impeded drainage boundaries. The results show that the time domain can be efficiently calculated using the numerical inversion of the Laplace transform. The semi-analytical solutions agree well with the different solutions and the solutions reported in the literature.

(16) Wei, P. et al., Characterization of Pores and Fractures in Soft Coal from the No. 5 Soft Coalbed in the Chenghe Mining Area [61].

This paper explored the pore structure characteristics of soft coal masses using a mercury intrusion method, low-temperature liquid nitrogen adsorption, and scanning electron microscopy (SEM). The results show that the mercury-injection and mercury-ejection curves of soft coals have significant differences, while those of hard coals are very close. This indicates that in the soft coal, the pores are mainly open while in the hard coal, the pores are commonly closed. Besides, the fractal dimension of soft coals, which is used for characterizing pore distributions, is larger than that of hard coals.

(17) Li, J. et al., Numerical Investigation of Hydraulic Fracture Propagation Based on Cohesive Zone Model in Naturally Fractured Formations [62].

This paper numerically investigated the hydraulic fracture propagation in naturally fractured formations on the basis of a cohesive zone model. The results show that the stress interference from multiple clusters plays an important role in the suppression and diversion of the fracture network. A lower stress difference is more effective to open the natural fractures than a larger stress difference.

(18) Sun, Z. et al., Simulating the Filtration Effects of Cement-Grout in Fractured Porous Media with the 3D Unified Pipe-Network Method [63].

This paper simulated the filtration effect of cement-grout in fractured porous media using a 3D unified pipe network method. The grout is assumed to exhibit two-phase flow and the filtration effect depends on the concentration and rheology of the grout, and porosity and permeability of the fractured porous media. The results show that the size of the grout penetration region has a limited effect due to the filtration.

(19) Hou, Y. et al., Study on the Preparation and Hydration Properties of a New Cementitious Material for Tailings Discharge [64].

This paper developed a new cement-based material with blast furnace slag to replace the ordinary Portland cement. The mechanical properties are tested using uniaxial compressive strength, X-ray diffraction, and thermal gravity, scanning electron microscope, and mercury intrusion porosimetry experiments. The results show that the mercury intrusion volume is decreased by 18% and 13%, and the most common pore size is decreased by 53% and 29%, for the samples manufactured using the new cement-based material and the ordinary Portland cement, respectively.

(20) Wang, C. et al., Experimental Study on the Shear-Flow Coupled Behavior of Tension Fractures Under Constant Normal Stiffness Boundary Conditions [65].

This paper carried out a series of tests to study the shear-flow coupled behaviors of tensile fractures under constant normal stiffness conditions using a sever-controlled direct shearing apparatus. The results show that with increasing the initial normal stress and/or the fracture surface roughness, the peak shear stress increases. The surface roughness of fractures also increases residual shear stress. The larger normal stiffness leads to a larger normal stress and a smaller normal displacement during shearing.

(21) Hao, J. et al., The Effects of Backfill Mining on Strata Movement Rule and Water Inrush: A Case Study [66].

This paper estimated the effect of backfill mining on strata movement and water inrush on the basis of a case study in Caozhuang coal mine, China. The results show that the backfill mining changes the deformation and amplitude of overburden and improves the ground pressure environment. The floor failure depth is much smaller in backfill mining, compared with that in traditional longwall mining.

(22) Hua, W. et al., Experimental Study on Mixed Mode Fracture Behavior of Sandstone under Water–Rock Interactions [67].

This paper studied the mechanical properties of sandstones under periodic water-rock interactions using central-cracked Brazilian disk specimens. The results show that the fracture resistance of sandstone is significantly influenced by the periodic water-rock interactions. The cyclic wetting-drying plays a more significant role on the fracture resistance of sandstones than the long-term immersion.

(23) Li, H. et al., Seepage Characteristics and Its Control Mechanism of Rock Mass in High-Steep Slopes [68].

This paper reviewed the up-to-date status of seepage characteristic evolutions of fractured rock masses from lab-scale to the engineering scale. The experimental findings and newly developed numerical techniques on the nonlinear flow regimes of fluid were introduced and discussed in detail. Finally, the authors outlined the scientific progress in fractured rock seepage control theory and optimal design technologies of high-steep slope engineering.

(24) Li, Z. et al., Experimental Study on Electric Potential Response Characteristics of Gas-Bearing Coal During Deformation and Fracturing Process [69].

This paper designed a test system to estimate the evolutions of multi-parameters of gas-bearing colas under loading, which was then used to investigate the electrical potential responses during the deformation and fracturing. The results show that as the gas pressure increases, the confining action and the erosion effect are promoted. When the coal is cracked/damaged, the electrical potential dramatically increases to its maximum value.

(25) Liu, R. et al., Model for the Patterns of Salt-Spray-Induced Chloride Corrosion in Concretes under Coupling Action of Cyclic Loading and Salt Spray Corrosion [70].

This paper experimentally studied the patterns of chloride ion erosion of unsaturated concrete, which subjected to a coupling effect of cyclic loading and salt spray corrosion. The Fick's second law was used to obtain the chloride diffusion coefficient by fitting the variation patterns of chloride concentration. The results show that the porosities of concrete with 30% fly ash and 20% mineral powder under three cyclic loading protocols of 50%, 65%, and 80% of the compression strength are 3.49%, 3.83%, and 4.71%, respectively.

(26) Qiao, S. et al., The Application of a Three-Dimensional Deterministic Model in the Study of Debris Flow Prediction Based on the Rainfall-Unstable Soil Coupling Mechanism [71].

This paper reported a three-dimensional deterministic model to study the debris flow prediction based on the rainfall-unstable soil coupling mechanism. The results show that the proposed method is suitable for forecasting debris flows at the regional scale, and the application of the Scoops3D model can more efficiently characterize the slope stability than traditional two-dimensional models.

(27) Yu, W. et al., Development of 3D Finite Element Method for Non-Aqueous Phase Liquid Transport in Groundwater as Well as Verification [72].

This paper developed a 3D finite element method to simulate non-aqueous phase liquid transport in groundwater and verified its validity. In the proposed model, the weak form of mass balance equations for water was discretized in space using the Galerkin finite element method. The authors also adopted the mixed form formulations using both pressure water head and water saturation. The developed model can be used for preliminary evaluation of the migration and diffusion of non-aqueous liquid pollution in the fields.

(28) Liu, D. et al., Experimental Investigation of Pore Structure and Movable Fluid Traits in Tight Sandstone [73].

This paper reported the variations in pore structures and movable fluid traits in tight sandstones through direction observations, indirect measurements, and imaging processing techniques. The results show that the decreasing rate of water saturation in macropores is larger than that in tiny pores. The sandstones having a poor reservoir quality do not exhibit a remarkable descending of lower limits of movable pore radius.

(29) Wang, X. et al., Experimental Development of Coal-Like Material with Solid-Gas Coupling for Quantitative Simulation Tests of Coal and Gas Outburst Occurred in Soft Coal Seams [74].

This paper developed coal-like materials with solid-gas coupling for quantitative simulation tests. The results show that the ratio of gypsum to petrolatum plays a dominant role on the density, compressive strength, elastic modulus, cohesion and deformation characteristics. The gas adsorption/desorption is controlled by the coal powder content. Finally, an empirical function is proposed to estimate the adsorption capacity.

(30) Zhang, S. et al., DEM Investigation of the Influence of Minerals on Crack Patterns and Mechanical Properties of Red Mudstone [75].

This paper studied the effect of minerals on crack propagation patterns and mechanical properties of red mudstones, based on numerical simulations using the discrete element method (DEM) modeling techniques. The results show that with the increases in circle size of the central mineral and/or mineral ratio, both the elastic modulus and uniaxial compression strength of the model increase. The mineral, which is located in the crack propagation paths, can effectively change the crack propagation direction.

(31) Huang, B. et al., Effect of Nitric Acid Modification on Characteristics and Adsorption Properties of Lignite [76].

This paper depicted the influence of nitric acid modification on pore structures and surface properties of lignite. The results show that nitric acid modification can help lignite adsorb Pb^{2+}. The nitric acid treatment increases the contents of polar oxygen-containing functional groups, including hydroxyl, carbonyl, and carboxyl groups on the surface of lignite.

Finally, we would like to thank all the authors for their valuable contributions, all the reviewers for their comments that greatly improved the quality of the papers and all the editors from the editorial office for their substantial support for us to handle this Special Issue.

Funding: This research, as well as the APC, was funded by National Natural Science Foundation of China, China (Grant No. 51709260) and the Fundamental Research Funds of State Key Laboratory for Geomechanics and Deep Underground Engineering, China University of Mining and Technology, China (Grant No. Z18011).

Conflicts of Interest: The authors declare no conflict of interest.

References

1. MacMinn, C.; Szulczewski, M.; Juanes, R. CO_2 migration in saline aquifers. Part 1. Capillary trapping under slope and groundwater flow. *J. Fluid Mech.* **2010**, *662*, 329–351. [CrossRef]
2. Wei, W.; Xia, Y. Geometrical, fractal and hydraulic properties of fractured reservoirs: A mini-review. *Adv. Geo-Energy Res.* **2017**, *1*, 31–38. [CrossRef]
3. Gerritsen, M.; Durlofsky, L. Modeling fluid flow in oil reservoirs. *Annu. Rev. Fluid Mech.* **2005**, *37*, 211–238. [CrossRef]
4. Haugen, A.; Fernø, M.; Graue, A.; Bertin, H. Experimental study of foam flow in fractured oil-wet limestone for enhanced oil recovery. *SPE Reserv. Eval. Eng.* **2012**, *15*, 218–228. [CrossRef]
5. Grant, M.; Donaldson, I.; Bixley, P. *Geothermal Reservoir Engineering*; Academic Press: New York, NY, USA, 1983.
6. Huang, N.; Liu, R.; Jiang, Y.; Cheng, Y.; Li, B. Shear-flow coupling characteristics of a three-dimensional discrete fracture network-fault (3D DFN-fault) model considering stress-induced aperture variations. *J. Hydrol.* **2019**, *571*, 416–424. [CrossRef]

7. Liu, R.; Li, B.; Jiang, Y.; Yu, L. A numerical approach for assessing effects of shear on equivalent permeability and nonlinear flow characteristics of 2-D fracture networks. *Adv. Water Resour.* **2018**, *111*, 289–300. [CrossRef]
8. Mora, P.; Wang, Y.; Alonso-Marroquin, F. Lattice solid/Boltzmann microscopic model to simulate solid/fluid systems—A tool to study creation of fluid flow networks for viable deep geothermal energy. *J. Earth Sci.* **2015**, *26*, 11–19. [CrossRef]
9. Long, J.; Remer, J.; Wilson, C.; Witherspoon, P. Porous media equivalents for networks of discontinuous fractures. *Water Resour. Res.* **1982**, *18*, 645–658. [CrossRef]
10. Yang, S.; Huang, Y.; Jing, H.; Liu, X. Discrete element modeling on fracture coalescence behavior of red sandstone containing two unparallel fissures under uniaxial compression. *Eng. Geol.* **2014**, *178*, 28–48. [CrossRef]
11. Yang, S.; Liu, X.; Jing, H. Experimental investigation on fracture coalescence behavior of red sandstone containing two unparallel fissures under uniaxial compression. *Int. J. Rock Mech. Min. Sci.* **2013**, *63*, 82–92. [CrossRef]
12. Yang, S.; Yang, D.; Jing, H.; Li, Y.; Wang, S. An experimental study of the fracture coalescence behaviour of brittle sandstone specimens containing three fissures. *Rock Mech. Rock Eng.* **2012**, *45*, 563–582. [CrossRef]
13. Yang, S. Crack coalescence behavior of brittle sandstone samples containing two coplanar fissures in the process of deformation failure. *Eng. Fract. Mech.* **2011**, *78*, 3059–3081. [CrossRef]
14. Chen, Y.; Hu, S.; Hu, R.; Zhou, C. Estimating hydraulic conductivity of fractured rocks from high-pressure packer tests with an Izbash's law-based empirical model. *Water Resour. Res.* **2015**, *51*, 2096–2118. [CrossRef]
15. Zhou, J.; Hu, S.; Fang, S.; Chen, Y.; Zhou, C. Nonlinear flow behavior at low Reynolds numbers through rough-walled fractures subjected to normal compressive loading. *Int. J. Rock Mech. Min. Sci.* **2015**, *80*, 202–218. [CrossRef]
16. Yin, Q.; Ma, G.; Jing, H.; Wang, H.; Su, H.; Wang, Y.; Liu, R. Hydraulic properties of 3D rough-walled fractures during shearing: An experimental study. *J. Hydrol.* **2018**, *555*, 169–184. [CrossRef]
17. Zhao, Z.; Jing, L.; Neretnieks, I. Evaluation of hydrodynamic dispersion parameters in fractured rocks. *J. Rock Mech. Geotech. Eng.* **2010**, *2*, 243–254. [CrossRef]
18. Liu, R.; Jiang, Y. Special Issue: Fluid Flow in Fractured Porous Media. *Processes* **2018**, *6*, 178. [CrossRef]
19. He, L.; Yin, Q.; Jing, H. Laboratory Investigation of Granite Permeability after High-Temperature Exposure. *Processes* **2018**, *6*, 36. [CrossRef]
20. Jin, Y.; Han, L.; Meng, Q.; Ma, D.; Han, G.; Gao, F.; Wang, S. Experimental Investigation of the Mechanical Behaviors of Grouted Sand with UF-OA Grouts. *Processes* **2018**, *6*, 37. [CrossRef]
21. Yan, S.; Liu, T.; Bai, J.; Wu, W. Key Parameters of Gob-Side Entry Retaining in a Gassy and Thin Coal Seam with Hard Roof. *Processes* **2018**, *6*, 51. [CrossRef]
22. Li, J.; Huang, Y.; Qiao, M.; Chen, Z.; Song, T.; Kong, G.; Gao, H.; Guo, L. Effects of Water Soaked Height on the Deformation and Crushing Characteristics of Loose Gangue Backfill Material in Solid Backfill Coal Mining. *Processes* **2018**, *6*, 64. [CrossRef]
23. Liu, J.; Wan, Z.; Xie, Q.; Li, C.; Liu, R.; Cheng, M.; Han, B. Investigation on Reinforcement and Lapping Effect of Fracture Grouting in Yellow River Embankment. *Processes* **2018**, *6*, 75. [CrossRef]
24. Ma, C.; Wang, P.; Jiang, L.; Wang, C. Deformation and Control Countermeasure of Surrounding Rocks for Water-Dripping Roadway below a Contiguous Seam Goaf. *Processes* **2018**, *6*, 77. [CrossRef]
25. Liu, R.; Liu, H.; Sha, F.; Yang, H.; Zhang, Q.; Shi, S.; Zheng, Z. Investigation of the Porosity Distribution, Permeability, and Mechanical Performance of Pervious Concretes. *Processes* **2018**, *6*, 78. [CrossRef]
26. Xie, X.; Liu, J.; Han, B.; Li, H.; Li, Y.; Li, X. Critical Hydraulic Gradient of Internal Erosion at the Soil–Structure Interface. *Processes* **2018**, *6*, 92. [CrossRef]
27. Cui, Y.; Lu, B.; Wu, M.; Luo, W. A New Pseudo Steady-State Constant for a Vertical Well with Finite-Conductivity Fracture. *Processes* **2018**, *6*, 93. [CrossRef]
28. Li, Z.; Li, L.; Zhang, Z.; Li, M.; Zhang, L.; Huang, B.; Tang, C. The Fracturing Behavior of Tight Glutenites Subjected to Hydraulic Pressure. *Processes* **2018**, *6*, 96. [CrossRef]
29. Chen, S.; Zhang, D.; Shang, T.; Meng, T. Experimental Study of the Microstructural Evolution of Glauberite and Its Weakening Mechanism under the Effect of Thermal-Hydrological-Chemical Coupling. *Processes* **2018**, *6*, 99. [CrossRef]
30. Zhu, D.; Jing, H.; Yin, Q.; Han, G. Experimental Study on the Damage of Granite by Acoustic Emission after Cyclic Heating and Cooling with Circulating Water. *Processes* **2018**, *6*, 101. [CrossRef]

31. Chen, Y.; Li, X.; Li, B. Coal Anisotropic Sorption and Permeability: An Experimental Study. *Processes* **2018**, *6*, 104. [CrossRef]
32. Zhang, Q.; Li, X.; Bai, B.; Hu, S.; Shi, L. Effect of Pore Fluid Pressure on the Normal Deformation of a Matched Granite Joint. *Processes* **2018**, *6*, 107. [CrossRef]
33. Chen, Y.; Qiu, Z.; Li, B.; Yang, Z. Numerical Simulation on the Dynamic Characteristics of a Tremendous Debris Flow in Sichuan, China. *Processes* **2018**, *6*, 109. [CrossRef]
34. Sha, Z.; Pu, H.; Li, M.; Cao, L.; Liu, D.; Ni, H.; Lu, J. Experimental Study on the Creep Characteristics of Coal Measures Sandstone under Seepage Action. *Processes* **2018**, *6*, 110. [CrossRef]
35. Zhou, L.; Wang, Y.; Feng, D. A High-Order Numerical Manifold Method for Darcy Flow in Heterogeneous Porous Media. *Processes* **2018**, *6*, 111. [CrossRef]
36. Wang, S.; Li, H.; Li, D. Numerical Simulation of Hydraulic Fracture Propagation in Coal Seams with Discontinuous Natural Fracture Networks. *Processes* **2018**, *6*, 113. [CrossRef]
37. Liu, W.; Zhao, J.; Nie, R.; Liu, Y.; Du, Y. A Coupled Thermal-Hydraulic-Mechanical Nonlinear Model for Fault Water Inrush. *Processes* **2018**, *6*, 120. [CrossRef]
38. Xu, S.; Li, C.; Liu, J.; Bian, M.; Wei, W.; Zhang, H.; Wang, Z. Deformation and Hydraulic Conductivity of Compacted Clay under Waste Differential Settlement. *Processes* **2018**, *6*, 123. [CrossRef]
39. Xue, D.; Zhou, J.; Liu, Y.; Zhang, S. A Strain-Based Percolation Model and Triaxial Tests to Investigate the Evolution of Permeability and Critical Dilatancy Behavior of Coal. *Processes* **2018**, *6*, 127. [CrossRef]
40. Li, Z.; Li, S.; Liu, H.; Zhang, Q.; Liu, Y. Experimental Study on the Reinforcement Mechanism of Segmented Split Grouting in a Soft Filling Medium. *Processes* **2018**, *6*, 131. [CrossRef]
41. Zhang, C.; Yu, L.; Feng, R.; Zhang, Y.; Zhang, G. A Numerical Study of Stress Distribution and Fracture Development above a Protective Coal Seam in Longwall Mining. *Processes* **2018**, *6*, 146. [CrossRef]
42. Wu, Q.; Wu, Q.; Xue, Y.; Kong, P.; Gong, B. Analysis of Overlying Strata Movement and Disaster-Causing Effects of Coal Mining Face under the Action of Hard Thick Magmatic Rock. *Processes* **2018**, *6*, 150. [CrossRef]
43. Zhao, C.; Zhang, R.; Zhang, Q.; Shi, Z.; Yu, S. Shear-Flow Coupled Behavior of Artificial Joints with Sawtooth Asperities. *Processes* **2018**, *6*, 152. [CrossRef]
44. Dou, Z.; Sleep, B.; Mondal, P.; Guo, Q.; Wang, J.; Zhou, Z. Temporal Mixing Behavior of Conservative Solute Transport through 2D Self-Affine Fractures. *Processes* **2018**, *6*, 158. [CrossRef]
45. Yang, X.; Jing, H.; Qiao, W. Numerical Investigation of the Failure Mechanism of Transversely Isotropic Rocks with a Particle Flow Modeling Method. *Processes* **2018**, *6*, 171. [CrossRef]
46. Jiang, D.; Cheng, X.; Luan, H.; Wang, T.; Zhang, M.; Hao, R. Experimental Investigation on the Law of Grout Diffusion in Fractured Porous Rock Mass and Its Application. *Processes* **2018**, *6*, 191. [CrossRef]
47. Qiu, Z.; Cao, T.; Li, Y.; Wang, J.; Chen, Y. Rheological Behavior and Modeling of a Crushed Sandstone-Mudstone Particle Mixture. *Processes* **2018**, *6*, 192. [CrossRef]
48. Yang, X.; Wang, J.; Hou, D.; Zhu, C.; He, M. Effect of Dry-Wet Cycling on the Mechanical Properties of Rocks: A Laboratory-Scale Experimental Study. *Processes* **2018**, *6*, 199. [CrossRef]
49. Zhang, Z.; Sui, W.; Wang, K.; Tang, G.; Li, X. Changes in Particle Size Composition under Seepage Conditions of Reclaimed Soil in Xinjiang, China. *Processes* **2018**, *6*, 201. [CrossRef]
50. Irfan, M.; Chen, Y.; Ali, M.; Abrar, M.; Qadri, A.; Bhutta, O. Geotechnical Properties of Effluent-Contaminated Cohesive Soils and Their Stabilization Using Industrial By-Products. *Processes* **2018**, *6*, 203. [CrossRef]
51. Zhang, M.; Cui, Y.; Cai, W.; Wu, Z.; Li, Y.; Li, F.; Zhang, W. High Mixing Efficiency by Modulating Inlet Frequency of Viscoelastic Fluid in Simplified Pore Structure. *Processes* **2018**, *6*, 210. [CrossRef]
52. Liu, S.; Liu, W. Experimental Development Process of a New Fluid–Solid Coupling Similar-Material Based on the Orthogonal Test. *Processes* **2018**, *6*, 211. [CrossRef]
53. Liu, L.; Li, L.; Elsworth, D.; Zhi, S.; Yu, Y. The Impact of Oriented Perforations on Fracture Propagation and Complexity in Hydraulic Fracturing. *Processes* **2018**, *6*, 213. [CrossRef]
54. Wang, F.; Liu, Y.; Hu, C.; Wang, Y.; Shen, A.; Liang, S. Experimental Study on Feasibility of Enhanced Gas Recovery through CO_2 Flooding in Tight Sandstone Gas Reservoirs. *Processes* **2018**, *6*, 214. [CrossRef]
55. Zhuo, Q.; Liu, W.; Xu, H.; Sun, X.; Zhang, H.; Liu, W. The Effect of Collision Angle on the Collision and Adhesion Behavior of Coal Particles and Bubbles. *Processes* **2018**, *6*, 218. [CrossRef]
56. Wang, H.; Jiao, J.; Wang, Y.; Du, W. Feasibility of Using Gangue and Fly Ash as Filling Slurry Materials. *Processes* **2018**, *6*, 232. [CrossRef]

57. Huang, S.; Wang, J.; Qiu, Z.; Kang, K. Effects of Cyclic Wetting-Drying Conditions on Elastic Modulus and Compressive Strength of Sandstone and Mudstone. *Processes* **2018**, *6*, 234. [CrossRef]
58. Zhang, Y.; Zhou, W.; Li, M.; Chen, Z. Experimental Study on Compression Deformation and Permeability Characteristics of Grading Broken Gangue under Stress. *Processes* **2018**, *6*, 257. [CrossRef]
59. Sun, X.; Li, G.; Zhao, C.; Liu, Y.; Miao, C. Investigation of Deep Mine Shaft Stability in Alternating Hard and Soft Rock Strata Using Three-Dimensional Numerical Modeling. *Processes* **2019**, *7*, 2. [CrossRef]
60. Huang, M.; Li, D. 2D Plane Strain Consolidation Process of Unsaturated Soil with Vertical Impeded Drainage Boundaries. *Processes* **2019**, *7*, 5. [CrossRef]
61. Wei, P.; Liang, Y.; Zhao, S.; Peng, S.; Li, X.; Meng, R. Characterization of Pores and Fractures in Soft Coal from the No. 5 Soft Coalbed in the Chenghe Mining Area. *Processes* **2019**, *7*, 13. [CrossRef]
62. Li, J.; Dong, S.; Hua, W.; Li, X.; Pan, X. Numerical Investigation of Hydraulic Fracture Propagation Based on Cohesive Zone Model in Naturally Fractured Formations. *Processes* **2019**, *7*, 28. [CrossRef]
63. Sun, Z.; Yan, X.; Han, W.; Ma, G.; Zhang, Y. Simulating the Filtration Effects of Cement-Grout in Fractured Porous Media with the 3D Unified Pipe-Network Method. *Processes* **2019**, *7*, 46. [CrossRef]
64. Hou, Y.; Ding, P.; Han, D.; Zhang, X.; Cao, S. Study on the Preparation and Hydration Properties of a New Cementitious Material for Tailings Discharge. *Processes* **2019**, *7*, 47. [CrossRef]
65. Wang, C.; Jiang, Y.; Luan, H.; Liu, J.; Sugimoto, S. Experimental Study on the Shear-Flow Coupled Behavior of Tension Fractures Under Constant Normal Stiffness Boundary Conditions. *Processes* **2019**, *7*, 57. [CrossRef]
66. Hao, J.; Shi, Y.; Lin, J.; Wang, X.; Xia, H. The Effects of Backfill Mining on Strata Movement Rule and Water Inrush: A Case Study. *Processes* **2019**, *7*, 66. [CrossRef]
67. Hua, W.; Li, J.; Dong, S.; Pan, X. Experimental Study on Mixed Mode Fracture Behavior of Sandstone under Water–Rock Interactions. *Processes* **2019**, *7*, 70. [CrossRef]
68. Li, H.; Tian, H.; Ma, K. Seepage Characteristics and Its Control Mechanism of Rock Mass in High-Steep Slopes. *Processes* **2019**, *7*, 71. [CrossRef]
69. Li, Z.; Niu, Y.; Wang, E.; Liu, L.; Wang, H.; Wang, M.; Ali, M. Experimental Study on Electric Potential Response Characteristics of Gas-Bearing Coal During Deformation and Fracturing Process. *Processes* **2019**, *7*, 72. [CrossRef]
70. Liu, R.; Yin, H.; Zhang, L.; Li, B.; Mao, X. Model for the Patterns of Salt-Spray-Induced Chloride Corrosion in Concretes under Coupling Action of Cyclic Loading and Salt Spray Corrosion. *Processes* **2019**, *7*, 84. [CrossRef]
71. Qiao, S.; Qin, S.; Chen, J.; Hu, X.; Ma, Z. The Application of a Three-Dimensional Deterministic Model in the Study of Debris Flow Prediction Based on the Rainfall-Unstable Soil Coupling Mechanism. *Processes* **2019**, *7*, 99. [CrossRef]
72. Yu, W.; Li, H. Development of 3D Finite Element Method for Non-Aqueous Phase Liquid Transport in Groundwater as Well as Verification. *Processes* **2019**, *7*, 116. [CrossRef]
73. Liu, D.; Sun, W.; Ren, D. Experimental Investigation of Pore Structure and Movable Fluid Traits in Tight Sandstone. *Processes* **2019**, *7*, 149. [CrossRef]
74. Wang, X.; Xie, W.; Su, Z.; Tang, Q. Experimental Development of Coal-Like Material with Solid-Gas Coupling for Quantitative Simulation Tests of Coal and Gas Outburst Occurred in Soft Coal Seams. *Processes* **2019**, *7*, 155. [CrossRef]
75. Zhang, S.; Zhang, D.; Zhao, Q.; Chi, M.; Zhang, W.; Yu, W. DEM Investigation of the Influence of Minerals on Crack Patterns and Mechanical Properties of Red Mudstone. *Processes* **2019**, *7*, 162. [CrossRef]
76. Huang, B.; Liu, G.; Wang, P.; Zhao, X.; Xu, H. Effect of Nitric Acid Modification on Characteristics and Adsorption Properties of Lignite. *Processes* **2019**, *7*, 167. [CrossRef]

processes

MDPI

Editorial

Special Issue: Fluid Flow in Fractured Porous Media

Richeng Liu [1,2] and Yujing Jiang [2,*]

1 State Key Laboratory for Geomechanics and Deep Underground Engineering, China University of Mining and Technology, Xuzhou 221116, China; liuricheng@cumt.edu.cn
2 School of Engineering, Nagasaki University, 1-14 Bunkyo-machi, Nagasaki 8528521, Japan
* Correspondence: jiang@nagasaki-u.ac.jp; Tel.: +81-95-819-2612

Received: 28 September 2018; Accepted: 28 September 2018; Published: 1 October 2018

The fluid flow in fractured porous media plays a significant role in the characteristic/assessment of deep underground reservoirs such as CO_2 sequestration [1–3], enhanced oil recovery [4,5] and geothermal energy development [6,7]. Many methods, including laboratory experiments, theoretical analysis and numerical simulation, and many models including the discrete fracture network (DFN) model [8–10], the channeling network (CN) model [11–13] and the dual-porosity model [14,15], have been employed to investigate fluid flow in fractured porous media. The early studies commonly assumed that fluid flow obeys the cubic law and that fluid flow is in the linear flow regime [16–23], whereas recent studies also investigated the nonlinear flow through rock fractures/fracture networks [24–33]. However, due to the complex and uncertain geometric properties of rock masses deep underground, extensive studies on the hydraulic properties of fractured porous media and engineering applications are still needed.

Therefore, we are handling a Special Issue (SI) of *Processes* on fluid flow in fractured porous media (http://www.mdpi.com/journal/processes/special_issues/porous_media), which is now ongoing until the deadline of 31 December 2018. This SI has published 27 high quality papers and another 8 papers are under review. The accepted papers include a wide variety of topics such as granite permeability investigation, grouting, coal mining, roadway, concrete and so on. The details of these accepted papers in the SI are summarized individually as follows:

(1) He, L. et al., Laboratory Investigation of Granite Permeability after High-Temperature Exposure. [34]

The very first accepted paper of the Special Issue on "Fluid flow in fractured porous media" presented a series of flow tests on granite after high temperature treatment of up to 800 °C. The results showed that higher temperatures induce a larger number of micro-fractures and the larger decrement ratios of both the P-wave velocity and the density of granite. The tested equivalent permeability increases with increasing temperature, following an exponential function.

(2) Jin, Y. et al., Experimental Investigation of the Mechanical Behaviors of Grouted Sand with UF-OA Grouts. [35]

This paper investigated the effect of a newly developed grouting material—which is a chemical material-ureaformaldehyde resin mixed with oxalate curing agent (UF-OA)—on the mechanical behaviors of sand specimens through uniaxial compression tests. The results showed that the presence of initial water can decrease both the uniaxial compressive strength (UCS) and the elastic moduli (E) of the grouted specimens; however, the peak strain (ε_c) does not change significantly after 14 days of curing.

(3) Yan, S. et al., Key Parameters of Gob-Side Entry Retaining in a Gassy and Thin Coal Seam with Hard Roof. [36]

This paper analyzed the gas concentration and air pressure in the gob and proposed a roof-cutting mechanical model of gob-side entry retaining (GER) with a roadside backfill body (RBB) to determine the key parameters of the GER-TCS. Here, TCS denotes the thin coal seam. They reported that a greater RBB width results in greater roof-cutting resistance and the floor heave of GER increases with increasing RBB width. Finally, the validity of the proposed model is verified by comparing the calculated results with in situ measurements.

(4) Li, J. et al., Effects of Water Soaked Height on the Deformation and Crushing Characteristics of Loose Gangue Backfill Material in Solid Backfill Coal Mining. [37]

This paper studied the effect of water soaked height in coal mines on the deformation and crushing characteristics of loose gangue backfill material (LGBM). They concluded that, with the increment of water socked height, both the crushing ratio and the maximum axial strain of LGBM specimens increase. The crushing ratio has a strong linear correlation with the maximum axial strain and a higher water soaked height leads to a lower deformation resistance and a more significant influence on the quality of backfilling.

(5) Liu, J. et al., Investigation on Reinforcement and Lapping Effect of Fracture Grouting in Yellow River Embankment. [38]

This paper carried out a series of laboratory and in situ experiments to characterize the reinforcement effect of fracture grouting on the Yellow River Embankment. The results showed that the optimum mixing proportion of the compound cement-silicate grout is 70% cement, 25% fly ash and 5% bentonite. The interval distance of the injection hole is suggested to be 1.2 m, in which the lapping effect of fracture grouting veins play an import role.

(6) Ma, C. et al., Deformation and Control Countermeasure of Surrounding Rocks for Water-Dripping Roadway below a Contiguous Seam Goaf. [39]

This paper studied the deformation of rocks surrounding the roadway using numerical simulation and proposed a mechanical model of water-dripping rock using a bolt support. They found that the erosion by water that is dripped on the roadway is substantial and changes significantly over time during roadway excavation. The suggested bolt line spacing is 0.7~0.9 m and the suggested tightening force should exceed 40 kN.

(7) Liu, R. et al., Investigation of the Porosity Distribution, Permeability and Mechanical Performance of Pervious Concretes. [40]

This paper investigated the performance of pervious concrete, such as its porosity, pore distribution, permeability coefficient and mechanical properties, with laboratory tests. The main conclusions are that the standard deviation of permeability is less than 0.03 under different hydraulic gradients. The permeability is closely correlated with the volumetric porosity (or effective pore size d_{50}), following polynomial functions.

(8) Xie, X. et al., Critical Hydraulic Gradient of Internal Erosion at the Soil–Structure Interface. [41]

This paper carried out orthogonal and single-factor tests to estimate the critical hydraulic gradient of internal erosion on soil-structure interfaces. They reported that the critical hydraulic gradient is mostly sensitive to degree of compaction, followed by roughness and clay content. The internal erosion is found to develop inside first and then extend to the outside along the interface.

(9) Cui, Y. et al., A New Pseudo Steady-State Constant for a Vertical Well with Finite-Conductivity Fracture. [42]

The authors proposed a new semi-analytical solution for the Pseudo Steady-State constant, which is defined as the difference between the dimensionless wellbore pressure and dimensionless average pressure and is an important parameter for describing Type Curves. They derived a new conductivity-influence function and a normalized conductivity-influence function, which are then utilized to derive the approximate solute of Pseudo Steady-State constant.

(10) Li, Z. et al., The Fracturing Behavior of Tight Glutenites Subjected to Hydraulic Pressure. [43]

Numerical simulations were performed to model the fracturing behaviors of tight glutenites, in which the glutenite heterogeneity has been taken into account. The main conclusions are that the hydraulic fractures propagate along the gravels, influenced by stress anisotropy and gravel strength. When the gravel distribution orientation, size and axial ratio are considered, the fractures propagate past the gravel, forming a bypass fracture.

(11) Chen, S. et al., Experimental Study of the Microstructural Evolution of Glauberite and Its Weakening Mechanism under the Effect of Thermal-Hydrological-Chemical Coupling. [44]

This paper focuses on the microsctructure evolution of glaubeite under different leaching conditions. Cylindrical specimens with a size of 3 mm in diameter and 10 mm in length were cored from a bedded salt deposit buried at over 1000 m depth in Yunying salt formation, China. The results show that those specimens immersed in a solution at a low temperature generate cracks at the center of the disc. With incremental increase of the temperature of the solution, the crack nucleation appears and the dissolution rate increases significantly.

(12) Zhu, D. et al., Experimental Study on the Damage of Granite by Acoustic Emission after Cyclic Heating and Cooling with Circulating Water. [45]

This paper experimentally studied the damage of granite using acoustic emissions after cyclic heating and cooling with circulating water. They found that the physical and mechanical parameter of granite exhibits exponential relationships with increasing number of cycles. The cumulative number of acoustic emission decreases with the increment of heating and cooling number and the temperature play a key role in the failure mode of samples.

(13) Chen, Y. et al., Coal Anisotropic Sorption and Permeability: An Experimental Study. [46]

This paper presented an experimental study on the anisotropic gas adsorption–desorption and permeability of coal. The results show that an increase in the bedding plane angle of the specimen expands the length and area of the contact surface, thereby increasing the speed and quantity of adsorption and desorption. With an increase in the bedding angle, the number of pores and cracks is found to increase together with the volumetric strain.

(14) Zhang, Q. et al., Effect of Pore Fluid Pressure on the Normal Deformation of a Matched Granite Joint. [47]

Laboratory experiments were carried out to estimate the relationship between pore fluid pressure and normal deformation of joints. They reported that during fluid injection, the joint opens and host rock dilates. The normal deformation of joints can be well described using the proposed exponential functions and the relative normal deformation of host rock during has a linear relationship with the pore fluid pressure.

(15) Chen, Y. et al., Numerical Simulation on the Dynamic Characteristics of a Tremendous Debris Flow in Sichuan, China. [48]

This paper reported the numerical simulation of Zoumaling debris flow based on computational fluid dynamics (CFD). All these results of the simulations and analyses can guide, and provide suggestions for, the design and construction of prevention engineering for Zoumaling debris flow.

(16) Sha, Z. et al., Experimental Study on the Creep Characteristics of Coal Measures Sandstone under Seepage Action. [49]

This paper investigated the creep characteristics of coal measures sandstone influenced by the effect of seepage. A seepage-creep coupling test system was established and scanning electron microscopy techniques were used. The results show that the maximum creep deformation gradually decreases with the increase in confining pressure. For a fixed confining pressure, the creep deformation, creep deformation rate, and accelerated creep deformation rate, increase dramatically with the increment of infiltration pressure.

(17) Zhou, L. et al., A High-Order Numerical Manifold Method for Darcy Flow in Heterogeneous Porous Media. [50]

The authors adopted a high-order numerical manifold method to improve the simulation accuracy of fluid flow in heterogeneous porous media. They developed a regular mathematical mesh, which is independent of the physical domain and is very efficient and flexible. The proposed new numerical manifold method model is confirmed to be more accurate and efficient, compared with the traditional numerical manifold method model.

(18) Wang, S. et al., Numerical Simulation of Hydraulic Fracture Propagation in Coal Seams with Discontinuous Natural Fracture Networks. [51]

This paper investigated the mechanism of hydraulic fracture propagation in coal seams with discontinuous natural fracture networks by using an innovative cohesive element method. The hydraulic fracture network characteristics, as well as the growth of secondary hydraulic fractures and the pore pressure distribution, were analyzed. The results depict that the hydraulic fracture network is spindle-shaped at a high stress difference condition. The ratio of total length of secondary fracture to that of main fracture is 2.11~3.62, which indicates that the secondary fracture plays a dominate role in coal seams. The process of hydraulic fracture propagation in deep coal seams shows very different behaviors with that in hand and/or tight rocks.

(19) Liu, W. et al., A Coupled Thermal-Hydraulic-Mechanical Nonlinear Model for Fault Water Inrush. [52]

A coupled thermal–nonlinear hydraulic–mechanical (THM) model for fault water inrush was proposed in this paper to study water–rock–temperature interactions and to predict the fault water inrush. The governing equations of the THM coupling model coupled the particle transport equation, the nonlinear flow equation, the mechanical equation and the heat transfer equation. Results show that the water velocity increases with the increasing water pressure. Temperature change in the fault zone is subjected to the interaction of the water pressure and the working face advanced distance. The temperature change in the fault zone can reflect the change in seepage field in the fault and confined aquifer, which can be used to predict the fault water inrush induced by coal mining activities.

(20) Xu, S. et al., Deformation and Hydraulic Conductivity of Compacted Clay under Waste Differential Settlement. [53]

This paper examined the effect of geogrid reinforcement on the deformation and conductivity of bentonite-sand mixtures subjected to differential settlement. When the hydraulic conductivity is 5×10^{-7} cm/s, the settlement for a 100 mm width bentonite-sand mixture with geogrid reinforcement is 100 mm and increases to 15 mm for the case without geogrid reinforcement. The obtained results are helpful for predicting the deformation and conductivity of a landfill cover system.

(21) Xue, D. et al., A Strain-Based Percolation Model and Triaxial Tests to Investigate the Evolution of Permeability and Critical Dilatancy Behavior of Coal. [54]

A strain-based percolation model was proposed to model the coupled process of strain and CH_4 seepage under triaxial compressions. Using a volume-covering method, it is found that the multiscale fracture network is fractal, which has a linear correlation with the porosity of matrix. The proposed percolation model, which links the axial strain to the permeability, is used to interpret the transitions of volumetric strain and CH_4 seepage.

(22) Li, Z. et al., Experimental Study on the Reinforcement Mechanism of Segmented Split Grouting in a Soft Filling Medium. [55]

A series of experiments were carried out to address the reinforcement by segmented split grouting in soft filling media. They establish a quantitative relationship between mechanical behaviors and hydraulic behaviors of soft filling media both before and after grouting. The results exhibit that three types of veins that are skeleton vein, cross-grid grouting vein and parallel dispersed grouting vein are formed with the increase in the distance from the grouting hole. After the skeleton support and compaction, the segmental split grouting shows a significant reinforcement effect on the soft filling media.

(23) Zhang, C. et al., A Numerical Study of Stress Distribution and Fracture Development above a Protective Coal Seam in Longwall Mining. [56]

This paper performed numerical simulations on stress distribution and fracture development above a protective coal seam in longwall mining in Jincheng, China. Five zones are divided for the rock mass above the protective coal seam, including: pre-mining zone, compression zone, expansion zone, recovery zone and re-compacted zone. The boreholes for gas drainage in the coal seam should be drilled in the volume expansion or the dilation zone, in which the gas concentration is high.

(24) Wu, Q. et al., Analysis of Overlying Strata Movement and Disaster-Causing Effects of Coal Mining Face under the Action of Hard Thick Magmatic Rock. [57]

This paper analyzed the overlying strata movement and studied the mechanical behaviors of breaking induced disaster of hard and thick magmatic rocks. The main results are that the magmatic rocks control the development of the bed separation. The distribution of the fracture zone alters with the advance of the working face. The gas and water are accumulated below the bed separation space during the breaking of magmatic rocks and tend to rush towards the working face.

(25) Zhao, C. et al., Shear-Flow Coupled Behavior of Artificial Joints with Sawtooth Asperities. [58]

This paper performed regular shear tests and shear-flow coupled tests on artificial joints that have sawtooth asperities. The influences of joint roughness, normal stress and seepage pressure were systematically analyzed. The results indicate that the shear strength is weakened after a seepage process due to the mutual reinforcing of offsetting and softening effect. The evolution of hydraulic aperture during the tests shows a four-stage characteristic: shear contraction, shear dilation, re-contraction and stability.

(26) Dou, Z. et al., Temporal Mixing Behavior of Conservative Solute Transport through 2D Self-Affine Fractures. [59]

The authors simulated the temporal mixing behavior of solute transport through self-affine fractures. The influences of Hurst exponent and Peclet number were investigated and the mixing was studied by scalar dissipation rate in fractures. For the parallel plate fracture, the Hurst exponent played a negligible role on the temporal evolution of the scalar dissipation rate when the Peclet number was small. The longitudinal scalar dissipation rate overestimated the global scalar dissipation rate for a large Peclet number.

(27) Yang, X. et al., Numerical Investigation of the Failure Mechanism of Transversely Isotropic Rocks with a Particle Flow Modeling Method. [60]

This paper numerically studied the failure mechanisms of transversely isotropic rocks using a particle flow modeling method. The results show that three modes are observed in the transversely isotropic rock models, including tensile failure across interfaces, shear failure along interfaces, tensile failure along interfaces. The bounded joint cohesion and joint friction angle significantly influence the uniaxial compressive strengths, whereas the joint coefficient of friction and joint tensile strength slightly influence the uniaxial compressive strengths.

Finally, we would like to thank all the authors for their valuable contributions, all the reviewers for their comments that greatly improved the quality of the papers and all the editors from the editorial office for their substantial support for us to handle this Special Issue.

Acknowledgments: This Special Issue on "Fluid flow in fractured porous media" has been partially funded by National Natural Science Foundation of China, China (No. 51709260) and the Fundamental Research Funds of State Key Laboratory for Geomechanics and Deep Underground Engineering, China University of Mining and Technology, China (No. Z18011). These supports are gratefully acknowledged.

Conflicts of Interest: The authors declare no conflict of interest.

References

1. Juanes, R.; Spiteri, E.; Orr, F.; Blunt, M. Impact of relative permeability hysteresis on geological CO_2 storage. *Water Resour. Res.* **2006**, 42. [CrossRef]
2. Juanes, R.; MacMinn, C.; Szulczewski, M. The footprint of the CO_2 plume during carbon dioxide storage in saline aquifers: Storage efficiency for capillary trapping at the basin scale. *Transp. Porous Media* **2010**, *82*, 19–30. [CrossRef]
3. MacMinn, C.; Szulczewski, M.; Juanes, R. CO_2 migration in saline aquifers. Part 1. Capillary trapping under slope and groundwater flow. *J. Fluid Mech.* **2010**, *662*, 329–351. [CrossRef]
4. Gerritsen, M.; Durlofsky, L. Modeling fluid flow in oil reservoirs. *Annu. Rev. Fluid Mech.* **2005**, *37*, 211–238. [CrossRef]
5. Haugen, Å.; Fernø, M.; Graue, A.; Bertin, H. Experimental study of foam flow in fractured oil-wet limestone for enhanced oil recovery. *SPE Reserv. Eval. Eng.* **2012**, *15*, 218–228.
6. Grant, M.; Donaldson, I.; Bixley, P. *Geothermal Reservoir Engineering*; Academic Press: New York, NY, USA, 1983.
7. Mora, P.; Wang, Y.; Alonso-Marroquin, F. Lattice solid/Boltzmann microscopic model to simulate solid/fluid systems—A tool to study creation of fluid flow networks for viable deep geothermal energy. *J. Earth Sci.* **2015**, *26*, 11–19. [CrossRef]
8. Long, J.; Remer, J.; Wilson, C.; Witherspoon, P. Porous media equivalents for networks of discontinuous fractures. *Water Resour Res.* **1982**, *18*, 645–658. [CrossRef]
9. Wei, W.; Xia, Y. Geometrical, fractal and hydraulic properties of fractured reservoirs: A mini-review. *Adv. Geo-Energy Res.* **2017**, *1*, 31–38. [CrossRef]
10. Hartley, L.; Roberts, D. *Summary of Discrete Fracture Network Modelling as Applied to Hydrogeology of the Forsmark and Laxemar Sites*; Report R-12-04; Swedish Nuclear Fuel and Waste Management Co.: Stockholm, Sweden, 2012.

11. Neretnieks, I. Channeling effects in flow and transport in fractured rocks-Some recent observations and models. In Proceedings of the GEOVAL-87, International Symposium, Stockholm, Sweden, 14–17 May 1987; pp. 315–335.
12. Dershowitz, W.; Fidelibus, C. Derivation of equivalent pipe network analogues for three-dimensional discrete fracture networks by the boundary element method. *Water Resour. Res.* **1999**, *35*, 2685–2691. [CrossRef]
13. Black, J.; Barker, J.; Woodman, N. *An Investigation of 'Sparse Channel Networks'—Characteristic Behaviours and Their Causes*; Report R-07-35; Swedish Nuclear Fuel and Waste Management Co.: Stockholm, Sweden, 2007.
14. Selroos, J.; Walker, D.; Ström, A.; Gylling, B.; Follin, S. Comparison of alternative modelling approaches for groundwater flow in fractured rock. *J. Hydrol.* **2002**, *257*, 174–188. [CrossRef]
15. Figueiredo, B.; Tsang, C.; Niemi, A.; Lindgren, G. Review: The state-of-art of sparse channel models and their applicability to performance assessment of radioactive waste repositories in fractured crystalline formations. *Hydrogeol. J.* **2016**, *24*, 1607–1622. [CrossRef]
16. Klimczak, C.; Schultz, R.; Parashar, R.; Reeves, D. Cubic law with aperture-length correlation: Implications for network scale fluid flow. *Hydrogeol. J.* **2010**, *18*, 851–862. [CrossRef]
17. Zhao, Z.; Jing, L.; Neretnieks, I. Evaluation of hydrodynamic dispersion parameters in fractured rocks. *Journal of Rock Mech. Geotech. Eng.* **2010**, *2*, 243–254. [CrossRef]
18. Cvetkovic, V.; Painter, S.; Outters, N.; Selroos, J. Stochastic simulation of radionuclide migration in discretely fractured rock near the Äspö Hard Rock Laboratory. *Water Resour. Res.* **2004**, 40. [CrossRef]
19. Zhao, Z. Gouge particle evolution in a rock fracture undergoing shear: A microscopic DEM study. *Rock Mech. Rock Eng.* **2013**, *46*, 1461–1479. [CrossRef]
20. Min, K.; Rutqvist, J.; Tsang, C.; Jing, L. Stress-dependent permeability of fractured rock masses: A numerical study. *Int. J. Rock Mech. Min. Sci.* **2004**, *41*, 1191–1210. [CrossRef]
21. Parashar, R.; Reeves, D. On iterative techniques for computing flow in large two-dimensional discrete fracture networks. *J. Comput. Appl. Math.* **2012**, *236*, 4712–4724. [CrossRef]
22. Latham, J.; Xiang, J.; Belayneh, M.; Nick, H.; Tsang, C.; Blunt, M. Modelling stress-dependent permeability in fractured rock including effects of propagating and bending fractures. *Int. J. Rock Mech. Min. Sci.* **2013**, *57*, 100–112. [CrossRef]
23. Huang, N.; Jiang, Y.; Li, B.; Liu, R. A numerical method for simulating fluid flow through 3-D fracture networks. *J. Nat. Gas Sci. Eng.* **2016**, *33*, 1271–1281. [CrossRef]
24. Zimmerman, R.; Al-Yaarubi, A.; Pain, C.; Grattoni, C. Non-linear regimes of fluid flow in rock fractures. *Int. J. Rock Mech. Min. Sci.* **2004**, *41*, 163–169. [CrossRef]
25. Zhang, Z.; Nemcik, J. Fluid flow regimes and nonlinear flow characteristics in deformable rock fractures. *J. Hydrol.* **2013**, *477*, 139–151. [CrossRef]
26. Javadi, M.; Sharifzadeh, M.; Shahriar, K.; Mitani, Y. Critical Reynolds number for nonlinear flow through rough-walled fractures: The role of shear processes. *Water Resour. Res.* **2014**, *50*, 1789–1804. [CrossRef]
27. Chen, Y.; Hu, S.; Hu, R.; Zhou, C. Estimating hydraulic conductivity of fractured rocks from high-pressure packer tests with an Izbash's law-based empirical model. *Water Resour. Res.* **2015**, *51*, 2096–2118. [CrossRef]
28. Zhou, J.; Hu, S.; Fang, S.; Chen, Y.; Zhou, C. Nonlinear flow behavior at low Reynolds numbers through rough-walled fractures subjected to normal compressive loading. *Int. J. Rock Mech. Min. Sci.* **2015**, *80*, 202–218. [CrossRef]
29. Zhou, J.; Wang, M.; Wang, L.; Chen, Y.; Zhou, C. Emergence of nonlinear laminar flow in fractures during shear. *Rock Mech. Rock Eng.* **2018**. [CrossRef]
30. Rong, G.; Yang, J.; Cheng, L.; Zhou, C. Laboratory investigation of nonlinear flow characteristics in rough fractures during shear process. *J. Hydrol.* **2016**, *541*, 1385–1394. [CrossRef]
31. Zou, L.; Jing, L.; Cvetkovic, V. Shear-enhanced nonlinear flow in rough-walled rock fractures. *Int. J. Rock Mech. Min. Sci.* **2017**, *97*, 33–45. [CrossRef]
32. Yin, Q.; Ma, G.; Jing, H.; Wang, H.; Su, H.; Wang, Y.; Liu, R. Hydraulic properties of 3D rough-walled fractures during shearing: An experimental study. *J. Hydrol.* **2018**, *555*, 169–184. [CrossRef]
33. Yin, Q.; Jing, H.; Ma, G.; Su, H.; Liu, R. Investigating the roles of included angle and loading condition on the critical hydraulic gradient of real rock fracture networks. *Rock Mech. Rock Eng.* **2018**. [CrossRef]
34. He, L.; Yin, Q.; Jing, H. Laboratory Investigation of Granite Permeability after High-Temperature Exposure. *Processes* **2018**, *6*, 36. [CrossRef]

35. Jin, Y.; Han, L.; Meng, Q.; Ma, D.; Han, G.; Gao, F.; Wang, S. Experimental Investigation of the Mechanical Behaviors of Grouted Sand with UF-OA Grouts. *Processes* **2018**, *6*, 37. [CrossRef]
36. Yan, S.; Liu, T.; Bai, J.; Wu, W. Key Parameters of Gob-Side Entry Retaining in a Gassy and Thin Coal Seam with Hard Roof. *Processes* **2018**, *6*, 51. [CrossRef]
37. Li, J.; Huang, Y.; Qiao, M.; Chen, Z.; Song, T.; Kong, G.; Gao, H.; Guo, L. Effects of Water Soaked Height on the Deformation and Crushing Characteristics of Loose Gangue Backfill Material in Solid Backfill Coal Mining. *Processes* **2018**, *6*, 64. [CrossRef]
38. Liu, J.; Wan, Z.; Xie, Q.; Li, C.; Liu, R.; Cheng, M.; Han, B. Investigation on Reinforcement and Lapping Effect of Fracture Grouting in Yellow River Embankment. *Processes* **2018**, *6*, 75. [CrossRef]
39. Ma, C.; Wang, P.; Jiang, L.; Wang, C. Deformation and Control Countermeasure of Surrounding Rocks for Water-Dripping Roadway below a Contiguous Seam Goaf. *Processes* **2018**, *6*, 77. [CrossRef]
40. Liu, R.; Liu, H.; Sha, F.; Yang, H.; Zhang, Q.; Shi, S.; Zheng, Z. Investigation of the Porosity Distribution, Permeability, and Mechanical Performance of Pervious Concretes. *Processes* **2018**, *6*, 78. [CrossRef]
41. Xie, X.; Liu, J.; Han, B.; Li, H.; Li, Y.; Li, X. Critical Hydraulic Gradient of Internal Erosion at the Soil–Structure Interface. *Processes* **2018**, *6*, 92. [CrossRef]
42. Cui, Y.; Lu, B.; Wu, M.; Luo, W. A New Pseudo Steady-State Constant for a Vertical Well with Finite-Conductivity Fracture. *Processes* **2018**, *6*, 93. [CrossRef]
43. Li, Z.; Li, L.; Zhang, Z.; Li, M.; Zhang, L.; Huang, B.; Tang, C. The Fracturing Behavior of Tight Glutenites Subjected to Hydraulic Pressure. *Processes* **2018**, *6*, 96. [CrossRef]
44. Chen, S.; Zhang, D.; Shang, T.; Meng, T. Experimental Study of the Microstructural Evolution of Glauberite and Its Weakening Mechanism under the Effect of Thermal-Hydrological-Chemical Coupling. *Processes* **2018**, *6*, 99. [CrossRef]
45. Zhu, D.; Jing, H.; Yin, Q.; Han, G. Experimental Study on the Damage of Granite by Acoustic Emission after Cyclic Heating and Cooling with Circulating Water. *Processes* **2018**, *6*, 101. [CrossRef]
46. Chen, Y.; Li, X.; Li, B. Coal Anisotropic Sorption and Permeability: An Experimental Study. *Processes* **2018**, *6*, 104. [CrossRef]
47. Zhang, Q.; Li, X.; Bai, B.; Hu, S.; Shi, L. Effect of Pore Fluid Pressure on the Normal Deformation of a Matched Granite Joint. *Processes* **2018**, *6*, 107. [CrossRef]
48. Chen, Y.; Qiu, Z.; Li, B.; Yang, Z. Numerical Simulation on the Dynamic Characteristics of a Tremendous Debris Flow in Sichuan, China. *Processes* **2018**, *6*, 109. [CrossRef]
49. Sha, Z.; Pu, H.; Li, M.; Cao, L.; Liu, D.; Ni, H.; Lu, J. Experimental Study on the Creep Characteristics of Coal Measures Sandstone under Seepage Action. *Processes* **2018**, *6*, 110. [CrossRef]
50. Zhou, L.; Wang, Y.; Feng, D. A High-Order Numerical Manifold Method for Darcy Flow in Heterogeneous Porous Media. *Processes* **2018**, *6*, 111. [CrossRef]
51. Wang, S.; Li, H.; Li, D. Numerical Simulation of Hydraulic Fracture Propagation in Coal Seams with Discontinuous Natural Fracture Networks. *Processes* **2018**, *6*, 113. [CrossRef]
52. Liu, W.; Zhao, J.; Nie, R.; Liu, Y.; Du, Y. A Coupled Thermal-Hydraulic-Mechanical Nonlinear Model for Fault Water Inrush. *Processes* **2018**, *6*, 120. [CrossRef]
53. Xu, S.; Li, C.; Liu, J.; Bian, M.; Wei, W.; Zhang, H.; Wang, Z. Deformation and Hydraulic Conductivity of Compacted Clay under Waste Differential Settlement. *Processes* **2018**, *6*, 123. [CrossRef]
54. Xue, D.; Zhou, J.; Liu, Y.; Zhang, S. A Strain-Based Percolation Model and Triaxial Tests to Investigate the Evolution of Permeability and Critical Dilatancy Behavior of Coal. *Processes* **2018**, *6*, 127. [CrossRef]
55. Li, Z.; Li, S.; Liu, H.; Zhang, Q.; Liu, Y. Experimental Study on the Reinforcement Mechanism of Segmented Split Grouting in a Soft Filling Medium. *Processes* **2018**, *6*, 131. [CrossRef]
56. Zhang, C.; Yu, L.; Feng, R.; Zhang, Y.; Zhang, G. A Numerical Study of Stress Distribution and Fracture Development above a Protective Coal Seam in Longwall Mining. *Processes* **2018**, *6*, 146. [CrossRef]
57. Wu, Q.; Wu, Q.; Xue, Y.; Kong, P.; Gong, B. Analysis of Overlying Strata Movement and Disaster-Causing Effects of Coal Mining Face under the Action of Hard Thick Magmatic Rock. *Processes* **2018**, *6*, 150. [CrossRef]
58. Zhao, C.; Zhang, R.; Zhang, Q.; Shi, Z.; Yu, S. Shear-Flow Coupled Behavior of Artificial Joints with Sawtooth Asperities. *Processes* **2018**, *6*, 152. [CrossRef]

59. Dou, Z.; Sleep, B.; Mondal, P.; Guo, Q.; Wang, J.; Zhou, Z. Temporal Mixing Behavior of Conservative Solute Transport through 2D Self-Affine Fractures. *Processes* **2018**, *6*, 158. [CrossRef]
60. Yang, X.; Jing, H.; Qiao, W. Numerical Investigation of the Failure Mechanism of Transversely Isotropic Rocks with a Particle Flow Modeling Method. *Processes* **2018**, *6*, 171. [CrossRef]

processes

MDPI

Article

Effect of Nitric Acid Modification on Characteristics and Adsorption Properties of Lignite

Bo Huang *, Guowei Liu, Penghui Wang, Xiang Zhao and Hongxiang Xu *

School of Chemical & Environmental Engineering, China University of Mining and Technology (Beijing), Beijing 100083, China; liuguowei7512@outlook.com (G.L.); wangpenghui008@outlook.com (P.W.); zhao928743213@163.com (X.Z.)

* Correspondence: huangbo1969@126.com (B.H.); xuhongxiang001@cumtb.edu.cn (H.X.); Tel.: +86-186-1013-3577 (B.H.); +86-130-5188-6977 (H.X.)

Received: 18 February 2019; Accepted: 18 March 2019; Published: 22 March 2019

Abstract: The objective of this research was to explore the changes of the pore structure and surface properties of nitric-modified lignite and base the adsorption performance on physical and chemical adsorbent characteristics. To systematically evaluate pore structure and surface chemistry effects, several lignite samples were treated with different concentrations of nitric acid in order to get different pore structure and surface chemistry adsorbent levels. A common heavy metal ion contaminant in water, Pb^{2+}, served as an adsorbate probe to demonstrate the change of modified lignite adsorption properties. The pore structure and surface properties of lignite samples before and after modification were characterized by static nitrogen adsorption, X-ray diffraction, Scanning electron microscope, Fourier transform infrared spectroscopy, zeta potential, and X-ray photoelectron spectroscopy. The experimental results showed that nitric acid modification can increase the ability of lignite to adsorb Pb^{2+}. The adsorption amount of Pb^{2+} increased from 14.45 $mg \cdot g^{-1}$ to 30.68 $mg \cdot g^{-1}$. Nitric acid reacted with inorganic mineral impurities such as iron dolomite in lignite and organic components in coal, which caused an increase in pore size and a decrease in specific surface areas. A hydrophilic adsorbent surface more effectively removed Pb^{2+} from aqueous solution. Nitric acid treatment increased the content of polar oxygen-containing functional groups such as hydroxyl, carbonyl, and carboxyl groups on the surface of lignite. Treatment introduced nitro groups, which enhanced the negative electrical properties, the polarity of the lignite surface, and its metal ion adsorption performance, a result that can be explained by enhanced water adsorption on hydrophilic surfaces.

Keywords: lignite; nitric acid modification; pore structure; surface characteristics; adsorption performance

1. Introduction

Lignite is a kind of low-rank coal with a low degree of coalification, and it is a macromolecular network structure compound that is composed of aromatic hydrocarbon rings and aliphatic chains, connected by bridge bonds. The molecular structure of lignite contains a large number of active oxygen-containing functional groups such as phenolic hydroxyl, carboxyl, ether bonds, and carbonyl, which impart the characteristics of weak acidity, hydrophilicity, and a cation exchange property to the coal [1]. Compared with other coal types, lignite has a more developed internal pore structure, a larger specific surface area, more rich oxygen-containing functional groups, and a higher porosity, which causes better adsorption properties of lignite and makes it useful as adsorbent to remove heavy metal ions and organic matter.

Shuying et al. [2] used Shulan lignite as adsorbent to handle wastewater that contained Ni^{2+}, Zn^{2+}, and Pb^{2+}. The results showed that lignite can remove Ni^{2+}, Zn^{2+}, and Pb^{2+} from an aqueous solution, and the adsorption capacities of Ni^{2+}, Zn^{2+}, and Pb^{2+} were 28.18, 28.36, and 29.29 $mg \cdot g^{-1}$, respectively. Lumin et al. [3] used Longkou lignite to adsorb mercury-containing wastewater. The mercury removal rate in wastewater was 99%, and the saturated adsorption capacity of lignite was 24.34 mg·g^{-1} under room temperature conditions and a pH = 5. On the other hand, natural porous lignite also shows good adsorption properties for phenol, p-nitrophenol, reactive dyes, and nonionic surfactants. Polat. et al. [4] adsorbed phenol-containing wastewater by Soma lignite. Experiments showed that the saturated adsorption capacity of lignite was 10 mg·g^{-1}, and its adsorption capacity per unit area (1.3 mg·m^{-2}) was much higher than that of activated carbon (0.05~0.3 mg·m^{-2}). The content of pollutants in the wastewater after adsorption was lower than the emission standard.

Although lignite has certain adsorption properties, in actual application there are some problems in the adsorption process, including low mechanical strength, high impurity content, poor chemical stability, and a tendency to shrink or swell. In order to overcome the aforementioned problems, some attempts have been done to prepare modified coal, including demineralization, nitration, sulphonation, pyrolysis, oxidation, granulation, and so on. Daocheng et al. [5] found that sulfonic acid groups can be introduced into the molecular structure of lignite coal by sulfonation with concentrated sulfuric acid. At the same time, more oxygen-containing functional groups, such as hydroxyl groups and phenolic hydroxyl groups, were formed on the surface of lignite, which made the lignite change from a weak acid-type ion exchange to a strong acid-type ion exchange, and greatly improved the adsorption capacity of lignite to metal ions. Choudhury et al. [6] studied the adsorption capacities of heavy metals on two low-rank Indian coals, which was oxidized by dilute nitric acid. Nitric acid oxidation of coal incorporates oxygen and nitrogen atoms into the coal matrix. The experimental data revealed good adsorption capacities of some heavy metal ions, such as Ni, Cu, Cd, and Pb. Kus et al. [7] studied the characteristics of oxidized coal, and found that coal oxidation lead to chemisorption of oxygen at the coal surface along with the formation of acid functional groups such as hydroxyl, carbonyl, carboxyl, and others, which led to subsequent thermal decomposition and a decrease in the aliphatic and alicyclic carbons. And during advanced stages of coal oxidation, vitrinite in coal will develop very extensive microcracks and microfissures accompanied by oxidation rims. During the modification process, the pore structure and surface chemistry of the adsorbent will change greatly and cause changes in adsorption performance. Thus, it is very important to study the changes of the pore structure and surface chemistry of modified coal to evaluate their effects on the adsorption properties of the adsorbent.

In this paper, different concentrations of nitric acid were used to modify lignite to get different pore structures and surface chemistry level samples. The pore structure, surface chemistry, and adsorption properties of different modified lignite samples were studied. Through the study we can explore the mechanism of lignite modification and the relationship between coal characteristics and adsorption performance. Furthermore, this study provides a basis for further research on the mechanism of lignite modification to improve the adsorption performance of lignite.

2. Experimental Parameters

2.1. Raw Materials and Experimental Coal Sample Preparation

The experimental lignite was obtained from Xilinhaote, Inner Mongolia. The raw coal was crushed and sieved, and the particles with a 45~75 μm size range were selected as the experimental samples. Moisture (M_{ad}), ash (A_{ad}), volatiles (V_{ad}), and fixed carbon (FC_{ad}) content based on air drying were tested according to GB/T 212-2008. The proximate analysis results of raw coal are listed in Table 1.

Table 1. Proximate analysis of the experiment lignite.

Content	M_{ad}	A_{ad}	V_{ad}	FC_{ad}
Wt/%	8.72	10.98	35.65	44.65

Ten raw coal samples with a mass of 10 g were weighed and added into 100 mL of nitric acid with concentrations of 0, 0.1, 0.3, 0.5, 0.7, 1, 3, 5, 7, and 10 mol·L^{-1}, marked as N0M (the raw coal), N0.1M,

N0.3M, N0.5M, N0.7M, N1M, N3M, N5M, N7M, and N10M, respectively. Then, the mixtures were placed in a shaking machine at 25 °C for 2 h. After the modification, the coal samples were washed with deionized water to a neutral state, and they were dried in a vacuum drying oven at 70 °C for subsequent use.

2.2. Coal Sample Mineral Composition Analysis

The mineral compositions of the brown coal before and after modification were measured by an X-ray diffractometer (SmartLab, Japan Science, Tokyo, Japan). The coal sample was sufficiently dried and ground before scanning with a speed of $15° \cdot min^{-1}$, in the range of 5° to 90°.

2.3. Coal Sample Pore Structure Characterization

The specific surface area and pore structure characteristics of coal samples under different nitric acid modification conditions were characterized by a static nitrogen adsorption instrument (JW-BK122W, Jingwei Gaobo, Beijing, China). The adsorption and desorption isotherms of the samples were determined by the static capacity method using nitrogen as the adsorption medium at −195.15 °C under the relative pressure of 0.001~1. All samples were degassed at 105 °C for 4 h before the measurement. The specific surface area, pore volume, and pore size distribution of the coal samples were calculated by the BET (Brunauer-Emmett-Teller), BJH (Barret-Joyner-Halenda) and HK (Horvaih-Kawazoe) models.

2.4. Coal Sample Surface Morphology Analysis

The surface morphology characteristics of coal samples under different nitric acid modification conditions were characterized by a scanning electron microscope (JSM-7800F, JEOL, Tokyo, Japan). In the operation, firstly, the conductive adhesive was pasted on the sample holder, and the coal powders were scattered on the conductive adhesive. Next, the coal powders that did not adhere to the conductive adhesive were blown away by a blowing device, and then they were scanned by a scanning electron microscope after the gold sputtering treatment. The acceleration voltage was set to 15 KV, the resolution was set to 0.8 nm, and the magnification was 5000×.

2.5. Coal Sample Surface Functional Group Analysis

The surface functional groups of lignite under different nitric acid modification conditions were determined using a Fourier transform infrared spectroscopy analyzer (Nicolet iS10, Thermo Fisher Scientific, Waltham, MA, USA). The coal samples to be tested were ground and mixed with KBr at a ratio of 1:160, and then they were pressed into thin slices. The coal samples were fully dried in a vacuum drying oven before compression, and the test range was 4000~400 cm^{-1}.

2.6. Coal Surface Potential Test

The surface potential of lignite samples under different nitric acid modification conditions were determined using a potential analyzer (Zetasizer Nano ZS90, Malvern, UK). One gram each of coal sample was added into 50 mL deionized water, mixed well, settled for 24 h, and the supernatant was used to determine results.

2.7. Analysis of the Existence of the Types of Coal Surface Elements

The types of coal elements existing on the surface of lignite samples were analyzed by an X-ray photoelectron spectrometer (ESCALAB 250Xi, Thermo Fisher Scientific, Waltham, MA, USA). The coal samples to be tested were ground and dried in a 50 °C drying oven for 1 h at a constant temperature, and they were then cooled in a desiccator to room temperature. After cooling, an X-ray photoelectron spectroscopy survey scan and narrow zone scan were performed.

2.8. Determination of the Adsorption Capacity of Pb^{2+} by Coal Samples

In this paper, Pb^{2+} was taken as the target pollutant to explore the adsorption efficiency of modified lignite. The $Pb(NO_3)_2$ stock solution was prepared in ultrapure water at a concentration of 200 $mg \cdot L^{-1}$. Coal samples weighing 0.6 g and a 100 mL $Pb(NO_3)_2$ solution were added into a 250 mL conical flask. The conical flask was then placed in a constant temperature shaking machine set at room temperature for 3 h. The mixture was filtered by a 45 μm membrane to obtain a clear liquid. The Pb^{2+} concentration in the filtrate was measured by a plasma emission spectrometer (ICPE-9000, Shimadzu, Kyoto, Japan) to calculate the Pb^{2+} removal efficiency.

3. Results and Discussion

3.1. Effects of Different Modified Concentrations of Nitric Acid on the Mineral Composition of Lignite

X-ray diffraction analysis of lignite before and after nitric acid modification was carried out. The analysis results are shown in Figure 1.

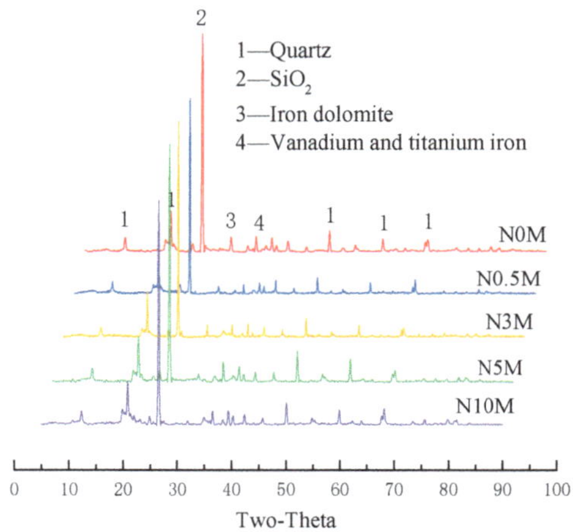

Figure 1. X-ray diffraction spectra of lignite and nitric acid-modified lignite.

Coal is mainly composed of two major components: organic carbon molecules and inorganic minerals [8]. As seen from Figure 1, the main mineral components in the experimental lignite were quartz crystals and silica. In addition, there was a small amount of iron dolomite and vanadium titanium iron minerals. After treatment with nitric acid, the peak of iron dolomite gradually weakened, while other mineral components remained unchanged. The content of iron dolomite in lignite decreased after treatment with 5 $mol \cdot L^{-1}$ nitric acid. Furthermore, under 10 $mol \cdot L^{-1}$ of nitric acid treatment, the peak of iron dolomite nearly disappeared. The chemical formula of iron dolomite is $Ca(Mg,Fe)(CO_3)_2$, a kind of alkali metal carbonate, consisting essentially of calcium carbonate and unequal amounts of iron, magnesium, and manganese. The results indicated that the alkali metal carbonate was easy to react with acid and was dissolved under acidic conditions. This left a dissolution hole in the coal body, resulting in changes in the pore structure of coal [9].

3.2. Effects of Different Modified Concentrations of Nitric Acid on the Specific Surface Area of Lignite

The change of specific surface area of lignite under different concentrations of nitric acid modification is shown in Figure 2.

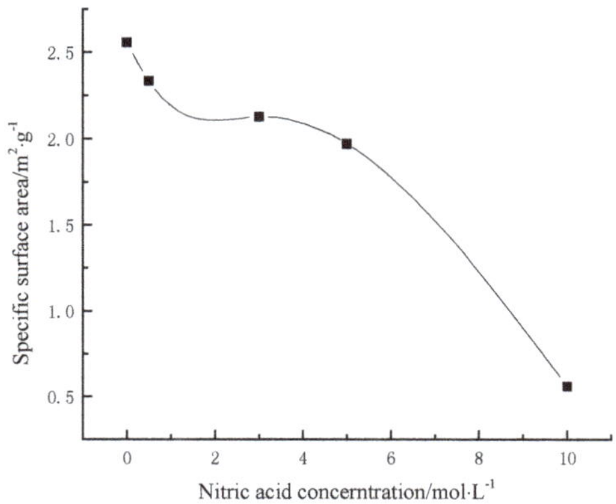

Figure 2. Specific surface area of lignite treated with different nitric acid concentrations.

It can be seen from Figure 2, in the lignite modification process, that the specific surface area gradually decreased with the increase of nitric acid concentration. And the higher the modified concentration of nitric acid was, the more the specific surface area decreased. After modification by nitric acid, some alkaline mineral impurities in the coal were dissolved to form new pores, and at the same time the original pores in the coal were destroyed. With the increase of nitric acid concentration, the dissolution reaction was more intense. At the same time, the organic matter that made up the molecular structure of coal also participated in the reaction, which caused the pores to change (explained in the discussion section). Therefore, the specific surface area of the coal samples after nitric acid treatment was lower than that of the lignite raw coal, and the higher the modified nitric acid concentration, the more the specific surface area decreased.

3.3. *Effects of Different Modified Concentrations of Nitric Acid on the Pore Shape of Lignite*

The nitrogen adsorption–desorption curves of lignite under different concentrations of nitric acid modification are shown in Figure 3.

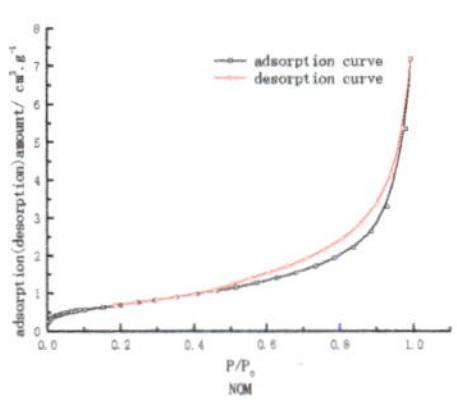

(**a**) N0M

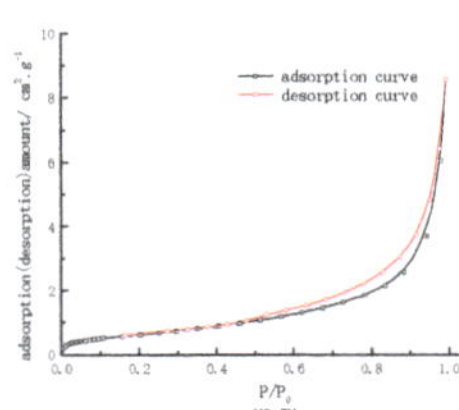

(**b**) N0.5M

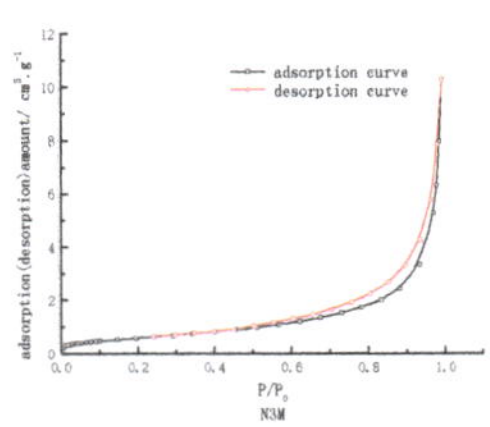

(**c**) N3M

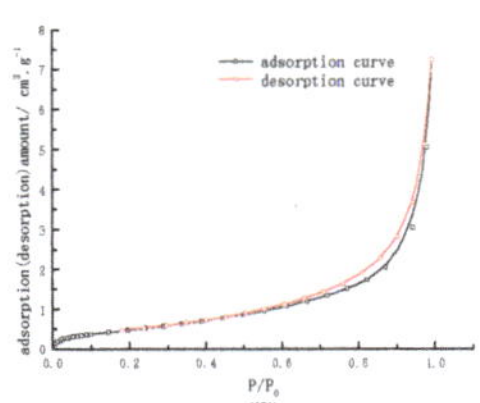

(**d**) N5M

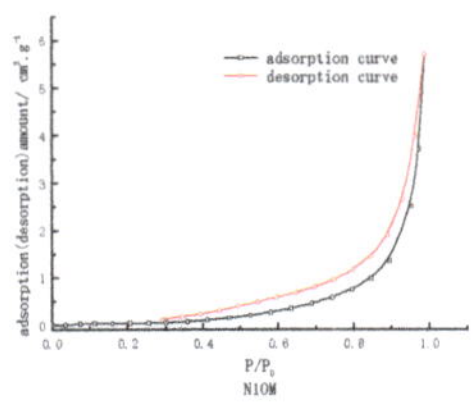

(**e**) N10M

Figure 3. Nitrogen adsorption–desorption curves of coal treated with different nitric acid concentrations.

It can be seen from Figure 3 that the shape of the adsorption–desorption curves of lignite before and after modification belonged to the type IV adsorption–desorption isotherm [10,11], meaning that when the relative pressure P/P_0 was less than 0.02, the nitrogen adsorption amount increased with P/P_0, which indicated that some micropores existed in the coal samples. When P/P_0 was greater than 0.4, the nitrogen adsorption amount in the coal samples increased rapidly with the increase of P/P_0, indicating that the coal samples contained abundant mesopores and macropores. In addition, the capillary condensation phenomenon of nitrogen in the mesopores led to the existence of hysteresis loops on the coal sample adsorption–desorption curve. The shape of the hysteresis loop reflected the pore shape in the coal samples. As seen from Figure 3, the shape of the hysteresis loop of each coal sample was uniform and conformed to the H1 type adsorption loop [12]. It can be judged from the shape of the hysteresis loop that the pore shapes in the coal samples were mostly narrow and long. At the beginning, as the modified concentration of nitric acid increased, the hysteresis loop gradually narrowed, indicating that the number of open pores decreased. This also indicated that the pore structure of lignite collapsed during nitric acid treatment, and the higher the concentration of nitric acid, the more severe the collapse of pores. When the concentration of nitric acid was 10 mol·L^{-1}, the width of the hysteresis loop became larger. At the same time, the nitrogen adsorption amount was significantly lower than that of other coal samples, indicating that the original pores in the coal sample collapsed and crosslinked. Looking at the P/P_0 range of the ring, the original pores in the coal samples mainly transformed into mesopores.

3.4. *Effects of Different Modified Concentrations of Nitric Acid on the Structural Characteristics of Lignite Pores*

The total pore volume and pore structure parameters of lignite treated with different concentrations of nitric acid are shown in Table 2. The distribution of pores in different coal samples are shown in Figure 4.

Table 2. The pore structure parameters of coal samples under different concentrations of nitric acid.

Samples	Total Pore Volume/(cm^3/g)	Average Pore Diameter/(nm)	Most Probable Pore Size/(nm)	Microporous Volume/(cm^3/g)	Microporous Most Probable Pore Size/(nm)
N0M	0.01110	17.36602	2.856	0.00089	1.07210
N0.5M	0.01321	22.62922	3.188	0.00084	1.07615
N3M	0.01588	23.89515	3.150	0.00079	1.01992
N5M	0.01219	25.85776	2.539	0.00065	1.03273
N10M	0.00882	68.80520	6.290	0.000098	1.01285

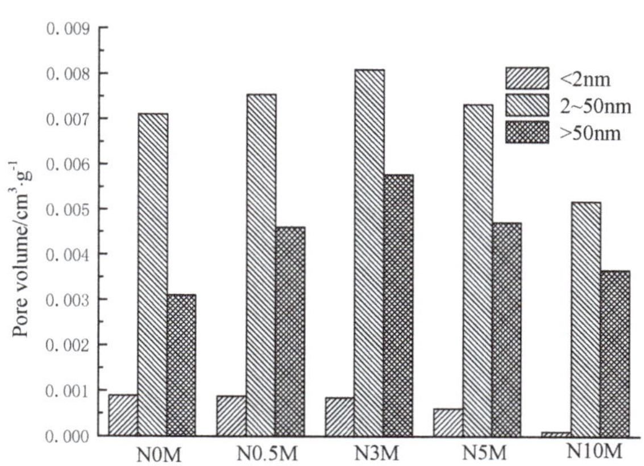

Figure 4. Pore volume of lignite treated with different concentrations of nitric acid.

It can be seen from Table 2 and Figure 4, after treatment with different concentrations of nitric acid, the total pore volume of lignite increased first and then decreased, and the micropore volume decreased with the increase of nitric acid concentration. Combined with the results of X-ray diffraction analysis, when the concentration of nitric acid was 0.5~3 mol·L^{-1}, the nitric acid reacted with the inorganic mineral impurities and the organic components in the coal, and it produced new pores, which increased the total pore volume of the coal samples. Then, when the concentration of nitric acid was higher than 3 mol·L^{-1}, the nitric acid entered the micropores and reacted violently with the mineral components in lignite. The physical and chemical structure of the coal body was destroyed, causing the pores in the coal to collapse. Thus, the pore volume of each type was reduced, and the total pore volume of the coal samples decreased.

3.5. *Effects of Different Modified Concentrations of Nitric Acid on the Surface Morphology of Lignite*

The changes in mineral composition, specific surface area, and pore structure can be more intuitively determined from the surface morphology of lignite. Scanning electron microscope images of lignite samples under different nitric acid concentrations were obtained using a scanning electron microscope, which is shown in Figure 5.

(**a**) N0M (**b**) N0.5M (**c**) N3M (**d**) N5M (**e**) N10M

Figure 5. Scanning electron microscope images of lignite treated with different concentrations of nitric acid.

It can be seen from Figure 5 that the surface of lignite raw coal was relatively flatter, and it gradually became rough after treatment with nitric acid. The higher the modified concentration of nitric acid, the more apparent this change was, which fully demonstrated the corrosive effect of nitric acid on the lignite structure.

3.6. *Effects of Different Concentrations of Nitric Acid on the Surface Functional Groups of Lignite*

The infrared scanning spectra results of lignite treated with different concentrations of nitric acid are shown in Figure 6.

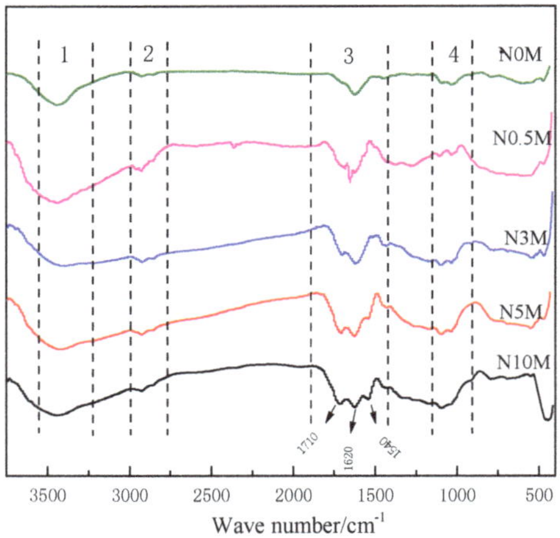

Figure 6. Fourier transform infrared spectra of lignite and nitric acid-modified lignite.

As shown in Figure 6, the peak range of the Fourier transform infrared spectra was divided into four main regions according to the peak positions of the Fourier transform infrared spectra of the coal samples before and after nitric acid treatment. The discussions on the functional group structure for each region follows.

(1) 3600~3200 cm^{-1}: The wide and strong absorption peak at this section was mainly the absorption band of phenolic hydroxyl (Ar-OH) in coal [13], and the characteristic absorption peak was around 3420 cm^{-1} [14]. It was found that the intensity of this peak in lignite raw coal was small. Compared with lignite raw coal, the width and strength of the peak were gradually enhanced with the increase of nitric acid concentration. The larger the nitric acid concentration was, the more apparent the increase was. The intensity of this peak after nitric acid modification was higher than that of raw lignite, indicating that the content of phenolic hydroxyl increased in the modified coal samples.

(2) 3000~2800 cm^{-1}: This section was mainly the stretching vibration region of aliphatic hydrocarbons. The two characteristic absorption peaks at 2920 cm^{-1} and 2850 cm^{-1} were attributed to symmetric and asymmetric stretching vibrations of C–H. In raw coal, the peak intensity of C–H was small, and the change of peak intensity under different nitric acid concentrations was not significant, indicating that the nitric acid modification had little effect on the content of C–H alkanes in coal.

(3) 1400~1800 cm^{-1}: The absorption peak in the range of 1700~1800 cm^{-1} was attributed to the C=O stretching vibration of the carboxylic acid, lactone, ester, and anthracene groups. The peak in the range of 1500~1660 cm^{-1} was the superposition zones of several functional groups, which were usually attributed to the stretching vibration peaks of C=C and C=O. In this region, lignite raw coal mainly had a strong absorption peak at 1620 cm^{-1} and a weak absorption peak at 1710 cm^{-1}. Generally speaking, the peak at 1620 cm^{-1} was caused by stretching vibration of the aromatic ring C=C and conjugated stretching of the benzene ring. The peak at 1710 cm^{-1} was the characteristic peak of the carboxyl group in coal [15]. The FTIR spectra of raw lignite and different nitric acid concentrations of modified coal showed that the peak intensity at 1710 cm^{-1} was obviously enhanced with the increase of modified nitric acid concentration, indicating that the content of carboxyl groups in the lignite structure increased during the process. It was found that a new peak at 1540 cm^{-1} appeared, and it had an increasing trend with the increase of nitric acid concentration. There should be an anti-symmetric absorption peak of -NO_2, in the forms of R-NO_2 and Ar-NO_2, indicating that a nitro group was introduced during the nitric acid modification treatment.

(4) 1200~900 cm^{-1}: The characteristic absorption peak of this region was complex. Some researchers think that there were carboxylic acid, acid anhydride, lactone, ester, ether, and stretching vibration peaks of C–O–C, C–O, and O–H in this region. It can be seen from the spectra in Figure 7

that there was no strong absorption peak, but two weak absorption peaks at 1090 cm^{-1} and 1030 cm^{-1}. This was attributed to the stretching vibration of C–O and C–O–C. Compared with the original coal, there was no new peak in this area, but the strength was slightly enhanced, indicating that the oxygen-containing functional groups increased in the nitric acid-modified lignite.

3.7. Effects of Nitric Acid Modification on the Types of Elements Existing on Coal Surfaces

3.7.1. Effects of Nitric Acid Modification on the Distribution of Chemical Functional Groups on the Lignite Surface

The survey XPS spectra of lignite and modified lignite is shown in Figure 7a,b.

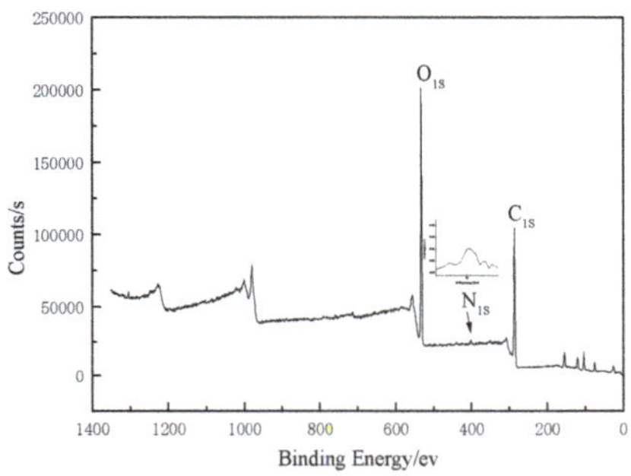

(**a**) N0M

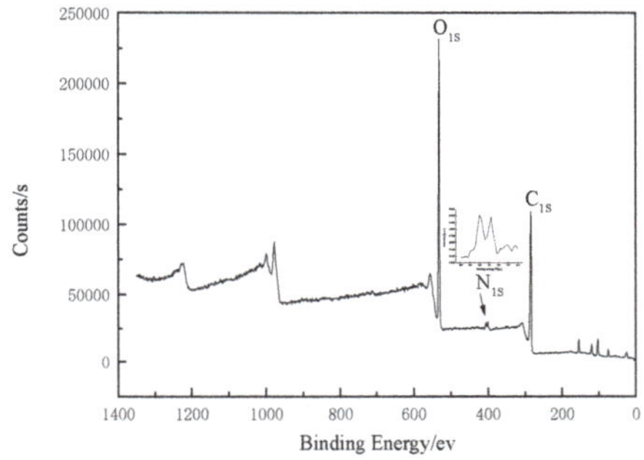

(**b**) N5M

Figure 7. Survey X-ray photoelectron spectra of raw lignite and the modified lignite sample.

It can be seen from Figure 7a,b that the lignite raw coal and the modified lignite had strong peaks of C and O elements, and gave a weaker peak of N element, indicating that the major elements in the brown coal before and after the modification were C and O. The XPS peak parameters of C, O, and N are shown in Table 3.

Table 3. The parameters of the C, N, and O XPS peaks for the raw lignite and modified lignite samples.

Samples / Parameter	N0M			N5M		
	C1S	N1S	O1S	C1S	N1S	O1S
Peak Height	11,670.18	298.92	22,296.43	11,796.75	641.8	25,943.64
Peak Area	29,752.88	1018.95	56,642.3	35,778.77	3648.17	65,990.71
Atomic %	58.68	1.57	39.75	55.24	3.59	41.17

It can be seen from Table 3 that the surface of raw lignite and modified lignite mainly contained C and O elements, and the content of N was small, indicating that the vast majority of the chemical functional groups of the raw lignite and modified lignite were oxygen-containing functional groups.

Comparing the nitric acid-modified lignite and the raw lignite samples, it can be seen that the C element peak area of the lignite surface increased by 20.25%, the oxygen element peak area increased by 16.5%, and the N element peak area increased by 258% after nitric acid modification. The data showed that the nitric acid acidification process had an effect on the surface element composition of lignite. This was mainly to increase the content of N and O elements on the surface of lignite.

3.7.2. Distribution of Carbon Chemical Functional Groups of Lignite and Modified Lignite Surfaces

The types of nitrogen and oxygen existing in coal are closely related to carbon atoms. When C atoms form chemical bonds with O or N, a large amount of surface chemical functional groups are created on the surface of the coal, and this leads to a change in binding energy.

In order to understand the change in the forms of C element existing on the surface of lignite before and after modification, XPS was carried out in a narrow range, and the peak of C1S was fitted by peakfit software. The peak-fitting results are shown in Figure 8a,b, and the peak parameters and attribution are shown in Table 4.

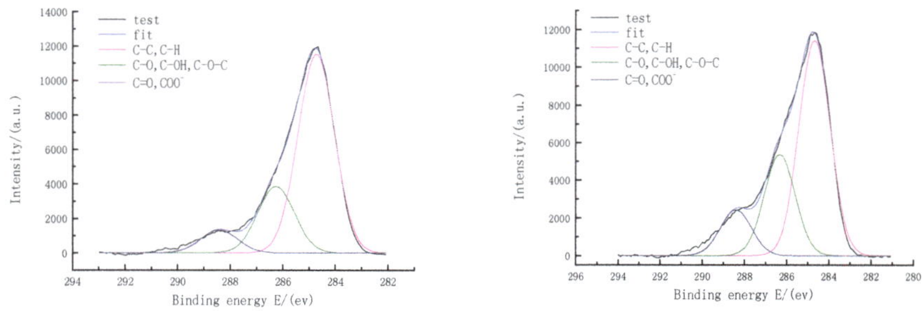

(**a**) N0M (**b**) N5M

Figure 8. XPS C1S spectra for lignite and nitric acid-modified lignite.

Table 4. XPS C1S synthesized data of lignite and nitric acid-modified lignite.

Peak	Position/eV	Area	Relative Content/%	Assignment
N0M 1	284.74	20,080.00	68.98	C-C, C-H
N0M 2	286.26	6701.38	23.02	C-O, C-OH, C-O-C
N0M 3	288.42	2327.59	8.00	C=O, COO^-
N5M 1	284.68	21,080.00	59.41	C-C, C-H
N5M 2	286.34	9909.68	27.93	C-O, C-OH, C-O-C
N5M 3	288.41	4490.42	12.66	C=O, COO^-

There are four types of carbon in coal: aromatic hydrocarbon or aromatic-substituted alkane, phenol carbon or ether carbon, the carbonyl group carbon, and the carboxyl group carbon. It can be seen from Figure 8 and Table 4 that the C-bonding energy of the raw lignite and modified lignite samples was mainly in the range of 282–290 eV. Both the lignite and the modified lignite were divided into three peaks under the best fitting state at almost the same position, which can be considered as the same species, and the peaks were aromatic hydrocarbon or aromatic-substituted alkane at 284.7 eV, the carbon–oxygen single bond at 286.3 eV, and the carbon–oxygen double bond at 288.4 eV [16,17].

Compared with the raw lignite, the peak area of aromatic carbon of the modified brown coal surface reduced by 4.98%, the carbon–oxygen single bond increased by 47.87%, and the carbon–oxygen double bond increased by 10.26%. This indicated that the nitric acid modification caused a decrease in the aromatic hydrocarbon or aromatic-substituted alkane on the lignite surface, and an increase in the oxygen-containing functional groups, such as the ether carbon or the phenol carbon, the carbonyl group, or the carboxyl group.

3.7.3. Distribution of Oxygen Chemical Functional Groups of Lignite and Modified Lignite Surfaces

The oxygen element in the coal mainly exists in the form of organic oxygen-containing groups [18]. The XPS scan of the lignite coal samples before and after nitric acid modification and the peak-fitting of the O1S spectra were carried out. The peak-fitting results are shown in Figure 9a,b. The peak parameters and attribution are shown in Table 5.

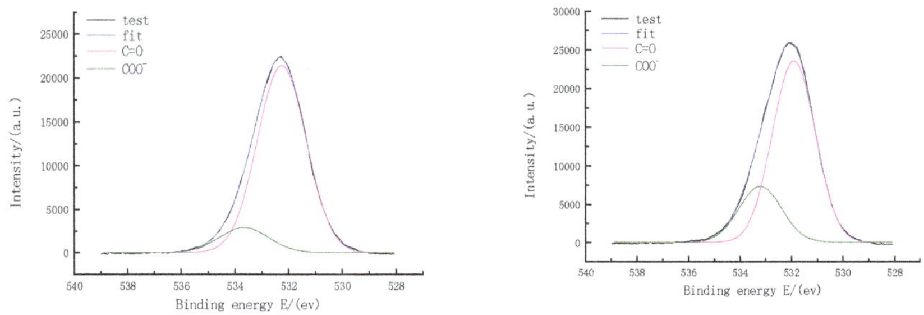

(**a**) N0M (**b**) N5M

Figure 9. XPS O1S spectra for lignite and nitric acid-modified lignite.

Table 5. XPS O1S synthesized data of lignite and nitric acid-modified lignite.

Peak	Position/eV	Area	Relative Content/%	Assignment
N0M 1	532.25	49,140.00	88.06	C=O
N0M 2	533.67	6661.53	11.94	COO^-
N5M 1	531.90	49,270.00	76.29	C=O
N5M 2	533.22	15,310.00	23.71	COO^-

It can be seen from Figure 9a,b that the peak of the O element ranged from 529 to 536 eV. The types of oxygen in the oxygen-containing functional groups on the coal surface can be mainly divided into three types: C=O bond in chemical functional groups of ketones, lactones, and carbonyls; carbonyl oxygen in the chemical functional groups of esters, amides, and acid anhydrides; and COO^- in the chemical functional groups of hydroxyl, ether oxygen, and carboxyl [19,20].

The XPS O1S spectra of lignite raw coal and nitric acid-modified lignite can be divided into two prominent peaks: C=O at 532 eV and COO^- at 533.5 eV. The change in peak area showed that the two oxygen types changed on the surface of lignite before and after nitric acid treatment, in which C=O increased by 0.26%, and COO^- increased by 129.82%. This indicated that nitric acid modification mainly changed the COO^- content of the lignite surface.

3.7.4. Distribution of Nitrogen Chemical Functional Groups of Lignite and Modified Lignite Surfaces

The XPS N1S spectra scan of the lignite before and after modification and the peak-fitting of the N1S spectra were carried out. The peak-fitting results are shown in Figure 10a,b. The peak parameters and attribution are shown in Table 6.

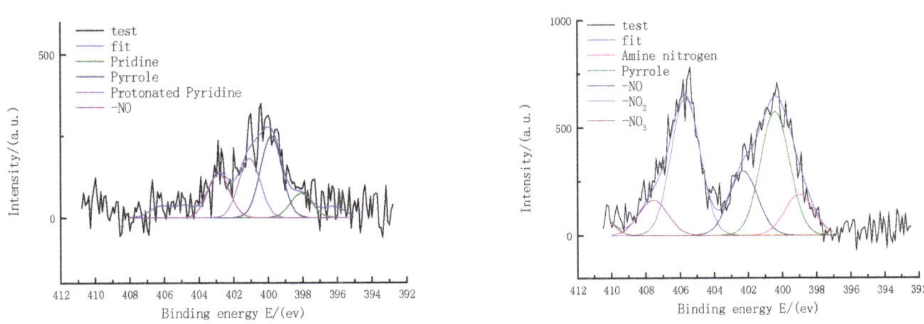

(**a**) N0M (**b**) N5M

Figure 10. XPS N1S spectra for lignite and nitric acid-modified lignite.

It can be seen from Figure 10a,b that the N1S peak position of the raw coal was between 395 and 405 eV, and the N1S spectrum had a single peak. The significant difference from the raw coal N1S spectrum was that the lignite N1S spectrum after nitric acid modification had two peaks at 400 eV and 406 eV, indicating that the nitrogen-containing functional groups on the surface of lignite had large changes before and after modification. The specific changes are shown in Table 5.

Table 6. XPS N1S peak synthesized data of lignite and nitric acid-modified lignite.

Peak	Position/eV	Area	Relative Content/%	Assignment
N0M 1	398.08	114.45	11.59	pyridine
N0M 2	399.81	385.53	39.03	pyrrole
N0M 3	401.07	278.90	28.24	protonated pyridine
N0M 4	402.84	208.84	21.14	-NO
N5M 1	399.00	404.37	10.21	amine nitrogen
N5M 2	400.41	1215.78	30.69	pyrrole
N5M 3	402.28	633.02	15.98	-NO
N5M 4	405.66	1360.00	34.33	$-NO_2$
N5M 5	407.55	348.64	8.80	NO_3^-

The nitrogen element in coal is mainly from plants in the coal-forming process; therefore, mainly in the forms of organic nitrogen. There are four main forms of nitrogen in coal: pyridine nitrogen 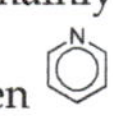 (398.0~398.1 eV), pyrrole nitrogen 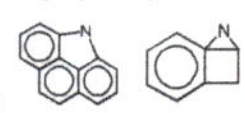 (400.0~400.4 eV), protonated pyridine  (401.6 eV), and nitrogen oxides 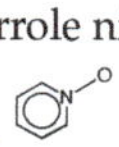 (402~405 eV) [21,22].

Four peaks were obtained by peak-fitting of lignite raw coal XPS N1S spectra, attributed to pyridine nitrogen, pyrrole nitrogen, protonated pyridine, and –NO. Therefore, the four organic nitrogen types existed in lignite raw coal. Among them, the content of pyrrole nitrogen and protonated pyridine were higher than the other two types, accounting for 39.03% and 28.24%, respectively. The contents of –NO was 21.14%, and the content of pyridine nitrogen was 11.59% of the total nitrogen content. The peak separation of N1S spectra of nitric acid-modified lignite obtained five peaks, attributed to amine nitrogen, pyrrole nitrogen, –NO, $–NO_2$, and NO_3, respectively. Among them, $–NO_2$ accounted for 34.33% of total nitrogen, pyrrole nitrogen content was 30.69%, amine nitrogen and –NO accounted for 10.21% and 15.98%, respectively, and the rest was a small amount of NO_3.

Compared with lignite raw coal, the types of nitrogen existing in nitric acid-modified lignite was quite different. The content of pyridine nitrogen and protonated pyridine on the surface of nitric acid-modified lignite was so little that it was not shown in the separation peak. The pyrrole nitrogen and –NO peak areas decreased, and new peaks of amine nitrogen, $–NO_2$ and NO_3 appeared, indicating that there was conversion of pyridine nitrogen and protonated pyridine to other types of nitrogen during the nitric acid modification process.

3.8. Effects of Different Modified Concentrations of Nitric Acid on the Surface Potential of Lignite

Zeta potential is an important aspect of adsorbent surface chemistry, which reflects the surface charge of the adsorbent. The electrostatic force plays an important role in the adsorption and purification of heavy metal ions. The negatively charged surface can provide electrostatic attraction to heavy metal ions, thereby enhancing the adsorption amount of heavy metal ions. At the same time, due to the dissociation of the chemical groups on the surface of the adsorbent, the change of the zeta potential can also reflect the change of the adsorbent surface chemical structure. The zeta potential of the lignite surface before and after nitric acid modification was determined. The measurement results are shown in Figure 11.

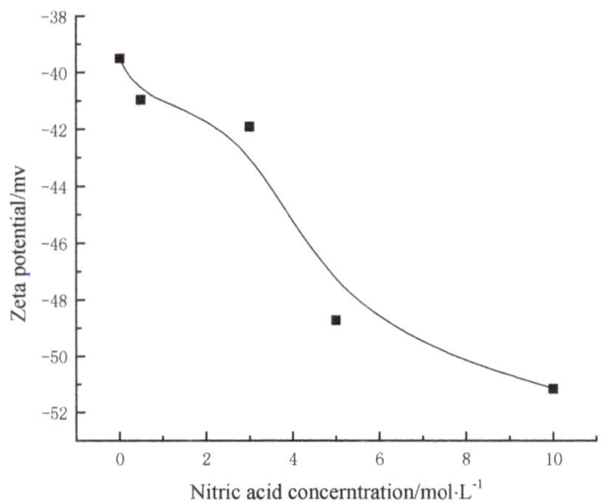

Figure 11. Zeta potential of lignite and nitric acid-modified lignite.

It can be seen from Figure 11 that raw lignite was negatively charged because of a large amount of oxygen-containing functional groups that were present, and the absolute value of zeta potential increased with the increase of nitric acid concentration. From the analysis of FTIR and XPS, it was found that the content of oxygen-containing functional groups on the surface of lignite increased after nitric acid treatment [23]. This can be confirmed from the change of zeta potential. The negative charge was enhanced, causing the electrostatic attraction between the metal ions and the surface of the lignite.

3.9. Adsorption Capacity of Pb^{2+} on Different Modified Coal Samples

To elucidate the effects of adsorbent surface chemistry and pore structure on the adsorption of Pb^{2+} from aqueous solution, a single-solute experiment was conducted with coal samples modified by different concentrations of nitric acid. To compare single-solute adsorption results in a quantitative manner, the adsorption experiment was conducted under the same conditions. The adsorption capacity changes of Pb^{2+} on different concentrations of nitric acid-modified lignite are shown in Figure 12.

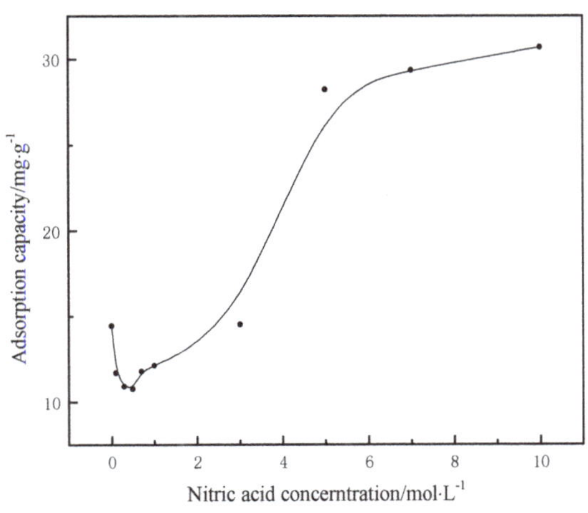

Figure 12. The change in Pb^{2+} adsorption capacities with the modified concentration of nitric acid.

It can be seen from Figure 12 that with the increase of the modified concentration of nitric acid, the adsorption amount of Pb^{2+} on lignite samples tended to decrease first and then increase to the minimum value of 0.5 mol·L^{-1}. In the previous analysis of the pore structure and surface properties of nitric acid-modified lignite, it was found that nitric acid modification increased the content of oxygen-containing functional groups on the lignite surface, while it decreased the specific surface area and pore volume of coal samples. When the modified concentration of nitric acid was less than 0.5 mol·L^{-1}, the adsorption amount of Pb^{2+} on lignite decreased with the increase in nitric acid concentration. At this time, the concentration of nitric acid was low, which mainly played an acidic role, causing a reduction in the specific surface area of lignite and a decreased adsorption amount. When the modified nitric acid concentration was above 0.5 mol·L^{-1}, the adsorption amount of Pb^{2+} on lignite increased with the increase in modified nitric acid concentration, and it increased rapidly in the range of 0.5 to 5.0 mol·L^{-1}. At this time, nitric acid showed strong oxidizing properties. The surface of lignite was rapidly oxidized to produce a large number of polar oxygen-containing functional groups, and the adsorption amount of Pb^{2+} was enhanced. When the concentration of modified nitric acid was in the range of 5.0–10 mol·L^{-1}, the surface chemistry functional group of lignite was close to oxidized saturation, and the growth rate of the Pb^{2+} adsorption amount slowed down. Before and after the modification, the adsorption amount of Pb^{2+} increased from 14.45 mg·g^{-1} to 30.68 mg·g^{-1}. It suggested that nitric acid modification can obviously improve the adsorption property of lignite, and at the same time, the effects of pore structure and surface chemistry on the adsorption performance of the adsorbent.

The adsorption of Pb^{2+} on the surface of lignite was the result of both the pore structure and the surface chemistry of the adsorbent. At the beginning, the adsorption amount of Pb^{2+} decreased with the decrease in lignite-specific surface area. Next, the adsorption amount of Pb^{2+} was not continually reduced, while it greatly increased due to the increase in oxygen-containing functional groups on the surface of lignite. It indicated that the surface chemical structure played a greater role than the pore structure in the process of lead ion adsorption. This indicated that the removal of metal ions from aqueous solution by coal primarily involved electrostatic adsorption, ion exchange, and subsequent metal complex formation processes, and those processes are associated with oxygen-containing functional groups.

4. Conclusions

To evaluate pore structure and surface chemistry effects on the adsorption of Pb^{2+} from aqueous solution, a matrix of lignite with different pore structure and surface chemistry levels was studied. From the relationship observed between the adsorbent characteristics and the contaminate adsorption data in ultrapure water, a number of conclusions can be drawn.

The adsorption of Pb^{2+} on the surface of lignite is the result of both the pore structure and the surface chemistry of the adsorbent. The adsorption capacity of lignite to Pb^{2+} increased from 14.45 mg·g^{-1} to 30.68 mg·g^{-1} after modification. From the X-ray diffraction, static nitrogen adsorption, and Scanning electron microscope characterizations of the nitric acid-modified lignite, it was found that nitric acid entered the pores of coal and reacted with the mineral components, and the pore structure of the coal body was destroyed. The pores in the coal collapsed in a high, modified concentration. Both of the specific surface areas and pore volumes of each sample were decreased. This can partly explain the decrease in adsorption capacity of lignite in the low, modified concentration.

From the analysis of Fourier transform infrared spectroscopy, X-ray photoelectron spectroscopy, and the zeta potential test, it was confirmed that nitric acid modification of coal incorporated oxygen and nitrogen atoms in coal matrix. The surface of lignite was rapidly oxidized to produce a large number of polar oxygen-containing functional groups, such as hydroxyl groups and carboxyl groups, which enhanced the surface-level, negative electrical properties of lignite. The nitrification reaction between nitric acid and the lignite organic components introduced nitro groups on the surface of lignite, which also enhanced the surface polarity of lignite and its adsorption properties to metal ions.

The physical parameters of lignite, such as specific surface areas, pore volume, and pore size, do not characterize the lignite adsorption properties well. The adsorption model of Pb^{2+}, the types of functional groups on the surface of adsorbent, and the quantity plays key roles in the adsorption of Pb^{2+} by lignite, which are determined between the adsorption capacity of Pb^{2+} and the number of oxygen-containing functional groups of lignite in the corresponding relationship.

Author Contributions: Data curation, G.L. and P.W.; Formal analysis, G.L. and H.X.; Investigation, G.L., P.W. and X.Z.; Methodology, B.H. and H.X.; Project administration, B.H.; Resources, H.X.; Software, X.Z.; Validation, X.Z.; Writing—original draft, P.W.; Writing—review & editing, G.L.

Funding: This work was supported by the National Natural Science Foundation of China (No. 51604280) and the Fundamental Research Funds for the Central Universities (No. 2016QH04).

Conflicts of Interest: The authors declare no conflict of interest.

References

1. Ping, Z. *Production and Application of Coal Humic Acid*; Chemical Industry Press: Beijing, China, 1991; p. 28.
2. Shuying, W.; Jinghong, L.; Yi, S. Adsorption Character of ShuLan Brown Coal for Heavy Metal Ions Such As Ni^{2+}, Zn^{2+} and Pb^{2+}. *J. Northeast China Inst. Electr. Power Eng.* **2004**, *24*, 85–91.
3. Lumin, W.; Changliang, D.; Shujing, Y. Study of Longkou Lignite to Purify of Mercury-containing Waste Water. *J. Yantai Univ. (Natl. Sci. Eng.)* **1998**, *2*, 153–156.
4. Polat, H.; Mlva, M.; Polat, M. Capacity and mechnism of phenol adsorption on lignite. *Int. J. Miner. Proc.* **2006**, *79*, 264–273. [CrossRef]
5. DaoCheng, L.; JunFeng, L.; ZhongYuan, L. Study on the Treatment of Chromium (VI)-containing Wastewater Use Sulfonated Lignite. *Coal Chem. Ind.* **2010**, *38*, 36–38.
6. Choudhury, D.; Srivastava, S.K.; Prasad, R. Removal of heavy metals in wastewater by its adsorption: On nitric acid oxidized coal. *Indian J. Environ. Prot.* **2013**, *33*, 804–815.
7. Kus, J.; Misz-Kennan, M. Coal weathering and laboratory (artificial) coal oxidation. *Int. J. Coal Geol.* **2017**, *171*, 12–36. [CrossRef]
8. Ward, C.R. Mineral matter in triassic and tertiary low-rank coals from South Australia. *Int. J. Coal Geol.* **1992**, *20*, 185–208. [CrossRef]
9. Aparna, M.; Srivastava, S.K.; Udaybhanu, G. Removal of cadmium ions from waste water using raw and treated coals as adsorbents. *Int. J. Chem. Sci.* **2011**, *9*, 265–280.
10. Rouquerol, J.; Avnir, D.; Fairbridge, C.W.; Everett, D.H.; Haynes, J.M.; Pernicone, N.; Ramsay, J.D.F.; Sing, K.S.W.; Unger, K.K. Physical and biophysical chemistry division commission on colloid and surface chemistry including catalysis. *Pure Appl. Chem.* **1994**, *66*, 1739–1758. [CrossRef]
11. Sing, K.S.W.; Everett, D.H.; Haul, R.A.W.; Moscou, L.; Pierotti, R.A.; Rouquerol, J.; Siemieniewska, T. Reporting Physisorption Data for Gas/Solid Systems with Special Reference to the Determination of Surface Area and Porosity. *Pure Appl. Chem.* **1985**, *57*, 603–619. [CrossRef]
12. Chunzhu, L. *Victoria Lignite Science Progress*; Chemical Industry Press: Beijing, China, 2009; p. 26.
13. Yuliani, G.; Garnier, G.; Chaffee, A.L. Utilization of raw and dried Victorian brown coal in the adsorption of model dyes from solution. *J. Water Proc. Eng.* **2017**, *15*, 43–48. [CrossRef]
14. Hong, Z.; Hulin, L.; Zeshen, O.; Wang, D.Z. Study on Surface Modification of Different Rank Coals by Using FTIR. *J. China Univ. Min. Technol.* **2001**, *30*, 365–369.
15. Xuedong, Z.; Zibin, Z.; Chongjia, H.; Chengfang, Z. Quantitative Determination of Oxygen-containing Functional Groups in Coal by FTIR Spectroscopy. *J. Fuel Chem. Technol.* **1999**, *4*, 335–339.
16. Xuqin, D.; Zune, W.; Chunbao, S. Surface Properties of Macerals from Shenfu Coal. *J. China Univ. Min. Technol.* **2007**, *36*, 630–635.
17. Chen, S.; Tang, L.; Tao, X.; Chen, L.; Yang, Z.; Li, L. Effect of oxidation processing on the surface properties and floatability of Meizhiyou long-flame coal. *Fuel* **2017**, *210*, 177–186. [CrossRef]
18. Xuqin, D.; Zuna, W. XPS analysis of oxygen contained functional group in coal macerals. *J. Liaoning Tech. Univ. (Natl. Sci.)* **2010**, *29*, 498–501.
19. Biniak, S.; Szymański, G.; Siedlewski, J.; Świątkowski, A. The characterization of activated carbons with oxygen and nitrogen surface groups. *Carbon* **1997**, *35*, 1799–1810. [CrossRef]
20. László, K.; Tombácz, E.; Josepovits, K. Effect of activation on the surface chemistry of carbons from polymer precursors. *Carbon* **2001**, *39*, 1217–1228. [CrossRef]

21. Mei, L.; Junhe, Y.; Qifeng, Z.; Hai-zhou, C.; Hui, S. XPS study on transformation of N- and S-functional groups during pyrolysis of high sulfur New Zealand coal. *J. Fuel Chem. Technol.* **2013**, *41*, 1287–1293.
22. Kelemen, S.R.; Gorbaty, M.L.; Kwiatek, P.J. Quantification of Nitrogen Forms in Argonne Premium Coals. *Energy Fuels* **1994**, *8*, 896–906. [CrossRef]
23. Qihui, L.; Guoguang, W.; Zhiqiang, S.; Guangyou, W.; Hongfeng, X.; Yongjian, Z. Study on the influence of coalification degree and particle size on zeta potential of coal surface. *Energy Technol. Manag.* **2007**, *3*, 81–82.

processes

Article

DEM Investigation of the Influence of Minerals on Crack Patterns and Mechanical Properties of Red Mudstone

Shuai Zhang, Dongsheng Zhang *, Qiang Zhao *, Mingbo Chi, Wei Zhang and Wei Yu

State Key Laboratory of Coal Resources and Safe Mining, School of Mines, China University of Mining and Technology, Xuzhou 221116, China; zhangshuai818@cumt.edu.cn (S.Z.); chimb@cumt.edu.cn (M.C.); zhangwei@cumt.edu.cn (W.Z.); yuwei2017@cumt.edu.cn (W.Y.)

* Correspondence: dshzhang123@cumt.edu.cn (D.Z.); 143302zhao@cumt.edu.cn (Q.Z.)

Received: 27 December 2018; Accepted: 2 March 2019; Published: 15 March 2019

Abstract: Rocks are natural heterogeneous materials. It is common for a rock to have several kinds of minerals, which will have a significant effect on its mechanical behavior. The purpose of the numerical simulation study in this paper is to explore the effects of minerals on the crack patterns and mechanical properties of rocks. First, the corresponding calculation model is established by using the discrete element method (DEM), whereby the mechanical parameters of the blocks and joints in the Tyson polygon procedure are fitted with the rock properties obtained in the laboratory. Then, various combination models of different mineral sizes and ratios are established to study the effects of mineral size, position, and ratio on the fracture distribution and mechanical properties of rock samples. The results indicate that with increased circle size of the center mineral and the mineral ratio, the elastic modulus and uniaxial compression strength (UCS) of the model gradually increase. The drop degree of post-peak stress decreases, and the integrity and bearing capacity increase. It is found that there is a quartic polynomial relationship between elastic modulus and mineral circle radius, with $R^2 \geq 0.94$. The minerals located in the crack propagation path will effectively block the crack and change the propagation direction. When the mineral position is close to the model boundaries, especially the lateral boundaries, it has a significant influence on the crack initiation position, causing crack initiation to occur at the upper boundary of the mineral circle and propagate to the middle of the model. With increased mineral percentage and small-size mineral circle proportion, the depth of extension of the crack from boundary to center is reduced, the crack has wide development in the boundary area, the number of central cracks in the rock specimen decreases, and the degree of fragmentation decreases.

Keywords: minerals; mechanical properties; uniaxial compressive strength; crack distribution characteristics; discrete element method

1. Introduction

Rocks are categorized as natural inhomogeneous materials that include different minerals, particles, holes, and fissures. These individual materials and defects usually have different physical and mechanical properties [1,2], which will cause different responses under the action of internal and external forces. Underground engineering (such as coal mining, chamber excavation, oil and gas extraction, etc.) will break the original stress balance, causing a redistribution of stress in the surrounding rock and achieving a new equilibrium state, which produces a more complex stress environment by the dual action of tectonic stress and mining stress. The response of heterogeneous rock under this stress environment will have a significant impact on geotechnical and coal mining work [3], therefore, it is of great theoretical and engineering value in many rock engineering problems to study the influence of mineral composition on the crack distribution and mechanical properties of rock specimens.

Under normal circumstances, the research methods of rock mechanics characteristics include field measurements, laboratory experiments and numerical simulations. The rock properties obtained by field measurement methods are usually reliable, but field measurement is not widely used because of complex geological conditions and high cost [4–6]. At present, most mechanical tests of rock samples are carried out in the laboratory; however, due to limitations in the current experimental equipment, it is difficult to accurately determine the mineral types and their locations within the rock samples, and it is impossible to quantify mineral ratios (the ratio of mineral area to the area of model) on the rock sample scale. In addition, due to the randomness of sampling and the influence of accidental factors, it is difficult to analyze the fracture modes and mechanical properties of different combinations of mineral sizes and ratios by experimental mechanical testing. In recent years, with the rapid development of computer performance, numerical simulation methods for the influence of minerals on rock properties have been used more. The most commonly used numerical simulation methods include finite element, discrete element, and finite-discrete hybrid models [7,8].

Considering the heterogeneity of rock, the granite model containing biotite, quartz, and feldspar was established by using FLAC finite element software [9]. The conventional Mohr–Coulomb and tensile stress failure criteria were used to study the tensile failure characteristics of granite, and the results show that the weakest mineral, biotite, was found to control the initiation and failure mode of cracks [10]. Tang et al. [11] used RFPA software to establish five models with the same heterogeneity and five models with different heterogeneity, and heterogeneity was found to play an important role in the deformation and strength characterization of specimens. More homogeneous specimens have stronger linear deformation characteristics than more nonuniform specimens before peak stress. Blair et al. [12] used a nonlinear, rule-based model to study the influence of microheterogeneity on the macroscopic mechanical behavior of rock samples and found that the heterogeneity of the local stress field caused by grain shape and loading had a significant effect on the macroscopic properties of rock. In the finite software, the material points in the research area remain in the same neighboring relation, which cannot simulate the fracture and separation process of the block; the continuous method has difficulties presenting the microscopic and macroscopic crack patterns [13].

Compared with the finite element method (FEM), the discrete element method (DEM) and hybrid finite/discrete methods can simulate movement and separation of the block. In the DEM method, the rock is simulated using polygon blocks that interact with neighboring blocks through contacts. Cracks can only occur along the polygon block boundaries. The initiation, expansion, and closure of joints between blocks can simulate the formation of fissures in rocks and the response of microscopic and macroscopic cracks under stress loads [14,15]. Lan et al. [16] established the GBM-UDEC model to simulate the microheterogeneity of rock samples caused by elastic variation and contact stiffness anisotropy, and found that microheterogeneity plays an important role in the micromechanical behavior and macroscopic response control under uniaxial compressive loading. Fu [17] used PFC software to establish different mineral compositions of particles with regular and irregular shapes to study the effect of particle size, stiffness, and friction coefficient on the microscopic and macroscopic behavior of granular materials.

In the course of rock foundation mechanical experiments (compressive, tensile, and shear, as shown in Figure 1), we found that when the mineral particles are in the propagation path of the crack, the extension direction of the original crack changes [18–20]. This causes the crack to extend along the edges of the blocks representing the mineral particles to other regions; during the failure process the cracks are not allowed to pass through the mineral particles. The experimental and numerical simulation methods mentioned above did not determine the influence of mineral size, position, and ratio on the mechanical properties of rock specimens. This shows that the gray-white minerals in red mudstone will have a great influence on the crack mode and macroscopic mechanical properties of rock specimens, and there have been no studies analyzing the mechanism of mineral influence on red mudstone to fill this gap. Considering the intrinsic advantages of DEM in simulating crack propagation, the UDEC polygon method is used to establish combination models of different mineral sizes and ratios to study the influence of mineral size, position, and ratio and different sizes of mineral ratios on rock crack distribution and mechanical properties.

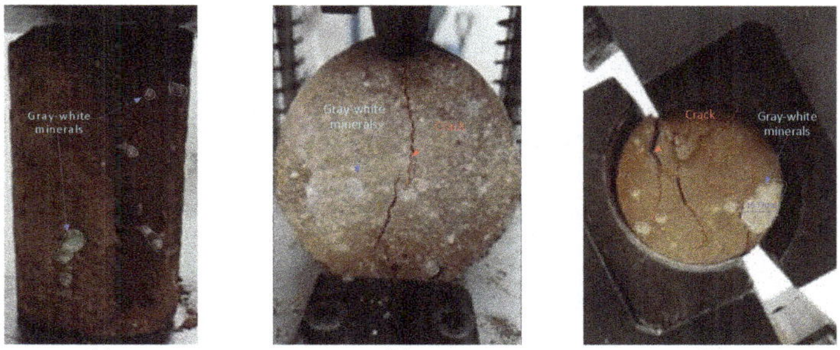

(a) (b) (c)

Figure 1. Rock foundation mechanical experiments: (**a**) compressive, (**b**) tensile, (**c**) shear.

2. Materials and Methods

2.1. Study Case

Red mudstone is continuously distributed on a large scale in the Laosangou field, located 15 km northwest of Xuejiawan town, Jungar banner, Erdos city, Inner Mongolia, as shown in Figure 2. Laosangou coal mine is located in the eastern part of the Loess Plateau of Ordos, the ground is high in the north and low in the south. The red mudstone surface is mainly red. Gray-white mineral particles can be seen in parts of the fading area, and there is a clear boundary between the fading area and the red rock. Gray-white minerals particles in the red mudstone appear grayish white and are mostly granular, lumpy and breccia, the maximum particle size is about 16.7 mm, as shown in Figure 1. During the grinding process of the rock samples, it was found that compared to the red mudstone, the gray-white minerals particles in the red mudstone are more difficult to grind into 200-purpose powdery. It indicates that the hardness of gray-white minerals is higher than red mudstone. This layer of red mudstone is of great significance for maintaining a reasonable ecological underground water level in this mine area. In order to improve the success rate of standard sample preparation, the rock samples extracted from the scene were cut and ground directly in the laboratory. The diameter of the rock samples was between 74 and 86 mm, and the height of the model was two times the corresponding diameter.

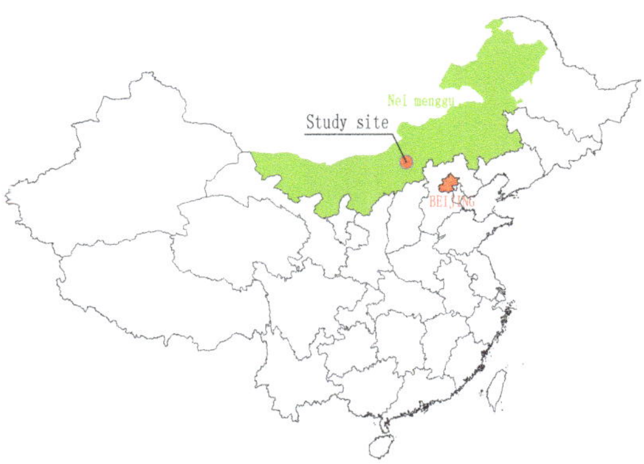

Figure 2. Geographical location of the study area.

2.2. Contact Constitutive Model

Multiple studies have shown that the 2D Voronoi model in UDEC can reliably simulate mechanical responses for both laboratory tests and field observations of rocks [21–26]. Both deformable and rigid polygon blocks can be produced in UDEC; polygon blocks cannot be destroyed, and a fracture can only be generated along the edges of a polygon. The generated polygon blocks are connected by joints, and the relationship between force and displacement at these joints is determined based on the normal joint stiffness and shear stiffness. The strength of a joint depends on its cohesion, internal friction angle, and tensile strength. Joints break when the stress between polygons is greater than the shear and tensile stress limits they can withstand. Therefore, the initiation, expansion, and closure of joints between polygons can be used to simulate the fracture generation process. The joint constitutive model is shown in Figure 3.

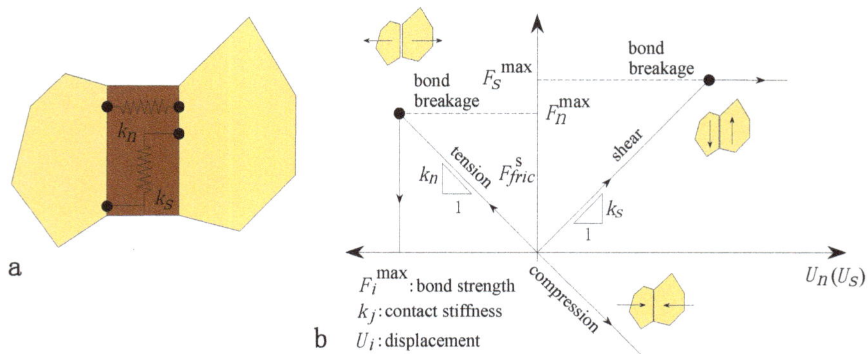

Figure 3. Joint constitutive model in UDEC. (**a**) Normal and shear stiffness between blocks. (**b**) Constitutive behavior in shear and tension (i = s, n) [7].

2.3. Model Construction and Parameter Fitting

2.3.1. Construction of Numerical Simulation Model

In this paper, considering the intrinsic advantages of UDEC in simulating the development of rock fissures, the UDEC polygon method was used to simulate the response of rock under uniaxial compression, with a model size of 160 mm height × 80 mm width. Using the Voronoi tessellation generator command, the blocks were segmented into polygons and joints by setting the number of seeds and usually had a relatively uniform size, as shown in Figure 4. The upper and lower loading plates were set as a rigid body to simulate the loading plate in the mechanical experiment. The other blocks were made deformable, the displacement control was applied to the upper loading plate, and the fixed constraints were applied to the bottom loading plate. The quasi-static loading method was used in rock mechanics experiments, and local damping (damping value is 0.8 by default) was chosen as the damping setting.

The size of the block will influence the mechanical properties and crack propagation of the model. In order to study the influence of block size on the mechanical properties of the model, nine models with an average block size of 1.5–4 mm were established to simulate the uniaxial compressive strength of the rock specimen, as shown in Figure 5. Kazerani and Yao [27,28] suggested that when the grain size is less than one-tenth of the model, the effect of block size on the mechanical properties is reduced. However, crystals that are too refined will significantly increase the model's establishment and calculation time. It takes 89 h using an ASUS computer (processor: Intel(R) Core(TM) i5-2450M CPU@2.50 GHz, RAM: 4 GB) to calculate a model with an average size of 1 mm. In order to reduce the influence of block size on the mechanical properties and improve the calculation efficiency of the model, the average size of a block was set at 2.25 mm, which is enough refinement to simulate the mechanical behavior.

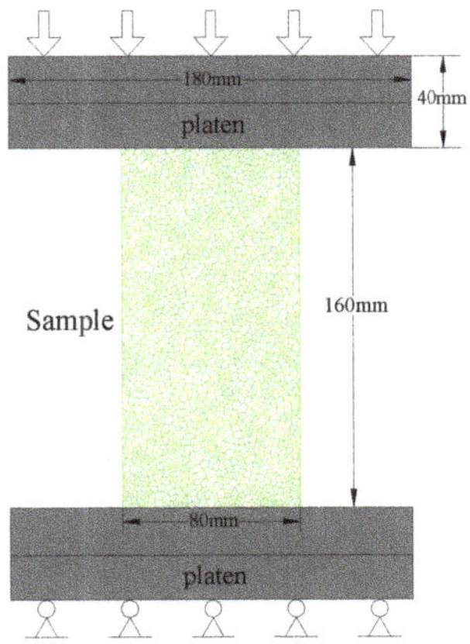

Figure 4. Numerical model of uniaxial compression test.

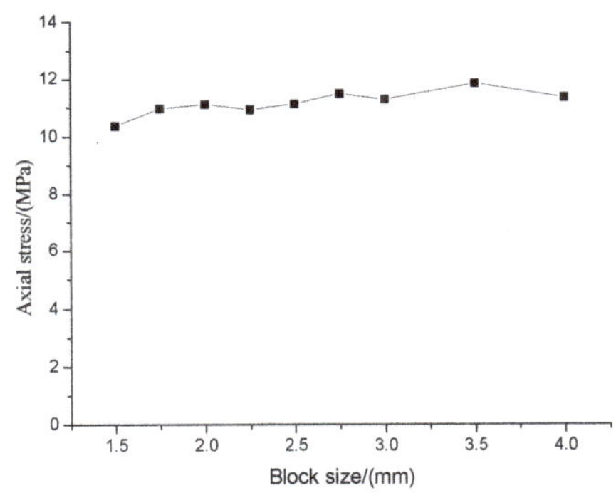

Figure 5. Effect of block size on uniaxial compression strength (UCS).

The displacement rate control applied by the upper loading plate has an important influence on the mechanical behavior of rock, and the loading speed must be slow enough to ensure that the rock specimen is in a quasi-static condition. In order to study the influence of loading rate on the mechanical properties of the model, different displacement rates (0.01, 0.02, 0.05, 0.10, 0.15.2 m/s) were exerted on the upper loading plate, and the axial stress was calculated by continuously recording the stress values in the contact area between the upper boundary and the loading plate, as shown in Figure 6. With increased loading rate, the uniaxial compressive strength and peak strain increased. The uniaxial compression strength with 0.01 m/s loading rate was as low as 9.974 MPa, and the maximum uniaxial compression strength with 0.2 m/s loading rate was 13.26 MPa. We used the 0.1 m/s loading rate, which has been validated to be low enough to ignore the effect of loading rate on mechanical properties [29].

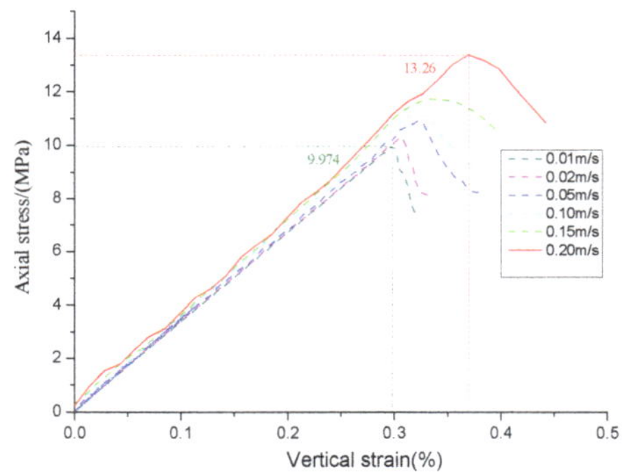

Figure 6. Stress–strain curves under different loading velocities.

Based on the findings of the previous rock foundation mechanical experiment (Figure 1), the cracks in the experiment did not penetrate the gray-white minerals in the red mudstone. In order to reproduce this phenomenon in numerical simulations, the contacts were joined in the specified range of the numerical model to simulate the mineral circles. Contacts that are joined cannot slide or open. The average size of the blocks of gray-white minerals in the model was 2.25 mm. The joint stiffness in the specified area was calculated automatically. The normal stiffness was based on an average zone stiffness multiplied by a factor, and the default value of the coefficient was 100. In the process of calculation, the maximum normal stiffness value that was calculated automatically was assigned to the joint, and the shear stiffness value was taken as half of the normal stiffness value [15]. Based on the grinding experiment of the rock samples, it was found that the hardness of gray-white minerals is higher than that of red mudstone. The bulk and shear modulus of the grains as well as normal and shear stiffness of the polygonal contacts control the sample's elastic constants, they have a relatively small effect on the sample's UCS [30]. In order to describe the difference between the mechanical properties of the gray-white minerals and red mudstone, it is assumed that the bulk modulus and shear modulus of the mineral circle were twice that of red mudstone.

2.3.2. Numerical Model Parameter Verification

In the process of numerical simulation calculation, selecting reasonable physical and mechanical parameters can effectively reproduce the mechanical behavior of rock. A previous method was used to fit data in this paper [31–33]. The mechanical parameters of the blocks and contacts were determined by fitting the uniaxial compressive strength of the rock. In order to obtain the physical and mechanical parameters of the blocks and contacts in the model to represent the mechanical behavior of red mudstone, a relatively uniform sample (compared to Figure 1a) was selected for the uniaxial compression text for parameter fitting. The blocks in the model was elastomer, and the mechanical strength was controlled by the joints. The FISH language was used to determine joint damage. Joints that produced tensile damage were the tensile damage group, and joints that produced shear damage were the shear damage group.

The physical and mechanical parameters applied to the numerical model are shown in Table 1. Figure 7 shows the numerical simulation results of the uniaxial compression test. As shown in Figure 7a. An X-shaped failure pattern occurs during the experimental uniaxial compression test. The same failure mode was obtained in the numerical simulation, as shown in Figure 7b. Figure 7c is the stress–strain curve of the model; the stress peak of uniaxial compression model was 11.39 MPa, the stress peak of the specimen was 11.44 MPa, and the error value was 0.44%. The numerical simulation results of uniaxial compression are well fitted to the experimental results. Before the stress peak was reached, the stress increased linearly with the increase in strain. The stress decreased rapidly after the peak stress. The stress value was reduced to the lowest value of 3.271 MPa and the uniaxial compression strength

(UCS) was 28.59%. The number of damaged joints was recorded during the simulation. The amount of damage was equal to the number of damaged joints divided by the total number of joints. Before reaching 51.1% of the peak stress, the joints were mainly affected by tensile damage, and when the stress value was exceeded, the joints began to produce shear damage and the quantity increased rapidly. When peak stress was reached, the percentage of shear and tensile damage was 14.47% and 17.22%, respectively. After the peak, the number of joints suffering shear damage increased and then stabilized at approximately 15.69%. The number of joints suffering tensile damage increased and then stabilized at approximately 27.1%. Nearly 42.79% of the joints were found to be cracked. The number of joints suffering tensile damage was approximately 1.73 times the number suffering shear damage.

Table 1. Mechanical parameters of sample fitting in numerical model.

Material Properties				
UCS		Young's modulus		
11.44 MPa		3.23 GPa		
Calibrated Micro-Parameters				
Bulk modulus	**Shear modulus**	**Joint normal stiffness**	**Joint shear stiffness**	**Joint cohesion**
8.54 GPa	3.636 GPa	2461 GPa	1453 GPa	5.8 MPa
Joint friction	Joint tension	Residual cohesion	Residual friction	Residual tension
10°	1.6 MPa	0 MPa	6°	0 MPa

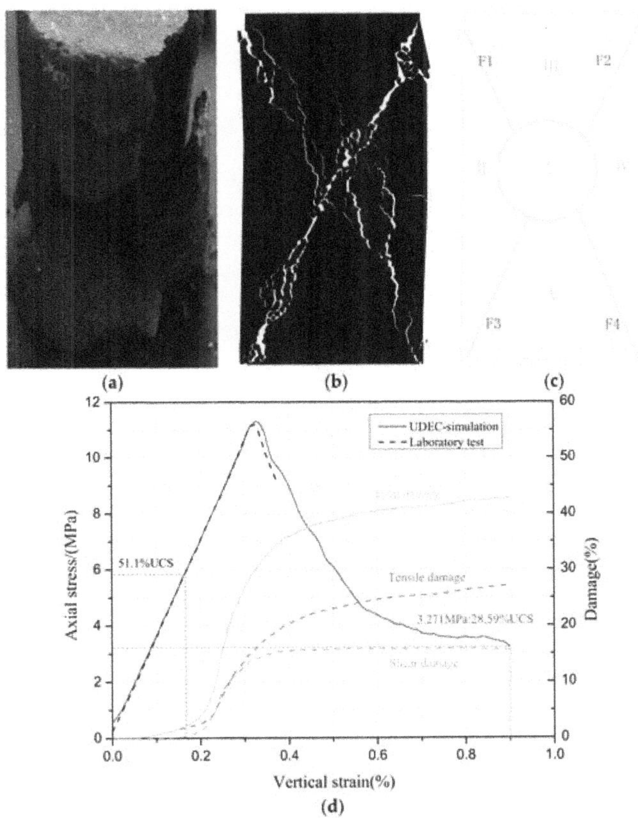

Figure 7. Experimental and numerical simulation results of uniaxial compression test. (**a**) failure pattern of the laboratory sample; (**b**) failure pattern of the numerical model; (**c**) Schematic diagram of failure pattern and (**d**) stress–strain curve of the numerical model.

The presence of minerals will change the stress distribution inside the rock specimen and cause a change in the position of stress concentration, resulting in a change in the crack initiation location and failure mode [11,18,19,34]. In order to describe the crack morphology, the broken fitting model (Figure 7b) was partitioned, as shown in Figure 7c. The model was divided into five regions: region I was the central crushing region, and regions II–V were the relatively complete zone of the model. The main macrocracks were named F1–F4 in the clockwise direction.

2.4. Determining the Numerical Simulation Scheme

Based on the Voronoi polygon method and mineral circle simulation technology, five numerical simulation schemes were determined and 42 calculation models were established. In scheme one, the mineral circle was arranged in the central coordinates (40 mm, 80 mm) of region I, and the radius of the mineral circle was 2.5–40 mm (C1–C9), as shown in Figure 8, to study the influence of different mineral sizes on rock specimen strength and failure mode. In scheme two, the mineral circle radius was 5 mm with a 0.6% mineral ratio, the upper left quarter of the rock sample was divided into three levels, and three positions were selected at each level to decorate the mineral circle (L1–L9), as shown in Figure 9. This was used to study the influence of mineral circle position on the mechanical properties of rock specimens. In scheme three, the mineral circle radius was 5 mm and four positions were selected in the rock sample, the central crushing region (I) and the relatively complete zone (II and III), and a mineral circle was arranged in the extension direction of the F1 crack. Then the sequence composition of $C_4{}^2$ was taken to analyze the influence of two mineral circle conditions on the mechanical properties of rock specimens, as shown in Figure 10. In scheme four, the mineral circle radius was 10, 5 and 2.5 mm (Figure 11a) and the mineral ratio was 2.45%, 4.9% and 9.8%. Then, according to the mineral circle radius, three computational models containing a single size mineral circle were established to explore the influence of different mineral ratios on the mechanical properties of rock samples, as shown in Figure 11. In scheme five, the mineral circle radius was 5 and 2.5 mm (Figure 12a) and the mineral ratio was 4.9%. The area ratio of the 5 and 2.5 mm mineral circles contained in the rock sample was 6:2, 4:4 and 2:6, and three numerical models were established for each ratio (Figure 12) to investigate the effects of different mineral circle proportions on the mechanical properties of rock samples. In schemes four and five, the mineral circles were randomly distributed in the model, and the FISH language built into UDEC was used to generate mineral circles and ensure that the circles were not embedded or produced at the edges of the model.

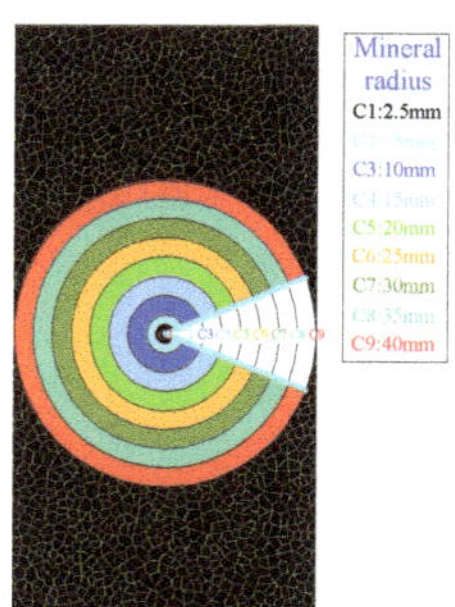

Figure 8. Mineral radii.

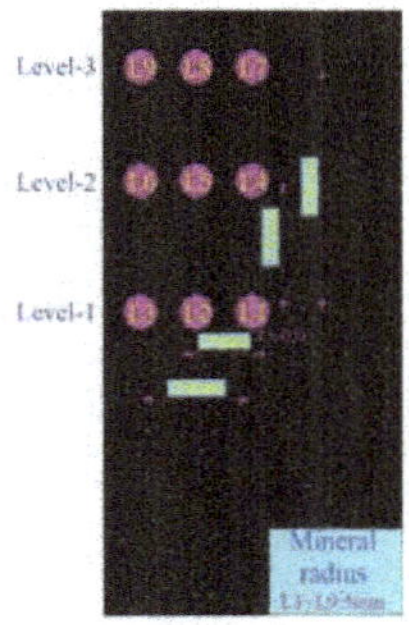

Figure 9. Mineral positions.

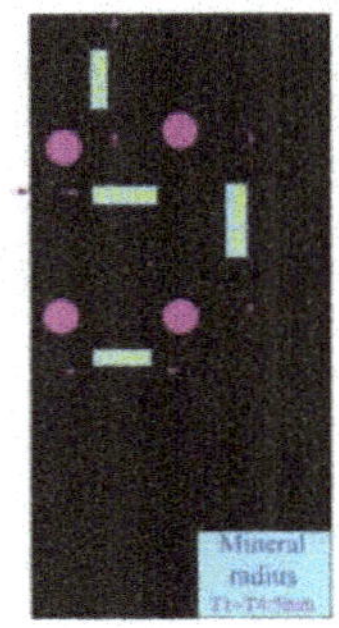

Figure 10. Two mineral circle ($C_4{}^2$) distribution diagram.

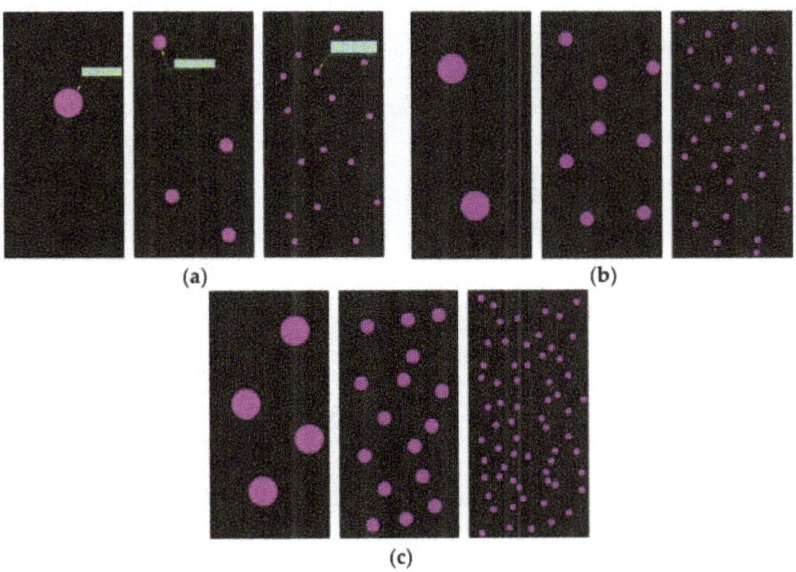

Figure 11. Mineral ratios for models under single circular dimensions 10, 5 and 2.5 mm: (**a**) S1–S3: 2.45%; (**b**) S4–S6: 4.9%; (**c**) S7–S9: 9.8%.

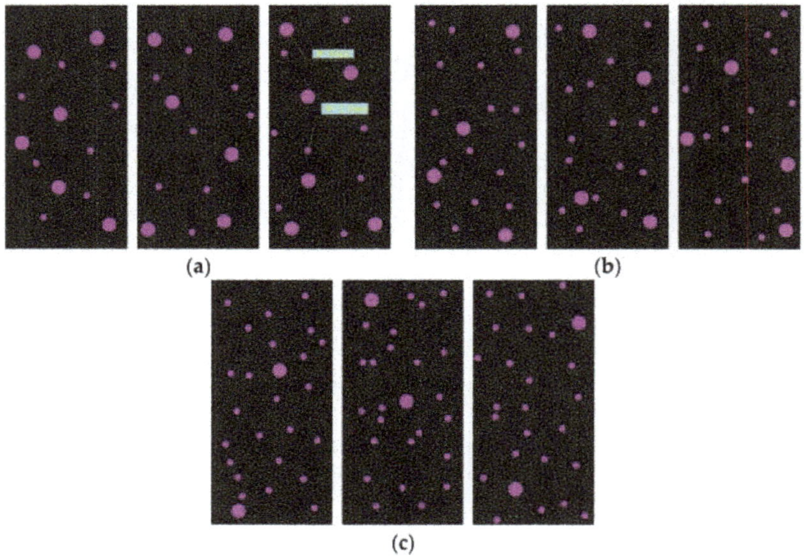

Figure 12. Proportion of 5 to 2.5 mm mineral circles under the same mineral ratio of 4.9%: (**a**) P1–P3: 6:2; (**b**) P4–P6: 4:4; (**c**) P7–P9: 2:6.

3. Numerical Simulation Results

In this section, the simulation results of the five experimental schemes are analyzed in detail. In order to study the mechanical characteristics of rock specimens in the whole stress–strain process, four characteristic values were selected as key objects: (1) elastic modulus, (2) uniaxial compressive strength, (3) maximum stress post-peak, and (4) minimum value post-peak. The crack patterns of rock specimens are studied. The UCS of the rock sample is marked on the bottom left of the model after it is broken, and the maximum value post-peak is marked on the bottom right. The maximum value is the value where the model goes through the post-peak descent stage before increasing to the maximum.

3.1. Mineral Circle Dimensions

In order to display the mechanical characteristics of the loading process of schemes C1–C9 more clearly and intuitively, Figure 13a is divided into three subfigures. Before the peak, stress increases linearly as strain increases. The elastic modulus of the model was calculated by selecting the linear segment with good stress and strain performance, as shown in Figure 13b. With increased mineral circle radius, the elasticity modulus of the model increases gradually. There is a quartic polynomial relationship between the elastic modulus and the mineral circle radius, with $R^2 \geq 0.94$. The stress value fluctuates near peak stress, because when stress is greater than 50% of the peak, the amount of tensile and shear damage of joints increases rapidly, then macroscopic cracks are formed gradually, and stress release and redistribution are caused by crack formation. The stress peak of each scheme is marked in Figure 13a1–a3. With increased mineral circle radius, the displacement becomes smaller when peak stress is reached. The stress peak shows different post-peak mechanical properties with increased mineral circle radius. The stress values of C1–C3 gradually decreased to the minimum as strain increased, and were respectively 41.95%, 47.37%, and 60.39% of their peaks. The stress of C4 and C5 decreased to the minimum value (49.95% and 54.36% of respective peaks) when the strain reached 0.57% and 0.69%, respectively, then fluctuated in a small range. The stress of C6 and C7 decreased to the minimum (63.20% and 61.97%) when the strain reached 0.56% and 0.51% of their peak values, then gradually increased to a maximum of 85.90% and 81.14% of the peak, respectively. C8 and

C9 maintained good stress levels after the peak, and the maximum value was 120.07% and 124.45%, respectively. This indicates that with increased central mineral circle size, post-peak axial stress values decreased monotonically (C1–C3), first decreased and then stabilized (C4, C5), first decreased and then increased (C6, C7), or first fluctuated slightly and then increased (C8, C9). In order to further analyze the influence of mineral circle size on the mechanical properties of the post-peak phase, two post-peak stress values were selected for analysis: the lowest stress value and the maximum value after the descent stage. The axial stress of different mineral sizes is shown in Figure 13c; the UCS of all models is greater than the UCS (11.44 MPa) of the homogeneous rock sample. As the mineral circle radius increases, the UCS as a whole tends to increase. The UCS of the 40 mm mineral circle model is the maximum, which is 1.3 times the UCS of homogeneous sample. With increased mineral circle radius, the lowest value of stress as a whole tends to increase. The minimum and maximum stress values after peak for C1–C7 are less than their respective UCS; the minimum stress value of C8 and C9 is less than their respective UCS, but the maximum stress value after peak is greater than their respective UCS.

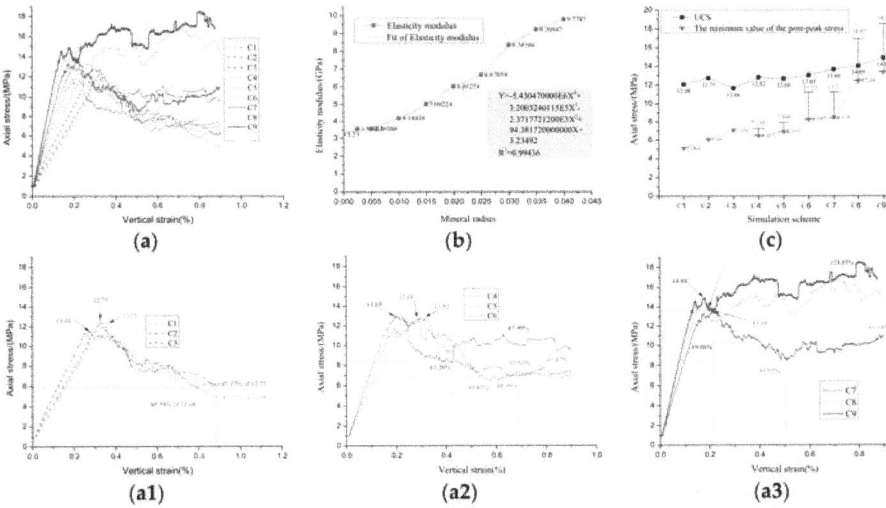

Figure 13. Mechanical characteristics of circular dimensions of different minerals: (**a**) stress–strain curve, (**b**) elastic modulus, (**c**) axial stress. a1, a2 and a3 are the stress-strain curves of C1-C3, C4-C6 and C7-C9 respectively.

The fracture distribution of specimens with a central mineral circle is shown in Figure 14. When the mineral circle radius is 2.5–10 mm (C1–C3), an X-shaped destruction form is present. When the rock specimen was destroyed, the central area gradually moved upward, and the cracks were widely distributed on the upper part of the specimen and both sides of the mineral circle. The degree of damage to the lower part decreased and the integrity improved significantly. When the mineral circle radius reached 15–20 mm (C4, C5), a V-shaped destruction form was present on the upper part of the specimen, and two cracks in the lower part first spread along the edge of the mineral circle and then extended to the lower left and lower right. When the mineral circle radius reached 25–30 mm (C6, C7), a V-shaped and inverted U-shaped destruction form was present on the upper and lower part of the specimen, respectively. The U opening increased as the mineral circle radius increased. This is because the two cracks first propagated along the edge of the mineral circle, and the two cracks at the lower part of C6 developed along the vertical direction. At 30 mm, the left lower crack was connected with the left edge and the right crack extended to the right edge. When the radius was greater than 30 mm (C8, C9), the crack was concentrated in the upper and middle part of the model and was not apparent at the bottom because the propagation path downward was blocked.

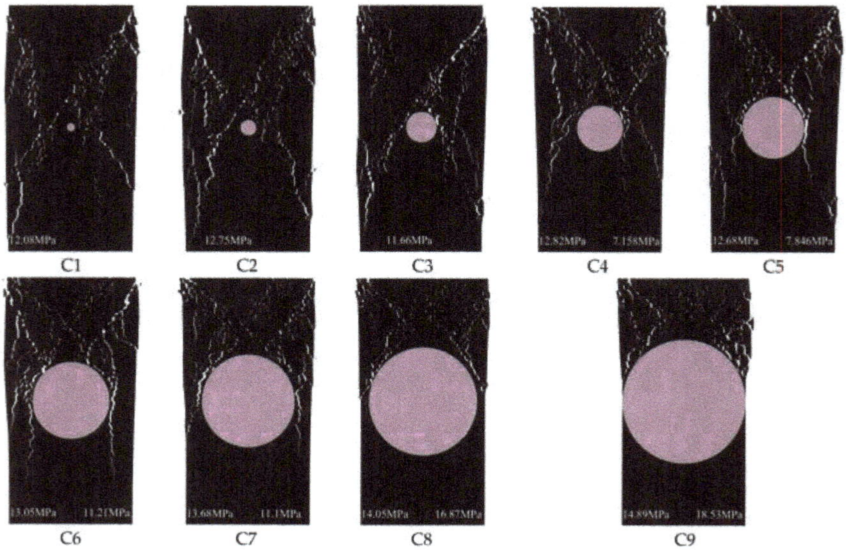

Figure 14. Crack pattern of mineral circle size after sample is broken.

3.2. One Mineral Circle in Different Positions

Before the peak, stress increases linearly as the strain increases (Figure 15a). The elastic modulus of the model was calculated by selecting the linear segment with good stress and strain performance, as shown in Figure 15b. When the stress reached the peak, the stress value of L1–l4 and L7–L9 gradually decreased to the minimum with increased strain. The stress value of L7 reduced to the minimum value, 4.579 MPa, which is 34.64% of UCS. The stress value of L5 and L6 decreased to the minimum when the strain reached 0.56% and 0.72%, respectively, which are 54.95% and 60.11% of their respective peaks. Subsequently, the stress value gradually increased to 72.05% and 71.54% of the peak, respectively. The elasticity modulus of the model is shown in Figure 15b, and the axial stress of one mineral circle at different positions is shown in Figure 15c. When the mineral circle was at level 1 (L1–L3), the elastic modulus and stress peak were lower than the average value, and minimum stress gradually decreased as the distance between the mineral circle and the center line increased. When the mineral circle was at level 2 (L4–L6), the elasticity modulus was greater than the average, the UCS of L5 and L6 was greater than the average, and the UCS of L4 was less than the average. Minimum stress increases as the distance between the mineral circle and the center line increases. When the mineral circle was at level 3 (L7–L9), the elastic modulus, UCS, and minimum stress gradually increased as the distance between the mineral circle and the center line increased.

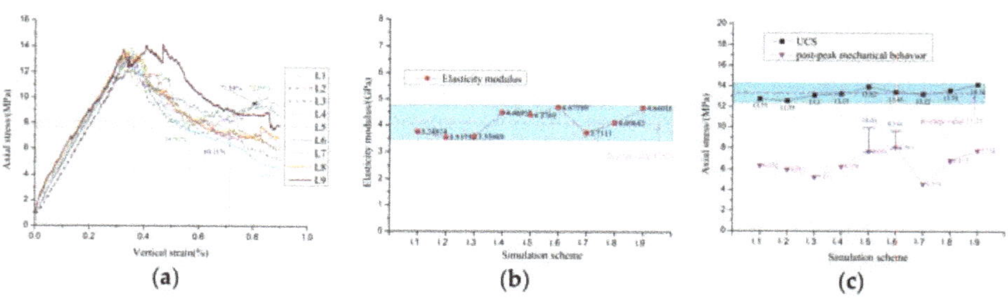

Figure 15. Mechanical characteristics of mineral circles at different positions: (**a**) stress–strain curve, (**b**) elastic modulus, (**c**) axial stress.

The fracture distribution of specimens with one mineral circle located in different positions is shown in Figure 16. When the mineral circle was at level 1 (L1–L3), the crack was widely distributed in the upper part of the model, and the degree of destruction of the upper part was significantly greater than the lower part, which indicates that the mineral circle changed the propagation direction of the crack and affected its distribution characteristics, protecting the integrity of the lower part. The order of integrity of the lower part of the model is L1 > L2 > L3. This shows that as the distance between the mineral circle and the left edge decreased, the effect of the mineral circle on decreasing the number of cracks and improving the integrity of the lower part was gradually reduced. When the mineral circle was located at level 2 (L4–L6), the upper part of L4 presented a V-shaped failure mode and the lower part presented a wormlike crack. The blocking effect of the mineral circle on crack propagation was not obvious. This is because the mineral circle was located in the upper part of region I and lower part of region III, and macroscopic cracks could not pass through the mineral circle. The mineral circles of L5 and L6 were located on the propagation path of the F1 crack, which effectively blocked the extension of the crack to region I. F1 cracks developed in the mineral circle and then extended along its edge to the lower left. It increased the overall integrity of the region below the mineral circle. This, together with region III, formed a relatively complete area in the upper left of the sample, which increased the bearing capacity of the rock specimen. This reasonably explains that in models L5 and L6, the stress value increased to 10.03 and 9.594 MPa, respectively, in the post-peak phase with the increased strain (Figure 15c). There were some microcracks in the relatively complete area of L6, indicating that as the distance between the mineral circle and left edge of the model decreased, the effect of the mineral circle on preventing the propagation of cracks gradually decreased. When the mineral circle was at level 3 (L7–L9), an X-shaped destruction form was present when the rock specimen was destroyed.

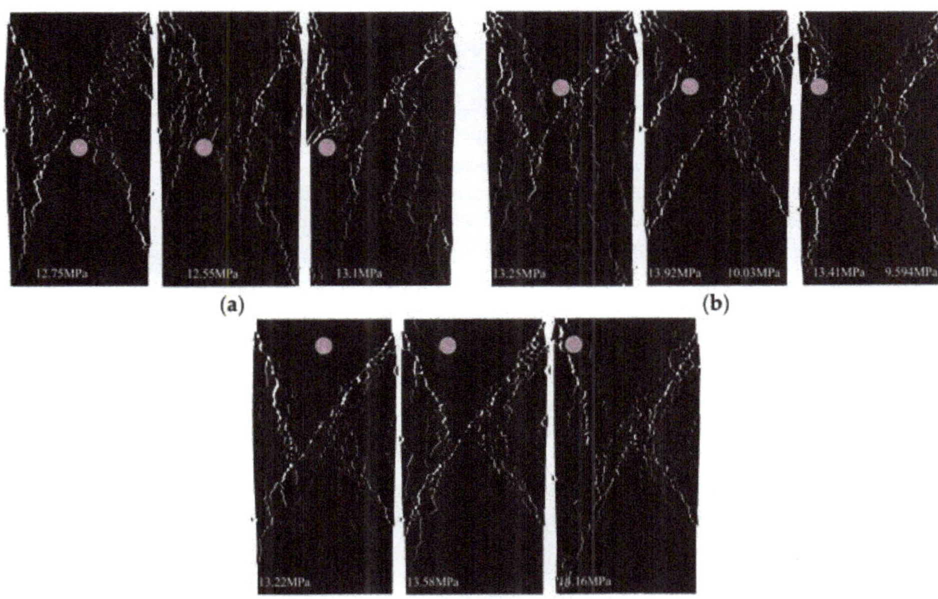

Figure 16. Crack pattern of rock specimen with one mineral circle in different positions after sample is broken: (**a**) level 1: L1–L3; (**b**) level 2: L4–L6; (**c**) level 3: L7–L9.

3.3. Mineral Circles in Two Positions

Before the peak, the stress first increases rapidly and then the growth rate slows down with increased strain (Figure 17a). The obtained elastic modulus is shown in Figure 17b. The average elastic modulus of the rapid and slow growth stages was 5.8235 and 3.8454 GPa, respectively. After the stress reaches the peak value, it decreases. The stress value of T2 reduced to the lowest value, 7.262 MPa, which was 53.28% of UCS. The stress value of T6 reduced by the least amplitude to 7.262 MPa, which was 75.24% of UCS. The stress decreases, then increases gradually. The axial stress of mineral circles in two different positions is shown in Figure 17c. The average UCS value was 13.43 MPA, and the stress peak values of T1 and T5 were less than the average, while those of other models were greater. After the peak stress, the decreased amplitude of T1, T2, and T5 was significantly greater than that of other models and the stress level was relatively low. The maximum value in the floating range after the peak was close to the minimum values of other models, which indicates that the post-peak stress bearing capacity of T1, T2, and T5 was smaller than that of other models.

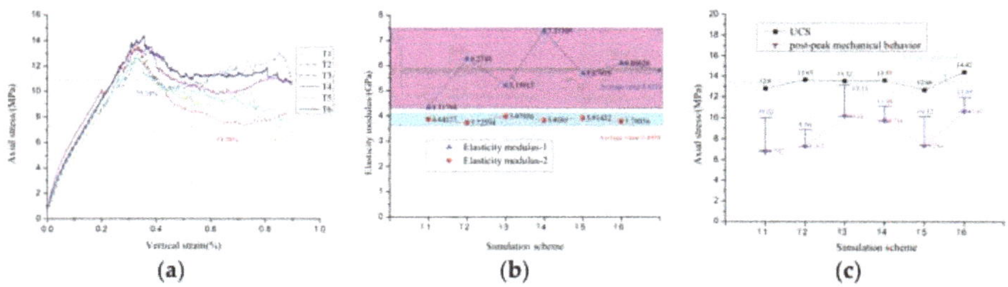

Figure 17. Mechanical characteristics of mineral circles in two positions: (**a**) stress–strain curve, (**b**) elastic modulus, (**c**) axial stress.

The fracture distribution of specimens with two mineral circles located in different positions is shown in Figure 18. The presence of a central mineral circle (T1, T2, T5) caused the central crushing area to move up, and the degree of breakage tended to increase. This reasonably explains why the degree of stress dropped after the peaks of T1 and T2, and T5 was greater than that of other models. Only when the second model was broken, an approximate X-shaped destruction form was present on the specimen. F1–F4 all cracked from the boundary of the model and successfully extended to the middle region. In other models, F2–F4 started from the boundary and extended successfully at the middle region; the propagation path of F1 was blocked. This blocking function can be roughly divided into two types: either the mineral circle is located on the propagation path of the crack, which blocks its extension direction (T3, T4, T5), or the mineral circle is near the boundary and the crack starts at the upper boundary and develops along the upper right of the mineral circle. F1 cracks start from the left boundary and extend to the central fracture zone, and the two cracks above will connect in the upper left (T1, T6). To some extent, such blocking can improve the integrity of the upper left part of the model and reduce the degree of breakage of the central region, which improves the bearing capacity of the rock sample after the peak.

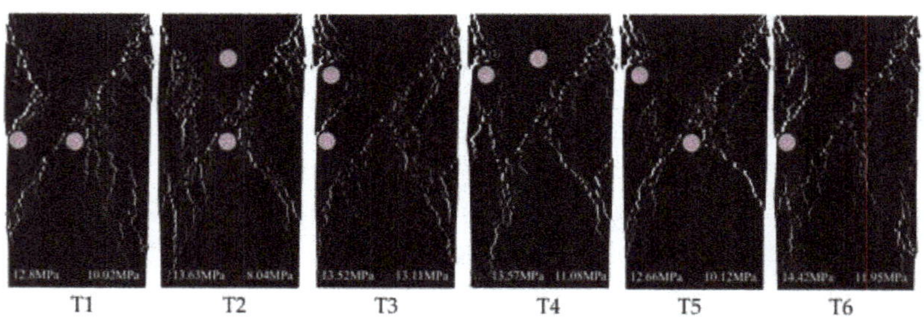

Figure 18. Crack pattern of two mineral circles located at different positions after sample is broken (T1–T6).

3.4. Mineral Ratios

Before the peak, the stress first increases rapidly and then the growth rate slows down with increased strain (Figure 19a). The obtained elastic modulus is shown in Figure 19b. The average elastic modulus of the rapid and slow growth stages was 8.8067 GPa (E1) and 4.8646 GPa (E2), respectively. The average values of elastic modulus E1 and E2 of the model with a mineral proportion of 2.5% were respectively 6.0479 and 4.0227 GPa, which are both lower than the average values of the overall elastic modulus. When the mineral accounted for 5%, elastic modulus E1 and E2 were 9.0225 and 4.8668 GPa, respectively, close to the average of overall elasticity modulus. Elastic modulus E1 and E2 of the model with a mineral ratio of 10% was 11.3498 and 5.7044 GPa, respectively, larger than the average of the overall elastic modulus. After reaching the peak, the stress value first reduced. The stress of S1 reduced to the lowest value (9.436 MPa, 69.48% of UCS). The stress value of S8 increased to a maximum (20 MPa), which is 120.99% of UCS. The axial stress of single mineral size with different mineral ratios is shown in Figure 19c. The stress peak values of S1–S3, with a mineral proportion of 2.5%, and S5, with a mineral proportion of 5%, were lower than the average value of UCS (14.6411 MPa), while the values of the other models were higher. With increased mineral proportion, the UCS of the rock sample tended to increase. The minimum value of post-peak strength was lower than the UCS of the respective models, and the maximum value was greater, except for S1.

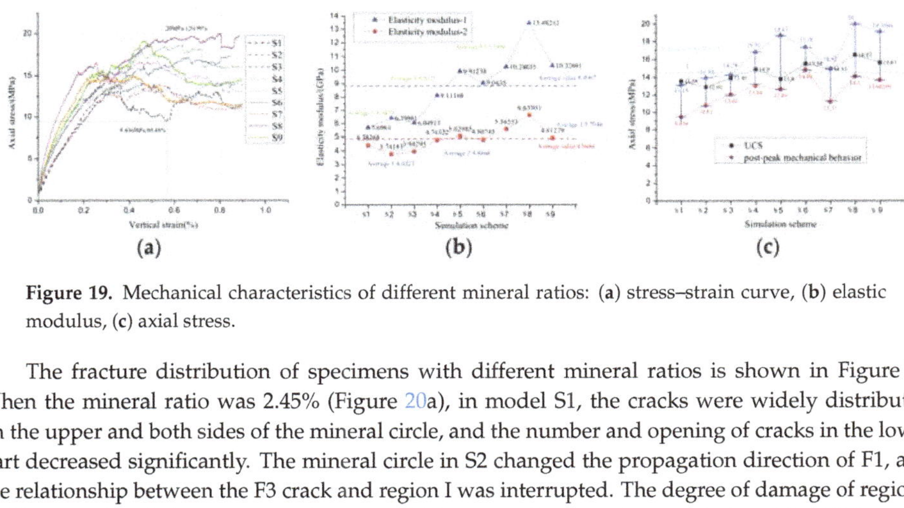

Figure 19. Mechanical characteristics of different mineral ratios: (**a**) stress–strain curve, (**b**) elastic modulus, (**c**) axial stress.

The fracture distribution of specimens with different mineral ratios is shown in Figure 20. When the mineral ratio was 2.45% (Figure 20a), in model S1, the cracks were widely distributed on the upper and both sides of the mineral circle, and the number and opening of cracks in the lower part decreased significantly. The mineral circle in S2 changed the propagation direction of F1, and the relationship between the F3 crack and region I was interrupted. The degree of damage of region I in S3 was reduced, the connection between F1 F2, and region I was interrupted, and the F3, F4 crack extended from the left and right edge to region I. The crack pattern was due to the formation of two protective structures, shown on the right side of Figure 20a, which changed the crack initiation and propagation pathways. When the mineral ratio was 4.9% (Figure 20b), in model S4, an X-shaped destruction form was present on the specimen. In S5, damage occurred in the upper left and upper right corner, and there was no obvious connection between these two failure regions and the failure of the lower part, which caused the crack initiation position of F3 and F4 to change from the lower edge to the right and left edge. This is because a protective structure, as shown in Figure 20b, formed, resulting in the crack propagating along its edge, successfully changing the propagation path and protecting the integrity of the lower region. In model S6, the crack widely developed in the upper left and upper right regions of the rock sample, and no obvious cracks were found in the middle and lower part. When the mineral ratio was 9.8% (Figure 20c), the main cracks developed on the middle and upper part of the model, which approximately presented a symmetrical hyperbolic shape. The maximum damage depth of the left and right sides of S7, S8, and S9 was 33.6 and 28.8 mm, 19.1 and 25.5 mm, and 18.3 and 21.8 mm, respectively. Comparing the results of Figure 20 horizontally (models with the same mineral proportion and different mineral radius), with decreased mineral radius, the number of cracks and the breaking range of rock samples tended to decrease. Comparing the results of Figure 20 vertically (models with the same mineral radius and different mineral proportions), with increased mineral proportion, the number of cracks and the breaking range in the rock sample tended to decrease, and the distribution area of the crack gradually shifted to the edge and upper part.

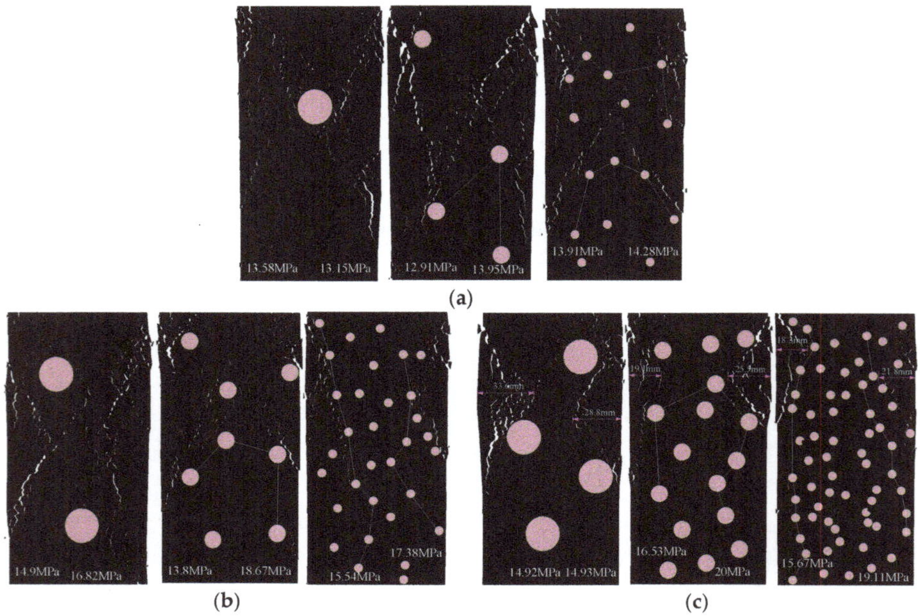

Figure 20. Crack patterns of different mineral ratios after sample is broken: (**a**) S1–S3: 2.45%; (**b**) S4–S6: 4.9%; (**c**) S7–S9: 9.8%.

3.5. Proportions of Different Size Minerals

Before the peak, the stress first increases rapidly and then the growth rate slows down with increased strain (Figure 21a). The obtained elastic modulus is shown in Figure 21b. The average elastic modulus of the rapid and slow growth stages was 9.8188 GPa (E1) and 4.48308 GPa (E2), respectively. The average elastic modulus (E1) of the models with 5 to 2.5 mm mineral circles with proportions 6:2, 4:4 and 2:6 was 9.18036, 10.8576 and 9.4185 GPa, respectively. With an increased proportion of 2.5 mm mineral, elastic modulus E1 first increased and then decreased. After reaching peak value, the stress value first decreased. The stress value of P8 reduced to the minimum, 13.45 MPa, which is 86.33% of UCS, and the value of P6 increased to the maximum, 20.49 MPa, which is 126.01% of UCS. The axial stress of different mineral circle proportions with the same mineral ratio is shown in Figure 21c. The peak stress of P1–P5 was slightly lower than the average value of UCS (15.2344 MPa), the peak stress of P6–P9 was slightly higher than the average value, the minimum value of post-peak strength was lower than the UCS of the respective models, while the maximum value of post-peak strength was greater.

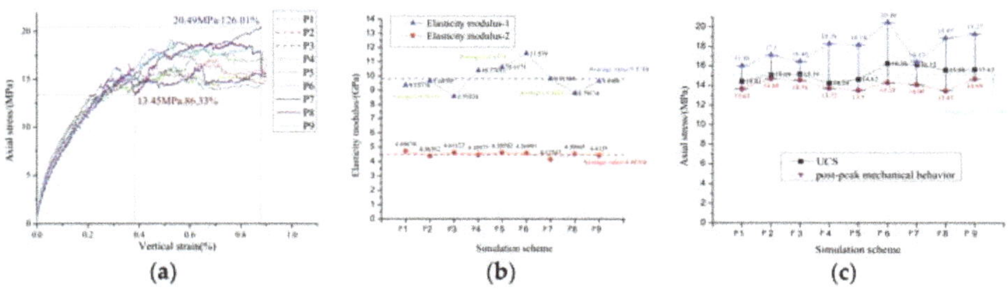

Figure 21. Mechanical characteristics of proportions of different size minerals with the same mineral ratio (4.9%): (**a**) stress–strain curve, (**b**) elastic modulus, (**c**) axial stress.

The fracture distribution of different mineral circle proportions with the same mineral ratio is shown in Figure 22. When the ratio of 5 and 2.5 mm was 6:2, in models P1 and P2, the mineral circle blocked the extension of F2 to the core region of the rock sample, and in F1, F3, and F4 cracks gathered at the center-left position. In model P3, F1–F4 converged approximately at the center of the model. After crack initiation, it mainly expanded to the center, and a crushing area was formed in the rock sample. When the ratio of 5 and 2.5 mm was 4:4, the relationship between cracks was weakened. After crack initiation, it not only expanded to the center, but also extended to both sides of the model, and the degree of brokenness in the middle was reduced. When the ratio of 5 and 2.5 mm was 2:6, after the main crack initiation, it did not successfully propagate to the midline, but was widely distributed at the left and right sides. The crushing depth of the left and right sides of P7, P8, and P9 was 26.3 and 19.7 mm, 24.5 and 30.7 mm, 20.1 and 19.3 mm, respectively. It is shown that the range of brokenness of rock samples in the model was mainly concentrated on the left and right sides, and there was a microcrack near the midline, but it did not form an obvious central crushing zone.

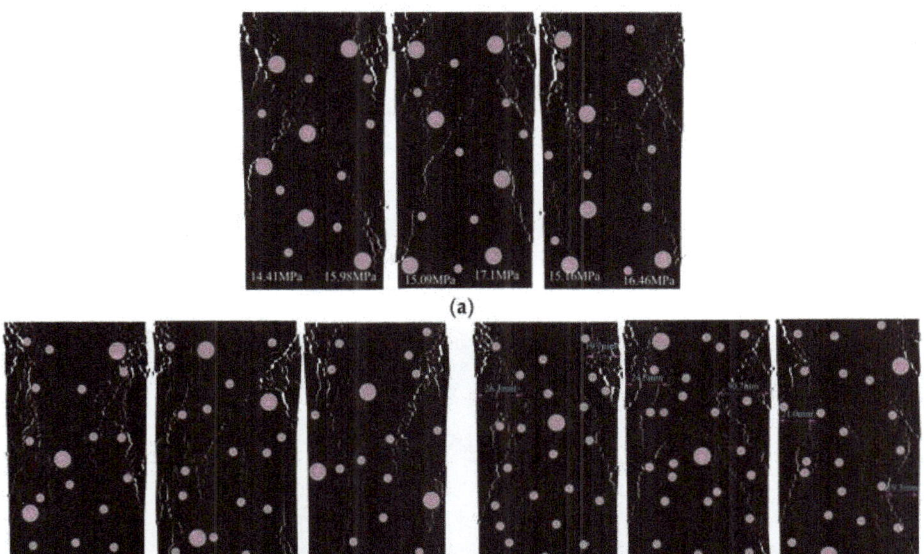

Figure 22. Crack patterns of proportions of different size minerals with the same mineral ratio (4.9%) after sample is broken: (**a**) P1–P3: 6:2; (**b**) P4–P6: 4:4; (**c**) P7–P9: 2:6.

4. Discussion

In this paper, a DEM model is established to study the crack development and mechanical properties of red mudstone containing mineral circles under uniaxial compression. The physical mechanical parameters of the calculation model are fitted with the UCS of red mudstone, which ensures that the parameters of the polygonal blocks and joints are reasonable and feasible.

Numerical simulation results show that mineral size, position, and quantity will affect crack patterns and macroscopic mechanical properties. In most cases, the crack initially extends from the boundary position to the central region of the model. When gray-white minerals particles appear in the model, the strength of the mineral is significantly greater than that of the red mudstone. The crack does not pass through the mineral; it first accumulates in the upper area, then extends along the sides to other areas, which increases the number of cracks and degree of brokenness in the upper part of the mineral circle and improves the integrity of the lower part, as shown in Figure 14. The mineral circle also influences the crack initiation position and extension direction. The intrinsic reason is that the existence of the mineral circle changes the original stress environment and the stress concentration position and leads to a different crack initiation position and fracture mode. It was found that when the mineral circle is located in the direction of crack extension, the crack is effectively blocked and the propagation direction is changed (L5 and L6). When the mineral circle is close to the boundary, cracks will start in the upper part of the mineral circle and propagate to the middle of the model (T1, T3, and T6).

In order to more intuitively show the influence of mineral circle location on the mechanical properties of rock samples, Surfer drawing software was used to plot distribution diagrams of elastic modulus, UCS, and minimum stress value post-peak with different mineral circle locations [35], as shown in Figure 23. When the mineral circle is located at the center of the model and the center of the upper and lower boundaries, especially at positions B1 and B2, the impact on the elastic modulus is minimal; the elastic modulus is most affected when the mineral circle is located at positions A1–A4. When the mineral circle is located at the center of the model, and especially at positions B1 and B2, the effect on UCS is minimal. When the mineral circle is located at the four corners of the model ((0,0), (0,0.16), (0.08,0.16) and (0.08,0)), the impact on UCS is greatest. When the mineral circle is at the center of the model's four boundaries ((0,0.08), (0.04,0.16), (0.08,0.08) (0.04,0)), it has the least influence on the minimum value of post-peak stress. When the mineral circle is located at position A1–A4, it has the greatest influence on the minimum stress.

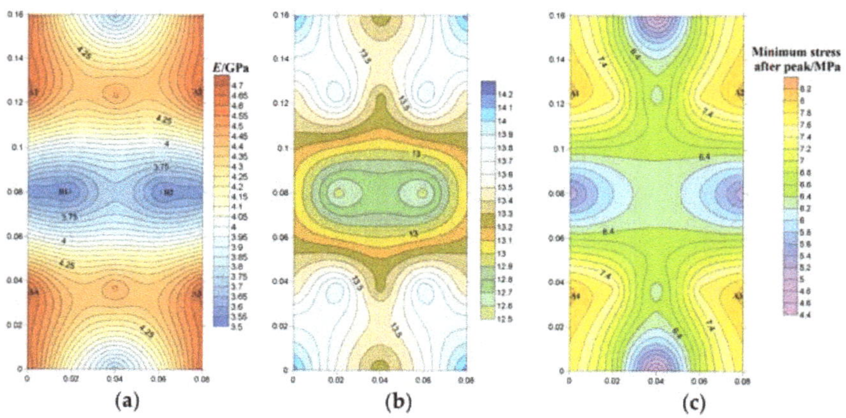

Figure 23. Mechanical properties of different mineral circular positions: (**a**) elasticity modulus, (**b**) UCS, (**c**) minimum stress post-peak.

For rock specimens with a single size mineral circle, the area of the 10 mm mineral circle is 2.45% of the area of the rock sample (160 × 80 mm) (Figure 20a). The number of 2.5, 5 and 10 mm mineral circles in the model with a mineral proportion of 9.8% and 4.9% is four and two times higher than that with a mineral proportion of 2.45%. For mineral circles of the same size, with an increased number of mineral circles, the area where the crack cannot pass increases. This is equivalent to increasing the area of the rock bridge in the sample to some extent; the rock bridge can stop the propagation of cracks, and this helps to increase the rock loading capacity. For mineral circles of different sizes, taking the particle composite structure of Figure 24 as an example, the radius of the mineral circles in Figure 24a,b is A and A/2, respectively. In the two-dimensional environment, the area of four mineral circles with radius A/2 is equal to the area of a mineral circle with radius A. However, compared with Figure 24a, the area where the crack cannot pass in Figure 24b was increased by $(1-\pi/4)A^2$, which is equivalent to increasing the area of the rock bridge in the sample. As the proportion of minerals or small size minerals with the same proportion of minerals increases, it is more likely to form a protective structure in the rock specimen (Figure 20), which can protect the lower and inner regions in the model. The effect is not to prevent rock samples from cracking in the protected area, but to reduce to a certain extent the cracking degree. It is reasonable to compare the results of Figure 20 vertically (the left of Figure 20a–c, the model (S1, S4, S7) with same mineral radius and different mineral proportion). With increased mineral proportion, the number of cracks and the breaking range in the rock sample tended to decrease, and the distribution area of the crack gradually shifted to the edge and the upper part of the model. The crack extension depth from the boundary to the center decreased, and the crack developed widely in the boundary region. The broken range of rock sample and the number of cracks in the center position tended to decrease. Comparing the results of Figure 20 horizontally (Figure 20a, the model (S1, S2, S3) with the same mineral proportion and different mineral radius), with decreased mineral radius, the number of cracks and the breaking range of rock samples tended to decrease. The phenomenon also occurred in the model containing two types of mineral circles. With a mineral proportion of 4.9% and a decreased proportion of 5 and 2.5 mm mineral circles, the crack started and developed widely in the boundary range of the model. The number of central cracks and the degree of brokenness in rock specimen decreased.

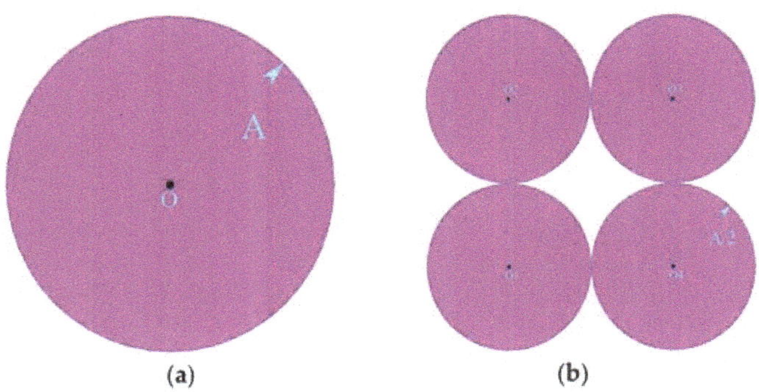

Figure 24. Mineral circles with the same area: (**a**) radius A, (**b**) radius A/2.

The area of the 5 mm mineral circle is about 0.6% of the area of the rock sample (160 × 80 mm), as shown in Figure 16. The mineral ratio of Figure 18 is 1.2%, and the mineral ratio of Figure 20a–c is 2.45%, 4.9% and 9.8%, respectively. Before the peak, for the mineral proportion of 0.6%, the stress increases linearly with increased strain, while for the mineral proportion over 1.2%, the stress first increases and then the growth rate slows down with increased strain. This is because a homogeneous sample has stronger linear deformation characteristics before peak stress than a nonhomogeneous sample [11,36]. The average elasticity modulus of different mineral ratios is shown in Figure 25, with the mineral ratio increasing, E-1 and E-2 gradually increasing, and the difference between the two also increasing. The average value of the axial stress of different mineral ratios is shown in Figure 26. When the mineral proportion is less than 2.45%, the maximum and minimum values post-peak are smaller than the UCS of the model. When the mineral proportion is greater than 2.45%, the maximum value post-peak is greater than the UCS and the minimum value is less. When the mineral ratio reaches 4.9%, with increased mineral proportion, UCS, the maximum and minimum stress post-peak increase rapidly. When the mineral ratio is more than 4.9%, the growth rate decreases significantly. The mineral circle can increase the elastic modulus and strength of the red mudstone, reduce the degree of stress drop, and increase the bearing capacity of the rock after the peak.

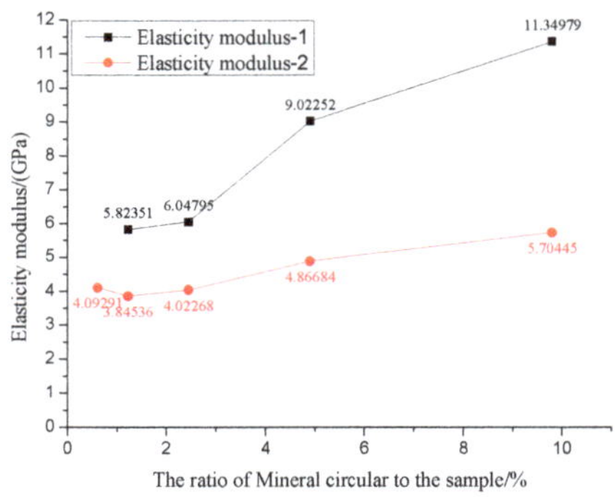

Figure 25. Mean value of elastic modulus for different mineral proportions.

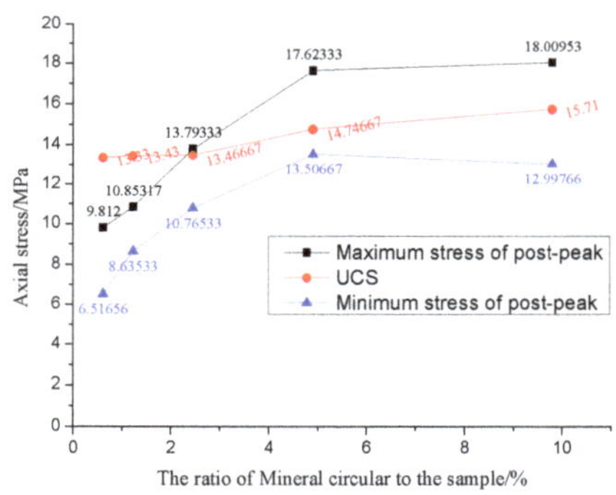

Figure 26. UCS and maximum and minimum stress post-peak for different mineral ratios.

For the mineral proportion of 4.9%, the proportion of 5 to 2.5 mm mineral circles is 6:2, 4:4, and 2:6, respectively. The mean values of elastic modulus and axial stress of proportions of different size minerals are shown in Figures 27 and 28, respectively. The mean values of elastic modulus and axial stress with a mineral proportion of 4.9% obtained earlier (Figure 19b) were used as reference values, shown as dotted lines in Figures 27 and 28. The biggest differences of E-1, E-2, maximum stress post-peak, UCS, and minimum stress post-peak are about 1.84 GPa, 0.52 GPa, 1.34 MPa, 1.03 and 0.78 MPa, which are 20.34%, 10.63%, 7.6%, 6.98%, and 5.75% of the reference values, respectively. The mean differences of E-1, E-2, maximum stress post-peak, UCS, and minimum stress post-peak are 0.80 GPa, −0.38 GPa, 0.26 MPa, 0.49 MPa, and 0.56 MPa, which are 8.83%, 7.89%, 1.46%, 3.31%, and 4.13% of the reference value, respectively. This indicates that the proportion of different size minerals influences the mechanical properties of rock specimens. However, the mean values of elastic modulus and axial stress of proportions of different size minerals fluctuate around the mean value without any obvious increase or decrease. This shows that the influence of the proportion of different size minerals on the mechanical properties is significantly less than that of the mineral proportion.

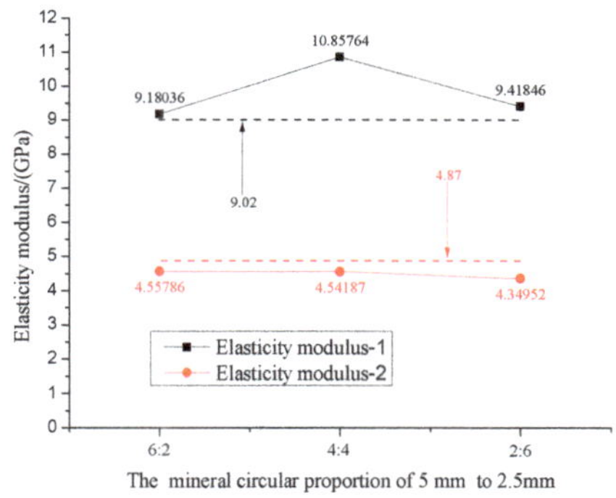

Figure 27. Mean value of elastic modulus for proportions of different size minerals.

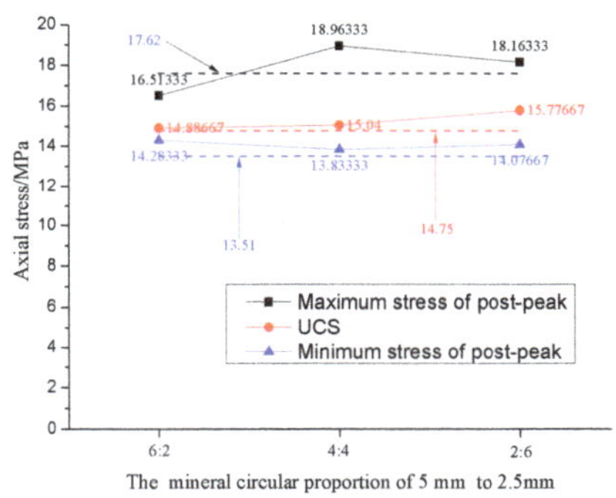

Figure 28. UCS and maximum and minimum stress post-peak for proportions of different size minerals.

The paper presents an application of a DEM particle method and this type of analysis in an excellent example how DEM can be adopted in order to further understand mechanical behavior of rock containing other kinds of minerals. The limitations of this study are as follows: (1) the findings presented in this paper do not represent the complexity of natural rocks. In most cases, minerals are embedded with each other and appear at the boundary, which will have an effect on the mechanical properties [37,38]. Due to the limitations of current equipment performance and numerical modelling, it is difficult to determine the shape, location and mineral ratio of gray-white minerals particle within red mudstone. Using the ideal circular to simulate mineral particles, there are still differences with the actual shape and size of mineral particles in the red mudstone. (2) Based on the fact that it is difficult to determine the mechanical parameters of gray-white mineral particles by method of parameter fitting. In the numerical mode, the joint stiffness in the mineral circle was calculated automatically, it is assumed that the bulk modulus and shear modulus of the mineral circle were twice that of red mudstone. With the simplification and assumption in computational models, some influencing factors and the resulting deviations were not considered. (3) This paper does not present the rock crack distribution and mechanical properties affected by sample thickness. Further work needs to be done to quantitatively evaluate the influence of specimen thickness on the UCS and fracture patterns of red mudstone containing gray-white minerals. The ideal shape of the mineral used here still provides some insight into how mineral size, position, and content influence crack initiation and the morphology and mechanical properties of the rock samples. These are useful for preparing rock samples and assessing the mechanical properties of uniaxial compression of rock samples containing minerals.

5. Conclusions

With an increased circle size of the central mineral and the mineral ratio, the elastic modulus and UCS of the model gradually increased, the drop degree of post-peak stress decreased, and the integrity and bearing capacity increased. It was found that there is a quartic polynomial relationship between the elastic modulus and mineral circle radius, with $R^2 \geq 0.94$. When the mineral ratio is more than 2.45%, the maximum post-peak stress is greater than the UCS, and the stress bearing capacity post-peak is greatly improved. When the mineral circle is in the center, its influence on the elastic modulus and UCS of the model is less; when the distance between the mineral circle and the corner decreases, the influence of the mineral circle on the elastic modulus and UCS gradually increases. When the mineral position is close to the model boundaries, especially the lateral boundaries, it has a significant influence on the crack initiation position, causing crack initiation at the upper boundary of the mineral circle and propagating to the middle of the model. When the mineral circle is located in the crack propagation path, it will effectively block the crack and change the direction of crack propagation. With increased mineral ratio, the blocking effect on the crack is enhanced, and it is more likely to form a protective structure in the rock sample, resulting in decreased extension depth of the crack from the boundary to the center. The crack is widely developed in the boundary region. With increased proportion of small size mineral circle, the crack starts to crack and develops widely in the boundary range, the number of central cracks and the degree of brokenness decreases, the integrity of the rock sample is improved, and the degree of post-peak stress is reduced. Based on the fact that it is difficult to determine the mechanical parameters of gray-white mineral particles by method of parameter fitting. The physical and mechanical parameters of mineral particles in the model are assumed and simplified, the precise quantification of block and joint parameters in the model is not realized. If it is to be extended to other models to study the influence of mineral particles on the mechanical properties of rock, further exploration is needed.

Author Contributions: S.Z. and D.Z. conceived and designed the research. S.Z. and M.C. analyzed the data and wrote the paper. Q.Z. and W.Y. verified the results and polished the article. Q.Z. and W.Z. collected the measured data. All of the authors have read and approved the final manuscript. The authors are also grateful for the helpful comments provided by the anonymous reviewers and the journal's editors.

Funding: The research was financially supported by the National Basic Research Program of China (2015CB251600), the Qing Lan Project (grant Sujiaoshi (2016)15), the National Natural Science Foundation (51874278), the Jiangsu Basic Research Program (Natural Science Foundation, BK20181357),the Fundamental Research Funds for the Central Universities of China (grant no. 2017XKQY022).

Acknowledgments: We also thank the Lao Sangou Coal Mine for their support. The authors are also grateful for the helpful comments provided by the anonymous reviewers and the journal's editors.

Conflicts of Interest: The authors declare no conflict of interest.

References

1. Martini, C.D.; Read, R.S.; Martino, J.B. Observations of brittle failure around a circular test tunnel. *Int. J. Rock Mech. Min. Sci.* **1997**, *34*, 1065–1073. [CrossRef]
2. Akesson, U. *Characterisation of Micro Cracks Caused by Coredisking*; SKB Rep P-08-103; Swed Nuclear Fuel and Waste Manage Co.: Stockholm, Sweden, 2008; 43p.
3. Li, S.J.; Li, D.; Wu, L.; Cao, L.J. Meso-simulation and fractal characteristics for uniaxial compression test of inhomogeneous rock. *J. China Coal Soc.* **2014**, *39*, 849–854.
4. Zhang, P.H.; Yang, T.H.; Yu, Q.L.; Xu, T.; Shi, W.H.; Li, S.C. Study of a Seepage Channel Formation Using the Combination of Microseismic Monitoring Technique and Numerical Method in Zhangmatun Iron Mine. *Rock Mech. Rock Eng.* **2016**, *49*, 1–10. [CrossRef]
5. Qu, Q.D.; Xu, J.L.; Wu, R.L.; Qin, W.; Hu, G.Z. Three-zone characterisation of coupled strata and gas behaviour in multi-seam mining. *Int. J. Rock Mech. Min. Sci.* **2015**, *78*, 91–98. [CrossRef]
6. Adhikary, D.P.; Guo, H. Modelling of Longwall Mining-Induced Strata Permeability Change. *Rock Mech. Rock Eng.* **2015**, *48*, 345–359. [CrossRef]

7. Lisjak, A.; Grasselli, G. A review of discrete modeling techniques for fracturing processes in discontinuous rock masses. *J. Rock Mech. Geotech. Eng.* **2014**, *6*, 301–314. [CrossRef]
8. Jing, L.; Hudson, J.A. Numerical methods in rock Mechanics. *Int. J. Rock Mech. Min. Sci.* **2002**, *39*, 409–427. [CrossRef]
9. Itasca Consulting Group, Inc. *FLAC3D User's Guide*; Itasca Consulting Group, Inc.: Minneapolis, MN, USA, 2011.
10. Chen, S.; Yue, Z.Q.; Tham, L.G. Digital image-based numerical modeling method for prediction of inhomogeneous rock failure. *Int. J. Rock Mech. Min. Sci.* **2004**, *41*, 939–957. [CrossRef]
11. Tang, C.A.; Liu, H.; Lee, P.K.K.; Tsui, Y.; Tham, L.G. Numerical studies of the influence of microstructure on rock failure in uniaxial compression—Part I: effect of heterogeneity. *Int. J. Rock Mech. Min. Sci.* **2000**, *37*, 555–569. [CrossRef]
12. Blair, S.C.; Cook, N.G.W. Analysis of compressive fracture in rock using statistical techniques Office of Scientific. *Tech. Inf. Tech. Rep.* **1994**, *35*, 849–861.
13. Bai, Q.S.; Tu, S.H.; Zhang, C. DEM investigation of the fracture mechanism of rock disc containing hole(s) and its influence on tensile strength Theoretical. *Appl. Fracture Mech.* **2016**, *86*. [CrossRef]
14. Ghazvinian, E.; Diederichs, M.S.; Quey, R. 3D random Voronoi grain-based models for simulation of brittle rock damage and fabric-guided micro-fracturing. *J. Rock Mech. Geotech. Eng.* **2014**, *6*, 506–521. [CrossRef]
15. Itasca Consulting Group Inc. *UDEC: Universal Distinct Element Code, Version 5.0*; ICG: Minneapolis, MN, USA, 2011.
16. Lan, H.; Martin, C.D.; Hu, B. Effect of heterogeneity of brittle rock on micromechanical extensile behavior during compression loading. *J. Geophys. Res. Solid Earth* **2010**, *115*. [CrossRef]
17. Fu, Y. *Experimental Quantification and DEM Simulation of Micro–Macro Behaviors of Granular Materials Using x-ray Tomography Imaging (Dissertation)*; Louisiana State University: Baron Rouge, LA, USA, 2005.
18. Eberhardt, E.; Stimpson, B.; Stead, D. The influence of mineralogy on the initiation of microfractures in Granite. In Proceedings of the 9th International Congress on Rock Mechanics, Paris, France, 25–28 August 1999; A.A. Balkema: Rotterdam, The Netherlands, 1999; pp. 1007–1010.
19. Eberhardt, E.; Stimpson, B.; Stead, D. Effects of grain size on the initiation and propagation of thresholds of stress induced brittle fractures. *Rock Mech. Rock Eng.* **1999**, *32*, 81–99. [CrossRef]
20. Yilmaz, N.G.; Karaca, Z.; Goktan, R.M.; Akal, C. Relative brittleness characterization of some selected granitic building stones: Influence of mineral grain size. *Constr. Build. Mater.* **2009**, *23*, 370–375. [CrossRef]
21. Cao, R.; Lin, H.; Cao, P. Strength and failure characteristics of brittle jointed rock-like specimens under uniaxial compression: Digital speckle technology and a particle mechanics approach. *Int. J. Min. Sci. Technol.* **2018**, *28*, 669–677. [CrossRef]
22. Damjanac, B.; Fairhurst, C. Evidence for a Long-Term Strength Threshold in Crystalline Rock. *Rock Mech. Rock Eng.* **2010**, *43*, 513–531. [CrossRef]
23. Kazerani, T.; Zhao, J. A Discrete Element Model for Predicting Shear Strength and Degradation of Rock Joint by Using Compressive and Tensile Test Data. *Rock Mech. Rock Eng.* **2012**, *45*, 695–709. [CrossRef]
24. Gao, F.Q. *Simulation of Failure Mechanisms Around Underground Coal Mine Openings Using Discrete Element Modelling (Dissertation)*; Simon Fraser University: Burnaby, BC, USA, 2013.
25. Liu, R.; Li, B.; Jiang, Y. A fractal model based on a new governing equation of fluid flow in fractures for characterizing hydraulic properties of rock fracture networks. *Comput. Geotech.* **2016**, *75*, 57–68. [CrossRef]
26. Liu, R.; Jiang, Y.; Li, B.; Wang, X. A fractal model for characterizing fluid flow in fractured rock masses based on randomly distributed rock fracture networks. *Comput. Geotech.* **2015**, *65*, 45–55. [CrossRef]
27. Kazerani, T.; Zhao, J. Micromechanical parameters in bonded particle method for modelling of brittle material failure International Journal for Numerical. *Anal. Methods Geomech.* **2010**, *34*, 1877–1895. [CrossRef]
28. Yao, C.; Jiang, Q.; Shao, J.; Zhou, C. A mesoscopic numerical model for simulation of rock fracturing. *Chin. J. Rock Mech. Eng.* **2013**, *32*. [CrossRef]
29. Gao, F.; Stead, D.; Kang, H.; Wu, Y. Discrete element modelling of deformation and damage of a roadway driven along an unstable goaf—A case study. *Int. J. Coal Geol.* **2014**, *127*, 100–110. [CrossRef]
30. Christianson, M.; Board, M.; Rigby, D. "UDEC simulation of triaxial testing of lithophysal tuff". In Proceedings of the 41st U.S. Symposium on Rock Mechanics (USRMS), Golden Rocks 2006, Golden, CO, USA, 17–21 June 2006.

31. Gao, F.Q.; Stead, D. The application of a modified Voronoi logic to brittle fracture modelling at the laboratory and field scale. *Int. J. Rock Mech. Min. Sci.* **2014**, *68*, 1–14. [CrossRef]
32. Zhang, C.; Tu, S.; Bai, Q. Evaluation of Pore Size and Distribution Impacts on Uniaxial Compressive Strength of Lithophysal Rock. *Arab. J. Sci. Eng.* **2018**, *43*, 1235–1246. [CrossRef]
33. Zhang, S.; Zhang, D.S.; Wang, Z.; Chen, M.W. Influence of Stress and Water Pressure on the Permeability of Fissured Sandstone Under Hydromechanical Coupling. *Mine Water Environ.* **2018**, *37*, 774–785. [CrossRef]
34. Kazerani, T. Effect of micromechanical parameters of microstructure on compressive; and tensile failure process of rock. *Int. J. Rock Mech. Min. Sci.* **2013**, *64*, 44–55. [CrossRef]
35. Barnes, R. *Variogram Tutorial*; Golden Software Inc.: Golden, CO, USA, 2004.
36. Dey, S.; Mohanta, M.K.; Singh, R. Mineralogy and textural impact on beneficiation of goethitic ore. *Int. J. Min. Sci. Technol.* **2017**, *27*, 445–450. [CrossRef]
37. Rigby, D.B. *Influence of Lithophysal Geometry on the Uniaxial Compression of Tuff-like Rock*; Department of Energy Technical Report TR-07-001, Task ORDFY04-013; System of Higher Education. U.S.: Las Vegas, NV, USA, 2007.
38. Lin, M.; Kicker, D.; Damjanac, B.; Board, M.; Karakouzian, M. Mechanical degradation of emplacement drifts at Yucca Mountain—A modeling case study—Part II: Lithophysal rock. *Int. J. Rock Mech. Min. Sci.* **2007**, *44*, 368–399. [CrossRef]

processes

MDPI

Article

Experimental Development of Coal-Like Material with Solid-Gas Coupling for Quantitative Simulation Tests of Coal and Gas Outburst Occurred in Soft Coal Seams

Xingkai Wang [1,2,*], Wenbing Xie [1,*], Zhili Su [2] and Qingteng Tang [2]

1 State Key Laboratory of Coal Resources and Safe Mining, China University of Mining and Technology, Xuzhou 221116, China

2 School of Mines, China University of Mining and Technology, Xuzhou 221116, China; suzhili1991@163.com (Z.S.); tangqingteng@126.com (Q.T.)

* Correspondence: tb16020019b2@cumt.edu.cn (X.W.); 1692@cumt.edu.cn (W.X.); Tel.: +86-183-5134-3101 (X.W.); +86-139-5220-6989 (W.X.)

Received: 28 January 2019; Accepted: 7 March 2019; Published: 13 March 2019

Abstract: Solid-gas coupling coal-like materials are essential for simulating coal and gas outbursts and the long-term safety study of CO_2 sequestration in coal. However, reported materials still differ substantially from natural coal in mechanical, deformation and gaseous properties; the latter two aspects are common not considered. There is a lack of a definite and quantitative preparation method of coal-like materials with high similarity for future reference. Here, 25 groups of raw material ratios were designed in the orthogonal experiment using uniaxial compression, shearing and adsorption/desorption tests. Experiment results indicated that the coal-like materials were highly similar to soft coals in properties mentioned above. And range analysis revealed the key influencing factors of each mechanical index. The gypsum/petrolatum ratio controls the density, compressive strength, elastic modulus, cohesion and deformation characteristic. The coarse/fine coal powder (1–2 and 0–0.5 mm) controls the internal friction angle and is the secondary controlling factor for compressive strength and elastic modulus. The effect of coal particle size on the sample strength was studied using scanning electron microscope (SEM). When the gypsum/petrolatum ratio increased, the deformation characteristics changed from ductile to brittle. The different failure modes in the samples were revealed. The coal powder content is a key in the gas adsorption/desorption properties and an empirical formula for estimating the adsorption capacity was established. Based on the range analysis of experimental results, a multiple linear regression model of the mechanical parameters and their key influencing factors was obtained. Finally, a composition closely resembling the natural coal was determined, which differs by only 0.47–7.41% in all parameters except porosity (11.76%). Possible improvements and extension to similar materials are discussed. The findings of this study can help for better understanding of coal and gas outburst mechanism and stability of CO_2 sequestration in soft coal seams.

Keywords: tectonically deformed coal; coal and gas outburst; coal-like material; mechanical properties; deformation feature; adsorption/desorption properties

1. Introduction

Coal and gas outburst accidents are among the most serious disasters affecting coal mining [1–4] and a problem in almost all major coal-producing countries [2–4]. Coal and gas outburst is an extremely complex gas dynamic phenomenon, in which large amounts of coal and gas are often ejected at a very fast rate from the coal rock seam to the mining space in a very short period of time (a few seconds to a few minutes). Such outbursts could destroy underground facilities, damage the ventilation system, cause a large number of casualties and even induce secondary accidents such as gas burning or explosion [2,4–7]. Today, more than half of the disasters due to coal and gas outbursts occur in China [8], causing major economic losses and casualties.

Many factors could cause coal and gas outbursts, including in-situ stress, gas pressure, geological structure, physical and mechanical properties of coal, mining methods and so forth. [6–10]. At present, there is no coherent model that can fully reveal their internal mechanism. In addition, the sudden, transient and dangerous nature makes it almost impossible to observe or study these processes on site. Therefore, simulation tests have been rapidly developed and used to explore the mechanism and process of coal and gas outbursts [5,9,11–15]. However, coal seams that are prone to coal and gas outbursts often have high gas content and low strength and so it is difficult to retrieve samples from the field for large-scale simulation tests. Hence, the development of similar solid-gas coupling coal-like materials in the lab is indispensable for studying the mechanism of coal and gas outbursts.

In addition, there is growing realization in recent years among researchers that coal seams with strong gas adsorption/desorption capacity may be able to capture and store CO_2 [16–21]. However, the key challenge here is the uncertain impact of CO_2 on the mechanical properties of coal and how this will affect the long-term safety and stability of storage. On the other hand, natural coal has extremely complex composition and physical structure, as well as strong heterogeneity–even two samples close to each other in the same coal seam could have significantly different mechanical properties. This seriously hinders the analysis of laboratory test results and the understanding of the above uncertainty. So, homogeneous and reproducible coal-like materials (also called reconstituted coal sample) will provide significant advantages for studies related to CO_2 sequestration in coal.

The key to solve the above two problems is to develop solid-gas coupling coal-like materials that closely resemble the target natural coal in their mechanical and gaseous properties and deformation features. Table 1 lists representative papers on preparing coal-like materials. There are mainly three types of methods. The first is to directly press the coal powder without any added substances [5,12,22]. For example, in 1953, a one-dimensional coal and gas outburst simulation test was completed using briquettes that were cold pressure formed [22]. Skoczylas [12] used fine coal powder to produce a series of briquettes (40 mm in diameter, 110 mm in height and with porosities of 11.2–32.0%) and the test results indicated that they had similar gaseous characteristics to natural coal. However, the coal samples prepared by only molding pressure had very low strength. The second type is pressing coal powders with water, oil or diesel but without any binder [11,20,23–27]. Jasinge et al. [20] prepared reconstituted brown coal sample by compacting 0–1 mm coal particles, in order to study the effect of coal swelling on its permeability in the laboratory. While there was a marked similarity in permeability between natural brown coal and reconstituted specimens, the strength of this type of briquette sample was still lower than natural coal and the gas adsorption/desorption characteristics were also quite different. The third type is pressing a powder mixture consisted of coal powder, cement, sand, lime and other substances [6,10,13,18,19,21,28–30]. Hu et al. [13] made coal-like materials using coal powder, cement, water, sand and activated carbon. They studied the effect of the proportion of each component on the density and mechanical properties of the samples. In 2017 and 2018, Wang et al. [10] and Zhao et al. [6] studied the gas and CO_2 adsorption characteristics of such coal-like materials. Zhang et al. [30] analyzed the influence of different material ratios on the elastic modulus. This last type of coal sample and its preparation method represent outstanding progress in the development of coal-like materials. Nevertheless, the deformation characteristics and failure modes of similar materials, which play an important role in the preparation, forming and developing stages during coal and gas outburst [13], are often not considered and rarely compared with those of natural coal. A clear discussion of related researches can be seen in Figure 1.

The abovementioned studies made important contributions to laboratory research on coal and gas outbursts and CO_2 sequestration. However, at present, there are still the following problems in the development of coal-like materials:

(1) The coal powders used in existing literature have different particle size and molding pressure, molding time and additives are also different. In future simulation test, repeated adjustments are usually needed to determine a suitable proportion of materials. There is still a lack of definite and quantitative preparation method of coal-like materials for future reference.

(2) The similarity remains low between coal-like materials and natural coal in mechanical and physicochemical properties, especially in that the former still have lower strength and very few indexes were measured. Their deformation features and adsorption and desorption characteristics are often not considered and still significantly different from those of natural coal [6,10,13,19,29]. As a result, most simulation studies only reported qualitative rather than quantitative results [13,31]. Today, the understanding of coal and gas outburst mechanism remained poor, the predictions were inaccurate and coal and gas outburst accidents still account for a large proportion of accidents in coal mines [14,32,33].

In this study, we propose a systematic and quantitative development method to determine a solid-gas coupling coal-like material that could closely resemble given natural coal in the mechanical properties, adsorption/desorption features and deformation characteristics. First, according to the physical and mechanical parameters of natural coal samples, raw materials were selected, that is, coarse and fine coal powder, gypsum, petrolatum and light calcium carbonate. Second, the orthogonal method was used to design 25 groups of schemes with different raw material ratios. Meanwhile, the sample preparation method was optimized to improve the homogeneity and reproducibility. Then, density, uniaxial compressive strength (UCS), elastic modulus (EM), cohesion, internal friction angle (IFA), deformation and failure modes, porosity and adsorption/desorption characteristics of samples were tested. Third, range analysis revealed the key influencing factors of each mechanical and physicochemical index and the levels of influence of each factor on a given index. A multivariate linear model was proposed to predict the mechanical parameters of the materials. Fourth, the effects of different particle sizes on the sample strength were examined by scanning electron microscope and the effects of gypsum/petrolatum ratio on the deformation evolution of samples were also analyzed. Finally, a material ratio scheme closely resembling the natural coal was identified. Related improvements and remaining problems in the preparation of similar material were also discussed. This study would contribute to the prevention and control of coal and gas outbursts, as well as other physical simulation tests related to soft coal seams.

Table 1. Representative literature methods for preparing coal-like materials.

No.	Coal Particle Size (mm)	Compaction Pressure (MPa)	Additives	Molding/Air Curing Time (h)	References
1	0–0.2	2.76–19.90	None	-	[5,12,22]
2	0.25–0.38	100	Water	Molding curing 0.33 h	[23–25]
3	0–1	4–13	Water	Molding curing 24 h	[20]
4	0.18–0.25	100	water	24 h drying in drying basin	[11]
5	0.18–0.25	100	Water	Molding curing 0.5 h	[26]
6	0–1.7	4	Water	-	[27]
7	0.1–1	5.7	Cement, water	Molding curing 24 h, air curing 96 h	[18,28]

Table 1. *Cont.*

No.	Coal Particle Size (mm)	Compaction Pressure (MPa)	Additives	Molding/Air Curing Time (h)	References
8	0–0.8	2–6	Cement, water	Molding curing 14.6–20 h, air curing 96–24 × 28 h	[19]
9	0.063–1	5	Phenol-Formaldehyde Resin, potassium hydroxide alkaline aqueous, distilled water	Molding curing 3–26 h, air curing time 144.5–598.5 h	[21]
10	0.18–0.25	100	Cement, water	Molding curing 0.17 h, air curing 24 × 28 h	[29]
11	0.18–0.38/0.38–0.83 mass ration 1:1	25	Cement, sand, activated carbon, water	Air cured 7 × 24 h	[6,10,13]

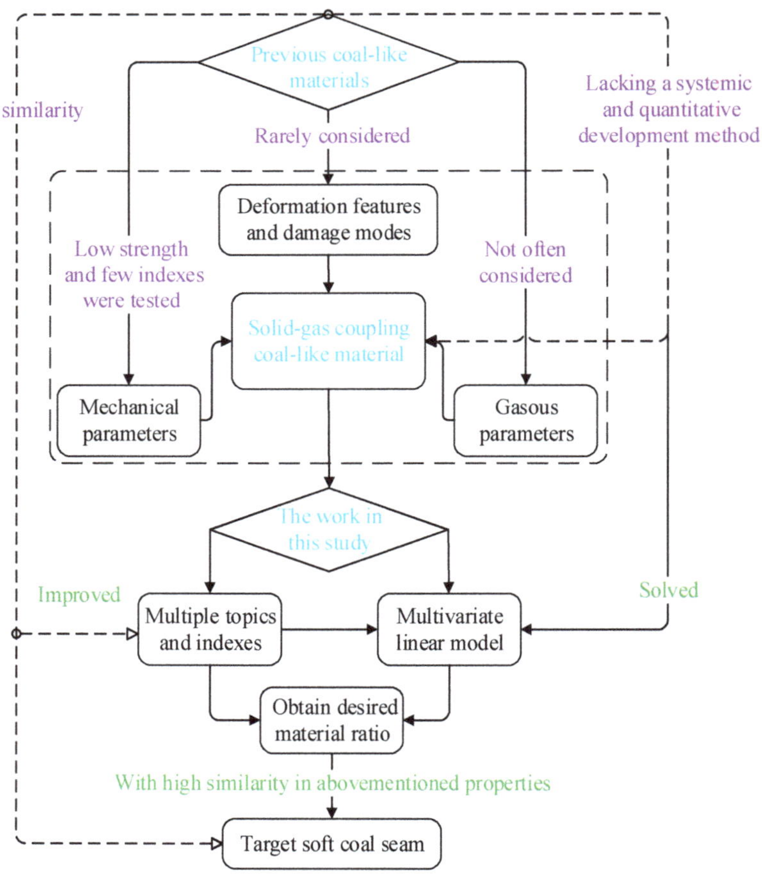

Figure 1. The difference between this work and the previous studies.

2. Materials and Methods

2.1. Preparation and Testing of Natural Coal Samples

Studies have shown that most of the coal and gas outburst occur in tectonically deformed coal (TDC) [34–36]. This is mainly because TDC is formed under mono- or multi-phase strong tectonic movements. Its original structure has been deformed or damaged, resulting in larger adsorption surface area and higher gas content than the primary coal [37]. In this paper, the natural coal samples were selected from a typical tectonically deformed coal seam from No. B-1 Coalbed, Zhengzhou, Western Henan, China, where there have been frequent coal and gas outburst disasters over the last 60 years [7]. According to the new structural-genetic classification system [38], the TDC samples belong to medium-rank and mylonitic structure coal and are characterized by low strength and strong gas adsorption/desorption [32].

Because this coal seam is soft and fragile, it is difficult to obtain large samples using the traditional drill core sampling method. The natural coal standard samples were obtained successfully only after many attempts and improvements. Specifically, a square iron sampler was used to cut the underground coal seam to obtain a sample of approximately 100 mm × 100 mm × 150 mm and the sample was quickly placed in a sealed bag packaging for preservation. The sample was then carefully polished in the lab using a grinding machine. Samples of two target sizes were prepared: 50 mm × 50 mm × 100 mm for uniaxial compression test and 50 mm × 50 mm × 50 mm for variable angle compression shear test. The sample standard requirements and test procedures are in accordance with methods for determining the physical and mechanical properties of coal and rock (Part 7 [39] and Part 11 [40]). The sample loading rate was 0.5 mm/min. The basic physical and mechanical parameters of coal sample are as follows: apparent density =1.280 g/cm^3, UCS = 1.72 MPa, elastic modulus = 126.35 MPa, cohesion = 0.17 MPa and internal friction angle = 27°. However, the success rate of natural coal sample preparation remained only approximately 5%. The proximate analysis results of this TDC are as follows: moisture content Mad = 0.95%, ash content Ad = 10.42%, volatile matter content Vad = 13.95%, porosity = 15.3%, gas adsorption constants a = 39.789 and b = 1.113 and index of initial velocity of gas diffusion Δp = 26–29.

The stress-strain curve in Figure 2 contains the typical stages of initial compaction, linear elastic deformation, strength hardening and softening. Notably, it includes a longer compaction phase than primary coal seam, as well as a plastic or shearing deformation stage. The peak-failure strain of TDC reached 1.73%, which was significantly larger than those of brittle coal or rock (approximately 0.5%) [41]. In addition, microstructure test (Figure 3) showed that the coal body was very broken. The original structure was almost completely destroyed and was composed of particles of different sizes and the texture was loose. This was the reason why the density and strength of TDC were both usually smaller than those of the primary coal seam.

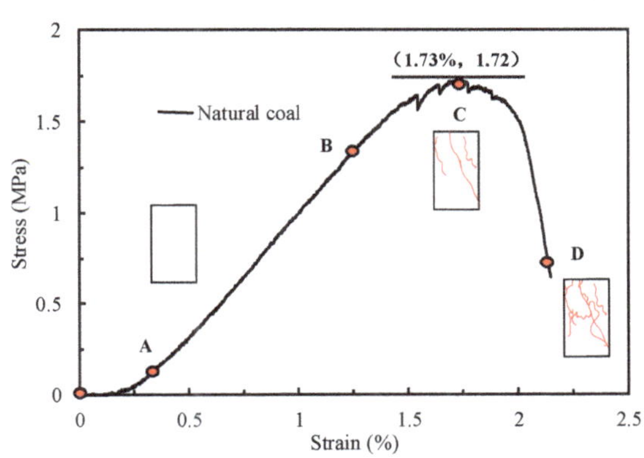

Figure 2. The stress-strain curve of target natural coal.

Figure 3. The structure of natural coal under scanning electron microscope.

2.2. Similarity Criteria and Index

Coal and gas outburst is a complex solid-gas coupling process. The currently accepted view is that it is mainly caused by the combined action of in-situ stress, gas pressure and physical and mechanical properties of coal [7–10]. The coal and gas outburst consists of three successive stages [13], among which the static deformation and failure of coal occur during the important outburst preparation stage, while the fracturing of gaseous coal and the movement of pulverized coal and gas occur during the forming and developing stage. Therefore, several similarity criteria (C_i in Equation (1) below) have been proposed to describe the whole process of coal and gas outburst, by using mechanical, deformation and energy models and considering the experimental conditions [6]:

$$\begin{cases} C_\gamma = C_n = C_\varphi = C_p = 1 \\ C_\sigma = C_E = C_c = C_l C_\gamma \end{cases} \tag{1}$$

where γ, n, φ, p, σ, E, c and l, are the volumetric weight, porosity, internal friction angle, gas pressure, compressive strength, elastic modulus, cohesion and length, respectively.

Judging from the above, the similarity index of the coal-like materials should include the physical and mechanical parameters (density, uniaxial compressive strength, elastic modulus, internal friction angle, etc.) and deformation properties and failure mode, as well as porosity, absorption constants (a, b) and desorption index (Δp) of the initial gas diffusion velocity.

2.3. Preparation of Coal-Like Material Samples

2.3.1. Composition Selection

Coal-like materials generally consist of aggregates, binders and additives. The raw materials should be selected according to the following principles: (1) similar to the natural coal material, (2) abundant and low-cost and (3) safe, non-toxic and environment-friendly. In the coal and gas outburst tests [6], the density of coal-like material should be the same as that of natural coal. So, the natural coal powder was selected as the aggregate. Another important reason for using coal powder was that it had good adsorption characteristics similar to the natural coal. However, the particle sizes of coal powders reported in Table 1 were different. Some studies [30,42] have shown that larger coal particles (greater than 3 mm) would undergo secondary crushing during the molding and pressing process, which would have an adverse effect on molding. Therefore, two types of coal powders with particle sizes of 1–2 mm and 0–0.5 mm were used respectively as the coarse and fine aggregates. The fine coal powder was the main aggregate to ensure the molding quality. Second, the binder was selected. Normally, the binder has the greatest influence on the strength and deformation parameters of the material. As the strength of sample obtained by pressing only coal powder was too low, gypsum was selected to adjust the strength parameters of coal-like materials. Portland cement was not used here, because the cement hardening time was too long (more than 28 days). Its strength would also change with time, which could lead to strength instability of coal-like materials [19] and poor homogeneity and reproducibility. On the other hand, petrolatum was selected to adjust the deformation characteristics of the coal-like materials.

The molding pressure has an important effect on the density and porosity of briquette [5,12]. Excessive molding pressure would damage aggregates with large particle size and affect the molding quality. Therefore, the molding pressure was designed based on the ground stress level of the coal seam. The vertical stress level of the coal seam was approximately 7.2 MPa. In addition, in order to allow the gypsum to play a bonding role, it was necessary to add an appropriate amount of water. After repeated tests, it was found that when the moisture exceeded 11%, the material was too wet and would affect the molding, while the raw material to water ratio of 10:1 was relatively reasonable. Because the density of gypsum (2.3 g/cm^3) is higher than that of natural coal, it was also necessary to use minuteness super white light calcium carbonate with very stable mechanical properties to match the target density while using different material ratios. The selected material compositions are shown in Table 2 and Figure 4.

Table 2. Selected raw materials.

Type	Name	Remarks
Aggregate	Coal powder	1–2 mm and 0–0.5 mm coal particles
Binder	Gypsum (The Group of TZU She Tang Gypsum, Taiwan, China)	Particle size 0.048 mm, density 2.3 g/cm^3, fast-hardening for 15–45 min
	Petrolatum (Dezhouchengze Co., LTD, Dezhou, in Shandong Province, China)	Medical grade, density 0.83 g/cm^3
Auxiliary material	Light Calcium carbonate (Darui Chemical Co., LTD, Gaoan, in Jiangxi Province, China)	Ultra-fine, ultra-white, light calcium carbonate, 0.1 μm < particle size ≤ 1 μm, stable performance, density 0.54 g/cm^3
Water		Tap water

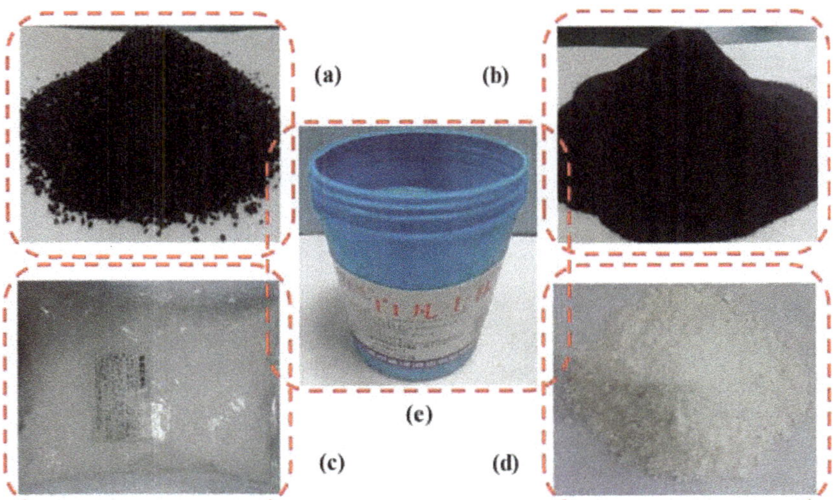

Figure 4. Photographs of the components. (**a**) The coarse aggregate, (**b**) fine aggregate, (**c**) gypsum, (**d**) Minuteness super white light calcium carbonate powder and (**e**) petrolatum.

2.3.2. Design Scheme

The components and their ratios both have an impact on the physical and mechanical properties of coal-like materials. The orthogonal test method was used to design 25 groups of ratios, as shown in Tables 3 and 4. The advantage of this method was that the schemes included a combination of any two factors at all levels and reduced the amount of experimental work [43].

Table 3. The 4 factors and 5 levels of the orthogonal test design.

Level \ Factor	A-Aggregate/ Binder	B-Coarse/Fine Aggregate	C-Gypsum/ Petrolatum	D-Calcium Carbonate (%)
1	6:1	1:9	5:5	2
2	7:1	2:8	6:4	3
3	8:1	3:7	7:3	4
4	9:1	4:6	8:2	5
5	10:1	5:5	9:1	6

Table 4. The 25 groups of material composition ratio schemes.

Group No. \ Factor	Aggregate/Binder	Coarse/Fine Aggregate	Gypsum/Petrolatum	Calcium Carbonate (%)
1	6:1	1:9	5:5	2
2	6:1	2:8	6:4	3
3	6:1	3:7	7:3	4
4	6:1	4:6	8:2	5
5	6:1	5:5	9:1	6
6	7:1	1:9	6:4	4
7	7:1	2:8	7:3	5
8	7:1	3:7	8:2	6
9	7:1	4:6	9:1	2
10	7:1	5:5	5:5	3
11	8:1	1:9	7:3	6
12	8:1	2:8	8:2	2
13	8:1	3:7	9:1	3
14	8:1	4:6	5:5	4
15	8:1	5:5	6:4	5
16	9:1	1:9	8:2	3
17	9:1	2:8	9:1	4
18	9:1	3:7	5:5	5
19	9:1	4:6	6:4	6
20	9:1	5:5	7:3	2
21	10:1	1:9	9:1	5
22	10:1	2:8	5:5	6
23	10:1	3:7	6:4	2
24	10:1	4:6	7:3	3
25	10:1	5:5	8:2	4

2.3.3. Sample Preparation Method

(1) The raw materials were weighed accurately using an electronic scale according to the given ratios based on a total amount of 1000 g. (2) The dry components (coal powder, gypsum powder and calcium carbonate powder) were mixed and stirred uniformly. (3) A mixture of water and petrolatum was heated to 45 to 50 °C to melt it into a liquid and then quickly mixed with the materials. (4) The well-mixed coal-like materials were divided into five equal parts by weight and put into the mold. Each layer was pressed by a servo press machine at the speed of 120 mm/min. The target pressure was 14.137 kN (the calculated stress was 7.2 MPa) and the load was maintained for 10 s. The actual load-time curve during the layer-by-layer compaction press is shown in Figure 5a. Finally, the whole sample was compacted by the press at the molding pressure of 7.2 MPa and maintained for 10 min, mainly to make the overall force on the whole sample consistent. The loading path on the whole sample can be seen in Figure 5b. It should be noted that the surface between two adjacent layers

should be roughened before filling in the next layer of material to increase the adhesion between layers. The steel mold had an inner diameter of 50 mm, a height of 120 mm and a wall thickness of 10 mm. The mold was approximately 20 mm higher than the sample, making it convenient for the fifth material to be added to the mold at one time. The inner surface of the mold had to be very smooth to reduce the friction. (5) With a self-designed demolding device, as shown in Figure 5d, the molded sample was pushed out at a speed of 5 mm/min by a press. The sample slowly entered the hollow steel pipe (wall thickness 10 mm, height 200 mm, inner diameter 100 mm, with soft material placed on the bottom to prevent sample damage). Then, the sample was cured at room temperature for 10 to 15 days. The water content was tested continuously in this process. When the water content was the same as that of the natural coal, the sample was sealed and stored in a plastic bag in time. In particular, it should be pointed out that many reconstituted coal samples were destroyed during the extrusion process in the past [18], while by using the designed demolding method here, the success rate of sample preparation was 100%. The use of a press machine with quantitative loads for compacting the materials was also an improvement over the manual pressing used in previous studies, which was difficult to quantify and lowered the homogeneity and reproducibility of samples.

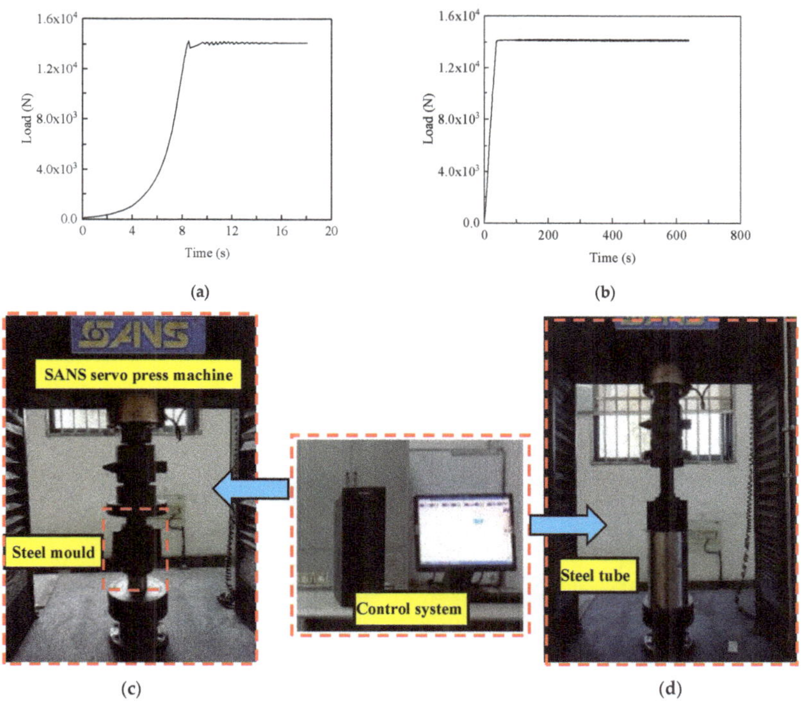

Figure 5. Sample molding and demolding process. (**a**) The load-time curve during the compaction process of each layer of material. (**b**) The load-time curve during the compaction process of the whole sample. (**c**) Sample compression molding process. (**d**) Sample demolding process.

2.3.4. Sample Tests

The sample preparation met the requirements suggested by International Society for Rock Mechanics [44]. The height of the sample was more than twice the diameter. For each group of formulation scheme, 6 samples of Φ 50 mm × 102 mm were made (a total of 150 samples). Two samples in each group were used for the uniaxial compression test (Figure 6a), three were cut into 6 pieces Φ 50 mm × 50 mm in size to perform the variable angle shear tests (Figure 6b). The spare sample would be used in the isothermal adsorption and desorption test.

The CSS-44300 universal testing machine (Changchun testing machine research institute, Changchun, China) with electro-hydraulic servo was used to perform the uniaxial compression and variable angle shear tests. The test steps referred to references [39,40]. The sample loading rate was 0.5 mm/min and the surface unevenness did not exceed 0.2 mm.

(a) (b)

Figure 6. Prepared coal-like samples of two sizes: (**a**) Φ 50 mm × 102 mm; (**b**) Φ 50 mm × 50 mm.

3. Result Analysis and Discussion

3.1. Mechanical Properties

3.1.1. Results of Orthogonal Test

The mechanical property test results are shown in Table 5. The property parameters of the 25 groups of coal-like materials covered a wide range: density of 1.171–1.305 g/cm^3, uniaxial compressive strength of 0.55–2.20 MPa, elastic modulus of 32.06–196.18 MPa, cohesion of 0.068–0.248 MPa and the internal friction angle of 16.8–35.6°. These samples could be used for physical simulation of most soft coal seams [13,19].

Table 5. The physical and mechanical parameters of each group.

No.	Density (g/cm^3)	UCS (MPa)	E (MPa)	Cohesion (MPa)	F (°)
1	1.235	0.67	39.76	0.131	17.4
2	1.248	1.24	76.22	0.153	23.8
3	1.251	1.68	91.90	0.192	27.2
4	1.253	1.82	141.45	0.201	32.4
5	1.242	1.74	152.75	0.248	35.6
6	1.228	0.91	35.11	0.132	18.2
7	1.229	1.49	108.22	0.164	20.4
8	1.231	1.94	163.66	0.185	27.8
9	1.296	2.05	181.85	0.203	32.8
10	1.176	0.55	32.06	0.089	28.2
11	1.223	1.16	63.90	0.152	18.7
12	1.289	1.72	111.81	0.173	21.1
13	1.305	2.20	196.18	0.201	30.1
14	1.188	0.89	63.11	0.073	23.7
15	1.184	0.79	49.05	0.095	28.9

Table 5. *Cont.*

No.	Density (g/cm^3)	UCS (MPa)	E (MPa)	Cohesion (MPa)	F (°)
16	1.256	1.39	64.43	0.164	21.6
17	1.252	1.94	163.59	0.213	24
18	1.172	0.99	60.69	0.068	23.4
19	1.171	1.13	84.24	0.089	25.6
20	1.212	1.04	55.07	0.119	29.8
21	1.238	1.61	97.41	0.191	22.4
22	1.181	0.77	49.24	0.061	16.8
23	1.218	1.23	74.83	0.078	22.2
24	1.211	1.38	93.58	0.108	27.4
25	1.209	1.27	77.25	0.128	30.6
TDC	1.280	1.72	126.35	0.17	27

3.1.2. Range Analysis

In addition, according to the orthogonal experimental theory, range analysis (also called visual analysis method) was used to reveal the influence of various factors on results from the orthogonal experiment [43]. That is, the results corresponding to the same level of each factor were averaged and the range was obtained by subtracting the minimum average value from the maximum average value for each level. The range value reflected the influence of different levels of a given factor on the index of interest. A large range indicated that different levels of this factor had strong and different influences on the test results. From the test results in Table 5, the average and range of each level of all factors affecting the density, compressive strength, elastic modulus, cohesion and internal friction angle could be obtained, as shown in Table 6. The effect of different levels of various influencing factors on each index could also be obtained, as shown in Figure 7.

The following observations can be found from Table 6 and Figure 7. (1) For density, the gypsum/petrolatum played the main controlling role. The calcium carbonate content was the secondary factor determining the material density. In Figure 7a, the density of coal-like material increased with increasing gypsum/petrolatum ratio but decreased linearly with increasing calcium carbonate content. (2) For UCS, the gypsum/petrolatum ratio had the highest degree of influence on the compressive strength. The coarse/fine aggregate ratio played a secondary controlling role, followed by the aggregate/binder ratio. Factor D (calcium carbonate content) had almost no influence. In addition, Figure 7b shows that (i) the compressive strength tends to be positively correlated with the gypsum/petrolatum ratio, indicating that increasing the gypsum content could improve the material strength significantly. (ii) The compressive strength first increased then decreased with increasing coarse/fine aggregate ratio, indicating that a reasonable ratio of aggregate particles of different sizes had an important influence on the material strength. When the fine aggregate/coarse aggregate content was approximately 70%, the aggregate had the highest strength. (iii) The compressive strength of coal-like material and aggregate/binder showed a negative correlation. (3) For ***E***, the influence of each factor on the elastic modulus was similar to that on the compressive strength, in the same order of C > B > A > D. (4) For cohesion, the influence of each factor was in the order of C > A > B > D. In Figure 7d, the cohesion of material increased with the increase of gypsum/petrolatum ratio but decreased with increasing aggregate/binder ratios. (5) Regarding IFA, the influence of each factor on internal friction angle was in the order of B > C > A > D. The coarse/fine aggregate had the most significant influence. As shown in Figure 7e, as the coarse aggregate content increased, the internal friction angle increased linearly. This indicated that the friction and bite force between coarse aggregates were significantly larger than those between fine aggregates. Under the condition of ensuring the molding quality, increasing the coarse aggregate content is beneficial to enhancing the internal friction angle of the material.

Table 6. Range analysis of the physical and mechanical indexes.

Level	Average Density (g/cm^3)				Average UCS (MPa)				Average EM (MPa)				Average Cohesion (MPa)				Average IFA (°)			
	A	B	C	D	A	B	C	D	A	B	C	D	A	B	C	D	A	B	C	D
1	1.246	1.236	1.190	1.250	1.43	1.15	0.77	1.34	100.41	60.12	48.97	92.66	0.185	0.154	0.084	0.141	27.28	19.66	21.90	24.66
2	1.232	1.240	1.210	1.239	1.39	1.43	1.06	1.35	104.18	101.82	63.89	92.49	0.155	0.153	0.109	0.143	25.48	21.22	23.74	26.22
3	1.238	1.235	1.225	1.226	1.35	1.61	1.35	1.34	96.81	117.45	82.53	86.19	0.139	0.145	0.147	0.148	24.50	26.14	24.70	24.74
4	1.213	1.224	1.248	1.215	1.30	1.45	1.62	1.34	85.60	112.84	111.72	91.36	0.131	0.135	0.170	0.144	24.88	28.38	26.70	25.50
5	1.211	1.205	1.267	1.210	1.25	1.08	1.91	1.35	78.46	73.23	158.36	102.76	0.113	0.136	0.211	0.147	23.88	30.62	28.98	24.90
Range	0.035	0.035	0.076	0.040	0.18	0.53	1.14	0.01	25.72	57.33	109.39	11.39	0.07	0.02	0.13	0.01	3.40	10.96	7.08	1.56

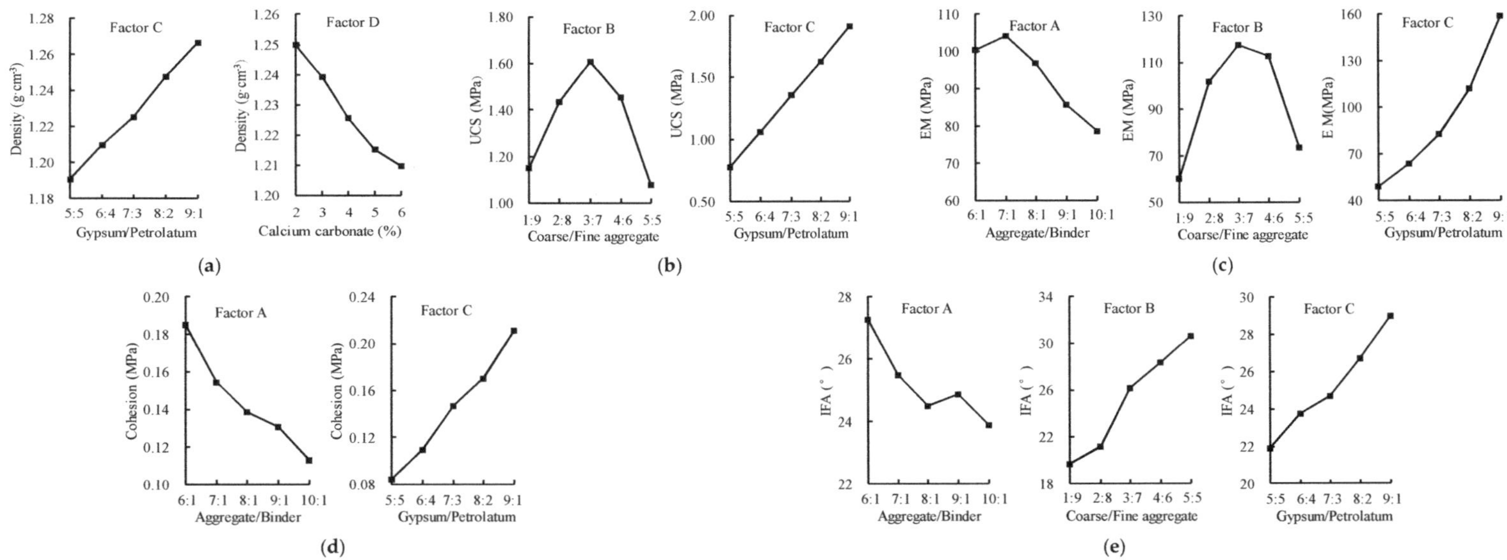

Figure 7. The effect of different levels of key influencing factors on the physical and mechanical indexes: (**a**) density; (**b**) UCS; (**c**) EM; (**d**) cohesion; (**e**) IFA.

3.2. Deformation Characteristics and Failure Modes of the Coal-Like Samples

The range analysis indicated that the gypsum/petrolatum ratio is the main factor affecting the strength indexes of the specimens. Since samples in the 25 groups used in orthogonal uniaxial compression tests showed different deformation processes, it was necessary to study the effect of gypsum/petrolatum ratio on the deformation features. To this end, coal-like materials with aggregate/binder = 6:1, coarse/fine aggregate = 3:7 and gypsum/petrolatum = 5:5, 6:4, 7:3, 8:2 and 9:1 were prepared and used in deformation tests. Figure 8 shows the stress-strain curves of the five types specimens with different levels of gypsum/petrolatum ratios in the uniaxial compression test. All samples contained the typical stages of initial compaction, linear elastic deformation, strength hardening and softening and a residual phase. Samples with low gypsum/petrolatum ratio had lower strength, longer plastic or ductile deformation phase before peak stress and a gentler drop after the peak value. While the sample with gypsum/petrolatum = 9:1 had the greatest strength and shortest plastic deformation phase, the stress declined linearly after the peak to reach a very small residual value, which corresponds to a brittle characteristic.

More importantly, the damage modes of the specimens with different gypsum/petrolatum ratios under uniaxial compression test could be divided into five main types, named as A, B, C, D and E. Figure 9 demonstrates the physical damage modes and corresponding schematic diagrams of the specimens. Type A is squeezing failure, which belongs to ductile damage and occurred in the sample with gypsum/petrolatum = 5:5, that is, a high content of petrolatum. The major features include apparent lateral convex deformation and the lack of obvious large fracture surfaces, as shown in Figure 9(A-1,A-2). Type B is wedge failure, which mainly occurred in the sample with gypsum/petrolatum = 6:4. As shown in Figure 9(B-1,B-2), large wedge-shaped coal blocks slipped from the sample, so that large fissure surfaces could be found in the specimens. Type C is plastic shear failure and mainly occurred in the sample with gypsum/petrolatum = 7:3. This type is characterized by shear cracks and damage everywhere, the absence of obvious large fissure surface, the coal body collapsing in the form of scales and noticeable dilative shear deformation (Figure 9(C-1,C-2)). Type D is brittle shear failure and occurred in the sample with gypsum/petrolatum = 8:2. In this case, first a major fissure appeared and then it propagated and extended through the whole sample. When the sample broke down, there was a large final fracture surface about 45–60° from the specimen axial direction (Figure 9(D-1,D-2)). Type E is brittle splitting failure and occurred in the sample with gypsum/petrolatum = 9:1, that is, high gypsum content. Such samples always split quickly after the peak stress. Figure 9(E-1,E-2) present the ultimate destruction characteristic of type E: one or several vertical cracks ran through the top and bottom of the sample and their directions were approximately parallel to the axis of the specimen.

In addition, the occurrence of coal and gas outbursts must be affected by the deformation and failure mechanism of the coal [13,37]. According to the different deformation mechanisms, different types of TDC are divided into three series of deformation and ten classes, that is, the brittle, the ductile and the brittle-ductile deformation [38]. In this study, upon decreasing the gypsum content or increasing the petroleum content, the deformation of the coal-like materials transitioned from ductile to brittle. The failure modes included typical squeezing damage, wedge splitting, plastic shear failure, brittle shear and brittle fracturing. These varied deformation features are useful in different coal and gas outburst simulation tests.

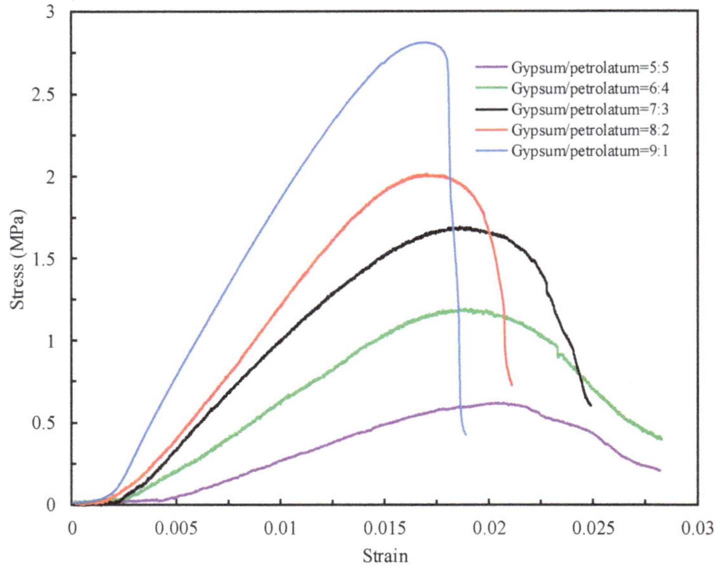

Figure 8. Stress-strain curves of coal-like material specimens with different gypsum/petrolatum ratios.

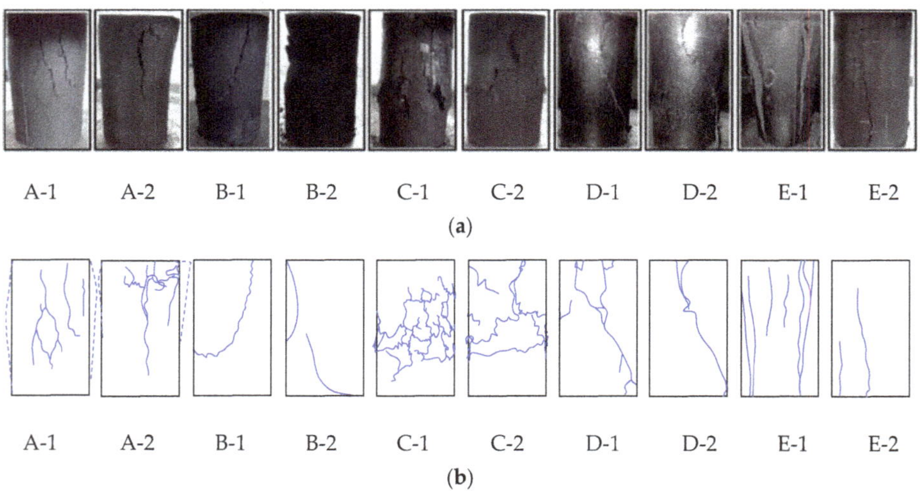

Figure 9. (**a**) The different physical damage modes and (**b**) the corresponding schematic diagrams of the specimens.

3.3. Physicochemical Properties

Samples in the four groups of 4, 8, 12 and 17 had similar mechanical parameters (compressive strength, elastic modulus, cohesion, internal friction angle, etc.) and deformation features to the natural coal. Thus, they were selected for further testing of porosity and gas adsorption/desorption.

3.3.1. Porosity

Porosity refers to the proportion of the volume of pores to the total volume. It can be calculated from the apparent density ρ_a and true density ρ_t of material according to the following equation:

$$k = \frac{1}{\rho_a} - \frac{1}{\rho_t} \tag{2}$$

After the sample was dried, its apparent density could be obtained by weighing using an electronic scale and size measurement using a Vernier caliper. The true density was measured by the automatic industrial analyzer. From the porosity of the three groups of samples (Table 7), it can be seen that their density and porosity were highly consistent with those of the natural coal. So, these coal-like materials could satisfy the porosity and volumetric weight similarity ratio shown in Equation (1) [6]:

Table 7. Porosity of the samples.

Sample	Apparent Density (g/cm^3)	True Density (g/cm^3)	Mass Ratio of Coal Powder (%)	Porosity (%)
TDC	1.268	1.573	-	15.3
No. 4	1.241	1.558	81.43	15.7
No. 8	1.277	1.623	82.25	16.1
No. 12	1.293	1.691	87.11	17.4
No. 17	1.240	1.591	86.40	16.9

3.3.2. Adsorption and Desorption Parameters

The gas adsorption constants a and b, used to quantify the adsorption capacity of coal, were determined by the isotherm adsorption test results using methane of 99.9% purity at 30 °C in combination with the Langmuir equation [6]. The parameter Δp represents the initial velocity of coal gas diffusion [45]. In this paper, the TDC and the samples 4, 8, 12 and 17 were crushed and grains with the particle size of 0.2 to 0.25 mm were sifted out. For each sample group, 6 portions of 3.5 g each were weighed. Each portion was placed in the WFC-2 initial gas diffusion velocity analyzer (Zhengzhou huazhi electronic technology co. LTD, Zhengzhou, China). Air tightness check and vacuum treatment were performed on the test instrument. The gas adsorption and desorption of samples were tested under the conditions of 99.9% pure methane, 0.1 MPa and 20 °C. The averages of the test results are shown in Table 8. It can be seen that the gas adsorption and desorption properties of coal-like materials were consistent with those of the natural coal and could meet the test requirements [6].

Table 8. Adsorption and desorption parameters of samples.

Sample	a (cm^3/g)	b (1/MPa)	Δp (mL/s)
TDC	39.789	1.113	27
No. 4	35.258	1.169	23
No. 8	35.463	1.163	23
No. 12	37.944	1.137	25
No. 17	37.562	1.146	24

Coal has a strong gas adsorption capacity mainly because of its good porosity and relatively large specific surface area [36,37]. Therefore, under the same molding pressure conditions, the coal mass ratio had the most impact on these characteristics. Tables 7 and 8 also verify this feature: the higher the coal content, the larger the porosity of the coal-like materials and the stronger the gas adsorption and desorption capacities. On the other hand, higher contents of gypsum, petrolatum and calcium carbonate reduce the porosity and the methane adsorption/desorption, which is consistent with existing research results [13,46]. Besides, previous results [47] presented that coal particles below 3.35 mm in size have almost no influence on the gas adsorption capacity. Therefore, the total amount of fine and coarse coal powder can be used to quantitatively estimate the adsorption constants *a*, as presented in Equation (3) and Figure 10 (with R^2 = 0.9967).

$$Y_a = 0.4841X - 4.25 \tag{3}$$

where Y_a is the value of gas adsorption constant *a* and *X* is the content of coal powder.

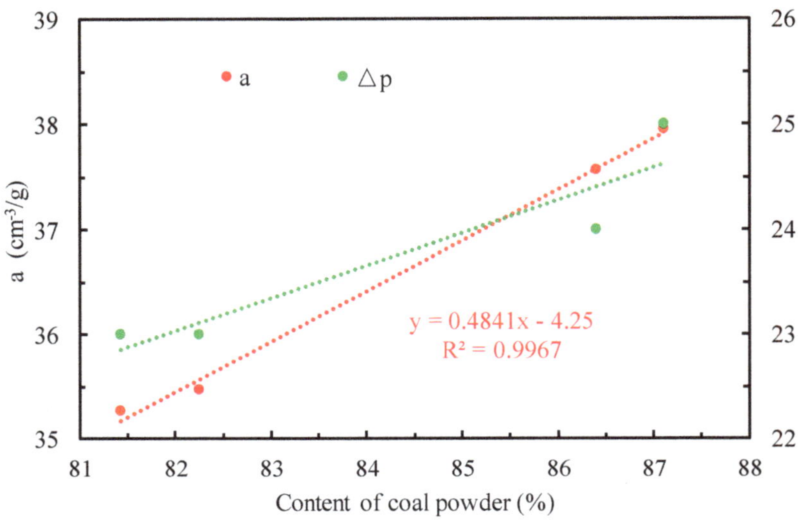

Figure 10. The quantitative relation between coal powder content and the gas absorption constant, *a* and the desorption index, Δp.

Combining Tables 4 and 7, it can be seen that the four coal-like material groups of 4, 8, 12 and 17 had a coal powder ratio of 81.43 to 87.11%, which almost coincides the overall range for groups 1 to 25 (80.57 to 89.09%). This indicated that the samples in group 1 to 25 all had similar porosity and gas adsorption/desorption characteristics as the natural coal.

3.4. Multivariate Linear Model for Predicting the Mechanical Parameters of Coal-Like Material

According to our range analysis (in Section 3.1.2), each mechanical index has multiple influencing factors. To determine the suitable proportion of materials used in physical simulation, repeated adjustments are usually needed. Thus, results from previous studies can hardly be used directly in future research and it is difficult to promote the rapid development of similar materials. Therefore, it is important to establish a multivariate linear model between the mechanical property indexes and their main influencing factors (as shown in Table 6 and Figure 7). The model can be described by the equations in (4):

$$\begin{cases} Y_{\sigma_c} = C_1 + a_1X_1 + b_1X_2 + c_1X_3 + d_1X_4 \\ Y_E = C_2 + a_2X_1 + b_2X_2 + c_2X_3 + d_2X_4 \\ Y_C = C_3 + a_3X_1 + b_3X_2 + c_3X_3 + d_3X_4 \quad (0 \le X_1, X_2, X_3, X_4, X_5 \le 100) \\ Y_F = C_4 + a_4X_1 + b_4X_2 + c_4X_3 + d_4X_4 \\ Y_D = C_5 + a_5X_1 + b_5X_2 + c_5X_3 + d_5X_4 + e_1X_5 \end{cases} \tag{4}$$

where C_i, a_i, b_i, c_i, d_i and e_i (*I* = 1, 2, 3, 4, 5) are the fitting constants. The contents of coarse aggregate, fine aggregate, gypsum, petrolatum and calcium carbonate were set as X_1(%), X_2(%), X_3(%), X_4(%) and X_5(%); and the indexes of compressive strength, elastic modulus, cohesion, internal friction angle and density were Y_{σ_c}, Y_E, Y_C, Y_F and Y_D, respectively.

It should be noted that in Figure 11, the elastic modulus of the material increased exponentially (with an excellent correlation of R^2 = 0.9144) against the compressive strength, which is consistent with previous studies [13]. More importantly, their similarity ratios are the same in the similarity criteria (as shown in Equation (1)). Therefore, Equation (4) could be simplified to (5),

$$\begin{cases} Y_{\sigma_c} = C_1 + a_1X_1 + b_1X_2 + c_1X_3 + d_1X_4 \\ Y_C = C_2 + a_2X_1 + b_2X_2 + c_2X_3 + d_2X_4 \quad (0 \leq X_1, X_2, X_3, X_4, X_5 \leq 100) \\ Y_F = C_3 + a_3X_1 + b_3X_2 + c_3X_3 + d_3X_4 \\ Y_D = C_4 + a_4X_1 + b_4X_2 + c_4X_3 + d_4X_4 + e_1X_5 \\ 100 = X_1 + X_2 + X_3 + X_4 + X_5 \end{cases} \tag{5}$$

Then, according to the orthogonal test result in Table 5, the multivariate Equation (5) were fitted using Origin program to identify the regression Formula (6). The R^2 values of Y_{σ_c}, Y_C, Y_F and Y_D were 0.838, 0.965, 0.945 and 0.937, respectively.

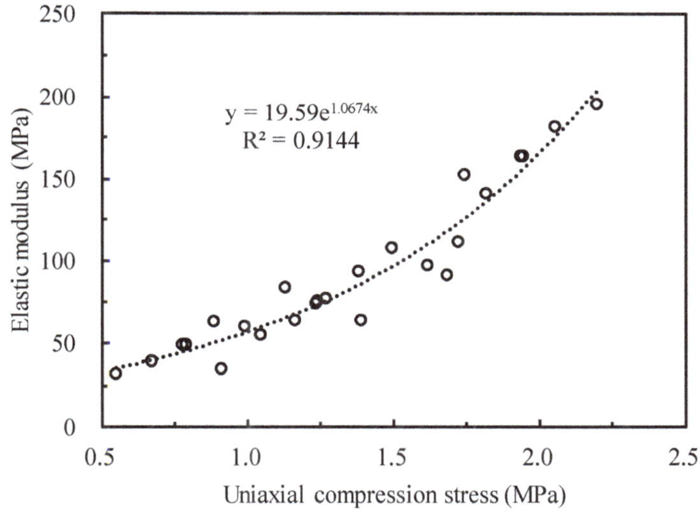

Figure 11. Relationship between modulus of elasticity and uniaxial compressive strength of samples.

$$\begin{cases} Y_{\sigma_c} = 0.01356X_1 + 0.0171X_2 + 0.1305X_3 - 0.13143X_4 - 0.58773 \\ Y_C = -0.0016X_1 - 0.00073X_2 + 0.02122X_3 - 0.00695X_4 + 0.08933 \quad (0 \leq X_1, X_2, X_3, X_4, X_5 \leq 100) \\ Y_F = 0.27831X_1 - 0.05061X_2 + 1.15883X_3 - 0.40811X_4 + 13.57658 \\ Y_D = -0.00024X_1 + 0.00083X_2 + 0.01278X_3 - 0.00487X_4 - 0.01045X_5 + 1.14431 \\ 100 = X_1 + X_2 + X_3 + X_4 + X_5 \end{cases} \tag{6}$$

According to the physical and mechanical parameters of the raw coal, the five-variable primary equations can be solved using the Mmult and Minverse functions in the Excel program to obtain the corresponding composition (26.6% coarse aggregate, 62.2% fine aggregate, 8.5% gypsum, 1.7% petrolatum and 1% calcium carbonate). The gas absorption capacity of this material was estimated by Equation (3) to be a = 38.738.

Then, samples were made according to the determined composition and their indexes were tested in the same conditions. Table 9 shows the differences of each index of mechanical, deformation and gaseous properties between the natural coal and developed coal-like material. The difference in mechanical and gas adsorption/desorption parameters was very small (approximately 0.47 to 11.76%). Figure 12 is the stress-strain curve of the coal-like material, showing evolution characteristics of strength and deformation (comprehensive damage mode of brittle shearing and plasticity) very consistent to that of the natural coal.

Table 9. Difference analysis between natural coal and determined coal-like material.

Sample	Apparent Density (g/cm^3)	UCS (MPa)	EM (MPa)	Cohesion (MPa)	IFA (°)	Porosity (%)	a	Δp
Coal-like material	1.286	1.78	131.85	0.18	25.8	17.1	38.343	25
TDC	1.280	1.72	126.35	0.17	27	15.3	39.789	27
Difference (%)	+0.47	+3.49	+4.35	−5.9	−4.44	+11.76	−3.63	−7.41

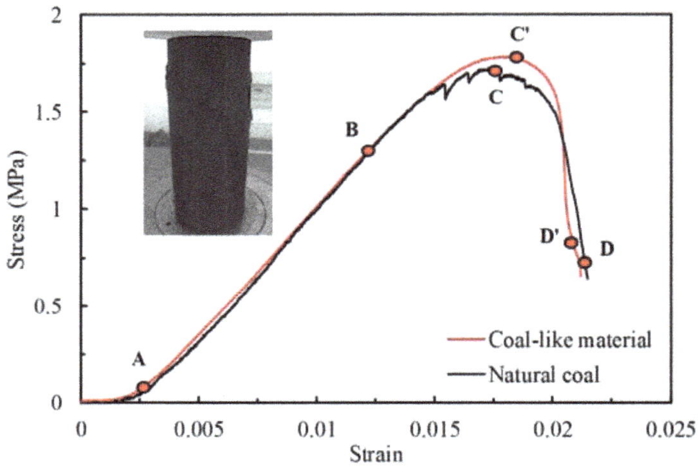

Figure 12. The stress-strain curves of coal-like material and natural coal.

3.5. Discussion

3.5.1. Microscopic Analysis of the Effect of Pulverized Coal Particle Size on Sample Strength

As stated in Section 3.1.2, a reasonable ratio of aggregate particles of different sizes has an important influence on the material strength and elastic modulus. Meanwhile, the particle sizes of coal powders reported in Table 1 were different. To study the effect of particle size on the preparation and strength of samples, coal-like materials with aggregate/gypsum = 10:1, coarse (1–2 mm)/fine (0–0.5 mm) coal particles = 1:9, 3:7 and 5:5 were prepared. Their microstructures were examined by SEM (Figure 13).

On the one hand, the morphology diagrams (Figure 13a) of the three cut specimens show that, when the proportion of fine aggregate was high (coarse/fine coal particles = 1:9 and 3:7), the sample was relatively flat in morphology and relatively dense in structure, while a high proportion of coarse aggregate (coarse/fine coal particles = 5:5) resulted in an uneven sample morphology, a looser structure and a low strength (shown in Figure 7b). On the other hand, in specimen No. 1 shown in Figure 13b, the gypsum (needle shapes) were stuck together when 90% of the coal had a particle size of 0–0.5 mm. This indicates that too much fine coal particles makes it difficult to mix the sample evenly with the cementing agent, resulting in poor strength stability and low homogeneity of the sample, which will have an important impact on the analysis of test results. However, in specimen No. 3 the gypsum was uniformly attached to the surface of coal particles and the very small coal particles were also firmly bonded to the larger ones. Therefore, a large proportion of coarse coal particles is beneficial to the uniform distribution of cementing agent. When the ratio of coarse/fine coal particles was about 3:7, the cement and aggregate were well mixed, and the specimen also had a compact structure.

In addition, it can be seen that gypsum with its extremely small particle size may reduce the gap between particles and the porosity would be reduced to a certain extent, which helps to reduce the permeability of reconstituted coal.

(a)

(b)

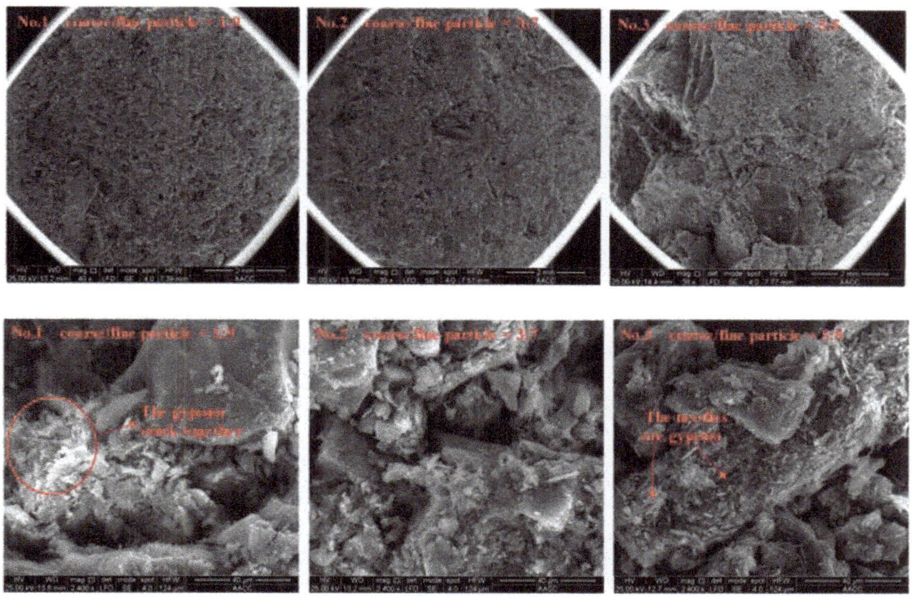

Figure 13. SEM images of three specimens composed with different coarse/fine coal particle ratios: (**a**) the morphology diagrams and (**b**) internal structures of the three cut specimens.

3.5.2. Optimization of Sample Preparation Method and Deformation Characteristic Controlling Binder

Samples made by directly pressing the coal powder have lower strengths larger deformation than the natural coal. Consequently, studies in the past [10,13,18,19,21,29] usually used Portland cement to improve the strength of coal-like materials. However, the mechanical properties of cement keep changing over time. The molding and air curing time (usually 28 days) also have significant influence on the mechanical strengths of the sample and should be considered as additional influencing factors in the preparation process. As a result, the sample preparation process becomes more complex, which makes it hard to ensure the homogeneity and reproducibility. In order to solve this problem, this study used gypsum instead of cement and improved the sample preparation method as discussed in Section 2.3.3 (i.e., using quantitative compaction samples and self-designed molding and demolding equipment). However, gypsum as a binder is less effective than cement in improving the mechanical parameters of materials and so it is more suitable for the preparation of low-strength (0 to 2.5 MPa) coal-like materials. More importantly, gypsum and petrolatum were selected as the compound binder to adjust the deformation characteristics of coal-like materials. Similar materials with ductile to brittle deformation features could be obtained by adjusting the mass ratio of gypsum/petrolatum.

3.5.3. Porosity Adjustment Method

Among the properties summarized in Table 9, the porosity difference was large (11.73%) while the difference in adsorption and desorption parameters was smaller (3.63–7.41%). This indicated that the specific surface area was not very different between the natural coal and coal-like material. Rather, the gap between particles was larger in the coal-like materials. Studies have shown that the porosity of coal-like materials decreases with increasing molding pressure [12,48]. Thus, the porosity could be adjusted by changing the molding pressure slightly but it could not be reduced by adding inorganic matter. For example, sand will decrease the coal powder content and gas adsorption/desorption capacity and the latter is the most important index in coal and gas outburst simulation tests [6].

3.5.4. Limitation and Future Direction to Improve the Similarity of Solid-Gas Coupling Coal-Like Materials

It can be seen from Table 9 that the difference in gaseous parameters between coal-like material and natural coal was larger than that that in the mechanical parameters. This was mainly because the coal-like material contained inorganic matters (gypsum and calcium carbonate) and petrolatum that do not have gas adsorption properties. Future new binders that could simultaneously increase the strength and adsorption characteristics of coal-like materials will further improve the similarity and broaden the application prospects.

3.5.5. Application Prospects of Solid-Gas Coupling Coal-Like Materials

Quantitative simulation tests of coal and gas outbursts and CO_2 sequestration need to ensure that the coal-like material has highly consistent mechanical, deformation and gas adsorption/desorption characteristics with the natural coal [13,19–21]. In the past, such similarity was difficult to achieve [6,10,13,18,19,29], mainly because the deformation characteristics and gas adsorption/desorption characteristics were less considered. This study tested multiple indexes including the density, compressive strength, elastic modulus, cohesion, internal friction angle, porosity and gas adsorption and desorption parameters. The deformation features and damage modes of coal-like materials were also considered. Finally, samples with highly consistent stress-strain curves to the natural coal were obtained.

Since the tectonically deformed coal seam is extremely soft and fragile, it remains difficult to obtain natural coal samples that meet the test requirements and the related triaxial/long-term mechanical properties are still unclear. Existing studies have shown that the strength and deformation characteristics of reconstituted coal-like materials and natural coal were similar [20,23,27]. In view of the highly consistent mechanical properties, the homogeneity and the reproducibility of the coal-like material developed in this study; the synthetic material could be used in place of natural coal sample to study the triaxial and rheological mechanical properties of soft tectonically deformed coal. The relevant experimental results will help guide the design of roadway support for soft coal seams, the stability control of the gas drainage borehole [49,50] and the prediction of coal and gas delayed outburst.

4. Conclusions

In this study, solid-gas coupling coal-like materials were developed according to the characteristics of low strength, strong plastic deformation and large gas adsorption capacities of natural TDC. Coal powder was used as the aggregate, gypsum and petrolatum as the composite binder, calcium carbonate as the additive and vertical ground stress as the molding pressure. The orthogonal method was used to design 25 groups of material ratio schemes. The physical and mechanical parameters, deformation features and gaseous constants were obtained by carrying out uniaxial compression, variable angle shearing and gas adsorption/desorption tests. The following conclusions were obtained.

(1) The range analysis indicates that the gypsum/petrolatum ratio played the main role in controlling the sample density, compressive strength, elastic modulus, cohesion and deformation characteristic. The coarse coal powder (1–2 mm)/fine coal powder (0–0.5 mm) ratio was mainly in control of the sample internal friction angle and also had an important influence on the compressive strength and elastic modulus. When the fine coal powder/aggregate ratio was about 70%, the sample had the highest strength. With the gypsum/petrolatum ratio increasing from 5:5, 6:4, 7:3, 8:2 to 9:1, the deformation characteristics of coal-like materials transitioned from ductile to brittle; and the

failure modes included typical squeezing damage, wedge splitting, plastic shear failure, brittle shear and brittle fracturing. The content of coal powder played a decisive role in the gas adsorption and desorption characteristics of coal-like materials. An empirical formula between coal powder content and gas adsorption index was also established.

(2) The microstructures of specimens indicate that an even mixing with the cementing agent is difficult when there is a large ratio of fine coal particles, resulting in poor strength stability and low homogeneity of the sample. When the ratio of coarse/fine coal particles was about 3:7, the binder and aggregate were relatively well mixed and the specimen had a compact structure.

(3) A multivariate linear model was established between the mechanical characteristic indexes and their main influencing factors. Then, the desired material ratio was determined using this model and the developed coal-like material showed high similarity in its mechanical and physicochemical properties to the natural coal sample. Meanwhile, developing new binders that simultaneously improve sample strength and gas adsorption/desorption performance would help further reduce the difference between coal-like material and natural coal.

(4) The method for preparing solid-gas coupling coal-like material was optimized and the developed samples had good reproducibility and homogeneity. This study lays the foundation for future quantitative work in the physical simulation of coal and gas outbursts, the long-term safety study of CO_2 sequestration with coal in the lab and other experiments on the mechanical properties of soft coal seams.

Author Contributions: X.W. prepared the manuscript and conducted the investigation, Z.S. was responsible for the methodology and software, Q.T. carried out the data processing, W.X. acquired the funding and wrote, reviewed and edited the manuscript.

Funding: This research was funded by State Key Laboratory of Coal Resources and Safe Mining, China University of Mining and Technology (Grant No. SKLCRSM15X01) and National Key Research and Development Plan (Grant No. 2017YFC0603001).

Acknowledgments: We wish to thank Wanxin Song for supporting this experiment, as well as Xiaoying Fan for her great encouragement during the research process. We would like to thank Editage [www.editage.cn] for English language editing.

Conflicts of Interest: The authors declare no conflict of interest.

References

1. Valliappan, S.; Zhang, W.H. Role of gas energy during coal outbursts. *Int. J. Numer. Meth. Eng.* **1999**, *44*, 875–895. [CrossRef]
2. Lama, R.D.; Bodziony, J. Management of outburst in underground coal mines. *Int. J. Coal. Geol.* **1998**, *35*, 83–115. [CrossRef]
3. Wang, F.; Zhao, X.; Liang, Y.; Li, X.; Chen, Y. Calculation Model and Rapid Estimation Method for Coal Seam Gas Content. *Processes* **2018**, *6*, 223. [CrossRef]
4. Beamish, B.B.; Crosdale, P.J. Instantaneous outbursts in underground coal mines: An overview and association with coal type. *Int. J. Coal. Geol.* **1998**, *35*, 27–55. [CrossRef]
5. Skoczylas, N.; Dutka, B.; Sobczyk, J. Mechanical and gaseous properties of coal briquettes in terms of outburst risk. *Fuel* **2014**, *134*, 45–52. [CrossRef]
6. Zhao, B.; Wen, G.C.; Sun, H.T.; Sun, D.L.; Yang, H.M.; Cao, J.; Dai, L.C.; Wang, B. Similarity criteria and coal-like material in coal and gas outburst physical simulation. *Int. J. Coal Sci. Technol.* **2018**, *5*, 167–178. [CrossRef]
7. Chen, L.; Wang, E.Y.; Ou, J.C.; Fu, J.W. Coal and gas outburst hazards and factors of the No. B-1 Coalbed, Henan, China. *Geosci. J.* **2018**, *22*, 171–182. [CrossRef]
8. Wang, G.; Wu, M.M.; Wang, H.Y.; Huang, Q.M.; Zhong, Y. Sensitivity analysis of factors affecting coal and gas outburst based on an energy equilibrium model. *Chin. J. Rock Mech. Eng.* **2015**, *34*, 238–248. (In Chinese) [CrossRef]
9. Yuan, L. Control of coal and gas outbursts in Huainan mines in China: A review. *J. Rock Mech. Geotech. Eng.* **2016**, *8*, 559–567. [CrossRef]

10. Wang, G.; Li, W.X.; Wang, P.F.; Yang, X.X.; Zhang, S.T. Deformation and gas flow characteristics of coal-like materials under triaxial stress conditions. *Int. J. Rock Mech. Min. Sci.* **2017**, *91*, 72–80. [CrossRef]
11. Peng, S.J.; Xu, J.; Yang, H.W.; Liu, D. Experimental study on the influence mechanism of gas seepage on coal and gas outburst disaster. *Saf. Sci.* **2012**, *50*, 816–821. [CrossRef]
12. Skoczylas, N. Laboratory study of the phenomenon of methane and coal outburst. *Int. J. Rock Mech. Min. Sci.* **2012**, *55*, 102–107. [CrossRef]
13. Hu, Q.T.; Zhang, S.T.; Wen, G.C.; Dai, L.C.; Wang, B. Coal-like material for coal and gas outburst simulation tests. *Int. J. Rock Mech. Min. Sci.* **2015**, *74*, 151–156. [CrossRef]
14. Sun, H.T.; Cao, J.; Li, M.H.; Zhao, X.S.; Dai, L.C.; Sun, D.L.; Wang, B.; Zhai, B.N. Experimental Research on the Impactive Dynamic Effect of Gas-Pulverized Coal of Coal and Gas Outburst. *Energies* **2018**, *11*, 797. [CrossRef]
15. Alexeev, A.D.; Revva, V.N.; Alyshev, N.A.; Zhitlyonok, D.M. True triaxial loading apparatus and its application to coal outburst prediction. *Int. J. Coal. Geol.* **2004**, *58*, 245–250. [CrossRef]
16. Clayton, J.L. Geochemistry of coalbed gas—A review. *Int. J. Coal. Geol.* **1998**, *35*, 159–173. [CrossRef]
17. Prusty, B.K. Sorption of methane and CO_2 for enhanced coalbed methane recovery and carbon dioxide sequestration. *J. Nat. Gas Chem.* **2008**, *17*, 29–38. [CrossRef]
18. Jasinge, D.; Ranjith, P.G.; Choi, S.K.; Kodikara, J. Mechanical Properties of Reconstituted Australian Black Coal. *J. Geotech. Geoenviron. Eng.* **2009**, *135*, 980–985. [CrossRef]
19. Jasinge, D.; Ranjith, P.G.; Choi, S.K. Development of a reconstituted brown coal material using cement as a binder. *Environ. Earth Sci.* **2011**, *64*, 631–641. [CrossRef]
20. Jasinge, D.; Ranjith, P.G.; Choi, X.; Fernando, J. Investigation of the influence of coal swelling on permeability characteristics using natural brown coal and reconstituted brown coal specimens. *Energy* **2012**, *39*, 303–309. [CrossRef]
21. Ranjith, P.G.; Shao, S.S.; Viete, D.R.; Jaysinge, D. Carbon Dioxide Storage in Coal: Reconstituted Coal as a Structurally Homogeneous Substitute for Coal. *Int. J. Coal Prep. Util.* **2012**, *32*, 265–275. [CrossRef]
22. Skochinski, A.A. *Communication of the Initiation of a Sudden Outburst of Gas and Coal in the Model in Outburst Laboratory of the Institute of Mining of AN, SSSR*; Ugol: Novokuznetsk, Russia, 1953; p. 10.
23. Yin, G.Z.; Huang, Q.X.; Zhang, D.M.; Wang, D.K. Test study of gas seepage characteristics of gas-bearing coal specimen during process of deformation and failure in geostress field. *Chin. J. Rock Mech. Eng.* **2010**, *29*, 336–343. (In Chinese)
24. Jiang, C.B.; Yin, G.Z.; Li, X.Q.; Cai, B. Experimental study of gas permeability of outburst coal briquettes in complete stress-strain process. *Chin. J. Rock Mech. Eng.* **2010**, *29*, 3482–3487. (In Chinese)
25. Cao, S.G.; Li, Y.; Guo, P.; Bai, Y.J.; Liu, Y.B. Comparative research on permeability characteristics in complete stress-strain process of briquettes and coal samples. *Chin. J. Rock Mech. Eng.* **2010**, *29*, 899–906. (In Chinese)
26. Chen, H.D.; Cheng, Y.P.; Zhou, H.X.; Li, W. Damage and Permeability Development in Coal During Unloading. *Rock Mech. Rock Eng.* **2013**, *46*, 1377–1390. [CrossRef]
27. Xu, J.; Geng, J.B.; Peng, S.J.; Liu, D.; Nie, W. Acoustic emission characteristics of coal and gas outburst under different moisture contents. *J. China Coal Soc.* **2015**, *35*, 1047–1054. [CrossRef]
28. Jasinge, D.; Ranjith, P.G.; Kodikara, J.; Al, E. A Comparision of Stress Strain Behavior of Reconstituted and Natural Black Coal. In Proceedings of the Congress of the 11th ISRM of the International Society of Rock Mechanics, Lisbon, Portugal, 9–13 July 2007.
29. Xu, J.; Ye, G.B.; Li, B.B.; Cao, J.; Zhang, M. Experimental study of mechanical and permeability characteristics of moulded coals with different binder ratios. *Rock Soil Mech.* **2015**, *36*, 104–110. (In Chinese) [CrossRef]
30. Zhang, Q.H.; Yuan, L.; Wang, H.P.; Kang, J.H.; Li, S.C.; Xue, J.H.; Zhou, W.; Zhang, D.M. Establishment and analysis of similarity criteria for physical simulation of coal and gas outburst. *J. China Coal Soc.* **2016**, *41*, 2773–2779. (In Chinese) [CrossRef]
31. Zhang, S.T.; Lin, D.C.; Wang, B.; Cao, J. Experimental study of mixture ratio of similar material for simulation of coal and gas outburst. *Coal Sci. Technol.* **2015**, *43*, 76–80. (In Chinese) [CrossRef]
32. Wang, J.L.; Li, M.; Xu, S.C.; Qu, Z.H.; Jiang, B. Simulation of Ground Stress Field and Advanced Prediction of Gas Outburst Risks in the Non-Mining Area of Xinjing Mine, China. *Energies* **2018**, *11*, 1285. [CrossRef]
33. Yin, W.T.; Fu, G.; Yang, C.; Jiang, Z.G.; Zhu, K.; Gao, Y. Fatal gas explosion accidents on Chinese coal mines and the characteristics of unsafe behaviors: 2000–2014. *Saf. Sci.* **2017**, *92*, 173–179. [CrossRef]

34. Wang, C.J.; Yang, S.Q.; Yang, D.D.; Li, X.W.; Jiang, C.L. Experimental analysis of the intensity and evolution of coal and gas outbursts. *Fuel* **2018**, *226*, 252–262. [CrossRef]
35. Li, H.Y. Major and minor structural features of a bedding shear zone along a coal seam and related gas outburst, Pingdingshan coalfield, northern China. *Int. J. Coal. Geol.* **2001**, *47*, 101–113. [CrossRef]
36. Cao, Y.X.; Mitchell, G.D.; Davis, A.; Wang, D.M. Deformation metamorphism of bituminous and anthracite coals from China. *Int. J. Coal. Geol.* **2000**, *43*, 227–242. [CrossRef]
37. Ju, Y.W.; Jiang, B.; Hou, Q.L.; Tan, Y.J.; Wang, G.L.; Xiao, W.J. Behavior and mechanism of the adsorption/desorption of tectonically deformed coals. *Chin. Sci. Bull.* **2009**, *54*, 88–94. [CrossRef]
38. Ju, Y.W.; Jiang, B.; Hou, Q.L.; Wang, G.L. The new structure-genetic classification system in tectonically deformed coals and its gelological significance. *J. China Coal Soc.* **2004**, *29*, 513–517. (In Chinese) [CrossRef]
39. China National Coal Association. *Methods for Determining the Physical and Mechanical Properties of Coal and Rock—Part 7: Methods for Determining the Uniaxial Compressive Strength and Counting Softening Coefficient*; GB/T 23561.7-2009; China National Coal Association: Beijing, China, 2009.
40. China National Coal Association. *Methods for Determining the Physical and Mechanical Properties of Coal and Rock—Part 11: Methods for Determining Shear Strength of Coal and Rock*; GB/T 23561.11-2010; China National Coal Association: Beijing, China, 2010.
41. Mansouri, H.; Ajalloeian, R. Mechanical behavior of salt rock under uniaxial compression and creep tests. *Int. J. Rock Mech. Min. Sci.* **2018**, *110*, 19–27. [CrossRef]
42. Gao, Y.J. Analysis of the Factors Influencing Briquette Forming and Design of Roll Briquetting Machine. Master's Thesis, Shanxi University, Taiyuan, China, 2009. (In Chinese with English abstract).
43. Liu, S.; Liu, W. Experimental Development Process of a New Fluid–Solid Coupling Similar-Material Based on the Orthogonal Test. *Processes* **2018**, *6*, 211. [CrossRef]
44. Brown, E.T. *Suggested Method for Determining the Uniaxial Compressive Strength of Rock Materials, Rock Characterization, Testing and Monitoring*; ISRM Suggested Methods; Pergamon Press: Oxford, UK, 1981.
45. Yang, D.D.; Chen, Y.J.; Tang, J.; Li, X.W.; Jiang, C.L.; Wang, C.J.; Zhang, C.J. Experimental research into the relationship between initial gas release and coal-gas outbursts. *J. Nat. Gas Sci. Eng.* **2018**, *50*, 157–165. [CrossRef]
46. Laxminarayana, C.; Crosdale, P.J. Role of coal type and rank on methane sorption characteristics of Bowen Basin, Australia coals. *Int. J. Coal. Geol.* **1999**, *40*, 309–325. [CrossRef]
47. Ruppel, T.C.; Grein, C.T.; Bienstock, D. Adsorption of methane on dry coal at elecated pressure. *Fuel* **1974**, *53*, 152–162. [CrossRef]
48. Jasinge, D.; Ranjith, P.G.; Choi, S.K. Effects of effective stress changes on permeability of latrobe valley brown coal. *Fuel* **2011**, *90*, 1292–1300. [CrossRef]
49. Tang, J.; Jiang, C.L.; Chen, Y.J.; Li, X.W.; Wang, G.D.; Yang, D.D. Line prediction technology for forecasting coal and gas outbursts during coal roadway tunnelling. *J. Nat. Gas Sci. Eng.* **2016**, *34*, 412–418. [CrossRef]
50. Xue, Y.; Gao, F.; Gao, Y.; Chen, H.M.; Liu, Y.K.; Hou, P.; Teng, T. Quantitative evaluation of stress-relief and permeability-increasing effects of overlying coal seams for coal mine methane drainage in Wulan coal mine. *J. Nat. Gas Sci. Eng.* **2016**, *32*, 122–137. [CrossRef]

processes

Article

Experimental Investigation of Pore Structure and Movable Fluid Traits in Tight Sandstone

Dengke Liu [1,*], Wei Sun [1] and Dazhong Ren [2]

1 Department of Geology, State Key Laboratory of Continental Dynamics, Northwest University, Xi'an 710069, China; sunwei5393901@sohu.com
2 College of Petroleum Engineering, Xi'an Shiyou University, Xi'an 710065, China; petro_gas@163.com
* Correspondence: liudengke02@126.com; Tel.: +86-1871-091-6008

Received: 6 February 2019; Accepted: 2 March 2019; Published: 8 March 2019

Abstract: Whether the variation of pore structures and movable fluid characteristics enhance, deteriorate, or have no influence on reservoir quality has long been disputed, despite their considerable implications for hydrocarbon development in tight sandstone reservoirs. To elucidate these relationships, this study systematically analyzes pore structures qualitatively and quantitatively by various kinds of direct observations, indirect methods, and imaging simulations. We found that the uncertainty of porosity measurements, caused by the complex pore-throat structure, needs to be eliminated to accurately characterize reservoir quality. Bulk water was more easily removed, while surface water tended to be retained in the pores, and the heterogeneity of pore structures was caused by the abundance of tiny pores. The rates of water saturation reduction in macropores are faster than those for tiny pores, and sandstones with poor reservoir quality show no marked descending of lower limits of movable pore radius, indicating that the movable fluid would advance exempted from the larger pores. This study suggests that the deterioration of reservoir quality is strongly affected by the reduction of larger pores and the aqueous phases tended to remain in the tiny pores in the forms of surface water.

Keywords: pore structure; movable fluid; tight sandstones; Ordos Basin

1. Introduction

With depletion of conventional hydrocarbon production and increasing demand for energy, unconventional reservoirs are expected to be a vital and realistic inventory in future energy supply [1,2]. As one of the major unconventional resources, tight sandstones reservoirs, to which many scholars have drawn attention, have been developed successfully in China [3–6]. The definition of tight sandstone in petroliferous basins is that its porosity is no more than 10% and in suit permeability generally less than 1 mD or air permeability less than 0.1 mD, whereupon it needs specific treatments (horizontal well fracturing, correct way of completion, advanced simulation jobs, etc.) to produce commercial hydrocarbon flow [7,8]. The complexity and irregularity of the pore structure are crucial microscopic factors for the deteriorated physical properties, poor reservoir quality, and complicated oil emplacement; therefore, comprehensive and accurate pore geometry evaluation has become a momentous issue in tight sandstones [9,10]. Meanwhile, due to the heterogeneity of pore structure in tight sandstones, the fluid that flows in the porous media is more complex than that of conventional reservoirs because of the intricate flow path [11]. Movable fluid parameters are significant in that they can provide a reliable evaluation of the characteristics of the fluid, and accurate movable fluid parameter estimation has been a point of interest because movable fluid parameters are valuable when considering the water retention and residual oil accumulation and development [12]. Therefore, the assessment of pore structure (porosity, pore size distribution (PSD), connectivity, tortuosity, heterogeneity, pore shape, etc.) and movable fluid features (movable fluid percentage, movable fluid porosity, etc.) are of paramount importance in evaluating the microscopic characterization of tight sandstones [13,14]. Accurate prediction of pore structure and movable fluid traits helps to assess storage capacity and evaluate percolation capacity and can provide insights into planning field exploration and development [15].

Currently, a series of optical techniques and experiments are being conducted to investigate the pore structure of tight sandstones, including radiation methods (e.g., thin section (TS), scanning electron microscope (SEM), small angle neutron scattering (SANS), computed tomography (CT), etc.), intrusive techniques (e.g., pressure-controlled mercury intrusion (PCMI), rate-controlled mercury intrusion (RCMI), etc.) and non-intrusive techniques (e.g., nuclear magnetism resonance (NMR), low-temperature N_2 adsorption (LTNA), etc.) [2,8,10–13,15–17]. With continuous increase in the resolution of optical microscopy and radiation methods, the field of views could be small and the costs could rise. Due to this trade-off between resolution and viewshed information, indirect techniques have been supplemented, which systematically compensate the limits of direct techniques in characterizing pore structure in tight sandstones [8,14]. Although many methods can be used to characterize pore structures, no sole direct and indirect experiment can be used to determine pore structures traits due to its own limitations and strengths [18]: TS and SEM can reveal pore morphology and occurrence qualitatively, but these techniques were unable to obtain quantitative pore structure-related data [19,20]. CT is an effective technique to quantitatively characterize the three-dimensional images of pore networks, but it has drawbacks in the calculation of pores larger than 3 μm and is often restricted by expense [2,21,22]. PCMI is understood to be a widely-used method of estimation of pore structures because it can detect a broad spectrum of PSD on account of relatively high injection pressure according to the Washburn equation [23]; however, this technique fails to acquire an exact count of large pores because of the shielding effect of small pores [24]. Unlike PCMI, RCMI injects mercury into pores at very low speed that keeps the interfacial tension and contact angle changeless and can partition the PSD into areas of pores and throats based on the periodic rising and falling of pressures, overcoming the pitfall in PCMI [2,15,25–27]. While it is impossible to detect pores with radius less than 0.12 μm due to the maximum injection pressure limitation (~900 psi/6.2 MPa), as a result, RCMI can be envisaged as a limited technique that exaggerates the proportion of large pores [10,28]. NMR has been established as an effective method to determine pore structure characterization [29,30], nevertheless, the calculation of pore radius requires calibration with other techniques (e.g., RCMI) to confirm surface relaxivity, namely, it would cause discrepancy between NMR-derived PSD and other independent pore size measurement if the process of surface relaxivity calibration is skipped and referred to an arbitrary pore size classification that is derived from other research areas [15,31,32]. Therefore, in order to better evaluate pore structure characterization, a combination of each technique is necessary [33,34].

Recently, many studies have documented that NMR is a fast and nondestructive method to determine flow characteristic in tight sandstones [11–13,35–37]. As for the characteristic parameters of fluid in tight sandstones, movable fluid saturation is commonly used NMR-derived parameters, which is related to the pore structures and can be regulated by T_2 cutoff values [12]. T_2 cutoff values, defined here as the decay time after the external magnetic simulation disappeared, correspond to threshold partitioning of the T_2 spectrum into two subcurves, namely, the free fluid areas (corresponding to T_2 higher than the threshold) and bound fluid areas (corresponding to T_2 lower than the threshold); hence, rigorous calculation of T_2 cutoff values is vital to exactly determine the percentage of free water [12]. The determination of T_2 cutoff values needs to compare the T_2 spectrum before and after centrifuging in a certain rotation speed because it can dramatically change among different specimens, even in the same research area; rather, empirically determined T_2 cutoff values ascertaining the proportion of movable fluid on the basis of predecessors is too arbitrary, and accurate determination of optimum centrifugal forces to obtain T_2 cutoff values requires an understanding of the calibration model [12,13,36,38,39]. Movable fluid percentage is a significant property used to predict estimated ultimate recovery (EUR), because the irreducible fluids were not productive at all [11]; thus, it is of great importance to quantitatively determine the movable fluid parameters, in order to provide insight into evaluating the fluid storage and percolation traits and help develop proper hydrocarbon exploitation strategies.

Considering the significance of pore structure and movable fluid for hydrocarbon development in tight sandstones, a comprehensive analysis of pore structures and evaluation of movable fluid are necessary to evaluate the mechanism of hydrocarbon storage and percolation in the investigated basin [40,41]. In our study, a suit of tight sandstone samples from the Upper Triassic in the Ordos basin was investigated by TS, SEM, PCMI, RCMI, NMR, and CT in order to reveal the characteristics of pore structure and movable fluid traits. We first lay out the spatial features of the pores needed for the qualitative evaluation of our work by TS and SEM. Then we propose an integrated method by a combination of PCMI, RCMI, and CT to describe PSD, and an improved PSD is elucidated. Further, we determine the optimum rotation speed and movable fluid traits and explain how pore structure and rock physical properties influence the movable fluid parameters by NMR-derived T_2 spectrum analysis before and after centrifuging at different centrifugal force, T_1-T_2 correlation spectra and magnetic resonance imaging (MRI). Finally, we investigate the effect of pore radius lower limits on reservoir quality. This multidisciplinary research unravels the impact of pore structure and movable fluid features on reservoir quality of tight sandstones, which is critical for the future production of the Upper Triassic tight sandstone reservoirs in Ordos Basin, China and offer technological merits in similar reservoirs elsewhere.

2. Materials and Methods

2.1. Tight Sandstones Specimens

All specimens were taken from the Upper Triassic Yanchang tight sandstones in the Yishan slope of the Ordos Basin in NW China (Figure 1a). It is a cratonic basin with an area of around 2.5×10^5 km^2, and Yishan slope is a hydrocarbon enrichment area, where dips at no more than 1° toward the west and internal faults are rarely seen [42,43]. The Yanchang formation, which is divided into ten members (Figure 1b), consists of sandstones intercalated with mudstones and shales, and Chang 2, Chang 4 + 5, Chang 6 and Chang 8 members are composed predominantly of sandstones [44,45]. The Yanchang formation is dominated by fine-to moderate-grained sands deposited in a fluvial, delta and lacustrine environment; compaction, cementation, and dissolution represent the most significant diagenetic process influencing pore structure and movable fluid traits [46,47]. In the present study, sandstones specimens from the Yanchang formation were collected from 14 wells, and the porosities of all specimens range from 5.53% to 14.24% with an average of 9.33%, while permeabilities vary from 0.02 mD to 2.70 mD with a mean value of 0.68 mD (Table 1), which indicates that the sandstones are typical tight sandstones [7]. The tight sandstones specimens of Yanchang formation were mainly arkose according to Folk's (1980) classification [47], and quartz is predominant in specimens (av. 37.38%) with a subdominant component of feldspar (av. 31.86%) and low rock fragments (average contents of metamorphic, volcanic and sedimentary rock fragments equal to 7.84%, 3.82%, and 4.56%, respectively) (Table 1). The clay minerals are dominated by chlorite (av. 4.81%) and illite (av. 4.72%), whereas kaolinite is less dominant (av. 1.50%) (Table 1). The mica and quartz overgrowth contents vary from 1.00% to 11.50% (av. 4.32%) and 1.00% to 4.00% (1.90%), respectively (Table 1).

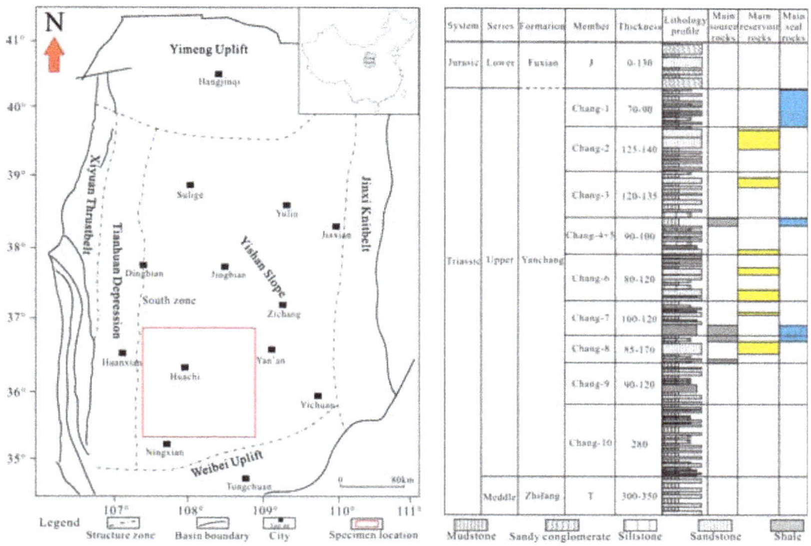

(a) (b)

Figure 1. (**a**) Location map that shows the research area in the southwestern Ordos Basin, China (modified after Gao and Li, 2015 [13]; Cui et al., 2017 [48]) and (**b**) generalized stratigraphic column of the research area (modified after Lai et al., 2016 [49]).

Table 1. Physical properties and petrological composition of Yanchang formation tight sandstones [1].

SN	φ, %	K, mD	Q, %	F, %	V, %	M, %	S, %	Mi, %	Ka, %	Il, %	Ch, %	Ca, %	QO, %
1#	9.60	0.59	44.00	26.00	3.00	9.00	6.00	4.00	-	5.00	-	3.00	-
2#	10.01	0.76	48.00	23.00	3.00	6.00	1.00	9.00	-	3.00	-	4.00	3.00
3#	10.47	1.56	21.00	52.00	3.00	10.00	3.00	3.00	-	-	7.00	1.00	-
4#	7.79	0.21	44.00	18.00	3.00	11.00	6.00	2.00	2.00	3.00	1.00	6.00	4.00
5#	7.68	1.41	32.50	32.00	2.00	4.50	3.00	8.00	-	2.00	2.00	6.00	2.00
6#	8.23	0.18	22.80	40.00	2.50	5.20	5.70	11.50	-	4.50	5.30	0.50	2.00
7#	7.03	0.11	48.00	19.00	4.00	7.00	9.00	2.00	1.00	3.00	1.00	3.00	3.00
8#	5.53	0.02	48.00	23.00	3.00	9.00	4.00	3.00	-	7.00	-	2.00	1.00
9#	8.95	0.12	21.00	50.00	3.00	8.00	6.00	2.00	-	-	8.00	2.00	-
10#	12.16	0.32	47.00	23.00	3.00	8.00	4.00	4.00	-	4.00	-	6.00	1.00
11#	14.24	2.70	40.00	33.00	8.00	9.00	-	1.00	-	-	7.00	1.00	1.00
12#	12.54	0.48	25.00	53.00	2.00	6.00	3.00	4.00	-	-	6.00	1.50	-
13#	5.68	0.05	46.00	22.00	3.00	7.00	4.00	4.00	-	11.00	-	2.00	1.00
14#	10.70	0.96	36.00	32.00	11.00	10.00	-	3.00	-	-	6.00	1.00	1.00

[1] SN—specimen number; φ—porosity; K—permeability; Q—quartz; F—feldspar; V—volcanic rock fragments; M—metamorphic rock fragments; S—sedimentary rock fragments; Mi—mica; Ka—kaolinite; Il—illite; Ch—chlorite; Ca—carbonate minerals; QO—quartz overgrowth.

2.2. Experimental Methods

First, the rock cores selected from wells were drilled parallel to the bedding plane and cylindrical core plugs with approximately 5 cm long and 2.54 cm in diameter were obtained. The specimens were polished and cleaned with solutions of alcohol and trichloromethane before the experiments in order to remove the bitumen or drilling mud that remained in pores from the specimens, and then each specimen was put into the vacuum systems and dried at 120 °C for 24 h. Subsequently, the specimens were put in the core holder of the FYK-I testing apparatus, then the helium-based porosity was repeatedly tested five times and averaged. After the porosity tests, the pressure-transient nitrogen pulse decay permeability test was carried out using the FYK-I instrument and the specimen was tautologically measured five times for an average.

After the physical property measurement, the specimen plugs were split into several chips and plugs for direct observations and indirect tests. The TS analyses were conducted on rock slices that were impregnated by red or blue epoxy resin, using ZEISS polarized light petrographic microscope to detect the pore geometry. SEM, which was used to observe the microscopic features of the tight sandstones, was conducted by a FEI Quanta 400 FEG scanning electron microscope with working accelerating voltage set at 20 kV and the samples were covered with 10 nm thick carbon films. An OXFORD IE 350 energy dispersive spectrometer was used to collect XRD data, and each specimen was crushed and ground into powder with a grain size of 300 mesh, then dried and mixed with ethylene glycol and heated to 550 °C for 3 h before the tests.

The PCMI test was conducted using a Micromeritic Autopore IV 9420 Instrument. Injection pressure was reached at 200 MPa, which corresponded to 3.6 nm based on the Washburn equation [23], then the pressure progressively decreased to zero and the intrusion and extrusion capillary pressure curves were acquired. The ASPE 730 mercury porosimeter was adopted for the purpose of RCMI measurements. Unlike the PCMI experiment, during the injection process in RCMI, mercury was injected into the specimen at a speed of 5×10^{-5} mL·min^{-1} in order to stabilize the interfacial tension and contact angle. The pressure increased as the mercury was injected into the throat, then an instantaneous drawdown was seen when the mercury entered the pores. The periodic rising and falling pressure was recorded by the software and real-time data were acquired to partition the void spaces into distributions of pores and throats [26,38]. Both mercury intrusion experiments followed the SY/T 5346-2005 Chinese Oil and Gas Industry Standard, and the contact angle and interfacial tension was 140° and 480 mN·m^{-1}, respectively.

Micro-CT scanning can provide a relatively high-resolution 3D reconstruction of the features of the pore network, when compared to other methods [2,50–52]. The 3D images of the pore characterization of tight sandstone were measured by a Phoenix Nanotom M scanner with a working voltage of 90 kV, and tomogram image spatial resolution of around 3 μm. In this method, an X-ray with conical beam penetrates the specimen and then attenuates depending on the sample density, mineral compositions and contents, and the thickness along the beam direction, and this attenuation is mainly associated with the decreasing sample density [22,53]. Finally, based on the threshold value segmentation method, rock matrix with high density and void space with low density can be determined [22].

The NMR is a nondestructive technique that provides information of pore features, and insights into the fluid and porous media can be acquired by this test [11,54]. The process that helped the magnetization vector recover from a non-equilibrium to equilibrium state is called relaxation, and the recovery time is called relaxation time [11]. The T_1 and T_2 relaxation time, which correspond to the time that the magnetization vector is parallel to and perpendicular to the external magnetic field returned to the equilibrium state, is referred to as longitudinal and transverse relaxation time, respectively [55]. This technique was performed in Niumag NMR spectrometer at a proton resonance frequency of 2 MHz and detected by a CPMG pulse sequence [55]. We set the echo number for 30 and 6000 for T_1 and T_2, respectively, and the echo spacing was varied from 10 μs to a few seconds for T_1 and the values of which were equal to 100 μs for T_2. Each specimen was scanned 64 times to get a good signal-to-noise ratio, and the wait time between successive scans was chosen as 5 s. The specimens were saturated in brine, which was a mixture of deionized water and Calcium Chloride ($CaCl_2$) with salinity of 2.5×10^4 mg·L^{-1} to prevent the clay minerals from swelling. It was used for NMR measurement to record initial relaxation time spectra and images. After the T_1 and T_2 spectrum and MRI for the saturated water specimens were determined, the centrifugal processing with a rotation speed of 2500 r·min^{-1}, 2900 r·min^{-1}, 3500 r·min^{-1}, 5000 r·min^{-1}, 7900 r·min^{-1} and 9100 r·min^{-1} for 1 h for each speed were conducted on each specimen, and the T_2 spectra distributions were tested after different rotation speed centrifugal processing was completed. Specifically, the T_1-T_2 correlation spectra distributions and MRI were generated after 9100 r·min^{-1} (corresponding to the maximum rotation speed) for centrifugal processing was implemented to determine the T_2 cutoff values, free and irreducible water distributions.

3. Results

3.1. Pore Structures Observed by TS and SEM

TS and SEM can be used to determine pore types and analyses pore geometry since they provide direct characterization of the pores [17,56]. Four main kinds of pore types are developed in the Upper Triassic Yanchang formation of tight sandstones, including intergranular pores, dissolution pores, intercrystalline pores, and microfractures. The intergranular pores are triangular or polygonal and remain after strong mechanical compaction and clay minerals recrystallization [20], have characteristics of relatively larger pore radius, pore-lining chlorite, and sometimes elongate euhedral quartz prisms (Figure 2a,b). The size of the intergranular pores was typically >40 μm, and the relative abundances of these pores by point counting based on TS observations were in the range from trace levels to 5.9% and averaged as 1.73%. The dissolution pores are generally a result of feldspar and unstable ferromagnesian rock fragments dissolution and show irregular geometry [2,57]; these pore systems are of three categories: intragranular dissolution pores (feldspar leaching generally developed along cleavage planes) (Figure 2c), intergranular dissolution pores (Figure 2d) and moldic pores (the pores derived from complete dissolution of grains and the shapes of the original framework were retained) (Figure 2e) [58]. Point counting results of the dissolution pores reveal a wide range—from trace amounts (due to feldspars leaching) to 1.0% (resulting from the connection of moldic pores and residual intergranular pores) with an average of 0.56%. The intercrystalline pores, which refers to pores within biotite planes, clay aggregates, quartz overgrowths, and carbonate minerals, are commonly small enough (often lower than 10 μm) to be hardly observed in TS; however, the examination under SEM suggested that lamellar chlorite and fibrous illite have widespread intercrystalline pores because of its abundance (Figure 2f,g). The extensive cementation of carbonate minerals can contribute little intercrystalline pores owing to the compacted voids in the lattice (Figure 2h), and pores within sheet like biotite and elongated authigenic quartz only in minor amounts (Figure 2i,j). The microfractures, which are dominated by diagenesis and scarcely observed in cores [59], offer features of microscale in length and nanoscale in width, and can only be detected under microscopic identifications. Microscopic observations of the specimens reveal that the microfractures along the edge of grains and through the grains are the predominate microfracture type in tight sandstones (Figure 2k,l), from which acid solution dissolution and mechanical compaction was derived, respectively [60]. Point counting results show that the contents of microfractures are minimal (<0.2%).

3.2. Pore Structures Characteristics Determined by PCMI

The PCMI-derived curves of mercury drainage and imbibition for the 14 specimens are presented in Figure 3 after Gane calibration [61] in order to make the curves smooth and continuous; the typical parameters for pore structures are listed in Table 2. The specimens could be categorized into four types based on intrusion and extrusion capillary curves. The first type (type α) shows low threshold pressure (P_d) (av. 0.07 MPa), low median pressure (P_{50}) (av. 0.81 MPa), high maximum mercury intrusion saturation (S_{Hgmax}) (av. 89.19%) and medium hysteresis between intrusion and extrusion curves, revealing that well-connected pores play a dominant role in this type of sandstone. Besides, partial percolations (plateau stage in mercury intrusion curves), corresponding to the distinct change in saturation over a small range of capillary pressure (gray boxes)), occurred in some specimens, indicating this type is better sorted (average sorting coefficient equal to 2.49) than other types (Figure 3a). For type β specimens, the average P_d and P_{50} rapidly increase, equal to 0.95% and 8.09%, respectively, and the average S_{Hgmax} decreases (av. 86.33%), while a relatively strong hysteresis is observed (Figure 3b). These phenomena suggest that some poorly-connected pores were developed in this type of sandstone. For the type γ specimens, the mercury intrusion curves display that the P_d for all the specimens exceed 1.16 MPa, P_{50} over 7.06 MPa, and S_{Hgmax} is no more than 87.57%, that is, mercury can hardly enter the pores and the proportion of non-connected pores is relatively high, when compared to type α and β. The intrusion-extrusion cycles show more moderate mercury saturation rise and fall, corresponding to weak hysteresis, and no percolation phenomenon. However, the specimen that has strong hysteresis may result from an abundance of mica, which is deformed in the process of mercury injection and leads to the change of pore structures (Figure 3c). In terms of type δ, the average P_d and P_{50} are the highest, equal to 3.67 MPa and 51.47 MPa, respectively, while the average S_{Hgmax} is no more than 83.85%. Strong hysteresis of these specimens is the evidence of poor-connected pores and minor amounts of clay minerals due to intense mechanical compaction (Figure 3d). According to the multitype pore space model proposed by Sakhaee-Pour and Bryant [40], medium hysteresis for type α suggests that intergranular pores mainly constitute the pore space, whereas the void spaces in type γ is predominantly intercrystalline pores, which corresponds to a relatively poor hysteresis; in addition, the dissolution pores may make the greatest contribution to the void spaces in type β because their diameters are midway between intergranular pores and intercrystalline pores, according to the description in Section 3.1. Figure 4 backs this and explains why more mercury was stranded in the spaces in intergranular and dissolution pore-dominated reservoirs, whereas the void spaces of intercrystalline pores can be thoroughly saturated with the wetting phase.

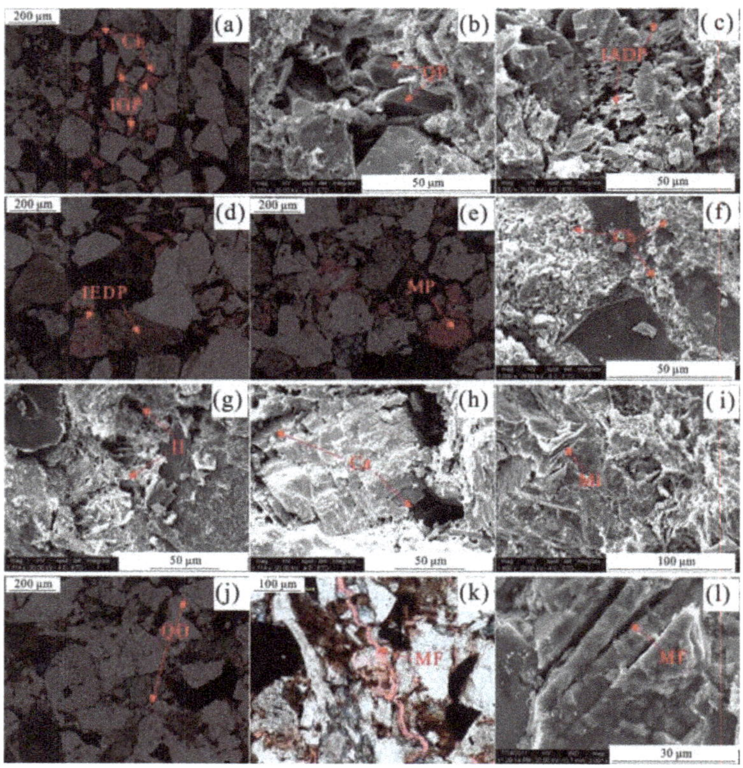

Figure 2. Typical pore types from Yanchang formation sandstones: (**a**) Plane-polarized light view of intergranular pore and presence of pore-lining chlorite; (**b**) SEM image of quartz prisms; (**c**) plane-polarized light view of intragranular dissolution pore; (**d**) plane-polarized light view of intergranular pore; (**e**) plane-polarized light view of moldic pore; (**f**) SEM image of pore-filling chlorite; (**g**) SEM image of illite; (**h**) SEM image of carbonate; (**i**) SEM image of mica; (**j**) plane-polarized light view of quartz overgrowth; (**k**) plane-polarized light view of microfracture; (**l**) SEM image of microfracture. IGP—intergranular pore; Ch—chlorite; QP—quartz prisms; IADP—intragranular dissolution pore; IEDP—intergranular dissolution pore; MP—moldic pore; Il—illite; Ca—carbonate; Mi—mica; QO—quartz overgrowth; MF—microfracture.

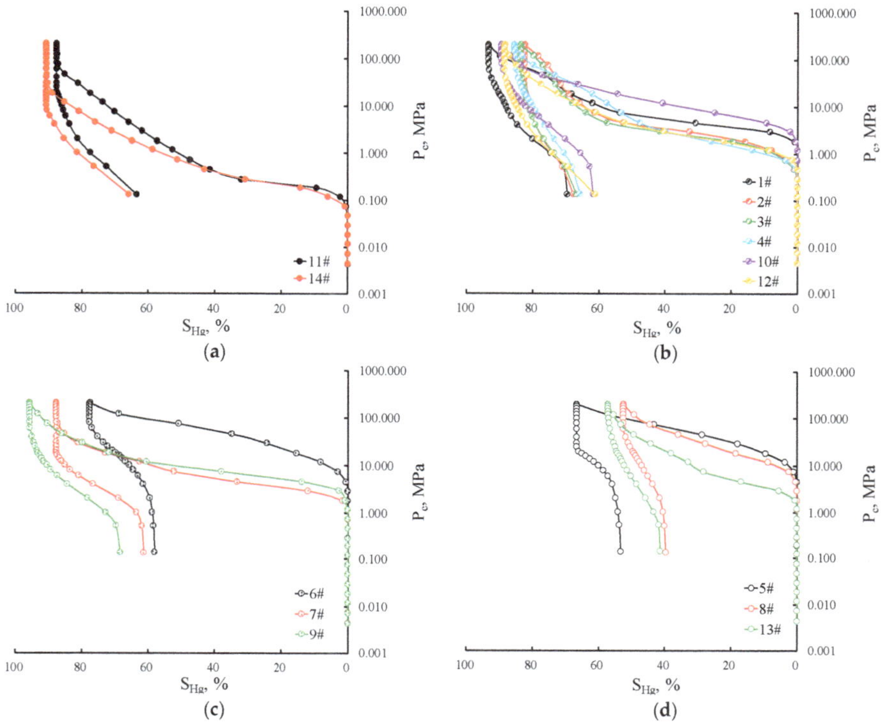

Figure 3. Comparison of four types of PCMI results: (**a**) Type α; (**b**) type β; (**c**) type γ; (**d**) type δ. The gray boxes represent a partial percolation stage.

Table 2. The parameters for pore structure characterization from PCMI [2].

Specimen	P_d, MPa	r_{max}, μm	P_{50}, MPa	r_{50}, μm	S_{Hgmax}, %	W_e, %	Sort	Type
11	0.07	9.94	0.94	0.78	87.61	27.48	2.75	α
14	0.07	9.94	0.68	1.09	90.78	27.39	2.24	α
1	1.80	0.41	14.45	0.05	88.40	34.27	1.99	β
2	0.45	1.62	4.29	0.17	82.47	18.09	2.64	β
3	0.45	1.62	3.75	0.20	83.63	20.34	2.55	β
4	0.45	1.62	5.72	0.13	85.56	23.33	2.67	β
10	1.80	0.41	16.12	0.41	89.46	31.04	1.92	β
12	0.73	1.01	4.21	0.17	88.49	30.93	2.20	β
6	4.49	0.16	72.99	0.01	77.39	25.12	3.24	γ
7	1.81	0.41	7.06	0.10	87.57	30.08	1.88	γ
9	1.16	0.63	10.88	0.07	79.65	23.41	2.94	γ
5	7.39	0.10	97.00	0.01	66.49	20.08	4.31	δ
8	1.81	0.41	7.22	0.10	83.85	19.44	2.39	δ
13	1.81	0.41	50.18	0.01	57.03	27.55	4.45	δ

[2] P_d—threshold pressure; r_{max}—maximum pore radius; P_{50}—median pressure; r_{50}—median pore radius; S_{Hgmax}—maximum mercury intrusion saturation; W_e—efficiency of mercury withdrawal; Sort—sorting coefficient.

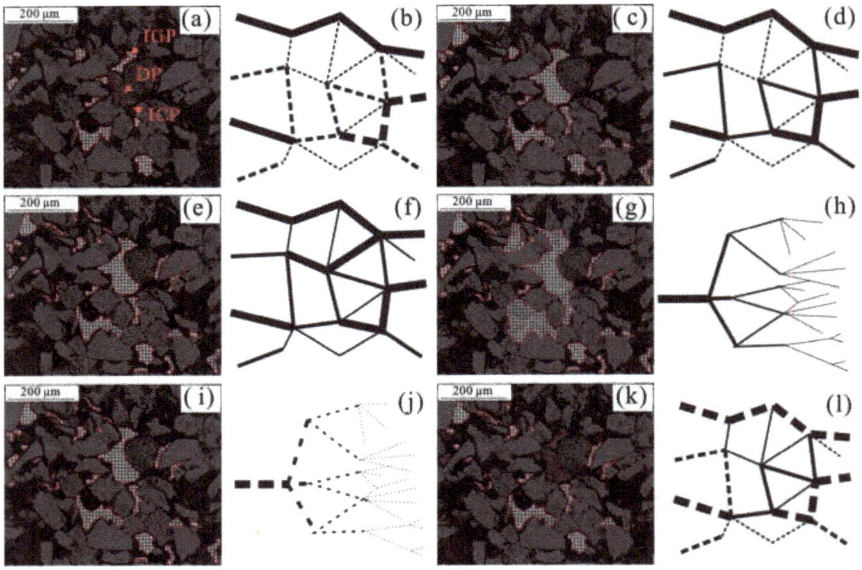

Figure 4. TS observations (net region with red rim represent the area that mercury occupies at this stage) and multitype model (the full line represents mercury fulfilling the network, the dashed line represents a wetting phase fulfilling the network and the line thickness represent the diameters of pore throats): (**a**) TS observation shows intergranular pores filled with mercury at the outset; (**b**) start of intrusion shows intergranular pores filled with mercury on the basis of a stick model; (**c**) TS observation show mercury invading the dissolution pores; (**d**) intrusion in this step shows that dissolution pores are occupied based on the stick model; (**e**) TS observation reveals that all connected pore throats are filled by mercury; (**f**) The end of invasion based on a stick model, mercury has occupied all the pore throats; (**g**) TS observation show that the pores among clay aggregates are filled with mercury; (**h**) The end of invasion based on a tree-like model, corresponding to the intercrystalline pores that are filled with mercury; (**i**) TS observation depicts the beginning of mercury withdrawal, reflecting mercury leaving the smallest pores; (**j**) Outset of mercury extrusion based on the tree-like model, intercrystalline pores did not leave a residual phase; (**k**) TS observation showing that mercury remains in weakly connected pores; (**l**) End of withdrawal, where the mercury remains in the tiny pores that are less unconnected to the outside, on the basis of a stick model. IGP—intergranular pore; DP—dissolution pore; ICP—intercrystalline pore.

The PCMI-derived PSD can be calculated on the basis of the Washburn equation [23]. The PSD of type α sandstones has a wide size range and possesses high peakedness over 1.0 μm with a long tail in the rests of pore radius, suggesting that specimens with high physical properties contain relatively larger pores and well-sorted PSD (Figure 5a). The PSD results of type β give two locations of the maxima, of which specimens 1# and 4# correspond to the wide range pore radius with bimodal, whereas other specimens show unimodal; the pores with radius larger than 1.0 μm are rare in this type (Figure 5b). The PSD of the type γ sandstones mainly varies from 0.01 μm to 0.4 μm, but the corresponding pore sizes of the peak pore volume values are different, from 0.008 μm to approximately 0.2 μm, indicating a heterogeneity of PSD with similar porosity and permeability as those of type γ (Figure 5c). Quantitative statistics performed on type δ sandstones indicate that the PSD mainly range from 0.003 μm to 0.15 μm, the main crests of these specimens being mainly centered at around 0.015 μm, whereas specimen 5# is accompanied by a peak larger than 0.1 μm, which might show evidence of microfractures (Figure 5d). The rapid increase of frequency in 0.003 μm represent the abundance of tiny pores (mainly the pores among clay minerals or the pores within the aluminosilicate layers); hence, porosity and permeability may be limited because of the rich clay minerals. The results of all the specimens show that the pore radius is distributed over a wide range and exhibits a multimodal PSD, and with the deterioration of physical properties, the PSD becomes more fluctuant, revealing that the pore structures are more complicated and heterogeneous. However, pores with radius over 20 μm are rare, which show a discrepancy between the PCMI and petrographic observations because many large pores are shielded by connected smaller pores. Hence, PCMI provides imperfect information of larger pores and can only uncover PSD of smaller ones, accordingly, the full range of PSD determination is required for PCMI in combination with other methods.

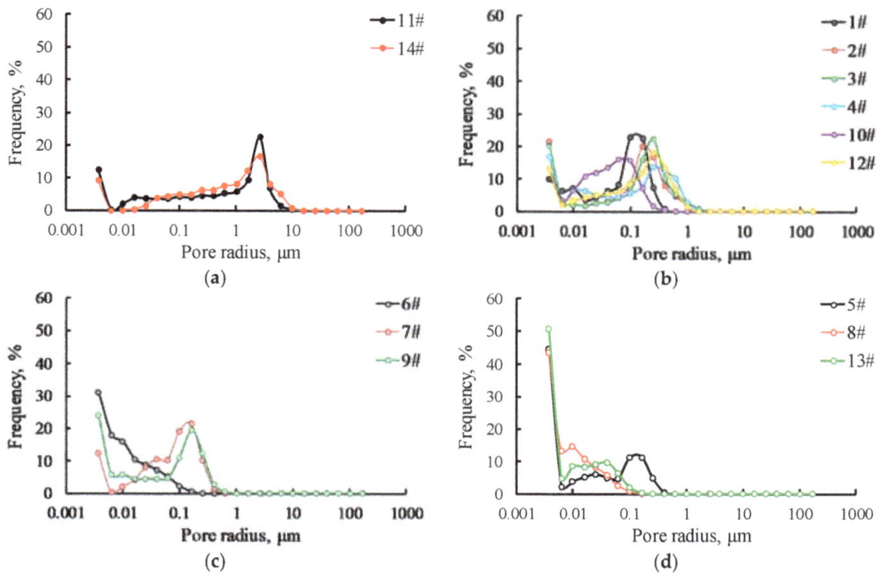

Figure 5. PCMI-derived PSD of four types of tight sandstones. (**a**) type α; (**b**) type β; (**c**) type γ; (**d**) type δ.

3.3. Pore Structures Characteristics Determined by RCMI

RCMI provides a reliable way to reflect detailed pores and throat structures. Since the void space can be divided into pore region and throat region, based on injection pressure fluctuation, the distribution of pores and throats radius are determined [15]. In order to investigate RCMI-derived PSD, six typical specimens with different physical properties were selected (Table 3). The capillary pressure curves of the specimens show that in the primary stage, the curves of total mercury intrusion saturation followed a trend of pore mercury intrusion saturation, while with the increasing pressure, the total mercury intrusion saturation was contributed to by the throat intrusion saturation and the intrusion pressure of the pore rapidly increased (Figure 6). Based on the trend of pore intrusion mercury saturation curves, the RCMI-derived PSD can be divided into a pore-dominated region corresponding to the stage that the pore mercury intrusion saturation increased exponentially, a throat-dominated region corresponding to the areas that the mercury saturation of pore remained unchanged, whereas the total mercury intrusion followed the trend of throat mercury intrusion, and a hybrid region by which both pores and throats are controlled (Figure 6). Furthermore, all specimens could be assigned to two groups based on the proportion of different regions. Group I exhibited a relatively lower threshold pressure (P_d) and medium pressure (P_{50}), while relatively higher maximum total mercury intrusion saturation (S_{Hgmax}); the percentage of pore-dominated and hybrid region exceeded that of throat-dominated, and shows a notable flat segment corresponding to the relatively good sorting of pores and throats (Table 3; Figure 6a,c,d). For Group II, the P_d increased while the S_{Hgmax} decreased and P_{50} disappeared. Moreover, the proportion of throat-dominated region surpasses that of the pore-dominated region in this group and no hybrid region existed, showcasing characteristics of low porosity and explaining why the storage of this group was relatively poor (Table 3; Figure 6e,f). Specimen 5# has relatively low P_d and S_{Hgmax}, indicating that it may contain microfractures that lead to the reduction of entry pressure (Table 3; Figure 6b).

Table 3. The parameters for pore structure characterization from RCMI. [3]

SN	P_d, MPa	r_{max}, μm	P_{50}, MPa	r_{50}, μm	r_t, μm	r_p, μm	η	r_m, μm	δ	S_{Hgmax}, %	S_p, %	S_t, %	Group
11	0.02	33.43	2.17	0.36	1.28	177.25	264.05	1.96	0.58	63.66	42.31	21.35	I
5	0.30	2.50	0.32	2.29	1.16	110.89	209.80	1.72	0.57	39.78	8.01	31.77	-
14	0.22	3.28	1.86	0.41	1.53	163.42	140.71	2.05	0.46	67.41	32.63	34.85	I
12	0.07	9.87	1.97	0.39	0.99	131.51	185.06	1.30	0.49	66.01	41.55	24.49	I
4	0.65	1.13	-	-	0.78	135.20	236.89	0.93	0.37	33.75	4.38	29.36	II
8	1.01	0.72	-	-	0.60	116.05	203.70	0.64	0.19	21.01	2.50	18.51	II

[3] P_d—threshold pressure; r_{max}—maximum pore radius; P_{50}—median pressure; r_{50}—median pore radius; r_t—average throat radius; r_p—average pore radius; η—pore throat radius ratio; r_m—mainstream throat radius; δ—sorting coefficient; S_t—throat maximum mercury intrusion saturation; S_p—pore maximum mercury intrusion saturation; S_{Hgmax}—total maximum mercury intrusion saturation.

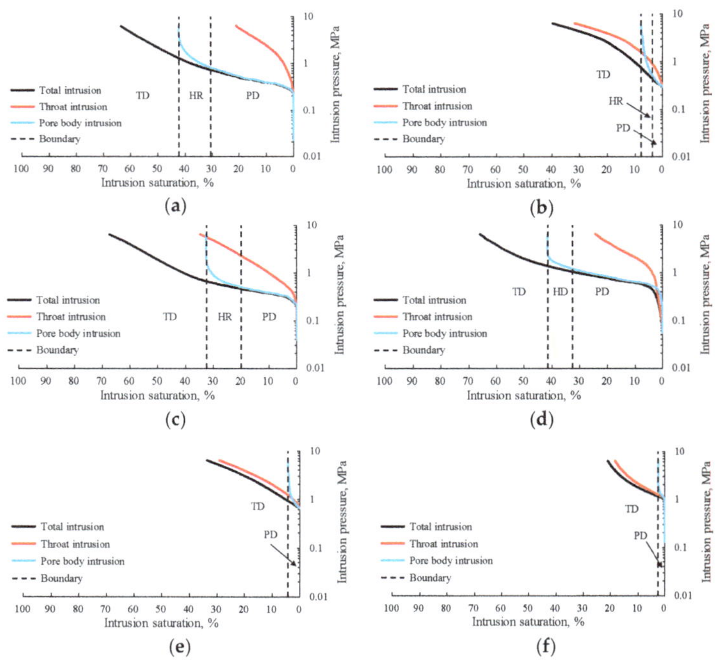

Figure 6. Capillary pressure curves measured by RCMI of typical tight sandstones specimens. (**a**) 11#; (**b**) 5#; (**c**) 14#; (**d**) 12#; (**e**) 4#; (**f**) 8#. PD—pore-dominated region; HR—hybrid region; TD—throat-dominated region.

The RCMI can test pores and throats separately; hence, the distribution of the pore radius, throat radius, and pore-throat radius ratio is recorded simultaneously [2]. The pores' radius mainly ranges from 88.23 μm to 230.15 μm with main crests of approximately 105 μm, showing no obvious discrepancy between different specimens (Table 3, Figure 7a). However, the throat radius demonstrates variations among specimens with distinctive features in spectral peaks and radius spectra, mainly ranging from 0.15 μm to 2.54 μm with an average of 1.06 μm (Table 3, Figure 7b). The distribution of pore-throat radius ratio is principally in the range of 55.21–430.78 with the average value of 206.70 (Table 3, Figure 7c). Unlike PCMI, RCMI can provide more reliable results due to the quasi-statics mercury injection process; however, this method generally leads to similar pore radius distributions among various specimens. In addition, the pore radius derived from RCMI shows that tons of pores larger than 100 μm existed, while pores with radius between 10 μm and 100 μm are seldom developed; nevertheless, it does not match with the TS and SEM observations (Figure 2). This comparison reveals that the RCMI may exaggerate the radius of the pores; therefore, necessary corrections for pore radius curves need to be made with the help of other methods.

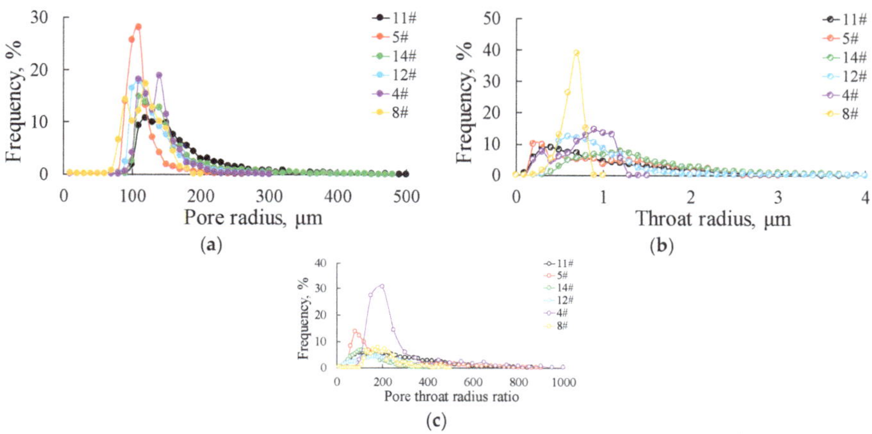

Figure 7. Pore-throat size distribution by RCMI of typical tight sandstones specimens: (**a**) pore size distribution; (**b**) throat size distribution; (**c**) pore throat radius ratio distribution.

3.4. Pore-Throat Microstructures Derived from Micro-CT

The pore-throat microstructures depicted by micro-CT allow accurate and detailed investigation of PSD; the pore-throat networks are illustrated in Figure 8. The matrix of the specimens is characterized by dark gray while the pore-throats are in color. The mutually inter-connected pores correspond to the same color, while the disconnected pores are presented by different colors or an isolated part. Due to the expense and time restriction, only two typical specimens with different permeability were selected for this measurement. Specimen 11# had relatively high porosity and permeability (φ: 14.24%, K: 2.70 mD); a mass of pores is mainly displayed in yellow and green, and the percentage of gray is relatively low, suggesting that this specimen had good pore-throat connectivity (Figure 8A,B). Many pores were found to be tubular or in larger spherical shapes, a few were isolated narrowed belt-like or small spherical pores, showcasing that intergranular and dissolution pores play a leading role in high permeability specimens (Figure 8A,B). For specimen 12#, which had relatively low porosity and permeability (φ: 12.54%, K: 0.48 mD), the pores were sparsely distributed, when compared to specimen 11#, and the same color was sporadically distributed, suggesting poor connectivity of these specimens. The pore radius distribution of these specimens show that a few narrow pores are found in 11#, whereas abundant pores with radius around 30 μm play important roles in forming major void spaces, which implies that micro-pores determined the storage and transport properties in tight sandstone (Figure 8C,F).

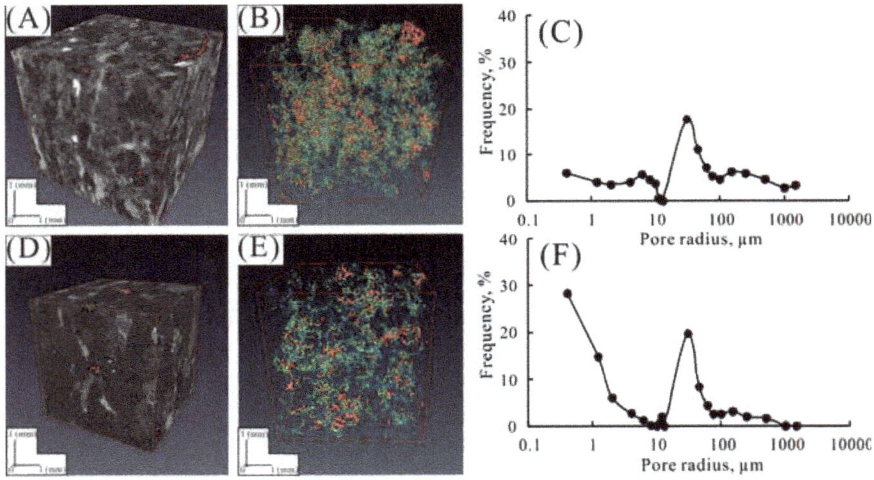

Figure 8. (**A**,**D**) Reconstructed 3D CT images and volume of (**B**,**E**) resolved pore-throat space and (**C**,**F**) micro-CT derived pore radius distribution of tight sandstone specimen of 11# and 12#, respectively. The pore-throat clusters are shown in different colors.

3.5. NMR Results

3.5.1. T_1 Relaxation Time before and after Centrifugal Processing

T_1 measurements in a fully saturated state and centrifuged state were taken for the specimens; these spectra are shown in Figure 9. The T_1 spectra of specimens show multimodal characteristics with the maximum crests being mainly distributed in the range of 51.79–719.69 ms and 4.64–64.49 ms in a fully saturated state and a centrifuged state, respectively. However, the major peaks show a broad massif-like peak in the fully saturated state and a narrow ridge-like peak in the centrifuged state; the corresponding T_1 relaxation time and amplitude moves from large to small (Figure 9). The spurious crests at high (>8000 ms) T_1 value in Figure 9a are the result of optimum values of the smoothening factors arising from the low error and unstable nature of the problem [62,63].

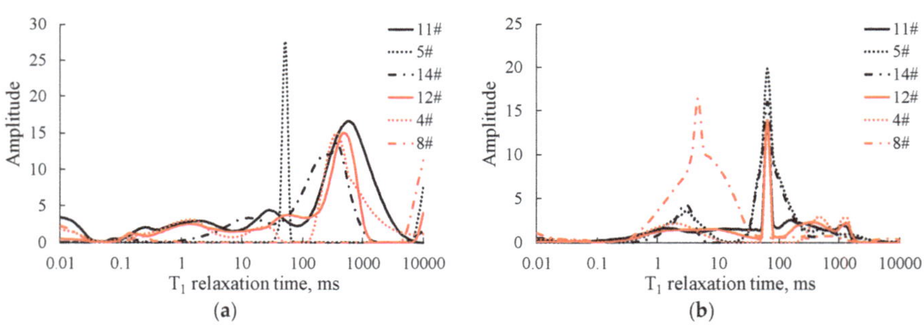

Figure 9. NMR T_1 distributions of specimens (**a**) before and (**b**) after centrifugations.

3.5.2. T_2 Relaxation Time before and after Centrifugal Processing

Unlike T_1 relaxation time, tests of T_2 relaxation time are preferred, as they are not time-consuming and can provide more accurate information about the void space by liquid-state NMR [32,64]. The T_2 distributions in the fully saturated and centrifugated state show bimodal or multimodal features with two crests; the short T_2 values represents micropores, mainly distributed in the range of 0.11–1.54 ms, while the long T_2 relaxation time corresponding to larger pores and cracks is mainly distributed in the range of 5.63–382.79 ms (Figure 10a). The T_2 spectra show similar shape before and after the centrifugal processing, while amplitude is narrowed in the irreducible water state (Figure 10). The left peak shows a relatively narrow peak distribution and larger amplitude than that of the right peaks in the fully saturated state, indicating that micropores have a higher percentage, while meso- and macropores have a wider pore radius range [65]. Besides, amplitude reduction with a slight drop in the left peak and distinct descending right peaks in the centrifugated state reveal that the irreducible water is mainly stranded in the micropores after centrifugal processing (Figure 10).

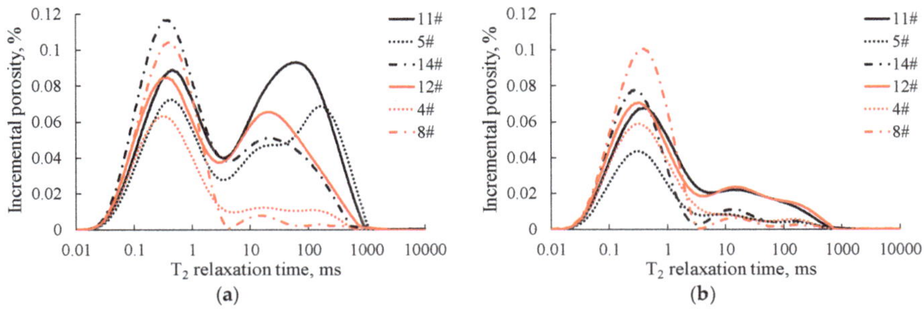

Figure 10. NMR T_2 distributions of specimens (**a**) before and (**b**) after centrifugations.

4. Discussion

4.1. Full PSD Calculated by Integration of PCMI, RCMI, and NMR

As mentioned above, PCMI, RCMI, micro-CT, and NMR tests were conducted on the same specimens; however, these tests do not lead to coincident results because of the limitations and drawbacks of these experiments. Thus, no single test can be used to investigate total PSD in tight sandstones; hence, the experimental data of these experiments may be modified for the sake of precise determination. PCMI can determine the characteristics of tiny pores but acquire defective larger pore radius spectra due to the shielding effect and mercury snap-off during extrusion [24,66]. RCMI could quantify the throat, but had a lower throat radius limit of 0.12 μm, while the pores' spectra detected by this method seems unreliable, because the overall pore radius of different specimens determined by RCMI are mainly in the range of 100–200 μm; observation of TS reveals few pores larger than 100 μm in the research area (Figure 2; Figure 7a). Besides, the pore radius distributions of the studied area are almost in accordance with that of other basins, such as the Songliao basin [15], suggesting that calibration needs to be done in order to acquire accurate pore radius spectra. Moreover, the RCMI-derived PSD shows that there is a "missing zone", namely, the pores radius range from 20 μm to 80 μm is rarely presented. These pores are commonly found in the TS and SEM micrographs (Figures 2 and 7), suggesting that this experiment may exaggerate the radius of mesopores. Micro-CT scanning may be a reasonable method to characterize PSD; however, it requires a trade-off between expense and representativeness. Moreover, the resolution of this method makes it unable to estimate the number of relatively tiny pores. The isolated pores, which have fewer contributions to storage capacity and no contributions to percolation ability, can be detected; thereupon, this test is inapplicable for the determination of PSD because commercial flow is often restricted by isolated pores [2,21,22,51]. The T_2 relaxation time, derived from the NMR test, can be used to describe PSD as a nondestructive method that does not destroy in-suit pore structures; however, this method requires calibration with an independent test to calculate surface relaxivity and specific surface area and then convert the relaxation time to pore radius [15,21,22,31,32].

In order to overcome these shortcomings, an integrated method is necessary. First, fitting procedures of T_2 relaxation time and mercury intrusion need to be done. As discussed above, the

PCMI-derived PSD is unreliable due to its shielding effect and induced cracks that resulted from high mercury injection pressure. There is a discrepancy between RCMI-derived pore radius distributions and optical observation results; hence, we use RCMI-derived throat radius curves to calibrate the T_2 spectra to PSD. According to previous research [16,65], the relationship between T_2 relaxation time of fluid in the porous materials and pore radius can be expressed as follows:

$$T_2 = Cr_t^n, \tag{1}$$

where r_t is the pore-throat radius, μm; C and n can be regarded as a constant parameter. In this way, T_2 relaxation time can be transformed to pore radius if C and n are determined. The procedures are listed below:

(1) The maximum throat intrusion saturation is multiplied by specimen-measured helium porosity φ to obtain throat porosity, φ_t. And then, the product of the incremental throat mercury intrusion saturation and φ_t can be calculated as incremental throat mercury intrusion porosity. (2) The incremental amplitude of T_2 relaxation time is multiplied by helium porosity φ to obtain incremental porosity, φ_{t2}. (3) The cumulative porosity according to φ_t and φ_{t2} should be calculated, and the comparison of both curves obtained. The results show that there is a good match between RCMI- and NMR-derived cumulative porosity, suggesting that these procedures and methods are reasonable and reliable for calibration (Figure 11a). Then the linear least square method is used to determine the most appropriate C and n (Figure 11b). At last, the T_2 relaxation time can be transformed to the pore radius, and the NMR-derived PSD is obtained.

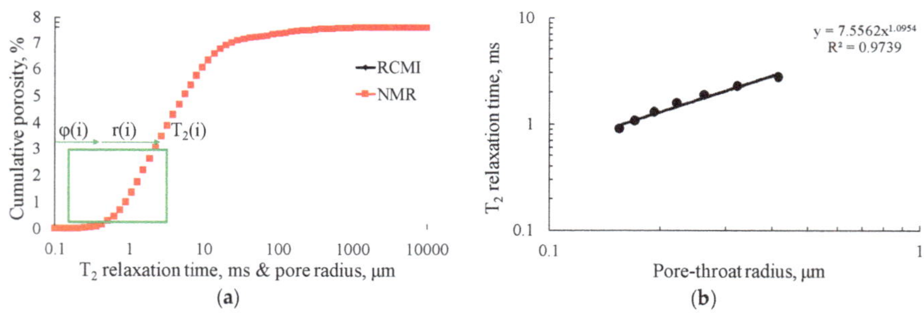

Figure 11. Methods for determining the improved-NMR PSD by the integration of NMR and RCMI, taking B116 specimen in the research area as an example: (**a**) data selection; (**b**) relationship between T_2 relaxation time and pore-throat radius.

The specimens were saturated with simulated formation water before it was put into the NMR apparatus; thus, some dead pores and ink-bottle pores saturated with fluid are inevitable. However, this non- or weak- connected pores contributed less to the percolation [65]. Besides, as mentioned above, the RCMI-derived throat distributions are reliable, although it cannot detect pores that are lower than 0.12 μm; hence, the integration of PCMI and NMR-derived PSD is regarded as a valid way to acquire the full range of PSD. The procedures are listed as follows: (1) the incremental porosity of improved NMR-derived PSD, which is more than 0.12 μm, is cumulated to obtain NMR cumulative porosity, φ_N. (2) Subtract the φ_N from helium measured porosity φ to get micropores porosity, φ_M. (3) The incremental mercury intrusion saturation of PCMI, which has radius less than 0.12 μm, is multiplied by φ to acquire the incremental PCMI-derived porosity. (4) The ratio of φ_M and φ is multiplied by the incremental PCMI-derived porosity to obtain calibrated PCMI-derived porosity. After the improved NMR-derived porosity and PCMI-derived porosity are determined, the full range PSD can be transformed to that of Figure 12a. The improved PSD matches well with the microscopic observations (Figure 2), demonstrating that this distribution is reasonable. In the same way, the calibration equations for all specimens can be acquired and displayed in Figure 12b.

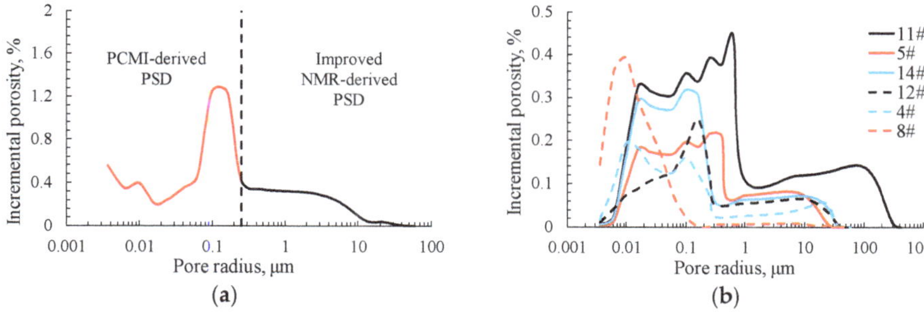

Figure 12. Full-range PSD by the integration of improved-NMR PSD and PCMI: (**a**) sample B116; (**b**) the PSDs of all specimens measured in this research.

4.2. T_2 *Cutoff Values*

T_2 cutoff is a significant value in the NMR experiment because the movable fluid saturation can be determined by this parameter [13]. This value is not a fixed constant, due to the variation of surface relaxivity and specific surface area of different specimens; centrifugal processing is an effective way to calculate this value. Figure 13 demonstrates how to calculate the T_2 cutoff: first, the cumulative porosity of T_2 spectrum in the fully saturated and irreducible water state is obtained, then a horizontal projection from the post centrifugation curve is drawn, and corresponding T_2 value of the intersection of this horizontal line and cumulative curve of the fully saturated state is the T_2 cutoff.

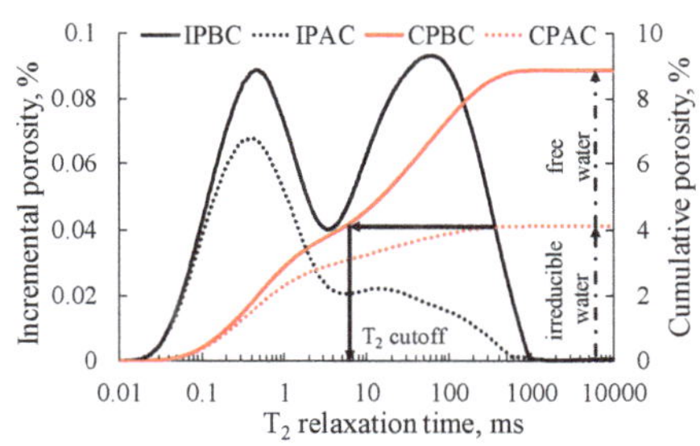

Figure 13. Specimen 11# shows a method to determine T_2 cutoff in 9100 $r{\cdot}min^{-1}$ rotation speed. IPBC—incremental porosity before centrifugation; IPAC—incremental porosity after centrifugation; CPBC—cumulative porosity before centrifugation; CPAC—cumulative porosity after centrifugation.

We repeat this centrifugal process with various rotation speeds. In order to acquire the most accurate movable fluid saturation and determine optimum rotation speed, Coates model [65,67] was used:

$$K_c = \left(\frac{1}{C_C}\right)^4 \varphi^4 \left(\frac{FFI}{BVI}\right)^2, \tag{2}$$

where K_C represents calculated permeability (mD), φ represents measured porosity (%), C_c is a constant, FFI represents movable fluid saturation, and BVI represents irreducible fluid saturation. Hence, the linear regressions between $\varphi^4(\frac{FFI}{BVI})^2$ under various rotation speeds and measured permeability (K_a) are performed to determine the optimum rotation speed. Figure 14 illustrate that 9600 $r{\cdot}min^{-1}$ can best fit Equation (2) with an R^2 value of 0.9428. The values of optimum T_2 cutoff are listed in Table 4.

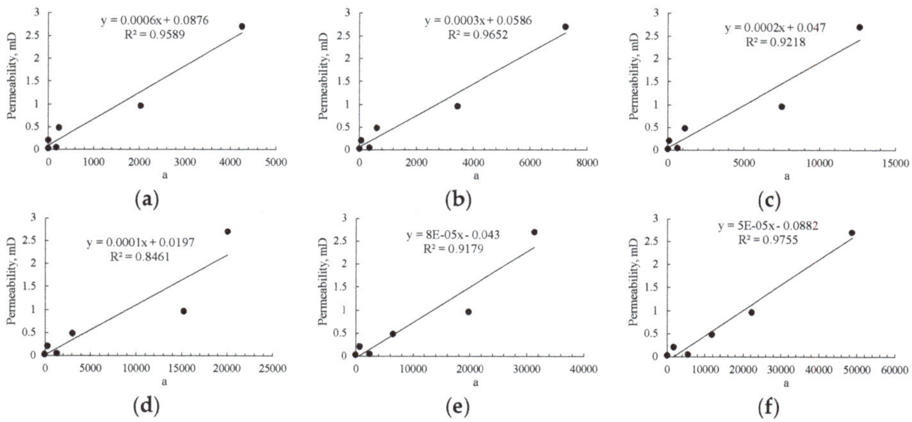

Figure 14. Linear regression analysis between $\varphi^4(\frac{FFI}{BVI})^2$ and measured permeability (a represents $\varphi^4(\frac{FFI}{BVI})^2$). (**a**) 2500 r·min^{-1}; (**b**) 2900 r·min^{-1}; (**c**) 3500 r·min^{-1}; (**d**) 5000 r·min^{-1}; (**e**) 7900 r·min^{-1}; (**f**) 9100 r·min^{-1}.

Table 4. The parameters for movable fluid parameters from NMR. [4]

SN	T_2, ms	S_m, %	S_p, %	Pore Radius Lower Limits in Different Rotation Speed, μm					
				2500 r·min^{-1}	2900 r·min^{-1}	3500 r·min^{-1}	5000 r·min^{-1}	7900 r·min^{-1}	9100 r·min^{-1}
11	6.83	52.12	7.42	1.17	0.83	0.55	0.39	0.24	0.14
5	0.74	69.75	3.96	1.35	0.67	0.34	0.15	0.04	0.01
14	0.42	64.24	6.87	0.19	0.10	0.03	0.02	0.01	0.01
12	7.32	40.93	5.13	1.91	1.17	0.83	0.44	0.26	0.15
4	13.67	18.61	1.45	1.10	0.59	0.39	0.36	0.27	0.15
8	1.96	10.53	0.58	0.44	0.26	0.22	0.19	0.16	0.04

[4] T_2-T_2 relaxation time; S_m—movable fluid saturation; S_p—movable fluid percentage.

4.3. Controls of Movable Fluid Traits on Reservoir Quality

4.3.1. Movable Fluid Parameters and Their Effects on Reservoir Quality

After determining the optimum rotation speed (9100 r·min^{-1}), the movable fluid parameters can be calculated. In NMR-derived relaxation time studies, the movable fluid saturation (MFS) and movable fluid porosity (MFP) reflect the volumetric fraction of movable fluid occurring in the voids and the practical fluid flow ability through the sandstones, respectively; thus, there is a significant parameter that needs to be considered in evaluating reservoir quality [12]. Reservoir quality is defined here as physical properties, including porosity and permeability. The results show that the relationship between the reservoir quality and movable fluid saturation is inferior to that of the movable fluid porosity; the correlation between movable fluid parameters and porosity are lower than that of permeability (Tables 1 and 4, Figure 15). These results demonstrate that the uncertainty of the porosity measurements, caused by the complex pore-throat structure, needs to be eliminated to characterize reservoir quality accurately; in addition, the movable fluid parameters are more closely aligned with permeability. Meanwhile, with decreasing reservoir quality and movable fluid saturation, chlorite decreases from 7% to 0%, while illite increases from 0% to 7%, indicating that illite exerts a significant impact on retaining water, whereas chlorite can retard compaction (Figure 16a–c). Besides, reduction rate of reservoir quality with movable fluid saturation tends to decrease in tight sandstones with intergranular pores-dominated reservoirs to those with clay minerals, which indicates that movable fluid parameters are sensitive to a higher content of clay minerals or the proportion of tiny throats, whereas the reduction of intergranular pores play a clearly important role in the descending permeability (Figure 16a,b,d).

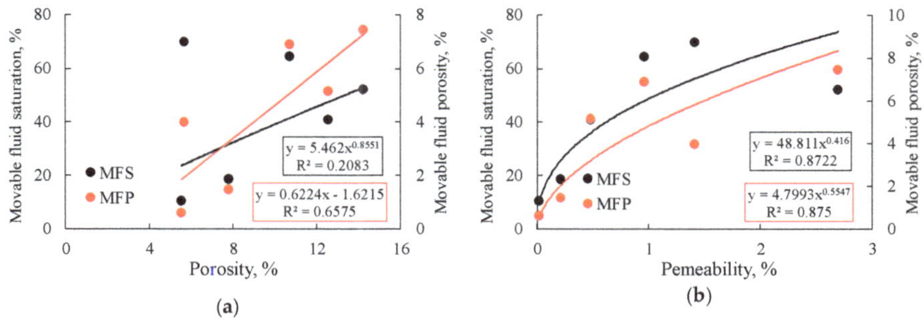

Figure 15. The plot of movable fluid parameters versus (**a**) porosity and (**b**) permeability. MFS-movable fluid saturation; MFP-movable fluid porosity.

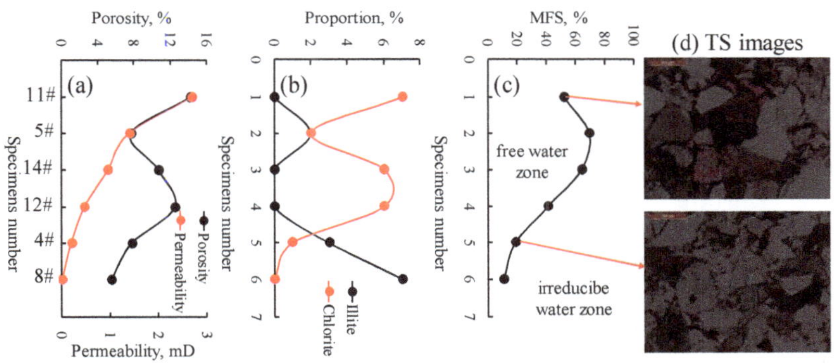

Figure 16. Graphs illustrating the impacts of pore types, movable fluid saturation and clay mineral contents on reservoir quality. (**a**) reservoir quality; (**b**) clay minerals; (**c**) movable fluid saturations (MFS); (**d**) thin section images record the pore types.

4.3.2. T_1-T_2 Correlation Spectra and Their Effects on Reservoir Quality

The multidimensional correlation functions of T_1 and T_2, which were used to determine the characteristics of PSD, are of great interest. Appendix A presents the T_1-T_2 spectrum for specimens of tight sandstones; the bright red represents high amplitude, while the light red represents low amplitude. Appendix A illustrates that in either good or poor reservoir qualities specimens, the range of T_1 and T_2 is approximately 3 to 4 orders of magnitude because of the wide range of PSD, and the figures showed a wide distribution before centrifugation and a relatively narrow ridge-like peak in the center of the spectra after centrifugation. The bumps along the axis are likely artifacts due to noise. According to Song et al. (2002) [55], the peaks with small T_1 and T_2 are dominated by surface water whereas the long T_1 and T_2 peaks are contributed by surface and bulk water, and these are consistent with the notion that the free water (bulk water) that is in the center of the pores is easily removed after centrifugal processing [55]. Besides, the zenith of the spectra before centrifugation moved from relatively high to low relaxation time in irreducible water conditions, suggesting that the aqueous phases have low mobility in tiny pores. Moreover, the ridges are more closely parallel to the line $T_1 = T_2$, meaning that they have similar T_1 and T_2 ratio corresponding to similar properties (Appendix A).

The effect of reservoir quality is further investigated. For the specimen with greatest reservoir quality (11#) (Appendix A (A,B)), the proportion of bulk water that corresponds to larger pores is relatively high, compared to poor reservoir qualities specimens. This manifests that mill metric voids play a predominant role in optimization of reservoir quality in tight sandstone reservoirs. With permeability decreasing, the reduction rate of the water covering spectra areas of all the specimens, except specimen 5#, are decline, suggesting that the pores that allowed water to immigrate freely (bulk water) were few in the low permeability specimen, even in trace amounts in extremely tight sandstones (Appendix A (K,L)). Specimen 5# is characterized as low porosity and high permeability due the presence of microcracks (Figure 2l), which may lead to the wide range of T_2 and narrow range of T_1 before centrifugal processing (Figures 9a and 10b, Appendix A (C,D)). However, the mechanism of this phenomenon needs further investigation. Besides, for good reservoir quality specimens, peaks and shoulders are continuously distributed, whereas poor ones show separate distributions before centrifugation, revealing that the pores are relatively homogeneous, distributed in tight sandstones with high physical properties (Table 1, Appendix A (A–D)). After centrifugation, the spectra tend to show discrete distributions because the heterogeneity of the PSD caused some bulk water, which is hard to remove, especially for medium permeability specimens (Table 1, Appendix A (E–H)). For the extremely tight sandstone, however, the PSD shows a homogeneous distribution that results in the continuous distributions of T_1-T_2 correlation spectra, because the very tiny pores generally show continuous arrangement (Table 1, Appendix A (I–L)).

4.3.3. NMR Imaging for Saturated and Irreducible Water Distributions

The images show distinct difference in simulated water saturation under fully saturated and irreducible water conditions; the red color represents oil and dark blue denotes deionized water, and the surrounding blue color may be caused by signal noise (Appendix B). The water saturation of the good reservoir quality specimens is always higher than that of poor reservoir quality, implying that brine can enter the majority of pore spectrum of high permeability specimens, whereas the low permeability specimens are dominated by tight section. Due to the hydrophilia of illite and I/S minerals [68] (low permeability specimens, which contain relatively abundant illite and I/S minerals), the change of water saturation before and after centrifugal processing is minor (Table 1, Appendix B). Besides, the percentage of high-saturated sections in high permeability specimens before and after centrifugal processing decline faster than that of the low permeability samples (slightly decrease) (Table 1, Appendix B). This implies that movable fluid mainly comes from a contribution of intergranular pores, while irreducible water is mainly stranded in intercrystalline pores.

4.4. *Effects of Movable Pore Radius Lower Limit on Reservoir Quality*

During NMR centrifugal processing, the specimens underwent different movable fluid loss, predominantly by centrifugal forces with minor gravity [65]. In terms of the PSD being converted from the T_2 spectra and RCMI data, a series of T_2 cutoff values, which is defined by the ratio of T_2 spectra before and after centrifugal processing with various centrifugal forces, were often used to determine the lower limit of movable pore radius [11,32]. In the case of high permeability specimens, a sharp decrease of macropores was observed corresponding to high T_2 relaxation time first, followed by a moderate and uniform reduction (Appendix A (A,B)). The correlation between lower limits of movable pore radius and movable fluid porosity show a distinct descending at the initial stage, followed by a moderate decrease (Figure 17). We can thus infer that the intergranular pores, which correspond to a larger pore radius, are the main result of the reduction of the lower limits of pore radius in higher permeability sandstones, and dissolution pores and intercrystalline pores are less important. For specimens with medium permeability, the amplitude of T_2 relaxation time drop uniformly (Appendix A (C,D)); however, the lower limits of movable pore radius show different changes: a distinct change in MFS and MFP over a small range of pore radius lower limits occurred in 14#, whereas the variance of 12 # is opposite (Figure 17), meaning that although these two specimens have similar reservoir quality, the percolation capacity of 14# is mainly dominated by heterogeneous pore structures with relatively low porosity, whereas the heterogeneity of the pore-throats in specimen

12# limits the improvement of permeability, even with relatively high porosity. Specimens with lower permeability resulted in no marked descending of T_2 and lower limits of movable pore radius (Appendix A (E,F)); the length of the curves for the tight specimen are generally short (except the first stage of 4# due to minor amounts of intergranular pores), indicating that pores in aqueous phases were hard to enter and were removed in poor reservoir quality sandstones (Figure 17). These results manifest that because the tight sandstones are highly heterogeneous, during the centrifugation process, rates of water saturation reduction in macropores are faster than those for tiny pores. For specimens with intergranular pores, the movable fluid advances, exempted from the larger pores.

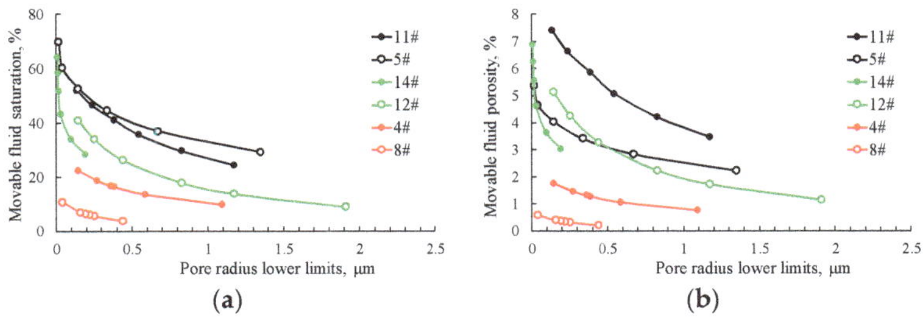

Figure 17. The plot of pore radius lower limits versus (**a**) movable fluid saturation and (**b**) movable fluid porosity.

5. Conclusions

A combination of TS, SEM, X-ray diffraction, PCMI, RCMI, CT, and NMR tests were conducted on typical tight sandstone specimens from Ordos Basin to investigate pore structures and movable fluid traits. The following observations were derived:

(1) Intergranular pores, dissolution pores, intercrystalline pores, and microfractures are four main kinds of pores in the research area. The PCMI-derived capillary pressure curves can be grouped into four types, and the RCMI-derived capillary pressure curves are divided into pore-dominated, throat-dominated and hybrid regions. Good pore-throat connectivity plays a vital role in enhancing physical properties based on CT images. The amplitude and incremental porosity distribution become small and narrow after centrifugal processing using NMR apparatus.

(2) The PSDs derived by RCMI show discrepancy with the direct observations, and the improved PSDs can be obtained by combining PCMI, RCMI and NMR and then reconstructing cumulative PSDs, which is consistent with the observation results of TS and SEM images.

(3) On the basis of Coates model, the linear regression results show that 9100 r·min^{-1} is the optimum rotation speed in the research area to determine appropriate T_2 cutoff values. MFP can characterize reservoir quality accurately, and a high proportion of chlorite with relatively abundant intergranular pores could improve MFS and reservoir quality.

(4) Bulk water makes great contributions to the movable fluid, whereas surface water is hard to be removed. Movable fluid mainly comes from the contribution of intergranular pores, while irreducible water is mainly stranded in intercrystalline pores, and the movable fluid could advance, exempted from the larger pores.

(5) Pore structures controls the lower limits of movable pore radius and hence affects reservoir quality. In sandstones with relatively high permeability, abundant intergranular pores can reduce the lower limits of pore radius; however, sandstones with complex pore structures have relatively poor reservoir qualities.

Author Contributions: Conceptualization, D.L. and D.R.; Data curation, D.L.; Formal analysis, D.L. and D.R.; Funding acquisition, W.S. and D.R.; Investigation, W.S.; Methodology, D.L.; Project administration, W.S.; Supervision, W.S.; Writing—original draft, D.L.; Writing—review & editing, D.R.

Funding: This research was co-funded by National Science and Technology Major Project, grant number 2016ZX05047-003-005; National Natural Science Foundation of China: grant number 41702146, 51874242, and 41802166; Doctoral Fund of Ministry of Education of China, grant number 2018M643554.

Acknowledgments: The authors appreciate the help and support from State Key Laboratory of Continental Dynamics, and State Engineering Laboratory for Exploration and Development of Low Permeability Oil and Gas Fields.

Conflicts of Interest: The authors declare no conflict of interest.

Appendix A

T_1-T_2 correlation spectra in this study:

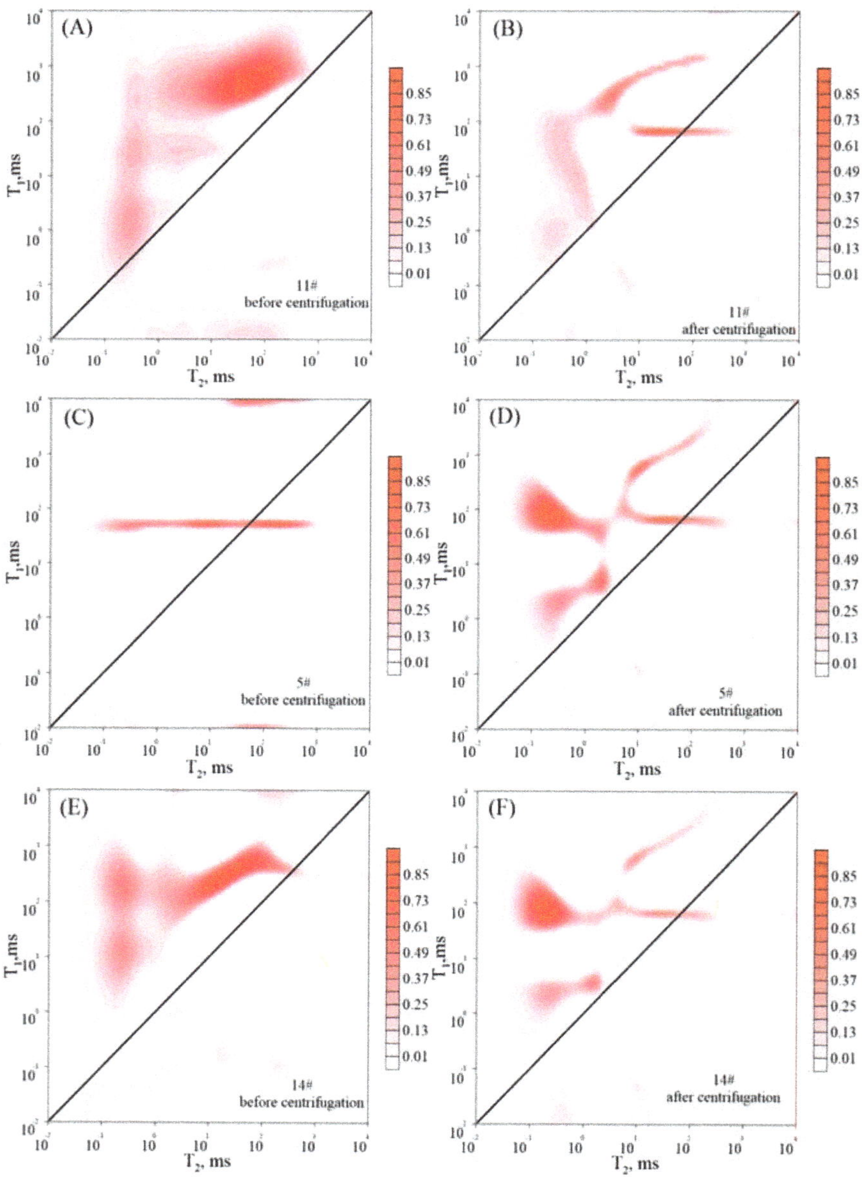

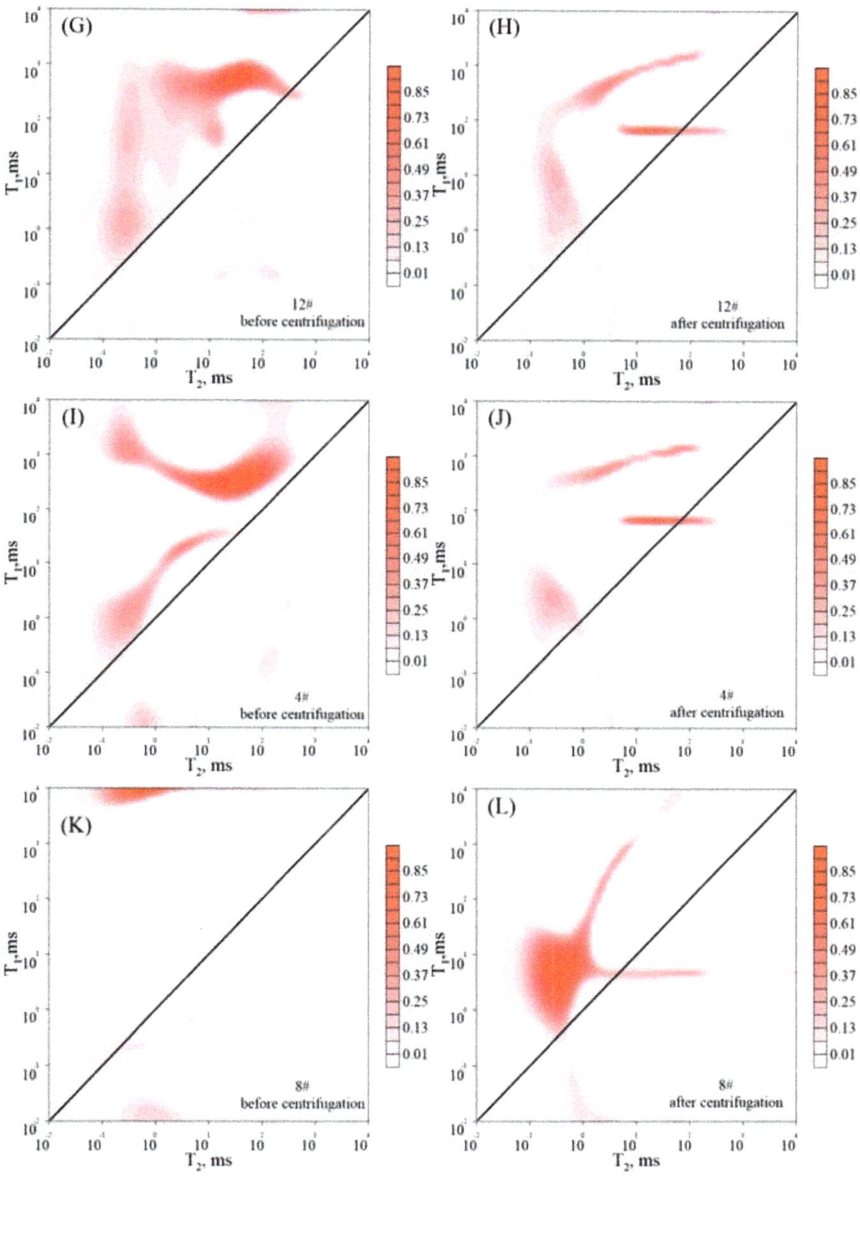
(G)
12#
before centrifugation
(H)
12#
after centrifugation
(I)
4#
before centrifugation
(J)
4#
after centrifugation
(K)
8#
before centrifugation
(L)
8#
after centrifugation
T1,ms
T2, ms
0.85
0.73
0.61
0.49
0.37
0.25
0.13
0.01

Appendix B

NMR imaging in this study:

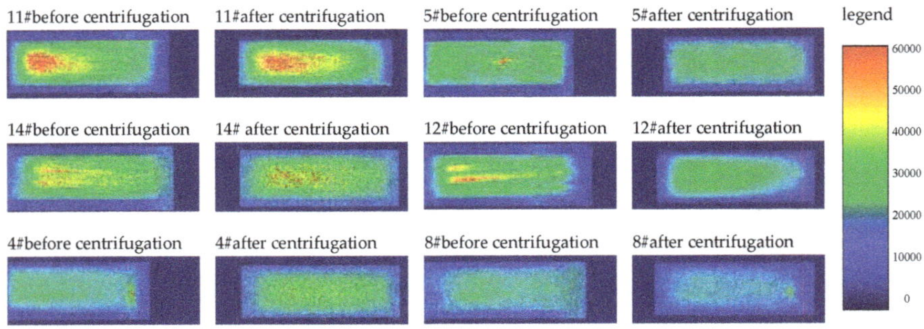

References

1. Hughes, J.D. Energy: A reality check on the shale revolution. *Nature* **2013**, *494*, 307–308. [CrossRef] [PubMed]
2. Wu, H.; Ji, Y.; Liu, R.; Zhang, C.; Chen, S. Insight into the Pore Structure of Tight Gas Sandstones: A Case Study in the Ordos Basin, NW China. *Energy Fuels* **2017**, *31*, 13159–13178. [CrossRef]
3. Hill, R.J.; Zhang, E.; Katz, B.J.; Tang, Y. Modeling of gas generation from the Barnett shale, Fort Worth Basin, Texas. *AAPG Bull.* **2007**, *91*, 501–521. [CrossRef]
4. Pollastro, R.M.; Cook, T.A.; Roberts, L.N.; Schenk, C.J.; Lewan, M.D.; Anna, L.O.; Gaswirth, S.B.; Lillis, P.G.; Klett, T.R.; Charpentier, R.R. *Assessment of Undiscovered Oil Resources in the Devonian-Mississippian Bakken Formation, Williston Basin Province, Montana and North Dakota*; No. 2008-3021; Geological Survey (US): Reston, VA, USA, 2008.
5. Zou, C.; Zhang, G.; Tao, S. Geological features and exploration discoveries of unconventional resources and the petroleum geological theory. *Pet. Explor. Dev.* **2010**, *37*, 129–145.
6. Kang, Y.L.; Luo, P.Y. Current status and prospect of key techniques for exploration and production of tight sandstone gas reservoirs in China. *Pet. Explor. Dev.* **2007**, *34*, 239–247.
7. Zou, C.; Zhu, R.; Liu, K.; Su, L.; Bai, B.; Zhang, X.; Yuan, X.; Wang, J. Tight gas sandstone reservoirs in China: characteristics and recognition criteria. *J. Pet. Sci. Eng.* **2012**, *88*, 82–91. [CrossRef]
8. Rezaee, R.; Saeedi, A.; Clennell, B. Tight gas sands permeability estimation from mercury injection capillary pressure and nuclear magnetic resonance data. *J. Pet. Sci. Eng.* **2012**, *88*, 92–99. [CrossRef]
9. Lai, J.; Wang, G. Fractal analysis of tight gas sandstones using high-pressure mercury intrusion techniques. *J. Nat. Gas Sci. Eng.* **2015**, *24*, 185–196. [CrossRef]
10. Xi, K.; Cao, Y.; Haile, B.G.; Zhu, R.; Jahren, J.; Bjørlykke, K.; Zhang, X.; Hellevang, H. How does the pore-throat size control the reservoir quality and oiliness of tight sandstones? The case of the Lower Cretaceous Quantou Formation in the southern Songliao Basin, China. *Mar. Pet. Geol.* **2016**, *76*, 1–15. [CrossRef]
11. Huang, H.; Sun, W.; Ji, W.; Zhang, R.; Du, K.; Zhang, S.; Ren, D.; Wang, Y.; Chen, L.; Zhang, X. Effects of pore-throat structure on gas permeability in the tight sandstone reservoirs of the Upper Triassic Yanchang formation in the Western Ordos Basin, China. *J. Pet. Sci. Eng.* **2018**, *162*, 602–616. [CrossRef]
12. Gao, H.; Li, H.A. Pore structure characterization, permeability evaluation and enhanced gas recovery techniques of tight gas sandstones. *J. Nat. Gas Sci. Eng.* **2016**, *28*, 536–547. [CrossRef]
13. Gao, H.; Li, H. Determination of movable fluid percentage and movable fluid porosity in ultra-low permeability sandstone using nuclear magnetic resonance (NMR) technique. *J. Pet. Sci. Eng.* **2015**, *133*, 258–267. [CrossRef]
14. Nelson, P.H. Pore-throat sizes in sandstones, tight sandstones, and shales. *AAPG Bull.* **2009**, *93*, 329–340. [CrossRef]
15. Xiao, D.; Jiang, S.; Thul, D.; Huang, W.; Lu, Z.; Lu, S. Combining rate-controlled porosimetry and NMR to probe full-range pore throat structures and their evolution features in tight sands: A case study in the Songliao Basin, China. *Mar. Pet. Geol.* **2017**, *83*, 111–123. [CrossRef]

16. Wang, L.; Zhao, N.; Sima, L.; Meng, F.; Guo, Y. Pore structure characterization of the tight reservoir: Systematic integration of mercury injection and nuclear magnetic resonance. *Energy Fuels* **2018**, *32*, 7471–7484. [CrossRef]
17. Anovitz, L.M.; Cole, D.R.; Rother, G.; Allard, L.F.; Jackson, A.J.; Littrell, K.C. Diagenetic changes in macro-to nano-scale porosity in the St. Peter Sandstone: An (ultra) small angle neutron scattering and backscattered electron imaging analysis. *Geochim. Cosmochim. Acta* **2013**, *102*, 280–305. [CrossRef]
18. Favvas, E.P.; Sapalidis, A.A.; Stefanopoulos, K.L.; Romanos, G.E.; Kanellopoulos, N.K.; Kargiotis, E.K.; Mitropoulos, A.C. Characterization of carbonate rocks by combination of scattering, porosimetry and permeability techniques. *Microporous Mesoporous Mater.* **2009**, *120*, 109–114. [CrossRef]
19. Loucks, R.G.; Ruppel, S.C. Mississippian Barnett Shale: Lithofacies and depositional setting of a deep-water shale-gas succession in the Fort Worth Basin, Texas. *AAPG Bull.* **2007**, *91*, 579–601. [CrossRef]
20. Loucks, R.G.; Reed, R.M.; Ruppel, S.C.; Jarvie, D.M. Morphology, genesis, and distribution of nanometer-scale pores in siliceous mudstones of the Mississippian Barnett Shale. *J. Sediment. Res.* **2009**, *79*, 848–861. [CrossRef]
21. Mayo, S.; Josh, M.; Nesterets, Y.; Esteban, L.; Pervukhina, M.; Clennell, M.B.; Maksimenko, A.; Hall, C. Quantitative micro-porosity characterization using synchrotron micro-CT and xenon K-edge subtraction in sandstones, carbonates, shales and coal. *Fuel* **2015**, *154*, 167–173. [CrossRef]
22. Liu, X.; Wang, J.; Ge, L.; Hu, F.; Li, C.; Li, X.; Yu, J.; Xu, H.; Lu, S.; Xue, Q. Pore-scale characterization of tight sandstone in Yanchang Formation Ordos Basin China using micro-CT and SEM imaging from nm-to cm-scale. *Fuel* **2017**, *209*, 254–264. [CrossRef]
23. Washburn, E.W. The dynamics of capillary flow. *Phys. Rev.* **1921**, *17*, 273–283. [CrossRef]
24. Kaufmann, J.; Loser, R.; Leemann, A. Analysis of cement-bonded materials by multi-cycle mercury intrusion and nitrogen sorption. *J. Colloid Interface Sci.* **2009**, *336*, 730–737. [CrossRef] [PubMed]
25. Huang, H.; Chen, L.; Sun, W.; Xiong, F.; Ji, W.; Jia, J.; Tang, X.; Zhang, S.; Gao, J.; Luo, B. Pore-throat structure and fractal characteristics of Shihezi Formation tight gas sandstone in the Ordos Basin, China. *Fractals* **2018**, *26*, 1840005. [CrossRef]
26. Yuan, H.H.; Swanson, B.F. Resolving pore-space characteristics by rate-controlled porosimetry. *SPE Form. Eval.* **1989**, *4*, 17–24. [CrossRef]
27. Gao, H.; Li, T.; Yang, L. Quantitative determination of pore and throat parameters in tight oil reservoir using constant rate mercury intrusion technique. *J. Pet. Explor. Prod. Technol.* **2016**, *6*, 309–318. [CrossRef]
28. Wang, R.; Chi, Y.; Zhang, L.; He, R.; Tang, Z.; Liu, Z. Comparative studies of microscopic pore throat characteristics of unconventional super-low permeability sandstone reservoirs: Examples of Chang 6 and Chang 8 reservoirs of Yanchang Formation in Ordos Basin, China. *J. Pet. Sci. Eng.* **2018**, *160*, 72–90. [CrossRef]
29. Liu, H.; d'Eurydice, M.N.; Obruchkov, S.; Galvosas, P. Determining pore length scales and pore surface relaxivity of rock cores by internal magnetic fields modulation at 2 MHz NMR. *J. Magn. Reson.* **2014**, *246*, 110–118. [CrossRef] [PubMed]
30. Yao, Y.; Liu, D. Comparison of low-field NMR and mercury intrusion porosimetry in characterizing pore size distributions of coals. *Fuel* **2012**, *95*, 152–158. [CrossRef]
31. Daigle, H.; Johnson, A. Combining mercury intrusion and nuclear magnetic resonance measurements using percolation theory. *Transp. Porous Media* **2016**, *111*, 669–679. [CrossRef]
32. Yao, Y.; Liu, D.; Che, Y.; Tang, D.; Tang, S.; Huang, W. Petrophysical characterization of coals by low-field nuclear magnetic resonance (NMR). *Fuel* **2010**, *89*, 1371–1380. [CrossRef]
33. Desbois, G.; Urai, J.L.; Kukla, P.A.; Konstanty, J.; Baerle, C. High-resolution 3D fabric and porosity model in a tight gas sandstone reservoir: A new approach to investigate microstructures from mm-to nm-scale combining argon beam cross-sectioning and SEM imaging. *J. Pet. Sci. Eng.* **2011**, *78*, 243–257. [CrossRef]
34. Zhao, H.; Ning, Z.; Wang, Q.; Zhang, R.; Zhao, T.; Niu, T.; Zeng, Y. Petrophysical characterization of tight oil reservoirs using pressure-controlled porosimetry combined with rate-controlled porosimetry. *Fuel* **2015**, *154*, 233–242. [CrossRef]
35. Guan, H.; Brougham, D.; Sorbie, K.S.; Packer, K.J. Wettability effects in a sandstone reservoir and outcrop cores from NMR relaxation time distributions. *J. Pet. Sci. Eng.* **2002**, *34*, 35–54. [CrossRef]
36. Li, P.; Sun, W.; Wu, B.; Gao, Y.; Du, K. Occurrence characteristics and influential factors of movable fluids in pores with different structures of Chang 6_3 reservoir, Huaqing Oilfield, Ordos Basin, China. *Mar. Pet. Geol.* **2018**, *97*, 480–492. [CrossRef]

37. Lai, J.; Wang, G.; Fan, Z.; Chen, J.; Wang, S.; Zhou, Z.; Fan, X. Insight into the pore structure of tight sandstones using NMR and HPMI measurements. *Energy Fuels* **2016**, *30*, 10200–10214. [CrossRef]
38. Hassanizadeh, S.M.; Celia, M.A.; Dahle, H.K. Dynamic effect in the capillary pressure–saturation relationship and its impacts on unsaturated flow. *Vadose Zone J.* **2002**, *1*, 38–57. [CrossRef]
39. Hossain, Z.; Grattoni, C.A.; Solymar, M.; Fabricius, I.L. Petrophysical properties of greensand as predicted from NMR measurements. *Pet. Geosci.* **2011**, *17*, 111–125. [CrossRef]
40. Sakhaee-Pour, A.; Bryant, S.L. Effect of pore structure on the producibility of tight-gas sandstones. *AAPG Bull.* **2014**, *98*, 663–694. [CrossRef]
41. Clarkson, C.R.; Freeman, M.; He, L.; Agamalian, M.; Melnichenko, Y.B.; Mastalerz, M.; Bustin, R.M.; Radliński, A.P.; Blach, T.P. Characterization of tight gas reservoir pore structure using USANS/SANS and gas adsorption analysis. *Fuel* **2012**, *95*, 371–385. [CrossRef]
42. Dai, J.; Li, J.; Luo, X.; Zhang, W.; Hu, G.; Ma, C.; Guo, J.; Ge, S. Stable carbon isotope compositions and source rock geochemistry of the giant gas accumulations in the Ordos Basin, China. *Organ. Geochem.* **2005**, *36*, 1617–1635. [CrossRef]
43. Guo, H.; Jia, W.; Peng, P.A.; Lei, Y.; Luo, X.; Cheng, M.; Wang, X.; Zhang, L.; Jiang, C. The composition and its impact on the methane sorption of lacustrine shales from the Upper Triassic Yanchang Formation, Ordos Basin, China. *Mar. Pet. Geol.* **2014**, *57*, 509–520. [CrossRef]
44. Xiao, X.M.; Zhao, B.Q.; Thu, Z.L.; Song, Z.G.; Wilkins, R.W.T. Upper Paleozoic petroleum system, Ordos Basin, China. *Mar. Pet. Geol.* **2005**, *22*, 945–963. [CrossRef]
45. Tang, X.; Zhang, J.; Wang, X.; Yu, B.; Ding, W.; Xiong, J.; Yang, Y.; Wang, L.; Yang, C. Shale characteristics in the southeastern Ordos Basin, China: Implications for hydrocarbon accumulation conditions and the potential of continental shales. *Int. J. Coal Geol.* **2014**, *128*, 32–46. [CrossRef]
46. Liu, D.; Sun, W.; Li, D.; Shi, J.; Ren, D. Pore structures characteristics and porosity evolution of tight sandstone reservoir: taking the Chang 6_3 tight sandstones reservoir of Huaqing Area in Ordos Basin as an instance. *FEB-Fresenius Environmental Bulletin* **2018**, 1043–1052.
47. Folk, R.L. *Petrology of Sedimentary Rocks*; Hemphill Publishing Company: Austin, TX, USA, 1980; Volume 21, pp. 714–727.
48. Cui, Y.; Wang, G.; Jones, S.J.; Zhou, Z.; Ran, Y.; Lai, J.; Li, R.; Deng, L. Prediction of diagenetic facies using well logs–A case study from the upper Triassic Yanchang Formation, Ordos Basin, China. *Mar. Pet. Geol.* **2017**, *81*, 50–65. [CrossRef]
49. Lai, J.; Wang, G.; Ran, Y.; Zhou, Z.; Cui, Y. Impact of diagenesis on the reservoir quality of tight oil sandstones: The case of Upper Triassic Yanchang Formation Chang 7 oil layers in Ordos Basin, China. *J. Pet. Sci. Eng.* **2016**, *145*, 54–65. [CrossRef]
50. Li, S.; Shi, Z.; Liu, X.; Yang, S.; Deng, X.; Liu, G.; Li, J. Quantitative analysis of the Mesozoic abnormal low pressure in Ordos Basin. *Pet. Explor. Dev.* **2013**, *40*, 566–571. [CrossRef]
51. Bai, B.; Zhu, R.; Wu, S.; Yang, W.; Gelb, J.; Gu, A.; Zhang, X.; Su, L. Multi-scale method of Nano (Micro)-CT study on microscopic pore structure of tight sandstone of Yanchang Formation, Ordos Basin. *Pet. Explor. Dev.* **2013**, *40*, 354–358. [CrossRef]
52. Shearing, P.R.; Bradley, R.S.; Gelb, J.; Lee, S.N.; Atkinson, A.; Withers, P.J.; Brandon, N.P. Using synchrotron X-ray nano-CT to characterize SOFC electrode microstructures in three-dimensions at operating temperature. *Electrochem. Solid-State Lett.* **2011**, *14*, 117–120. [CrossRef]
53. Krakowska, P.; Dohnalik, M.; Jarzyna, J.; Wawrzyniak-Guz, K. Computed X-ray microtomography as the useful tool in petrophysics: A case study of tight carbonates Modryn formation from Poland. *J. Nat. Gas Sci. Eng.* **2016**, *31*, 67–75. [CrossRef]
54. Wang, X.; Xiao, L.; Xie, R.; Zhang, Y. Study of NMR porosity for terrestrial formation in China. *Sci. China Ser. G* **2006**, *49*, 313–320. [CrossRef]
55. Song, Y.Q.; Venkataramanan, L.; Hürlimann, M.D.; Flaum, M.; Frulla, P.; Straley, C. T_1–T_2 correlation spectra obtained using a fast two-dimensional Laplace inversion. *J. Magn. Reson.* **2002**, *154*, 261–268. [CrossRef] [PubMed]
56. Ziarani, A.S.; Aguilera, R. Pore-throat radius and tortuosity estimation from formation resistivity data for tight-gas sandstone reservoirs. *J. Appl. Geophys.* **2012**, *83*, 65–73. [CrossRef]
57. Loucks, R.G.; Reed, R.M.; Ruppel, S.C.; Hammes, U. Spectrum of pore types and networks in mudrocks and a descriptive classification for matrix-related mudrock pores. *AAPG Bull.* **2012**, *96*, 1071–1098. [CrossRef]

58. Huang, W.; Lu, S.; Hersi, O.S.; Wang, M.; Deng, S.; Lu, R. Reservoir spaces in tight sandstones: Classification, fractal characters, and heterogeneity. *J. Nat. Gas Sci. Eng.* **2017**, *46*, 80–92. [CrossRef]
59. Anders, M.H.; Laubach, S.E.; Scholz, C.H. Microfractures: A review. *J. Struct. Geol.* **2014**, *69*, 377–394. [CrossRef]
60. Lai, J.; Wang, G.; Cai, C.; Fan, Z.; Wang, S.; Chen, J.; Luo, G. Diagenesis and reservoir quality in tight gas sandstones: the fourth member of the Upper Triassic Xujiahe Formation, Central Sichuan Basin, Southwest China. *Geol. J.* **2018**, *53*, 629–646. [CrossRef]
61. Gane, P.A.; Kettle, J.P.; Matthews, G.P.; Ridgway, C.J. Void space structure of compressible polymer spheres and consolidated calcium carbonate paper-coating formulations. *Ind. Eng. Chem. Res.* **1996**, *35*, 1753–1764. [CrossRef]
62. Butler, J.P.; Reeds, J.A.; Dawson, S.V. Estimating solutions of first kind integral equations with nonnegative constraints and optimal smoothing. *SIAMJ Numer. Anal.* **1981**, *18*, 381–397. [CrossRef]
63. Gallegos, D.P.; Smith, D.M.A. NMR technique for the analysis of pore structure: Determination of continuous pore size distributions. *J. Colloid Interface Sci.* **1988**, *122*, 143–153. [CrossRef]
64. Kleinberg, R.L.; Straley, C.; Kenyon, W.E.; Akkurt, R.; Farooqui, S.A. Nuclear magnetic resonance of rocks: T_1 vs. T_2. In Proceedings of the SPE Annual Technical Conference and Exhibition, Houston, TX, USA, 3–6 October 1993.
65. Lyu, C.; Ning, Z.; Wang, Q.; Chen, M. Application of NMR T_2 to Pore Size Distribution and Movable Fluid Distribution in Tight Sandstones. *Energy Fuels* **2018**, *32*, 1395–1405. [CrossRef]
66. Gane, P.A.; Ridgway, C.J.; Lehtinen, E.; Valiullin, R.; Furo, I.; Schoelkopf, J.; Paulapuro, H.; Daicic, J. Comparison of NMR cryoporometry, mercury intrusion porosimetry, and DSC thermoporosimetry in characterizing pore size distributions of compressed finely ground calcium carbonate structures. *Ind. Eng. Chem. Res.* **2004**, *43*, 7920–7927. [CrossRef]
67. Coates, G.R.; Xiao, L.; Prammer, M.G. *NMR Logging: Principles and Applications*; Haliburton Energy Services: Houston, TX, USA, 1999; Volume 344.
68. Dékány, I.; Nagy, L.G. Immersional wetting and adsorption displacement on hydrophilic/hydrophobic surfaces. *J. Colloid Interface Sci.* **1991**, *147*, 119–128. [CrossRef]

processes

MDPI

Article

Development of 3D Finite Element Method for Non-Aqueous Phase Liquid Transport in Groundwater as Well as Verification

Wei Yu and Hong Li *

School of Civil Engineering, Dalian University of Technology, Dalian 116024, China; yuwei2017@mail.dlut.edu.cn
* Correspondence: hong.li@dlut.edu.cn; Tel.: + 86-138-894-72105

Received: 30 December 2018; Accepted: 18 February 2019; Published: 25 February 2019

Abstract: Groundwater contamination previously occurred at a broad range of locations in present-day China. There are thousands of kinds of contaminants which can be divided into soluble and insoluble categories in groundwater. In recent years, the non-aqueous phase liquid (NAPL) pollution that belongs to the multi-phase seepage flow phenomenon has become an increasingly prominent topic due to the challenge brought by groundwater purification and its treatment. Migrating with seepage flow and moving into the potable water sources, these contaminants directly endanger people's health. Therefore, it is necessary to research how these contaminants not only migrate, but also are then accordingly remedied. First, as an analysis means, an effective numerical method is necessary to be built. A three-dimensional finite element method program for analyzing two-phase flow in porous media, which can be applied to the immiscible contaminant transport problem in subsurface flow has been developed in this paper. The fundamental theory and numerical discretization formulations are elaborated. The numerical difficulty brought about by the distinct non-linearity of the temporal evolution of saturation-dependent variables is overcome by the mixed-form formulation. The effectiveness of simultaneous solution (SS) method and its improvement in efficiency are explained. Finally, two computational examples are given for verifying the correctness and demonstrating the preliminary applicability. In addition, the function of two-phase immiscible flow, especially in Fast Lagrangian Analysis of Continua (FLAC) is used to simulate the same examples and the results are compared to further verify the correctness of the numerical development.

Keywords: non-aqueous phase liquid; finite element method; two-phase flow; mixed-form formulation; FLAC

1. Introduction

Different from the traditional of subsurface contaminant problems dominated by diffusion dissolved in water and convection miscible in fluid transport, in recent years, pollution of non-aqueous phase liquids (NAPLs), which belongs to multi-phase seepage flow phenomenon, is becoming increasingly prominent and complicated in the groundwater purification and treatment. Such liquid pollutants are immiscible in water. Most of them are organic pollutants such as hydrocarbons, petroleum products, chlorinated organic solvents, etc. According to the density of NAPLs compared with water, they are divided into light non-aqueous liquids (LNAPLs, lighter than water) and heavy non-aqueous liquids (DNAPLs, heavier than water). During actual industry production, they are commonly seen and exposed in many places [1–3]: Strong organic chlorine solvents, such as trichloroethylene and tetrachloroethylene, are widely applicable to dry cleaning of metal, integrated circuits, and electronic components to remove paints, oil, and fat stains [4]. Additionally, a large number of gas stations and oil storage facilities have been newly built in many places in the world since the development of petroleum energy and chemicals play an important role in human industrial production activities. Due to improper disposal or poor management, or accidental breakage of storage containers or pipelines, they leak out and seep into the ground [5,6]. As shown in Figure 1, these leaking petroleum products will pass through surface soil to aquifers to the aquitard, and they will sneak into the groundwater and expand laterally on the aquitard top boundary. Additionally, the volatile component will quickly spread far away, and even spillover out of the surface and pollute the air. NAPLs can be chronically bounded in the pores of media, and then accumulated there in large quantities. They are extremely difficult to degrade, finally becoming a persistent source of contamination. It retains in the environment, able to biologically accumulate in ecosystems, has substantial negative impacts on human health and the environment, and even directly affects the price of land.

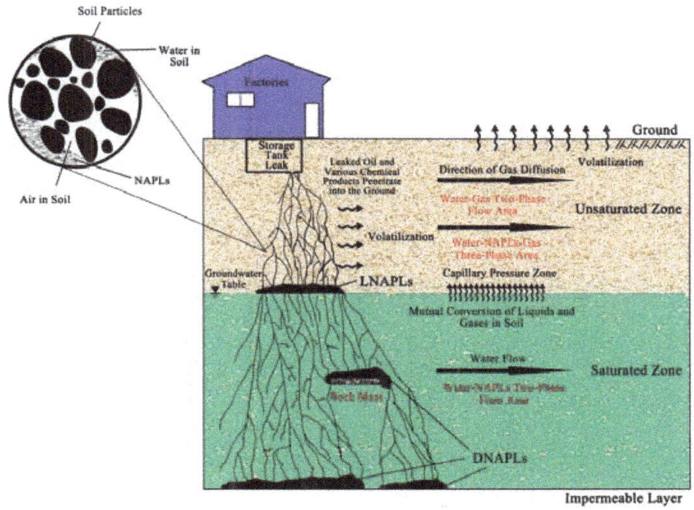

Figure 1. Schematic illustration of the subsurface transport mechanism of NAPL contaminants.

In these cases, it is necessary to analyze the mutual dependence and simultaneous flowing of the non-miscible phase liquid and the ground water in the infiltration field. Compared with the miscible fluid, the difference of NAPL in subsurface water is that the different liquids are separated by the interfaces existing in the pores of the medium. The liquid with higher wettability tends to adhere to pore walls. Due to the interface between the liquids is bent by interfacial tension, it is recessed to the liquid with the higher wettability. To balance this effect, the less humidified liquid has a higher pressure in the pores than the more humidified liquid. This usually not negligible pressure difference is called capillary pressure. When the pore size is kept constant, capillary pressure depends on the percentage of the pore space occupied by different liquid phase contents, that is, saturation. The effective relative permeability of each different liquid also depends on the degree of the phase corresponding saturation. These saturation-dependent parameters exhibit nonlinear characteristics in time evolution, and sometimes even relatively strong nonlinearities, posing a challenge to the rigor of numerical solutions.

1.1. Numerical Developing Status on NAPL Contaminant Analysis

The migration of incompatible pollutants in groundwater was only noticed internationally in the late 1980s [7–9]. In the past 40 years, one scholar after another conducted comparatively systematic research. For most cases of the soil contaminated with NAPLs, generally speaking, the problem of two-phase flow is involved in the water-saturated domain of interest, while that of three-phase flow considering condensation and gasification phenomena should be taken into account in the vadose zone. Conceptually, NAPLs are presented in small pores, pore wedges, bypassed pores as films or lenses on water or solid surfaces. They might not be drained from pores in vadose zones after long drainage periods in a strongly water-wet porous medium [10]. White, Oostrom, and Lenhard implemented a numerical model on the flow of a nonvolatile nonaqueous phase liquid (NAPL) and aqueous phases that accounts for mobile, entrapped, and residual NAPL in variably-saturated water-wet porous media. The results were also compared against those from detailed laboratory experiments [11]. Comparisons between the numerical simulations and experiments demonstrated the necessity to include the residual NAPL formation process in multiphase flow simulators. Xue et al. established a coupling model for analyzing the transport of organic contaminants in soil and water environments [12]. Wu and Wang researched the relationship between the riser oil holding rate and the volume fraction with regards to oil measured by the coaxial conductivity sensor, and designed an oil content measuring system based on coaxial conductivity sensor [13]. Chen, Yang, and Tian conducted a study on microscopic mechanism of non-aqueous phase liquids (NAPLs) migration in porous media [14]. Kleinknecht and Braun undertook not only a large column experiment, but also its numerical simulation to study the density-driven migration of DNAPL gas (carbon disulfide (CS_2) vapor) in the vadose zone [15]. Kikumoto and Nakamura developed a comprehensive numerical method for simulating transport of non-aqueous phase liquids (NAPLs) in unsaturated subsurface domains [16]. Javanbakht et al. conducted a study where X-ray microtomography experiments were performed to investigate the impact of surfactants and microemulsions on the mobilization and resolvability of NAPL in heterogeneous rocks [17]. Xie et al. investigated the dynamic behavior, such as flow rate and multi-scale time irreversibility, of different flow patterns based on the measurement signals obtained from oil-gas-water three-phase and oil- water two-phase flow experiments [18,19]. Tan et al. studied the flow patterns of horizontal oil-water two-phase pipe flow with water holdup fluctuations provided by a set of conductivity and capacitance sensors [20]. Picchi and Battiato, tackled the problem of the limitations of Darcy's law in properly modeling the flow at the continuum scale by proposing a set of upscaled equations based on pore-scale flow regimes, that is, the topology of flowing phases [21]. Balasuriya et al. performed a study on detailed field characterization of elemental mercury DNAPL distribution with depth, together with two-phase flow modelling by using STOMP [22]. Li et al. developed an energy demodulation algorithm for flow velocity measurement of oil-gas-water three-phase flow [23]. Kacem and Benadda built a model to simulate the multiphase extraction (MPE) applied to soil polluted by toluene. The transport and transfer between three phases by using the capillarity equations were simulated [24]. These studies involve various aspects of multiphase flows in subsurface water purification. However, in most practical problems, because of the heterogeneity of the considered domain, the irregular shapes of its boundaries, the existence of flow turbulence and phase interfacial interactions, and the liquid mixture in porous medium often exhibit complex behaviors. It is not possible to solve these mathematical models analytically. Instead, the mathematical model is transformed into a numerical one that can be solved by means of computer programs [25]. Although there is a large amount of literature on finite difference or finite volume methods for multi-phase flow, the flexible geometry and intrinsic boundary adaptation ability were usually not considered sufficiently. The literature on finite element methods is particularly needed when conducting such issues. However, among these studies and studies, those that discuss the finite element method in this field were rarely seen, especially in the three-dimensional FEM literature, which introduce numerical programs for their counterparts. Descriptions in such detail with operable procedures is hardly found.

This paper introduces the program development of the three-dimensional finite element method for non-miscible and incompressible two-phase flow and preliminary verification with two examples, which belong to LNAPL (oil) and DNAPL (trichloroethylene), respectively.

2. Fundamental Theory of Two-Phase Flow in Porous Media

In order to use mathematical means to depict the motion disciplines of fluids in porous media, the continuum medium method is introduced [26]. It is needed to define the parameters of fluids and porous medium at arbitrary point of domain in this method, such as the velocity of flow, porosity, driving pressure, etc. Therefore, it is hypothesized that a medium is continuously filled over the entire field of research and the actual multiphase seepage porous flow microscopic structure is replaced by average macroscopic meaning, that is, the continuum medium mechanics which at least contains two phases—fluid phase and porous media phase. Both medium and fluid are continuously filled with the entire domain of interest. The parameters of the porous medium, fluid and motion such as porosity, permeability, density, flow rate, concentration, etc., can be defined at any point. This method to describe the migration of fluids and contaminants in porous media is called the continuum medium method, and it has avoided the difficulty of conducting the law of fluid mass point motion in a single pore, studying its parameters using their macroscopic average values instead. These obtained parameters such as flow rate, pressure and concentration are the best approximation of the actual flow, which meets the actual demands.

2.1. Governing Equations

In a porous medium, an infinitesimal elemental volume centered at a point of media is taken, which is a volume that can represent the average physical properties near the point. It is called a representative elemental volume (REV, sometimes called the control volume). Similar to the partial differential equation for establishing single-phase seepage flow of porous media, as shown in Figure 2, a cuboid REV is considered.

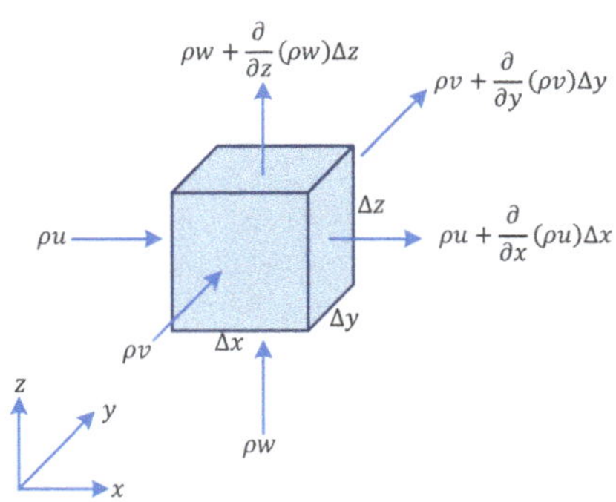

Figure 2. Mass conservation of flows passing through a small rectangular parallelepiped.

In the concept of continuum mechanics, it must be small enough to approximate macroscopic continuity and large enough to be equivalent to microscopic statistical averaging.

The flow rate (Q_i) flowing into the elemental volume can be expressed as:

$$Q_i = \rho u \Delta y \Delta z + \rho v \Delta x \Delta z + \rho w \Delta x \Delta y, \tag{1}$$

where ρ is the mass density of the liquid; $u, v,$ and w are the average flow rates through the left, front, and bottom of the element (positive and negative, respectively, represent in and out).

In direction x, if the mass flux flowing into the element is ρu, after the same time interval Δx, the mass flux flow out of the element can be expressed as Taylor series.

$$(\rho u)_{x+\Delta x} = \rho u + \frac{\partial}{\partial x}(\rho u)\Delta x + \frac{\partial^2}{2!\partial x^2}(\rho u)(\Delta x)^2 + \frac{\partial^3}{3!\partial x^3}(\rho u)(\Delta x)^3 + \ldots, \tag{2}$$

If high-order terms of a small variable are omitted, the corresponding flow (Q_t) flowing out of the cuboid per unit time can be expressed as:

$$Q_t = \left[\rho u + \frac{\partial}{\partial x}(\rho u)\Delta x\right]\Delta y\Delta z + \left[\rho v + \frac{\partial}{\partial y}(\rho v)\Delta y\right]\Delta x\Delta z + \left[\rho w + \frac{\partial}{\partial z}(\rho w)\Delta z\right]\Delta x\Delta y, \tag{3}$$

On the other hand, the cumulative storage amount (Q_a) in the cuboid after per unit time can be expressed by the following formula:

$$Q_a = \frac{\partial}{\partial t}(\rho\phi S)\Delta x\Delta y\Delta z, \tag{4}$$

where ϕ is the porosity of the media, $\phi = V_v/V$, V_v and V are the volume of the pores and medium, respectively; S is the saturation of the flow phase, $S = V_f/V_v$, V_f is the flow phase volume.

For a flow phase, according to the mass conservation principle $Q_i - Q_t = Q_a$, combining Equation (1), (3), and (4) yields:

$$-\frac{\partial}{\partial x}(\rho u) - \frac{\partial}{\partial y}(\rho v) - \frac{\partial}{\partial z}(\rho w) = -\frac{\partial}{\partial t}(\rho\phi S), \tag{5a}$$

Introducing Einstein's summation convention, Equation (5a) can be written as follow:

$$-\frac{\partial}{\partial x_i}(\rho u_i) = -\frac{\partial}{\partial t}(\rho\phi S), \tag{5b}$$

As the simplified form of conservation of momentum in the flow phase, Darcy's law applies and can be also expressed as:

$$u_i = -\frac{k}{u}K_{ij}\frac{\partial}{\partial x_j}(p + \rho g H_i) = -\frac{k}{u}K_{ij}\left(\frac{\partial p}{\partial x_j} + \gamma\delta_{ji}\right), \tag{6a}$$

where, k is relative permeability co-efficient of flow phase, μ is dynamic viscosity of flow phase, K_{ij} is absolutely permeability tensor, p is the pressure of flow phase, g is gravitational acceleration, H_i is the coordinate in the direction of each coordinate axis, γ is relative density of fluid, and δ_{ji} is the sign of the Kronecker delta.

The item $\rho g H_i \neq 0$ only when H_i along the gravity direction (assumed direction 3). Thus, Equation (6a) can be also written as:

$$u_i = -\frac{k}{u}K_{ij}\frac{\partial}{\partial x_j}(p + \rho g H_3) = -\frac{k}{u}K_{ij}\left(\frac{\partial p}{\partial x_j} + \gamma\delta_{j3}\right), \tag{6b}$$

As already mentioned in the above, the flow phases discussed in this paper are all approximated as incompressible, and the pores of the porous media are also approximated as rigid.

Substituting Equation (6b) into Equation (5b), then phase control equations yield as follows:

$$\nabla\cdot\frac{k_\alpha}{\mu_\alpha}\kappa\nabla(p_\alpha + \rho_\alpha g z) = \frac{\partial}{\partial t}(\phi S_\alpha) \quad (\alpha = w, o), \tag{7}$$

where $\nabla\cdot$ is the divergence operator of a vector field; ∇ is the gradient operator of a scalar field; p_α, k_α, ρ_α, S_α, and μ_α are pressure, relative permeability coefficient, mass density, saturation, and dynamic viscosity of each phase; κ is the intrinsic or absolute permeability. z is the vertical coordinate of the porous medium. In this paper, the intrinsic or absolute permeability and porosity are only functions of position. "w" and "o" are representing water and non-aqueous liquids.

As for two-phase flow that contains water and non-aqueous only, there are:

$$S_w + S_o = 1, \tag{8a}$$

and:

$$\frac{\partial S_o}{\partial t} = -\frac{\partial S_w}{\partial t}, \tag{8b}$$

Noticing Equation (8a), Equation (7) can be rewritten as the following equation group:

$$\left.\begin{array}{l} \phi\frac{\partial S}{\partial t} = \nabla\cdot k_w K\nabla(\psi_w + z) \\ -\phi\frac{\partial S}{\partial t} = \nabla\cdot\frac{k_o}{\mu_r}K\nabla(\psi_o + \rho_r z) \end{array}\right\}, \tag{9}$$

where S is saturation of water, $S = S_w$; μ_r is relative viscosity of water; K is hydraulic conductivity, $K = \kappa\rho_w g/\mu_w$; $\psi_\alpha(\alpha = w, o)$ is pressure expressed in the water head of phase α, $\psi_\alpha = p_\alpha/(\rho_w g)$; ρ_r is the relative density, $\rho_r = \rho_o/\rho_w$.

2.2. Finite Differential Discretization in Time

The finite difference discretization is used for time evolution of Equation (9) and the following equations are obtained:

$$\left.\begin{array}{l} \phi\frac{S^{n+1,\,v+1}-S^n}{\Delta t} - \nabla\cdot k_w^{n+1-\theta,\,v}K\nabla[(1-\theta)\psi_w^n + \\ \quad\theta\psi_w^{n+1,\,v+1}] - \frac{\partial k_w^{n+1-\theta,\,v}K}{\partial z} = 0 \\ \phi\frac{S^n-S^{n+1,\,v+1}}{\Delta t} - \nabla\cdot\frac{k_w^{n+1-\theta,\,v}}{\mu_r}K\nabla[(1-\theta)\psi_o^n + \\ \quad\theta\psi_o^{n+1,\,v+1}] - \frac{\partial}{\partial z}\left(\rho_r\frac{k_o^{n+1-\theta,\,v}K}{\mu_r}\right) = 0 \end{array}\right\}, \tag{10}$$

where the superscript indicates "n" is the known value of the variable at the end of the previous time step, $\psi^n = \psi(t)$, "n + 1" indicates the value to be evaluated at the end of the current time step, $\psi^{n+1} = \psi(t + \Delta t)$; and the superscript indicates "v" is the value obtained by the current time step in the previous iteration, "v + 1" indicates the value obtained by the current time step variable at the completion of this iteration; θ is a parameter that indicates the type of the difference method; when $\theta = 1$, the backward difference format is adopted, and when $\theta = 1/2$, the central difference format is adopted.

In order to solve the equations, the pressure head of each phase $\psi_w = \psi_w(x, y, z, t)$ and $\psi_o = \psi_o(x, y, z, t)$. are selected to be the basic variables, and the following equations are used:

$$\left.\begin{array}{l} \psi_\alpha^{n+1,\,v=0} = \psi_\alpha^n\ (n = 0) \\ \psi_\alpha^{n+1-\theta,\,v} = \psi_\alpha^n + \theta\frac{(\Delta t)^{n+1}}{(\Delta t)^n}\left(\psi_\alpha^n - \psi_\alpha^{n-1}\right) \\ \qquad (n > 0, v = 0) \\ \psi_\alpha^{n+1-\theta,\,v} = (1-\theta)\psi_\alpha^n + \theta\psi_\alpha^{n+1,\,v} \\ \qquad (n > 0, v > 0) \end{array}\right\}, \tag{11}$$

Based on these, the saturation of water can be expressed as the function of ψ_w or ψ_o. In two-phase flow, the void space is completely filled by the two fluids. One of the fluids (the wetting fluid) wets the porous medium more than the other (the non-wetting fluid). As a result, the pressure in the non-wetting fluid will be higher than the pressure in the wetting fluid. The pressure difference is the capillary pressure ψ_c, which is a function of saturation. As shown in Figure 3a, the saturation of water phase can be expressed as a function of the capillary pressure head, $S = S(\psi_c)$, where capillary pressure is defined as $\psi_c = \psi_o - \psi_w$. Water phase saturation can be obtained by the following finite-difference equation:

$$\left.\begin{array}{r}S^{n+1,\,v+1}=S\left(\psi_o^{n+1,\,v+1}-\psi_w^{n+1,\,v+1}\right)\\ S^n=S(\psi_o^n-\psi_w^n)\end{array}\right\}, \quad (12)$$

As shown in Figure 3b, the relative permeability coefficient of each phase is a function of water phase saturation, and it can also be further expressed as a function of ψ_w and ψ_o. For example, permeability coefficient of water phase can be expressed as:

$$K_w^{n+1-\theta,\,v}=K_w\left(S\left(\psi_o^{n+1-\theta,\,v}-\psi_w^{n+1-\theta,\,v}\right)\right)=K_w\left(S\left(\psi_o^{n+1-\theta,\,v}-\psi_w^{n+1-\theta,\,v}\right)\right), \quad (13)$$

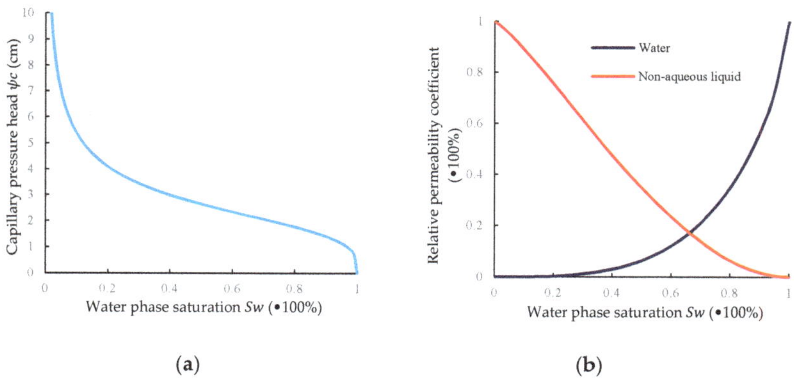

Figure 3. Nonlinear character of water and NAPL two-phase flow: (**a**) Two-phase flow characteristic curve of the water-non-aqueous liquid; and (**b**) the relative permeability coefficient of each phase depends on the change of the water phase saturation.

The Taylor series of water phase saturation expands to:

$$S^{n+1,\,v+1}=S^{n+1,\,v}+\left(\frac{\partial S}{\partial \psi}\right)^{n+1,\,v}\left(\psi^{n+1,\,v+1}-\psi^{n+1,\,v}\right)+O\left[\left(\psi^{n+1,\,v+1}-\psi^{n+1,\,v}\right)^2\right], \quad (14a)$$

where:

$$\psi=\psi_c,$$

This expansion corresponds to the backward difference format, which has a first-order precision after omitting the high-order terms. If the Taylor series of saturation expands as follows:

$$\left.\begin{array}{r}S^{n+1,\,v+1}=S^{n+1,\,v+\frac{1}{2}}+\left(\frac{\partial S}{\partial \psi}\right)^{n+1,\,v+\frac{1}{2}}\varepsilon\\ +\frac{1}{2}\left(\frac{\partial S}{\partial \psi}\right)^{n+1,\,v+\frac{1}{2}}\varepsilon^2+O(\varepsilon^3)\\ S^{n+1,\,v}=S^{n+1,\,v+\frac{1}{2}}-\left(\frac{\partial S}{\partial \psi}\right)^{n+1,\,v+\frac{1}{2}}\varepsilon\\ +\frac{1}{2}\left(\frac{\partial S}{\partial \psi}\right)^{n+1,\,v+\frac{1}{2}}\varepsilon^2-O(\varepsilon^3)\end{array}\right\}, \quad (14b)$$

where:

$$\varepsilon=\psi^{n+1,\,v+1}-\psi^{n+1,\,v+1/2},$$

adding the first and the second formula of Equation (14b), the follow equation can be obtained:

$$S^{n+1,\,v+1}=S^{n+1,\,v}+\left(\frac{\partial S}{\partial \psi}\right)^{n+1,\,v+1/2}\left(\psi^{n+1,\,v+\frac{1}{2}}-\psi^{n+1,\,v}\right)+O\left[\left(\psi^{n+1,\,v+1}-\psi^{n+1,\,v}\right)^3\right], \quad (15)$$

Equation (15) corresponds to central difference format, which has second-order precision after omitting higher-order terms.

If the specific water content function is defined as $C(\psi) = \phi \partial S / \partial \psi$, which represents the change in water content caused by per unit change in capillary pressure head. Then, the evolution of saturation can be expressed as:

$$S^{n+1,\, v+1} - S^{n+1,\, v} = \frac{1}{\phi} C^{n+1,\, w} \left(\psi^{n+1,\, v+1} - \psi^{n+1,\, v} \right) = \frac{1}{\phi} C^{n+1,\, w} (\varepsilon_o - \varepsilon_w), \tag{16}$$

where:

$$\varepsilon_\alpha = \psi_\alpha^{n+1,v+1} - \psi_\alpha^{n+1,v} \quad (\alpha = o, w),$$

It should be pointed out that $\Delta_t S = S^{n+1,v+1} - S^{n+1,v}$ should be able to converge to 0 in the gradual iteration. In theory, Equation (16) can be rewritten into:

$$\Delta_t S = S^{n+1,\, v+1} - S^n = \frac{1}{\phi} C^{n+1,\, w} \left(\psi^{n+1,\, v+1} - \psi^n \right) = \frac{1}{\phi} C^{n+1,\, w} \Delta_t \psi, \tag{17}$$

2.3. Mixed-Form Formulation Adopted in FEM Program Development

However, Equation (17) does not consistently meet the requirements for the conservation of mass. As shown in Figure 4, because of the nonlinearity of the water content characteristic curve, there exists such a case: where the actual saturation change is large corresponding to a small capillary pressure head increment. Additionally, if Equation (17) is substituted into Equation (10) to eliminate the saturation term, and equations with the pressure head as the unknown quantity will be directly formed. These equations whose linear values correctly approximate the saturation require unusually high convergence accuracy, so the efficiency is extremely low or not feasible at all. If this is not recognized and general convergence accuracy is set, the calculation process may be unstable or the results will be totally unacceptable.

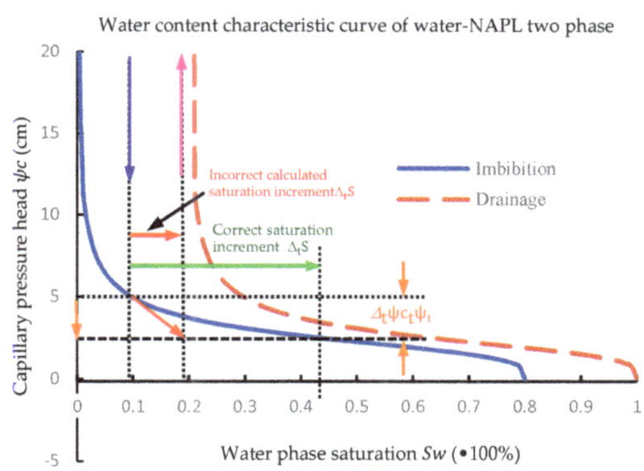

Figure 4. Illustration of inability of using single form formulation to model immiscible flow.

The main performance is that the conservation of mass is not guaranteed, and the position of the depth of the infiltration is estimated incorrectly. Therefore, it is necessary to adopt the mixed-form formula advocated by Celia et al. [27]. Equation (10) is expressed as follows:

$$\left.\begin{array}{c}\frac{1}{\Delta t}C^{n+1,\ v+\ w}\varepsilon_o-\frac{1}{\Delta t}C^{n+1,\ v+\ w}\varepsilon_w-\\ \nabla\cdot k_w^{n+1,\ v+w}K\nabla\ (\theta\varepsilon_w)=\\ \nabla\cdot k_w^{n+1,\ v+w}K\nabla\left[(1-\theta)\psi_w^n+\theta\psi_w^{n+1,\ v}\right]\\ +\frac{\partial k_w^{n+1,\ v+w}K}{\partial Z}-\phi\frac{S^{n+1,\ v+w}-S^n}{\Delta t}\\ \frac{1}{\Delta t}C^{n+1,\ v+\ w}\varepsilon_w-\frac{1}{\Delta t}C^{n+1,\ v+\ w}\varepsilon_o-\\ \nabla\cdot\frac{k_w^{n+1,\ v+w}}{u_r}K\nabla\ (\theta\varepsilon_o)=\\ \nabla\cdot\frac{k_w^{n+1,\ v+w}}{u_r}K\nabla\left[(1-\theta)\psi_o^n+\theta\psi_o^{n+1,\ v}\right]\\ +\frac{\partial}{\partial Z}\left(\rho_r\frac{k_w^{n+1,\ v+w}}{u_r}K\right)+\phi\frac{S^{n+1,\ v+w}-S^n}{\Delta t}\end{array}\right\}, \tag{18}$$

Equation (10) and Equation (18) are in hybrid form because some items are based on the pressure head, and some items are based on the saturation. In Equation (18), the right side of the equations are known values of the variable at the last time step or the end of the previous iteration. The left items of the equations are expressed as the pressure head that will increase in the current iteration, and the quantity is the unknown value to be sought.

3. Numerical Discretization Formulations Using the Finite Element Method

In space, the shape of the variable to be sought is discretized as follows:

$$\psi_\alpha(x,y,z,t)=N_\xi(x,y,z)[\Psi_\alpha(t)]_\xi, \tag{19a}$$

We adopt an eight-node hexahedron iso-parametric element, where:

$$N_\xi(x,y,z)=1/8(1+x_\xi x)(1+y_\xi y)(1+z_\xi z) \tag{19b}$$

is an interpolation function. $P_0(x_\xi,y_\xi,z_\xi)$ represents the local coordinates of the corner points in the eight-node hexahedral element; ξ is the local grid point number for the corresponding element, $\xi=1,\ 2,\ \ldots,\ 8$.

According to the standard Galerkin finite element method format, the discrete residual function should be equal to 0 in the global domain, so there are:

$$\left.\begin{array}{c}\sum_{e=1}^{N}\int\ V_e\left\{\frac{1}{\Delta t}C^{n+1,\ v+\ w}\varepsilon_o-\frac{1}{\Delta t}C^{n+1,\ v+\ w}\varepsilon_w-\right.\\ \nabla\cdot k_w^{n+1,\ v+w}K\nabla\ (\theta\varepsilon_w)-\\ \nabla\cdot k_w^{n+1,\ v+w}K\nabla\left[(1-\theta)\psi_w^n+\theta\psi_w^{n+1,\ v}\right]\\ \left.-\frac{\partial k_w^{n+1,\ v+w}K}{\partial Z}+\phi\frac{S^{n+1,\ v+w}-S^n}{\Delta t}\right\}N_m dV_e=0\\ \sum_{e=1}^{N}\int\ V_e\left\{\frac{1}{\Delta t}C^{n+1,\ v+\ w}\varepsilon_w-\frac{1}{\Delta t}C^{n+1,\ v+\ w}\varepsilon_o-\right.\\ \nabla\cdot\frac{k_w^{n+1,\ v+w}}{u_r}K\nabla\ (\theta\varepsilon_o)=\\ \nabla\cdot\frac{k_w^{n+1,\ v+w}}{u_r}K\nabla\left[(1-\theta)\psi_o^n+\theta\psi_o^{n+1,\ v}\right]\\ \left.-\frac{\partial}{\partial Z}\left(\rho_r\frac{k_w^{n+1,\ v+w}}{u_r}K\right)-\phi\frac{S^{n+1,\ v+w}-S^n}{\Delta t}\right\}N_m dV_e=0\end{array}\right\}, \tag{20}$$

where e. is element sequence number, $e=1,\ 2,\ \ldots,\ N$; m is sequence number for the grid point in the whole domain, $m=1,\ 2,\ \ldots,\ NP$.

Applying partial integral and Gaussian divergence theorem to Equation (20), after settlement it yields:

$$\begin{bmatrix} \theta(A_w)_{m(\xi)} - \frac{F_m}{\Delta t}\delta_{m(\xi)} & \frac{F_m}{\Delta t}\delta_{m(\xi)} \\ \frac{F_m}{\Delta t}\delta_{m(\xi)} & \theta(A_w)_{m(k)} - \frac{F_m}{\Delta t}\delta_{m(\xi)} \end{bmatrix} \cdot \begin{Bmatrix} (\varepsilon_w)_{(\xi)} \\ (\varepsilon_o)_{(\xi)} \end{Bmatrix} = \begin{Bmatrix} (B_w)_m \\ (B_o)_m \end{Bmatrix}, \tag{21a}$$

where:

$$(B_w)_m = (Q_w)_m - (G_w)_m - (A_w)_{m(\xi)} \cdot \left[\theta(\Psi_w)_{(\xi)}^{n+1,\, v+w} + (1-\theta)(\Psi_w)_{(\xi)}^{n}\right] - \frac{1}{\Delta t}\zeta_t\Theta_m, \tag{21b}$$

$$(B_o)_m = (Q_o)_m - (G_o)_m - (A_o)_{m(\xi)} \cdot \left[\theta(\Psi_o)_{(\xi)}^{n+1,\, v+w} + (1-\theta)(\Psi_o)_{(\xi)}^{n}\right] + \frac{1}{\Delta t}\zeta_t\Theta_m, \tag{21c}$$

$$(A_w)_{m(\xi)} = \sum_{e=1}^{N} \int_{V_e} \frac{\partial N_m}{\partial x_i} k_w K_{ij} \frac{\partial N_{(\xi)}}{\partial x_j} dV_e \tag{21d}$$

$$(A_o)_{m(\xi)} = \sum_{e=1}^{N} \int_{V_e} \frac{\partial N_m}{\partial x_i} \frac{k_o}{\mu_r} K_{ij} \frac{\partial N_{(\xi)}}{\partial x_j} dV_e \tag{21e}$$

$$(G_w)_{m(\xi)} = \sum_{e=1}^{N} \int_{V_e} \frac{\partial N_m}{\partial x_i} k_w K_{i3} dV_e \tag{21f}$$

$$(G_o)_{m(\xi)} = \sum_{e=1}^{N} \int_{V_e} \frac{\partial N_m}{\partial x_i} \frac{k_o}{\mu_r} K_{i3} dV_e \tag{21g}$$

$$(Q_w)_{m(\xi)} = \sum_{e=1}^{N} \int_{\Gamma_e} k_w K_{ij} \frac{\partial}{\partial x_j}(\psi_w + H_3) N_m n_i d\Gamma_e \tag{21h}$$

$$(Q_o)_{m(\xi)} = \sum_{e=1}^{N} \int_{\Gamma_e} \frac{k_o}{\mu_r} K_{ij} \frac{\partial}{\partial x_j}(\psi_w + \rho_r H_3) N_m n_i d\Gamma_e \tag{21i}$$

$$F_m = \sum_{e=1}^{N} \int_{V_e} C N_m dV_e \tag{21j}$$

$$\zeta_t\Theta_m = \sum_{e=1}^{N} \int_{V_e} \phi(S_m - S_m^n) N_m dV_e \tag{21k}$$

$$\left.\begin{array}{c} k_w = k_w(\psi_w^{n+1,\, v+w}),\ k_o = k_o(\psi_o^{n+1,\, v+w}) \\ \mu_r = \dfrac{\mu_o(\psi_o^{n+1,\, v+w})}{\mu_w(\psi_w^{n+1,\, v+w})},\ \rho_r = \dfrac{\rho_o(\psi_o^{n+1,\, v+w})}{\rho_w(\psi_w^{n+1,\, v+w})} \\ C = C(\psi^{n+1,\, v+w}),\ S = S(\psi^{n+1,\, v+w}) \end{array}\right\}, \tag{21l}$$

when $m = (\xi)$, $\delta_{m(\xi)} = 1$, when $m \neq (\xi)$, $\delta_{m(\xi)} = 0$; (ξ) represents the global node number corresponding to the local node number of the node in the element. n_i is normal direction vector

The number of equations or variables to be solved in Equation (21a) is double to the total number of nodes because the two-phase flow itself has two degrees of freedom. Noticing that in this form the coefficient matrix will remain symmetrical, an unsteady iterative solution can be used to solve the linear equations formed during each iteration of each time step, such as the conjugate gradient method, which is very efficient, stable, fast, and accurate. As for the iterative steps for solving a system of nonlinear equations, as expressed by Equation (18) and (21), if an incremental form of the unknown pressure head is used, the general iterative method is equivalent to the Newton–Raphson iterative solution when the incremental form is not used.

4. Verifying the Numerical Solution with Two Examples

4.1. Water Flooding Oil in the Pillar Rock Sample (LNAPL)

The first example is to simulate a rock sample water displacing oil experiment to test the numerical method presented in this paper. As shown in Figure 4, the static rock sample is filled with oil, such as the initial saturation condition of the water is zero. The initial value of the capillary pressure of the sample for oil and water is measured in advance, for example, the result is water head pressure 25 cm. The side is sealed and the top boundary is kept as a certain fixed pressure. At the beginning of the experiment, water is injected from the bottom boundary with a certain, but slightly higher, pressure to gradually displace most of the oil from the top. The specific pressure values at the top and bottom boundaries are shown in Figure 5.

The capillary pressure head and the relative permeability coefficient depend on water saturation using the formulas of Van Genucheten [28] and Parker et al. [29–31]:

$$\psi_c = \lambda^{-1}\left(S_{we}^{-1/r} - 1\right)^{1/n}, \tag{22}$$

$$k_w = \sqrt{S_{we}}\left[1 - \left(1 - S_{we}^{-1/r}\right)^r\right]^2, \tag{23}$$

$$k_o = \sqrt{1 - S_{we}}\left(1 - S_{we}^{-1/r}\right)^{2r}, \tag{24}$$

where, S_{we} is the effective water saturation, $S_{we} = (S_w - S_{wr})/(1 - S_w - S_{wr}), S_{wr}$ and S_{or} are the residual saturations of water and non-aqueous liquid, respectively; k_w and k_o are the relative permeability coefficients of water and non-aqueous liquid; α and n are both van Genuchten constants determined by experimental data; and $r = 1 - 1/n$. In this example, $\lambda = 0.431\ cm^{-1}$, $n = 3.7$.

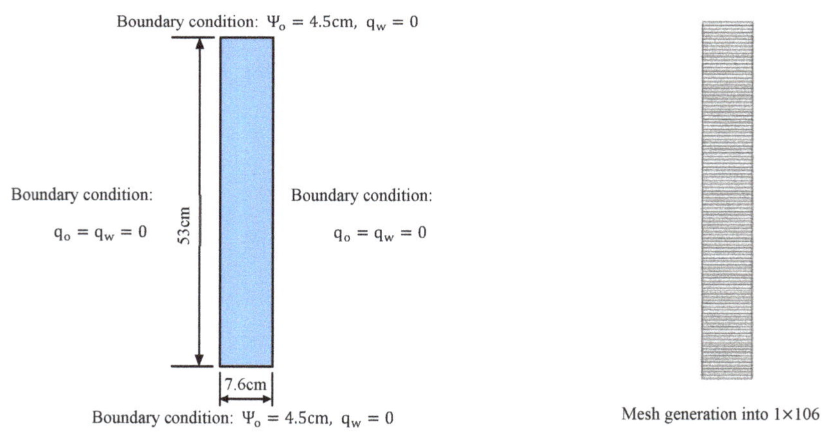

Figure 5. Displacing test thorough a rock sample: (**a**) Model size and boundary conditions; and (**b**) model meshing.

The residual saturation of oil, and water are both 0. The required samples and other physical parameters of the liquids are shown in Table 1.

Table 1. Physical properties of rock sample and fluids (24 °C).

Density/g·cm^{-3}		Dynamic Viscosity/cp		Sample	
Water	Oil	Water	Oil	Porosity	Saturated permeability coefficient/cm · s^{-1}
1.00	0.86	0.914	38.560	0.37	3.243×10^{-2}

Different from the numerical method above, FLAC2D, version 7.0, is calculating with finite-difference method. Noticed that the sample is a quadrangular, and FLAC considers only a plane model (the third direction is processed by unit length). In order to testify the influence of the sample thickness in this example, a unit thickness was set to the quadrangle and it was found that this is a complete one-dimensional problem. Therefore, the difference in thickness can be ignored. In FLAC, the increments of the nodal pore pressure and saturation are expressed as the following equations [28,32]:

$$\Delta\psi_w = -\frac{\Delta t}{\phi VD}\left[Q_w\left(1-\frac{S_o\psi_c'}{K_o}\right)+Q_o\right]-\beta\frac{\Delta V}{\phi VD}\left[S_w\left(1-\frac{S_o\psi_c'}{K_o}\right)+S_o\right], \tag{25}$$

$$\Delta S_w = -\frac{\Delta t}{\phi VD}\left[Q_w\frac{S_o}{K_o}-Q_o\frac{S_w}{K_w}\right]-\beta\frac{\Delta V}{\phi VD}S_wS_o\left[S_w\left(1-\frac{S_o\psi_c'}{K_o}\right)+S_o\right], \tag{26}$$

$$\Delta\psi_o - \Delta\psi_w = \psi_c'\Delta S_w\,, \tag{27}$$

$$\Delta S_w + \Delta S_o = 0, \tag{28}$$

where, by definition:

$$D = \frac{S_w}{K_w}+\frac{S_o}{K_o}-\frac{S_w}{K_w}\frac{S_o\psi_c'}{K_o}, \tag{29}$$

and ψ_w, ψ_o, S_w, S_o are nodal pressure, and saturation of water and non-wetting fluid. K_w, K_o are fluid bulk moduli, V is the nodal volume, Q is the nodal flow rate, β is the undrained coefficient, which is a constant and equal to one for mechanical coupling and zero for a stand-alone fluid flow calculation. ψ_c' is the derivative of the capillary curve.

Differentiating f equations with respect to S_w gives the following Equation (30):

$$S_o\psi_c' = -\psi_o\frac{1-r}{r}(1-S_{we})S_{we}{}^{-1/r}\left[1-S_{we}{}^{1/r}\right]^{-r} \tag{30}$$

Visualizing the calculated date results from self-development by visualization software, and then generating the saturation distribution contours, the water saturation distribution in the rock sample after 5000 s is shown in Figure 6. The more detailed water distribution information is shown in Figure 7. The water content distribution at that time is shown as Figure 8. The specific water capacity refers to a change in the volumetric water content corresponding to the unit change of pressure head. The distribution of the relative permeability coefficients of water and oil after 5000 s is shown in Figure 9.

As shown in Figures 6–9, the results which were calculated by self-developed program are essentially consistent with those correspondingly by FLAC. The discrepancy yet existing between the outcomes from source code and FLAC is due to the difference in numerical formulations between the implicit FEM and explicit finite difference method. Concretely speaking, the latter is an overlaid quadrilateral (four triangular elements for each) 'element' as its feature. Additionally, the striking divergence between the theory described before and mentioned in this chapter will also contribute to this subtle difference. It lies in whether the compressibility of fluids is considered.

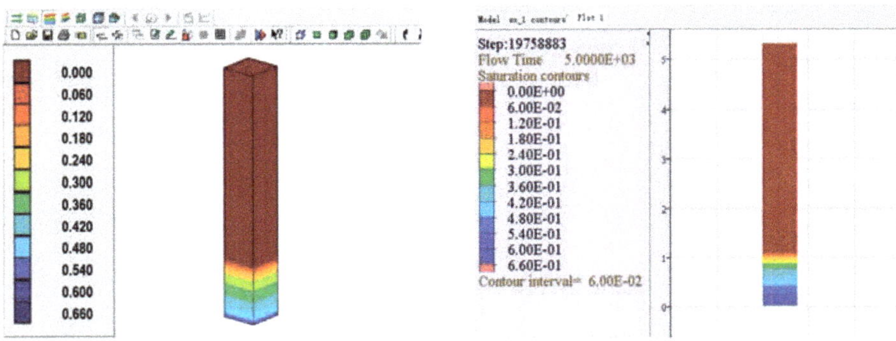

Figure 6. Contrast of simulated water saturation distribution after 5000s: (**a**) Instantaneous water saturation distribution calculated by self-developed source codes; and (**b**) instantaneous water saturation distribution calculated by FLAC.

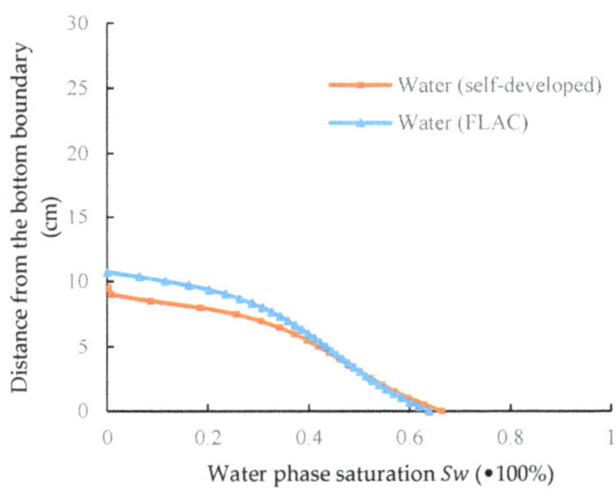

Figure 7. Simulated water-phase saturation distribution after 5000 s.

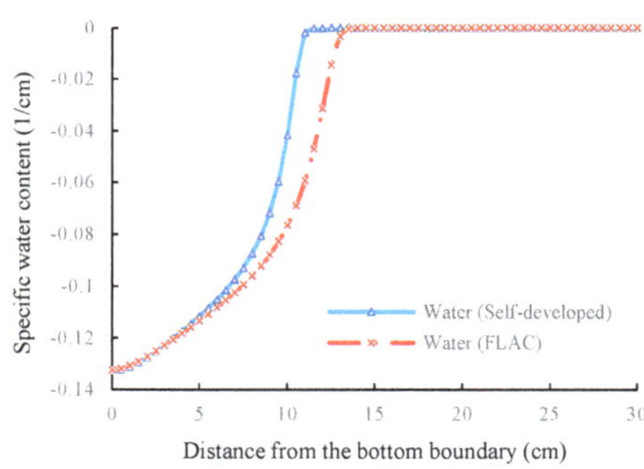

Figure 8. Simulated specific volumetric water content distribution after 5000 s.

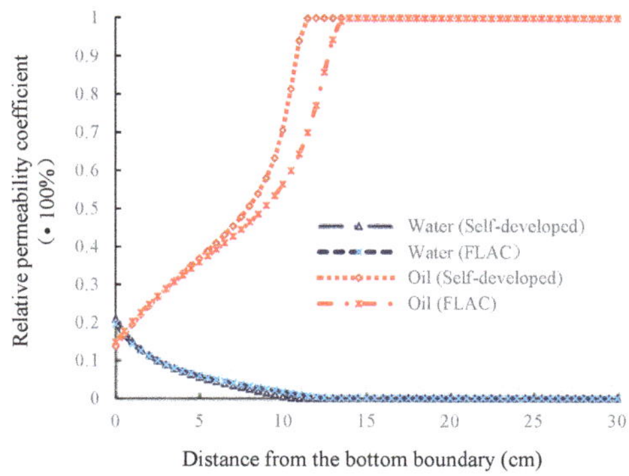

Figure 9. Simulated relative permeability distribution after 5000 s.

Therefore, before explaining the specific reason of this discrepancy, it is necessary to clarify the impact of the difference based on the theory first. To simplify the comparison process, a two-dimensional Darcy's flow was elaborated. In a saturated fluid steady horizontal flow in the ground, if considering the compressibility of the fluid, the governing equations are the follows:

$$-\left(\rho u_i\right)_{,i} = -\frac{\partial}{\partial t}(\rho\phi S_w) \quad i = x, y\,, \tag{31a}$$

$$-\frac{\partial}{\partial x}(\rho u_x) - \frac{\partial}{\partial y}\left(\rho u_y\right) = -\frac{\partial}{\partial t}(\rho\phi\cdot 1), \tag{31b}$$

$$-\frac{\partial}{\partial x}\left(\rho\frac{k_r}{\mu}K_{xx}\frac{\partial\psi}{\partial x}\right) - \frac{\partial}{\partial y}\left(\rho\frac{k_r}{\mu}K_{yy}\frac{\partial\psi}{\partial y}\right) = -\frac{\partial}{\partial t}(\rho\phi\cdot 1), \tag{31c}$$

$$\frac{k_r}{\mu}K_{xx}\cdot\frac{\partial}{\partial x}\left(\rho\frac{\partial\psi}{\partial x}\right) + \frac{k_r}{\mu}K_{yy}\cdot\frac{\partial}{\partial y}\left[\rho\cdot\left(\frac{\partial\psi}{\partial y} + \rho g y\right)\right] = -\frac{\partial}{\partial t}(\rho\phi\cdot 1), \tag{31d}$$

$$\tfrac{k_r}{\mu}K_{xx}\cdot\left(\tfrac{\partial\rho}{\partial x}\cdot\tfrac{\partial\psi}{\partial x} + \rho\tfrac{\partial^2\psi}{\partial x^2}\right) + \tfrac{k_r}{\mu}K_{yy}\cdot\left[\left(\tfrac{\partial\psi}{\partial y} + \rho g y\right)\tfrac{\partial\rho}{\partial y} + \rho\tfrac{\partial^2\psi}{\partial y^2} + g y\cdot\tfrac{\partial\rho}{\partial y} + \rho g\right] = -\tfrac{\partial}{\partial t}(\rho\phi\cdot 1), \tag{31e}$$

while ignoring compressibility of fluid:

$$\frac{k_r}{\mu}K_{xx}\cdot\frac{\partial^2\psi}{\partial x^2} + \frac{k_r}{\mu}K_{yy}\cdot\left(\frac{\partial^2\psi}{\partial y^2} + g\right) = -\frac{\partial}{\partial t}(\phi\cdot 1), \tag{32}$$

Comparing Equation (31) with Equation (32), it can be noticed that, in fact, FLAC's built-in algorithm has considered redundant items $\partial\rho/(\partial x)$ and $\partial\rho/\partial y$ (fluid density ρ can be regarded as the function of moduli K_α, while a constant in self-development code), which contributes to the main difference. So as to cut down the deviations to the lowest, the most direct and simple solution is to increase the value K_α. However, in realistic calculations, oversized K_α (almost the actual value) will result in a significant increase in the calculation time, as shown in Equation (33), and even terminating the process in order to ensure the convergence for transit seepage flow:

$$\Delta_t = L_z^2 \min\left(\frac{1}{k_w K_w}, \frac{1}{k_o K_o}\right), \tag{33}$$

where L_z is the smallest zone size in the simulation, and k_o is the saturated mobility coefficient for the non-wetting fluid ($k_o = k_w\mu_w/\mu_o$).

Although it is difficult to precisely position the interface between water and oil, it is believed that the required accuracy can still be attained by finer gridding and longer computational time.

4.2. Simulation of Trichloroethylene Transport in the Aquifer (DNAPL)

The second example is described below. It is used to verify the correctness of the developed program by simulating a typical contamination transport problem. As shown in Figure 10, there is a non-aqueous liquid (trichloroethylene) pressure head with a 0.1 m source at the midpoint of the aquifer model and a low permeability rock formation in the center of the aqueous model (0.667 m × 0.100 m). The top and bottom of the model are impervious borders, while the left and right sides are fixed hydrostatic pressure boundaries based on initial saturation conditions. In the horizontal direction, the water is driven by a 2% pressure gradient and slowly seeps from right to left.

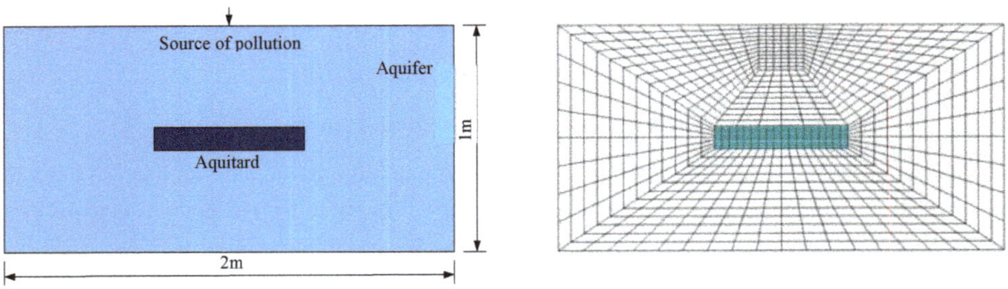

(a) (b)

Figure 10. Aqueous model test: (**a**) Model size and boundary conditions; and (**b**) model meshing.

The physical properties of the medium are shown in Table 2. For model aquifers and low-permeability aquifers, the residual saturation of water is 5%, while the residual saturation of non-aqueous liquids is 0.5.

Table 2. Physical properties of media (24 °C).

Media	Density ($g{\cdot}cm^{-2}$)	Dynamic Viscosity (cp)	Surface Tension ($N{\cdot}m^{-1}$)	
Water NAPL Aquifer	999.1 1462.0	0.914 0.55	0.0728 0.0293	
Media	**Porosity**	**Hydraulic Conductivity** ($m{\cdot}s^{-1}$)	**Van Genuchten Parameters**	
			r/m^{-1}	n
Water NAPL Aquifer	0.35	1.0×10^{-5}	10	2

The calculation results can be obtained by discretizing the fields and performing numerical simulations using a computer program written in accordance with the finite element formula described above. Meanwhile, the same parameters according to Table 2 in FLAC were set and then this simulation ran. To further verify the correctness of the source program, the numerical simulation result contour figures of the two algorithms after 360 min, 720 min, 1080 min, 1440 min, 1800 min, 2160 min, 2520 min, and 2880 min are compared. As an example, the distribution and variation of trichloroethylene saturation over time in the model domain is shown in Figure 10.

During the infiltration process, under the action of gravity, DNAPL (trichloroethylene) moves in three directions in this model simultaneously and generates an approximately circular pollution area, the closer the place is away from the injection source, the higher the concentration of contaminant. Meanwhile, horizontal flow driving force makes the circular area gradually transform into an elliptical shape, as shown in Figure 11a.

As the infiltration process continues, DNAPL has reached the aquitard and accumulated at the top surface of there, shown in Figure 11b. With the infiltration time increased, DNAPL migration states are shown in the next series of contour figures (Figure 11c–h).

According to these comparison results of the above saturation contours, it can be concluded that either calculated by the self-developed source code or FLAC, the range scale of transport of trichloroethylene in the aquifer after the same time interval is basically regarded as consistent.

These two typical examples have been well simulated using two difference methods: the self-developed finite element method and the advanced finite differential method in FLAC as the existing commercial simulator. The calculation results had both been obtained ideally. They suggest that the comparisons were essentially in consistency, which verifies the correctness of the developed source program, and also show its preliminary practical applicability.

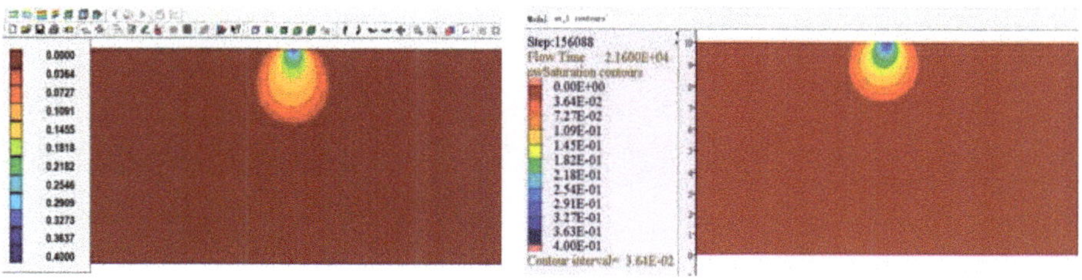

(**a**) At 360 min.

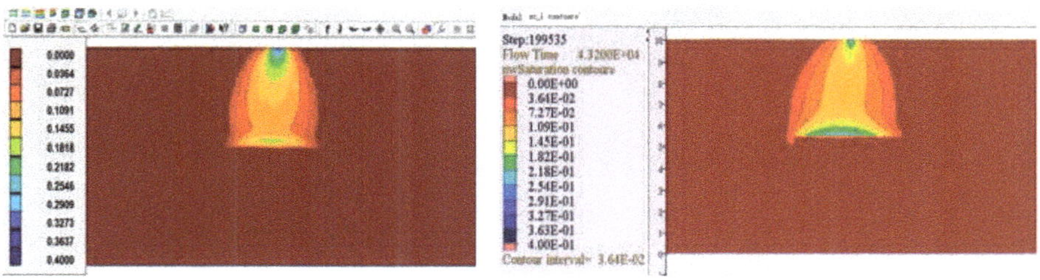

(**b**) At 720 min.

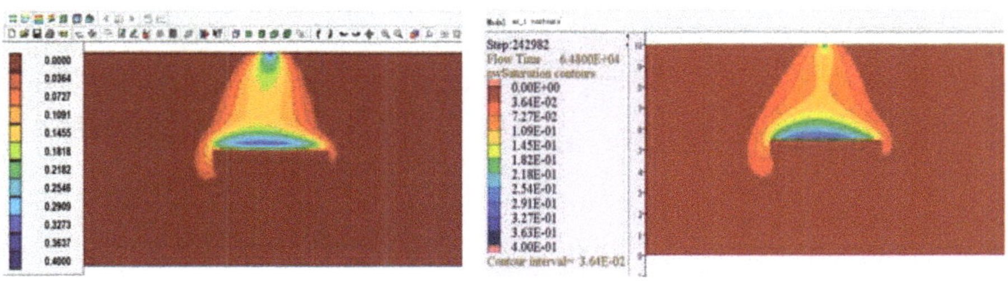

(**c**) At 1080 min.

Figure 11. *Cont.*

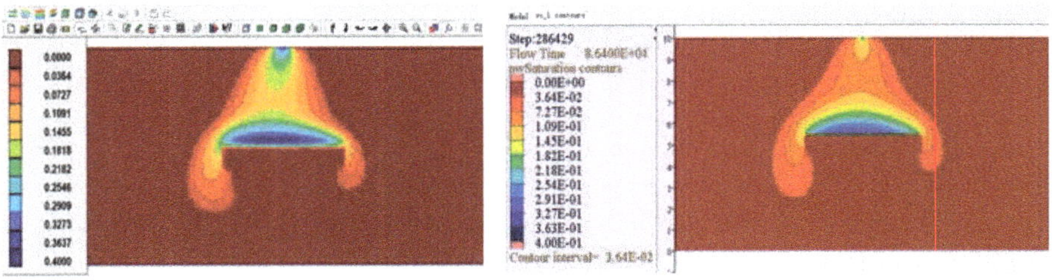

(**d**) At 1440 min.

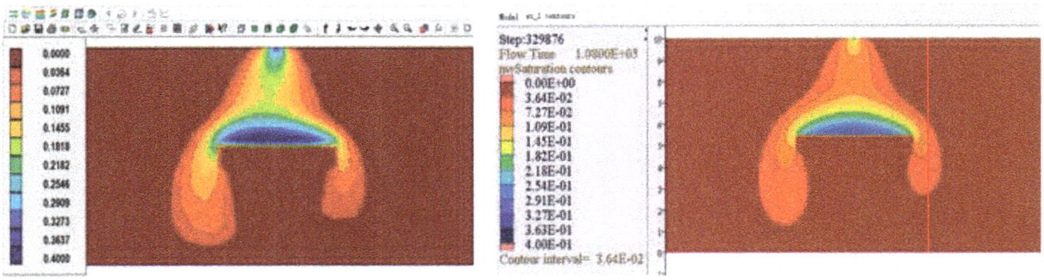

(**e**) At 1800 min.

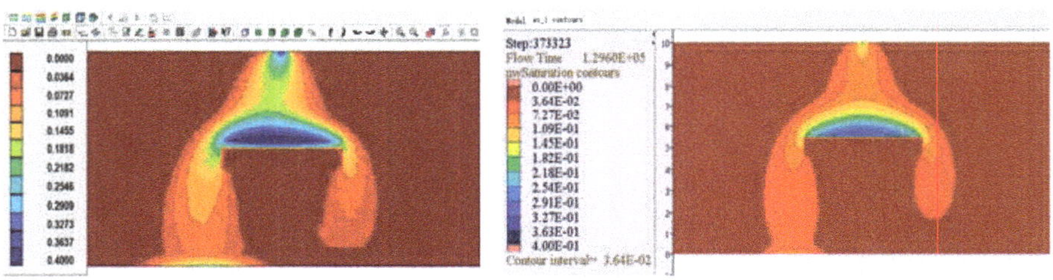

(**f**) At 2160 min.

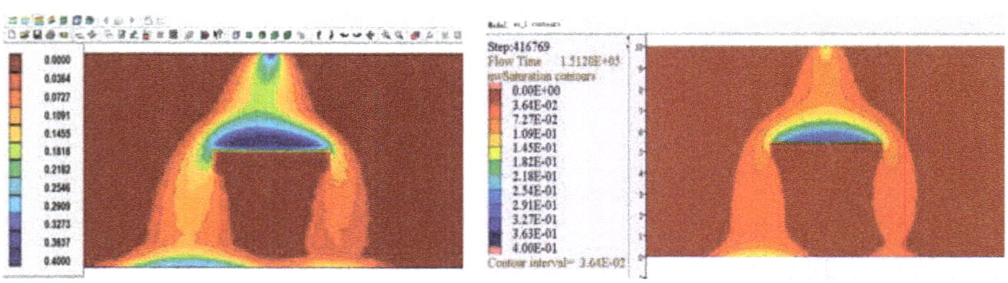

(**g**) At 2520 min.

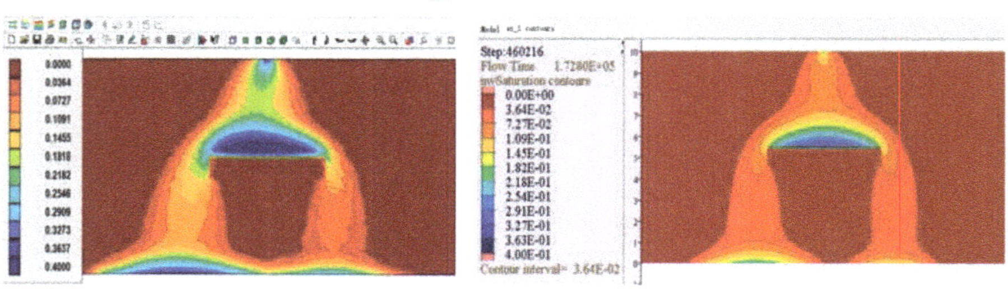

(**h**) At 2880 min.

Figure 11. Distribution of saturation with regards to trichloroethylene and its change with time simulated using self-developed source code (**left**) and FLAC (**right**).

5. Conclusions

In this paper, the procedure of developing three-dimensional finite element method source code programs for the numerical simulation of non-miscible two-phase flows in subsurface water saturated soil is expounded. In the proposed numerical method, the weak forms of mass balance equations for water, NAPLs were discretized in space using the Galerkin finite element method, and mixed-form formulations using both pressure water head and water saturation were adopted. Based on this particular calculation method, two typical examples which belong to LNAPL and DNAPL problems were the modelled cases, respectively. In addition, their outcomes were compared with the results simulated by FLAC. The comparisons demonstrated that they were approximately in coincidental agreement. Therefore, it is convincing that the correctness of the source program is obtained. Moreover, these simulations, having been successfully applied to these problems, also indicates the potential auxiliary analysis potentials of the algorithms developed in this paper on the rock sample displacing experiment and the preliminary evaluation on the migration and diffusion of non-aqueous liquid pollution in the fields. However, it should be noted that the program software developed here is still in its infancy and needs to be further developed and continuously verified based on practical problems in the laboratory and on-site so as to gradually attain consecutive improvements.

Author Contributions: Conceptualization, H.L.; writing—original draft preparation, W.Y.; methodology, formal analysis, H.L.; data curation, visualization, validation, W.Y.; writing—review and editing, W.Y. and H.L.; supervision, H.L.;

Funding: This research was funded by National Basic Research Program of China, grant number 2011CB013503.

Acknowledgments: The authors acknowledge the support by Ministry of Science and Technology of China for its funding to the project 'Seepage Characteristics and Seepage Control Mechanisms of Rock Mass in High-steep Slopes' (No. 2011CB013503) and National Key Research and Development Plan of China (2017YFC1503103, 2018YFC1505301). The authors would also like to thank the National Natural Science Foundation of China (grant No. 51774064), the State Key Laboratory of Coastal and Offshore Engineering (grant No. LY1805), and the State Key Laboratory of Water Resource Protection and Utilization in Coal Mining (grant No. SHJT-17-42.15).

Conflicts of Interest: The authors declare no conflicts of interest.

Abbreviations

FLAC	Fast Lagrangian Analysis of Continua
Q_i (Q_t)	Flow rate flowing into (out of) the infinitesimal element
Q_a	Cumulative storage amount in the infinitesimal element ($Q_i - Q_o = Q_a$)
Q_w	Water volumetric flow rate
Q_o	NAPL volumetric flow rate
ρ_w	Mass density of water
ρ_o	Mass density of NAPL
V	Volume of the porous media
V_f	The fluid volume
V_v	Volume of the pores contained in the porous media
ϕ	Porosity ($\phi = V_v/V$)
S_w	Water saturation
S_o	NAPL saturation
k_w	The relative permeability coefficient of water
k_o	The relative permeability coefficient of NAPL
K_{ij}	Absolutely permeability tensor
p_w	Pore pressure of water phase
p_o	NAPL pressure of NAPL phase
g	Gravitational acceleration
γ	Relative density of NAPL to water ($\gamma = \gamma_o/\gamma_w$)
δ_{ji}, δ_{j3}, $\delta_{m(\xi)}$	Signs of Kronecker delta
κ	Intrinsic or absolute permeability

μ	Dynamic viscosity of flow phase
μ_w	Dynamic viscosity of water
μ_o	Dynamic viscosity of NAPL
μ_r	Relative viscosity of NAPL to water ($\mu_r = \mu_o/\mu_w$)
K	Hydraulic conductivity
ψ_w	Pressure in water head of water ($\psi_w = p_w/(\rho_w g)$)
ψ_o	Pressure in water head of NAPL ($\psi_o = p_o/(\rho_w g)$)
ψ_c	Capillary pressure in water head ($\psi_c = \psi_o - \psi_w$)
C	Specific water content
S_{we}	Effective water saturation
S_{wr}	Residual saturations of water
S_{or}	Residual saturations of NAPL
β	Undrained coefficient
K_w	Water bulk modulus
K_o	NAPL bulk modulus

References

1. Bernaus, A.; Gaona, X.; Ree, D.V.; Valiente, M. Determination of mercury in polluted soils surrounding a chlor-alkali plant: Direct speciation by X-ray absorption spectroscopy techniques and preliminary geochemical characterisation of the area. *Anal. Chim. Acta* **2006**, *565*, 73–80. [CrossRef]
2. Bollen, A.; Wenke, A.; Biester, H. Mercury speciation analyses in HgCl2-contaminated soils and groundwater—Implications for risk assessment and remediation strategies. *Water Res.* **2008**, *42*, 91–100. [CrossRef] [PubMed]
3. Arbestain, M.C.; Rodriguez-Lado, L.; Bao, M.; Macías, F. Assessment of Mercury-Polluted Soils Adjacent to an Old Mercury-Fulminate Production Plant. *Appl. Environ. Soil Sci.* **2009**, *2009*. [CrossRef]
4. Lynge, E.; Tinnerberg, H.; Rylander, L.; Romundstad, P.; Johansen, K.; Lindbohm, ML.; Heikkilä, P.; Westberg, H.; Clausen, L.B.; Piombino, A.; et al. Exposure to tetrachloroethylene in dry cleaning shops in the Nordic countries. *Ann. Occup. Hyg.* **2011**, *55*, 387–396. [PubMed]
5. Kvenvolden, K.; Cooper, C. Natural seepage of crude oil into the marine environment. *Geo-Mar. Lett.* **2003**, *23*, 140–146. [CrossRef]
6. Das, N.; Chandran, P. Microbial degradation of petroleum hydrocarbon contaminants: An overview. *Biotechnol. Res. Int.* **2011**, 1–13. [CrossRef] [PubMed]
7. Bear, J. *Dynamics of Fluids in Porous Media*; Li, J.S., Chen, C.X., Eds.; China Architecture and Building Press: Beijing, China, 1983. (In Chinese)
8. Abriolal, M.; Pinder, G.F. A multiphase approach to the modeling of porous media contamination by organic compounds 1. Equation development. *Water Resour. Res.* **1985**, *21*, 11–18. [CrossRef]
9. Abriolal, M.; Pinder, G.F. A multiphase approach to the modeling of porous media contamination by organic compounds 2. Numerical simulation. *Water Resour. Res.* **1985**, *21*, 19–26. [CrossRef]
10. Lenhard, R.J.; Oostrom, M.; Dane, J.H. A constitutive model for air-NAPL-water flow in the vadose zone accounting for immobile, non-occluded (residual) NAPL in strongly water-wet porous media (Journal of Contaminant Hydrology (2004) 71 (261-282) doi:10.1016/j.jconhyd.2003.10.014). *J. Contam. Hydrol.* **2004**, *73*, 281.
11. White, M.D.; Oostrom, M.; Lenhard, R.J. A practical model for mobile, residual, and entrapped NAPL in water-wet porous media. *Ground Water* **2004**, *42*, 734–746. [CrossRef] [PubMed]
12. Xue, Q.; Liang, B.; Liu, X.L. Study on environmental prediction model of organic contaminants transport in soil-water environment. *J. Hydraul. Eng.* **2003**, *6*, 48–55. (In Chinese)
13. Wu, D.Y.; Wang, C. Analysis of the Relationship Between the Oil Holdup Measured by Coaxial Conductivity Sensor and the Oil Volume Fraction. *J. Tianjin Univ.* **2013**, *46*, 276–280. (In Chinese)
14. Chen, J.J.; Yang, J.; Tian, L. Advances in NAPLs transport experiment in porous media based on pore network model. *Adv. Earth Sci.* **2007**, *22*, 997–1004. (In Chinese)
15. Kleinknecht, S.M.; Class, H.; Braun, J. Density-driven migration of heavy NAPL vapor in the unsaturated zone. *Vadose Zone J.* **2015**, *14*. [CrossRef]

16. Kikumoto, M.; Nakamura, K. Simulation of seepage flow of non-aqueous phase liquid in vadose zone. *Environ. Geotech.* **2017**, *4*, 171–183. [CrossRef]
17. Javanbakht, G.; Arshadi, M.; Qin, T.; Goual, L. Micro-scale Displacement of NAPL by Surfactant and Microemulsion in Heterogeneous Porous Media. *Adv. Water Resour.* **2017**, *105*, 173–187. [CrossRef]
18. Joun, W.; Lee, S.-S.; Koh, Y.E.; Lee, K.-K. Impact of water table fluctuations on the concentration of borehole gas from NAPL sources in the vadose zone. *Vadose Zone J.* **2016**, *15*. [CrossRef]
19. Xie, B.; Kong, D.; Kong, L.; Kong, W.; Li, L. Analysis of vertical upward oil-gas-water three-phase flow based on multi-scale time irreversibility. *Flow Meas. Instrum.* **2018**, *62*, 9–18. [CrossRef]
20. Tan, C.; Li, P.; Dai, W.; Dong, F. Characterization of oil-water two-phase pipe flow with a combined conductivity/capacitance sensor and wavelet analysis. *Chem. Eng. Sci.* **2015**, *134*, 153–168. [CrossRef]
21. Picchi, D.; Battiato, I. The Impact of Pore-Scale Flow Regimes on Upscaling of Immiscible Two-Phase Flow in Porous Media. *Water Resour. Res.* **2018**, *54*, 6683–6707. [CrossRef]
22. Balasuriya, S. Transport between two fluids across their mutual flow interface: The Streakline Approach. *SIAM J. Appl. Dyn. Syst.* **2017**, *16*, 1015–1044. [CrossRef]
23. Li, Y.; Gao, J.; Liu, X.; Xie, R. Energy demodulation algorithm for flow velocity measurement of oil-gas-water three-phase flow. *Math. Probl. Eng.* **2014**, *5*, 1–13. [CrossRef]
24. Kacem, M.; Benadda, B. Mathematical Model for Multiphase Extraction Simulation. *J. Environ. Eng. USA* **2018**, *144*. [CrossRef]
25. Bear, J.; Cheng, H.D. *Modeling Groundwater Flow and Contaminant Transport*; Springer: Berlin, Germany, 2010.
26. Bear, J. *Dynamics of Fluids in Porous Media*; American Elsevier: Amsterdam, The Netherlands, 1972; Volume 764.
27. Celia, M.A.; Bouloutas, E.T.; Zarba, R.L. A general mass-conservative numerical solution for the unsaturated flow equation. *Water Resour. Res.* **1990**, *26*, 1483–1496. [CrossRef]
28. Van Genucheten, M.T. A closed-form equation for predicting the hydraulic conductivity of unsaturated soils. *Soil Sci. Soc. Am. J.* **1980**, *44*, 892–898. [CrossRef]
29. Parker, J.C.; Lenhard, R.J.; Kuppusamy, T. A parametric model for constitutive properties governing multiphase flow in porous media. *Water Resour. Res.* **1987**, *23*, 618–624. [CrossRef]
30. Parker, J.C.; Lenhard, R.J. A model for hysteretic constitutive relations governing multiphase flow: 1. Saturation–pressure relations. *Water Resour. Res.* **1987**, *23*, 2187–2196. [CrossRef]
31. Parker, J.C.; Lenhard, R.J. A model for hysteretic constitutive relations governing multiphase flow: 2. Permeability–saturation relations. *Water Resour. Res.* **1987**, *23*, 2197–2206. [CrossRef]
32. Nikolaevskii, V.N. *Mechanics of Porous and Fractured Media*; World Scientific: Singapore, 1990.

processes

MDPI

Article

The Application of a Three-Dimensional Deterministic Model in the Study of Debris Flow Prediction Based on the Rainfall-Unstable Soil Coupling Mechanism

Shuangshuang Qiao, Shengwu Qin *, Junjun Chen, Xiuyu Hu and Zhongjun Ma

College of Construction Engineering, Jilin University, Changchun 130026, China;
qiaoss17@mails.jlu.edu.cn (S.Q.); junjunchen90@163.com (J.C.); xiuyu17@mails.jlu.edu.cn (X.H.);
mzjgeology@gmail.com (Z.M.)

* Correspondence: qinsw@jlu.edu.cn; Tel.: +86-135-0441-8387

Received: 15 December 2018; Accepted: 7 February 2019; Published: 15 February 2019

Abstract: As debris flow is one of the most destructive natural disasters in many parts of the world, the assessment and management of future debris flows with proper forecasting methods are crucial for the safety of life and property. So increasing attention has been paid to the forecasting methods on debris flows. A debris flow forecasting method based on the rainfall-unstable soil coupling mechanism (R-USCM) is presented in the current study. This method is based on the debris flow formation mechanism. The density of sediment is introduced as an evaluation index to determine the susceptibility of debris flow occurrence. The forecasting method includes two phases: (1) rainfall and soil coupling and (2) runoff and unstable soil coupling. Scoops3D, a three-dimensional (3D) model for analyzing slope stability, was introduced into the debris flow forecasting method. In order to test the forecasting accuracy of this method, Jiaohe County was selected as a research area, and the serious debris flow disasters attributed to strong rainfall on 20 July 2017 were taken as the research case. By comparing the forecasting results with the debris flow distribution map for Jiaohe County, the method based on the R-USCM is feasible for forecasting debris flows at the regional scale. The application of the Scoops3D model can more reasonably analyze the slope stability than the traditional two dimensional (2D) method and improve the forecasting ability of debris flows.

Keywords: debris flow; forecasting; rainfall-unstable soil coupling mechanism(R-USCM); scoops3D; Jiaohe

1. Introduction

Rainfall-induced debris flow is a mixture of unconsolidated sediment and is one of the most important of all natural hazards, occurring in many areas [1]. Debris flows cause severe damage to both life and property every year worldwide, occurring at different intervals and with varying durations [2]. To reduce debris flow-related disasters, the assessment and management of future debris flows that can be achieved through appropriate forecasting methods cannot be overlooked [3].

The current debris flow forecasting methods mostly establish the critical threshold triggering debris flow formation in the study area based on commonly used precipitation parameters [4]. Aleotti et al. took Piedmont Region in the northwest of Italy as the study area and determined the precipitation threshold leading to debris flows by studying the statistical relationship between precipitation events and debris flows occurrence [5]. However, as a region changes, so does the threshold of the rainfall [6]. A few practical forecast models based on long-term observation in a debris flow valley were obtained [7]. However, such statistics-based debris flow forecasting methods are not always economically and practically suitable for satisfying the demand for disaster mitigation. With the development of debris flow forecasting research, the combination of rainfall parameters (precipitation duration, intensity and cumulative precipitation, etc.) is used replace the single rainfall parameter as the determination factor of debris flows. Bacchini et al. proposed a Rainfall Intensity-Duration curve for debris flow forecasting in the Los Angeles of the United States [4]. Scholars have also carried out analyses and research on the stability of soil on slope under the condition that soil mechanical properties change. And the debris flow initiation models have been established. Iverson deduced the debris flow initiation model based on Mohr-Coulomb criterion, and made a discussion on the role of pore water pressure in the process of the debris flow formation by using the 100-meter flume test conducted by the United States Geological Survey (USGS) [8]. Additionally, in recent years, the debris flow susceptibility model has become one of the most important models for assessing areas susceptible to debris flows [9]. Weighted integration methods are used to synthesize multiple debris flow-causing factors to delineate debris flow-prone areas. However, one of the main challenges of the weighted integration method is quantification of the impact of individual factors on debris flow susceptibility mapping [10]. Another challenge is the availability of layer data for specific factors affecting the debris flow. Furthermore, with the change in climate and other variables, the spatial relationship between causative factors and the evaluation of debris flow has also changed [11]. Therefore, in order to solve these issues, a mechanism-based prediction method was proposed in this paper.

A landslide can easily be converted into a debris flow when it contains enough water, especially the loose deposits on its surface under the action of raindrop impact or runoff [12]. Large amounts of deposits are drawn into the runoff and continue to move with the runoff [13]. The interaction between the runoff and large amounts of loose deposit can lead to debris flows [14]. The formation process of debris flows can be constructed based on two rainfall-unstable soil coupling processes. Firstly, precipitation is coupled with slope soil, which makes the soil on the slope unstable; secondly, debris flow is formed by the coupling of runoff and unstable soil [6]. Based on the mechanism of debris flow formation, this forecast method established a coupling relation between the rainfall and unstable soil of the underground surface. The density of the mixture reaching a certain threshold is a necessary condition for the debris flow to form. Therefore, the density of the debris flow is introduced as an evaluation index to determine the susceptibility of debris flow occurrence. To calculate the density of the mixture, the first phase of the forecast method based on the rainfall-unstable soil coupling mechanism (R-USCM) needs to determine the volume of unstable soil, and the second phase needs to determine the volume of the runoff.

An accurate calculation of the amount of unstable soil failure is crucial in the first phase of the forecast method based on the R-USCM. It is also an indispensable factor for a number of hazard assessments, sediment budgets, and initial conditions for landslide run-out models. The 2D model has been widely used to analyze slope stability. However, a traditional 2D slope stability analysis (SSA) cannot consider the direction of the slip surfaces, and failure bodies are forced to move in the presupposed direction. Furthermore, the traditional 2D model assumes that the sliding surface is parallel to the slope, which may not be completely consistent with the actual slope instability [15]. To estimate the sum of the volumes of unstable soil, detailed data about the soil thickness are essential, which is hard to acquire for a large regional scale. Thus, it is still a huge challenge to obtain the relevant complete data essential to use a traditional 2D model to calculate the sum of the volumes of the unstable soil for forecasting debris flows.

Therefore, the 3D deterministic model Scoops3D is adopted in this paper to evaluate and predict the stability of unstable soil. The model uses the circular arc method to search for sliding surfaces with different curvatures in the study area in 3D space and uses the limit equilibrium method to analyze the stability of each grid based on the digital elevation model (DEM) SSA data. This solves the problem where the sliding surface is parallel to the slope surface, which is inconsistent with the actual situation. Furthermore, Scoops3D computes the volume of each column in the potential failure mass and adds it to the total. It also does not ignore the columns along the margin of the potential failure mass, which may be only partly contained within the scopes of the sphere. Thus, the volume of the unstable soil is calculated more accurately. In addition, the 3D SSA usually provides more stable results than the one-dimensional (1D) and 2D methods [16–18], and it is demonstrated that the Scoops3D model has the potential to overcome the problem of over-prediction. According to the above, using the Scoops3D model to determine the volume of unstable soil can improve the accuracy of the forecast method based on the R-USCM.

In this paper, the forecast method based on the R-USCM is used to predict debris flows in Jiaohe County, Jilin Province. The serious debris flow disaster attributed to strong rainfall in Jiaohe County on 20 July 2017, was taken as the research case. The Scoops3D model and traditional 2D method are used to analyze the slope stability and calculate the volume of unstable soil. The soil conservation service (SCS) curve number (CN) method is used to calculate the volume of the runoff. The Scoops3D model combined with the SCS-CN method is called the S-CN forecast method, and the traditional 2D model combined with the SCS-CN method is called the T-CN forecast method. In order to test the forecasting ability of the two methods, the forecasting results of the two methods were compared with the debris flow distribution map for Jiaohe County and tested by the receiver operating characteristic (ROC) curve method.

2. Forecast Method of Debris Flow Based on the Rainfall-Unstable Soil Coupling Mechanism

2.1. Forecast Theory

Soil is unsaturated before persistent or heavy rainfall infiltration, being in a relatively stable state. However, the gradual increase in the soil moisture caused by rainfall infiltration will reduce the matrix suction and apparent cohesion of the soil, thereby reducing the shear strength of the soil. Therefore, in the case of short-term heavy or continuous rainfall, runoff and unstable soil can cause debris flows. According to the R-USCM, the process of debris flow formation can be simplified into two phases: The first phase is the soil on the slope becomes unstable due to rainfall, which provides a sufficient source basis for the generation of debris flows; the second phase is that the unstable soil will form a debris flow when it encounters runoff.

However, so far the processes of the debris flows formation are too complex to be simulated by numerical and physical models. Debris flow can be considered to consist of water and sediment. The density of the sediment in the debris flow reaching a certain threshold is a necessary condition for debris flow to form. Therefore, the density of the debris flow is introduced as an evaluation index to determine the susceptibility of debris flow occurrence. However, the density of debris flow changes continuously during the process of formation. Therefore, the complex dynamic rainfall-unstable soil coupling process is expressed quantitatively with the static coupling results. Specifically, this means using runoff and unstable soil caused by rainfall to calculate the density of the mixture. The susceptibility of debris flow occurrence has been judged by the density we have calculated.

$$\rho = \frac{\rho_r v_r + \gamma_{us} v_{us}/g}{v_r + v_{us}} \tag{1}$$

where ρ is the density of the debris flow, ρ_r is the density of the runoff, and γ_{us} is the unit weight of the unstable soil. g is the acceleration of gravity, v_r is the volume of the runoff, and v_{us} is the volume of the unstable soil which can provide the material source basis for the debris flow. Therefore, according to Equation (1), the accurate calculation of v_r and v_{us} is an important part of evaluating the accuracy of the forecast method based on the R-USCM.

According to the previous research [19], the density of the standard debris flow fluctuates between 1.1×10^3 and 2.3×10^3 kg/m^3. We divided the density of the mixture into five reference intervals, as shown in Table 1. They are used as a criteria for evaluating the susceptibility of debris flow occurrence from low to high. Based on these density intervals, five levels of debris flow warnings were determined (Table 1). Figure 1 shows the flowchart of the debris flow prediction method based on the R-USCM.

Table 1. The susceptibility of debris flow occurrence based on the water-soil mixture density.

The Density of Debris Flows	Class	Susceptibility of Debris Flow	Color of Warning
$\rho < 1.2$	1	Very low	Blue
$1.2 < \rho < 1.5$	2	Low	Green
$1.5 < \rho < 1.8$	3	Moderate	Yellow
$1.8 < \rho < 2.0$	4	High	Orange
$2.0 < \rho < 2.3$	5	Very high	Red

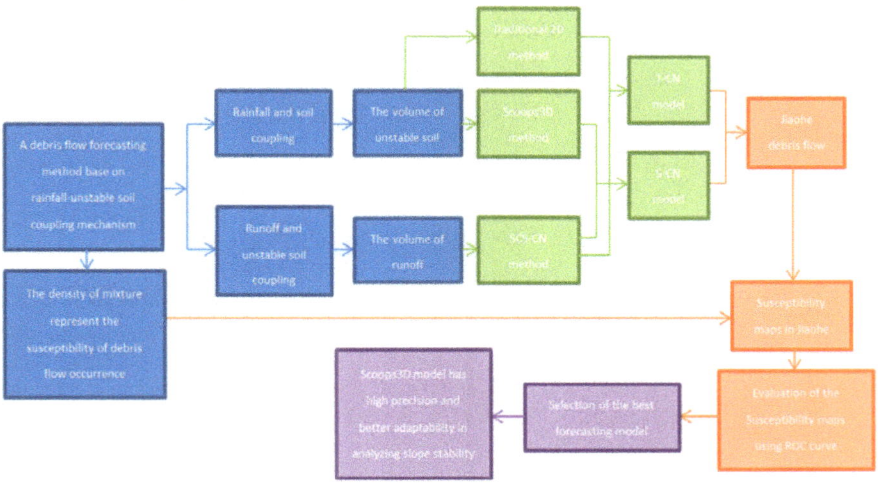

Figure 1. The flowchart of the debris flow prediction method based on the rainfall-unstable soil coupling mechanism (R-USCM). (SCS-CN: soil conservation service-curve number).

2.2. The Volume of Unstable Soil Calculated by the Scoops3D Model

The 3D slope stability assessment has not been widely used due to its intricacy, its computational requirements, and the lack of relevant data. The majority of methods use 1D (infinite slope model on a raster cell-by-cell basis) [20,21] or 2D (based on a series of obtained by DEM) [22] models to evaluate the slope stability. Some scholars have noted that 1D and 2D models are receivable in SSA; not only are their calculations easier, but they also provide more cautious results than 3D models [23]. However, they may not simulate the practical mechanism of the motion process of unstable soil [15]. Compared to other models, the Scoops3D model has the advantage of considering the terrain described by DEM to identify various potential sliding bodies in the whole region, not just for a single landslide on the pre-defined hill-slope. If materials have no sharp contrast or discontinuity, the failure is usually curved. According to this feature, Scoops3D uses the potential failure or slip surface made by parts of a sphere and presumes a rotational slip. Through a 3D analysis, the Scoops3D model provides direct estimates of the potential failure volume. The result of the calculation is more accurate than that of traditional models. This can also improve the accuracy of the forecast method based on the R-USCM.

Scoops3D analyzes the stability of each grid based on the DEM data. First, the sliding surface is determined in order to calculate the weight of the potential slide block, and then the stability of the object is calculated by the 3D limit equilibrium method. In the process of calculation, a 3D retrieval region is established based on the grid of the DEM. A mass of search points are identified in the region. Each search point is taken as the center of the ball in turn, and the search is expanded according to a certain radius. The grid element that is finally delineated by the sphere is the sliding block [24]. Since the bottom of each grid element is part of the sphere, the surface is approximately treated as a plane with an angle of ε to the horizontal plane [24].

Scoops3D calculated the shear strength through the linear Coulomb-Terzaghi instability criterion [25].

$$s = c + (\sigma_n - u)\tan\varphi \tag{2}$$

where c and φ are the apparent cohesive force and angle of internal friction of the soil, respectively. σ_n is the normal stress of the sliding block, and u is the pore water pressure acting on the shear surface.

Limit equilibrium models define F as the ratio of the shear resistance (s) to the shear stress (τ). $F < 1$ shows the slope is in an unstable state in theory.

We choose the 3D expansion method of Bishop's simplified model for the limit equilibrium model because it provides a simple and effective solution for a large number of practical problems [26]. In the Scoops3D model, the main force from the downward sliding of soil comes from gravity on the slider, in addition to the action of a seismic load ($K_{eq}W$). K_{eq} is the horizontal acceleration coefficient, and the following formulas are used to calculate the safety factor using the 3D simplified Bishop method [24].

$$F_{3D} = \frac{\Sigma R_{i,j}\left[c_{i,j}A_{h_{i,j}} + \left(W_{i,j} - u_{i,j}A_{h_{i,j}}\right)\tan\varphi_{i,j}\right]/m_{\alpha_{i,j}}}{\Sigma W_{i,j}\left(R_{i,j}\sin\alpha_{i,j} + k_{eq}e_{i,j}\right)} \tag{3}$$

$$m_{\alpha_{i,j}} = \cos\varepsilon_{i,j} + \left(\sin\alpha_{i,j}\tan\varphi_{i,j}\right)/F_{s3D} \tag{4}$$

The calculation of unstable soil volume is a key step in SSA, Scoops3D computes the volume (V_c) of each column which may be unstable and adds it to the total. Some of the columns distributed on the margin of potential failure surfaces may only be partially contained within spherical boundaries. When two or three corners of columns are contained within the spherical trial surface, Scoops3D can recognize them and take the incomplete volume of the columns into account, rather than full columns that other approaches use exclusively. Therefore, Scoops3D can accurately give the estimates of the volume of the unstable soil [24].

The volume of each column (partial or full), which is above the trial surface, can be approximately calculated by the following equation:

$$V_c = (1/6)\Delta x(S_0 + 4S_1 + S_2) \tag{5}$$

where Δx represents the DEM mesh spacing, S_0 and S_2 are the superficial areas of the two parallel sides of the column, and S_1 is the superficial area of a vertical cross section through the middle of the column [24].

For full column, its area calculation becomes the calculation of the quadrilateral area, and its volume calculation becomes the calculation of the cube volume. For those partial columns with two nodes, its area calculation becomes the calculation of the triangular area, and its volume calculation becomes the calculation of the pyramid volume. For those partial columns with three nodes, the shape of one side is roughly a quadrilateral, whereas the other side is a triangle [24].

2.3. Runoff Volume

The runoff depth plays an important role in water space redistribution. To estimate the runoff depth and thus obtain the total volume of the runoff, the SCS-CN method is applied [27]. The runoff depth is used to assess the potential water supply during runoff. Its most recent calculated equation was published by the Natural Resource Conservation Service (NRCS) in 2004:

$$Q = \frac{(P - I_a)^2}{(P - I_a) + R} \tag{6}$$

where Q is the runoff depth (mm), P is rainfall (mm), R is the potential maximum retention after the runoff begins (mm), and I_a is an initial abstraction (mm) that includes all losses before the beginning of runoff, evaporation, infiltration, and water interception by plants. It is defined as:

$$I_a = \lambda R \tag{7}$$

According to the reference of NRCS, the value of λ in this study is 0.2 [28]. Equation (6) can therefore be expressed as:

$$Q = \frac{(P - 0.2R)^2}{(P + 0.8R)} \tag{8}$$

R can be calculated by the following equation through the relationship between R and CN:

$$R = \frac{25,400}{CN} - 254 \tag{9}$$

CN provided by the Soil Conservation Service is an important indicator of the runoff depth. CN ranges from 0 to 100, representing the runoff response to a given rainfall event. The higher the value of CN is, the more rainfall is converted to surface runoff [27,29]. The values of CN are determined by the effect of runoff on soil and land cover. Therefore, the value of CN for each pixel in the study area can be estimated according to the soil-texture and land cover maps. In other words, the antecedent runoff conditions (ARCs) determine the selection of the CN value. ARC level is determined by the actual pre-rainfall of the debris flows. The ARC can be divided into three levels: ARC I represents dry, ARC III represents wet, and ARC II accounts for the antecedent runoff condition between dry and wet [27]. After determining the ARC level, the corresponding value of CN can be obtained from the data provided by NRCS.

Then, the sum of the volumes of runoff v_r which resulted from the rainfall can be calculated using the following formula:

$$Sum_{v_r} = \sum_{i=1}^{N} S_i \times Q_i \tag{10}$$

where Sum_{v_r} is the total volume of the runoff, Q_i is the runoff depth of each grid cell, S_i is the area of each grid cell, and N is the total number of grid cells.

3. Debris Flow Prediction and Test in the Study Area

Jiaohe County is prone to geological disasters and is one of the regions with the most severe geological catastrophes in Jilin Province. Because debris flow has several divisions within one zone, more hazardous geological disasters occur [30]. In the past several years, debris flows frequently occurred in the area due to strong rainfall, which has caused a great loss of local residents' lives and properties. Until 2017, a geological survey of Jilin province at a 1:50,000 scale found 162 debris flow disaster locations (Figure 2), with 23,231 square kilometers of crops affected, 1766 houses damaged, more than 59 km of roads damaged, 91 bridges and 7 culverts damaged, and 79 landslides and other geological disasters produced. The direct economic loss was approximately 202.62 million yuan (Figure 3). On 20 July 2017, Jiaohe County suffered strong rainfall which caused serious debris flow disasters. The debris flow disasters and rainfall event was used here as a study case. The forecast method based on R-USCM was developed for Jiaohe County, including the S-CN and T-CN forecast methods. The forecast results were compared with the debris flow hazard points discovered by the 2017 geological survey of Jilin province at a 1:50,000 scale, and then a preliminary assessment of the accuracy was made.

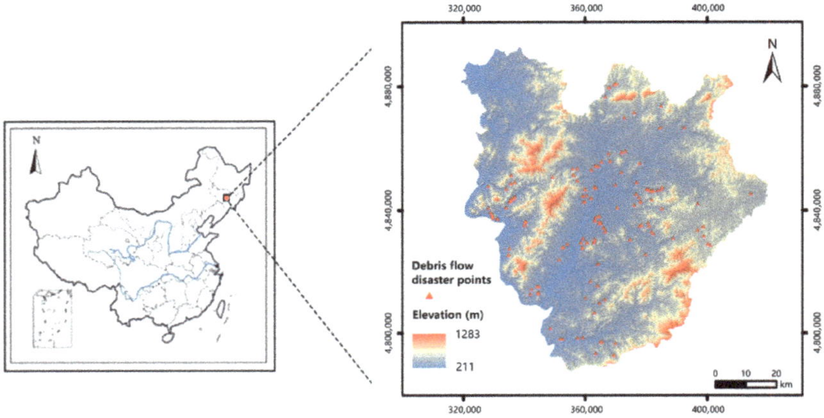

Figure 2. Debris flow locations in Jiaohe County.

Figure 3. Debris flow disasters identified in the field: (**a**) debris flow deposit and destroyed wall; (**b**) erosional surface; (**c**) destroyed road; and (**d**) destroyed farmland.

3.1. Calculate the Volume of Failure Soil Mass Caused by Rainfall in Jiaohe County

Scoops3D predicted the soil stability and calculated the volume of failure soil based on the DEM and soil mechanical parameters in Jiaohe County. The DEM of Jiaohe County with a spatial resolution of 7×7 m was obtained from Google Earth by using the software of 91 graphic assistant (v16.9.9.2, Beijing Qianfanyunlian technology co. LTD, Beijing, China, 2018) (Figure 4). The mechanical parameters of the soil include soil cohesion c, internal friction angle of soil φ and soil unit weight γ. Lithology data were extracted from the geological map of Jilin Province at a 1:50,000 scale, and the mechanical parameters were derived from the rock mechanics manual. According to the rainfall data from the two observation stations in Jiaohe County (Figure 5), the debris flows occurred on 20 July and continuous rainfall preceded the day that debris flow occurred. Concentrated and continuous heavy rainfall will significantly increase the water content and reduce the strength of the soil, and the stability of the slope will decrease. Therefore, the mechanical parameters of soil under wetting conditions with 30% gravimetric water content were selected as the input parameters of debris flow forecast methods (Figure 6). According to the rock mechanics manual, the unit weight of the soil is 1.75×10^4 kN/m^3 when the soil is under wetting conditions with 30% gravimetric water content. The Scoops3D model can directly generate the volume of unstable soil according to the above parameters set (Figure 7).

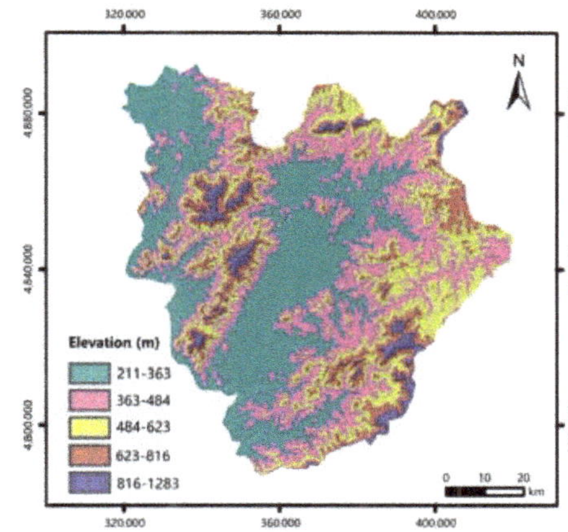

Figure 4. Digital Elevation Model (DEM) of Jiaohe County (m).

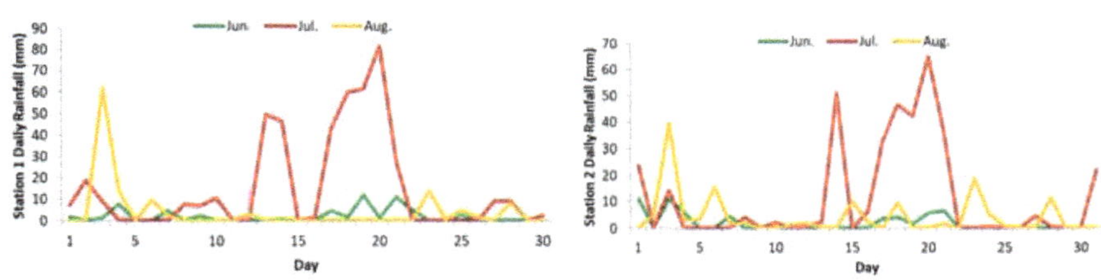

Figure 5. Daily rainfall data from June to August (2017) of the Jiaohe County weather stations.

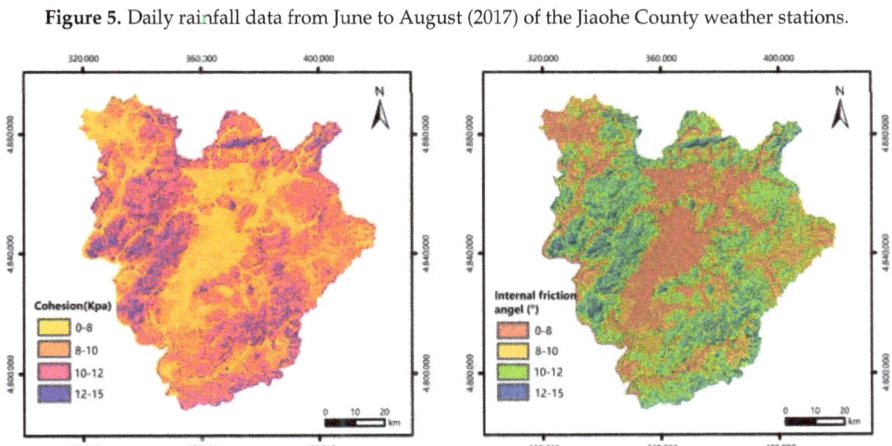

(a) (b)

Figure 6. Distribution of mechanical parameters of soil in Jiaohe County. (a) Distribution of cohesion (kPa); (b) Distribution of internal friction angle (°).

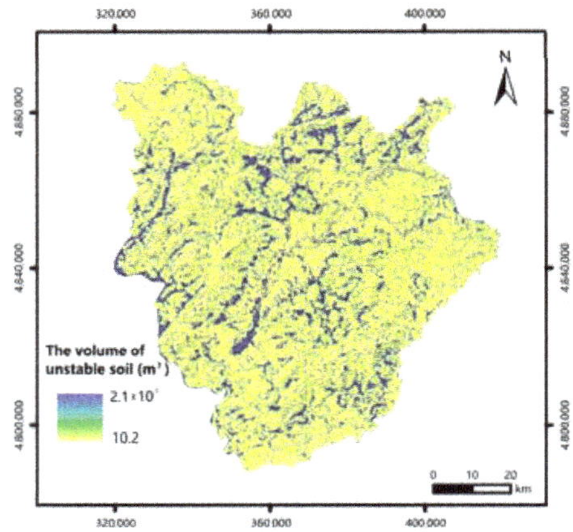

Figure 7. The volume of unstable soil mass in Jiaohe County using the Scoops3D model.

To calculate the volume of the unstable soil using a traditional 2D method, the thickness of the unstable soil should first be determined. In the current research, the soil thickness was obtained by using the relation between the slope and soil thickness proposed by Veit et al [31]. It can be approximated that there is a linear relationship between soil depth and slope angle [23]. The slope was directly extracted from the DEM (Figure 8). In this way, the soil depth of an any pixel (y) can be calculated by the function $y = -0.0486x + 3.5$. After determining the total thickness of the soil, several layers were subdivided, and the safety factors of each layer were calculated. More details about soil thickness are provided in Table 2. The mechanical parameters of the soil required for the traditional 2D method are the same data used in the Scoops3D model. It is essential to resample these data to ensure the grid cells are of the same size as the DEM. Then, the sum of the volumes of the unstable soil in the study area was calculated by Equation (11) and is shown in Figure 9.

$$Sum_{v_s} = \sum_{i=1}^{N} S_i \times D_i \tag{11}$$

where Sum_{v_s} is the sum of the volumes of the unstable soil, D_i is the unstable soil thickness of each grid cell, S_i is the grid cell area, and N is the sum of the numbers of the unstable grid cells.

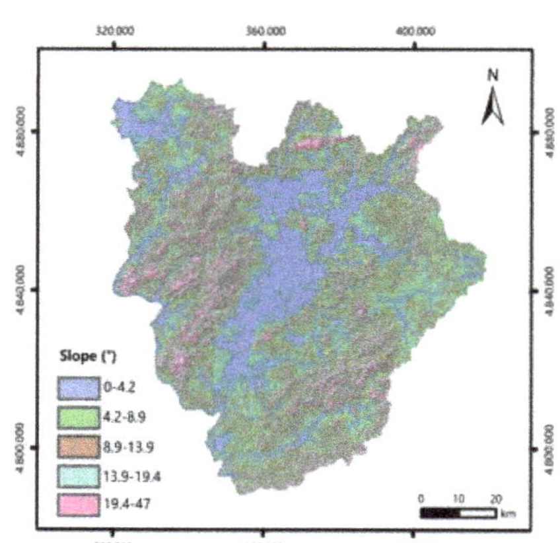

Figure 8. Slope of Jiaohe County (°).

Table 2. Soil thickness in Jiaohe County, Jilin Province.

Class	Thickness (m)	Relative Area (%)	Thickness of Subdivided Soil Layers						
			1	2	3	4	5	6	7
I	3.25–3.30	7.10%	0.1	0.2	0.3	0.4	0.5	0.75	0.95
II	3.30–3.35	39.06%	0.1	0.2	0.3	0.4	0.5	0.75	1.00
III	3.35–3.40	24.59%	0.1	0.2	0.3	0.4	0.5	0.75	1.05
IV	3.40–3.45	29.25%	0.1	0.2	0.3	0.4	0.6	0.75	1.10

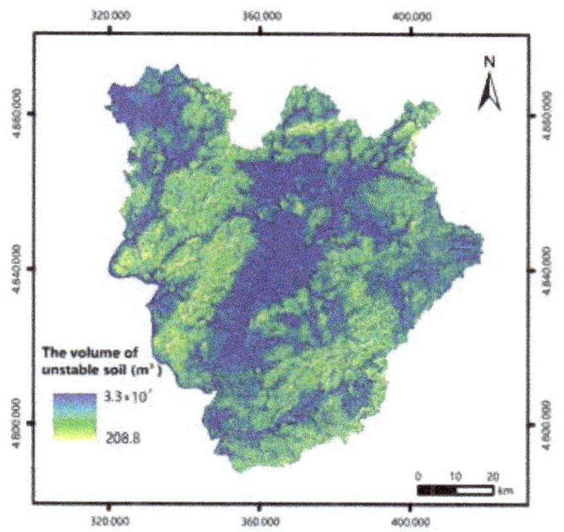

Figure 9. The volume of unstable soil mass using the traditional 2D model.

3.2. Calculate the Runoff Volume in Jiaohe County

The volume of runoff was calculated by the SCS-CN method, which required the CN values and rainfall data. The CN values are determined by land use. Land use in Jiaohe was determined from a geological map of Jilin province at a 1:50,000 scale, including agriculture land, forest, irrigation canals and ditches, and other scarce vegetation (Figure 10). The CN values are provided in Table 3 [27].

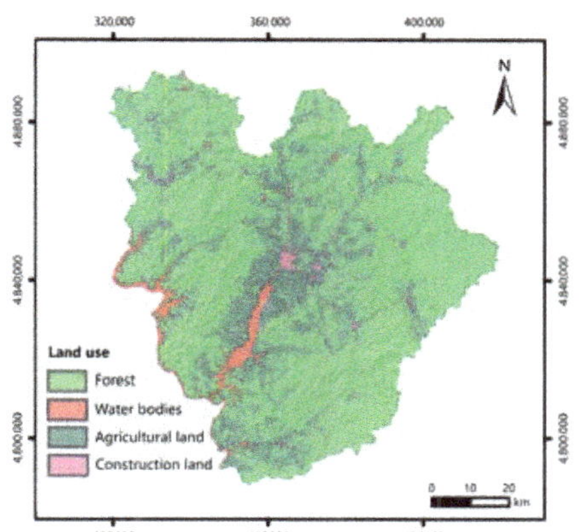

Figure 10. Land use of Jiaohe County.

There are two weather stations in Jiaohe County. Based on the rainfall data from weather stations, the annual precipitation is unevenly distributed, primarily concentrated from June to August. The daily rainfall curves from July to August (2017) are shown in Figure 5. According to different types of data sources, such as historical records, field surveys, and interviews with local residents, it can be determined that the large-scale debris flow caused by rainfall occurred on 20 July 2017. Daily rainfall around the north and south of the weather stations reached 81.6 and 65 mm, respectively. Therefore, debris flow in Jiaohe is predicted using the forecast method based on the R-USCM under the same rainfall conditions as 20 July 2017.

Table 3. Curve number (CN) values reflecting antecedent runoff condition III (ARC III).

Land Use	CN
Agriculture land	95
Forest	89
Irrigation canals and ditches	88
Other scarce vegetation	97

Combined with the rainfall data and CN values in Jiaohe County, the total volume of runoff can be obtained according to Equation (10) (Figure 11).

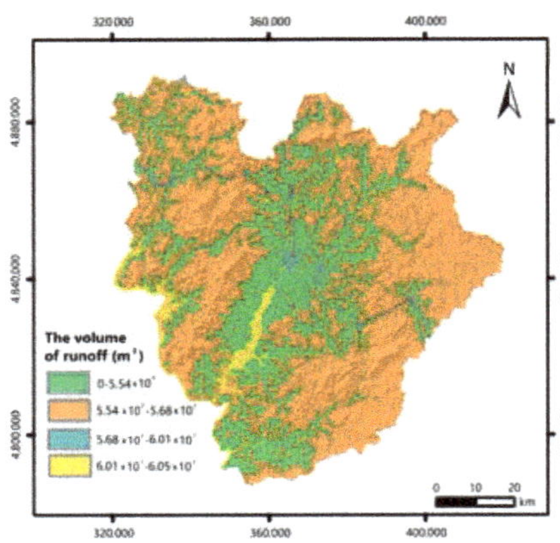

Figure 11. The volume of runoff in Jiaohe County.

3.3. Calculate the Density of the Water-Soil Mixture in Jiaohe County

According to Equation (1), the maps of the water-soil mixture density calculated by the S-CN method and T-CN method are shown in Figure 12.

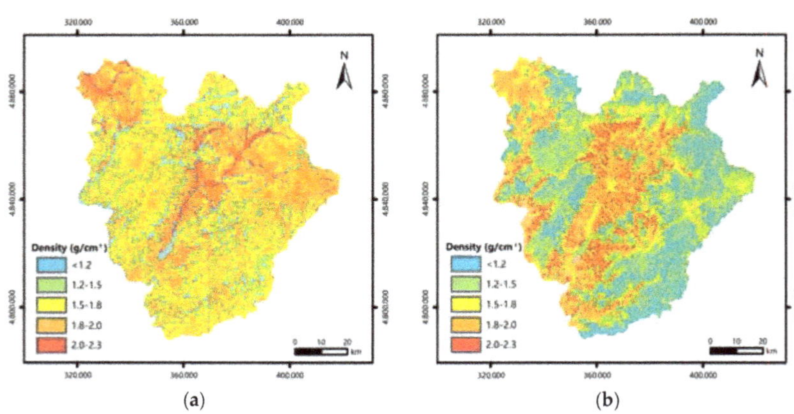

Figure 12. The predicted density of the water-soil mixture: (**a**) S-CN method (**b**) T-CN method.

4. Forecast Results and Discussion

In order to forecast the areas susceptible to debris flows, the density of the water-soil mixture and the corresponding susceptibility must be referenced (Table 1). Therefore, early alerts for areas susceptible to debris flows can be derived from the maps of forecast results (Figure 13). This is important for providing an effective theoretical basis for geological disaster prevention, control planning and risk management. As a result, loss of property and life can be reduced when debris flow occurs.

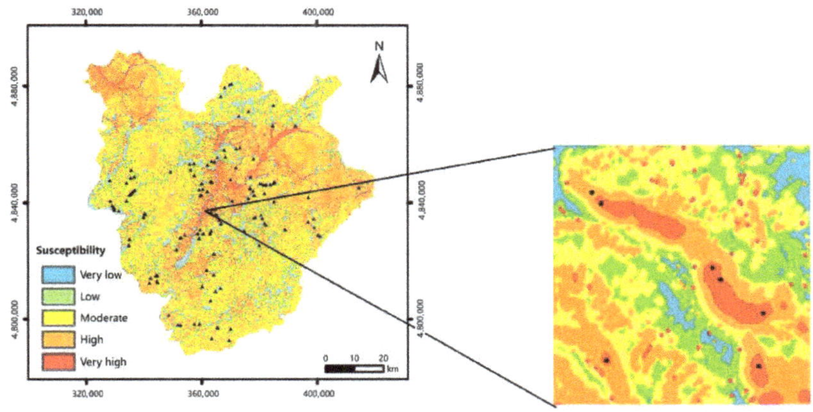

(a)

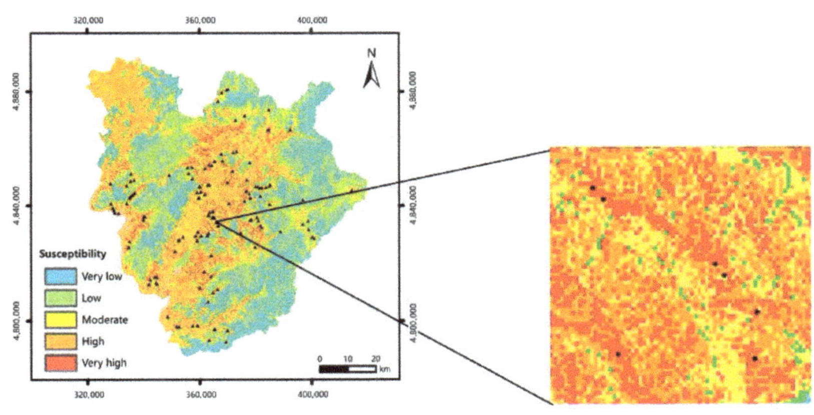

(b)

Figure 13. The maps of forecast results: (**a**) S-CN method and (**b**) T-CN method.

We analyzed the predicted results provided by the S-CN and T-CN forecast methods under the same rainfall conditions as 20 July 2017. This makes it possible for us to compare the prediction results with the field survey debris flow disaster points and evaluate the prediction ability macroscopically based on the R-USCM. In general, according to the predicted results by the S-CN and T-CN forecast methods, there are 22,391 and 33,024 km^2 in the very high and high susceptibility areas which accounted for 34.39% and 50.72% of the total study area, respectively. In addition, very low and low susceptibility areas only accounted for 32.58% and 27.25% of the total study area. It is shown that the study area belongs to the areas susceptible to debris flows, which is consistent with the actual investigation.

In this paper, the ROC curve method which is an effective tool to determine the quality of forecast methods, is also used to evaluate the prediction results of the two methods. The area under the ROC curve (AUC) represents the accuracy of a reliable probabilistic model for forecasting the occurrence or non-occurrence of debris flows. The reference line serves as the standard for judging the degree of model fitting. When AUC is greater than 0.5, the model fits well, and when AUC is less than 0.5, the model fits randomly. The most common method is to use the success rate and prediction rate curves to verify the model. The success rate curve is used to determine how well the maps of forecasting results have classified the areas of existing debris flows. The prediction rate curve shows how well the model and predictor variables predict debris flows. Both success and prediction rate curves are shown in Figure 14. The AUC values for the S-CN model and the T-CN model are 0.889 (88.9%) and 0.768 (76.8%), respectively. The prediction rate curves show that the AUC values for the S-CN model and T-CN model are 0.880 (88.0%) and 0.806 (80.6%), respectively, demonstrating that the S-CN model and T-CN model have a good performance for debris flow forecasting.

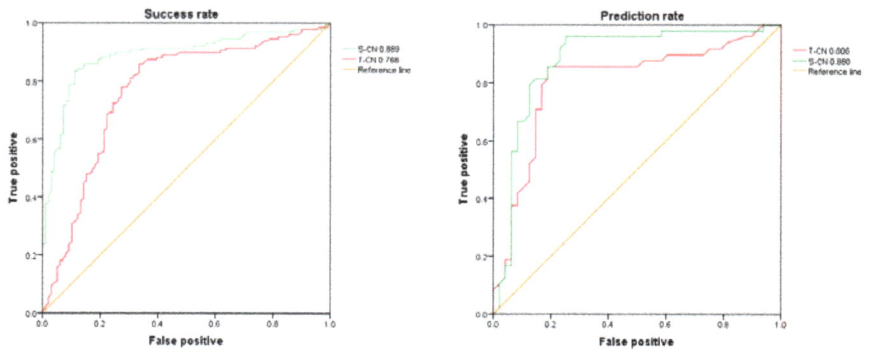

(**a**) (**b**)

Figure 14. Validation of debris flow prediction results: (**a**) success rate curve and (**b**) prediction rate curve.

Furthermore, under the same conditions of runoff, it can be seen that the prediction accuracy can be improved by using the Scoops3D model to determine the volume of the unstable soil mass. The Scoops3D model uses a circular arc search sliding surface method to determine the unstable slope surface, overcoming the shortcomings of the 2D method, which assumes that the failure surface is parallel to the slope surface. Scoops3D conducts SSA based on DEM of the study area. Due to the fact that the topographic features are an important factor affecting the slope stability, it can be preliminarily confirmed that the forecast results of the Scoops3D model have a high fit with the actual distribution of the unstable slopes. Beyond that, the Scoops3D computes the volume of each column in the potential failure mass and adds it to the total. This includes partial columns in the volume computation if two or three corners of the column at the ground surface are contained within the spherical trial surface, rather than full columns that other approaches use, especially when the potential failure mass includes only a small number of full columns. The Scoops3D model can calculate the volume of unstable soil more accurately.

According to the prediction rate curve, the prediction rate of the S-CN model is higher than that of the T-CN model. As seen from Figure 13, although the susceptibility area predicted by the T-CN method is larger, The Scoops3D model usually provides more stable results than the 2D methods in SSA. This is because in the analysis stage of Scoops3D model, although most safety factors of the searched unstable sliding surface are low, the stability of the partial region is improved due to the influence of local micro-geomorphology. Therefore, it is shown that the S-CN method has the ability to overcome the over-prediction problems in debris flow forecasting.

Figure 13 shows that the distribution map of debris flow obtained by the S-CN method is more realistic than that by the T-CN method in terms of the distribution characteristics of debris flow. It can be seen from Figure 13a that the debris flow tends to occur in small individual watershed areas instead of scattered points. The reason for this distinction is that in the analysis of slope stability, each soil column has its individual Fs value, which is separate from other cells in the traditional 2D method. There are some scattered unstable cells even in places far from the sliding sites. However, in the Scoops3D model, slip surfaces tend to occur in a single block, and a hypothesis must be made in order to aggregate unstable pixels together and define landslide blocks, which makes them readily observable. It also makes the result more reasonable.

According to the predicted results by the S-CN and T-CN forecast methods, there are 135 and 116 debris flow disaster locations in the very high and high susceptibility areas which accounts for 83.33% and 71.60% of the total debris flow disaster locations, respectively. No trace of debris flows has been found in partial very high and high susceptibility areas, which may be due to the occurrences of small-scale debris flows in these locations in the past. These traces may be destroyed or covered by plants. Field investigations can only determine the recent development of debris flows. In order to further validate, it is necessary to conduct detailed surveys on the spatial and temporal scales of the study area in the later stage. There are 15 and 24 debris flow disaster locations in the very low and low susceptibility areas which accounts for 9.26% and 14.81% of the total debris flow disaster locations, respectively. This is probably because in the actual conditions, climatic conditions, ground fissures, earthquakes, and other effects will lead to the instability of the slope. However, the current forecast methods cannot incorporate all factors into the SSA. Slope instability caused by factors not included in the SSA cannot be predicted. Therefore, it will affect the accuracy of the debris flow forecast method based on R-USCM. Another influence factor is due to the limited number of rainfall observation stations and the finiteness of the method to collect rainfall data, which makes it very difficult to accurately obtain the daily accumulated rainfall in each location of Jiaohe County. Based on the data of two rainfall observation stations, all of Jiaohe County is divided into two parts. In each part of Jiaohe County, different locations may have different amounts of rainfall over the same time period, so it is not reasonable to use the rainfall station data to represent the rainfall values for the entire region. Therefore, calculating the runoff by using the SCS-CN method in this paper may lead to errors. Further research should use a reasonable method for determining the accurate rainfall values of each location.

Finally, we emphasize that the use of the 3D method to find slip surfaces based on a high-precision DEM is usually time-consuming. The Scoops3D model cannot run in a geographic information system (GIS) environment nor can it realize parallel computing on multicore machines. Allowing very large areas to be simulated in a short time is the next step that requires improvement.

5. Conclusions

In the current study, a debris flow forecasting method based on the R-USCM is proposed. The key to the R-USCM is calculating the volume of unstable soil and the volume of runoff. The Scoops3D model was applied in calculating the volume of the unstable soil because it can analyze the stability of the slope according to the actual situation and calculate the volume of the unstable soil more accurately. The SCS-CN method was also applied in calculating the volume of runoff. The accuracy and applicability of the S-CN were proved in the case study in Jiaohe County. The results are compared with those of the T-CN method. Through the detailed analysis of the forecast results, the following conclusions can be made:

(1) The prediction model of the debris flow based on the R-USCM is due to the formation mechanism of debris flow. The coupling relations between the rainfall and unstable soil from the slope is established. As long as the unstable soil and rainfall information are obtained at the regional scale, a debris flow forecasting method can be used to predict the susceptibility of the debris flows at the regional scale. Thus, the debris flow forecasting method based on the R-USCM has strong applicability.

(2) The use of this S-CN method based on the R-USCM can successfully forecast debris flows at the regional scale. According to the results from the ROC curve, the method had a higher success and prediction rate than the T-CN method. It can be determined that the accuracy of calculating the

volume of unstable soil when runoff conditions are the same can also influence the forecasting results. Therefore, compared with the traditional 2D model, the Scoops3D model has a higher precision and better adaptability in analyzing the slope stability and calculating the volume of the unstable soil.

(3) The distribution characteristics of debris flow are largely determined by the distribution characteristics of unstable soil. It can be seen from Figure 13 that the S-CN method provides a more realistic result than the T-CN method. The debris flow tends to occur in small individual watershed areas instead of scattered points. The main reason is that the Scoops3D identifies the unstable sliding surface instead of some scattered unstable cells. Slip surfaces also tend to occur in individual blocks. Therefore, the risk map generated by the S-CN method is better applied in debris flow control and early warning.

(4) The results show that the T-CN method has the problem of excessive prediction of debris flow. The prediction of a dangerous area is too large to properly measure for early warning. It is certain that the over-prediction problems of the traditional 2D method of calculating the unstable soil can also influence the results. Because of the use of the Scoops3D models, the S-CN method has the ability to overcome the over-prediction problems in debris flow forecasting to provide a reliable basis for geological disaster prevention, control planning and risk management.

In conclusion, it is convincing that the Scoops3D model combined with the SCS-CN method is a practical tool for a 3D, spatially-distributed assessment and forecast of debris flow base on the R-USCM. At present, this method can only predict the debris flow susceptibility zone and cannot realize the prediction of the size of the debris flow trigger area. Further development of this work should focus on drawing orthophotographic maps of the research area to identify the actual scale and hazard scope of each debris flow. This reforming will help to evaluate the size of the debris flow triggering area in detail, thus allowing case-by-case comparisons between the model predictions and field reality. In addition, future research should also use a more reasonable method to determine the rainfall in each location of Jiaohe County and consider more comprehensive factors in SSA. More accurate forecasting results can be obtained.

Author Contributions: All authors were responsible for difference parts of this paper, S.Q. (Shuangshuang Qiao) was responsible for the research ideas, numerical studies and writing the whole paper; J.C. analyzed the simulated result; Z.M. and X.H. contributed the revision of the manuscript; S.Q. (Shengwu Qin) was responsible for the research grant and the revision of the manuscript.

Funding: This research was funded by the National Natural Science Foundation of China (Grant no 41202197), the key Projects of the National Natural Science Foundation of China (Grant no 41330636), Jilin Provincial Science and Technology Department (No.20170101001JC), and China Postdoctoral Science Foundation (Grant no 20100471265).

Conflicts of Interest: The authors declare no conflict of interest.

Abbreviations

R-USCM	Rainfall-unstable soil coupling mechanism
SCS	Soil conservation service
CN	Curve number
1D	One-dimensional
2D	Two-dimensional
3D	Three-dimensional
ROC	Receiver operating characteristic
SSA	Slope stability analysis
DEM	Digital elevation model
S-CN	Scoops3D model combined soil conservation service-curve number method
T-CN	Traditional 2D model combined soil conservation service-curve number method
NRCS	Natural Resource Conservation Service
GIS	geographic information system

List of symbol

v	Volume
R	Radius
x	Direction of the coordinate axis
y	Direction of the coordinate axis
z	Direction of the coordinate axis
s	Shear resistance
c	Cohesive force
u	Pore water pressure
F	safety factor
W	Weight
K_{eq}	Horizontal acceleration coefficient
A_h	Horizontal area
e	Horizontal driving force moment arm
S	Surface areas
Q	Runoff depth
P	Rainfall
R	Potential maximum retention
I_a	Initial abstraction
Sum_v	Total volume
N	Number
D	Unstable soil depth

Greek letters

ρ	Density
ε	Inclusion angle
α	Slip direction
φ	Internal friction angle
σ_n	Normal stress
τ	Shear stress
λ	Coeffcient of proportionality

References

1. Wang, J.; Yang, S.; Ou, G.; Gong, Q.; Yuan, S. Debris flow hazard assessment by combining numerical simulation and land utilization. *Bull. Eng. Geol. Environ.* **2018**, *77*, 13–27. [CrossRef]
2. Hungr, O.; McDougall, S.; Wise, M.; Cullen, M. Magnitude–frequency relationships of debris flows and debris avalanches in relation to slope relief. *Geomorphology* **2008**, *96*, 355–365. [CrossRef]
3. Cloke, H.L.; Pappenberger, F. Ensemble flood forecasting: A review. *J. Hydrol.* **2009**, *375*, 613–626. [CrossRef]
4. Bacchini, M.; Zannoni, A. Relations between rainfall and triggering of debris-flow: Case study of Cancia (Dolomites, Northeastern Italy). *Nat. Hazards Earth Syst.* **2003**, *3*, 71–79. [CrossRef]
5. Aleotti, P. A warning system for rainfall-induced shallow failures. *Eng. Geol.* **2004**, *73*, 247–265. [CrossRef]
6. Zhang, S.; Wei, F.; Liu, D.; Yang, H.; Jiang, Y. A Regional-Scale Method of Forecasting Debris Flow Events Based on Water-Soil Coupling Mechanism. *J. Mt. Sci.-Engl.* **2014**, *11*, 1531–1542. [CrossRef]
7. Wei, F.; Gao, K.; Cui, P.; Hu, K.; Xu, J.; Zhang, G.; Bi, B. Method of debris flow prediction based on a numerical weather forecast and its application. In *Wit Transactions on Ecology and the Environment*; Lorenzini, G., Brebbia, C.A., Emmanouloudis, D., Eds.; WIT Press: Southampton, UK, 2006; Volume 90, pp. 37–46. [CrossRef]
8. Iverson, N.R.; Mann, J.E.; Iverson, R.M. Effects of soil aggregates on debris-flow mobilization: Results from ring-shear experiments. *Eng. Geol.* **2010**, *114*, 84–92. [CrossRef]
9. Bhandary, N.P.; Dahal, R.K.; Timilsina, M.; Yatabe, R. Rainfall event-based landslide susceptibility zonation mapping. *Nat. Hazards* **2013**, *69*, 365–388. [CrossRef]
10. Al-Abadi, A.M.; Shahid, S.; Al-Ali, A.K. A GIS-based integration of catastrophe theory and analytical hierarchy process for mapping flood susceptibility: A case study of Teeb area, Southern Iraq. *Environ. Earth Sci.* **2016**, *75*, 687. [CrossRef]

11. Melelli, L.; Taramelli, A. An example of debris-flows hazard modeling using GIS. *Nat. Hazards Earth Syst.* **2004**, *4*, 347–358. [CrossRef]
12. Gabet, E.J.; Mudd, S.M. The mobilization of debris flows from shallow landslides. *Geomorphology* **2006**, *74*, 207–218. [CrossRef]
13. Korup, O.; McSaveney, M.J.; Davies, T. Sediment generation and delivery from large historic landslides in the Southern Alps, New Zealand. *Geomorphology* **2004**, *61*, 189–207. [CrossRef]
14. Cui, P.; Zhou, G.G.D.; Zhu, X.H.; Zhang, J.Q. Scale amplification of natural debris flows caused by cascading landslide dam failures. *Geomorphology* **2013**, *182*, 173–189. [CrossRef]
15. Bromhead, E.N.; Ibsen, M.L.; Papanastassiou, X.; Zemichael, A.A. Three-dimensional stability analysis of a coastal landslide at Hanover Point, Isle of Wight. *Q. J. Eng. Geol. Hydrogeol.* **2002**, *35*, 79–88. [CrossRef]
16. Arellano, D.; Stark, T.D. Importance of Three-Dimensional Slope Stability Analyses in Practice. In *Slope Stability*; ASCE: Reston, VA, USA, 2000.
17. Xie, M.; Esaki, T.; Qiu, C.; Wang, C. Geographical information system-based computational implementation and application of spatial three-dimensional slope stability analysis. *Comput. Geotech.* **2006**, *33*, 260–274. [CrossRef]
18. Chakraborty, A.; Goswami, D. State of the art: Three Dimensional (3D) Slope-Stability Analysis. *Int. J. Geotech. Eng.* **2016**, *10*, 493–498. [CrossRef]
19. Kang, Z.; Lee, C.; Law, K.; Ma, A. *Debris Flow Research in China*; Science Press: Beijing, China, 2004; pp. 21–22. Available online: http://ir.imde.ac.cn/handle/131551/3663 (accessed on 15 December 2018). (In Chinese)
20. An, H.; Tran, T.V.; Lee, G.; Kim, Y.; Kim, M.; Noh, S.; Noh, J. Development of time-variant landslide-prediction software considering three-dimensional subsurface unsaturated flow. *Environ. Model. Softw.* **2016**, *85*, 172–183. [CrossRef]
21. Baum, L.; Savage, W.Z.; Godt, J.W. *Trigrs-A Fortran Program for Transient Rainfall Infiltration and Grid-Based Regional Slope-Stability Analysis*; Version 2.0, US Geological Survey Open-File Report, 2008. Available online: http://pubs.usgs.gov/of/2008/1159 (accessed on 15 December 2018).
22. Gu, T.; Wang, J.; Fu, X.; Liu, Y. GIS and limit equilibrium in the assessment of regional slope stability and mapping of landslide susceptibility. *Bull. Eng. Geol. Environ.* **2015**, *74*, 1105–1115. [CrossRef]
23. The, V.T.; Alvioli, M.; Lee, G.; An, H.U. Three-dimensional, time-dependent modeling of rainfall-induced landslides over a digital landscape: A case study. *Landslides* **2018**, *15*, 1071–1084.
24. Reid, M.; Christian, S.; Brien, D.; Henderson, S. Scoops3d-Software to Analyze Three-dimensional Slope Stability throughout a Digital Landscape. In *U.S. Geological Survey*; Volcano Science Center: Menlo Park, CA, USA, 2015.
25. Brien, D.L.; Reid, M.E. Modeling 3-D Slope Stability of Coastal Bluffs Using 3-D Ground-Water Flow, Southwestern Seattle, Washington. In *Scientific Investigations Report*; Geological Survey (US): Reston, VA, USA, 2007.
26. Hungr, O. An extension of Bishop's simplified method of slope stability analysis to three dimensions. *Géotechnique* **1987**, *37*, 113–117. [CrossRef]
27. NRCS. *National Engineering Handbook, Part 630 Hydrology*; United States Department of Agriculture, Natural Resources Conservation Service: Washington, WA, USA, 2004.
28. Melesse, A.M.; Shih, S.F. Spatially distributed storm runoff depth estimation using Landsat images and GIS. *Comput. Electron. Agric.* **2002**, *37*, 173–183. [CrossRef]
29. Krois, J.; Schulte, A. GIS-based multi-criteria evaluation to identify potential sites for soil and water conservation techniques in the Ronquillo watershed, northern Peru. *Appl. Geogr.* **2014**, *51*, 131–142. [CrossRef]
30. Li, M.; Zhang, T. The Formation and Prevention of Debris Flow in Qiupi of Ji'an City. In *Advanced Materials Research*; Zhang, X., Zhang, B., Jiang, L., Xie, M., Eds.; Trans Tech Publications: Zurich, Switzerland, 2014; Volume 838, p. 2153.
31. Tran, T.V.; Lee, G.; Trinh, M.T.; An, H.U. Effect of Digital Elevation Model Resolution on Shallow Landslide Modeling Using TRIGRS. *Nat. Hazards Rev.* **2017**, *18*, 04016011.

processes

Article

Model for the Patterns of Salt-Spray-Induced Chloride Corrosion in Concretes under Coupling Action of Cyclic Loading and Salt Spray Corrosion

Ruixue Liu [1,2], Huiguang Yin [2], Lianying Zhang [2], Bing Li [2] and Xianbiao Mao [1,*]

[1] State Key Laboratory for Geomechanics & Deep Underground Engineering, China University of Mining and Technology, Xuzhou 221116, China; ruixue1964@163.com

[2] School of Civil Engineering, Xuzhou Institute of Technology, Xuzhou 221008, China; yhg@xzit.edu.cn (H.Y.); zhanglianying@126.com (L.Z.); Libing_libing_2007@163.com (B.L.)

* Correspondence: xbmaocumt@163.com; Tel.: +86-151-6218-3074

Received: 15 December 2018; Accepted: 2 February 2019; Published: 7 February 2019

Abstract: In this study, the patterns of chloride ion erosion of unsaturated concrete subjected to the coupling action of cyclic loading and salt spray corrosion were experimentally studied, and Fick's Second Law was used to fit the variation patterns of chloride concentration to obtain the chloride diffusion coefficient. Accordingly, we have established a mathematical model that describes chloride transport in unsaturated concrete and accounts for the effects of gas flow, water migration, convection diffusion, and capillary action. This model is composed of three equations—the gas flow equation, the solution flow equation, and the solute convection–diffusion equation. The COMSOL numerical analysis software was subsequently used to obtain solutions for this model, based on parameters such as porosity and the chloride diffusion coefficient. Subsequently, the saturation, relative permeability, and the chloride ion concentration during the first corrosion cycle were analyzed. The numerical results were consistent with the experimental values and were therefore superior to the values obtained using Fick's Second Law.

Keywords: Unsaturation; chloride; concrete; coupling model; numerical calculation

1. Introduction

Concrete structures in tidal zones (or splash zones) are prone to damage owing to various physical, chemical, and biological factors [1–4]. One of the main causes of damage in reinforced concrete structures is chloride intrusion, which usually leads to steel-bar corrosion and concrete cracks [5,6]. There have been many studies on the pattern and mechanism of chloride intrusion in tidal zones (or splash zones). Most of these studies have focused on the transport of chloride in unsaturated concrete. It is generally accepted that chloride intrusion in a tidal or splash zone primarily occurs via capillary adsorption on the concrete surface and diffusion within the concrete [7–9]. Based on Fick's Second Law, Jin et al. studied the effects of water on chloride intrusion and established a convection–diffusion model for chlorides in unsaturated concrete [7]. Pan et al. established a convection–diffusion equation for chlorides and obtained solutions for this equation, based on the theoretical understanding that the transport processes of chlorides in unsaturated concrete are jointly determined by diffusion and convection effects [10]. Using dual-porosity models, Yan et al. established equations that describe the transport of chlorides in concrete and the fractures owing to the coupling of concentration gradients and pore-fluid saturation gradients [11]. Li et al. proposed a model for chloride transport based on water transport mechanisms on concrete surfaces that are alternatingly wet and dry, and they experimentally validated this model [12,13]. Polder studied chloride transport patterns (in wet–dry cycles) in unsaturated concrete using an electrochemically accelerated experimental approach [14].

Several scholars have also constructed water–chloride transport models based on the effects of water migration in concrete pores [15–18]. For example, Zanden et al. adopted the diffusion coefficient to describe the water transport processes in concrete, and they established the transport equations for water and chlorides in concrete samples subjected to wet–dry cycles using the convection–diffusion model [19].

Currently, most studies on the patterns of chloride transport in unsaturated concrete are based on Fick's Second Law; these studies comprehensively account for the effects of capillary action, convection–diffusion, and water migration [20–22]. However, the transmission of chlorides in unsaturated concrete is an extremely complex process and is influenced by a variety of factors. In actual systems, gas flow is also an important factor for the transport of chlorides. Therefore, the objective of this study was to establish a chloride transport model that accounts for the effects of gas flow, water migration, capillary action, and convection–diffusion. The COMSOL numerical analysis program was used to obtain solutions for this model. The results were compared with experimental data to validate the model.

2. Experimental Research on Chloride Transport in Concrete

2.1. Materials

The cementitious materials used in the experiment were Portland cement, mineral powder (G), and fly ash (F) that were supplied by two local companies. The fine aggregate was river sand with a fineness modulus of 2.8 and an apparent density of 2769 kg/m^3. The coarse aggregate was crushed limestone with a size of 5–20 mm and an apparent density of 2719 kg/m^3. The slushing agent for the concrete was carboxylic acid provided by Wulong, and the mix proportion of the concrete is summarized in Table 1.

Table 1. Mix ratio of concrete.

Water–Cement Ratio	Admixture	Content (%)	Cement (kg/m^3)	Fly Ash (kg/m^3)	Mineral Powder (kg/m^3)	Sand (kg/m^3)	Stone (kg/m^3)	Water (kg/m^3)
0.32	F + G	15 + 35	226	68	158	740	1112	145

The specimens were kept in molds for about 24 h at room temperature (20 ± 5 °C) after casting. Then, they were remolded and placed in a standard curing room of a controlled temperature (20 ± 2 °C) and a relative humidity of more than 95% for 28 days before they were tested.

2.2. Test Methods

The dimensions of the concrete samples were 100 × 100 × 400 mm. The serial numbers and quantity of the samples are shown in Table 2, whereas Figure 1 shows the flowchart of the tests. The experimental procedures are described below:

1. Uniaxial compression tests were performed on the concrete samples that had been cured for 28 days to measure their strength (σ_c).
2. Three cyclic loading protocols (i.e., 50%, 65%, and 80% σ_c) were used to obtain concrete samples with three different porosities and were numbered as #1, #2, and #3, respectively. The detailed experimental steps of applying the cyclic loads are as follows: firstly, each specimen was loaded at the loading rate of 5 kN/s using force control, until the load reached the specified load setting value (here, the set value was 50%, 65%, or 80% of the compressive strength limit (c)), and then the load was maintained for 120 s. After that, each specimen was unloaded at the unloading rate of 5 kN/s, until the load value was 0 kN, and then the completely unloaded state was maintained for 120 s. This procedure was repeated for 5 cycles in total, according to the above loading and unloading operations. Finally, the sample was completely unloaded (load was 0 kN).

3. After cyclic loading, the samples with different porosities were placed in a drying oven and dried at 40 °C for 48 h. Subsequently, the samples were weighed, and the initial weights were recorded. Subsequently, the samples were soaked in water for 48 h. The weights of the samples at saturation were measured, and their porosities were calculated (Table 3).
4. One of the 100 × 400 mm surfaces of the samples was used as the corrosion surface, whereas the other sides were sealed using epoxy resins. Subsequently, the concrete samples were placed in an artificial marine climate simulation laboratory for salt spray corrosion tests. The relative humidity was 75–80%, and the duration of the salt spray was 12 h, after which the samples were dried for another 12 h—this cycle was continuously repeated for the duration of the test. The concentration of the NaCl solution used in this test was 5%.
5. When the corrosion time reached 30, 110, 190, and 270 days, concrete powders were obtained from the samples at depths of 5, 10, 15, 20, 25, 30, 35, and 40 mm, which were sealed in plastic bags. Subsequently, 4 g of the concrete powder was mixed with 40 g of distilled water in a reagent bottle. The bottle was shaken for 30 min in a solution shaker before being allowed to settle for 24 h.
6. The concentration of the chloride ions was measured using a DY-2501A Chloride Meter.
7. After testing, the holes in the concrete samples were filled with silicon sealant, and the samples were returned to the laboratory for further corrosion until the subsequent time point, and subsequently, step (6) was repeated. This was continued until the end of the experiment.

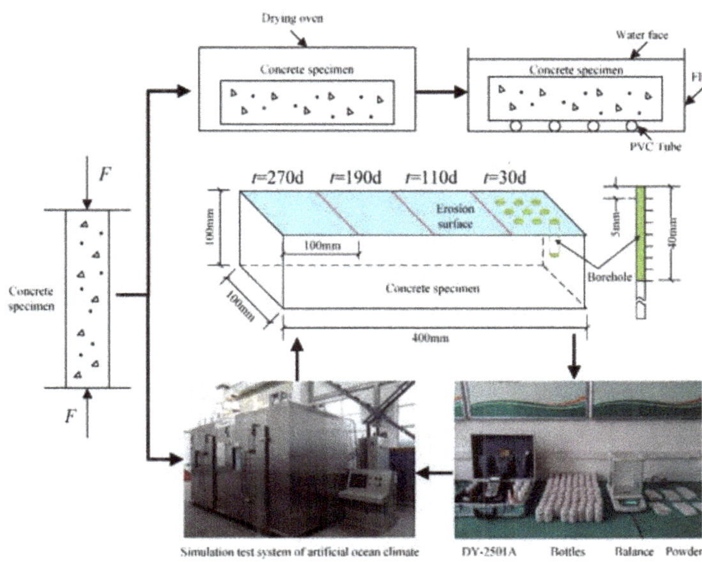

Figure 1. Flow chart of the test.

Table 2. Number of specimens.

Sample NO.	1#	2#	3#
NO./piece	3	3	3

Table 3. Porosity of concrete.

Sample NO.	Porosity ϕ (%)			
	Sample 1	Sample 2	Sample 3	Mean
1	3.92	3.22	3.30	3.49
2	3.90	4.21	3.38	3.83
3	4.73	4.13	5.26	4.71

2.3. Results and Discussion

Figure 2 shows the variational curve of c against l in the concrete samples (c is the concentration of chloride ions in the concrete samples, and l is the depth from the specimen surface). Figure 2a–d correspond to the time points at 30, 110, 190, and 270 days, respectively. It was observed that c decreased with an increase in l. When $l < 15$ mm, c decreased rapidly, with an average decrease of 80.80%; when $l > 15$ mm, c decreased slowly, and the average decrease was only 39.10%. As stated previously, cyclic salt sprays were used as the mechanism of corrosion in this study. The intrusion of chloride ions was mainly driven by capillary actions and concentration gradients. Since the surface of the concrete was the first area that came into contact with the external chlorides, large quantities of chloride accumulated on the surface, resulting in a large chloride concentration gradient. By comparison, fewer chloride ions accumulated deep inside the concrete samples, and, consequently, the chloride ion concentrations were relatively small. It was also observed that, for the same l, c increased with an increase in the porosity; this is because a high porosity increases water and gas permeability, which promotes chloride accumulation.

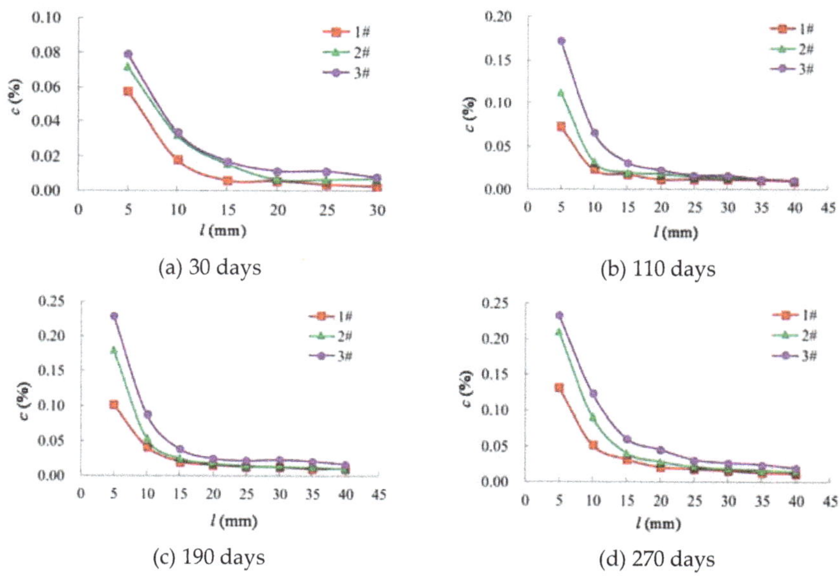

Figure 2. The change curves of chloride ion concentration, c, with depth, l.

Fick's Second Law was used to fit the pattern of chloride corrosion.

Fick's Second Law is expressed as

$$C(l,t) = C_0 + (C_s - C_0)(1 - erf\frac{l}{2\sqrt{Dt}}) \tag{1}$$

where C_0 denotes the initial mass concentration of chloride in an arbitrary section of the concrete, C_s is the mass concentration of chloride ions on the concrete surface at an arbitrary time (t), D denotes the diffusion coefficient of chloride ions, t represents time, l represents the depth, and *erf* represents the error function.

By re-arranging Equation (1), the diffusion coefficient of chloride ions can be calculated by

$$D = \frac{1}{t}\left\{l/2erf^{-}\left[1 - \frac{C(l,t) - C_0}{C_s - C_0}\right]\right\}^2 \tag{2}$$

where C_s is calculated as follows:

$$c_s = \frac{\phi\rho_w\omega}{\rho_s + \phi\rho_w\omega} \times \frac{35.5}{58.5} \times 100\% \tag{3}$$

where ϕ denotes the concrete porosity, ρ_w is the density of the NaCl solution, ρ_s is the concrete density, and ω is the mass percentage of NaCl in the solution. Further, ϕ assumed the values of the average porosities in Table 3, which were 3.49%, 3.83%, and 4.71%, ρ_s was 2420.67 kg/m^3, ρ_w was 1050 g/m^3, and ω was 5%.

The diffusion coefficient of the chlorides when l = 5 mm was calculated using MATLAB with Equation (3) (Table 4). The average value was used in the subsequent numerical calculations.

Table 4. Chloride diffusion coefficient, D.

Sample NO.	Chloride Diffusion Coefficient (D) ($\times10^{-12}$ m^2/s)				
	t = 30 d	t = 110 d	t = 190 d	t = 270 d	Mean
1	5.243	1.592	1.126	0.979	2.235
2	5.804	2.089	2.046	1.957	2.974
3	6.123	3.314	3.066	2.667	3.793

3. Model for Coupled Water–Gas–Chloride Transport

3.1. Basic Assumption

The model for coupled water–gas–chloride transport in unsaturated concrete was established with the following assumptions: (a) concrete is a single porosity medium, (b) the liquids and gases in concrete are ideal liquids and gases, (c) the influence of temperature on parameters such as dynamic viscosity is negligible, (d) only the transmission of a single solute (chloride ions) in concrete is considered, (e) in unsaturated concrete, the liquid phase is the chloride solution, whereas the gas phase is air, (f) there is no exchange of matter between the gas and the liquid phases over the entirety of the process, and (g) it is assumed that the solute (chloride ions) does not evaporate throughout the process, i.e., there is no exchange of matter between the gas phase and the solute [23–26].

In this paper, the gas–liquid two-phase flow model in a single-porosity medium was deduced first, and then the diffusion and gas–liquid two-phase flow were linked using the mass continuous equation according to the relationship between the saturation and the diffusion model.

3.2. Gas Flow Model

The equation of the continuity of gaseous masses without considering the source term is as follows:

$$\frac{\partial m_g}{\partial t} + \nabla \cdot \left(\rho_g \vec{q_g}\right) = 0 \tag{4}$$

where m_g denotes the mass of air, ρ_g is the air density, and $\vec{q_g}$ is the Darcy velocity vector.

In the following subsection, explanations are provided for each of the terms in Equation (4).

(1) Expression of the Gaseous Mass

The gaseous mass is expressed as

$$m_g = \rho_g \phi (1 - S_w) \tag{5}$$

where ϕ denotes the porosity of concrete and S_w is the water saturation.

According to the state equation of ideal gases, $\rho_g = \frac{M_g}{RT} P_g$, where M_g denotes the molecular mass of the gas, R is the gas constant, and T is the absolute temperature, Equation (5) can be written as

$$m_g = \frac{M_g}{RT} P_g \phi (1 - S_w) \tag{6}$$

(2) Equation of Porosity

When subjected to small loads or no load, the deformations in the porous media are mainly driven by pore pressure; hence, pore deformation occurs without significant changes in the particle volume.

In order to facilitate subsequent derivations, α was introduced to represent the pore deformation caused by changes in pore pressure:

$$\alpha = \frac{1}{U_w} \frac{dU_w}{dP} \tag{7}$$

where U_w denotes the volume of the pores. Thus, porosity may be expressed by

$$\phi = \frac{f(P)}{1 + f(P)} \tag{8}$$

where $f(P) = \frac{\phi_0}{1+\phi_0} e^{\alpha(P-P_0)}$, ϕ_0 is the initial porosity, and P_0 is the initial pore pressure.

(3) Darcy Velocity Vector of the Gas Phase

Darcy's law in the gas phase for a unit volume of porous concrete can be expressed as

$$\vec{q_g} = -\frac{k_{ge}}{\mu_g} \nabla P_g \tag{9}$$

where k_{ge} is the effective permeability of the gas phase (phase permeability) and μ_g is the dynamic viscosity of the gas phase.

(4) Expression for Effective Permeability

The expression for effective permeability is

$$k_{ge} = k_{rg} \cdot k \tag{10}$$

where k_{rg} is the relative permeability of the gas phase and k is the absolute permeability of the porous concrete.

(5) Expression for Absolute Permeability

Absolute permeability is the permeability when only one fluid (single phase) exists in the pores of the concrete, the fluid does not react with the concrete, and the fluid flows in accordance with the Darcy linear percolation law.

Absolute permeability is an important parameter for characterizing porous media, since it is closely related to the porosity of the medium. In this study, the relationship between the permeability and porosity in porous media, as reported by Chilingar, was defined as [27]

$$k = \frac{d^2 \phi^3}{72(1 - \phi^2)} \tag{11}$$

where k denotes the permeability of porous media, ϕ is the porosity of porous media, and d is the effective diameter of the particles in porous media. Based on Equation (8), the following equation was obtained:

$$k = k_0 \frac{(1-\phi_0)^2}{\phi_0{}^3} \frac{f(P)^3}{1-f(P)} \tag{12}$$

(6) Saturation and Relative Permeability

Fluid saturation is defined as the ratio of the volume of a fluid in a pore of an object to the pore volume, expressed as a percentage or fraction.

Based on the theories related to oil reservoir models [15], the relationship between the relative permeability, water pressure, gas pressure, and saturation [28,29] is derived in the following.

The capillary pressure is expressed as

$$P_c = P_e (S_w^*)^{-\frac{1}{\lambda}} = P_g - P_w \tag{13}$$

where P_e denotes the constant capillary pressure at the entrance, S_w^* is the saturation of the standard liquid phase, and λ is the pore size distribution index.

The relative permeability is expressed as

$$k_{rw} = (S_w^*)^{\frac{2+3\lambda}{\lambda}} \tag{14}$$

$$k_{rg} = (1 - S_w^*)^2 [1 - (S_w^*)^{\frac{2+\lambda}{\lambda}}] \tag{15}$$

The relationship between the standard liquid phase saturation and water–gas saturations is expressed as

$$S_w^* = \frac{S_w - S_{wr}}{1 - S_{gr} - S_{wr}} \tag{16}$$

$$S_w + S_g = 1 \tag{17}$$

where S_{wr} and S_{gr} are relevant constants.

By re-arranging Equation (13), we obtain

$$S_w^* = \left(\frac{P_e}{P_g - P_w} \right)^{\lambda} \tag{18}$$

By substituting Equation (18) into Equations (14) and (15), the relationships between the relative permeability, water pressure, and gas pressure are derived as

$$k_{rw} = \left(\frac{P_e}{P_g - P_w} \right)^{2+3\lambda} \tag{19}$$

$$k_{rg} = \left[1 - \left(\frac{P_e}{P_g - P_w} \right)^{\lambda} \right]^2 \left[1 - \left(\frac{P_e}{P_g - P_w} \right)^{2+\lambda} \right] \tag{20}$$

The relationship between saturation, water pressure, and gas pressure can be obtained by rearranging Equation (16):

$$S_w = (1 - S_{gr} - S_{wr}) \left(\frac{P_e}{P_g - P_w} \right)^{\lambda} + S_{wr} \tag{21}$$

(7) Flow Model for the Gaseous Phase

By substituting Equations (6), (8), (9), and (19)–(21) into Equation (4), the flow model for the gaseous phase can be obtained as

$$\begin{aligned}&\left[\phi(1-S_w)[1+\tfrac{\alpha P_g}{1+f(P_g)}]+\phi P_g\lambda(1-S_{gr}-S_{wr})\tfrac{P_e^\lambda}{(P_g-P_w)^{(1+\lambda)}}\right]\cdot\tfrac{\partial P_g}{\partial t}\\&-\left[\phi P_g\lambda(1-S_{gr}-S_{wr})\tfrac{P_e^\lambda}{(P_g-P_w)^{(1+\lambda)}}\right]\cdot\tfrac{\partial P_w}{\partial t}+\nabla\cdot\left(-\tfrac{k_{rg}k}{\mu_g}P_g\nabla P_g\right)=0\end{aligned} \tag{22}$$

3.3. Flow Model for the Liquid Phase

(1) Continuity Equation of Liquid Masses

The continuity equation of liquid masses is expressed as

$$\frac{\partial m_w}{\partial t}+\nabla\cdot(\rho_w\vec{q_w})=0 \tag{23}$$

where m_w denotes the mass of the solution in a unit volume of porous concrete, ρ_w is the density of the solution, and $\vec{q_w}$ refers to the Darcy velocity vector of the solution.

(2) Mass of the Liquid Phase

The mass of the liquid phase is expressed as

$$m_w=\rho_w\phi S_w \tag{24}$$

where ρ_w is the density of the solution, ϕ is the porosity, and S_w is the saturation of the solution.

During chloride transport in concrete, the pore pressure and the solution concentration change, which leads to changes in the solution density. Therefore, ρ_w is a function of P and c:

$$\rho_w=\rho_w(P,c) \tag{25}$$

In order to characterize the variation of density with the pore pressure and the solution concentration, we introduced a compressible pressure coefficient, β_P, and a compressible concentration coefficient, β_c:

$$\begin{aligned}\beta_P&=\tfrac{1}{\rho_w}\tfrac{d\rho_w}{dP}\\\beta_c&=\tfrac{1}{\rho_w}\tfrac{d\rho_w}{dc}\end{aligned} \tag{26}$$

where β_p denotes the ratio between the change in density caused by the unit pressure changes and the initial density, whereas β_c denotes the ratio between the density changes caused by the changes in the unit concentration and the initial density.

Subsequently, the relationship between the variations in solution density with the pore pressure and the solution concentration can be obtained as follows:

$$\rho_w=\rho_{w0}e^{[\beta_P(P-P_0)+\beta_c(c-c_0)]} \tag{27}$$

where ρ_{w0} denotes the initial solution density, P_0 is the initial pore pressure, and c_0 is the initial concentration of the chloride ions.

(3) Darcy Velocity Vector of the Liquid Phase

Darcy's law for the liquid phase in a unit volume of porous concrete can be expressed as

$$\vec{q_w}=-\frac{k_{we}}{\mu_w}\nabla P_w \tag{28}$$

where k_{we} denotes the effective permeability of the liquid phase (phase permeability) and μ_w is the dynamic viscosity of the liquid phase.

The effective permeability can be expressed as

$$k_{we} = k_{rw} \cdot k \tag{29}$$

where k_{rw} denotes the relative permeability of the liquid phase and k is the absolute permeability of the porous concrete media. Therefore, we obtain

$$\vec{q_w} = -\frac{k_{rw} \cdot k}{\mu_w} \nabla Pw \tag{30}$$

(4) Flow Model for the Liquid Phase

The flow model of the liquid phase is

$$\begin{aligned} &-\nabla \cdot (\tfrac{\rho_w k_{rw} k}{\mu_w} \nabla P_w) - \left[\lambda (1 - S_{gr} - S_{wr}) \frac{\rho_w \phi P_e^\lambda}{(P_g - P_w)^{(1+\lambda)}} \right] \frac{\partial P_g}{\partial t} \\ &+ \left\{ \rho_w \phi S_w (\beta_P + \beta_c) + \tfrac{S_w \rho_w \alpha \phi}{1+f(P_w)} + \left[\lambda (1 - S_{gr} - S_{wr}) \frac{\rho_w \phi P_e^\lambda}{(P_g - P_w)^{(1+\lambda)}} \right] \right\} \tfrac{\partial P_w}{\partial t} + [\rho_w \phi_0 (\beta_P + \beta_c)] \tfrac{\partial c}{\partial t} \end{aligned} \tag{31}$$

3.4. Convection–Diffusion Models for Chlorides

The transport of chlorides in concrete is a complex process, and it may be treated as the transport of solutes (chlorides) in solution. Solute transport in porous media is influenced by various factors, such as convection, molecular diffusion, mechanical dispersion, precipitation–dissolution combined effects, and chemical adsorption. In this study, only convection, mechanical dispersion, and chemical adsorption were considered in the convection–diffusion model for chlorides in unsaturated concrete.

When the concrete pores are filled with the solution, the pore water inside the concrete does not migrate as a whole. Under the condition that the chloride ion and the hydration product of the concrete do not react chemically, the process of chloride ions migrating from the outside of the concrete to the interior of the concrete, depending on the concentration gradient difference inside and outside of the concrete, can be considered as a pure diffusion process. Convection refers to the phenomenon of the whole substance migrating with a carrier solution. The main cause of the convection of chloride ions in concrete is the directional seepage of the pore liquid under pressure, capillary adsorption, and electric field forces [30].

Accordingly, S_w was introduced into the continuity equation for the mass of chlorides in saturated concrete, from which the continuity equation for chloride masses in unsaturated porous media was obtained as follows:

$$\frac{\partial (m_c)}{\partial t} - \nabla \cdot (S_w \phi D \cdot \nabla c) + \nabla \cdot (\vec{v} c) = 0 \tag{32}$$

where the mass of chloride ions in the solution is expressed as $m_c = S_w \phi c - (1 - \phi) \rho_s K_L \frac{C_{max} c}{1 + K_L c}$ and $\vec{v_w}$ is the Darcy velocity vector of the solution.

By substituting the relevant equations into Equation (32), the convection–dispersion model for chlorides in unsaturated concrete was obtained as follows:

$$\begin{aligned} &-\left[(1 - S_{gr} - S_{wr}) \frac{\phi c \lambda P_e^\lambda}{(P_g - P_w)^{(1+\lambda)}} \right] \tfrac{\partial P_g}{\partial t} + \left[(c + \rho_s K_L \tfrac{C_{max} c}{1+K_L c}) \tfrac{S_w \alpha \phi}{1+f(P_w)} + (1 - S_{gr} - S_{wr}) \frac{\phi c \lambda P_e^\lambda}{(P_g - P_w)^{(1+\lambda)}} \right] \tfrac{\partial P_w}{\partial t} \\ &+ \left[S_w \phi + S_w (\phi - 1) \rho_s K_L \frac{C_{max}}{(1+K_L c)^2} \right] \tfrac{\partial c}{\partial t} - \nabla \cdot (S_w \phi D \nabla c) + \nabla \cdot (\tfrac{k_{rw} k}{\mu_w} \nabla P_w c) = 0 \end{aligned} \tag{33}$$

3.5. Model for Coupled Water–Gas–Chloride Transport

The model for coupled water–gas–chloride transport in unsaturated concrete consists of three parts: the flow model for the gaseous phase, the flow model for the liquid phase, and the convection–diffusion model for chloride ions, as shown in Figure 3a–c. This model accounts for the effects of convection, diffusion, chemical adsorption, and capillary action and reflects on the coupling effects between gas–liquid flows and solute transport in unsaturated porous media. The "gas phase term" represents the influence of the gas phase on the solute.

Capillary action term; Liquid-gas interaction term

$$\left[\phi(1-S_w)\left[1+\frac{\alpha P_g}{1+f(P_g)}\right]+\phi P_g\lambda(1-S_{gr}-S_{wr})\frac{P_e^{\lambda}}{(P_g-P_w)^{(1+\lambda)}}\right]\cdot\frac{\partial P_g}{\partial t}-\left[\phi P_g\lambda(1-S_{gr}-S_{wr})\frac{P_e^{\lambda}}{(P_g-P_w)^{(1+\lambda)}}\right]\cdot\frac{\partial P_w}{\partial t}+\nabla\cdot\left(-\frac{k_{rg}k}{\mu_g}P_g\nabla P_g\right)=0 \qquad (a)\ \text{Gas phase}$$

Gas-liquid interaction term; Solute-liquid interaction term

$$-\nabla\cdot\left(\frac{\rho_w k_{rw}k}{\mu_w}\nabla P_w\right)+\left[\lambda(1-S_{gr}-S_{wr})\frac{\rho_w\phi P_e^{\lambda}}{(P_g-P_w)^{(1+\lambda)}}\right]\frac{\partial P_g}{\partial t}+\left\{\rho_w\phi S_w(\beta_P+\beta_c)+\frac{S_w\rho_w\alpha\phi}{1+f(P_w)}+\left[\lambda(1-S_{gr}-S_{wr})\frac{\rho_w\phi P_e^{\lambda}}{(P_g-P_w)^{(1+\lambda)}}\right]\right\}\frac{\partial P_w}{\partial t}+[\rho_w\phi_0(\beta_P+\beta_c)]\frac{\partial c}{\partial t}=0 \qquad (b)\ \text{Liquid phase}$$

Gas-solute interaction term; Liquid-solute interaction term; Adsorption term; Diffusion term; Convection

$$\left[(1-S_{gr}-S_{wr})\frac{\phi c\lambda P_e^{\lambda}}{(P_g-P_w)^{(1+\lambda)}}\right]\frac{\partial P_g}{\partial t}+\left[\left(c+\rho_s K_L\frac{C_{max}c}{1+K_Lc}\right)\frac{S_w\alpha\phi}{1+f(P_w)}+(1-S_{gr}-S_{wr})\frac{\phi c\lambda P_e^{\lambda}}{(P_g-P_w)^{(1+\lambda)}}\right]\frac{\partial P_w}{\partial t}+\left[S_w\phi+S_w(\phi-1)\rho_s K_L\frac{C_{max}}{(1+K_Lc)^2}\right]\frac{\partial c}{\partial t}-\nabla\cdot(S_w\phi D\nabla c)+\nabla\cdot\left(\frac{k_{rw}k}{\mu_w}\nabla P_w c\right)=0 \qquad (c)\ \text{Solute phase}$$

Figure 3. Water–gas–chloride coupling transport model in unsaturated concrete.

4. Model Calculations and Validation

4.1. Calculation Scheme

(1) Numerical model

The geometrical dimensions of the numerical model were 100 × 100 mm. The left boundary was subjected to chloride corrosion, whereas the remaining boundaries were non-seepage boundaries. Point A was slightly above the central line and was 5 mm from the left boundary, as shown in Figure 4.

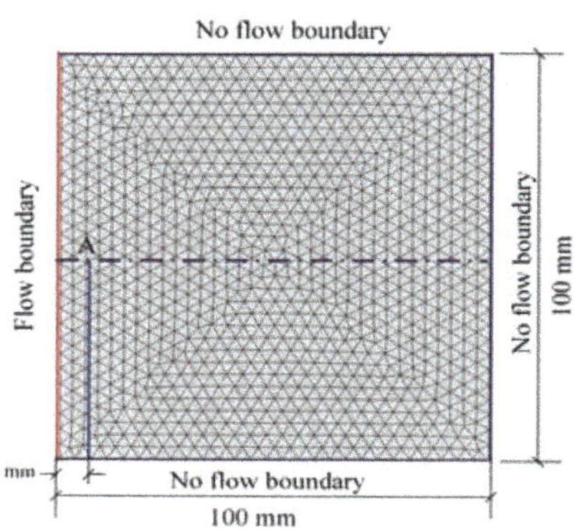

Figure 4. Numerical calculation model.

(2) Configuration of boundary conditions

According to the fluctuation of the sea level in the tidal zone during 24 h, the variation of the chloride concentrations in the left boundary during a single cycle (24 h) was described using Equation (34), as shown in Figure 5a. The variation of the pressure was described using Equation (35), as shown in Figure 5b.

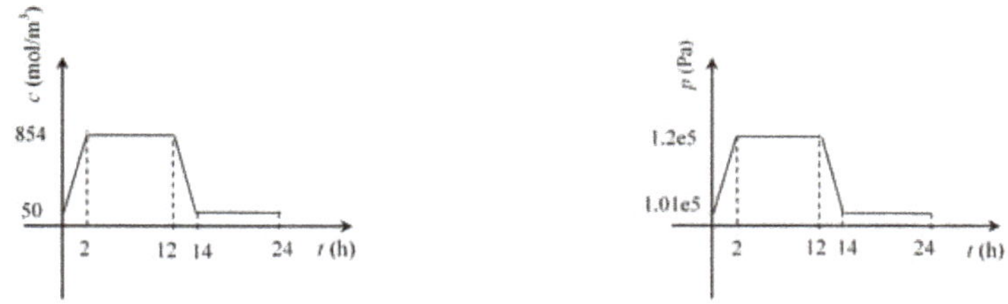

(a) the chloride concentrations in the left boundary **(b) the pressure**

Figure 5. Boundary conditions.

In Equation (34), n = 0, 1, 2, 3, ... , t represents the time (s), and c (t) denotes the concentration of chloride ions (mol/m^3):

$$c(t) = \begin{cases} 50 + \frac{804}{7200} \times (t - 86400 \times n) & 0 + 86400 \times n \leq t < 7200 + 86400 \times n \\ 854 & 7200 + 86400 \times n \leq t < 43200 + 86400 \times n \\ 854 - \frac{804}{7200} \times (t - 43200 - 86400 \times n) & 43200 + 86400 \times n \leq t < 50400 + 86400 \times n \\ 50 & 50400 + 86400 \times n \leq t < 86400 + 86400 \times n \end{cases} \quad (34)$$

$$p(t) = \begin{cases} 1.01e5 + \frac{0.19e5}{7200} \times (t - 86400 \times n) & 0 + 86400 \times n \leq t < 7200 + 86400 \times n \\ 1.2e5 & 7200 + 86400 \times n \leq t < 43200 + 86400 \times n \\ 1.2e5 - \frac{0.19e5}{7200} \times (t - 43200 - 86400 \times n) & 43200 + 86400 \times n \leq t < 50400 + 86400 \times n \\ 1.01e5 & 50400 + 86400 \times n \leq t < 86400 + 86400 \times n \end{cases} \quad (35)$$

In Equation (35), n = 0, 1, 2, 3 ... , t is the time (s), and p (t) is the boundary pressure (Pa).

(3) Scheme of the numerical calculation

Numerical simulation scheme is shown in Table 5.

Table 5. Numerical simulation scheme.

Scheme	Parameter	
	ϕ_0 %	D_0 ($\times 10^{-12}$ m^2/s)
Scheme 1	3.49	2.235
Scheme 2	3.83	2.974
Scheme 3	4.71	3.793

(4) Parameters for the numerical calculation

Since it is very difficult to obtain solutions for coupled models with complicated boundary conditions, the model in this work was further simplified prior to the calculations. For example, ϕ, ρ_w, and D were set to constant values, based on the values described by previous studies (Table 6).

Table 6. Numerical simulation parameters.

Parameter	Value
Initial porosity ϕ_0/%	ϕ_0
Pore deformation coefficient α/(1/Pa)	1e-6
Pressure compressibility β_p /(1/Pa)	5e-8
Concentration compressibility β_c/ (m^3/mol)	1e-5
Water permeability k_w/(m^2)	1e-15
Langmuir adsorption constant K_L/(m^3/mol)	0.001
Langmuir maximum adsorbing capacity C_{max}/(mol/m^3)	100
Initial concentration C_0/(mol/L)	50
Solution density ρ_w/(kg/m^3)	1025
Gas density ρ_g/(kg/m^3)	1.23
Concrete density ρ_c/(kg/m^3)	2450
Chloride diffusion coefficient D_0/(m^2/s)	D_0
Liquid dynamic viscosity u_w/(Pa·s)	0.00101
Gas dynamic viscosity u_g/(Pa·s)	0.00085
Liquid initial pressure P_0/(Pa)	1.01e5
Liquid initial saturation S_{wr}/%	30
Gas initial saturation S_{gr}/%	70
Concrete permeability K_0/(m^2)	1e-14

4.2. Results and Analysis

(1) Variation of model parameters

In this section, the variation of parameters such as saturation, permeability, relative permeability, and chloride ion concentration during the first cycle were analyzed.

(1-1) Variations in saturation

Figure 6 shows the variation of saturation, *S*, at Point A with time, *t*, during the first cycle.

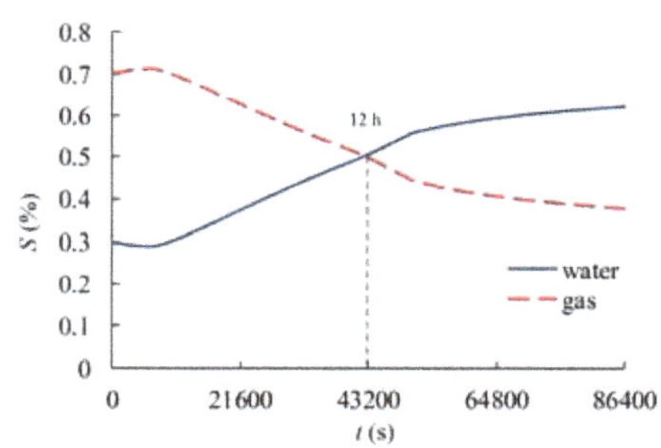

Figure 6. The change curves of saturation, *S*, with time, *t*.

It can be observed that the chlorides were transmitted into the concrete within the first 12 h, thus increasing the solution saturation and decreasing the gas saturation. During the period between the 12th and the 24th hour, surface water evaporation led to a decrease in the solution saturation and an increase in the gas saturation.

(1-2) Variations in fluid permeability

Figure 7 shows the variation of the ratio between the solution permeability, k_w, and the initial solution permeability, k_{w0}, (k_w/k_{w0}) at Point A with time, *t*, during the first cycle.

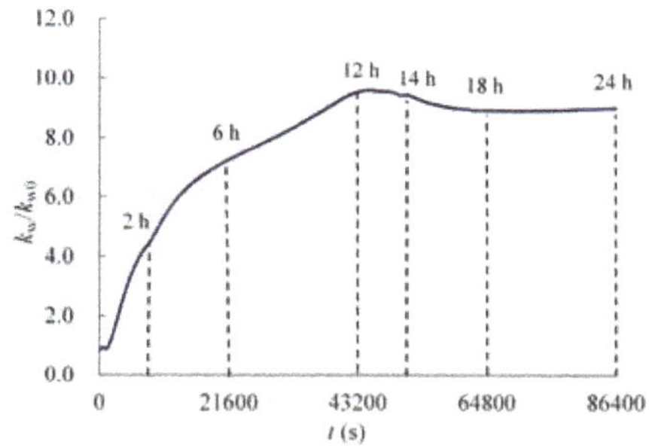

Figure 7. The change curves of the solution permeability ratio, k_w/k_{w0}, with time, t.

At t = 12 h, k_w reached its maximum value, which was 10 times the value of k_{w0}; at t = 24 h, the value of k_w was 9 times that of k_{w0}. Further, the curve can be divided into six segments. When t = 0–2 h, the solution permeability ratio, k_w/k_{w0}, grew at its most rapid rate, at 1.50/h. This suggests that the solution rapidly entered the concrete during this period. When t = 2–6 h, the rate of k_w/k_{w0} growth decreased to 0.75/h. When t = 2–12 h, k_w/k_{w0} continued to increase, but with a lower, stabilized rate of 0.33/h. When t = 12–14 h, k_w/k_{w0} began to decrease slowly, indicating that the solution on the concrete surface had begun to evaporate. When t = 14–18 h, k_w/k_{w0} continued to decrease, and eventually stabilized, such that k_w/k_{w0} = 9.0 when t = 18 h. When t = 18–24 h, the solution permeability ratio, k_w/k_{w0}, was maintained at a steady value.

(1-3) Variations in relative permeability

Figure 8 shows the variation of the relative permeability of the solution, k_{rw}, and the relative permeability of the gas phase, k_{rg}, at point A with time, t, during the first cycle.

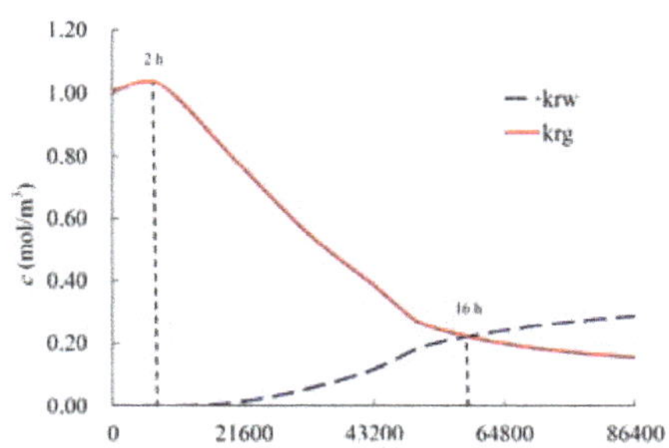

Figure 8. The change curves of gas and liquid relative permeability, k_{rg} and k_{rw}, respectively, with time, t.

When t = 0–2 h, k_{rg} increased, whereas k_{rw} was nearly zero. Most of the pores in the concrete were initially stable and filled with gas. When the boundary liquid pressure increased, the gas within the concrete changed from a stable state to a flow-dominated state. Consequently, the relative permeability of the gas increased slightly. Since the solution did not reach Point A, k_{rw} was zero. When t = 2–16 h, k_{rw} increased slowly and k_{rg} declined rapidly. When t = 16 h, k_{rw} = k_{rg}, indicating that most of the pores in the concrete were filled with water. When t = 12–24 h, k_{rw} continued to increase, albeit quite slowly, whereas k_{rg} continued to decrease slowly.

(1-4) Variations in chloride concentration within 24 h.

Figure 9 illustrates the variation of chloride concentrations, c, with depth, l, on the central line during the first cycle.

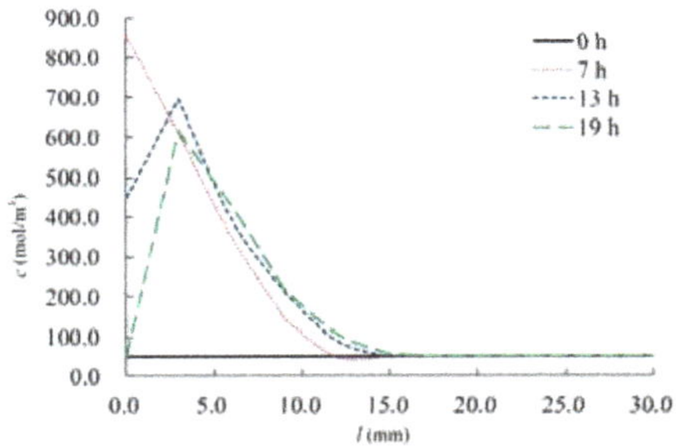

Figure 9. The change curves of chloride ion concentration, c, with depth, l.

At t = 0 h, the initial concentration, c, was 50 mol/m^3 at each of the points on the line. When t = 0–12 h, c decreased with an increase in l. When l reached a certain value, c reached its minimum value and remained steady. A peak value was not observed. When t = 12–24 h, c increased initially, and subsequently decreased with an increase in l. When l reached a certain value, the value of c reached a plateau. Therefore, there was a peak value of c. When the solution evaporated on the concrete surface, the chlorides dispersed and were transported internally and externally owing to the concentration gradients and capillary action. Consequently, a peak could be observed at a certain distance from the concrete surface.

(2) Variation of chloride concentrations

Numerical calculations were performed according to the schemes listed in Table 5. The initial porosities of each scheme were 3.49%, 3.83%, and 4.71%. The molar concentrations were converted to mass concentration to facilitate the experimental validation of the numerical results. The conversion formula is

$$c = \frac{c_1 M_{CL}\phi}{\rho_s} \tag{36}$$

where c_1 is the molar concentration of chloride ions, M_{CL^-} is the molar mass of chloride ions, ϕ is the porosity, and ρ_s is the density of the concrete samples.

The results of the coupling model presented in this study were compared with the experimental data and fitted results of Fick's Second Law, as shown in Figure 10. Figure 10a–d correspond to the results at 30, 110, 190, and 270 days, respectively. In the figure, 'test' corresponds to the experimental value, 'Fick' corresponds to the result obtained using Fick's Second Law, and 'couple' corresponds to the results of the proposed model. The following conclusions were drawn from the analysis of the results.

The results of the coupling model proposed in this study were consistent with the experimental results, indicating that the model provides a reasonable description of chloride transport in concrete. Nonetheless, a few deviations were observed between the model and experimental results because the equations were highly nonlinear and the calculations attenuated some of these expressions. Furthermore, the model did not account for the heterogeneity of the concrete.

As the depth increased, the fits obtained by Fick's Second Law gradually became smaller than the experimental results and their final values approached zero. The results of the model proposed in this work are thus superior to those of Fick's Second Law, since the proposed model produced a closer approximation of the experimental values. Since Fick's Second Law only considers the diffusion of chlorides and ignores the effects associated with gaseous flows, water migration, capillary actions, etc., it cannot provide an accurate description of chloride transport processes. These deficiencies have been addressed in the model proposed in this work.

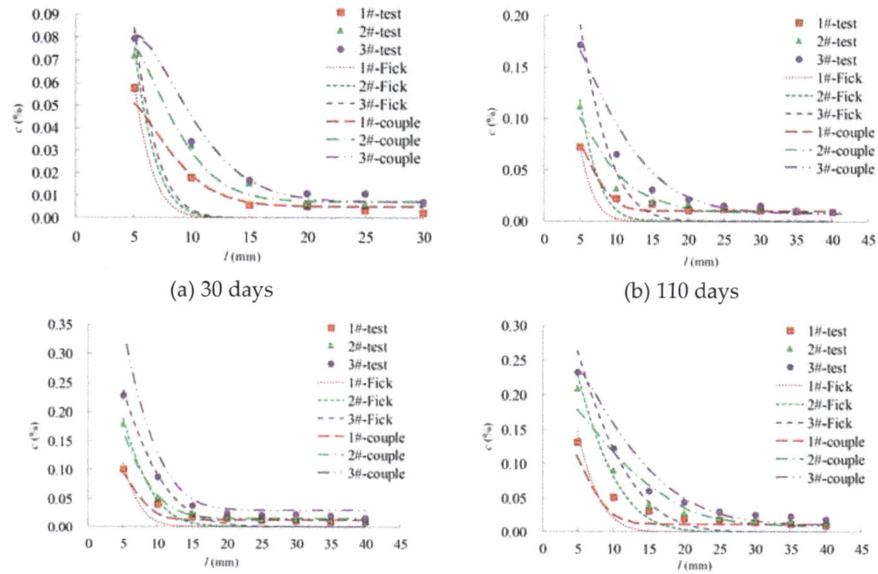

(a) 30 days (b) 110 days (c) 190 days (d) 270 days

Figure 10. The comparison curves of chloride ion concentration, c, with depth, l.

5. Conclusions

(1) The porosities of concrete with 30% fly ash and 20% mineral powder were calculated under the action of three cyclic loading protocols (i.e., 50%, 65%, and 80% σ_c) and were 3.49%, 3.83%, and 4.71%, respectively.

(2) The chloride ion transport law of concrete was obtained by a salt spray corrosion test under the action of three cyclic loading protocols, and the average chloride ion diffusion coefficients of concrete were 2.235, 2.974, and 3.793 $\times 10^{-12}$ m^2/s, respectively, which were calculated by using Fick's Second Law.

(3) We have constructed a model for coupled water–gas–chloride transport in unsaturated concrete using gas flow models, liquid flow models, and solute convection–diffusion models. This model accounts for the effects of gas flows, water migration, convection–diffusion, and capillary actions and reflects on the coupling between gas–liquid flows and solute transport in unsaturated porous media.

(4) The coupled water–gas–chloride transport model was solved based on cyclic boundary conditions. The variations in parameters such as saturation, relative permeability, and chloride concentration during the first corrosion cycle were also analyzed. Through calculations performed on concretes with different initial porosities, this study has determined that the chloride concentrations calculated using the proposed model are consistent with the experimental results and are therefore superior to those obtained using Fick's Second Law.

Author Contributions: R.L. conceived and designed the experiments. X.M. and B.L. contributed to the theoretical calculation and numerical simulation parts of the work and composed the paper manuscript. H.Y. and L.Z. analyzed the data. All of the authors contributed to the writing of the paper.

Funding: The authors gratefully acknowledge the financial support for this work provided by the Outstanding Youth Science Foundation of China (No.51322401), the Science and Technology Project of the Chinese Ministry of Housing and Urban–Rural Construction of China (No.2013-K4-22, 2014-K4-042 and 2016-K4-033), the Science and Technology Project of Housing and Urban-Rural Construction Department of Jiangsu Province (No.2017ZD163), the Natural Science Foundation of the Jiangsu Higher Education Institutions of China (No.16KJB560017 and 18KJB560019), the General Program of Chinese National Building Materials Industry Technology Innovation

Program of China (No. 2016-Z-2, 2016-Z-5, 2016-Z-7 and 2016-Z-8), and the Research Program of Xuzhou Institute of Technology (XKY2016228).

Conflicts of Interest: The authors declare no conflicts of interest.

References

1. Lv, J.F.; Tab, Z.J.; Ba, H.J. Micro-structure and performance analysis of marine concrete after extended exposure to marine environment. *J. China Univ. Min. Technol.* **2010**, *39*, 530–531.
2. Rita, B.F. Electrochemical sensors for monitoring the corrosion conditions of reinforced concrete structures: A review. *Appl. Sci.* **2017**, *7*, 1157.
3. Geng, J.; Easterbrook, D.; Liu, Q.F.; Li, L.Y. Effect of carbonation on release of bound chlorides in chloride contaminated concrete. *Mag. Concr. Res.* **2015**, *68*, 353–363. [CrossRef]
4. Yu, B.; Ning, C.; Li, B. Probabilistic durability assessment of concrete structures in marine environments: Reliability and sensitivity analysis. *China Ocean Eng.* **2017**, *31*, 63–73. [CrossRef]
5. Dai, H.J.; Wang, F.C.; Wu, Y.H.; Nan, F. Durability of concrete used for oil-platform in tidal zone. *J. Building Mater.* **2000**, *3*, 135–141.
6. Li, B.; Yin, H.G.; Mao, X.B.; Li, Y.; Zhang, L.Y.; Liu, R.X.; Qiu, P.T. Macroscopic and microscopic fracture features of concrete used in coal mine under chlorine salt erosion. *Int. J. Min. Sci. Technol.* **2016**, *26*, 455–459. [CrossRef]
7. Jin, W.L.; Zhang, Y.; Lu, Z.Y. Mechanism and mathematic modeling of chloride permeation in concrete under unsaturated state. *J. Chin. Ceram. Soc.* **2008**, *36*, 1362–1369.
8. Jin, W.L.; Jin, L.B.; Yan, Y.D.; Yao, C.J. Field inspection on chloride ion-intrusion effect of seawater in dry-wet cycling zone of concrete structures. *J. Hydraul. Eng.* **2009**, *40*, 364–365.
9. Zhang, Q.Z.; Gu, X.L.; Zhang, W.P.; Huang, Q.H. Model on capillary pressure saturation relationship for concrete. *J. Tong Ji Univ. Nat. Sci.* **2012**, *40*, 1753–1755.
10. Pan, Z.C.; Chen, A.R. Numerical simulation of chloride transportation process in unsaturated concrete. *J. Tong Ji Univ. Nat. Sci.* **2011**, *39*, 314–315.
11. Yan, Y.D.; Jin, W.L.; Wang, H.L.; Lu, C.H. Chloride transport in cracked concrete under wet-dry cycling condition. *J. Tong Ji Univ. Nat. Sci.* **2013**, *44*, 2060–2062.
12. Li, C.Q.; Li, K.F. Chloride ion transport in cover concrete under drying-wetting cycles: Theory, experiment and modeling. *J. Chin. Ceram. Soc.* **2010**, *38*, 581–582.
13. Li, C.Q.; Li, K.F. Moisture transport in cover concrete under drying-wetting cycles: Theory, experiment and modeling. *J. Chin. Ceram. Soc.* **2010**, *38*, 1151–1153.
14. Polder, B.P.; Peelen, W.H.A. Characterisation of chloride transport and reinforcement corrosion in concrete under cyclic wetting and drying by electrical resistivity. *Cem. Concr. Compos.* **2002**, *24*, 427–435. [CrossRef]
15. Thiery, M. Modelisation du sechage des betons: Analyse des differents modes de transfert hydrique. *Revue Européenne de Génie Civil* **2007**, *11*, 541–578. [CrossRef]
16. Ma, D.; Cai, X.; Li, Q.; Duan, H.Y. In-situ and numerical investigation of groundwater inrush hazard from grouted karst collapse pillar in longwall mining. *Water* **2018**, *10*, 1187. [CrossRef]
17. Li, K.F.; Li, C.Q.; Chen, Z.Y. Influential depth of moisture transport in concrete subject to drying-wetting cycles. *Cem. Concr. Compos.* **2009**, *31*, 693–698. [CrossRef]
18. Ma, D.; Duan, H.Y.; Liu, J.F.; Li, X.B.; Zhou, Z.L. The role of gangue on the mitigation of mining-induced hazards and environmental pollution: An experimental investigation. *Sci. Total Environ.* **2019**. [CrossRef]
19. Zanden, A.; Taher, A.; Arends, T. Modelling of water and chloride transport in concrete during yearly wetting/drying cycles. *Constr. Build. Mater.* **2015**, *81*, 120–123. [CrossRef]
20. Zhao, Y.X.; Chen, C.; Gao, X.J.; Jin, W.L. Seasonal variation of surface chloride ion content and chloride diffusion coefficient in a concrete doc. *Adv. Struct. Eng.* **2013**, *16*, 395–403. [CrossRef]
21. Zhou, Z.L.; Cai, X.; Ma, D.; Du, X.; Chen, L.; Wang, H.; Zang, H. Water saturation effects on dynamic fracture behavior of sandstone. *Int. J. Rock Mech. Min. Sci.* **2019**, *114*, 46–61. [CrossRef]
22. Wang, L.C.; Wang, J.Z. Mesoscale simulation of chloride diffusion in concrete subjected to flexural loading. *Adv. Struct. Eng.* **2014**, *17*, 561–571. [CrossRef]
23. Bourbatache, K.; Millet, O.; Aït-Mokhtar, A. Ionic transfer in charged porous media. Periodic homogenization and parametric study on 2D microstructures. *Int. J. Heat Mass Transf.* **2012**, *55*, 5979–5991. [CrossRef]

24. Ma, D.; Duan, H.Y.; Cai, X.; Li, Z.; Li, Q.; Zhang, Q. A global optimization-based method for the prediction of water inrush hazard from mining floor. *Water* **2018**, *10*, 1618. [CrossRef]
25. Wu, Y.; Liu, J.S.; Elsworth, D. Dual Poroelastic Response of coal Seam to CO_2 Injection. *Int. J. Greenh. Gas Control* **2010**, *4*, 668–678. [CrossRef]
26. Ma, D.; Rezania, M.; Yu, H.-S.; Bai, H.-B. Variations of hydraulic properties of granular sandstones during water inrush: Effect of small particle migration. *Eng. Geol.* **2017**, *217*, 61–70. [CrossRef]
27. Chilingar, G.V. Relationship between porosity, permeability, and grain size distribution of sands and sandstones. *Dev. Sedimentol.* **1964**, *1*, 71–74.
28. Zhou, Z.; Cai, X.; Ma, D.; Chen, L.; Wang, S.; Tan, L. Dynamic tensile properties of sandstone subjected to wetting and drying cycles. *Constr. Build. Mater.* **2018**, *182*, 215–232. [CrossRef]
29. Ma, D.; Cai, X.; Zhou, Z.L.; Li, X.B. Experimental investigation on hydraulic properties of granular sandstone and mudstone mixtures. *Geofluids* **2018**, *2018*, 9216578. [CrossRef]
30. Ge, Y.F. *Fine Numerical Simulation and Experimental Study of Chloride Corrosion of Self-Compacting Concrete*; School of Civil Engineering Southeast University: Nanjing, China, 2017; pp. 3–4.

processes

Article

Experimental Study on Electric Potential Response Characteristics of Gas-Bearing Coal During Deformation and Fracturing Process

Zhonghui Li [1,2,3,4,†,*], Yue Niu [1,2,3,4,†], Enyuan Wang [1,2,3,4,*], Lanbo Liu [5], Honghao Wang [6], Mingfu Wang [7] and Muhammad Ali [4,8]

1 Key Laboratory of Gas and Fire Control for Coal Mines (China University of Mining and Technology), Ministry of Education, Xuzhou 221116, China; ahsyls69@cumt.edu.cn
2 Engineering Research Center for Coal Gas Control, China University of Mining and Technology, Xuzhou 221116, China
3 State Key Laboratory of Coal Resources and Safe Mining, China University of Mining and Technology, Xuzhou 221116, China
4 School of Safety Engineering, China University of Mining and Technology, Xuzhou 221116, China; muhammad.ali@cumt.edu.cn
5 Department of Civil & Environmental Engineering, University of Connecticut, Storrs, CT 06268, USA; lanbo_liuuconn@163.com
6 Yanghe Coal Industry Limited Company, Zhengzhou Coal Industry (Group), Zhengzhou 452382, China; whh19901110@163.com
7 Xinzheng Coal Power Limited Company, Zhengzhou 450000, China; yuer_new@126.com
8 Department of Mining Engineering, Balochistan University of Information Technology, Engineering and Management Sciences, Quetta 87300, Pakistan
* Correspondence: lzhemr@126.com (Z.L.); weytop@cumt.edu.cn (E.W.); Tel.: +86-13814443973 (Z.L.); +86-15150037723 (E.W.)
† These authors contributed equally to this work.

Received: 23 December 2018; Accepted: 26 January 2019; Published: 1 February 2019

Abstract: Coal mass is deformed and fractured under stress to generate electrical potential (EP) signals. The mechanical properties of coal change with the adsorption of gas. To investigate the EP response characteristics of gas-bearing coal during deformation and fracture, a test system to monitor multi-parameters of gas-bearing coal under load was designed. The results showed that abundant EP signals were generated during the loading process and the EP response corresponded well with the stress change and crack expansion, and validated this with the results from acoustic emission (AE) and high-speed photography. The higher stress level and the greater the sudden stress change led to the greater EP abnormal response. With the increase of gas pressure, the confining action and erosion effect are promoted, causing the damage evolution impacted and failure characteristics changes. As a result, the EP response is similar while the characteristics were promoted. The EP response was generated due to the charge separation caused by the friction effect etc. during the damage and deformation of the coal. Furthermore, the main factors of the EP response were different under diverse loading stages. The presence of gas promoted the EP effect. When the failure of the coal occurred, EP value rapidly rose to a maximum, which could be considered as an anomalous characteristic for monitoring the stability and revealing failure of gas-bearing coal. The research results are beneficial for further investigating the damage-evolution process of gas-bearing coal.

Keywords: gas-bearing coal; electrical potential; charge separation; gas adsorption; damage evolution

1. Introduction

As the basic energy used in the world, coal resources play an important part in industrial production and economic life [1]. During mining activities, coal and gas outburst disasters greatly threaten safe production in coal mines, leading to serious casualties and property loss. Examples of accidents include coal and gas outburst and rock burst [2–4]. For example, in 2014, a coal burst damaged mine equipment seriously and resulted in the death of two miners in the Austar coal mine in Australia [5]. Therefore, the research on revealing the initiation and generation mechanism of coal and gas outburst disasters is of great significance for monitoring the stability of coal-rock mass while mining, and predicting the generation of coal and gas outburst disaster [6].

Under the combined effect of ground stress, mining-induced stress, and gas pressure, the internal damage of coal-rock mass constantly changes to finally trigger structural instability and dynamic failure, which causes the coal and gas outburst disaster [7]. Monitoring the stress state and damage-evolution process of coal-rock mass is the key to monitoring and early-warning of coal and rock outburst disasters [8]. At present, with the gradual reduction of coal resources in the shallow underground coal seams, coal mining shifts to deep underground coal seams [9]. Deep coal seams basically contain abundant high-pressure gas and the coupling effect of stress and gas gradually imposes increasing influences on the state of the coal mass [10]. When the gas is sufficiently adsorbed by the coal, the gas is stored in the pores of the coal mass at adsorbed state and free state to form gas-solid coupling system with the coal mass [11]. The formation of this gas-solid coupling system changes the physical and mechanical properties of the coal (including mechanical property, deformation and fracture process, and failure mode). The formation of the system further influences the occurrence of various disasters, such as coal and gas outburst and rock burst [12].

Energies are released during the initiation and occurrence process of coal and gas outburst disasters. The released forms include the elastic energy, acoustic energy, electromagnetic energy, etc. [13]. Therefore, numerous geophysical methods (such as acoustic emission (AE) and electromagnetic radiation) are developed to monitor the stability and predict the failure of the coal mass [14,15]. The methods have been widely explored in the laboratory, and corresponding technologies also have been applied in mining activities in the field [16–20]. Previous studies indicated that coal-rock mass could be electrically charged during deformation under load. It triggers the electrical potential (EP) response on the material surface. The EP response is closely related to the deformation and fracture of the coal mass, which characterizes the stress state and damage-evolution process of the coal mass. The abnormal change of the EP can be considered as the precursor of failure of coal-rock materials. Yoshida [21] found that the EP changed markedly just prior to dynamic rupture. Takeuchi et al. [22,23] studied the electrokinetic properties of quartz and granite and demonstrated that the fracture surface and sliding friction surface were charged with the density up to 10^{-4}~10^{-2} C/m^2. By exploring the EP effect of the coal-rock not containing gas under load, Wang et al. [24] showed that the change of the EP was highly correlated with loads and rates of load change. The distribution of the EP field corresponded well to that of strain field. Additionally, they discussed different mechanisms of the EP response by referring to the electromagnetic radiation (EMR) [25]. Archer et al. [14] found EP signals were stimulated during linear-elastic and -inelastic deformation, which was associated with micro-cracking. As a conclusion, the article provided an effective and advanced method for structural health monitoring of rocks. Niu et al. [26] presented the EP response characteristics and its mechanism in a similar simulation of coal-mining activities. The above research provides a favorable theoretical basis for monitoring the failure of coal-rock mass by testing the EP response [27–29].

Previous research on the EP effect of the coal-rock mass was mostly carried out on the coal-rock mass not containing gas. However, research on the EP response characteristics of coal dynamic failure under the stress-gas coupling effect and the influence of gas is rare. To solve the problem, the influences of the stress-gas coupling effect on the deformation and fracturing process of the coal mass were investigated. Furthermore, a multi-parameter test system for gas-bearing coal under load was designed to carry out compressive tests under different gas pressures. During the loading process, the various responses (including EP and AE) during the deformation process of gas-bearing coal under load were synchronously recorded. Simultaneously, the evolution process of coal fractures is subjected to high-speed photography in real time by using an industrial camera. The AE count can reflect the micro-fracturing happening during the damage and deformation of the coal mass while the amplitude of the signals can reflect the intensity of micro-cracks [30]. The high-speed photography can catch the instantaneous state of crack evolution of the coal mass during different loading stages [31]. This method allows us to analyze the EP response and the damage evolution of the coal mass. Therefore, the research result is beneficial to further reveal the initiation mechanism of coal-rock dynamic disasters. In this way, the attempt is carried out to provide valuable information about the damage evolution under the stress-gas coupling effect and monitor the failure of the coal mass by using the EP response.

2. Experimental Materials and Methods

2.1. Specimen System

The experiment was conducted in the Faraday shielded room in China University of Mining and Technology. It could effectively prevent the electric noise from affecting the recorded signals [15]. The experimental system was designed and established individually to acquire multiple types of information during loading process of the gas-bearing coal (see Figure 1). The subsystem parts are shown as follows.

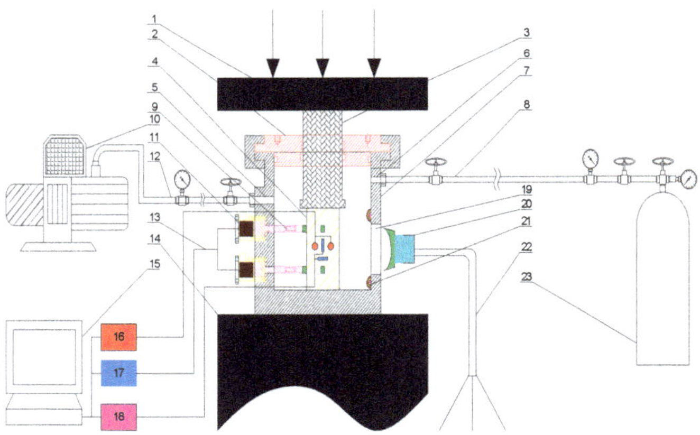

Figure 1. Experimental system of multi-data acquisition during loading process of gas-bearing coal.

Number of parts in experimental system of Figure 1.

1	loading plate	13	data transmission line
2	chamber cover	14	supporting base
3	loading shaft	15	data acquisition control host
4	specimen	16	EP controller
5	exhausting hole	17	AE controller
6	inflating hole	18	loading controller
7	chamber	19	visible window
8	gas transmission line	20	industrial camera
9	waveguide rod	21	illumination source
10	vacuum pump	22	triangular bracket
11	AE sensor	23	gas cylinder
12	exhaust line		

It includes a hermetically sealed chamber subsystem, an axial compressive loading subsystem, an EP monitoring subsystem, an AE monitoring subsystem, a high-speed camera shooting subsystem, and a gas inflating-extracting subsystem

(1) The hermetically sealed chamber subsystem, including a chamber cover, a loading shaft, gas spiracles, waveguide rods, and a visible window (see Figure 1). The gas can be pumped into the chamber with spiracles and kept at a certain pressure after closing the valve.

(2) The EP monitoring subsystem, including the electrode, data transmission line, an amplifier, A/D converter and data storage. When the EP signals were measured by the electrode on the surface of specimen, and then translated by the data transmission line. The host computer can analyze the signal and display it in real time. The maximum sampling frequency is 100 kHz.

(3) The AE monitoring subsystem, reaching a series of functions, such as the parameter setting, signal acquisition, data storage, graphics display, waveform acquisition, and spectral analysis. During the loading process of the specimen, it can acquire simultaneously the counts and energy of AE signals and locate the damage position inside the specimen.

Conventional AE sensors fail to endure high gas pressure. However, in previous monitoring methods, sensors are distributed on the surface of gas-bearing containers, which causes loss and inaccurate measurements of AE signals [32]. To solve the problem, the specialized waveguide rods are designed, as shown in Figure 2. The one end of the waveguide rods is connected to the specimen while the other end is connected to the AE sensor. Moreover, coupling agents favorable for conduction of acoustic waves are uniformly painted on the two ends of the waveguide rods. Composite materials are filled between the waveguide rods and the wall of the gas cylinder, to prevent the transmission of acoustic waves from waveguide rods to the gas cylinder. In this way, AE signals generated by the specimen can be completely transmitted to the AE sensors through waveguide rods as far as possible. The composite materials are connected to the waveguide rods through the screw joints so that the waveguide rods can move to the wall of the cylinder relatively and the specimen can be readily dismantled. Besides, the space between the waveguide rods and the composite materials is sealed through a flexible seal ring. It is an innovative design to transmit the AE signals of coal fracture in the high-gas environment through the waveguide rods.

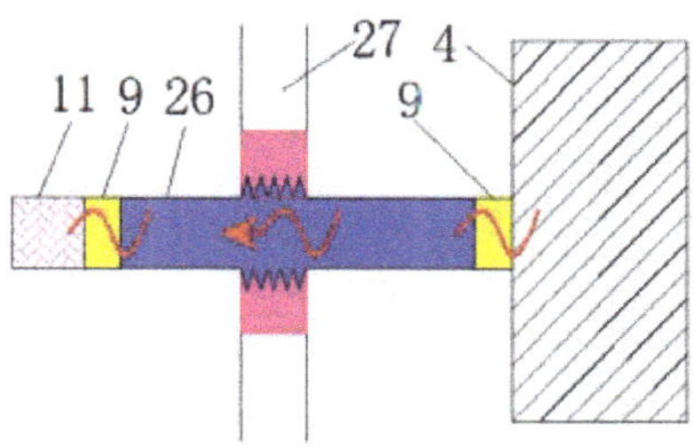

Figure 2. Schematic diagram of AE signal from waveguide wave. 11—AE sensor; 9—coupling agent; 26—waveguide rod; 27—chamber wall; 4—specimen.

(4) The axial compressive loading subsystem, including a loading frame, a hydraulic drive device, a controller host, and a program (see Figure 3). It can record the data of displacement, load, and time synchronously during the loading process. The oil source is placed outside the faraday shielded room, to reduce the interferon on the experiment.

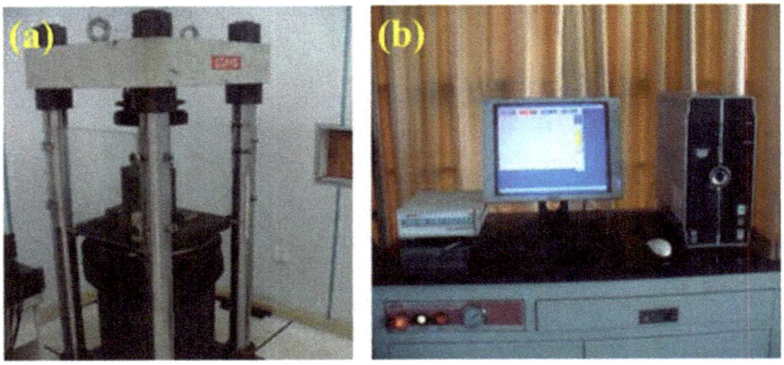

Figure 3. Axial compressive loading subsystem. (**a**) Loading frame. (**b**) Hydraulic drive device and controller host.

(5) The high-speed camera shooting subsystem. It includes the industrial camera, the illumination source, and the triangular bracket. The instant crack expansion on the specimen surface during loading process can be shot clearly by using the industrial camera through a visible window. The resolution ratio of photograph is up to 1920 × 1080 pixels with 40 frames per second.

(6) The gas inflating-extracting subsystem, including the vacuum pump, gas transmission, exhaust lines, and gas cylinder. The gas also can be inflated into the chamber by the gas transmission line and extracted after the experiment.

Under the coupled action of loading stress and gas pressure, the system could be used to conduct the compression experiment of coal. During the process, the signals of EP and AE and crack photograph could be obtained in real time to be analyzed.

2.2. Specimen Preparation

The coal mass was derived from No. 5 coal seam in Yangzhuang coal mine, Huaibei City, China. Then the specimens were prepared according to the standard size of 50 (width) × 50 (height) × 100 (length) mm (see Figure 4).

Figure 4. Coal specimen. With size of 50 (width) × 50 (height) × 100 (length) mm.

2.3. Experimental Scheme

The experiment was carried out according to the following steps.

(1) First, all specimens were placed in the sealed dry container for 24 h before the experiment, to prevent them from absorbing excessive water.

(2) The airtightness of the cylinder was tested to ensure that the cylinder has a favorable sealing effect. The various parts of the test system were connected and kept a turn-on state.

(3) Electrode and waveguide rods (the other end connected to AE detectors) were distributed on the surface of the coal mass and sealed in the cylinder. By using a vacuum pump, the cylinder was vacuumized. To provide electrical isolation, two thin teflon-plates were placed on the top and bottom of specimen.

(4) Gas was injected into the cylinder. After reaching a certain pressure, the pressure was stabilized for 8 h so that the coal mass could fully adsorb gas.

(5) Experimental parameters were set.

(6) The load was applied on the specimen by using the press to further synchronously measure various data such as EP and AE.

(7) After the specimen was damaged, the experiment was ended and gas in the gas cylinder was released.

3. Experimental Results and Analyses

3.1. Test Results of Multi-parameters during the Damage of Gas-Bearing Coal under Load

In the series, experiments under different gas pressures were carried out. As the example, the experimental results under 2.0 MPa were analyzed as follows. The experimental parameters were set as Table 1.

Table 1. Setting of experimental parameters.

Shielding Effect for Electromagnetic Signal	Threshold Value for AE Signal Acquisition	Axial Loading Rate of the Press	Gas Pressure in Chamber
85 dB	45 dB	50 N/s	2 MPa

(1) Responses of strain and stress

During loading process, the gas pressure was kept at 2 MPa after reaching adsorption equilibrium. The loading stress on the top surface of specimen by loading system is defined as loading stress (LS). Owing to gas pressure, the initial LS was not zero. The curves of stress and strain with respect to loading time are shown in Figure 5.

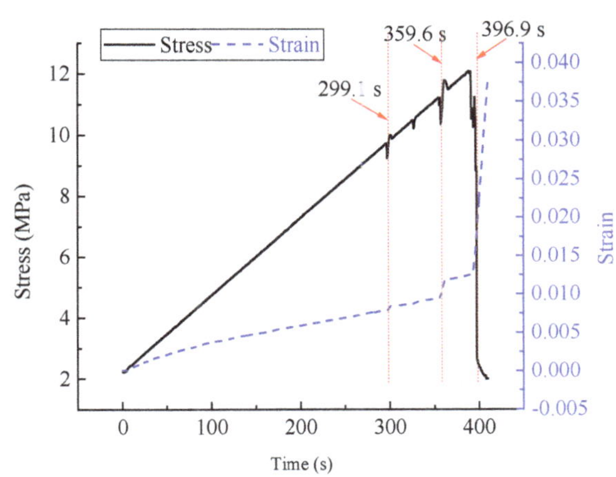

Figure 5. Curves of stress and stain with respect to loading time. The red dotted lines indicate the mutation time of curves.

It can be seen from Figure 5 that with increasing loading time, the stress and strain on the specimen gradually rose. Stress and strain suddenly changed at small amplitudes both at 299.1 s and 359.6 s, implying that the great damage appeared in the local zone of the specimen. Moreover, the local crack expansion resulted in the instant increase of the specimen deformation while stress fluctuated. With the constant loading, stress and strain constantly rose. At 396.9 s, the primary crack was found and therefore the structure of the specimen lost stable. In this case, stress dramatically and suddenly changed and rapidly reduced while strain rapidly increased.

(2) Responses of AE signals

Previous research shows that under the effect of external load, the coal mass is damaged to thus lead to the initiation and expansion of cracks, consequently triggering numerous AE events [30]. Dislocation and slippage between particles in coal matrix can also induce the fracturing of bridge bonds between coal molecules, consequently generating AE phenomenon [33]. AE counts characterize the times of micro-cracking happening during damage and fracturing of the coal mass while the amplitude of the signals reflects the strength of micro-cracking. The parameters related to AE can be used to describe the evolution process of damage and crack expansion and release process of energies in the coal mass, to further judge the deformation and fracturing of the coal specimen and predict the occurrence of failure of the coal specimen [20]. Figure 6 shows the curves of AE counts and MAS (mean amplitude strength) with respect to loading time under gas pressure of 2 MPa.

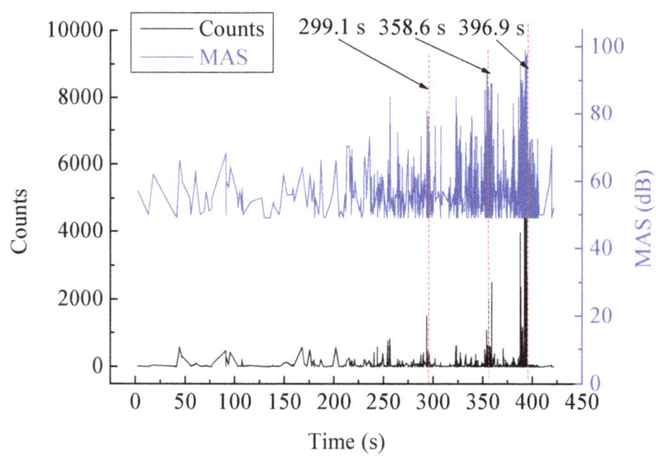

Figure 6. Curves of AE counts and MAS with respect to loading time.

As shown in Figure 6, in the early stage of loading, there were few AE signals with low AE count and amplitude. AE signals were mainly generated due to the mutual slippage of primary cracks in compacted gas-bearing coal and the friction between particles. With the constant growth of stress, the coal damage was constantly aggravated, gradually transiting to plastic damage stage from elastic stage. In the process, new cracks initiated, split, and expanded along the weak structural plane to generate new fracturing signals thus further. Therefore, the AE counts and amplitude both increased during which the amplitude became more concentrated. At 299.1 s and 358.6 s, the specimen was greatly damaged and therefore cracks rapidly expanded, resulting in many fracturing events. In this context, fractured zones were formed in local areas and high-energy gas constantly impacted the coal mass along the weak structural plane, also exhibiting a significant friction effect. AE counts rose in an impulse type, with a high amplitude, reflecting the sudden increase in micro-cracks in the coal mass and the large crack strength, thus releasing huge elastic energies. The result confirmed to the phenomenon that stresses suddenly changed at corresponding time moment in Figure 5, which can be taken as a favorable complementary evidence. When the primary cracks occurred at 396.9 s, AE count and amplitude both reached the maximum. It indicated that the number and strength of micro-cracks in the coal mass dramatically rose to thus result in the great expansion and cut-through of cracks, consequently triggering the failure of the specimen. After the primary crack was found, AE signals rapidly reduced and almost disappeared.

(3) Responses of crack expansion

The evolution processes of crack expansion at different time moments were recorded by applying high-speed photography, as shown in Figure 7.

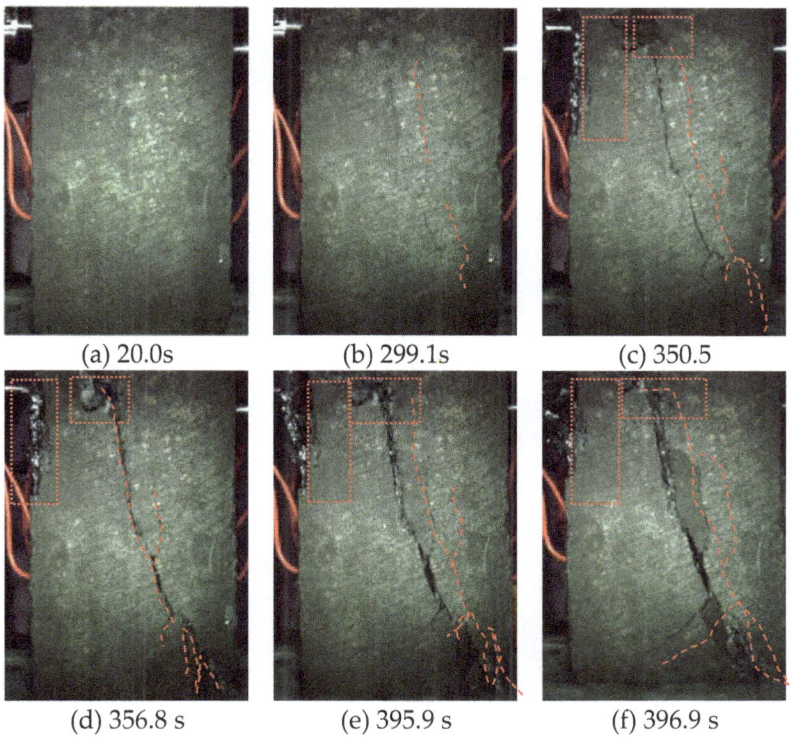

Figure 7. Pictures of fracture expansion in gas-bearing coal under load at different time moments; red imaginary lines refer to the tracks of crack expansion while red wireframes denote the centralized zone with cracks of specimen.

As shown in Figure 7, in the early stage of loading (for example, at 20.0 s), although accumulated damage of the specimen was present, it was insignificant, and no micro-crack was formed on the surface of the specimen. With increasing stress, the micro-defects and micro-cracks in the coal-rock mass gradually aggregated and connected in local zones after undergoing constant extension and expansion in the early stage, showing certain self-organization, thus triggering the generation of micro-cracks in the coal-rock mass. At 299.1 s, stress suddenly changed at a small amplitude and therefore two separated thin and long micro-cracks were formed on the surface of the specimen. At 350.5 s, the damage and fracturing of the coal mass aggravated and two previous micro-cracks mutually cut through and further expanded to thus form multiple secondary cracks. Additionally, microscopic breakage appeared at the left side of the specimen and the top inclined to the left. At 356.8 s, the stress approximated to the maximum, and the width and length of the primary cracks further rose, and the number of secondary cracks increased and constantly expanded. The damage zones at the left side and the top of the specimen were expanded and therefore the damage degree increased. At 395.9 s, the cracks in the middle part in the frontage of the specimen expanded and gradually cut through while the cracks at the bottom were staggered. The structure of the coal mass became increasingly unstable. At 396.9 s, the primary crack occurred in the specimen while the cracks in the middle part were totally connected. Moreover, the width and length of cracks in the lower part witnessed great increases and the cracks at the bottom were connected. Additionally, secondary cracks developed into secondarily primary cracks. The through-running cracks showed the combination of tensile and shear failure. Repetition

(4) Responses of EP signals

As shown in Figure 8, abundant EP signals can be generated during the damage of the gas-bearing coal under load. The changing trend of the EP was consistent with that of stress and there was a favorable corresponding relationship between EP and stress. With the increase of stress, damage in the specimen evolved. As a result, new fracturing constantly happened (shown as the growth of AE count and amplitude in Figure 6) and cracks constantly expanded (see Figure 7). During the above damage evolution, the EP intensity constantly rose. During the loading of the specimen, the EP intensity rose at a ladder type at 299.1 s and 359.6 s, with a significant increase amplitude. The phenomenon conformed to the time moments when stress and strain suddenly changed in Figure 5 and the time moments when AE count and amplitude varied suddenly in Figure 6. At these moments, the EP damage dramatically rose and therefore micro-cracks rapidly increased, showing growing strength. As a result, huge energy was released, which was also verified in Figure 7. At 396.9 s, as stress maximum appeared, the specimen was subjected to primary micro-cracks to thus be damaged due to loss of bearing capacity. In this case, the specimen was greatly and rapidly damaged. Moreover, the EP dramatically fluctuated and rapidly increased to a maximum, showing an extremely significant response. Afterward, the EP intensity rapidly reduced and stabilized at a low level. After the specimen was damaged, stress rapidly declined and the EP rapidly reduced and stabilized at a low level.

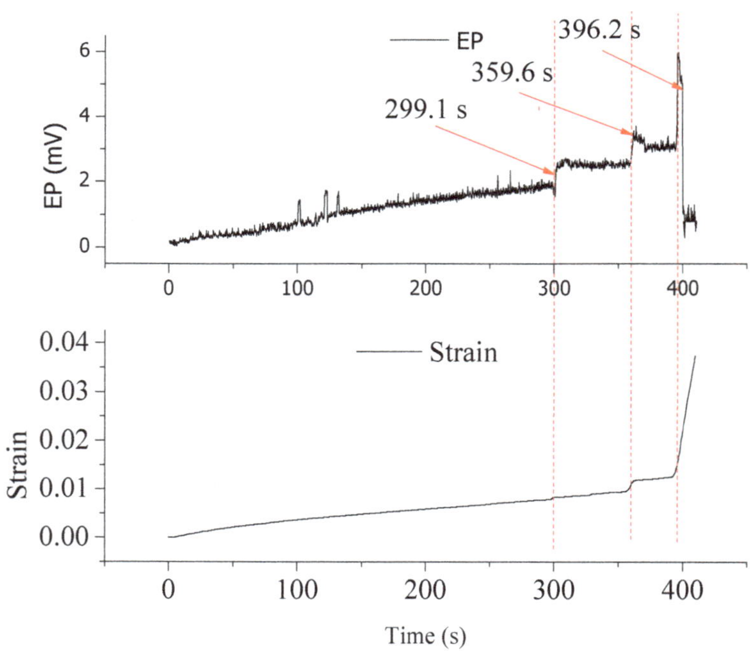

Figure 8. Curves of strain and EP with respect to loading time.

The strain's response characteristics were similar to the stress response. The difference was that the large stress drop occurring at 299.1 s. Currently, since the specimen was still in the elastic loading stage, the strain changes little even the EP response was significant. When loading phase of sample becomes to be plastic phase, especially around the failure, the strain response became more significant. It indicated that the strain monitoring is more sensitive when the specimen is plastic, especially for the main rupture.

The representative time points were selected to calculate corresponding LS levels. On this basis, the abnormal response characteristics of EP, AE, and crack expansion are attained, as shown in Table 2 and their datum statistics is displayed in Figure 9.

Table 2. Statistics on characteristics of EP, AE and crack expansion under different loading time moments (stress levels)

Time (s)	LS Level (%)	Stress Response	EP Response	Crack Expansion
100.1	37.1	Steady increase	Recover after the sudden change	No significant cracks
299.1	80.3	Recover after the sudden change	Rise after the sudden change	Two thin and long cracks
359.6	93.7	Recover after the sudden change	Rise after the sudden change	Cracks running from the top to the bottom
396.9	100	Cliff-type reduction	Sudden change to the maximum and then reduce	Increases in crack width and length

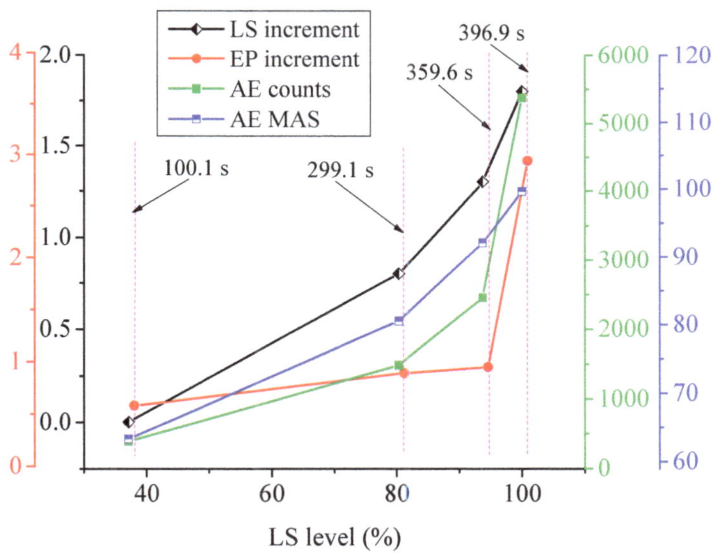

Figure 9. Datum statistics on EP and AE responses at different stress levels.

As shown in Table 2 and Figure 9, on the condition that stress suddenly changed, EP and AE tended to vary suddenly, showing significant abnormal response characteristics, and cracks greatly expanded and constantly aggravated. Correspondingly, the higher the stress level was and the larger the sudden change of stress was, the more significant the damage of the coal mass and thus the more dramatic the abnormal responses of EP and AE. It meant that the larger the increment of EP was, the more the AE counts and the higher the intensity. Additionally, crack expansion enhanced to thus aggravate the damage of the coal mass.

Therefore, based on the changes of stress and AE and analysis result of crack expansion, the change trend and response characteristics of EP can reflect the stress state of the specimen and reveal its damage-evolution process.

3.2. The EP Response Results under Different Gas Pressures

The EP responses under different gas pressures are shown in Figure 10. It is worth noting that gas pressure of 0 MPa means that gas is not injected into the cylinder and previous air in the cylinder is not extracted. Therefore, the air press is kept as the standard atmospheric pressure.

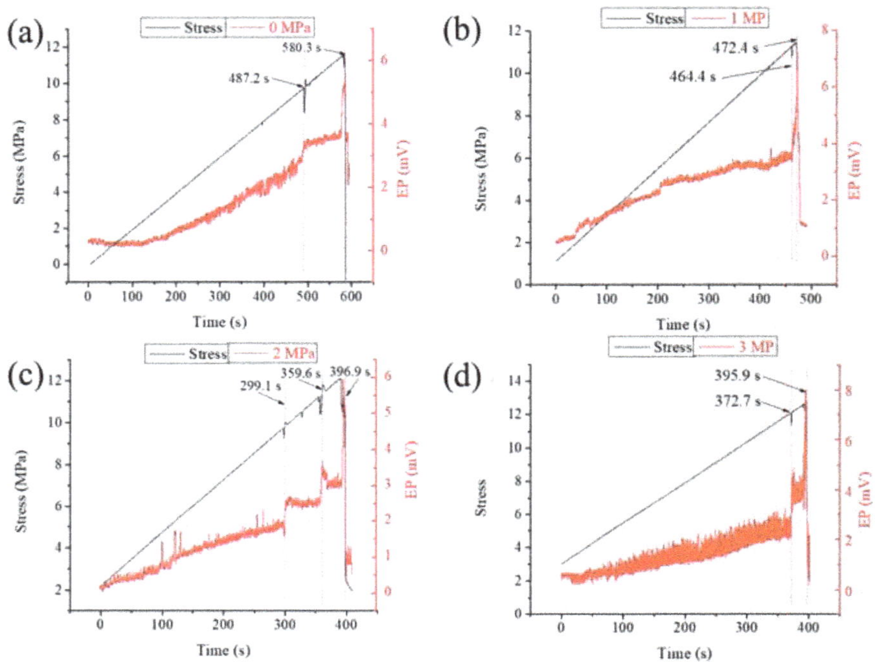

Figure 10. Changes of stress and EP during damage of coal mass under load under different gas pressures. Black and red actual lines represent the changes of stress and EP with loading time, respectively. (**a**) 0 MPa, (**b**) 1 MPa, (**c**) 2 MPa, (**d**) 3 MPa.

As shown in Figure 10, the EP responses under different gas pressures exhibited a basically same change law, which was similar to the results shown in Section 3.1 (4) With the increase of stress, the damage of the specimen under load constantly exacerbated and EP intensity gradually grew. When the specimen was subjected to a serious structural damage, stress suddenly changed and therefore the EP tended to vary abruptly (see 487.2 s in Figure 10a, 464.4 s in Figure 10b, 299.1 s and 359.6s in Figure 10c, and 372.7 s in Figure 10d). The specimen lost its bearing capacity after being subjected to primary crack and therefore stress declined at a cliff type and the EP also rapidly rose to the maximum, with the largest sudden change (see 580.3 s in Figure 10a, 472.4 s in Figure 10b, 396.9 s in Figure 10c, and 395.9 s in Figure 10d).

The broken strength, the strain maximum, the EP maximum, and the EP variation coefficient of specimens under different gas pressures were computed, as shown in Figure 11.

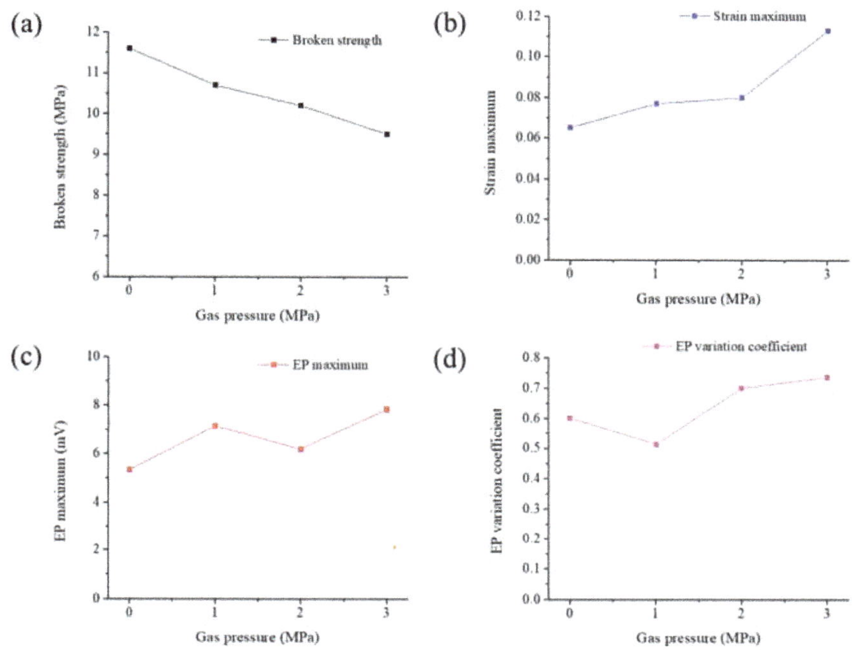

Figure 11. Statistical results of EP responses under different gas pressures. (**a**) Broken strength of specimen, (**b**) strain maximum of specimen, (**c**) EP maximum, (**d**) EP variation coefficient.

(1) The effective stress on the gas-bearing coal can be expressed as follows [34]:

$$\sigma_e = \sigma_L - \varnothing P - 2a\rho_s RT\ln(1 + bP)/3PV_m \quad (1)$$

where, σ_e, σ_L, $\varnothing$, P, a, ρ_s, R, T, b, and V_m refer to effective stress, LS, equivalent pore coefficient, gas pressure, ultimate gas adsorption of unit mass of rock mass at experimental temperature, apparent density of coal, molar gas constant, absolute temperature, adsorption constant, and molar volume, respectively.

Owing to the other conditions are unchanged, the aforementioned formula can be simplified if only the broken strengths (effective maximum stress on the specimen) of specimens under gas pressures were compared. It meant that relative broken strength ($\sigma_e\prime$) can be simplified and expressed by using the difference between LS maximum and gas pressure:

$$\sigma_e\prime = \sigma_t - P \quad (2)$$

As shown in Figure 11a, with the growth of gas pressure, the relative broken strength of the specimen gradually declined. Under the gas pressure of 3 MPa, the relative broken strength of the specimen significantly reduced, decreasing by 18.1% compared with that without gas pressure.

(2) The deformation of the specimen can be represented by using axial mean strain. As shown in Figure 11b, the deformation of the specimen greatly increased with increasing gas pressure. When the gas pressure was 3 MPa, the maximum strain rose by 112.3% compared with that on the condition of having no gas pressure.

(3) As shown in Figure 11c, the EP maximum tended to occur before or after primary micro-cracks of specimens, which was the most significant characteristic of the EP abnormal response. With the increase of gas pressure, although the EP maximum slightly fluctuated, it generally rose, implying that

gas pressure promoted the EP response and therefore EP maximum was at a high level, which was more valuable for analysis.

(4) The EP variation coefficient is the ratio of mean to standard deviation of whole EP data, which can objectively describe the fluctuating response of EP. As shown in Figure 11d, similar to the EP maximum, the EP variation coefficient also generally grew with the increase of gas pressure.

The aforementioned results showed that gas pressure promoted the EP response. The EP response characteristics can monitor the evolution process of damage and fracturing of gas-bearing coal (especially coal containing high content of gas).

4. Discussion

4.1. Damaging and Fracturing Process of Gas-Bearing Coal

The coal belongs to a typical heterogeneous structure, which contains a great number of micro-defects including pores, cracks, and dislocation, that is, Griffith defects [35]. The result can be clearly verified by observing the microstructures of gas-bearing coal by using the scanning electron microscope (SEM) (see Figure 12).

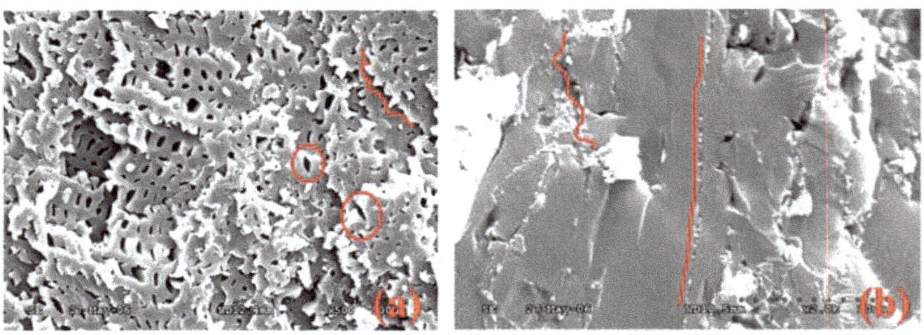

Figure 12. SEM images of microstructures of coal mass before loading. (**a**) Primary pores are abundant and crack was even, mostly appeared as spongy shape under magnification of 500 times; (**b**) Numerous cracks were found in pore clusters under magnification of 5000 times. Red curves refer to the tracks of cracks while red circles represent the location of cracks.

Cracks first initiated at the edge of primary cracks. After reaching the critical breaking strength, cracks expanded along a certain angle to thus generate new cracks. With the constant increase of stress, the damage of the coal mass exacerbated. Therefore, many micro-cracks appeared, expanded, split, closed and mutually connected along the direction with a weak strength to thus form a failure zone of cracks with a certain width. The phenomenon resulted in the generation of micro-cracks in the specimen finally (see Figure 7). Figure 13 displays the images of different fracture surfaces in the damaged gas-bearing coal at different magnification times. According to the pictures, the generation and expansion of cracks during the fracturing of coal as well as the inflection, bending, and splitting during the expansion can be clearly observed.

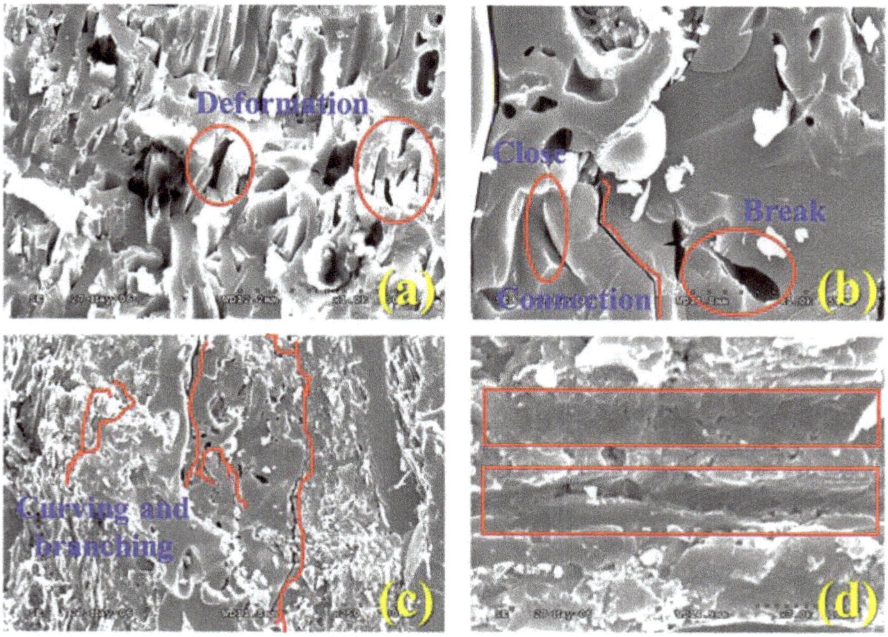

Figure 13. SEM images of microstructures of damaged gas-bearing coal; (**a**) Primary pores were deformed under magnification of 1000 times. (**b**) Pores were damaged, closed and then connected under magnification of 2000 times. (**c**) Tensile cracks were bent and split, which were densely distributed under magnification of 250 times. (**d**) Fracture surface appeared as step-shaped and river-shaped fractures under magnification of 100 times. Red rectangles denote the location of fractures.

4.2. The EP Response Analyses

As a micro-molecular mixture, coal is composed of multiple atom groups. The interiors of micro-molecules are connected through multiple bridge bonds, such as covalent bond, hydrogen bond and Van der Waals' force. The atom groups carry non-uniformly distributed charges so that they show polarity outward. As a result, the micro-surface of the rock mass shows weak electrical property [36] and the surface charge density exhibits the magnitude of 10^{-5}~10^{-4} C/m^2 [37], endowing the rock mass with a certain conductivity. Under the effect of external factors, the charged groups on the surface can lead to charge separation to form electric field. When the gas-bearing coal is damaged and deformed, the particles of coal matrix, mineral particles, and cement in internal structure of coal are subjected to relative slippage and dislocation to generate free charges due to the friction effect. The combined effect between triboelectrification and thermionic emission effect induces the EP response. However, the initiation and expansion of cracks can cause the fracturing of cementitious chemical bonds (even covalent bonds) between coal molecules to generate dangling bonds. It consequently leads to charge separation. Additionally, when stress is applied to micro-cracks, stress concentration effect occurs at the tip of cracks, which causes the energy in coal molecules to dramatically rise at the crack tip. Furthermore, molecular structures are subjected to distortion while outer electrons of molecules escape [16,19,24,26,28].

The charge separation and charge accumulation happening in the aforementioned process can lead to electrostatic charge field, which can be regarded as a point source of the surface EP effect. The interior of the coal mass is composed of heterogeneous coal matrixes and pores filled with gas molecules and coal matrix is greatly different from gas molecules in dielectric constant. Therefore, the polarization electric field can be formed at the interfaces of different dielectrics. Thus, it can be considered that EP at a point within the coal mass is formed due to the superposition of constantly generated charges under the combined effect of variable static electric field and polarization electric field [38]. To simplify the solution process of EP, the EP value can be calculated with imaging method [39]. As shown in Figure 14, it is assumed that the dielectric constants of two semi-infinite dielectrics (coal matrix and gas) are ε_1 and ε_2 and their interface is set as M. Moreover, it is supposed that a point charge appears at a point O, with the electric charge quantity of q. Additionally, it is assumed that the symmetrical location of O in the mirror image corresponding to the interface M is $O\prime$, with the electric charge of $q\prime$.

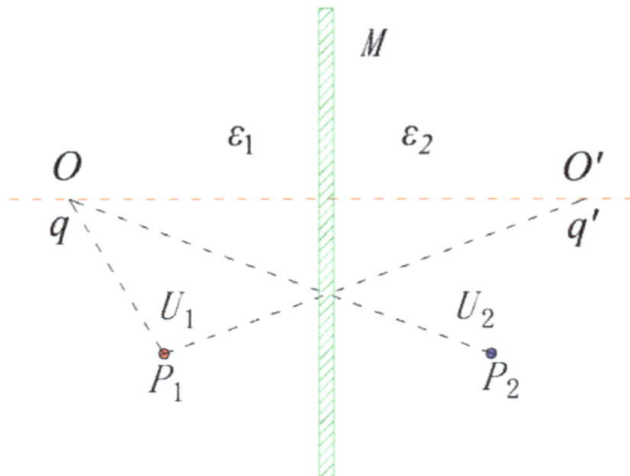

Figure 14. The sketch map of the imaging method for solving EP based on a single boundary. Where $O\prime$ and O refer to the true and imaging charges while U_1 and U_2 denote the EPs at points P_1 and P_2, respectively.

Therefore, the EP at the point P_1 with the same dielectric location with O is expressed as U_1. Similarly, the EP at the point P_2 with the same dielectric point with $O\prime$ is calculated as U_2. Thus,

$$U_1 = q(1/r + K_{12}/r_1)/(4\pi\varepsilon_1) \tag{3}$$

$$U_2 = q(1/r_2 - K_{12}/r_2)/(4\pi\varepsilon_2) \tag{4}$$

where, r, r_1, r_2, and $K_{12} = (\varepsilon_1 - \varepsilon_2)/(\varepsilon_1 + \varepsilon_2)$ refer to the distance of P_1 to point O, the distance of $O\prime$ to the point P_1, the distance of O to the point P_2, and the reflection coefficient of the dielectric ε_1 to ε_2, respectively.

The aforementioned model is popularized to the finite boundary. As shown in Figure 15, the point charge q is bounded within a finite space by four boundaries (A, B, C, and D) and isolated from different dielectrics of external environment. The initial imaging charges of q generated in the four boundaries are q_1, q_2, q_3, and q_4, respectively. Similarly, existing imaging charges can generate new imaging charges along the other boundaries. In this way, there are infinite such imaging charges. However, the farther the new imaging charges distancing to the point charge is, the less significant their effect on the EP of the measurement point [40].

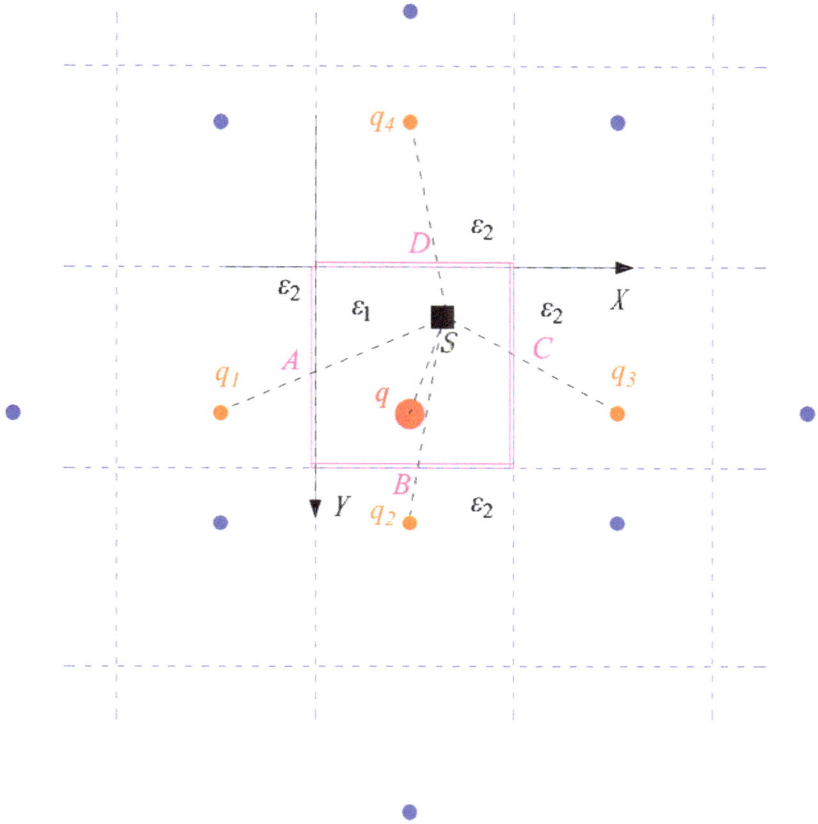

Figure 15. Schematic map of imaging method for solving EPs under finite boundaries. The red solid circle and the pink box refer to actual point charge and four boundaries, respectively. Brown solid circles denote the initial imaging charge of actual point charge at the aforementioned boundaries while blue solid circles represent new imaging charges generated based on the initial imaging charges at the corresponding boundaries. The solid box refers to the location of the point S.

Through simplification, the EP at the measurement point S can be expressed as follows:

$$U_S = q/(4\pi\varepsilon_1 r) + \sum_{j=0}^{m} K_{12}{}^{j} \sum_{i=0}^{n} K_{12}{}^{i}\left[1/r_{1ij} + 1/r_{2ij} + 1/r_{3ij} + 1/r_{4ij}\right]/(4\pi\varepsilon_2) \quad (5)$$

where, $K_{ab}{}^{i}$ and $K_{ab}{}^{j}$ separately denote reflection coefficients while i and j. are superscripts representing the reflecting interfaces separately parallel to axes x and y. Moreover, n and m denote corresponding mirror imaging reflections and r_{1ij}, r_{2ij}, r_{3ij}, and r_{4ij} refer to the distances of imaging charges to the measurement point P.

4.3. The Influence of Gas on EP Effect

As the pore structure is well developed in coal, a large amount of gas is adsorbed on the coal mass after the complete adsorption. Non-adsorbed gas freely moves in coal pores at a free state. A gas-solid coupling system is formed with the adsorbed and free-state gas with coal pores. The formation of the system changes the mechanical properties as well as the damage and fracturing process of the coal mass to thus further influence of response characteristics of EP [6].

(1) The influence of gas on mechanical properties of the coal mass

After gas is fully absorbed by the coal mass, surface free energy of pores reduces [41]. As a result, the attractive force between coal molecules on the fracture surface decreases and the capability of matrix for restricting coal molecules weakens, triggering the expansion-induced deformation of coal matrixes. Microscopically, the change is reflected by the reduced cohesion between coal matrix particles, which finally reduces the force and energy required during failure of the coal mass, and therefore the failure strength declines while the deformation amount increases. This has been verified in the results of Section 3.2, as shown in Figure 10a,b. Additionally, the free-state gas enters large microscopic fractures in the coal mass under the effect of pore pressure to therefore strengthen the effective normal stress, and the structure of fractures will be split and expanded. Additionally, the frictional resistance of cracked surfaces will also be weakened. Moreover, the pore structure is changed, and the mechanical strength of pore structures reduced [42].

(2) The influence of gas on the damage and fracturing process of the coal mass

The process of deformation and fracturing of the coal mass is discontinuous and non-uniform and sometimes local zone is subjected to expansion or shrinkage. On the one hand, gas can deteriorate the structure of the coal mass and promote the damage and fracturing of the coal mass. On the other hand, gas provides confining pressure to change the stress state of the interior of the specimen [42,43]. Under the effect of axial compression stress, the coal mass is subjectged to transverse deformation to induce transverse tensile stress. Under the effect of axial compression stress, transverse tensile stress, and transverse gas stress (confining pressure), the specimens are characterized by the combination of tensile and shear failure. Then the fracturing direction of the specimen shows a certain included angle with the axial direction. With the increase of gas pressure, the difference between confining pressure and peak loading stress leading to failure of the coal mass reduces, and such failure characteristic of the specimen becomes more significant, with a larger included angle (see Figure 16).

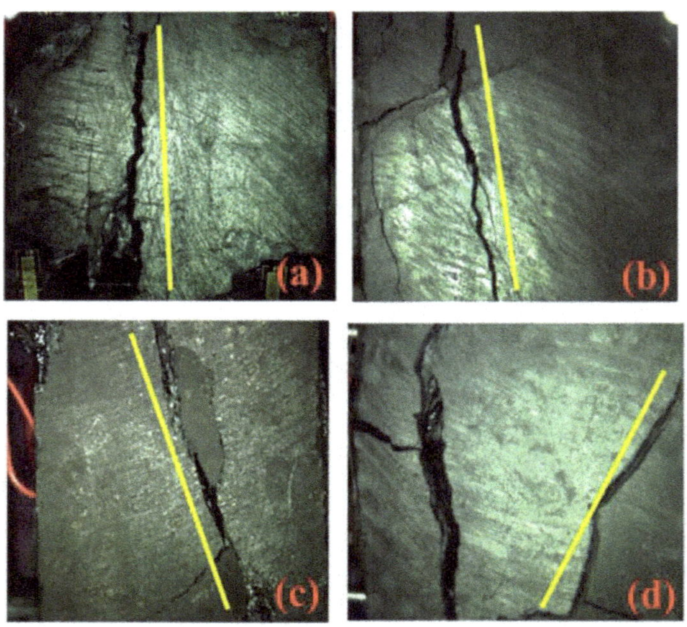

Figure 16. Pictures of real coal mass under load under different gas pressures. (**a**) 0 MPa, (**b**) 1 MPa, (**c**) 2 MPa, (**d**) 3 MPa.

(3) The influence of gas on the EP response

The presence of gas promoted the evolution of damage and fracturing of the coal mass. In particular, the expansion effect and the friction effect of cracks promoted the EP response. Additionally, free-state gas migrated and diffused in coal pores to constantly lead to collision and energy exchange with the edge of micro-cracks and to generate streaming EP with a certain intensity as well.

The EP response was triggered by cracks in the coal mass under load and abnormal sudden change of EP can reveal the failure of the coal mass. For this reason, during the fracturing of gas-bearing coal under load, with the increase of gas pressure, cracks were likely to initiate and propagate in the coal mass and friction effect was strengthened. As a result, the EP effect was more significant, and EP maximum and EP variation coefficient both gradually rose, which are also varied in Section 3.2, as shown in Figure 10c,d.

4.4. Analysis of the EP Response of Gas-Bearing Coal in Different Loading Stages

By taking 2.0 MPa of gas pressure as an example, the damage process of the gas-bearing coal under load can be approximately divided into five stages (see Figure 17) [32].

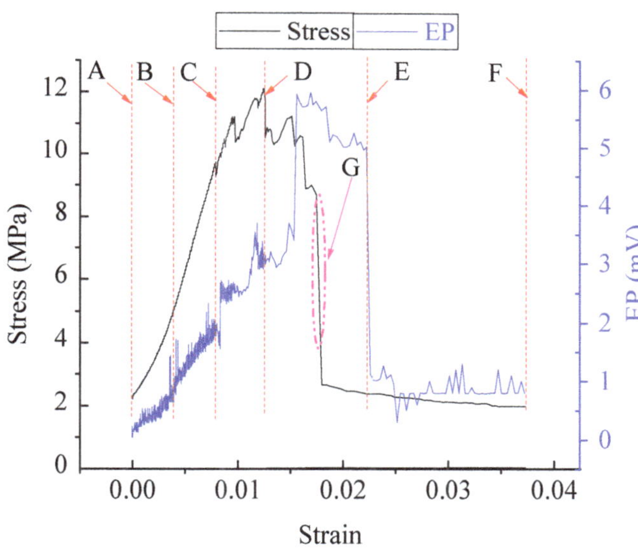

Figure 17. Complete stress-strain curves during the damage of gas-bearing coal under load.

(1) AB stage: Under the effect of stress, the original structural plane and micro-cracks were compacted and closed gradually. Therefore, the EP intensity in the initial stage was generally low while greatly fluctuated. In terms of the EP effect in the stage, the friction effect generated due to the closure of primary crack surfaces mainly appeared while new cracks hardly occurred.

(2) BC stage: In the elastic stage, the coal matrix was mainly subjected to linear-elastic deformation while suffered from little plastic damage. With the increase of strain, stress steadily rose and therefore new primary cracks constantly reached the critical strength, thereby causing the initiation, expansion, and splitting of cracks and steady rise of the EP intensity. In terms of the EP response in the stage, the friction effect between particles of coal matrix and crack surface and little crack expansion effect were mainly found. The presence of gas aggravated the damage and deformation of the coal mass, with a significant promotion effect on the EP response.

(3) CD stage: In the yielding stage, after stress reached the yield point, some irreversible deformations appeared in the specimen. Micro-cracks greatly expanded and resulting charge separation was taken as the dominant mechanism of the EP response. Moreover, the EP response was relatively active, and the EP intensity constantly rose.

(4) DE stage: Plastic deformation mainly happened in the stage. The damage of the specimen constantly exacerbated and micro-cracks and secondary cracks rapidly expanded and connected, which resulted in the occurrence of a primary crack. In the process, a great quantity of intermolecular and even intramolecular chemical bonds in the coal was fractured to thus further generate many charge separations. Additionally, the rapid expansion of cracks also led to the generation of friction effect and a great number of charges were accumulated under the two effects. Owing to the local zone was seriously damaged, the electron-escaping effect caused by stress concentration was also enhanced. The combined effect of multiple factors caused many charges to instantaneously accumulate. As a result, dramatic fluctuation appeared, and EP signals rapidly rose to a maximum. The promotion effect of gas was extremely significant (as shown in Figures 10 and 11).

(5) EF stage: Primary cracks appeared, and cracks fully connected, depriving bearing capacity of the specimen. After the specimen was completely damaged, generated friction effect, crack expansion, etc. of EP signals also basically ended. Therefore, the EP intensity also rapidly reduced and stabilized.

Combined with the analysis in Section 3.1., the EP response was closely related to loading state, AE response, and crack expansion, which can express the evolution of cracks and damage during the loading process of the gas-bearing coal. The presence of gas showed a promotion effect on the EP response. As shown in Figure 16, during DE stage, the specimen was greatly damaged and fractured and therefore failure happened to the coal. Moreover, the abnormal response of EP was extremely significant. The specimen was damaged, which meant that bearing capacity of the specimen rapidly reduced, that is, LS rapidly declined, which can be seen in G zone of Figure 17. In this case, EP rapidly rose to a maximum, which can be taken as the abnormal characteristic for the failure of the gas-bearing coal to monitor the failure of the coal mass.

4.5. Research Significance of the EP Effect of Gas-Bearing Coal

In the mining field of gas-bearing coal, dynamically monitoring the damage and fracturing process of gas-bearing coal is the premise of warning the coal and gas outburst disasters. In addition, EP monitoring can further reflect the stress level and damage state of the coal mass under different loading stages. Therefore, if the stress state of the coal mass cannot be tested directly, monitoring the EP response provides a favorable reference for monitoring the damage-evolution process of gas-bearing coal. Further investigating the EP effect and the mechanism of gas-bearing coal is conducive to further exploring the damage-evolution process of the coal mass. It is beneficial for the study on disaster-causing mechanism of coal failure under the stress-gas coupling effect. It also provides a new idea for monitoring the stability (nondestructive detection) of gas-bearing coal based on the EP response. Moreover, it provides a new idea for exploring the initiation and occurrence of rock and gas outburst disasters based on the EP response.

Compared with traditional geophysical information such as electromagnetic radiation and AE, the EP response is more accurate. Its monitoring process shows a low requirement for the shielding of environmental-noise and non-contact electromagnetic interference. Moreover, signal screening is not complex. Therefore, the EP response exhibits a favorable superiority in engineering application [26]. It possesses the important significance to be used to monitor the damage evolution of gas-bearing coal seams and provide an indication prior to forecasting dynamic disasters in mines.

5. Conclusions

During the loading and damaging process of gas-bearing coal, multi-information was measured and analyzed to obtain the research results as follow:

(1) Abundant EP signals are generated during the damage of gas-bearing coal under load. With the growth of stress, the damage of the specimen was aggravated, and EP was strengthened. Moreover, AE counts and amplitude increased, and crack expansion were exacerbated. When the specimen was subjected to local fracturing and stress suddenly changed, EP and AE tended to vary suddenly, and crack expansion was significant and constantly aggravated. The higher the stress level was and the greater the sudden change of stress was, the more dramatic the damage of the coal mass and therefore the greater the abnormal response of EP and AE. The EP response showed similar characteristics under different gas pressures and the presence of gas promoted the EP response. The changing trend and response characteristics of EP exhibit the stress state and reveal the damage-evolution process of the specimen.

(2) Under the coupling effect of stress and gas, the damage of the coal mass constantly aggravated. It caused internal cracks to constantly initiate, then propagate and finally converge, and connect, triggering the fracturing of the coal mass. Charge separation happened under the effects of crack expansion, friction effect between crack surface and coal matrixes, electron emission induced by stress concentration, etc. As a result, the EP response was triggered. Furthermore, the calculation method for EP is simplified with imaging method.

(3) After gas was fully absorbed by the coal mass, surface free energy of pores reduced, which caused the decline of intramolecular and intermolecular attractive forces. It led to expansion-induced deformation of coal and reduction of cohesion. This reduced the broken strength of the coal mass and increased deformation. Additionally, free-state gas entered large fractures in the coal mass under the effect of pore pressure. It had a splitting and expansion effect on fracture structure and weakened the friction resistance of crack surface. Therefore, the presence of gas promoted the crack expansion of the coal mass and friction effect to strengthen the EP effect. In addition, the electrokinetic effect generated due to the flow of free-state gas in pores also exerted a certain influence on the EP effect.

(4) At different loading stages, different factors dominated the EP response of gas-bearing coal. In the early stage of loading, the friction effect played a dominant part while crack expansion mainly appeared in the later period of loading. The electron emission was caused by stress concentration and the electrokinetic effect induced by gas flow both exhibited a certain effect during the whole loading stage. During the failure of the specimen, the EP rapidly rose to a maximum, so did the AE count. Moreover, signals showed a high amplitude and cracks rapidly expanded and ran through from the top to the bottom of the specimen. It led to the failure of gas-bearing coal finally. After the specimen was completely damaged, EP signals rapidly reduced and then stabilized. The abnormal characteristic of EP can be taken as an index for monitoring the stability of gas-bearing coal and warning the failure of the coal mass.

Author Contributions: Conceptualization and methodology: Z.L., E.W. and L.L.; analysis and investigation: Z.L. and Y.N.; original draft reparation, writing: Z.L. and Y.N.; review, editing and funding acquisition: Z.L., Y.N., E.W., L.L., H.W., A.M. and M. W.

Funding: This work was supported by National Natural Science Foundation of China (51674254, 51504244, 51514140), the State Key Research Development Program of China (Grant No. 2016YFC0801404, 2016YFC0801401), State Key Laboratory of Coal Resources and Safe Mining, CUMT (SKLCRSM15X03), and the Project Funded by the Priority Academic Program Development of Jiangsu Higher Education Institutions (PAPD).

Acknowledgments: We thank anonymous reviewers for their comments and suggestions to improve the manuscripts.

Conflicts of Interest: The authors declare no conflict of interest.

Abbreviations

a	Ultimate gas adsorption	r_{3ij}	Distances of imaging charges to the measurement point P
b	Molar gas constant	r_{4ij}	Distances of imaging charges to the measurement point P
K_{12}	Reflection coefficient of the dielectric ε_1 To ε_2	T	Absolute temperature
${K_{ab}}^i$	Reflection coefficients while i represent the superscripts of reflecting interface separately parallel to axes x and y	U_1	EP at the point P_1
${K_{ab}}^j$	Reflection coefficients while j represent the superscripts of reflecting interface separately parallel to axes x and y	U_2	EP at the point P_1
m	Another corresponding times of mirror imaging reflection	U_S	EP at the measurement point S
n	Corresponding times of mirror imaging reflection	V_m	Molar volume
P	Gas pressure	ε_1	Semi-infinite dielectrics of coal
q	Charge quantity of initial charge	ε_2	Semi-infinite dielectrics of gas
R	Molar gas constant	$\varnothing$	Equivalent pore coefficient
r	Distance of P_1 To point O	ρ_s	Apparent density
r_1	Distance of $O\prime$ to the point P_1	σ_e	Effective stress
r_2	Distance of O to the point P_2	$\sigma_e\prime$	Relative broken strength
r_{1ij}	Distances of imaging charges to the measurement point P	σ_L	Loading stress
r_{2ij}	Distances of imaging charges to the measurement point P		

References

1. Song, X.; Li, X.; Li, Z.; Zhang, Z.; Cheng, F.; Chen, P.; Liu, Y. Study on the characteristics of coal rock electromagnetic radiation (EMR) and the main influencing factors. *J. Appl. Geophys.* **2018**, *148*, 216–225. [CrossRef]
2. Yuan, L. Theory and practice of integrated coal production and gas extraction. *Int. J. Coal Sci. Technol.* **2015**, *2*, 3–11. [CrossRef]
3. Lu, S.; Li, L.; Cheng, Y.; Sa, Z.; Zhang, Y.; Yang, N. Mechanical failure mechanisms and forms of normal and deformed coal combination containing gas: Model development and analysis. *Eng. Fail. Anal.* **2017**, *80*, 241–252. [CrossRef]
4. Ma, C.; Wang, P.; Jiang, L.; Wang, C. Deformation and Control Countermeasure of Surrounding Rocks for Water-Dripping Roadway Below a Contiguous Seam Goaf. *Processes* **2018**, *6*, 77. [CrossRef]
5. Hebblewhite, B.; Galvin, J. A review of the geomechanics aspects of a double fatality coal burst at Austar Colliery in NSW, Australia in April 2014. *Int. J. Min. Sci. Technol.* **2017**, *27*, 3–7. [CrossRef]
6. Tang, Z.; Yang, S.; Wu, G. Occurrence Mechanism and Risk Assessment of Dynamic of Coal and Rock Disasters in the Low-Temperature Oxidation Process of a Coal-Bed Methane Reservoir. *Energy Fuels* **2017**, *31*, 3602–3609. [CrossRef]
7. Freund, F.; Sornette, D. Electro-magnetic earthquake bursts and critical rupture of peroxy bond networks in rocks. *Tectonophysics* **2007**, *431*, 33–47. [CrossRef]
8. Marland, S.; Merchant, A.; Rowson, N. Dielectric properties of coal. *Fuel* **2001**, *80*, 1839–1849. [CrossRef]
9. Lamich, D.; Marschalko, M.; Yilmaz, I.; Bednářová, P.; Niemiec, D.; Kubečka, K.; Mikulenka, V. Subsidence measurements in roads and implementation in land use plan optimisation in areas affected by deep coal mining. *Environ. Earth Sci.* **2016**, *75*, 69. [CrossRef]

10. Lu, P.; Li, P.; Chen, J.; Zhang, C.; Xue, J.; Yu, T. Gas drainage from different mine areas: Optimal placement of drainage systems for deep coal seams with high gas emissions. *Int. J. Coal Sci. Technol.* **2015**, *2*, 84–90. [CrossRef]
11. Pillalamarry, M.; Harpalani, S.; Liu, S. Gas diffusion behavior of coal and its impact on production from coalbed methane reservoirs. *Int. J. Coal Geol.* **2011**, *86*, 342–348. [CrossRef]
12. Mishchuk, N.; Ralston, J.; Fornasiero, D. Influence of dissolved gas on van der Waals forces between bubbles and particles. *J. Phys. Chem. A* **2002**, *106*, 689–696. [CrossRef]
13. Karacan, C.; Okandan, E. Adsorption and gas transport in coal microstructure: Investigation and evaluation by quantitative X-ray CT imaging. *Fuel* **2001**, *80*, 509–520. [CrossRef]
14. Archer, J.; Dobbs, M.; Aydin, A.; Reeves, H.; Prance, R. Measurement and correlation of acoustic emissions and pressure stimulated voltages in rock using an electric potential sensor. *Int. J. Rock Mech. Min. Sci.* **2016**, *89*, 26–33. [CrossRef]
15. Steffanson, M.; Bone, D.J. Apparatus and Method for Electromagnetic Radiation Sensing. U.S. Patent 9,851,256, 26 December 2017.
16. Enomoto, Y.; Shimamoto, T.; Tsutumi, A. Rapid electric charge fluctuation prior to rock fracturing: Its potential use for an immedi-ate earthquake precursor. In *Proceedings of International Workshop on Electromagneticphenomenarelat to Earthquake Prediction*; Hayakawa, M., Fujinawa, Y., Eds.; Terra Scientific Publishing Co.: Tokyo, Japan, 1993; pp. 64–65.
17. Freund, F.T.; Takeuchi, A.; Lau, B.W. Electric currents streaming out of stressed igneous rocks—A step towards understanding pre-earthquake low frequency EM emissions. *Phys. Chem. Earth Parts A/B/C* **2006**, *31*, 389–396. [CrossRef]
18. Vallianatos, F.; Triantis, D.; Tzanis, A.; Anastasiadis, C.; Stavrakas, I. Electric earthquake precursors: From laboratory results to field observations. *Phys. Chem. Earth Parts A/B/C* **2004**, *29*, 339–351. [CrossRef]
19. He, X.; Nie, B.; Chen, W.; Wang, E.; Dou, L.; Wang, Y.; Liu, M.; Hani, M. Research progress on electromagnetic radiation in gas-containing coal and rock fracture and its applications. *Saf. Sci.* **2012**, *50*, 728–735. [CrossRef]
20. Wang, X.; Liu, X.; Wang, E.; Li, X.; Zhang, X.; Zhang, C.; Kong, B. Experimental research of the AE responses and fracture evolution characteristics for sand-paraffin similar material. *Constr. Build. Mater.* **2017**, *132*, 446–456. [CrossRef]
21. Yoshida, S.; Clint, O.C.; Sammonds, P.R. Electric potential changes prior to shear fracture in dry and saturated rocks. *Geophys. Res. Lett.* **1998**, *25*, 1577–1580. [CrossRef]
22. Takeuchi, A.; Lau, B.W.; Freund, F.T. Current and surface potential induced by stress-activated positive holes in igneous rocks. *Phys. Chem. Earth Parts A/B/C* **2006**, *31*, 240–247. [CrossRef]
23. Wang, E.; Li, Z.; Liu, Z.; Li, Y.; Song, X. An Experimental Study on Surface Electric Potential of Loaded Coal. *Chin. J. Geophys.* **2009**, *52*, 641–649. [CrossRef]
24. Wang, E.Y.; Liu, X.F.; Li, Z.H.; Liu, Z.; He, X.Q. *Application of Electromagnetic Radiation (EMR) Technology in Monitoring and Warning of Coal and Rock Dynamic Disasters*; Springer International Publishing: Berlin, Germany, 2014; p. 4.
25. Li, Z.H.; Wang, E.Y.; He, X.Q. *Study on Theory and Mechanism of Surface Potential during Coal Fracture*; China University of Mining and Technology Press: Xuzhou, China, 2013.
26. Niu, Y.; Li, Z.; Kong, B.; Wang, E.; Lou, Q.; Qiu, L.; Kong, X.; Wang, J.; Dong, M.; Li, B. Similar simulation study on the characteristics of the electric potential response to coal mining. *J. Geophys. Eng.* **2017**, *15*, 42. [CrossRef]
27. Stavrakas, I.; Anastasiadis, C.; Triantis, D.; Vallianatos, F. Piezo stimulated currents in marble specimen: Precursory and concurrent-with-failure signals. *Nat. Hazards Earth Syst. Sci.* **2003**, *3*, 243–247. [CrossRef]
28. Aydin, A.; Dobbs, M.R.; Reeves, H.J.; Kirkham, M.P.; Graham, C.C. *Stress Induced Electric Field Measurements of Different Rock Lithology Using the Electric Potential Sensor*; American Rock Mechanics Association: New York, NY, USA, 2013.
29. Haas, A.K.; Revil, A.; Karaoulis, M.; Frash, L.; Hampton, J.; Gutierrez, M.; Mooney, M. Electric potential source localization reveals a borehole leak during hydraulic fracturing. *Geophysics* **2013**, *78*, D93–D113. [CrossRef]
30. Cai, Y.; Liu, D.; Mathews, J.P.; Pan, Z.; Elsworth, D.; Yao, Y.; Li, J.; Guo, X. Permeability evolution in fractured coal—Combining triaxial confinement with X-ray computed tomography, acoustic emission and ultrasonic techniques. *Int. J. Coal Geol.* **2014**, *122*, 91–104. [CrossRef]

31. Hou, W. Identification of coal and gangue by feed-forward neural network based on data analysis. *Int. J. Coal Prep. Util.* **2019**, *39*, 33–43. [CrossRef]
32. Kong, X.; Wang, E.; Hu, S.; Li, Z.; Liu, X.; Fang, B.; Zhan, T. Critical slowing down on acoustic emission characteristics of coal containing methane. *J. Nat. Gas Sci. Eng.* **2015**, *24*, 156–165. [CrossRef]
33. Lomas, H.; Roest, R.; Wells, A.; Wu, H.; Jiang, Z.; Sakurovs, R.; Stuart, R.; North, L.; Thorley, T.; Mahoney, M.R. Estimating coke fracture toughness using acoustic emissions and changes in coefficient of friction during scratch testing. *Fuel* **2018**, *226*, 564–572. [CrossRef]
34. Yin, G.Z.; Wang, D.K.; Zhang, D.M.; Huang, G. Solid-gas coupling dynamic model and numerical simulation of coal containing gas. *Chin. J. Geotech. Eng.* **2008**, *30*, 1430–1436.
35. Polesek-Karczewska, S. Estimation of the structure-related share of radiation heat transfer in a carbonised packed coal bed. *Fuel* **2017**, *195*, 243–252. [CrossRef]
36. Dokukin, M.; Olac-Vaw, R.; Guz, N.; Mitin, V.; Sokolov, I. Addressable photocharging of single quantum dots assisted with atomic force microscopy probe. *Appl. Phys. Lett.* **2009**, *95*, 173105. [CrossRef]
37. He, X.; Liu, X.; Nie, B.; Song, D. FTIR and Raman spectroscopy characterization of functional groups in various rank coals. *Fuel* **2017**, *206*, 555–563. [CrossRef]
38. Revil, A.; Ehouarne, L.; Thyreault, E. Tomography of self-potential anomalies of electrochemical nature. *Geophys. Res. Lett.* **2001**, *28*, 4363–4366. [CrossRef]
39. Patella, D. Introduction to ground surface self-potential tomography. *Geophys. Prospect.* **1997**, *45*, 653–681. [CrossRef]
40. Long, H.L.; Hao, J.Q. The Theoretical and Experimental Research on Self-Potential Tomography. *Chin. J. Geophys.* **2005**, *48*, 1408–1415. [CrossRef]
41. He, X.; Wang, E.; Lin, H. Coal deformation and fracture mechanism under pore gas action. *J. China Univ. Min. Technol.* **1996**, *25*, 6–11.
42. Liu, X.; Wang, X.; Wang, E.; Kong, X.; Zhang, C.; Liu, S.; Zhao, E. Effects of gas pressure on bursting liability of coal under uniaxial conditions. *J. Nat. Gas Sci. Eng.* **2017**, *39*, 90–100. [CrossRef]
43. Su, F.; Itakura, K.; Deguchi, G.; Ohga, K. Monitoring of coal fracturing in underground coal gasification by acoustic emission techniques. *Appl. Energy* **2017**, *189*, 142–156. [CrossRef]

processes

Article

Experimental Study on Mixed Mode Fracture Behavior of Sandstone under Water–Rock Interactions

Wen Hua [1,2], Jianxiong Li [1], Shiming Dong [1,2,*] and Xin Pan [1]

1 MOE Key Laboratory of Deep Earth Science and Engineering, College of Architecture and Environment, Sichuan University, Chengdu 610065, China; wenhua@scu.edu.cn (W.H.); jianxiongml@126.com (J.L.); pan-px@foxmail.com (X.P.)

2 Failure Mechanics and Engineering Disaster Prevention and Mitigation Key Laboratory of Sichuan Province, Sichuan University, Chengdu 610065, China

* Correspondence: smdong@scu.edu.cn; Tel.: +86-28-8541-6486

Received: 24 December 2018; Accepted: 26 January 2019; Published: 1 February 2019

Abstract: Water–rock interactions can significantly deteriorate the physical and mechanical properties of rocks, and it has been identified as one of the significant factors influencing the stability and safety of structures in rock–soil engineering. In this study, the fracture mechanical properties of sandstone under periodic water–rock interactions and long-term immersion have been studied with central cracked Brazilian disk specimens. The degradation mechanism of water–rock interactions was also studied using a scanning electron microscope (SEM). Finally, the generalized maximum tangential stress and generalized maximum tangential strain criteria were adopted to evaluate the experimental results. The results show that periodic water–rock interactions can remarkably affect the fracture resistance of sandstone. With the increase in the number of cycles, the pure mode I, pure mode II, and mixed mode fracture toughness decreases greatly, however, the values of K_{If}/K_{IC} and K_{IIf}/K_{IC} decrease slightly. Furthermore, the fracture resistance of sandstone influenced by cyclic wetting–drying is more significant than long-term immersion. Moreover, the fracture criteria, which considers the effect of T-stress, can reproduce the test results very well.

Keywords: water–rock interaction; degradation mechanism; mixed mode fracture resistance; fracture criteria; T-stress

1. Introduction

In engineering practices, such as tunneling, mining, oil and gas exploration, and excavation, the deformation and failure of rock materials is generally accompanied with water. These failures are strongly associated with water–rock interactions, such that many geological disasters like earthquakes, debris flow, landslides, and ground subsidence, have occurred in recent years. Water–rock interactions can significantly deteriorate the physical and mechanical properties of rocks, and they have been identified as one of the significant factors influencing the stability and safety of structures in rock–soil engineering and tunnel engineering [1–3].

In recent years, several studies on water content affecting the mechanical properties of different types of rocks, including uniaxial compressive strength, cohesion resistance, tensile strength, and fracture resistance, have been conducted by researchers [4–12]. The results showed that even if the water content had a small increment, it might make the strength and stiffness of the rocks reduce remarkably, which means that the water content could significantly degrade the mechanical properties of rock materials. Furthermore, the subcritical crack growth of sandstone and igneous rock affected by the relative humidity was studied by Nara et al. [13,14]. Tang et al. [15] performed numerous fracture tests with three-point bending samples so as to investigate the change of rules of the mode I fracture resistance of different types of rocks after soaking in a hydrochemical solution. Li et al. [16] also studied the deformation and strength properties of rock in an acid solution, and proposed a chemical damage model to predict the degradation of the rock strength and deformability.

In addition to the short or long term immersion, rock masses may also be subjected to periodic drying and wetting because of rain or changes in the groundwater level. Therefore, several experimental studies on periodic wetting–drying affecting the physical and mechanical properties of rocks have been performed in recent decades [17–26]. In addition, Yuan and Ma [27], and Zhou et al. [28,29] studied the dynamic mechanical behavior of selected sandstones under cyclic wetting–drying. The cyclic wetting–drying and freezing–thawing effects on some of the mechanical parameters of the selected rock materials were studied by Hale and Shakoor [30], Khanlari and Abdilor [31], and Özbek [32]. Moreover, Li and Zhang [33] proposed a damage model, which considered the wetting and drying effect, to characterize the variation in the uniaxial compressive stress–strain for sandstone. Liu et al. [34] analyzed the strength deterioration and the variations in the micro-structures of the shaly sandstone samples under wetting–drying cycles by using the PFC2D (particle flow code in two-dimensions) software. More recently, Zhao et al. [35] performed Brazilian disk splitting tests to study the influence of long-term soaking and periodic wetting and drying on the tensile strength of a selected rock.

It was found that the deterioration effects of periodic wetting–drying on different mechanical parameters for the same rock were significantly different. However, there was a common viewpoint that the strength of the rock would degrade after periodic wetting–drying, and the damage to the rocks was irreversible and progressive. The mechanical parameters in the aforementioned studies were mainly compressive strength, cohesion resistance, tensile strength, shear strength, and so on. However, studies on the fracture mechanical properties of the rocks affected by periodic water–rock interactions and long-term immersion are scarce [1–3,9,15,36]. It has not yet been incompletely understood how the mixed mode fracture behavior of rocks is influenced by periodic water–rock interactions and long-term immersion. In particular, the differences between the degradation effects of these two occurrences have not been entirely clarified.

The purpose of this study is to examine how the mixed mode fracture resistance of sandstone is influenced by periodic water–rock interactions and long-term immersion using central cracked Brazilian disk (CCBD) specimens. The degradation mechanism of water-rock interactions was also studied using a scanning electron microscope (SEM) in this research. Furthermore, the generalized maximum tangential stress (GMTS) and generalized maximum tangential strain (GMTSN) criteria were adopted to evaluate the experimental results.

2. Materials and Methods

There are many experimental methods and test configurations to determine the fracture resistance of brittle or quasi-brittle materials. Some well-known test configurations are the CCBD specimen, semi-circular bend specimen, compact tension shear specimen, edge cracked four-point bend beam specimen, notched three-point bend specimen, inclined center crack plate, and so on. Among these specimens, the CCBD specimen is frequently used to conduct mixed mode fracture tests on rock materials, owing to the existence of analytical expressions for stress intensity factors and the easy realization of complete mode combinations of mode I and mode II [37–39]. In addition, the cracked chevron notched Brazilian disc specimen, which is one of the CCBD specimens, has been suggested by the International Society for Rock Mechanics (ISRM) for determining the pure mode I fracture toughness of rocks [40]. In this research, the CCBD specimen is employed for measuring the mixed mode fracture resistance of sandstone under cyclic wetting–drying and long-term immersion conditions.

2.1. Stress Intensity Factors for the CCBD Specimens

Figure 1 represents a diagram of the CCBD specimen with thickness t and diameter D subjected to radial load P. The initial crack length in the Brazilian disk is represented by $2a$, and the loading angle is β. The analytical expressions of the mode I and mode II stress intensity factors K_I and K_{II} were deduced by Dong et al. [41], as follows:

$$K_I = \sigma\sqrt{\pi a} \cdot Y_I = \sigma\sqrt{\pi a}\left[f_{11} + 2\sum_{i=1}^{n} A_{1i} f_{1i} \alpha^{2(i-1)}\right] \quad (1)$$

$$K_{II} = \sigma\sqrt{\pi a} \cdot Y_{II} = \sigma\sqrt{\pi a} \cdot \left[2\sum_{i=1}^{n} A_{2i} f_{2i} \alpha^{2(i-1)}\right] \quad (2)$$

where $\alpha = a/R$; $\sigma = P/(\pi t R)$; and Y_I, Y_{II} are the normalized forms of K_I and K_{II}, respectively. The coefficients f_{ji} and A_{ji} (i = 1, 2, 3, $\cdots n$; j = 1, 2) in the above expressions can be found in the paper by Dong et al. [41]. According to the authors of [1,41], the sum coefficient n = 100 is adopted to ensure the calculating precision of stress intensity factors in this paper.

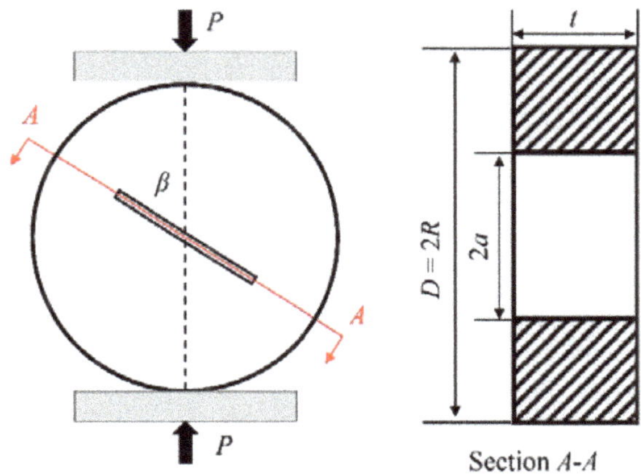

Figure 1. Diagram of a central cracked Brazilian disk (CCBD) specimen under compression.

2.2. Sample Preparation and Testing Procedure

In this study, all of the sandstone specimens were extracted from the same rock mass with fine particles and good homogeneity, which came from Ziyang City, Sichuan Province, China. The main mineralogical compositions of the selected sandstone were quartz, clinochlore, albite, and gordonite, according to the X-ray diffraction (XRD, PANalytical B.V., Almelo, The Netherlands) analyses. The sandstone had a moisture content of 1.20%, dry density of 2.24 g/cm^3, water absorption of 6.18%, Poisson's ratio of 0.21, and tensile strength of 2.68 MPa. In addition, the microstructure of the intact sandstone was observed using SEM (JEOL, Tokyo, Japan). It showed many pores, faults, and micro-cracks in the intact sandstone (Figure 2), which enabled the occurrence of water–rock interactions [3].

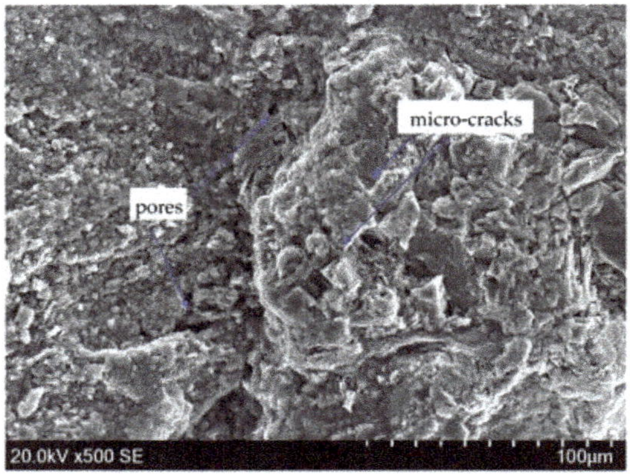

Figure 2. The typical microstructure photograph of the intact sandstone sample.

The CCBD specimen was employed for measuring the mixed mode fracture resistance of sandstone under water–rock interactions. The nominal diameter and thickness of the CCBD specimen were 75 mm and 25 mm, respectively. According to the analysis of Dong [42], a relative crack length a/R ranging from 0.4 to 0.6, is recommended to conduct a fracture test with a CCBD specimen. Therefore, a relative crack length of 0.5 for all the CCBD specimens was considered. According to Dong et al. [41], the pure mode I crack could be achieved when the loading angle was 0°. Moreover, the loading angle was 23° for a pure mode II crack when a/R = 0.5. Hence, when adjusting the loading angle from 0° to 23°, the complete mode combinations of mode I and mode II can be conveniently obtained. In this study, four different scenarios with the CCBD specimens loaded by 0°, 10°, 15°, and 23° were considered for the fracture tests.

Water–rock interactions were performed under the following two different treatment conditions: cyclic wetting–drying and long-term immersion. In each wetting–drying cycle, the sandstone samples were completely submerged in water for two days in order to reach the saturated state at atmospheric conditions. The specimens were then removed and heated in an oven at 105 °C for one day, and subsequently, the samples were cooled to an atmospheric temperature [3,22,30]. However, for the long-term immersion, it was considered as a single cycle when the sandstone samples were completely soaked in water for three days. The sandstone specimens were subjected to three and seven cycles with these two treatment approaches, respectively. After the samples underwent the required number of cycles, they were once again completely submerged in water for two days for fracture tests. It should be noted that the intact sandstone specimens soaked in water for only two days were considered to have undergone no cycle. In addition, the intact sandstone samples that did not undergo any treatment were also measured in this study.

The fracture experiments were performed with an electronic universal material testing machine (Shenzhen Reger Instrument, Shenzhen, China) for loading at a rate of 0.05 mm/min for all of the samples [43–46]. At least four CCBD specimens of sandstone in each group were successfully measured. During each test, the load–deformation behavior of the sandstone specimen from initial loading to final failure were completely recorded by the testing system. Figure 3 represents the failure modes of some of the sandstone specimens after undergoing water–rock interactions. It is observed that the crack propagation path is always along the initial crack line when the loading angle is 0° (pure mode I). However, when the CCBD specimens are under pure mode II or mixed mode loading, the crack extends in the direction deviating from the initial crack.

Figure 3. Failure modes of some sandstone specimens.

3. Experimental Results and Discussion

3.1. Water–Rock Interactions Affecting Fracture Resistance of Sandstone

Many fracture tests have been performed on sandstone under periodic water–rock interactions and long-term immersion with CCBD specimens. The fracture toughness of the sandstone samples determined from Equations (1) and (2) are listed in Table 1. Moreover, the averaged values of the pure mode I, pure mode II, and mixed mode fracture toughness of this sandstone are shown in Figure 4. It is observed that the mean values of the pure mode II fracture toughness are always greater than those of pure mode I for the sandstone samples under the same treatment conditions. This is very similar to the experimental results obtained with the CCBD specimens reported in the literature [43,47–49]. In addition, the pure mode I, pure mode II, and mixed mode fracture toughness of the intact sandstone are always larger than those of the sandstone after undergoing water–rock interactions. This is mainly because the sandstone samples are in a water-saturated state after undergoing water–rock interactions, and the water content within the sandstone is very high, which has been identified as an important factor for lowering the strength of rocks [8–10].

As observed in Figure 4, the pure mode I, pure mode II, and mixed mode fracture toughness of the sandstone samples decrease with the increase in the number of cycles under both cyclic wetting–drying and long-term immersion conditions. For the sandstone samples under cyclic wetting–drying, when the samples undergo none, three, and seven cycles, the mean values of the pure mode I and pure mode II fracture toughness are 0.232, 0.174, and 0.148 MPa $m^{0.5}$; and 0.341, 0.246, and 0.204 MPa $m^{0.5}$, respectively. The pure mode I fracture toughness decreases by 25.0% for the three-cycle case and 36.2% for the seven-cycle case compared with those of the saturated sandstone (no cycle). The pure mode II fracture toughness decreases by 27.9% and 40.2% after cyclic wetting–drying for the three-cycle and seven-cycle cases, respectively. This indicates that the pure mode II fracture toughness influenced by cyclic wetting–drying is more significant than the pure mode I fracture toughness.

For the sandstone samples under long-term immersion conditions, when the samples undergo three and seven cycles, the pure mode I fracture toughness is 0.187 and 0.173 MPa $m^{0.5}$, respectively, which decrease by 19.4% and 25.4%, respectively, after water–rock interactions. However, the pure mode II fracture toughness corresponding to three and seven cycles are 0.267 and 0.247 MPa $m^{0.5}$, respectively, which decrease by 21.7% and 27.6%, respectively. These values show that the effect on the pure mode II fracture toughness under long-term immersion is more significant, which is similar to the treatment condition of cyclic wetting–drying. After three cycles, the pure mode I and pure mode II fracture toughness of sandstone under long-term immersion decrease by 19.4% and 21.7%, respectively, which are smaller than the corresponding values of 25.0% and 27.9%, respectively, under cyclic wetting–drying. A similar phenomena is also observed when the number of cycles is seven. In conclusion, the fracture resistance of sandstone is more significantly affected by cyclic wetting–drying than long-term immersion.

Table 1. Test values of fracture toughness for sandstone under different water–rock interaction conditions.

Treatment Conditions	Number of Cycles	Fracture Toughness K/MPa m$^{0.5}$					
		K_{IC} ($\beta = 0°$)	K_{If} ($\beta = 10°$)	K_{IIf} ($\beta = 10°$)	K_{If} ($\beta = 15°$)	K_{IIf} ($\beta = 15°$)	K_{IIC} ($\beta = 23°$)
Intact	-	0.332	0.264	0.298	0.188	0.443	0.502
		0.327	0.275	0.312	0.171	0.404	0.509
		0.348	0.272	0.308	0.185	0.434	0.498
		0.344	0.257	0.291	0.177	0.415	0.485
Saturation	0	0.224	0.181	0.206	0.119	0.285	0.337
		0.231	0.174	0.199	0.127	0.304	0.372
		0.245	0.199	0.228	0.122	0.291	0.327
		0.229	0.191	0.218	0.113	0.268	0.326
Cyclic wetting and drying	3	0.173	0.136	0.152	0.089	0.211	0.246
		0.180	0.136	0.155	0.085	0.203	0.239
		0.164	0.141	0.158	0.088	0.212	0.255
		0.177	0.133	0.151	0.092	0.223	0.245
	7	0.145	0.115	0.131	0.077	0.185	0.198
		0.143	0.114	0.130	0.072	0.172	0.210
		0.156	0.111	0.127	0.080	0.190	0.215
		0.148	0.117	0.133	0.073	0.174	0.193
Long-term immersion	3	0.180	0.148	0.168	0.091	0.215	0.250
		0.183	0.144	0.162	0.097	0.232	0.290
		0.195	0.134	0.151	0.094	0.221	0.258
		0.188	0.144	0.162	0.099	0.236	0.268
	7	0.180	0.133	0.151	0.086	0.204	0.235
		0.184	0.123	0.139	0.082	0.197	0.251
		0.159	0.126	0.143	0.093	0.224	0.245
		0.171	0.140	0.159	0.089	0.212	0.255

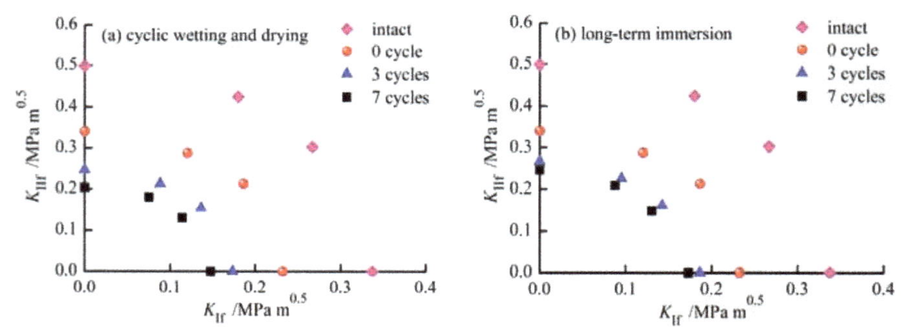

Figure 4. Fracture toughness of sandstone under different water–rock interaction conditions.

In order to compare the mixed mode fracture toughness of sandstone conveniently, the concept of effective fracture toughness (K_{eff}) is introduced, and it can be calculated by the following formula [3,50,51]:

$$K_{eff} = \sqrt{K_I^2 + K_{II}^2} \tag{3}$$

Figure 5 shows the effective fracture toughness of the sandstone samples under long-term immersion and cyclic wetting–drying conditions. As seen from Figure 5, the values of K_{eff} for sandstone under four different loading angles always decrease with the increasing number of cycles in both cyclic wetting–drying and long-term immersion conditions. When the sandstone samples with a loading angle of 10° undergo none, three, and seven wetting–drying cycles, the effective fracture toughness values K_{eff} are 0.283, 0.206, and 0.173 MPa $m^{0.5}$, respectively, which are reduced by 27.3% and 38.8%, respectively, compared with the saturated sandstone. However, for the sandstone samples loaded by 15°, the values of K_{eff} decrease by 26.0% and 37.3% after undergoing three and seven cycles, respectively. Furthermore, after seven wetting–drying cycles, when the loading angle changes from 0° (pure mode I), 10°, 15°, and 23° (pure mode II), the effective fracture toughness K_{eff} decreases by 36.2%, 38.8%, 37.3%, and 40.2%, respectively. It shows that the degradation effect on pure mode II and mixed mode fracture toughness is more significant, which has also been reported by Hua et al. [3] for another type of rock.

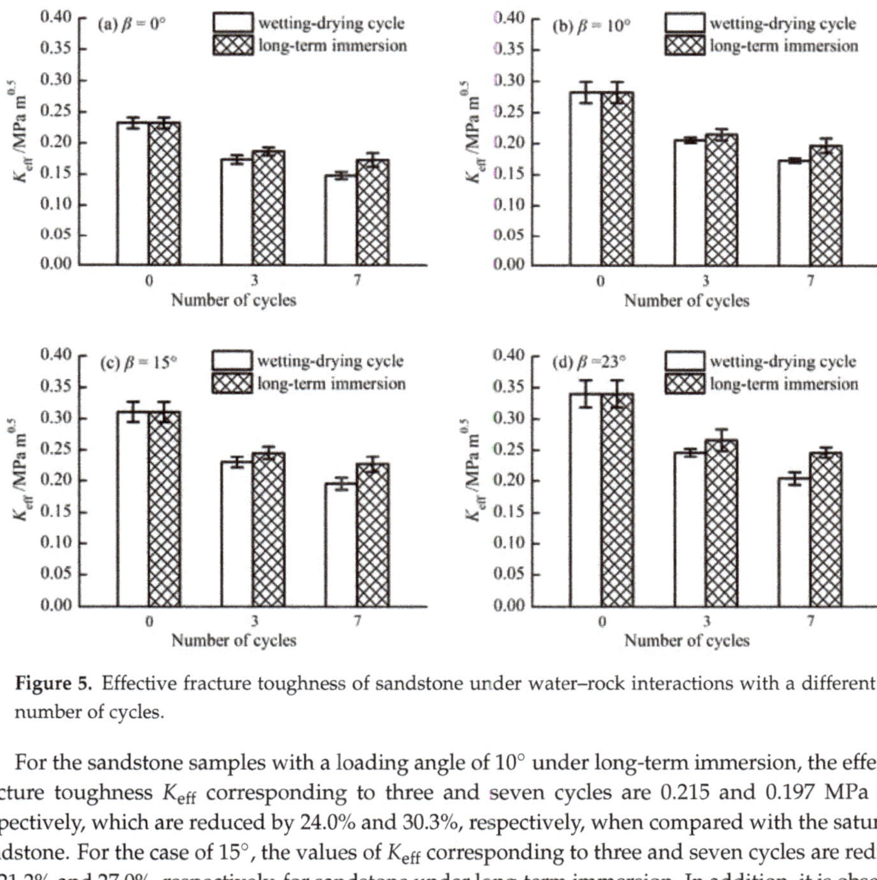

Figure 5. Effective fracture toughness of sandstone under water–rock interactions with a different number of cycles.

For the sandstone samples with a loading angle of 10° under long-term immersion, the effective fracture toughness K_{eff} corresponding to three and seven cycles are 0.215 and 0.197 MPa $m^{0.5}$, respectively, which are reduced by 24.0% and 30.3%, respectively, when compared with the saturated sandstone. For the case of 15°, the values of K_{eff} corresponding to three and seven cycles are reduced by 21.2% and 27.0%, respectively, for sandstone under long-term immersion. In addition, it is observed that the degradation effect of long-term immersion on pure mode I fracture toughness is the weakest, which is very similar to the cyclic wetting–drying process.

Moreover, in Figure 5, the values of K_{eff} for sandstone under long-term immersion are always larger than that under cyclic wetting–drying for the same number of cycles (except for no cycle). For example, when the sandstone is loaded by 10°, the values of K_{eff} for sandstone under long-term immersion and cyclic wetting–drying are 0.215 and 0.206 MPa $m^{0.5}$, respectively, after undergoing three cycles each. Furthermore, for the case of 15°, the reduction of K_{eff} for sandstone undergoing seven wetting–drying cycles is approximately 37.3%, which is larger than the corresponding value of 27.0% under long-term immersion. This behavior is similar for the other loading angles of 0° and 23°.

Hence, we can conclude that the fracture resistance of sandstone influenced by cyclic wetting–drying is more significant when compared with the case of long-term immersion.

3.2. The Degradation Mechanism of Water–Rock Interactions

As we known, the mechanical properties of rocks are closely associated with the mineralogical compositions and microstructure [3,28,34]. SEM is used to observe the changing regularity of the microstructure of the sandstone samples under the two treatment conditions of long-term immersion and cyclic wetting–drying. Figure 6 shows the photomicrographs at ×500 magnification for the sandstone samples undergoing three and seven cycles under the two treatment conditions. As shown in Figure 6a, the pores between the mineral particles increased gradually, and some particle structures tended to be loose. The new secondary micro-cracks appeared along the boundary of the mineral particles after three cycles. Furthermore, the exfoliated particles increased and the pore size of the sandstone sample increased gradually after seven cycles, which is observed in Figure 6b. When the samples were immersed in an aqueous solution, the water molecules moved gradually towards the interior of the rock mass owing to the pores, micro-cracks, faults, and the contact surfaces between the mineral particles. Furthermore, the aqueous solution could dissolve some soluble minerals and cements of the sandstone, increase the porosity (or water content), and consequently reduce the cohesive force between the mineral particles.

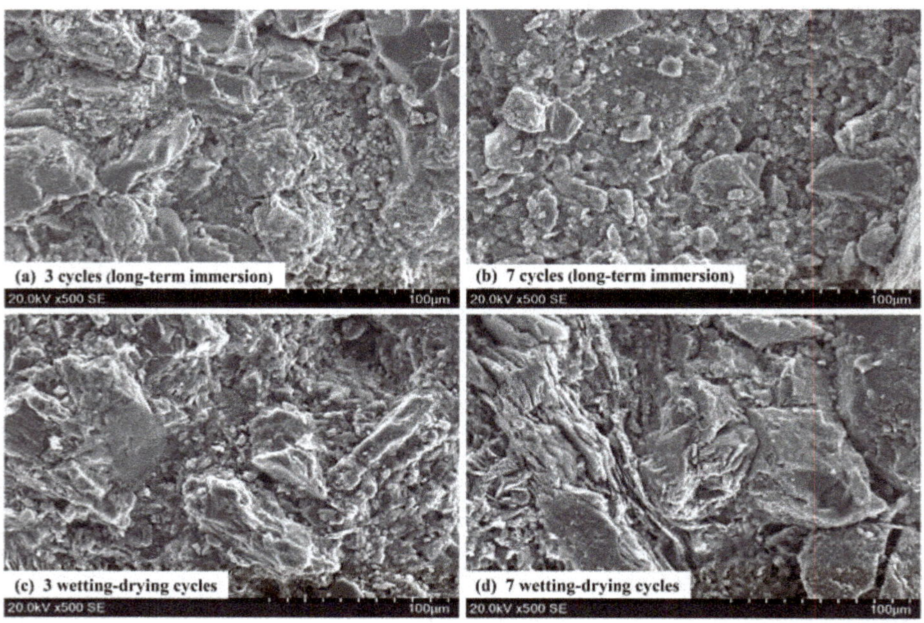

Figure 6. Scanning electron microscope (SEM) photomicrographs of sandstone samples after water–rock interactions.

The treatment procedure of cyclic wetting–drying in this experiment actually included not only wetting and drying, but also heating and cooling processes. After soaking, the sandstone samples were heated in an oven. Heating the sandstone samples directly might induce thermal damage to the sandstone, and this thermal damage to the rock was very large because of the higher water content in the sandstone after the wetting procedure. The presence of water in the rocks is considered a precondition for the occurrence of heating and cooling degradation [30]. The efficacy of heating degradation of the moisture-laden rock stems, in part, from the ability of the heated water to mobilize the soluble minerals towards the rock surface, resulting in the increased porosity of the rock. In addition, owing to the different thermal expansion coefficients of various minerals of sandstone, the mineral particles expanded and squeezed with each other when the sandstone samples were heated, however, the mineral particles contracted during the cooling process [3]. These processes led to the propagation of the micro-cracks and pores between the mineral particles, or produced new secondary micro-cracks (see Figure 6c,d).

It can be also found from Figure 6 that the size and quantity of the micro-cracks (or pores) between the mineral particles under the treatment condition of cyclic wetting–drying increased more significantly when compared with long-term immersion with the same number of cycles. This indicates that the degradation effect of cyclic wetting–drying was more remarkable than the long-term immersion, which is completely consistent with the test results. In short, the water–rock interaction was a gradual process of physical and chemical damage accumulation, which dissolved some soluble minerals and cements, reduced the adhesive forces between the mineral particles, made the micro-cracks and pores between the mineral particles propagate or produce new secondary micro-cracks, and consequently changed the microstructure of the sandstone. These changes in the microstructure characteristics were the basic reasons for the weakening of the fracture resistance of the sandstone.

4. Analysis According to the GMTSN Criterion

The GMTSN criterion was developed by considering the influence of T-stress on the maximum tangential strain (MTSN) criterion [52], which was primitively put forward by Ayatollahi and Abbasi [53]. Mirsayar [54] proposed a similar strain-based criterion and employed it to evaluate the fracture mechanical behavior of polymethyl methacrylate. Hua et al. [50] extended the GMTSN criterion by considering both plane strain and plane stress conditions, and made use of it to evaluate the brittle fracture behaviors of the CCBD specimens made of graphite and rock materials. More recently, the GMTSN criterion has been frequently employed to evaluate the onset of the fracture of different materials [51,55–57]. Based on the GMTSN criterion [50,53,54], the crack initiation angle (θ_0) can be calculated using the following formula:

$$Y_{\mathrm{I}}[(5-8\kappa)\sin\frac{\theta_0}{2}-3\sin\frac{3\theta_0}{2}]+Y_{\mathrm{II}}[(5-8\kappa)\cos\frac{\theta_0}{2}-9\cos\frac{3\theta_0}{2}]+8\sqrt{\frac{2r_0}{a}}T^*\sin 2\theta_0=0 \tag{4}$$

Here, Y_{I} and Y_{II} are the normalized forms of K_{I} and K_{II}, respectively; $T^* = T/\sigma$ is the normalized T-stress; a is half of the crack length; and r_0 is the critical distance around the crack tip. Moreover, κ is a coefficient for distinguishing the plane strain or plane stress condition, which is associated with the Poisson's ratio ν. For the plane stress problems, the value of the coefficient κ is equal to $1/(1+\nu)$, whereas its value is $1-\nu$ for the plane strain problems [50]. Similarly, based on the GMTSN criterion [50], the onset of brittle fracture can be determined from the following equations:

$$\frac{K_{\mathrm{If}}}{K_{\mathrm{IC}}}=\frac{(2\kappa-1)+\frac{T_{\mathrm{I}}^*}{Y_{\mathrm{I0}}}\sqrt{\frac{2r_0}{a}}(\kappa-1)}{f_1(\theta_0)+\frac{Y_{\mathrm{II}}}{Y_{\mathrm{I}}}f_2(\theta_0)+\frac{T^*}{Y_{\mathrm{I}}}\sqrt{\frac{2r_0}{a}}f_3(\theta_0)} \tag{5}$$

$$\frac{K_{\mathrm{IIf}}}{K_{\mathrm{IC}}}=\frac{(2\kappa-1)+\frac{T_{\mathrm{I}}^*}{Y_{\mathrm{I0}}}\sqrt{\frac{2r_0}{a}}(\kappa-1)}{\frac{Y_{\mathrm{I}}}{Y_{\mathrm{II}}}f_1(\theta_0)+f_2(\theta_0)+\frac{T^*}{Y_{\mathrm{II}}}\sqrt{\frac{2r_0}{a}}f_3(\theta_0)} \tag{6}$$

with

$$\begin{cases} f_1(\theta)=\frac{1}{4}[(8\kappa-5)\cos\frac{\theta}{2}+\cos\frac{3\theta}{2}] \\ f_2(\theta)=-\frac{1}{4}[(8\kappa-5)\sin\frac{\theta}{2}+3\sin\frac{3\theta}{2}] \\ f_3(\theta)=(\kappa-\cos^2\theta) \end{cases} \tag{7}$$

where Y_{I0} and $T_I{}^*$ are the normalized stress intensity factor and T-stress for pure mode I loading, respectively. For the CCBD specimens, the values of Y_I and Y_{II} can be easily calculated according to Equations (1) and (2), and the T-stress can also be obtained by an analytical formula developed by Hua et al. [58]. In this study, when the CCBD specimens with a/R = 0.5 are loaded by 0°, 10°, 15°, and 23°, the corresponding dimensionless T-stresses are −5.790, −4.634, −3.465, and −1.483, respectively. If the values of Y_I, Y_{II}, T^*, r_0, and ν are known, one can obtain the crack initiation angle (θ_0) from Equation (4), and the corresponding ratios of K_{If}/K_{IC} and K_{IIf}/K_{IC} can be also calculated according to Equations (5) and (6). It is noteworthy that the strain-based GMTSN criterion can be simplified to the stress-based GMTS criterion when the Poisson's ratio is v = 0 [50].

In order to use the GMTSN criterion, it is necessary to know the value of the critical distance around the crack tip r_0. This value is associated with the tensile strength σ_t and pure mode I fracture resistance K_{IC}, and can be calculated as follows [3,48,59]:

$$r_0 = \frac{1}{2\pi}\left(\frac{K_{IC}}{\sigma_t}\right)^2 \tag{8}$$

For obtaining the critical distance r_0, the tensile strength of this sandstone was measured by the Brazilian splitting method. Before measurement, the treatment process of the Brazilian disk samples was completely consistent with that of the CCBD specimens. At least four Brazilian disk samples were successfully tested in each group, and the averaged values of the tensile strength σ_t are summarized in Table 2. Furthermore, the values of the critical distance r_0 determined according to Equation (8) are also shown in Table 2, which would be used to compute the values of θ_0, K_{If}/K_{IC}, and K_{IIf}/K_{IC} according to Equations (4)–(6).

Table 2. The values of critical distance r_0 for sandstone under different water–rock interaction conditions.

Treatment Conditions	Number of Cycles	Tensile Strength σ_t/MPa	Mode I Fracture Toughness K_{IC}/MPa m$^{0.5}$	Critical Distance r_0/mm
Intact	-	2.677	0.338	2.54
saturation	0	1.897	0.232	2.38
Cyclic wetting and drying	3	1.483	0.174	2.17
	7	1.312	0.148	2.03
Long-term immersion	3	1.621	0.187	2.12
	7	1.536	0.173	2.02

Figure 7 represents the comparison of the experimental values and theoretical predictions based on the fracture criteria. As observed from Figure 7, the test values are always larger than the theoretical predictions according to the conventional maximum tangential stress (MTS) and the MTSN criteria. This is because the MTSN and MTS criteria only consider the stress intensity factors and neglect the influence of T-stress. Moreover, the test results agree well with the theoretical predictions according to the GMTSN and GMTS criteria, which consider the influence of T-stress. For example, when the sandstone samples with a loading angle of 10° undergo none, three, and seven wetting-drying cycles, the test values of K_{If}/K_{IC} are 0.803, 0.784, and 0.772, and the values of K_{IIf}/K_{IC} are 0.917, 0.885, and 0.880, respectively. Based on the GMTSN criterion, the theoretical predictions of K_{If}/K_{IC} are 0.815, 0.807, and 0.802, and those of K_{IIf}/K_{IC} are 0.935, 0.926, and 0.919, corresponding to none, three, and seven wetting–drying cycles, respectively. According to the GMTS criterion, the theoretical predictions of K_{If}/K_{IC} are 0.783, 0.776, and 0.772, and those of K_{IIf}/K_{IC} are 0.898, 0.890, and 0.885, corresponding to none, three, and seven wetting–drying cycles, respectively. The relative errors between the experimental values and the theoretical predictions based on the GMTSN and GMTS criteria are less than 5.0%. Furthermore, for the sandstone samples subjected to long-term immersion for three cycles, the measured values of K_{If}/K_{IC} and K_{IIf}/K_{IC} are 0.762 and 0.860, respectively. According to the GMTSN and GMTS criteria, the theoretical values of K_{If}/K_{IC} and K_{IIf}/K_{IC} are

0.806 and 0.775, and 0.924 and 0.889, respectively. These results show that both the GMTSN and GMTS criteria can evaluate the experimental results very well. In addition, the experimental results indicate that the values of $K_{\mathrm{If}}/K_{\mathrm{IC}}$ and $K_{\mathrm{IIf}}/K_{\mathrm{IC}}$ decrease slightly with the increasing number of cycles under the conditions of cyclic wetting–drying and long-term immersion.

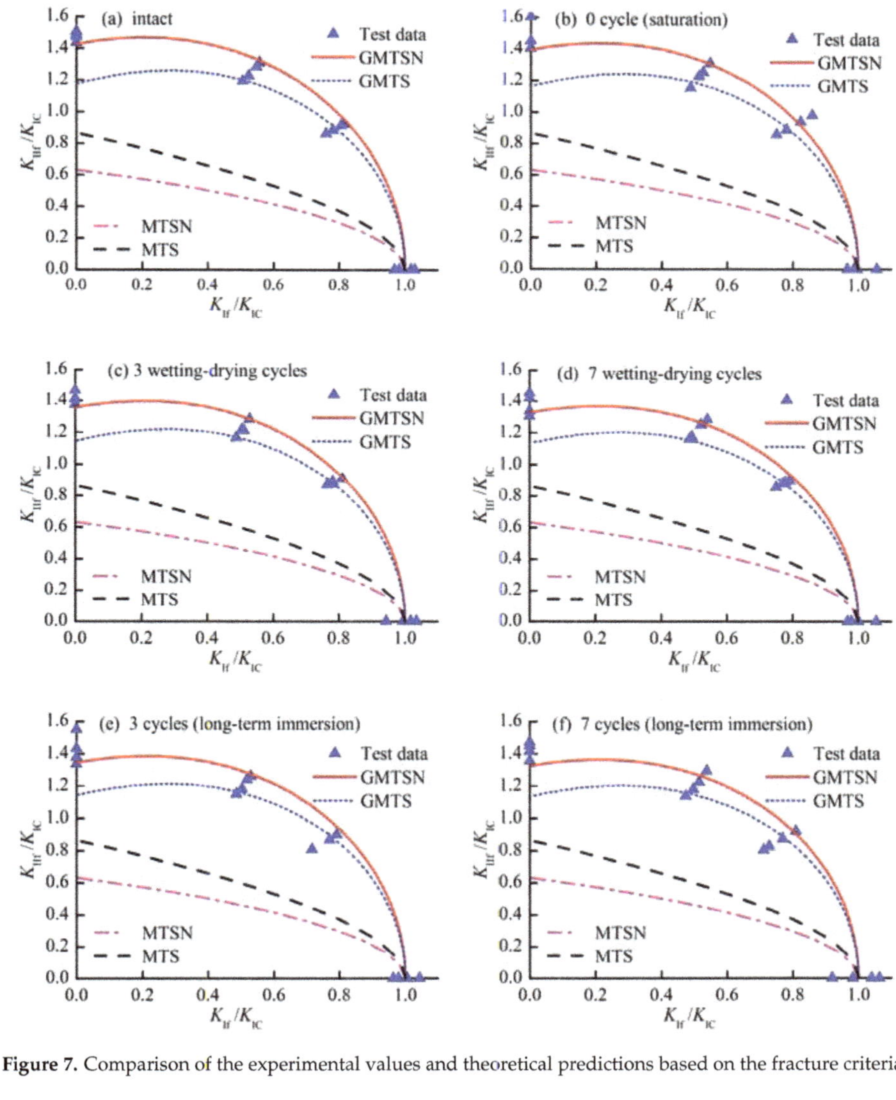

Figure 7. Comparison of the experimental values and theoretical predictions based on the fracture criteria.

As shown in Figure 7, variations still exist between the experimental values and theoretical predictions of $K_{\mathrm{IIC}}/K_{\mathrm{IC}}$ based on the GMTS criterion when approaching the pure mode II loading condition. To discuss the pure mode II case, the theoretical and experimental values of $K_{\mathrm{IIC}}/K_{\mathrm{IC}}$ are summarized in Table 3. We observed that the theoretical predictions according to the GMTSN criterion are closer to the testing values as compared with the GMTS criterion, which indicates that the GMTSN criterion can provide a better prediction for pure mode II. For instance, the ratio of $K_{\mathrm{IIC}}/K_{\mathrm{IC}}$ is 1.468 for the water-saturated sandstone, and 1.395 and 1.163 according to the GMTSN and GMTS criteria, respectively. This phenomenon can be also observed for sandstone under other treatment conditions.

Further study shows that the maximum relative error between the experimental values and theoretical predictions according to the GMTSN criterion is less than 7.0%. Compared with the GMTS criterion, the GMTSN criterion considers not only the influences of multiple stress components (σ_r, σ_θ, and σ_z) around the crack tip, but also the Poisson's ratio. Nevertheless, the GMTS criterion only takes into account the tangential stress component at the crack tip and ignores the effect of Poisson's ratio. Aliha and Saghafi [60] reported that the stress intensity factors and T-stress are significantly affected by Poisson's ratio. Moreover, Poisson's ratio is one of the important parameters for evaluating the onset of the fracture of brittle materials in the GMTSN criterion, and is the key factor to distinguish between stress- or strain-based criteria [50].

Table 3. The theoretical and experimental values of K_{IIC}/K_{IC} for sandstone under different treatment conditions. GMTSN—generalized maximum tangential strain; GMTS—generalized maximum tangential stress; MTSN—maximum tangential strain; MTS—maximum tangential stress.

Treatment Conditions	Number of Cycles	K_{IIC}/K_{IC}				
		Test Data	GMTSN	GMTS	MTSN	MTS
Intact	-	1.475	1.424	1.176	0.630	0.866
saturation	0	1.468	1.395	1.163	0.630	0.866
Cyclic wetting and drying	3	1.415	1.354	1.147	0.630	0.866
	7	1.378	1.327	1.136	0.630	0.866
Long-term immersion	3	1.425	1.344	1.143	0.630	0.866
	7	1.424	1.324	1.135	0.630	0.866

5. Conclusions

(1) Water–rock interactions can significantly influence the mixed mode fracture resistance of sandstone. The pure mode I, pure mode II, and mixed mode fracture toughness of sandstone decrease with the increase in the number of cycles.

(2) The values of K_{If}/K_{IC} and K_{IIf}/K_{IC} decrease slightly with the increasing number of cycles under the conditions of cyclic wetting–drying and long-term immersion.

(3) The degradation effect of cyclic wetting–drying on the fracture resistance of rock materials is more significant when compared with long-term immersion.

(4) Both the GMTSN and GMTS criteria, which consider the effect of T-stress, can evaluate the experimental results very well. Moreover, the GMTSN criterion may be better than the GMTS criterion for pure mode II loading.

Author Contributions: W.H. and S.D. conceived the experiments; W.H., J.L., and X.P. designed and performed the experiments; W.H. and J.L. analyzed the data; W.H. and S.D. wrote the manuscript.

Funding: This research was funded by the National Natural Science Foundation of China (grant No. 11872042), the Sichuan Science and Technology Program (No. 2018JY0024 & 2019YJ0156), and the Postdoctoral Research Foundation of Sichuan University.

Conflicts of Interest: The authors declare no conflict of interest.

References

1. Hua, W.; Dong, S.; Li, Y.; Xu, J.; Wang, Q. The influence of cyclic wetting and drying on the fracture toughness of sandstone. *Int. J. Rock Mech. Min. Sci.* **2015**, *78*, 331–335. [CrossRef]
2. Hua, W.; Dong, S.; Li, Y.; Wang, Q. Effect of cyclic wetting and drying on the pure mode II fracture toughness of sandstone. *Eng. Fract. Mech.* **2016**, *153*, 143–150. [CrossRef]
3. Hua, W.; Dong, S.; Peng, F.; Li, K.; Wang, Q. Experimental investigation on the effect of wetting-drying cycles on mixed mode fracture toughness of sandstone. *Int. J. Rock Mech. Min. Sci.* **2017**, *93*, 242–249. [CrossRef]
4. Erguler, Z.A.; Ulusay, R. Water-induced variations in mechanical properties of clay-bearing rocks. *Int. J. Rock Mech. Min. Sci.* **2009**, *46*, 355–370. [CrossRef]

5. Török, Á.; Vásárhelyi, B. The influence of fabric and water content on selected rock mechanical parameters of travertine, examples from hungary. *Eng. Geol.* **2010**, *115*, 237–245. [CrossRef]
6. Vásárhelyi, B.; Ván, P. Influence of water content on the strength of rock. *Eng. Geol.* **2006**, *84*, 70–74. [CrossRef]
7. Wang, M.; Cao, P.; Li, R.; Fan, X. Effect of water absorption ratio on tensile strength of red sandstone and morphological analysis of fracture surfaces. *J. Cent. South Univ.* **2017**, *24*, 1647–1653. [CrossRef]
8. Wong, L.N.Y.; Maruvanchery, V.; Liu, G. Water effects on rock strength and stiffness degradation. *Acta Geotech.* **2016**, *11*, 713–737. [CrossRef]
9. Zhou, Z.; Cai, X.; Ma, D.; Cao, W.; Chen, L.; Zhou, J. Effects of water content on fracture and mechanical behavior of sandstone with a low clay mineral content. *Eng. Fract. Mech.* **2018**, *193*, 47–65. [CrossRef]
10. Vasarhelyi, B. Statistical analysis of the influence of water content on the strength of the miocene limestone. *Rock Mech. Rock Eng.* **2004**, *38*, 69–76. [CrossRef]
11. Roy, D.G.; Singh, T.N.; Kodikara, J.; Das, R. Effect of water saturation on the fracture and mechanical properties of sedimentary rocks. *Rock Mech. Rock Eng.* **2017**, *50*, 2585–2600. [CrossRef]
12. Kim, E.; Changani, H. Effect of water saturation and loading rate on the mechanical properties of red and buff sandstones. *Int. J. Rock Mech. Min. Sci.* **2016**, *88*, 23–28. [CrossRef]
13. Nara, Y.; Morimoto, K.; Hiroyoshi, N.; Yoneda, T.; Kaneko, K.; Benson, P.M. Influence of relative humidity on fracture toughness of rock: Implications for subcritical crack growth. *Int. J. Solids Struct.* **2012**, *49*, 2471–2481. [CrossRef]
14. Nara, Y.; Morimoto, K.; Yoneda, T.; Hiroyoshi, N.; Kaneko, K. Effects of humidity and temperature on subcritical crack growth in sandstone. *Int. J. Solids Struct.* **2011**, *48*, 1130–1140. [CrossRef]
15. Tang, L.; Zhang, P.; Wang, S. Testing study on effects of chemical action of aqueous solution on crack propagation in rock. *Chin. J. Rock Mech. Eng.* **2002**, *21*, 822–827.
16. Li, N.; Zhu, Y.; Su, B.; Gunter, S. A chemical damage model of sandstone in acid solution. *Int. J. Rock Mech. Min. Sci.* **2003**, *40*, 243–249. [CrossRef]
17. Zhang, Z.; Jiang, Q.; Zhou, C.; Liu, X. Strength and failure characteristics of jurassic red-bed sandstone under cyclic wetting–drying conditions. *Geophys. J. Int.* **2014**, *198*, 1034–1044. [CrossRef]
18. Qin, Z.; Chen, X.; Fu, H. Damage features of altered rock subjected to drying-wetting cycles. *Adv. Civ. Eng.* **2018**, *2018*, 5170832. [CrossRef]
19. Razouki, S.S.; Salem, B.M. Frequency effect of cyclic soaking and drying on the resilient modulus of gypsum-rich roadbed sand. *Transp. Geotech.* **2016**, *7*, 13–20. [CrossRef]
20. Nara, Y.; Tanaka, M.; Harui, T. Evaluating long-term strength of rock under changing environments from air to water. *Eng. Fract. Mech.* **2017**, *178*, 201–211. [CrossRef]
21. Jia, X.; Zhao, E. Experimental study on the influence of wetting-drying cycle on the strength characteristics of expansive rock. In Proceedings of the International Conference on Energy and Environmental Protection, Zhuhai, China, 29–30 June 2017; Advances in Engineering Research: Zhuhai, China, 2017; pp. 242–245.
22. Sumner, P.D.; Loubser, M.J. Experimental sandstone weathering using different wetting and drying moisture amplitudes. *Earth Surf. Proc. Land.* **2008**, *33*, 985–990. [CrossRef]
23. Zhao, Y.; Ren, S.; Jiang, D.; Liu, R.; Wua, J.; Jiang, X. Influence of wetting-drying cycles on the pore structure and mechanical properties of mudstone from simian mountain. *Constr. Build. Mater.* **2018**, *191*, 923–931. [CrossRef]
24. Zhu, D.; Jing, H.; Yin, Q.; Han, G. Experimental study on the damage of granite by acoustic emission after cyclic heating and cooling with circulating water. *Processes* **2018**, *6*, 101. [CrossRef]
25. Huang, S.; Wang, J.; Qiu, Z.; Kang, K. Effects of cyclic wetting-drying conditions on elastic modulus and compressive strength of sandstone and mudstone. *Processes* **2018**, *6*, 234. [CrossRef]
26. Sun, Q.; Zhang, Y. Combined effects of salt, cyclic wetting and drying cycles on the physical and mechanical properties of sandstone. *Eng. Geol.* **2019**, *248*, 70–79. [CrossRef]
27. Yuan, P.; Ma, Q. Energy analyses of uniaxial compressive tests for coalmine sandstone after cyclic wetting and drying. In Proceedings of the 2014 ISRM European Regional Symposium on Rock Engineering and Rock Mechanics: Structures in and on Rock Masses, EUROCK 2014, London, UK, 27–29 May 2014; pp. 101–105.
28. Zhou, Z.; Cai, X.; Chen, L.; Cao, W.; Zhao, Y.; Xiong, C. Influence of cyclic wetting and drying on physical and dynamic compressive properties of sandstone. *Eng. Geol.* **2017**, *220*, 1–12. [CrossRef]

29. Zhou, Z.; Cai, X.; Ma, D.; Chen, L.; Wang, S.; Tan, L. Dynamic tensile properties of sandstone subjected to wetting and drying cycles. *Constr. Build. Mater.* **2018**, *182*, 215–232. [CrossRef]
30. Hale, P.A.; Shakoor, A. A laboratory investigation of the effects of cyclic heating and cooling, wetting and drying, and freezing and thawing on the compressive strength of selected sandstones. *Environ. Eng. Geosci.* **2003**, *9*, 117–130. [CrossRef]
31. Khanlari, G.; Abdilor, Y. Influence of wet–dry, freeze–thaw, and heat–cool cycles on the physical and mechanical properties of upper red sandstones in central Iran. *Bull. Eng. Geol. Environ.* **2014**, *74*, 1287–1300. [CrossRef]
32. Özbek, A. Investigation of the effects of wetting–drying and freezing–thawing cycles on some physical and mechanical properties of selected ignimbrites. *Bull. Eng. Geol. Environ.* **2014**, *73*, 595–609. [CrossRef]
33. Li, K.; Zhang, Y. Damage constitutive model for rock considering cyclic drying-wetting effect and its experimental validation. *Electron. J. Geotech. Eng.* **2016**, *21*, 277–286.
34. Liu, X.; Jin, M.; Li, D.; Zhang, L. Strength deterioration of a shaly sandstone under dry–wet cycles: A case study from the three gorges reservoir in China. *Bull. Eng. Geol. Environ.* **2017**, *S3*, 1–15. [CrossRef]
35. Zhao, Z.; Yang, J.; Zhang, D.; Peng, H. Effects of wetting and cyclic wetting–drying on tensile strength of sandstone with a low clay mineral content. *Rock Mech. Rock Eng.* **2016**, *50*, 485–491. [CrossRef]
36. Hou, P.; Gao, F.; Gao, Y.N.; Yang, Y.G.; Cai, C.Z. Effect of pore pressure distribution on fracture behavior of sandstone in nitrogen fracturing. *Energ. Explor. Exploit.* **2017**, *35*, 609–626. [CrossRef]
37. Atkinson, C.; Smelser, R.E.; Sanchez, J. Combined mode fracture via the cracked Brazilian disk test. *Int. J. Fract.* **1982**, *18*, 279–291. [CrossRef]
38. Hua, W.; Xu, J.; Dong, S.; Song, J.; Wang, Q. Effect of confining pressure on stress intensity factors for cracked Brazilian disk. *Int. J. Appl. Mech.* **2015**, *7*, 1550051. [CrossRef]
39. Meng, T.; Bao, X.; Zhao, J.; Hu, Y. Study of mixed mode fracture toughness and fracture characteristic in gypsum rock under brine saturation. *Environ. Earth Sci.* **2018**, *77*. [CrossRef]
40. Fowell, R.J. Suggested method for determining mode I fracture toughness using cracked chevron notched Brazilian disc (CCNBD) specimens. *Int. J. Rock Mech. Min. Sci. Geomech. Abstr.* **1995**, *32*, 57–64. [CrossRef]
41. Dong, S.; Wang, Y.; Xia, Y. Stress intensity factors for central cracked circular disk subjected to compression. *Eng. Fract. Mech.* **2004**, *71*, 1135–1148. [CrossRef]
42. Dong, S.M. Theoretical analysis of the effects of relative crack length and loading angle on the experimental results for cracked Brazilian disk testing. *Eng. Fract. Mech.* **2008**, *75*, 2575–2581. [CrossRef]
43. Hua, W.; Dong, S.; Xu, J. Experimental study on the fracture toughness of rust stone under mixed mode loading condition. *Mater. Res. Innov.* **2015**, *19*, 531–536. [CrossRef]
44. Khan, K.; Al-Shayea, N.A. Effect of speciment geometry testing method on mixed mode fracture toughness of limestone rock from Saudi Arabia. *Rock Mech. Rock Eng.* **2000**, *33*, 179–206. [CrossRef]
45. Aliha, M.R.M.; Ayatollahi, M.R. Mixed mode I/II brittle fracture evaluation of marble using SCB specimen. *Proc. Eng.* **2011**, *10*, 311–318. [CrossRef]
46. Hua, W.; Dong, S.; Fan, Y.; Pan, X.; Wang, Q. Investigation on the correlation of mode II fracture toughness of sandstone with tensile strength. *Eng. Fract. Mech.* **2017**, *184*, 249–258. [CrossRef]
47. Al-Shayea, N.A.; Khan, K.; Abduljauwad, S.N. Effects of confining pressure and temperature on mixed-mode (I–II) fracture toughness of a limestone rock. *Int. J. Rock Mech. Min. Sci.* **2000**, *37*, 629–643. [CrossRef]
48. Hou, C.; Wang, Z.; Liang, W.; Li, J.; Wang, Z. Determination of fracture parameters in center cracked circular discs of concrete under diametral loading: A numerical analysis and experimental results. *Theor. Appl. Fract. Mech.* **2016**, *85*, 355–366. [CrossRef]
49. Ayatollahi, M.R.; Sistaninia, M. Mode II fracture study of rocks using brazilian disk specimens. *Int. J. Rock Mech. Min. Sci.* **2011**, *48*, 819–826. [CrossRef]
50. Hua, W.; Dong, S.; Pan, X.; Wang, Q. Mixed mode fracture analysis of CCBD specimens based on the extended maximum tangential strain criterion. *Fatigue Fract. Eng. Mater. Struct.* **2017**, *40*, 2118–2127. [CrossRef]
51. Mirsayar, M.M. On fracture analysis of dental restorative materials under combined tensile-shear loading. *Theor. Appl. Fract. Mech.* **2018**, *93*, 170–176. [CrossRef]
52. Chang, K.J. On the maximum strain criterion-a new approach to the angled crack problem. *Eng. Fract. Mech.* **1981**, *14*, 107–124. [CrossRef]

53. Ayatollahi, M.R.; Abbasi, H. Prediction of fracture using a strain based mechanism of crack growth. *Build. Res. J.* **2001**, *49*, 167–180.
54. Mirsayar, M.M. Mixed mode fracture analysis using extended maximum tangential strain criterion. *Mater. Des.* **2015**, *86*, 941–947. [CrossRef]
55. Wei, M.D.; Dai, F.; Xu, N.W.; Liu, Y.; Zhao, T. A novel chevron notched short rod bend method for measuring the mode I fracture toughness of rocks. *Eng. Fract. Mech.* **2018**, *190*, 1–15. [CrossRef]
56. Mirsayar, M.M.; Razmi, A.; Aliha, M.R.M.; Berto, F. EMTSN criterion for evaluating mixed mode I/II crack propagation in rock materials. *Eng. Fract. Mech.* **2018**, *190*, 186–197. [CrossRef]
57. Ghouli, S.; Ayatollahi, M.R.; Bushroa, A.R. Fracture characterization of ceria partially stabilized zirconia using the gmtsn criterion. *Eng. Fract. Mech.* **2018**, *199*, 647–657. [CrossRef]
58. Hua, W.; Li, Y.; Dong, S.; Li, N.; Wang, Q. T-stress for a centrally cracked brazilian disk under confining pressure. *Eng. Fract. Mech.* **2015**, *149*, 37–44. [CrossRef]
59. Ayatollahi, M.R.; Aliha, M.R.M. On the use of brazilian disc specimen for calculating mixed mode I–II fracture toughness of rock materials. *Eng. Fract. Mech.* **2008**, *75*, 4631–4641. [CrossRef]
60. Aliha, M.R.M.; Saghafi, H. The effects of thickness and poisson's ratio on 3D mixed-mode fracture. *Eng. Fract. Mech.* **2013**, *98*, 15–28. [CrossRef]

processes

Article

The Effects of Backfill Mining on Strata Movement Rule and Water Inrush: A Case Study

Jian Hao [1,2], Yongkui Shi [1,*], Jiahui Lin [1,*], Xin Wang [1,*] and Hongchun Xia [3]

1 State Key Laboratory of Mining Disaster Prevention and Control Co-founded by Shandong Province and the Ministry of Science and Technology, Shandong University of Science and Technology, Qingdao 266590, China; mkaqjs@163.com

2 Shandong Tai'an Huaining Group, Taian 271400, China

3 College of Civil and Architectural Engineering, Dalian University, Dalian 116622, China; xhch@dlu.edu.cn

* Correspondence: shiyongkui@163.com (Y.S.); m17854258373_2@163.com (J.L.); woxingwangao@163.com (X.W.); Tel.: +86-178-542-58373 (J.L.)

Received: 9 December 2018; Accepted: 22 January 2019; Published: 29 January 2019

Abstract: Backfill mining is widely used to control strata movement and improve the stress environment in China's coal mines. In the present study, the effects of backfill mining on strata movement and water inrush were studied based on a case study conducted in Caozhuang Coal Mine. The in-situ investigation measured abutment pressure distribution (APD), roof floor displacement (RFD), and vertical stress in the backfill area. Results are as follows: (i) The range and peak of APD, RFD, and vertical stress in the backfill area are smaller than in traditional longwall mining with the caving method. (ii) Backfill mining could change the movement form and amplitude of overburden and improve the ground pressure environment. (iii) Floor failure depth (FFD) is much smaller in backfill mining. Backfill mining can be an effective method for floor water inrush prevention.

Keywords: water inrush prevention; backfill mining; strata movement; ground pressure; floor failure depth

1. Introduction

In past decades, there has been a growing demand for coal resources in China. Hence, the depth at which coal deposits are extracted has increased considerably [1,2]. Deep coal mining has inherent risks associated with it due to in-situ stress and hydraulic damage conditions produced by overburden pressure, tectonic movements, and pressurized water in aquifers underlying the floor strata [3–5]. Historically, longwall mining with the caving method has been widely adopted in China. After removal of the mined panel, the overburden is divided into three zones: The caving zone, the fracture zone, and the slow subsidence zone [4,6]. These zones are demarcated according to differences in the magnitude of strata movement. Meanwhile, abutment pressure forms around the coal panel due to the movement of the overburden. Abutment pressure and strata movement may result in floor failure, which can induce water inrush.

Backfilling, an effective way to control strata movement induced by underground mining activities has been widely adopted in China [7–9]. In backfill mining, the goaf is filled with backfill materials, thus supporting part of the overburden stress and changing its movement rule and stress distribution. Therefore, there are great differences in strata movement and stress transfer between backfill mining and traditional longwall mining with the caving method. The problem of strata movement and stress transfer has been well studied in the past. Zhang et al. [10] described ground pressure behaviors in gauge backfill panels by theoretical analysis. Liu et al. [11] studied strata movement with a similar material experiment. Stresses in backfilled stopes were evaluated using numerical simulations [12,13].

However, only a few previous studies have investigated the deformation and stress response in the backfill area due to a lack of monitoring equipment.

To find out the strata movement rule and its influence on water inrush, in the present study, the surrounding rock's response to backfill mining was measured through an in-situ investigation conducted in Caozhuang Coal Mine. The backfill system, materials, and process were introduced. In field monitoring, multiple factors related with strata movement in backfill mining were considered, including abutment pressure distribution (APD), roof-floor displacement (RFD), vertical stress, and floor failure depth (FFD). Results indicate that backfill mining might improve the stress environment and decrease water inrush risks.

2. Study Area

The Feicheng mining area, located in Taian, Shandong Province, China, is famous for water inrush disasters and abundance of Ordovician limestone (Figures 1 and 2). Since the 1960s, 283 mine accidents have occurred in the Feicheng mining area. It is thought that 65% of recoverable reserves are lying above Ordovician water but are still seriously threatened by Ordovician water [14,15].

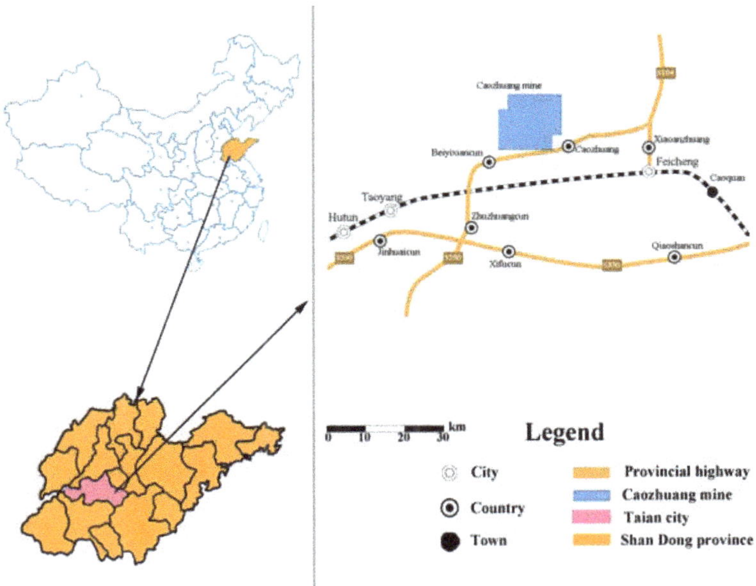

Figure 1. The location of Caozhuang Coal Mine.

The target coal seam is the #8 coal seam in Caozhuang Coal Mine CO., LTD of Feicheng Bureau of Shandong Energy Group CO., LTD. The #8 coal seam, excavated by coal panel 81006, has a depth of 550 m and belongs to the lower coal groups of the Northern China coal field. The coal seam has a mean thickness of 1.96 m, dips at an angle of 12–29°, and has a protodyakonov coefficient of 1.5. The immediate roof is the No. 4 limestone with a thickness of 5.3 m, and the floor is fine sandstone with a thickness of 4.78 m. The stratigraphic column of the mine field is shown in Figure 3. There are two water aquifers beneath the floor. The No. 5 limestone aquifer, with a bursting coefficient of 0.18 Mpa/m before mining, is 38 m away from the floor. The Ordovician limestone, with a bursting coefficient of 0.11 Mpa/m, is 61 m away from the floor. Additionally, faulted structures in the mine field are very complicated. The coal panel is severely threatened by the No. 5 and Ordovician limestone aquifers. Hence, the water-inrushing risks should not be overlooked.

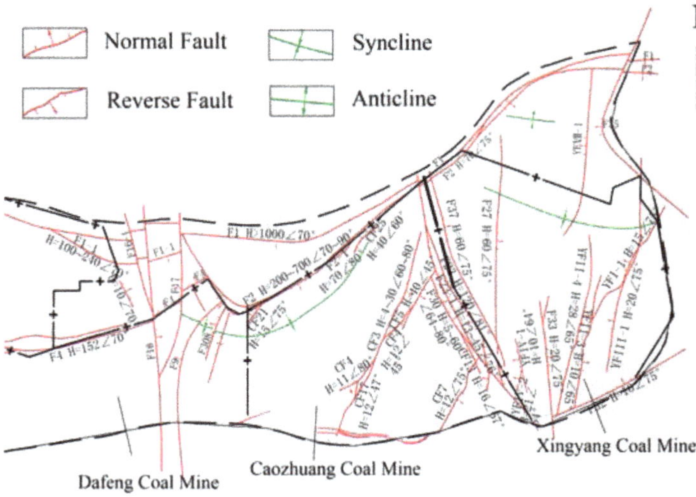

Figure 2. Geological map of Caozhuang Coal Mine.

Lithology	Thickness(m)
No. 4 limestone	5.3
No. 8 coal	1.4
Fine sandstone	4.68
marlstone	8.12
No. 9 coal	1.40
Siltstone	1.00
No. 10-1 coal	0.48
Siltstone	1.10
No. 10-2 coal	1.96
Clay rock	9.60
Limestone	1.50
No. 11 coal	0.46
Fine sandstone	7.60
No. 5 limestone	10.20
Clay rock	13.60
Ordovician limestone	800

Figure 3. Strata histogram of the mine field.

Water inrush disasters have had a great impact on coal extraction, and are not easily controlled by conventional water prevention and control methods. To prevent water inrushing disasters, non-pillar backfill mining is adopted in coal panel 81006 to pre-control the failure degree of the floor and prevent fracture due to water inrush. Longwall panels and backfill mining were designed respectively in coal panel 81006. For best performance of the grout-resistance wall and assuring successful backfill, the coal panel was divided into 4 sections from the inner to outer sections.

3. The Backfill System and Process

3.1. Materials and Methods

Paste-like backfill material is utilized in coal panel 81006 of Caozhuang Coal Mine. The material is comprised of cement, gangue powder, and coal ash, with a ratio of 1.8:6:6. The mass percentage of the backfilling slurry is 60%, and the specific gravity is 1.56. The backfill uniaxial compressive strength with time is shown in Figure 4. The backfill material production process is shown in Figure 5.

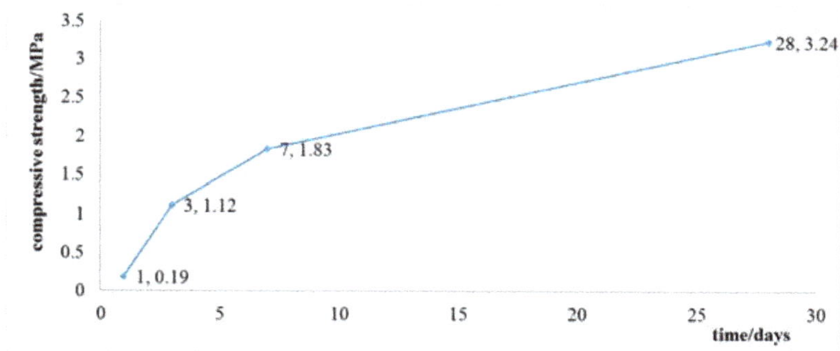

Figure 4. Backfill mass strength with time.

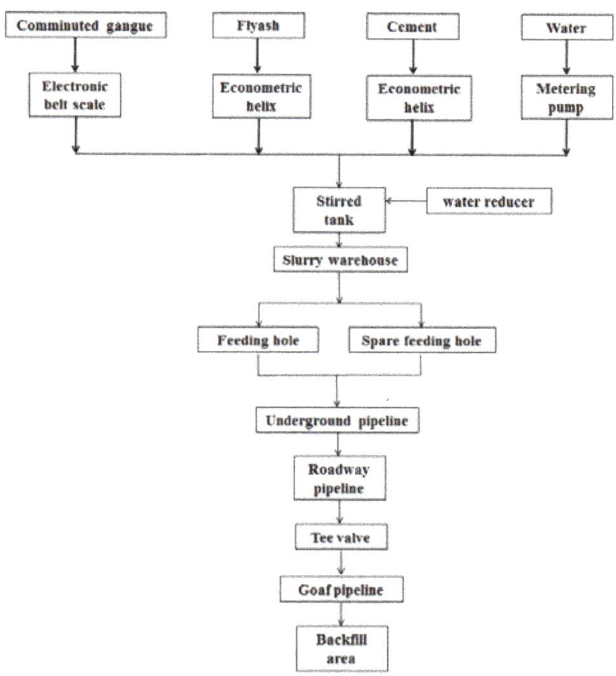

Figure 5. Diagram of the backfill material production process.

While the backfill system runs, the backfill slurry flows into the goaf behind the coal panels through feeding holes and a filling pipeline. The elevations of the feeding hole openings on the ground surface and the tail underground are +114.4 m and −240 m, respectively. The hole depth is 354.4 m. The elevation of the coal panel is −397.7 m to about −360 m, and the length of the backfill pipeline is 1850 m. The altitude difference from the feeding hole head to coal panels is 512 m, and the ratio of the pipeline length to altitude difference is 3.6. The diagram of the backfilling pipelines is shown in Figure 6.

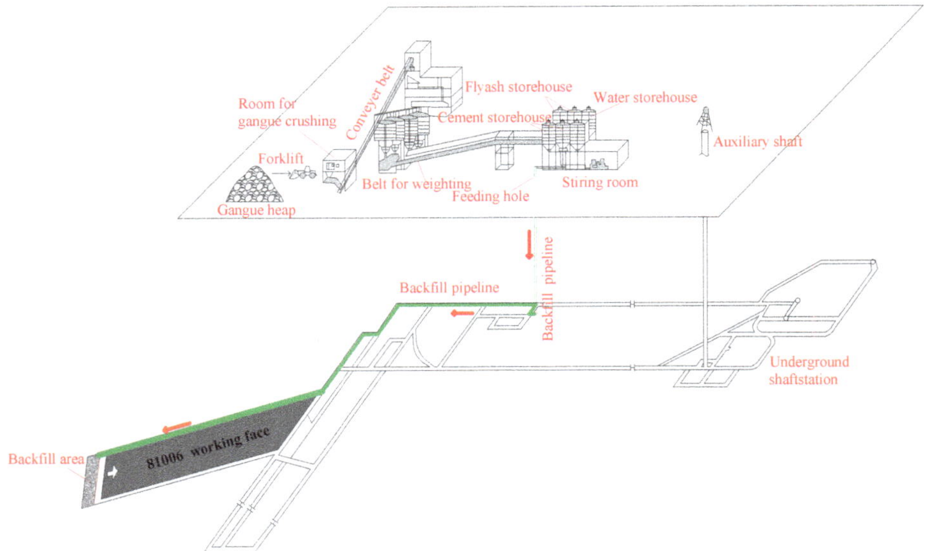

Figure 6. Diagram of backfilling pipelines.

3.2. Backfill System

The system consists of surface parts and underground parts. The backfill system has the characteristics of simple operation, real-time monitoring, tube-plugging reduction, and suitable transportation intensity. Material feeding, stirring, and transferring are controlled automatically in the backfill system. The surface system has 6 parts: The material production system, the material storage system, the water, and power supply system, the automatic control and metering system, the monitoring and communicating system, and the emergency system. The underground system has 2 parts, including the pipeline delivery and slurry-resistance systems in the coal panel. Upward mining was adopted in the longwall coal panel. Therefore, while the backfill system ran, the backfill material could flow down to the goaf. The coal panel layout is shown in Figure 7.

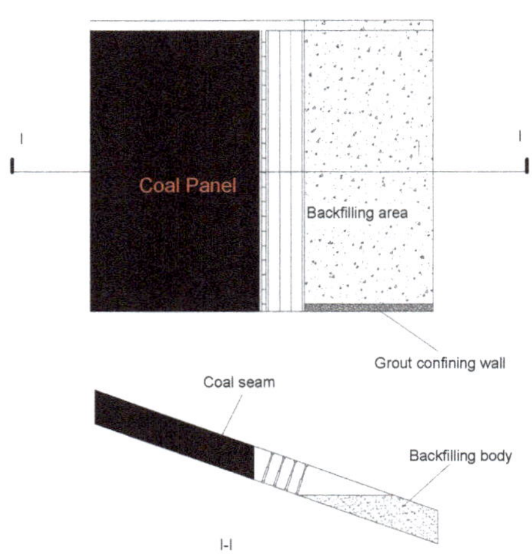

Figure 7. Overview of mining and filling in the coal panel.

4. Field Monitoring Equipment and Layout

4.1. Surrounding Rock Responding Monitoring Equipment in Backfill Coal Panel

4.1.1. Introduction

To evaluate the effect of backfill mining, a dynamic monitoring system consisting of several meters was installed in the coal and backfill area. These meters were the MA15Z coal mass stress meter, MA650 roof displacement meter, and MA60 backfill stress meter (all shown in Figure 8), manufactured by Xian Xinyuan Company in China. Monitoring data consisted of APD beyond coal panels, the RFD inside backfill areas, and the vertical stress inside backfill areas. The MA15Z mine-specialized coal mass stress meters have a wide measuring range of 10 MPa and high accuracy of 0.01 MPa. They were able to monitor APD in front of the coal wall. For best performance, the MA15Z should be installed in a borehole ahead of the working face prior to mining.

(**a**) The MA15Z mine-specialized coal stress meter.

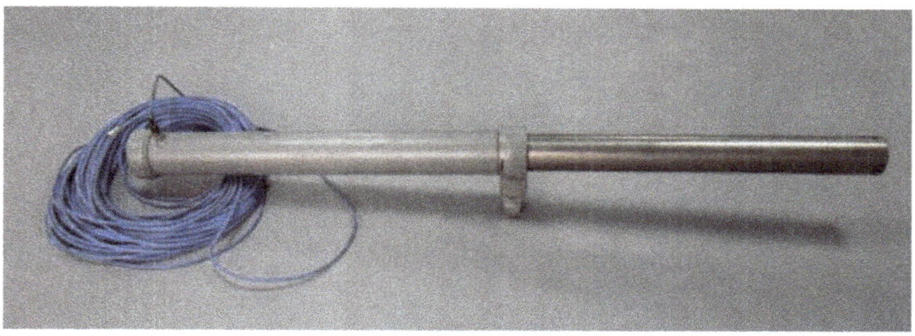

(**b**) The MA650 mine-specialized roof displacement meter.

(**c**) The MA60 backfill stress meter.

Figure 8. The ground pressure monitoring sensor in backfill mining.

The MA650 roof displacement meter was installed in the backfill area and measured RFD there. As the roof compressed, the MA650's internal resistance bridge circuit calculated the compression and sent the data to a collector in real time. The MA60 backfill stress meter measured the vertical stress inside the backfill area and then sent data to a collector through communication lines.

4.1.2. Monitoring Equipment Layout

As the working face advanced, the monitoring system received data during the whole process. The MA15Z mine-specialized coal stress meters were installed into the coal mass before mining; the others were installed in the backfill area. The meters' layout is shown in Figure 9.

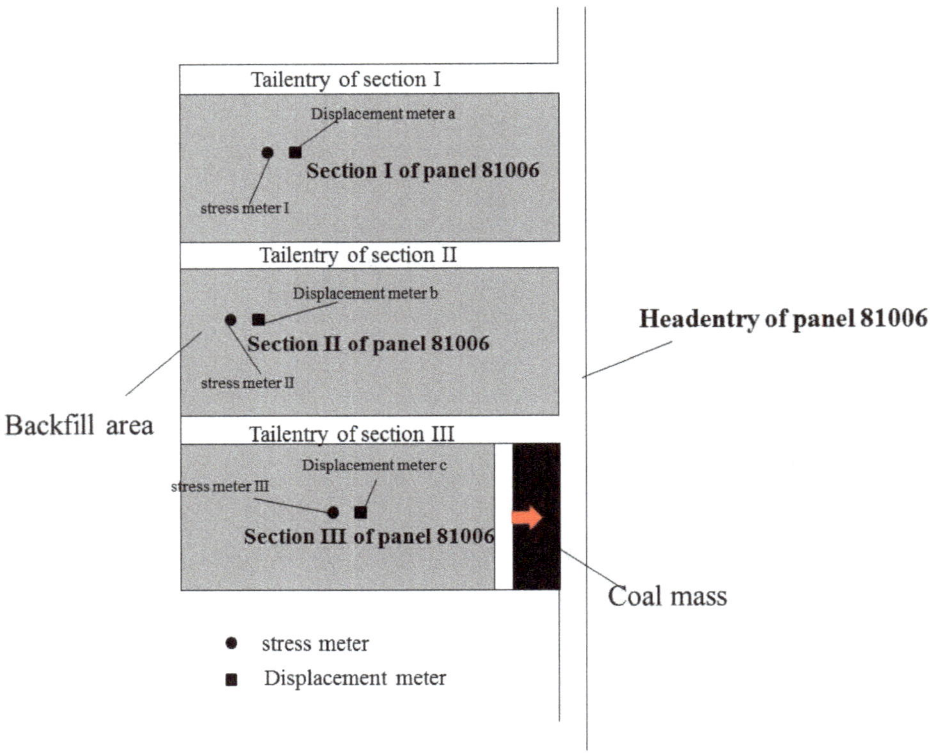

Figure 9. Meters' layout.

4.2. *Floor failure Depth (FFD) Detecting Equipment and Detecting Method*

4.2.1. Introduction

Isolated borehole section flow testing is used in many coal mines due to its simple construction/ installation and data analysis as well as low cost. The detection system consists of a testing device, a plugging control device, a water injection control device, and connector pipes as shown in Figure 10. It has two separate loops, including a plugging loop and a water injection loop. The packers at the ends of the testing device are used to isolate sections of the boreholes under the control of the plugging control device, and the leakage rate in the isolated section can be used to learn the failure condition of the strata.

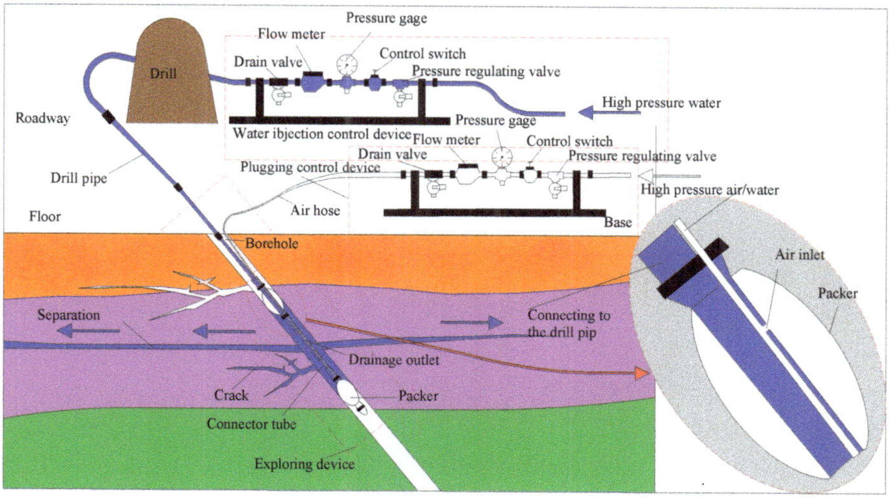

Figure 10. Diagram of an isolated borehole section flow testing system.

The exploring tube has two connected packers which stayed in a contracted state when not working. It can be pushed to any depth in a drill hole by the drilling tubes. By injecting air into the packers through a pressure regulating valve, the packers expanded into a spherical shape and formed an embolism at both ends of the exploring tube. A borehole blockage about 1m long then formed. Water was injected at a constant pressure through the drilling tubes, the regulating valve, and the pressure gauge. Then, the volume of water leakage per unit time through fractures in the hole wall was measured.

Theoretically, it can be proven that under a certain water injection pressure, the water injection flow is determined by the permeability of the rocks and the size of the fractures. Thus, the water-injection flow should increase with the increasing permeability coefficient and growth of the fractures. Results from the actual measurement demonstrated that the water-injection flow value is less than 1 L/min and even approaches 0 in unbroken rocks with a water-injection pressure of 0.1 MPa/min in every 1 m hole segment. In highly fractured rock, the flow value can be up to 30 L/min.

4.2.2. Detecting Method

In order to probe floor failure depth (FFD) of coal panel 81006, two drilling areas (A and B) were designed in the tail entry of section II. Drilling area A contains probing holes D_1 and D_2. Hole D_1 had an azimuth angle of 321°, a dip angle of −1°, and a depth of 91.8 m. Hole D_2 had an azimuth angle of 321°, dip angle of −8°, and depth of 77 m. Drilling area B contains probing holes D_3 and D_4. Hole D_3 had an azimuth angle of d 240°, a dip angle of 3°, and depth of 91.8 m. Hole D_4 has an azimuth angle of 240°, a dip angle of −3°, and a depth of 87 m.

5. Results and Discussion

5.1. Abutment Pressure Distribution Characteristics

A coal stress meter was installed in the tail entry of section 1 of panel 81006 in front of the working face. A hole was drilled in the coal face and the meter was injected into it prior to mining. These meters monitored APD as the working face advanced. The coal stress meters monitored the variation of relative stress inside the coal mass which reflected the APD as the working face advanced. The monitoring results are shown in Figure 11.

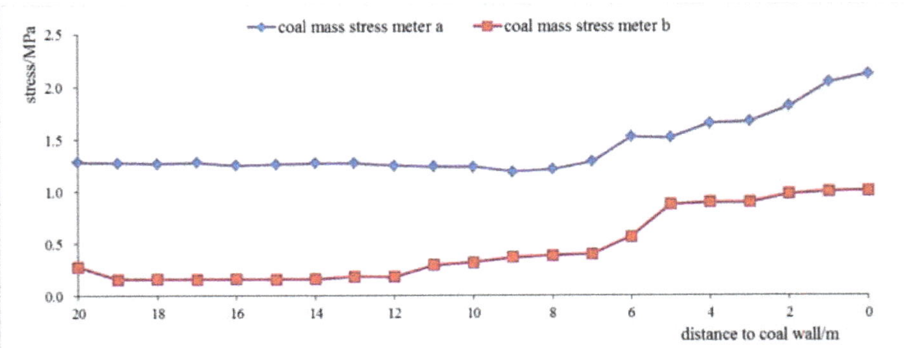

Figure 11. Variations of coal stress in the meters.

Figure 11 shows the relative stress in the coal mass as the working face advanced. The horizontal axis shows the distance from the installation location of the stress meter to the coal wall. As the working face advances, the relative stress in the coal mass increased gradually until reaching a maximum when the working face arrives at the installation location. It can be inferred that APD range, with no plastic zone, is approximately 12 m. Abutment pressure peaks at the wall edge and declines from there, indicating that abutment pressure concentration factor was small and the coal mass beyond the working face is in an elastic state.

The buried depth of the 81006 coal panel is about 550 m, and the uniaxial compressive strength of the coal is 15 MPa as provided by the mine. The abutment pressure concentration factor would be 2–3 normally [16–18]. We selected 2. If traditional longwall mining technology were adopted, the maximum supporting pressure would be as follows:

$$2\gamma H = 2.0 \times 2.5\ \text{t/m}^3 \times 550\ \text{m} = 27.5\ \text{MPa} \tag{1}$$

where γ is the strata's average density, 2.5 t/m^3 and H is the coal body's buried depth, 550 m.

Through theoretical analysis, the abutment pressure's peak in a traditional working face is higher than the uniaxial compressive strength of the coal body, creating a plastic zone at the coal wall edge. Monitoring data in other coal panels of this coal mine could prove this point. In backfill mining, according to the monitoring data, there is no plastic zone in front of the working face. This is because the backfill area supports the overburden, so less ground pressure is transferred to the coal mass in front of the working face. The ground pressure near the working face is therefore smaller.

5.2. Roof-Floor Displacement (RFD) Monitoring in the Backfill Area

The displacement meter monitored the amount of RFD in the backfill area, which reflected the deformation rule and the backfill compression ratio underground pressure. These two parameters are important to evaluate the backfill effect. The graph in Figure 12 was drawn according to the data gathered from the backfill displacement meter. From this graph, the greatest RFD of displacement meter was 104 mm. The curve's slope is relatively high at first, which indicates that the displacement between the roof and the floor happens quickly; then the slope decreases gradually, and the roof-floor displacement also tends to be stable and slow as the roof and floor reaches stable conditions.

It took about 70 days until the roof and the floor were stable, and the greatest RFD was 104 mm. The mining height is calculated as 2 m and the compression ratio of the backfill is 5.2%. The greatest RFDs at b and c are 93 mm and 129 mm, and the compression ratios are 4.74% and 6.45%, respectively. The curve variation rules of b and c are basically the same as that of displacement meter a.

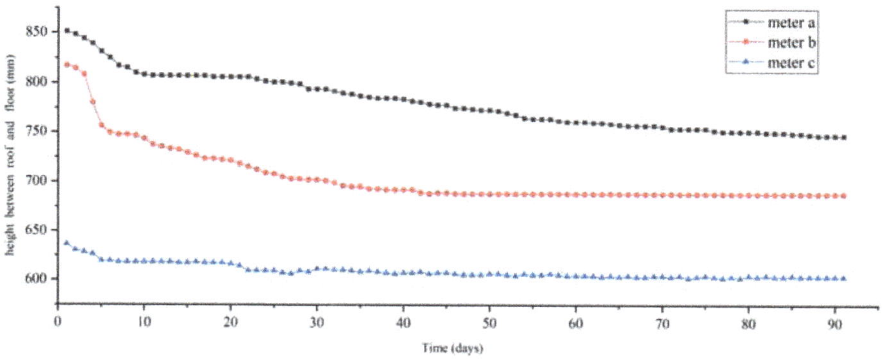

Figure 12. Variation of displacement in meter a.

5.3. Variations of Vertical Stress in the Backfill Area

Inside the backfill area, several backfill stress meters were installed to monitor the vertical stress variations with the advancement of the working face. As shown in Figure 13, the curve graph was drawn based on the data obtained by monitoring.

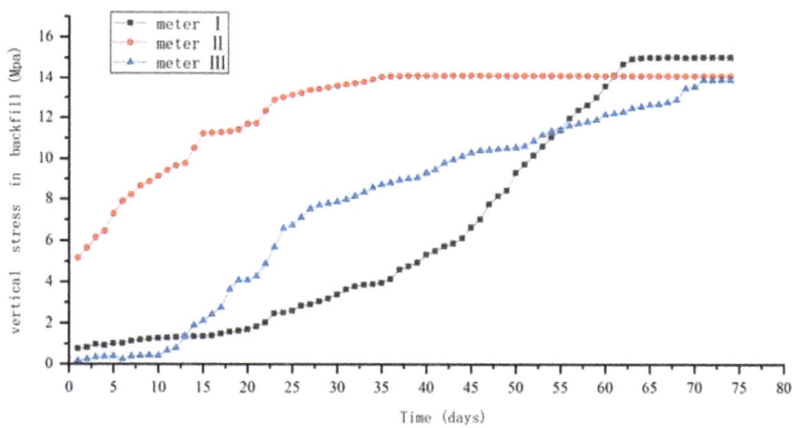

Figure 13. Variation of stress meter I.

From Figure 13, initially, ground pressure transferred to the backfill area is low and increases slowly. Then, vertical stress in the backfill area increases gradually until it reaches stability. The final stress value is 15.06 MPa, almost equal to the initial vertical stress (γH = 13.75 MPa), indicating that the backfill area has adequate capacity to support the overlying strata. This is also why the abutment pressure distribution's range is relatively narrow. It took around 70 days for vertical stress to stabilize, consistent with the displacement monitoring result.

Backfill stress monitored by meters II and III stabilized at 13.02 MPa (data before transmission line was destroyed) and 13.92 MPa, respectively. Their stress histories are shown in Figures 14 and 15, which had much of the same variation rule as stress meter I.

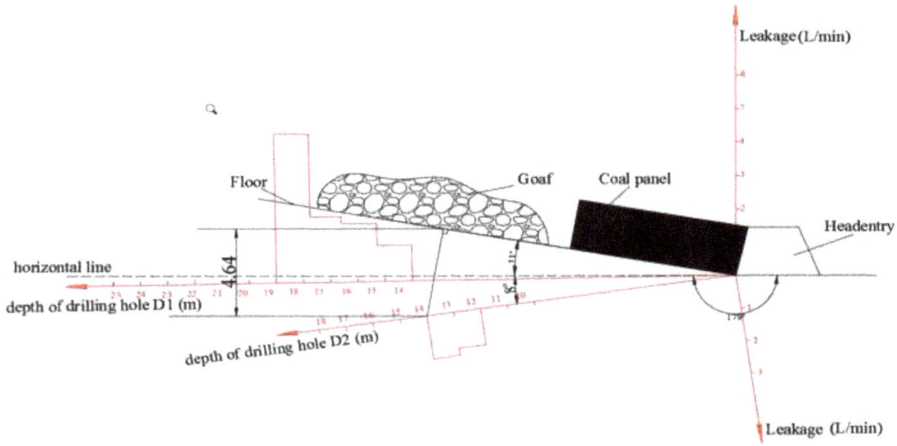

Figure 14. The leakage rates of drilling holes d1 and d2.

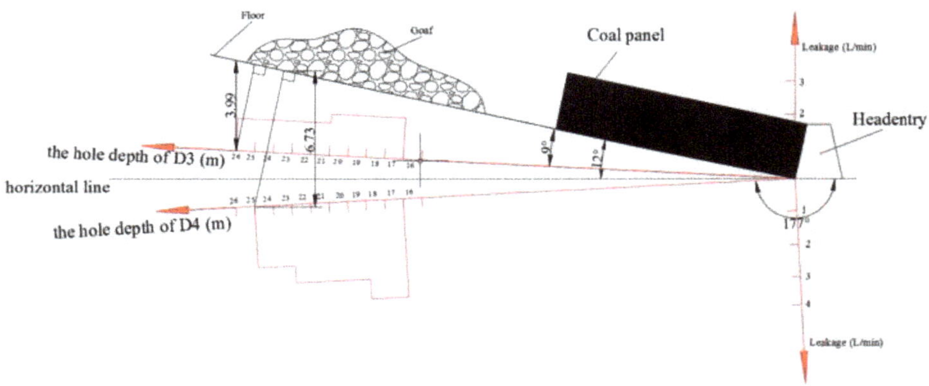

Figure 15. The leakage rates of drilling holes d3 and d4.

5.4. FFD in Backfill Mining

According to the probing data, the graph of the leakage rate of all drill holes was drawn as shown in Figure 14. From Figure 14, drill hole D_1 has a certain leakage amount from depths of 14–19 m. The leakage rate of other segments in D_1 is zero, indicating that rock fractures developed in this segment, which is the floor failure zone. Drill hole D_2 has a certain leakage amount from depths of 12–14 m, indicating that rock fractures developed in this segment, which is the floor failure zone. Analyzing the probing results of the two drill holes, the FFD is 4.64 m. From Figure 15, D_1 has a certain leakage amount above a depth of 25 m, while the leakage amounts of other segments are zero, indicating that rock fractures developed in this segment. D_2 has a certain amount of leakage above a depth of 27 m, indicating that rock fractures developed in this segment. Therefore, after analyzing the probing results of the two drill holes, the greatest FFD is 6.73 m.

5.5. Differences Between Strata Movement Rule of Backfill Mining and Traditional Longwall Mining and its Effect on Water Inrush

According to the above-mentioned field investigations, the abutment pressure did not surpass the strength limit of the coal, which remained in an elastic state. The range and peak of abutment pressures are relatively small compared to traditional longwall mining. The abutment pressure caused by strata movement was an important power resource to surrounding rock failure, and also to floor failure depth. Obviously, backfill mining could change the abutment pressure range and intensity. Therefore, through backfilling mining, the stress environment can be improved, and surrounding rock failure could be reduced to some extent.

Vertical stress in the backfill area reached the original vertical stress computed by theoretical analysis, indicating that the weight of the overburden could be transferred to the backfill area after the working face moved forward a certain distance. The weight of the overburden will never disappear, but it will transfer to the surrounding rock. When vertical stress in the goaf is higher, stress in the surrounding rock (reflected by abutment pressure) will be lower. When the stress in the backfill approaches the original vertical stress, there will be little weight transfer to surrounding rock; therefore, the abutment pressure will be smaller.

RFD values in the backfill area of coal panel 81006 are 104 mm, 93 mm, and 129 mm, respectively, for sections I, II and III. The compression ratios are 5.2%, 4.74%, and 6.45%, respectively. We can infer that in backfill mining, RFD is much smaller than with the traditional caving method. The form of roof movement was changed in backfill mining. The roof did not cave, but rather it bent and subsided.

From the comprehensive analysis of the above characteristics, it is concluded that backfill mining with the paste-like material is quite different from traditional longwall mining. In traditional longwall mining, coal extraction results in the lower strata losing their original balanced state, bending, falling, and then creating rooms for the upper strata's motion. As the coal face advances, the overburden moves upward from the bottom. According to the differences in the magnitude of strata movement, the overburden is normally divided into three zones: The caved zone, fractured zone, and continuous bending zone (abbreviated as "three zones") [16,17,19], as shown in Figure 16. As the working face advances, the overburden's weight that is no longer supported by coal mass due to coal extraction moves to the surrounding rock in which the concentrated stress, called abutment pressure, forms. Generally, the concentrating coefficient K of the abutment pressure can reach 2–3, thus forming the plastic zone in front of the coal face.

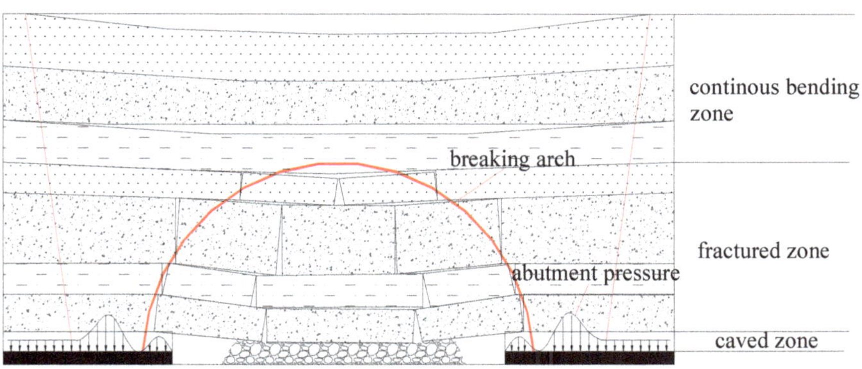

Figure 16. Strata movement characteristics in longwall mining.

In backfill mining, the goaf is backfilled before the immediate roof caves, leaving little room for the overburden's movement. The empty space that causes caving and subsidence no longer exists, which differs from longwall mining [8–10]. Due to the supporting effect of the backfill area, the lower strata obtain a new balanced state after the coal is replaced. The immediate roof does not cave or break. Furthermore, most of the overburden's weight is transferred to the backfill area. High-stress concentrations do not form in the coal. Commonly, no plastic zone appears in front of the coal wall. The abutment pressure peaks at the coal wall's edge. The concentration coefficient K of the abutment pressure in backfill mining is lower than that in traditional longwall mining, as shown in Figure 17.

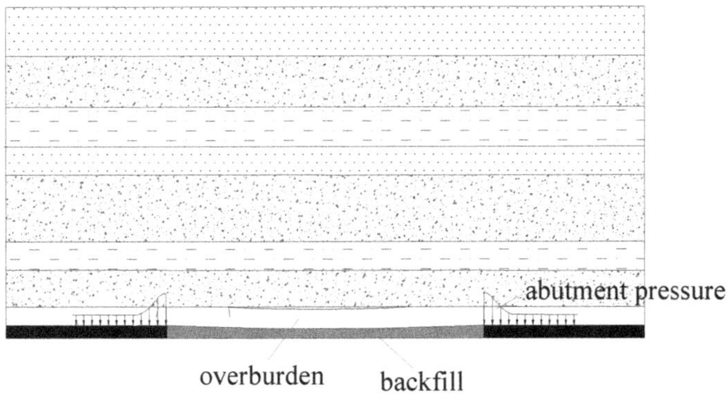

Figure 17. A structural model of strata movement in backfill mining.

In summary, the major differences between traditional longwall mining and backfill mining are given in Table 1.

Table 1. Strata movement characteristics of longwall caving mining and backfill mining.

Serial Number	Index	Traditional Longwall Mining	Filling Mining
1	Overburdens movement scope	"Three zones" exist	"Three zones" do not exist
2	Overburdens movement form	Immediate roof caves	Immediate roof does not cave
3	Roof subsidence	Great	Small
4	Abutment pressure	High	Low
5	Plastic zone	Wide plastic zone	Narrow or no plastic zone

Cracks in a floor disturbed by mining could provide a water inrush channel, and, therefore, FFD is an important index by which to evaluate coal panel safety. The greater is the FFD, the more dangerous the threat of a water inrush disaster. The FFD has a great relationship with strata movement and ground pressure, which can be reflected by APD, RFD, and vertical stress in backfill. The monitoring mentioned above showed that APD, RFD, and vertical stress in backfill are much smaller than in traditional longwall mining, implying that the form and aptitude of strata movement and ground pressure were improved significantly. It was not difficult to deduce that FFD would be relatively small, which may have an influence on water inrush disaster to some extent.

Monitoring results showed that the greatest FFD in the 81006 coal panel was 6.7 m, which agrees with our deduction above. According to previous monitoring data in other coal panels of the same coal seam extracted by traditional longwall mining, the FFD is more than 20 m. Thus, FFD in backfill mining is much smaller than in traditional longwall mining. Backfill mining can prevent water inrush channels from forming and developing and may be useful for water inrush prevention especially in mines with high floor pressure water.

6. Conclusions

A case study on a trial test of water inrush prevention technology with backfill mining was carried out in Caozhuang Coal Mine. According to field observations and monitoring, the trial test showed good control of ground pressure and floor fracture development by employing backfill mining.

(1) In-situ measurements of abutment pressure show that its range is about 10 m, and it peaks at the edge of the coal wall. Compared to traditional longwall mining, the range and peak of the abutment pressure with backfill technology are much smaller. Abutment pressure caused by strata movement was an important power resource to surrounding rock failure and to floor failure depth. Obviously, backfill mining could change the abutment pressure range and intensity, and therefore,

the stress environment of surrounding rock was improved implying that surrounding rock failure could be decreased to some extent, which can decrease the danger of water inrush.

(2) RFDs monitored in the backfill area of coal panel 81006 are 104 mm, 93 mm, and 129 mm, and the compression ratios are 5.2%, 4.74%, and 6.45%. Roof displacement with backfill technology is much smaller than with the traditional caving method. Roof displacement could reflect strata movement intensity, and roof displacement is very small without caving in backfill mining, indicating that backfill mining can change the form of the strata movement.

(3) Vertical stresses monitored in backfill areas I, II, and III were respectively 15.06 MPa, 13.02 Mpa, and 13.92 Mpa. These values were roughly equal to the original vertical stress (13.75 MPa), indicating that the weight of the overburden has been transferred to the backfill area. The weight of the overburden will never disappear, but it will transfer to the surrounding rock. When vertical stress in the goaf is higher, stress in surrounding rock (reflected by abutment pressure) will be lower. When the stress in backfill approaches the original vertical stress, there will be little weight transfer to surrounding rock, and therefore, the abutment pressure will be smaller. The results of vertical stress and abutment pressure indicated mutual verification.

(4) The case study indicates that backfill mining could change the form and aptitude of strata movement. Meanwhile, backfill mining could improve the stress environment, which could decrease FFD. The "three zones" do not appear in backfill mining. Rather, the ground pressure is much smaller with backfill technology, and FFD is much smaller in backfill mining, implying that backfill technology can be helpful for preventing floor water inrush. This study demonstrated that through backfill mining, FFD can be significantly decreased, and water inrush disasters may be eliminated. Backfilling mining may be an optional method for floor water inrush prevention and control.

Author Contributions: Conceptualization, Y.S.; Data curation, Y.S., X.W. and H.X.; Formal analysis, J.H.; Investigation, J.H. and J.L.; Methodology, H.X.; Resources, X.W.; Supervision, J.H.; Visualization, J.L.; Writing—review & editing, J.H.

Funding: This study was supported by the National Natural Science Foundation of China (Grant No.51804180, No.51574055), the key research plan of Shandong province (Grant No.2018GSF116007), the Natural Science Foundation of Shandong province (Grant No. ZR2017BEE033), Postdoctoral Innovation Foundation of Shandong province (201703076), "Top Disciplines" Special Project of Mining and Safety Engineering College(1SY04602).

Conflicts of Interest: The authors declare no conflict of interests.

References

1. Please, C.P.; Mason, D.P.; Khalique, C.M.; Ngnotchouye, J.M.T.; Hutchinson, A.J.; van der Merwe, J.N.; Yilmaz, H. Fracturing of an euler–bernoulli beam in coal mine pillar extraction. *Int. J. Rock Mech. Min. Sci.* **2013**, *64*, 132–138. [CrossRef]
2. Singh, G.S.P.; Singh, U.K. A numerical modeling approach for assessment of progressive caving of strata and performance of hydraulic powered support in longwall workings. *Comput. Geotech.* **2009**, *36*, 1142–1156. [CrossRef]
3. He, M.; Xie, H.; Peng, S. Study on rock mechanics in deep mining engineering. *Chin. J. Rock Mech. Eng.* **2005**, *24*, 2803–2813.
4. Peng, S.S. *Coal Mine Ground Control*, 3rd ed.; Peng SS Publisher: Morgantown, WV, USA, 2008; pp. 229–237.
5. Surinaidu, L.; Rao, V.V.S.G.; Ramesh, G. Assessment of groundwater inflows into Kuteshwar limestone mines through flow modeling study, Madhya Pradesh. *India Arab. J. Geosci.* **2013**, *6*, 1153–1161. [CrossRef]
6. Tan, Y.L. *Ground Pressure and Strata Control*; China Coal Industry Publishing: Beijing, China, 2011.
7. Zhang, X.; Lin, J.; Liu, J.; Li, F.; Pang, Z. Investigation of hydraulic-mechanical properties of paste backfill containing coal gangue-fly ash and its application in an underground coal mine. *Energies* **2017**, *10*, 1309. [CrossRef]
8. Deng, X. Strata behavior in extra-thick coal seam mining with upward slicing backfilling technology. *Int. J. Min. Sci. Technol.* **2016**, *26*, 587–592. [CrossRef]
9. Wu, D.; Sun, G.; Liu, Y. Modeling the thermo-hydro-chemical behavior of cemented coal gangue-fly ash backfill. *Constr. Build. Mater.* **2016**, *111*, 522–528. [CrossRef]

10. Zhang, J.X.; Zhou, N.; Huang, Y.L.; Zhang, Q. Impact law of the bulk ratio of backfilling body to overlying strata movement in fully mechanized backfilling mining. *J. Min. Sci.* **2011**, *47*, 73–84. [CrossRef]
11. Liu, C.Y.; Yang, P.J.; Hou, C.J. Movement law and stability analysis of overlaying strata under the condition of mining with filling. *J. China Univ. Min. Technol.* **2004**, *3*, 166–169.
12. Liu, G.; Li, L.; Yang, X.; Guo, L. Numerical analysis of stress distribution in backfilled stopes considering interfaces between the backfill and rock walls. *Int. J. Geomech.* **2016**, *17*, 06016014. [CrossRef]
13. Sivakugan, N.; Widisinghe, S.; Wang, V.Z. Vertical stress determination within backfilled mine stopes. *Int. J. Geomech.* **2014**, *14*, 06014011:1–06014011:5. [CrossRef]
14. Liu, X.L.; Wang, S.Y. Mine water inrush forecasting during the mining under waters. *Disaster Adv.* **2012**, *5*, 876–881.
15. Wang, X.Y.; Li, S.F.; Xu, G.Q.; Zhao, Y.Q.; Gao, Z.J.; Dong, D.L. *Applied Hydrogeology*; China University of Mining Press: Xuzhou, China, 2011. (In Chinese)
16. Song, Z.Q. *Practical Ground Pressure Controlling*; China University of Mining & Technology Press: Xuzhou, China, 1998.
17. Shi, Y. *Simulation and Application of the Dynamic Structural Mechanics Model of Working Face*; Shandong University of Science and Technology: Taian, China, 2001. (In Chinese)
18. Tan, Y.L.; Liu, X.S.; Ning, J.G.; Tian, C.L. Front Abutment Pressure Concentration Forecast by Monitoring Cable-Forces in the Roof. *Int. J. Rock Mech. Min. Sci.* **2015**, *77*, 202–207. [CrossRef]
19. Song, Z.; Hao, J. Study on water inrush from fault's prevention and control theory. *J. China Coal Soc.* **2013**, *38*, 1511–1515.

processes

Article

Study on the Preparation and Hydration Properties of a New Cementitious Material for Tailings Discharge

Yunbing Hou, Pengchu Ding *, Dong Han, Xing Zhang and Shuxiong Cao

College of Resources & Safety Engineering, China University of Mining & Technology (Beijing), D11 Xueyuan Road, Haidian District, Beijing 100083, China; houyunbing2000@163.com (Y.H.); tsp1600101007@student.cumtb.edu.cn (D.H.); tsp1600101008@student.cumtb.edu.cn (X.Z.); tsp1600101006@student.cumtb.edu.cn (S.C.)

* Correspondence: tbp1600101004@student.cumtb.edu.cn; Tel.: +86-178-8882-6795

Received: 27 November 2018; Accepted: 14 January 2019; Published: 17 January 2019

Abstract: Blast furnace slag (BFS) is often used as a cement-based raw material for underground filling and surface cemented paste discharge of tailings during mining processes. This paper studied a new cement-based material (NCM) with BFS to replace ordinary Portland cement (OPC). A uniaxial compressive strength (UCS) experiment was used to test the mechanical strength of samples; X-ray diffraction and thermal gravity experiments were used to test the crystalline phases and amount of hydration products by samples; a scanning electron microscope experiment was used to observe the influence of the hydration products morphology by samples; mercury intrusion porosimetry experiment was used to analyze the pore size distribution of samples. The samples with NCM had an optimum UCS; the crystalline phases of the hydration products were similar in OPC and NCM. However, the amount of product formed in OPC was less than that in NCM at the same curing time; more ettringite and calcium silicate hydrate were produced in samples with NCM, which filled the pores and enhanced the UCS of the samples. The final mercury intrusion volume of the samples with NCM were lower than the samples with OPC at the same curing time, which showed that samples with NCM had lower porosities. For the samples with NCM and OPC cured from 7 days to 28 days, the mercury intrusion volume was reduced by 18% and 13%, and the most common pore size of the samples reduced by 53% and 29%, respectively. This showed after 21 days curing time, the pores of all the samples getting smaller; however, the samples with NCM were more compact. The main ingredients of the NCM were clinker, lime, gypsum and BFS, and its ratio was 14:6:10:70. The content of additives to NCM was 0.4%, and the ratio of sodium sulfate: alum: sodium fluorosilicate was 2:1:1.

Keywords: new cementitious material; cement-based paste discharge; XRD; TG/DTG; SEM; MIP; mechanical behaviors

1. Introduction

After the ore has been recovered from underground, tailings are usually treated in two ways [1]. One method is filling the tailings in the underground goafs. This treatment can reduce the accumulation of tailings on the ground, and ensure the safety of underground operations. This method has been widely promoted in underground mining operations [2]. The filling method can remove about 50% of the tailings, and the remaining part is usually discharged in the tailings pond [3–5]. However, this method is prone to dam breaks, environmental pollution, and cause casualties [6,7]. For example, on 8 September 2008, a major tailings dam broke in Xinta mining Co., Ltd., of Lixian county, Linfen city, Shanxi province, causing 281 deaths [8]. According to a study by the Clarke University's pollution assessment team, the damage caused by tailings pond accidents ranked 18th among 93 types of accidents and hazards worldwide [9]. In view of the disadvantage of tailings pond, some scholars combined both methods and proposed the technology of cemented paste discharge of tailings

(CPDT) [10,11]. The method involved adding some cement to the tailings and discharging them to valleys or subsidence areas; the tailings need to be transported by pipes or belts. This method guarantees the stability of the tailings for the cemented piles. In addition, mines no longer need to build tailings pond and can also achieve safe pile-up. Because the cemented tailings pile can be built up to a high height, the limited footprint can pile up more tailings than tailings ponds. Presently, this technology has been applied and developed in some metal mines [11].

During the cemented paste backfill, the amount of cement was 2% to 7% of the total weight of the tailing slurry, and it may be up to 10% to remain independent across pillar extraction process [12]. Cement costs account for more than 60–80% of the filling costs, as discussed by Li et al. [13]. The idea of CPDT was derived from cemented paste backfill, thus, the process of CPDT was similar to this, and the cost of cement also occupied a large proportion of CPDT costs. Therefore, the search for cheap, cementitious industrial by-products to replace cement and reduce the cost of CPDT is of great significance. Blast furnace slag (BFS) is one of the by-products released from blast furnace [14,15]. At present, China produces 0.3 billion tons of BFS every year [16], which causes a great waste of resources. The major chemical compositions consist of CaO, SiO_2 and Al_2O_3. Moreover, it has a small amount of MgO, FeO and sulfide. The higher proportions of CaO, Al_2O_3 and MgO, the better activity of the BFS [17]. It was an alternative material to replace ordinary Portland cement (OPC), which could improve some properties and bring economic and environmental benefits [18]. Previous works have already studied the hydration activity of BFS and its excitation mechanisms such as alkali activation and mechanical activation [19–21]. He et al. [22] investigated the effect of BFS (both dry-separation and wet-milling) on the properties of cement slurry indicated that as the increase of dry-separation BFS, the initial and final setting time gradually decreased, while the wet-grinding BFS system showed the opposite trend. Yin et al. [23] investigated the environmental perspectives of recycling various combustion ashes in cement production and indicated that there was a range (5–10% by weight) of ashes for mixed cement production, beyond which may cause significant changes in cement composition. Cihangir et al. [24] studied the use of BFS in paste backfill by using scanning electron microscope (SEM) and X-ray diffraction (XRD) analysis, and drew the conclusion that the BFS has great potential for use as an alternative binder in cement paste backfill.

However, there is limited research on the performance of tailings under the condition of adding some additives when BFS, clinker, lime and gypsum were mixed according to a certain proportion. Therefore, the main purpose of this paper was to study the results of a new cement material's effect on the strength and pore volume development for tailings. Clinker plus lime and gypsum were used as stimulants, and a small amount of sodium sulfate, alum and sodium fluorosilicate were used as additives to make a new type of tailings cementitious material (NCM). The effect of different amounts of compound activator on the uniaxial compressive strength (UCS) of CPDT samples with NCM was explored. Thermal gravity (TG) and differential thermal gravimetric (DTG) were carried out to assess the amount of hydration products of different cementitious materials on microstructural development [25]. The SEM was used to analyze the composition and microcosmic morphology of the hydration products of the samples with OPC and NCM at the same curing time [26]. Finally, the microscope pore structures of the samples were studied through the mercury intrusion porosimetry (MIP) experiment [27].

2. Materials and Methods

2.1. Materials

The materials consisted of water, OPC 42.5$^{\#}$ (Jilong Cement Co., Ltd., Tangshan, China), BFS (Tangshan Iron and Steel Group Co., Ltd., Tangshan, China), lime (Yuanfeng Calcium Industry Co., Ltd., Xinxiang, China), gypsum (Longsheng Hengtong Gypsum Co., Ltd., Xingtai, China), clinker (Longfeng Cement Co., Ltd., Tangshan, China), sodium sulfate (Sinopharm Chemical Reagent Co., Ltd.,

Beijing, China), alums (Sinopharm Chemical Reagent Co., Ltd., Beijing, China), sodium fluosilicate (Sinopharm Chemical Reagent Co., Ltd., Beijing, China) and tailings (Lilou Iron Mine, Huoqiu, China).

2.1.1. Water and Binders

The water used for the experiment was tap water. OPC was the most familiar binder used in the disposing of the tailings [28]. NCM was made up of BFS, lime, gypsum, clinker, sodium sulfate, alums and sodium fluosilicate. The performance results of the NCM to CPDT were compared with OPC.

2.1.2. Tailings

The tailings taken back from the site were precipitated, dehydrated and dried, then, the particle size characteristics were analyzed. The results are shown in Table 1. The d_{10}, d_{30}, d_{50}, d_{60} and d_{90} represent the cumulative content of the particle composition curve, with corresponding particle size of (volume fraction) 10%, 30%, 50%, 60% and 90%, respectively. The main chemical composition of the tailings was shown in Table 2. According to non-uniform coefficient and curvature coefficient, the tailings belong to the graded bad material category. The median particle size was 38.3 μm, which is characteristic of fine tailings.

Table 1. Physical properties of the materials.

Element Unit	d_{10}/μm	d_{30}/μm	d_{50}/μm	d_{60}/μm	d_{90}/μm	Cu	Cc
Tailings	14.55	26.61	38.32	54.27	82.33	3.73	0.89

Table 2. Main chemical properties of the materials (OPC: ordinary Portland cement).

Element Unit	MgO (wt.%)	Al_2O_3 (wt.%)	SiO_2 (wt.%)	CaO (wt.%)	SO_3 (wt.%)	Fe_2O_3 (wt.%)	Total
Tailings	2.41	3.85	82.05	2.46	0.18	8.01	98.96
Slag	8.38	14.79	33.81	36.95	0.28	0.89	95.09
Clinker	2.45	4.47	22.01	64.31	2.45	3.45	99.14
Gypsum	2.14	0.12	0.98	45.85	42.45	0.11	91.66
Lime	0.56	0.23	0.38	72.29	0.13	0.26	73.84
OPC	2.19	15.49	21.86	63.59	2.42	2.66	96.97

2.1.3. NCM Ingredients

The specific surface area of the BFS was 475 m^2/kg and the OPC was 312 m^2/kg. The chemical composition of the materials used in this test is shown in Table 2.

2.1.4. Preparation of CPDT samples

In this study, the concentration of the CPDT samples was 78%, and the mix proportion adopted included a water-to-cement (w/c) ratio of 5.6 and a binder (NCM) content of 4.5 wt.% of the tailings. The w/c ratio and cement ratio in all mixtures remained constant. The basic principle of using this binder content was to balance cost and strength. According to the ratio of the tests, BFS, lime, gypsum and clinker were mixed to produce the new cement-based material [29–31]. The weighed tailings and dry materials were mixed in a container until a uniform color was obtained. Then, the weighed water was added to the container and mixed until a uniform paste was obtained. They were then stirred with a mixer for 7 min. The prepared CPDT samples were poured into plastic containers with a diameter of 5 cm and a height of 10 cm. They were cured with a controlled temperature of 20 ± 2 °C and more than 95% relative humidity for different curing time (3 days, 7 days and 28 days). Nearly 124 CPDT samples were prepared for the UCS, SEM MIP experiments.

Furthermore, samples for XRD and TG/DTG analysis were prepared according to the procedure above. The w/c ratio of all the samples was 2 (to represents the high volume of water in CPDT). Various tests were subjected to the NCM and the OPC samples after the required curing time. Nearly eight CPDT samples were prepared in this study.

2.2. Experimental Methods

2.2.1. The UCS Experiment

The mechanical strength or the stability of the CPDT samples was usually evaluated using UCS [12,32] in accordance with TYE-300D (Wuxi Jianyi Instrument Machinery Co., Ltd., Wuxi, China), as illustrated in Figure 1. UCS tests were carried out on the CPDT samples with different amount of NCM and OPC and curing 3, 7 and 28 days. The load was executed at a relatively slow rate (0.1 kN/s).

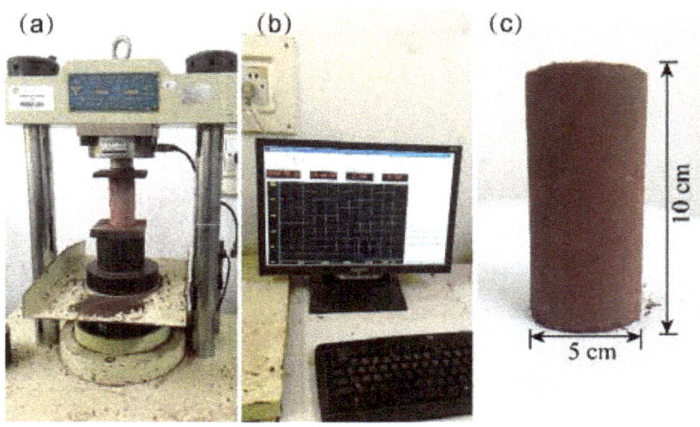

Figure 1. The TYE-300D and sample. (**a**) The TYE-300D test instrument; (**b**) the experiment bench; (**c**) the sample.

2.2.2. The XRD Experiment

XRD was a common measurement for crystal phases structure identification in cement-based materials slurry [21]. By using Empyrean Diffractometer, the analyses were carried out under the 2-θ range of 5–60° and 0.02° step to study the crystalline phases of the samples, as illustrated in Figure 2. XRD (PANalytical B.V., Almelo, The Netherlands) analyses were performed on dried and powdered cement-based slurry.

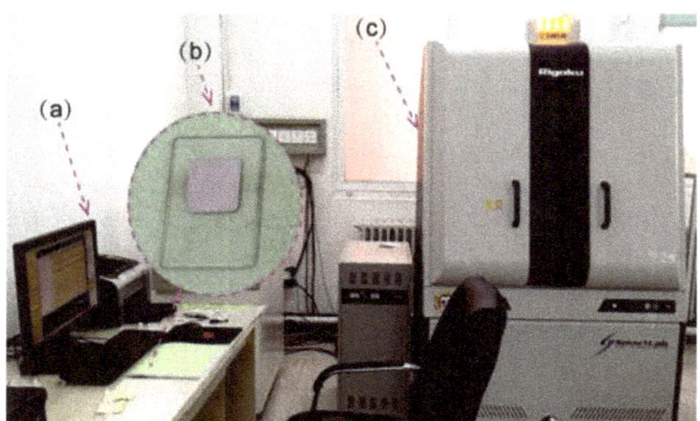

Figure 2. The Empyrean Diffractometer and sample. (**a**) The experiment bench; (**b**) the sample; (**c**) the X-ray diffraction (XRD) analyzer.

2.2.3. The TG/DTG Experiment

TG and DTG tests were performed to assess the amount of hydration products of the different cement-based materials on the microstructural development of the sample. This was done using a STA449F3 TG analyzer (NET Scientific Instruments Trading (Shanghai) Co., Ltd., Shanghai, China) that can raise temperatures up to 1200 °C, as illustrated in Figure 3. The temperature was increased from room temperature to 900 °C at a rate of 10 °C/min N_2 purge. The reason why the temperature was not raised to 1200 °C was that the mass quality does not change when the temperature was higher than 900 °C. The weight change sensitivity 1 μg/min, the test standard has been followed GB/T 27761-2011. The samples used for TG and DTG experiment were the same that with XRD experiment.

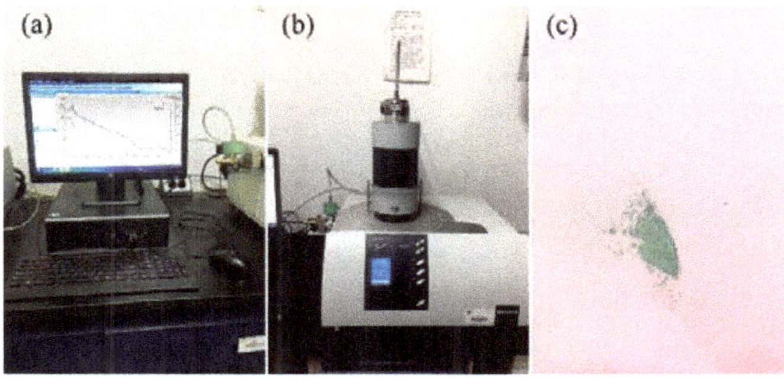

Figure 3. STA449F3 and samples. (**a**) The experiment bench; (**b**) the thermal gravity (TG) analyzer; (**c**) the samples.

2.2.4. The SEM Experiment

SEM analysis was used to observe the influence of the hydration products morphology on the samples with NCM and OPC [33]. The SEM observations were carried out under a 7001F analyzer (Japan Electronics Corporation, Shanghai, China), as illustrated in Figure 4. The magnification levels were in the range of 10–500 k.

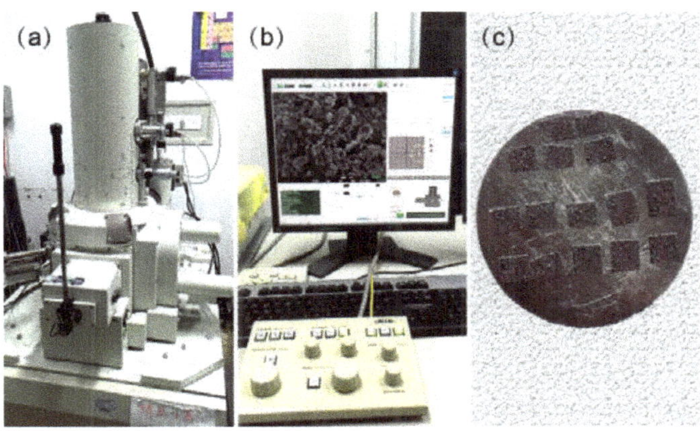

Figure 4. The JSM-7001F and samples. (**a**) The scanning electron microscope (SEM); (**b**) the experiment bench; (**c**) the samples.

2.2.5. The MIP Experiment

MIP is a high precision method for analyzing the micropore structure of materials [34]. Under continuous pressure, the volume of mercury invading the sample could be determined according to the external pressure, and the pore size distribution of the sample could be obtained. It was measured using a Micromeritics Auto Pore IV-9500 analyzer (Micromeritics Instrument (Shanghai) Ltd., Shanghai, China) with a range of 33,000 psia (228 MPa), as shown in Figure 5. The instrument can detect diameters ranging from 0.005 μm to 360 μm, while mercury injection volumes can be as accurate as 0.1 μL. The center samples were obtained from cylindrical specimens, weighing approximately 1.3 g. All of the samples were dried in a 50 °C oven until the quality was no longer reduced, the pore volume will not change at this drying temperature [35]. The samples were then immersed in mercury under gradually increasing pressure. The relationship between the test pressure and the diameter applies to Washburn–Laplace's law, as the following equation [36]:

$$d = \frac{-4\sigma \cos\theta}{P} \tag{1}$$

where, P was the applied pressure; d was the pore diameter of the sample, σ was the surface tension (N/m) and θ was the contact angle between the pore wall and the mercury. The contact angle was assumed 140° in this study [37,38].

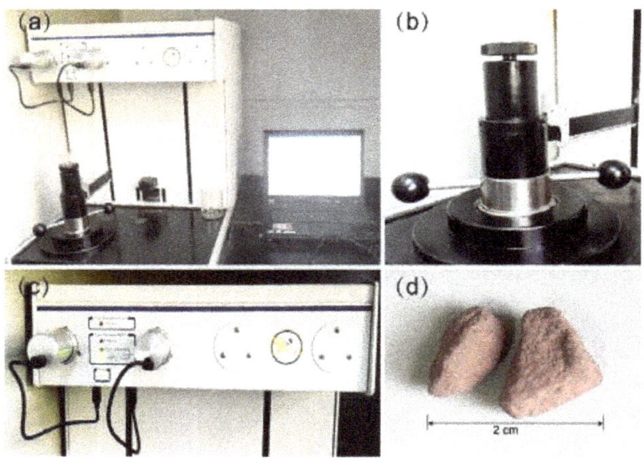

Figure 5. The AutoPore IV and sample. (**a**) Mercury intrusion porosimetry; (**b**) the high-pressure test pipeline; (**c**) the low-pressure test pipeline; (**d**) the sample.

3. Results and Discussion

3.1. Compound Activator Dosage Optimization

The hydration activity of BFS can only be stimulated under certain conditions. Common excitation methods included mechanical excitation, acid excitation and alkali excitation. In this test, the specific surface area of the BFS was controlled by 475 m^2/kg. Lime and gypsum combined with clinker were used to carry out compound excitation. The influence of different activators on the UCS of the samples with NCM was studied to optimize the compound activator dosage.

According to the test analysis, the three levels of clinker were 12%, 14% and 16%; the three levels of lime were 4%, 6% and 8% and the three levels of gypsum were 8%, 10% and 12%. A three-level, four-factor orthogonal design test $L_9(3^4)$ was conducted in this study. The orthogonal test plan and test results were shown in Table 3. All proportions in this paper were mass percentages.

Table 3. Orthogonal test table and uniaxial compressive strength (UCS) results.

Numbering	Concentration (%)	w/c	Cementitious Material Mass Fraction (%)				Compressive Strength (Mpa)		
			X_1	X_2	X_3	X_4	3 Day	7 Day	28 Day
H-1	78	5.6:1	12	4	8	76	0.607	1.384	2.536
H-2	78	5.6:1	12	6	10	72	0.861	1.879	2.623
H-3	78	5.6:1	12	8	12	68	0.811	1.835	2.466
H-4	78	5.6:1	14	4	10	72	0.863	1.626	2.502
H-5	78	5.6:1	14	6	12	68	0.896	1.941	2.673
H-6	78	5.6:1	14	8	8	70	0.707	1.826	2.549
H-7	78	5.6:1	16	4	12	68	0.724	1.273	1.774
H-8	78	5.6:1	16	6	8	70	0.726	1.611	2.333
H-9	78	5.6:1	16	8	10	66	0.782	1.766	2.303

Where X_1 denotes clinker, X_2 deontes lime, X_3 denotes gypsum and X_4 denotes blast furnace slag (BFS).

Range analysis is often used to analyze the most influential factors in experimental data. In this test, the UCS data were treated with this method; the results are shown in Table 4.

Table 4. Range analysis results for the UCS data.

Ages/Day	Factor		Clinker	Lime	Gypsum	Influence Sequence	Optimum Combination
3	Average of UCS/MPa	Level 1	0.760	0.731	0.680	$X_3 > X_2 > X_1$	$X_{1(Level\ 2)}$ $X_{2(Level\ 2)}$ $X_{3(Level\ 2)}$
		Level 2	0.869	0.875	0.835		
		Level 3	0.744	0.767	0.810		
	R [1]/MPa		0.125	0.144	0.178		
7	Average of UCS/MPa	Level 1	1.699	1.428	1.607	$X_2 > X_1 > X_3$	$X_{1(Level\ 2)}$ $X_{2(Level\ 2)}$ $X_{3(Level\ 2)}$
		Level 2	1.798	1.810	1.757		
		Level 3	1.550	1.809	1.683		
	R/MPa		0.248	0.382	0.150		
28	Average of UCS/MPa	Level 1	2.542	2.271	2.473	$X_1 > X_2 > X_3$	$X_{1(Level\ 2)}$ $X_{2(Level\ 2)}$ $X_{3(Level\ 2)}$
		Level 2	2.575	2.543	2.476		
		Level 3	2.137	2.439	2.304		
	R/MPa		0.438	0.272	0.172		

[1] R = the difference between maximum value and minimum value.

It can be seen from Table 4 that the UCS measures of 3 days, 7 days and 28 days for the activator optimum formula represented 14% of clinker, 6% of lime and 10% of gypsum. At these ratio conditions, the UCS of the samples with the NCM was the biggest. The reason for this was that under this ratio, BFS was hydrated to the fullest extent, and the remaining of lime and gypsum were hydrated the least, while the structure of the samples was more compact.

The range analysis cannot meticulously reflect the degree which the measured values agree with each other and data fluctuations caused by test errors during the experiment. It is also impossible to give an accurate quantitative estimate of the importance of the influence of the factors. To compensate for the lack of intuitive analysis, the variance was analyzed for the data above; the results are shown in Table 5.

Table 5. Variance analysis results.

Ages/Day	Source of Variance	Deviation	Degrees of Freedom	Mean Square Deviation	F	Significance	Closure Probability
3	X_1	0.028	2	0.014	8.225	$X_3 > X_2 > X_1$	0.108
	X_2	0.034	2	0.017	9.884		0.092
	X_3	0.056	2	0.028	16.518		0.057
	Error	0.003	2	0.002			
7	X_1	0.093	2	0.047	38.231	$X_2 > X_1 > X_3$	0.025
	X_2	0.292	2	0.146	119.578		0.008
	X_3	0.034	2	0.017	13.829		0.067
	Error	0.002	2	0.001			
28	X_1	0.357	2	0.178	6.544	$X_1 > X_2 > X_3$	0.133
	X_2	0.113	2	0.057	2.078		0.325
	X_3	0.058	2	0.029	1.060		0.485
	Error	0.055	2	0.027			

In Table 5, we find that the results of the variance and range analyses were consistent. For the UCS of 3 days, X_3(gypsum) had the greatest impact, X_2(lime) was second and X_1(Clinker) had minimal impact. For the UCS of 7 days, X_2 had the greatest impact, X_1 was second and X_3 had minimal impact. For the strength of 28 days, X_1 had the greatest impact, X_2 was second and X_3 had minimal impact. This showed that for the CPDT samples at the time of 3 days, gypsum had the greatest effect on the hydration reaction, while the role of gypsum was gradually reduced from 7 to 28 days. The reason for this phenomenon was that a large amount of SO_4^{2-} in gypsum can promote the hydration reaction of cementitious materials in the early stage. The absorption of SO_4^{2-} due to the hydration products of the C-S-H gels resulted in the decrease of SO_4^{2-} in the solution, which has a negative effect on the development of the UCS. Lime contained a large amount of OH^-, and excitation agents need to work in alkaline solution; thus, from 3 to 7 days, the effect of lime on the UCS of the samples gradually increased. From 7 to 28 days, due to the progress of the hydration reaction and self-drying, the moisture in the samples gradually decreased. Additionally, the effect of OH^- on the hydration reaction gradually decreased, which was consistent with the findings of Cui and Fall [39]. The influence of BFS on the UCS increased from 3 to 28 days. This showed that with the progress of the hydration reaction, BFS was the main component of the cementitious materials.

Based on the analysis above, the optimal ratio was 14% for clinker, 6% for lime, 10% for gypsum and 70% for BFS. The UCS of the test samples gathered by adding this ratio of materials was compared with that adding OPC; the results were shown in Table 6. The UCS of the samples with NCM was higher than that with OPC across all aging time.

Table 6. The results of UCS with OPC and new cement-based material (NCM).

Cementitious Materials	The UCS		
	3 Day	7 Day	28 Day
Samples with NCM	0.545	1.453	2.181
Samples with OPC	0.344	0.853	1.943

The same method was used to optimize the addition of sodium sulfate, alum and sodium fluorosilicate. The results were obtained when the materials ratio of clinker: lime: gypsum: BFS was 14:6:10:70, and the amount of admixture added was 0.4% of NCM. The ratio of sodium sulfate: alum: sodium fluorosilicate was 2:1:1, and the samples had the highest UCS across all ages. 0.831 MPa at 3 days, 2.019 MPa at 7 days, 3.307 MPa at 28 days with NCM were 2.4, 2.4 and 1.7 times higher as compared with OPC. The reason for this was that the fineness of the slag was larger than with OPC, and the finer of the slag, the larger the specific surface area and surface energy, significantly improving the activity of the slag. The fine particles of the slag filled in smaller pores to lower the porosity of the samples, which could also increase the strength of the samples (the conclusion will be confirmed in part 3.4). In addition, the slag had been stimulated at the beginning because of the addition of admixtures and more hydration products were produced in the earlier time. From the discussion above, it seems the NCM can replace OPC with regards to strength.

3.2. The Crystalline Phases and Amount of Hydration Products of NCM and OPC

The samples of NCM and OPC were prepared and cured at 20 ± 2 °C with relativity humidity greater than 95% for 7 and 28 days. The hydration reaction was then dried in a 50 °C oven until the quality was no longer reduced. The coupling excitation of the alkaline environment and mechanical grinding promotes the pozzolanic reaction of BFS, thus contributing to the development of the UCS for the samples. The crystalline phases of cement-based materials hydration products of NCM and OPC was demonstrated by the experimental evidence presented in Figures 6 and 7 for curing time of 7 and 28 days. The increasing amount of them were presented in Figures 8 and 9. These figures present the results of XRD, TG and DTG analyses of the powdered cemented materials. It can be seen from Figures 6 and 7 that the crystalline phases of the cement materials hydration products and the materials that do not participate in reaction like SiO_2 were almost the same. However, from 7 to 28 days, the diffraction peaks of $CaCO_3$ and C-S-H were stronger than those at 7 days, and the diffraction peaks of Et and calcium hydroxide (CH) became weak, whether it was OPC or NCM. It indicated that the phases produced by NCM was similar to that produced by OPC, however, the amount of each product changed.

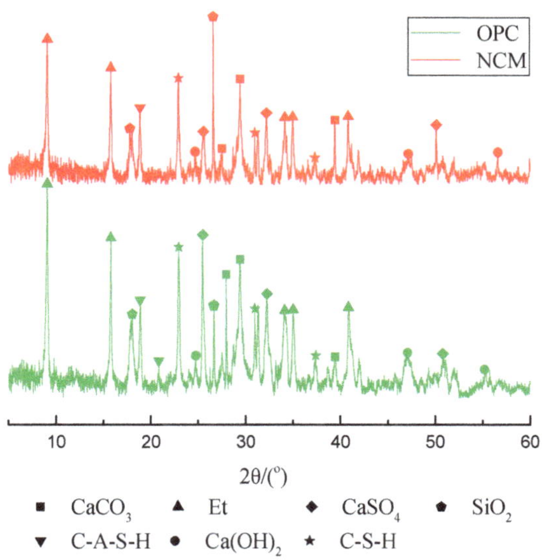

Figure 6. XRD images of NCM and OPC for 7 days of curing time (NCM: new cement-based material; OPC: ordinary Portland cement).

From Figures 8 and 9, the valleys at these different temperatures range (50 °C to 800 °C) showed the existence of C-S-H, ettringite, calcium hydroxide (CH) and calcite as demonstrated in several studies [40–42]. The weight loss between 50 °C and 105 °C was due to the disappearance of free water and bound water. The weight loss between 110 °C and 200 °C was due to the dehydration of ettringite, gypsum and C-S-H. The weight loss between 450 °C and 500 °C was attributed to the decomposition of CH, and the significant weight loss at 650–750 °C results from the decomposition of calcite. The comparison of the TG curves of the samples with OPC and NCM showed that the weight loss of water evaporation was almost the same, however, the weight loss caused by the dehydration of the cement-based hydration products (110 °C–200 °C) of NCM was greater than OPC. This indicated that more cement-based hydration products were generated in NCM. It can also be seen from Figures 8 and 9 that the amount of CH decreased, indicating that more CH participated in the hydration reaction, generating stronger hydration products and enhancing the strength of the CPDT samples. Both figures showed that the amount of bound water and the residual quality of OPC were greater than in NCM, indicating that there were more unstable hydration products in OPC; NCM produces more stable hydration products.

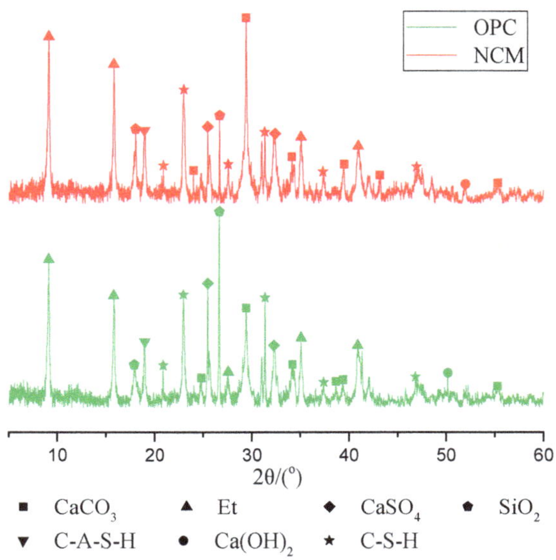

Figure 7. XRD images of NCM and OPC for 28 days of curing time.

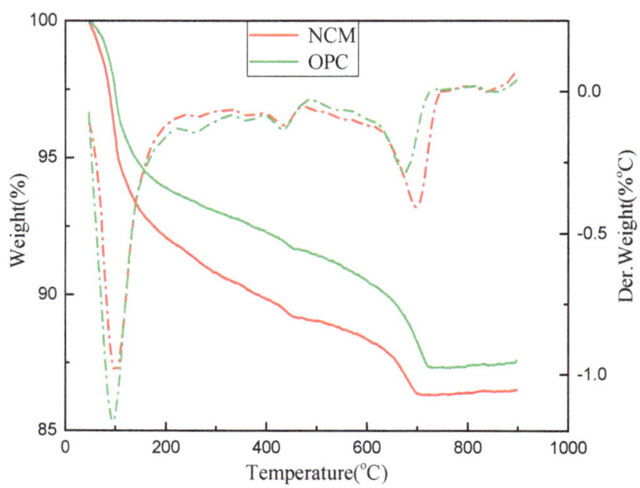

Figure 8. TG/ differential thermal gravimetric (DTG) images of NCM and OPC for 7 days of curing time.

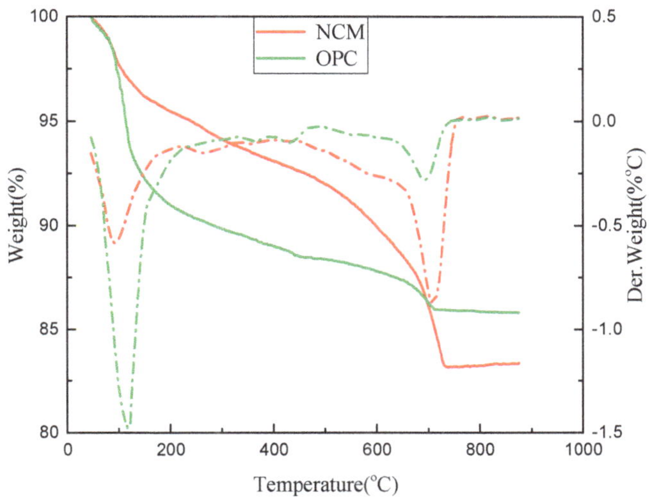

Figure 9. TG/DTG images of NCM and OPC for 28 days of curing time.

3.3. *The Hydration Mechanism of NCM and OPC*

The CPDT samples made of the optimum ratio of the NCM were cured at 20 ± 2 °C with relative humidity greater than 95% for 3, 7, and 28 days. The hydration reaction was then terminated with anhydrous ethanol and dried in a 50 °C oven until the quality no longer reduced. The samples were sprayed with gold and placed in an SEM experimental setup for microanalysis. The hydration products composition and microstructure of the samples with NCM determined the macroscopic strength of the CPDT samples. Figures 10–12 show the SEM images of the hydration products morphologies at 3, 7 and 28 days, respectively.

(**a**) Sample with NCM for 3 days (**b**) Sample with OPC for 3 days

Figure 10. SEM images of samples with NCM and OPC for 3 days of curing time.

In Figure 10a, the CPDT samples with NCM which had been hydrated for 3 days generated a large amount of ettringite and a small amount of C-S-H gels. While Figure 10b showed the sample with OPC hydrated for 3 days, the hydration product was only ettringite. In Figure 11a, the hydration reaction of the sample with NCM was observed for 7 days. The ettringite was reduced, which a large amount of C-S-H gels encapsulated the ettringite and the porosity decreased. The UCS of the sample with NCM showed great improvement, as seen in Table 6. On the other hand, in Figure 11b, the amount of C-S-H produced in the sample with OPC was less than that with NCM at 7 days, a large amount of ettringite was exposed on the surface, and the degree of envelopment for the ettringite was not as high as that with the NCM. In Figure 12a, the sample with NCM at 28 days produced a large amount of C-S-H gels and the ettringite was completely encapsulated and existed in entirety, which greatly increased the UCS of the CPDT sample. At the same time, in Figure 12b, the external surface of the sample with OPC was uneven. A large amount of C-S-H gels were generated too, however, some of the ettringite was still exposed on the surface, which showed that the amount of C-S-H generated was relatively less. This affects the UCS of the CPDT sample.

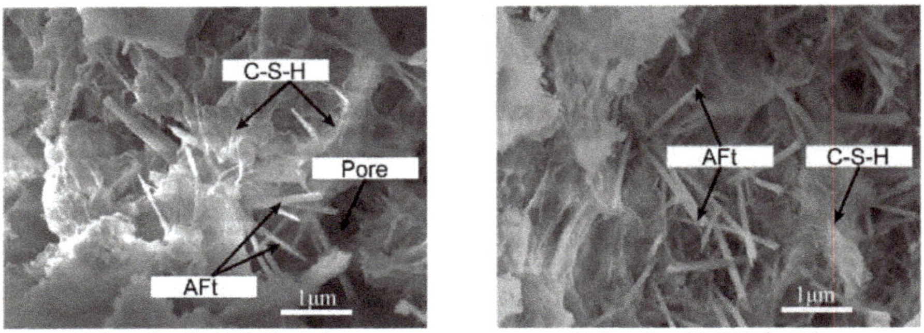

(**a**) Sample with NCM for 7 days (**b**) Sample with OPC for 7 days

Figure 11. SEM images of samples with NCM and OPC for 7 days of curing time.

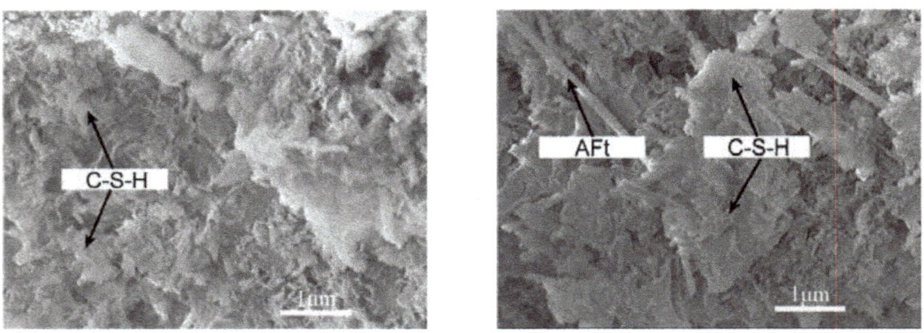

(**a**) Sample with NCM for 28 days (**b**) Sample with OPC for 28 days

Figure 12. SEM images of NCM and OPC samples for 28 days of curing time.

The reason why the microscopic structure images of the sample with NCM at the same ages were more compact than that with OPC can be explained as follows: when gypsum and lime are added to NCM, the lime provides Ca^{2+} (a raw material for hydration reactions) and OH^-. In the alkaline environment of the reaction, calcium sulfate provides Ca^{2+} and SO_4^{2-}, and SO_4^{2-} reacts with C_3A to produce secondary ettringite. The resulting ettringite was filled in the pores, and the strength of the samples was enhanced. The SO_4^{2-} was absorbed by the C-S-H so that the amount of ettringite was produced and thus the pore structure was coarsened. This was not only related to OH^- concentration, but also related to SO_4^{2-} concentration. A high concentration of SO_4^{2-} can increase the amount of secondary C-S-H and ettringite.

The BFS itself was not active, however, in the calcium hydroxide solution, a significant hydration reaction occurred, thereby inspiring its pozzolanic effect. Hydration was faster in the saturated calcium hydroxide solution. The hydration reaction was generally considered to be the following:

$$xCa(OH)_2 + SiO_2 + mH_2O \rightarrow xCaO{\cdot}SiO_2 \cdot nH_2O \tag{2}$$

$$xCa(OH)_2 + Al_2O_3 + mH_2O \rightarrow xCaO \cdot Al_2O_3 \cdot n\,H_2O \tag{3}$$

where the value of x depends on the type of mixture, the ratio of lime and active silica, the ambient temperature and the duration of the action. The value was generally equal to 1 or was slightly larger, and the value of n generally between 1 and 2.5.

The process of the CH interacting with SiO_2 was such that the amorphous silicic acid absorbs Ca^{2+} to form an indefinite component of the adsorption system, and an amorphous hydrated calcium silicate was then formed. This transformed into microcrystalline or crystalline calcium silicate gels after a long time period. CH interacts with Al_2O_3 to form calcium aluminate hydrate ($3CaO{\cdot}Al_2O_3{\cdot}6H_2O$). Hydrated calcium sulfoaluminate is a water-insoluble needle crystal that precipitated around the clinker particles and impeded moisture. Therefore, entry has played an important role in delaying the coagulation of cement-based materials. When gypsum was present in the liquid phase, it reacted with the calcium hydrated aluminate to form high-sulfur hydration calcium sulfoaluminate ($3CaO{\cdot}Al_2O_3{\cdot}3CaSO_4{\cdot}31H_2O$, ettringite). When the gypsum was completely consumed, part of it became monosulfur-type hydrated calcium sulfo-aluminate ($3CaO{\cdot}Al_2O_3{\cdot}3CaSO_4{\cdot}12H_2O$), which has a relatively high strength.

$$3CaO \cdot Al_2O_3 \cdot 6H_2O + 3(CaSO_4 \cdot 2H_2O) \rightarrow 3CaO \cdot Al_2O_3 \cdot 3CaSO_4 \cdot 31H_2O \quad (4)$$

$$3CaO \cdot Al_2O_3 \cdot 3CaSO_4 \cdot 31H_2O + 2(3CaO \cdot Al_2O_3 \cdot 6H_2O) \rightarrow 3(3CaO \cdot Al_2O_3 \cdot 3CaSO_4 \cdot 12H_2O) \quad (5)$$

3.4. The Pore Size Distribution of CPDT Samples with NCM and OPC

For research the effects of the new cement-based material on the pore size distribution of the tailings, the MIP tests on CPDT samples were carried out after curing 7 and 28 days. The volumes of invading mercury obtained by the MIP tests were compared with the pore diameters of the CPDT samples, which was plotted in Figure 13. According to the functional relationship of Equation (1), the pressure applied by the device was inversely proportional to the pore size. The intrusion volume of mercury increases as the intrusive pressure increases.

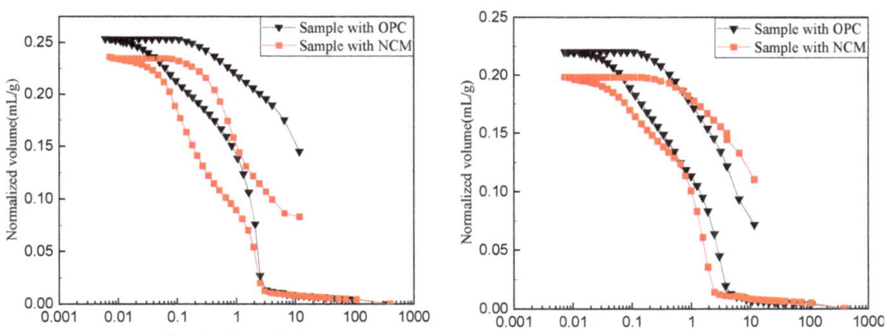

(**a**) Samples for 7 days (**b**) Samples for 28 days

Figure 13. The intrusion and extrusion curves for samples with different cement materials and curing time.

In Figure 13, the normalized volume increased slowly at the beginning of the lower pressure, where, primarily, pores larger than 7.2 μm were intruded. It increased rapidly at pore sizes of 4.9–6 μm. When the pressure caused the tailings particles rearrange, the normalized volume will rise to a larger value under smaller pressure. Thereafter, pores of less than 0.01 μm would be filled with mercury. However, even under nearly the highest pressure condition, it was difficult for mercury to enter the smallest and closed pores in the sample. Therefore, the mercury intrusion curve eventually tended to flatten.

In Figure 13a, for the normalized volume of the samples after curing for 7 days, the final normalized volume of the sample with NCM was about 8.0% larger than that with OPC. In Figure 13b, for the final normalized volume of the sample after curing for 28 days, the sample with NCM was about 11% larger than that with OPC. It can be seen from Figures 13a and 13b that after curing for 28 days, the porosity of the sample with OPC was reduced by 13%, and with NCM it was reduced by 18% compared with samples cured for 7 days. This showed that as the curing time increased, the porosity of samples with OPC and NCM decreased; however, the porosity of the sample with NCM decreased more, and the sample with OPC had a higher porosity than that with NCM, whether it was 7 days or 28 days. The low porosity indicated that the hydration reaction of the sample with NCM was quicker and more complete than that with OPC at 7 days and 28 days. More hydration products were generated and filled in the interparticle pores; the results made the samples more compact and stronger.

Figure 14 showed the Log-derivative mercury volume curves of the samples after curing for 7 and 28 days. For the 7 days sample, in Figure 14a, the most common pore diameters of the samples with OPC and NCM were 2.49 μm and 2.48 μm, respectively, representing a decrease of 0.4%. For the 28 days samples, in Figure 14b, the most common pore sizes of the samples with OPC and NCM were 1.93 μm and 1.62 μm, respectively, representing a decrease of 19%. From 7 days to 28 days, the most common pore size of the sample was reduced. The sample with OPC reduced by 29%, and with NCM reduced by 53%. This showed that during this time, the hydration reaction was kept on going, however, the hydration reaction of the sample with NCM was more complete, which was consistent with the analysis in Figure 13. In addition, for either the 7 days or the 28 days log-derivative mercury volume curves, the samples with NCM had a left shift relative to that with OPC, which indicated that the most probable pore size decreased.

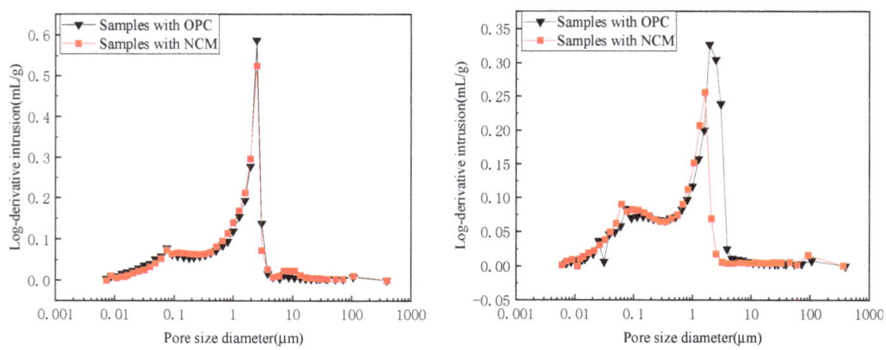

(**a**) Samples for 7 days (**b**) Samples for 28 days

Figure 14. The log-derivative pore volume curves for samples with different cement materials and curing time.

When the mercury intrusion of the sample with OPC occurred at 7 days and the pore size ranged from 0.6 μm to 1.9 μm, the value was larger than that with NCM, which means that in this range, the pore volume of the sample with OPC was relatively larger. From 0.09 to 1.9μm, the curves of the sample with NCM were higher than that with OPC, and from 0.008 to 0.09 μm, the curve of the sample with OPC was higher than that with NCM. This showed that the sample with NCM had more macropores transformed into mesopores relative to that with OPC, and some small pores turned into dense parts. When the mercury intrusion of the samples occurred at 28 days, and the pore size ranged from 0.5 to 1.5 μm, the curves of the sample with NCM were higher than that with OPC, and from 0.2 to 0.5 μm, the sample with OPC were higher than that with NCM. When it was less than 0.2 μm, the curves of the sample with NCM were higher than that with OPC. The reason for this was that the curve of the sample with NCM shifted to the left too much after 28 days; additionally, there were more macropores converted to small and medium pores.

4. Conclusions and Future Work

In this paper, a new type of cementitious material was produced and compared with OPC 42.5# with the hydration reaction characteristics after curing for 3, 7 and 28 days. The analyses of crystalline phases and amount of hydration products were carried out with XRD and TG/DTG. The SEM was used to analyze the formation of hydration products at different curing time. The MIP method was used to analyze the changes in pore size distribution after 7 days and 28 days. The major findings of this study included the following:

(1) The compound activator composed of gypsum, lime and clinker, together with some additives, had a good hydration activation effect on BFS. The ratio of the NCM was clinker: lime: gypsum: BFS equal to 14:6:10:70, the amount of admixture added was 0.4% of NCM and the ratio of sodium sulfate: alum: sodium fluorosilicate was 2:1:1. The UCS of the CPDT samples was 0.831 MPa at 3 days, 2.019 MPa at 7 days and 3.307 MPa at 28 days, which were 2.4, 2.4 and 1.7 times higher than those found with OPC. This indicates that the samples with NCM had the optimum UCS across all aging time.

(2) The hydration products crystalline phases of NCM and OPC were similar, however, a greater amount of hydration products formed in NCM at the same curing time, which resulted in a higher strength of CPDT samples with NCM. This showed NCM had a higher hydration reactivity.

(3) The NCM produced a large amount of ettringite and some C-S-H gels after 3 days. A large amount of C-S-H gels formed after 7 days. The ettringite was gradually wrapped and pores were filled. More C-S-H gels were formed after 28 days. The complete wrapping of ettringite greatly enhanced the UCS of the CPDT samples.

(4) Comparing the pore size distribution of the samples after curing for 7 and 28 days, the pore volume of the sample with OPC reduced by 13% and with NCM was reduced by 18%. The most probable pore size of the sample with OPC was reduced by 29%, and with NCM reduced by 53%. This means the porosity of all the samples reduced, however, the samples with NCM reduced more.

The study of the NCM in this paper was focused on CPDT, however, the NCM also applicable to cemented paste backfill. However, the experiments used in this paper were iron tailings, thus, the NCM may not necessarily be suitable for other metal or coal mines. In other mines, the ratio of cementitious materials needs to be re-tested. This paper provides ideas and methods for the study of cement-based materials, which was applicable to all mines.

Author Contributions: Y.H. and P.D. conceived and designed the experiments; P.D., X.Z. and D.H. performed the experiments; S.C. and D.H. collected and analyzed the data; Y.H. contributed materials/analysis tools; Y.H. and P.D. wrote the paper.

Funding: This work is supported by the National Natural Science Foundation of China (no.51674263) and the Fundamental Research Funds for the Central Universities (2011YZ02).

Conflicts of Interest: The authors declare no conflict of interest.

References

1. Li, W.; Fall, M. Sulphate effect on the early age strength and self-desiccation of cemented paste backfill. *Constr. Build. Mater.* **2016**, *106*, 296–304. [CrossRef]
2. Kesimal, A.; Ercikdi, B.; Yilmaz, E. The effect of desliming by sedimentation on paste backfill performance. *Miner. Eng.* **2003**, *16*, 1009–1011. [CrossRef]
3. Khademi, H.; Abbaspour, A.; Martínez-Martínez, S.; Gabarrón, M.; Shahrokh, V.; Faz, A.; Acosta, J.A. Provenance and environmental risk of windblown materials from mine tailing ponds, Murcia, Spain. *Environ. Pollut.* **2018**, *241*, 432–440. [CrossRef] [PubMed]
4. He, Z.; Wu, C. Research of Risk Assessment System on Tailings Pond Water Pollution. *Procedia Eng.* **2011**, *26*, 1788–1797. [CrossRef]

5. Wang, T.; Zhou, Y.; Lv, Q.; Zhu, Y.; Jiang, C. A safety assessment of the new Xiangyun phosphogypsum tailings pond. *Miner. Eng.* **2011**, *24*, 1084–1090. [CrossRef]
6. Milanović, P.T. *Dam Engineering and Its Environmental Aspects*; Springer: New York, NY, USA, 2012.
7. Zhang, L.; Huang, Y.; Wu, X.; Skibniewski, M.J. Risk-based estimate for operational safety in complex projects under uncertainty. *Appl. Soft Comput.* **2017**, *54*, 108–120. [CrossRef]
8. Mei, G.D. Quantitative Assessment Method Study Based on Weakness Theory of Dam Failure Risks in Tailings Dam. *Procedia Eng.* **2011**, *26*, 1827–1834. [CrossRef]
9. Men, Y.; Cai, J. Safety Status and Accident Control Measures of Tailings Storage in China. *Chin. J. China Safe Prod. Sci. Technol.* **2009**, *5*, 48–52. [CrossRef]
10. He, Z.; Tian, S.; Sui, L.; Xie, W. Mine Tailings Emission Status and Effective Ways of Disposal. *Chin. J. Min. Technol.* **2008**, *8*, 78–80. [CrossRef]
11. Hou, Y.; Tang, J.; Wei, S. Study on Tailings Consolidation Emissions Technology. *Chin. J. Met. Mine* **2011**, *06*, 59–62.
12. Fall, M.; Pokharel, M. Coupled effects of sulphate and temperature on the strength development of cemented tailings backfills: Portland cement-paste backfill. *Cem. Concr. Compos.* **2010**, *32*, 819–828. [CrossRef]
13. Li, W.; Zhou, X.; Liao, M.; Lin, Q. Research and Practice of Cement Filling Substitute Materials. *Chin. J. Min. Technol.* **2011**, *11*, 19–21. [CrossRef]
14. Kim, G.; Khalid, H.R.; Kim, H.J.; Lee, H.K. Alkali activated slag pastes with surface-modified blast furnace slag. *Cem. Concr. Compos.* **2017**, *76*, 39–47. [CrossRef]
15. Bong, S.; Hoon, H.; Yong, C. Effects of aluminate rich slag on compressive strength, drying shrinkage and microstructure of blast furnace slag cement. *Constr. Build. Mater.* **2017**, *140*, 293–300. [CrossRef]
16. Zhu, J. High Temperature Resistance of Alkali Slag Cementitious Materials and Its Application in Engineering. Ph.D.Thesis, Harbin Institute of Technology, Harbin, China, 2014.
17. Fu, X.; Hou, W.; Yang, C.; Li, D.; Wu, X. Studies on Portland cement with large amount of slag. *Cem. tConcr. Res.* **2000**, *30*, 645–649. [CrossRef]
18. Escalante, J.I.; Gómez, L.Y.; Johal, K.K.; Mendoza, G.; Mancha, H.; Méndez, J. Reactivity of blast-furnace slag in Portland cement blends hydrated under different conditions. *Cem. Concr. Res.* **2001**, *31*, 1403–1409. [CrossRef]
19. Agyei, N.M.; Strydom, C.A.; Potgieter, J.H. The removal of phosphate ions from aqueous solution by fly ash, slag, ordinary Portland cement and related blends. *Cem. Concr. Res.* **2002**, *32*, 1889–1897. [CrossRef]
20. Motz, H.; Geiseler, J. Products of steel slags an opportunity to save natural resources. *Waste Manag.* **2001**, *21*, 285. [CrossRef]
21. Vuk, T.; Tinta, V.; Gabrovšek, R.; Kaučič, V. The effects of limestone addition, clinker type and fineness on properties of Portland cement. *Cem. Concr. Res.* **2001**, *31*, 135–139. [CrossRef]
22. He, X.; Ma, M.; Su, Y.; Lan, M.; Zheng, Z.; Wang, T.; Strnadela, B.; Zeng, S. The effect of ultrahigh volume ultrafine blast furnace slag on the properties of cement pastes. *Constr. Build. Mater.* **2018**, *189*, 438–447. [CrossRef]
23. Yin, K.; Ahamed, A.; Lisak, G. Experimental investigation on the relationship between pore characteristics and unconfined compressive strength of cemented paste backfill. *Waste Manag.* **2018**, *78*, 401–416. [CrossRef]
24. Cihangir, F.; Ercikdi, B.; Kesimal, A.; Turan, A.; Deveci, H. Utilisation of alkali-activated blast furnace slag in paste backfill of high-sulphidemill tailings: Effect of binder type and dosage. *Miner. Eng.* **2012**, *30*, 33–43. [CrossRef]
25. Haha, M.B.; Lothenbach, B.; Saout, L.; Winnefeld, F. Influence of slag chemistry on the hydration of alkali-activated blast-furnace slag—Part I: Effect of MgO. *Cem. Concr. Res.* **2011**, *41*, 955–963. [CrossRef]
26. Yang, S.; Arvanitis, A.; Cao, Z.; Sun, X.; Dong, J. Synthesis of Silicalite Membrane with an Aluminum-Containing Surface for Controlled Modification of Zeolitic Pore Entries for Enhanced Gas Separation. *Processes* **2018**, *6*, 13. [CrossRef]
27. To, A. Characterization of Microscopic Pore Structures of Rock Salt through Mercury Injection and Nitrogen Absorption Tests. *Geofluids* **2018**, *2018*, 9427361. [CrossRef]
28. Pokharel, M.; Fall, M. Combined influence of sulphate and temperature on the saturated hydraulic conductivity of hardened cemented paste backfill. *Cem. Concr. Compos.* **2013**, *38*, 21–28. [CrossRef]
29. Bentz, D.P. A review of early-age properties of cement-based materials. *Cem. Concr. Res.* **2008**, *38*, 196–204. [CrossRef]

30. Tian, B.; Cohen, M.D. Does gypsum formation during sulfate attack on concrete lead to expansion? *Cem. Concr. Res.* **2000**, *30*, 117–123. [CrossRef]
31. Zhu, G.; Zheng, H.; Zhang, Z.; Tshukudu, T.; Zhang, P.; Xiang, X. Characterization and coagulation–flocculation behavior of polymeric aluminum ferric sulfate (PAFS). *Chem. Eng. J.* **2011**, *178*, 50–59. [CrossRef]
32. Yang, X.; Wang, J.; Hou, D.; Zhu, C.; He, M. Effect of Dry-Wet Cycling on the Mechanical Properties of Rocks: A Laboratory-Scale Experimental Study. *Processes* **2018**, *6*, 199. [CrossRef]
33. You, Z.; Lai, Y.; Zhang, M.; Liu, E. Quantitative analysis for the effect of microstructure on the mechanical strength of frozen silty clay with different contents of sodium sulfate. *Environ. Earth Sci.* **2017**, *76*, 143. [CrossRef]
34. Wu, S.; Yang, J.; Yang, R.; Zhu, J.; Liu, S.; Wang, C. Investigation of microscopic air void structure of anti-freezing asphalt pavement with X-ray CT and MIP. *Constr. Build. Mater.* **2018**, *178*, 473–483. [CrossRef]
35. Ghirian, A.; Fall, M. Coupled Behavior of Cemented Paste Backfill at Early Ages. *Geotech. Geol. Eng.* **2015**, *33*, 1141–1166. [CrossRef]
36. Zhang, Z.L.; Cui, Z.D. Effects of freezing-thawing and cyclic loading on pore size distribution of silty clay by mercury intrusion porosimetry. *Cold Reg. Sci. Technol.* **2018**, *145*, 185–196. [CrossRef]
37. Cui, Z.; Tang, Y. Microstructures of different soil layers caused by the high-rise building group in Shanghai. *Environ. Earth Sci.* **2011**, *63*, 109–119. [CrossRef]
38. Zhang, L.M.; Li, X. MicroPorosity Structure of Coarse Granular Soils. *J. Geotech. Geoenviron. Eng.* **2010**, *136*, 1425–1436. [CrossRef]
39. Cui, L.; Fall, M. An evolutive elasto-plastic model for cemented paste backfill. *Comput. Geotech.* **2016**, *71*, 19–29. [CrossRef]
40. Zhou, Q.; Glasser, F. Thermal Stability and Decomposition Mechanisms of Ettringite at <120 °C. *Cem. Concr. Res.* **2001**, *31*, 1333–1339. [CrossRef]
41. Alarconruiz, L.; Platret, G.; Massieu, E.; Ehrlacher, A. The use of thermal analysis in assessing the effect of temperature on a cement paste. *Cem. Concr. Res.* **2005**, *35*, 609–613. [CrossRef]
42. Fall, M.; Célestin, J.C.; Pokharel, M.; Touré, M. A contribution to understanding the effects of curing temperature on the mechanical properties of mine cemented tailings backfill. *Eng. Geol.* **2010**, *114*, 397–413. [CrossRef]

processes

MDPI

Article

Simulating the Filtration Effects of Cement-Grout in Fractured Porous Media with the 3D Unified Pipe-Network Method

Zizheng Sun [1], Xiao Yan [2], Weiqi Han [3], Guowei Ma [1,*] and Yiming Zhang [1]

1 School of Civil and Transportation Engineering, Hebei University of Technology, 5340 Xiping Road, Beichen District, Tianjin 300130, China; zizheng.sun@hebut.edu.cn (Z.S.); yiming.zhang@hebut.edu.cn (Y.Z.)
2 State Key Laboratory for Geo-Mechanics and Deep Underground Engineering, China University of Mining and Technology, 1 Daxue Road, Xuzhou 221116, China; Xiao.Yan@ruhr-uni-bochum.de
3 School of Public Health, Peking University, 38 Xueyuan Road, Beijing 100191, China; Weiq.Han@durham.ac.uk
* Correspondence: guowei.ma@hebut.edu.cn; Tel.: +86-022-6043-8005

Received: 8 December 2018; Accepted: 14 January 2019; Published: 16 January 2019

Abstract: In grouting process, filtration is the retention and adsorption of cement-grout particles in a porous/fractured medium. Filtration partly/even completely blocks the transportation channels in the medium, greatly decreasing its permeability. Taking into account filtration effects is essential for accurately estimating the grout penetration region. In this paper, the 3D unified pipe-network method (UPM) is adopted for simulating 3D grout penetration process in a fractured porous medium, considering filtration effects. The grout is assumed to exhibit two-phase flow, and the filtration effects depend on not only the concentration and rheology of the grout but also the porosity and permeability of the fractured porous medium. By comparing the model with the experimental results, we firstly verify the proposed numerical model. Then sensitivity analysis is conducted, showing the influences of grout injection pressures, the water–cement ratios of grout (W/C) and the grout injection rates on filtration effect. Finally, the grout filtration process in a complex 3D fractured network is simulated, indicating that the size of the grout penetration region is limited due to filtration.

Keywords: filtration effects; grout penetration; unified pipe-network method; two-phase flow; fractured porous medium

1. Introduction

Cement grouting is a widely used method for sealing underground water and improving the mechanical characteristics of surrounding media, such as fractured rocks, in geotechnical engineering. Over the past few decades, many on-site experiments have been conducted to assess the grouting quality [1–6] and to improve the strength and penetration distance. Some researchers remarked that cement particles can accumulate and deposit on the surfaces of pores/fractures [7–9], thereby blocking transport channels within the porous/fractured medium and terminating any further penetration of the grout. This phenomenon is called filtration, an irreversible process that significantly reduces the permeability of the porous/fractured medium [10–12]. Hence, assessments of the distance over which grout has penetrated will be greatly overestimated if the effects of filtration are ignored.

Researchers have experimentally investigated the influences of filtration on grouting with one-dimensional column injection tests considering constant grout injection pressures, injection rates, or cement grout particle concentrations. These researchers found that the cessation of cement grout penetration depends not only on the physical characteristics of the grout [13–15], such as the concentration of the cement grout particles and rheology of the grout, but also on the properties of the porous/fractured medium [16,17], such as the pore shape and fracture distribution. Furthermore, the permeability of the grouted medium decreases substantially in penetration zones [18–20]. Nevertheless, while the results of experimental investigations make it possible to verify proposed analytical and numerical models, their main disadvantage is that the influencing factors are very difficult to control as planned [21].

Many numerical and analytical methods, including phenomenological models [22], continuum models [23], trajectory analysis models [24], stochastic models [25], and network models [26–30], have been proposed to analyse the effects of filtration in porous but not fractured media. With these methods, some important conclusions have been drawn; for example, Saada et al. [19] and Maghous et al. [31] found that small variations in the porosity can lead to high variations in the permeability, as the injection of grout towards a void can be stopped by cement particle filtration, after which no significant reduction in the porosity will occur. Researchers have also proposed single [32] and multi-phase [19] models that are all applicable, but they have shown different precisions.

Moreover, research has been presented on the simulation of grout flow considering filtration effects in a fractured, porous medium [13,33]. Generally, fractures exist naturally within rock-like materials; thus, the initiation and propagation of such fracture need to be modelled by sophisticated numerical tools, such as the eXtended Finite Element Method (XFEM) [34–36], the strong discontinuity embedded approach [37–42], the mixed-mode FEM [43–46], and the phase-field model [47–49]. However, when focusing only on mass/energy transport within existing fractures and not their mechanical evolution, much simpler models can be used instead of these sophisticated techniques. The unified pipe-network method (UPM) is one of these simplified methods; it uses equivalent pipe networks to simulate the mass/energy transport processes within a porous/fractured medium. With the UPM, 2D and 3D mass/energy transport problems can be transformed into equivalent 1D problems, similar to the lattice element method (LEM) [50–52]. The fractures and porous medium can then be discretized using the same pipe-network system [53,54], which represents a significant advantage that provides a higher computing efficiency and greater numerical stability than other methods.

In this work, the grout filtration process within a 3D porous and fractured rock mass is simulated with a 3D UPM model. The grout flow is assumed to be an incompressible liquid [13,15,27]. In comparison with our former work [55], which focused on grout flow with a considerable time-dependent viscosity, we focus on the blocking effects of cement particles correlated with the microscopic structures of fractured, porous media. The remainder of this paper is organized as follows. In Section 2, the 3D two-phase filtration model is clearly introduced, and both the fluid flow and the mass transportation equations are given. In Section 3, the discretized equations of the filtration model based on the 3D UPM are established, thereby covering the interactions between the grout flow and the rock matrix and fractures. In Section 4, the developed numerical model is verified through a comparison with the experimental results given in [56] in tandem with a sensitivity analysis of the filtration process, including the grout injection pressure, grout injection rate and grout water–cement ratio. In Section 5, grout penetration in a porous and fractured rock is simulated considering different fracture inclination angles and random fracture networks. In Section 6, concluding remarks are provided.

2. Mathematical Model

We consider two-phase (fluid and solid particles) flow to study the filtration processes in a fractured and porous rock. This model is based on the interaction of Darcy's flow and mass transportation. The cement-based grout is considered to be a Newtonian fluid. The saturated single-phase grout flow in both the fractures and rock matrix obeys Darcy's law and can be calculated based on a mass conservation equation:

$$\frac{\partial}{\partial t}(\phi^{\tau}\rho) + \nabla \cdot (\rho \cdot \vec{u}) = \rho q, \tag{1}$$

where τ is a sign expressing the different media in the model domain, for example, $\tau = m$ expressing a matrix and $\tau = f$ expressing a fracture, where ϕ^m is the porosity of the rock matrix and ϕ^f is the porosity of a fracture; ρ is the grout density; q is the source term; and $\vec{u}$ is the flow velocity vector, which can be expressed as

$$\vec{u} = \frac{k^\tau}{\mu}(\nabla P - \rho\vec{g}), \tag{2}$$

in which μ is the grout fluid viscosity, P is the grout pressure, $\vec{g}$ is the gravitational acceleration and k^τ is the rock matrix and intrinsic fracture permeability tensor. The aperture of the fractures in the current simulation is assumed to be smooth, and the laminar flows of the grout in the fractures can be described by a cubic law; thus, the intrinsic fracture permeability can be calculated as $k^f = a^2/12$, where a is the fracture aperture.

For analyzing the transport of cement grout particles through a porous medium, the advection–dispersion equation is used with an additional term $(\partial\sigma/\partial t)$, which is defined as the rate of suspended particle mass transfer between the grout and the pore skeleton. Although the hydrodynamic dispersion coefficient is typically non-zero for multiphase flows in porous media [57], researchers have noted that the hydrodynamic dispersion tensors of porous media and fractures can be ignored since the effect of hydrodynamic dispersion on the dilution of grout is generally small [10,19,58–60]. Hence, the following relation is obtained for a porous medium:

$$\frac{\partial}{\partial t}(\phi^m C) + \vec{u}\cdot\nabla C = -\frac{\partial\sigma}{\partial t} = \lambda C, \tag{3}$$

where C is the concentration of transported particles, D^m is the hydrodynamic dispersion tensor of the porous medium, λ is the constant deposition coefficient and σ is the mass of grout particles being deposited per initial unit pore volume. In the above equation, the inhibition of particle deposition by inter-particle forces is assumed to be minimal [61], and the filtration process is modeled through a kinetic equation for mass transfer and represented by a first-order relationship [10,18,56].

Similarly, when assuming that filtration cannot proceed after cement grout particles have migrated into the fractures, the following relation is obtained for fractures:

$$\frac{\partial}{\partial t}(\phi^f C) + \vec{u}\cdot\nabla C = 0. \tag{4}$$

The filtration of cement grout particles is enhanced by a decrease in the volume of the void space. When the grout is injected into the porous medium, the porous medium becomes plugged, and the porosity decreases with space and time. The porosity is thus assumed to decrease in proportion to the mass of filtered particles as follows:

$$\phi^m = \phi_0^m - \frac{\sigma}{\rho_s}, \tag{5}$$

where ϕ_0^m is the initial porosity and ρ_s is the rock density.

The intrinsic permeability of the porous rock decreases with the clogging of the pore space. The experimental work performed by [19] proved that the permeability is dramatically influenced by the porosity, which means that very low variations in the porosity may induce large variations in the permeability. However, the well-known Kozeny–Carman equation expresses that the change in the permeability is consistent with the change in the porosity. Thus, a hyperbolic function of the (filtration-induced) porosity change is adopted in this paper to express the relationship between the porosity and permeability:

$$k^m = \frac{k_0^m}{1 + b(\phi^m - \phi_0^m)}, \tag{6}$$

where k_0^m is the initial intrinsic permeability and b is a negative scalar.

The concentration of cement grout particles has a considerable effect on its rheological properties [62]. Except for the porosity ϕ^τ and intrinsic permeability k^τ, the viscosity μ and the grout density ρ vary with time because of the cement solid filtration. The effective viscosity and density of the grout can be expressed as

$$\mu = \mu_w(1 + 2.5\delta), \tag{7}$$

$$\rho = \rho_w + (\rho_c - \rho_w)\delta, \tag{8}$$

where μ_w is the viscosity of water, ρ_w is the density of water, δ is the volume fraction of cement particles, which can be calculated as $\delta = C/\rho_s$, and ρ_c is the density of the cement particles, where $\rho_c = \rho_s$.

3. The Discretization Model for the 3D Unified Pipe-Network Method

In the framework of the 3D UPM, the equivalent hydraulic parameters and diffusion parameters of the fracture and matrix pipes are derived based on discrete unstructured elements.

Matrix Pipe-Network Model

The 3D porous medium is discretized using tetrahedral elements in this model, where each tetrahedron has six edges, and these elements are regarded as the flow pipes of the matrix, as shown in Figure 1a. When assuming that the edges of the tetrahedral elements are equivalent pipes, the mass/energy transport problem in a fractured and porous medium is transformed into an equivalent transport problem in the pipe network system. To obtain the equivalent parameters of the flow matrix pipes, the center of the sphere circumscribed around the tetrahedron (point o) and the circumcenters of the four faces of the tetrahedron (points cn, $(n = 1,2,3,4)$) are determined as division points (see Figure 1b). Linking the point o to the points c_n and to the midpoint of their joint edge in sequence encloses a partition face of the tetrahedron. The tetrahedron will form six partition faces in total, each of which is perpendicular to its respective intersecting edges. The tetrahedron is thus divided into four parts by these six partition faces, and the total pore volume of each part is assigned to its corresponding node, see Figure 1a.

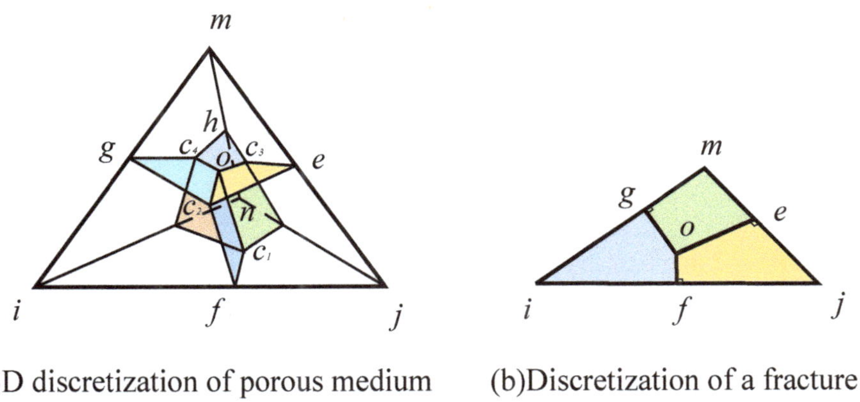

Figure 1. Discrete elements within fractured and porous media.

The equations for grout flow and cement grout particle transport through the control volume of a node are derived by formal integration methods, such as the finite volume method (FVM), expressed as follows:

$$\int_V \frac{\partial}{\partial t}(\phi^\tau \rho)dV + \int_A \vec{n} \cdot (\rho \cdot \vec{u})dA = \int_V \rho q dV, \tag{9}$$

$$\int_V \phi^m \frac{\partial C}{\partial t}dV + \int_A \vec{n} \cdot (\vec{u} \cdot \nabla C)dA = -\int_V \frac{\partial \sigma}{\partial t}dV, \tag{10}$$

$$\int_V \phi^f \frac{\partial C}{\partial t} dV + \int_A \vec{n} \cdot (\vec{u} \cdot \nabla C) dA = 0, \tag{11}$$

where V is the nodal control volume and A is the nodal control area generating the 3D Voronoi element.

The pressures and concentrations within a tetrahedral element can be written using the following interpolations:

$$P(x, y, z) = \sum N_k p_k \qquad (k = i, j, m, n), \tag{12}$$

$$C(x, y, z) = \sum N_k C_k \qquad (k = i, j, m, n), \tag{13}$$

where N_k is the linear shape function as in the FEM and P_k and C_k denote the pressures and grout concentrations, respectively. Here, we would like to emphasize that hydro-mechanical effects are ignored in the UPM; these effects require higher-order shape functions for the displacement field and lower-order shape functions for the pressure/concentration fields, such as the hybrid formulations used in [63–65], to avoid numerical instabilities [66,67]. When using the same shape functions for the pressure/concentration fields in our examples, we do not encounter any numerical problems.

The grout fluid flow Q_{ij}^m in pipe ij is equal to the flow through the area $oc1fc2$, see Figure 1a, and can be calculated as

$$Q_{ij}^m = Q_{oc1fc2}^m = \int_{Aoc1fc2} \vec{n}_{oc1fc2} \cdot \vec{u} = K_{ij}^m (P_i - P_j), \tag{14}$$

where $\vec{n}_{oc1fc2}$ is the unit normal vector, A_{oc1fc2} is the area of the face $oc1fc2$, P_i and P_j are the pressures at nodes i and j, respectively, and K_{ij}^m is the equivalent conductance coefficient of pipe ij, which can be written as [68–70]

$$K_{ij}^m = \frac{A_{oc1fc2} k^m}{l_{ij} \mu}, \tag{15}$$

where l_{ij} is the length of pipe ij. Furthermore, fractures in the 3D domain are represented as 2D disks. The equivalent hydraulic parameters of the pipes are derived from unstructured triangular elements, as shown in Figure 1b. The conductance coefficient K_{ij}^f for fracture pipe ij in the 3D domain is derived as [55]

$$K_{ij}^f = \frac{A_{of} k^f}{l_{ij} \mu} = \frac{l_{of} a^3}{l_{ij} \mu}. \tag{16}$$

The grout flow in each node obeys Darcy's law. Thus, within the framework of the UPM, for each node i, the governing equation for single-phase saturated flow is

$$\frac{\partial(\phi_i^m V_i^m)}{\partial t} + \sum_{j=1}^{n_i} Q_{ij}^m = Q_{s_i}, \tag{17}$$

where ϕ_i^m is the porosity of node i, V_i is the control volume of node i, the subscript n_i denotes the total number of connected pipes, and Q_{s_i} is the source term of node i.

The effective diffusion coefficient in a 3D porous medium can be derived using the same discretization method used for the aforementioned equivalent conductance coefficient. The mass of the grout particles $\dot{m}_{ij}^m$ transported in pipe ij is also equal to the mass being transported through the area $oc1fc2$ and can be calculated as

$$\dot{m}_{ij}^m = \dot{m}_{oc1fc2}^m = \int_{Aoc1fc2} \vec{n}_{oc1fc2} \cdot \vec{u} \cdot \nabla C = Q_{ij}^m \left(\frac{C_i + C_j}{2} \right), \tag{18}$$

where C_i and C_j are the concentrations at nodes i and j, respectively. The grout particles being transported through each node obey the nodal laws. For each node i, the governing equation for mass transport is

$$\frac{\partial(\phi_i^m V_i^m C_i)}{\partial t} + \sum_{j=1}^{n_i} \dot{m}_{ij}^m + \lambda V_i C_i = 0. \tag{19}$$

In the 3D fractured and porous rock, the fractures and rock matrix are reconstructed using 1D pipes. The pipes with the start and end nodes belonging to both the matrix and fracture pipes are defined as interface pipes, which means that both the fracture and matrix pipes will possess the same nodes at the interface. The nodal control volumes and the equivalent conductance coefficient and diffusion coefficients of the interface pipes are calculated by the superposition of the respective values of the fracture pipes and matrix pipes. In addition, the pore pressures and particle concentrations are assumed to be continuous at the interface without interaction between fractures and the matrix. Finally, the governing equations for each node are written as

$$\frac{\partial(\phi_i^\tau V_i^\tau)}{\partial t} + \sum_{j=1}^{n_i} Q_{ij}^\tau = Q_{s_i}, \tag{20}$$

$$\frac{\partial M_i^\tau}{\partial t} + \sum_{j=1}^{n_i} \dot{m}_{ij}^\tau + \lambda V_i^m C_i = 0. \tag{21}$$

4. Model Verification

4.1. Experimental Verification

Bouchelaghem [56] conducted a series of experiments on the flow and transport of microscopic to fine cement in sand using 1D column tests. These experiments use a well-graded and medium to coarse sand; thus, the porous medium can be considered to be homogeneous. A diagram of the grout injection experiment is given in Figure 2a. Following Bouchelaghem's experiments, the current numerical simulation uses a column with dimensions of 35 × 35 × 820 mm for a comparison with Bouchelaghem's results.

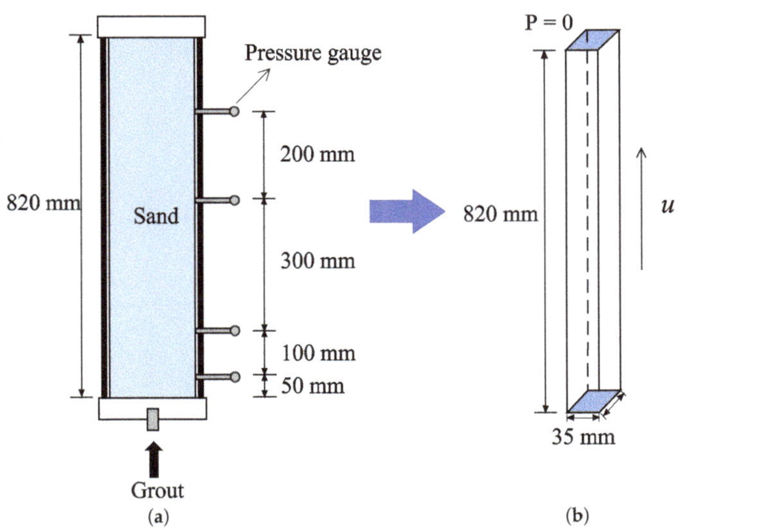

Figure 2. A schematic of grouting in sand. (**a**) 1D column injection test [56]; (**b**) the numerical verification model.

The coupled model of fluid flow and mass transport is used to simulate the grout filtration process. The parameters are listed in Table 1. In this model, the grout injection rate and grout concentration are constant at the inlet boundary, while the pressure is a constant at the outlet boundary and is defined as zero. The initial conditions at the four other transverse boundaries are such that $\vec{n} \cdot \nabla P = 0$ and $\vec{n} \cdot \nabla C = 0$.

Table 1. Parameters for the experimental verification simulation.

Parameter	Symbol	Unit	Value
Injection grout velocity	u	m/s	0.00117
Injection grout concentration	C	kg/m^3	1150
Water viscosity	μ_w	$Pa \cdot s$	0.001
Water density	ρ_w	kg/m^3	1000
Rock density	ρ_s	kg/m^3	3000
Initial matrix porosity	ϕ_0^m	–	0.34
Intrinsic matrix permeability	k_0^m	m^2	2×10^{-11}
Hydrodynamic dispersion tensor of	D_m	m^2/s	0
Constant deposition coefficient	λ	1/s	6×10^{-5}
Constant negative scalar	b	–	−120
Gravitational acceleration constant	g	m/s^2	9.8

Figure 3 shows the pore pressures at four elevations at different times under a constant injection velocity from both the experimental and the simulation results. The pore pressure evidently increases with time, and the pore pressures closer to the injection point are larger than those farther from the injection point. The pressure changes slowly far away from the injection point. The numerical results are in good agreement with the experimental results for the four elevations and are consistent with the numerical results obtained by Kim et al. [59] in comparison with the same experimental results. In summary, the key parameter causing the differences between the numerical simulations and experiments is the filtration rate, λ.

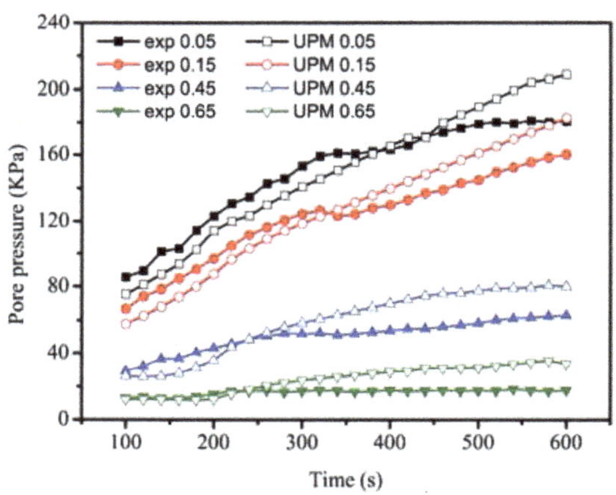

Figure 3. Comparison between the UPM model and the results obtained by Bouchelaghem [56].

4.2. Sensitivity Analysis of Grout Filtration

The grout injection pressure, injection rate and water–cement ratio of grout are three significant engineering parameters influencing the grouting effect, and all three are usually considered before engineering projects are initiated. In this sensitivity study, the grout filtration process with different grouting operations is modeled. The model is a column with dimensions of 5 × 2 × 2 m, as shown in Figure 4. The meshes are generated by self-developed mesh techniques [71], and the total number of grids in this model is 103,792. The transient flow pressure, mass concentration, porosity and permeability of rock are obtained and compared in time and space. The results are compared by analyzing the points' values along line AB, as shown in Figure 4, which is defined as the central line of the simulation model.

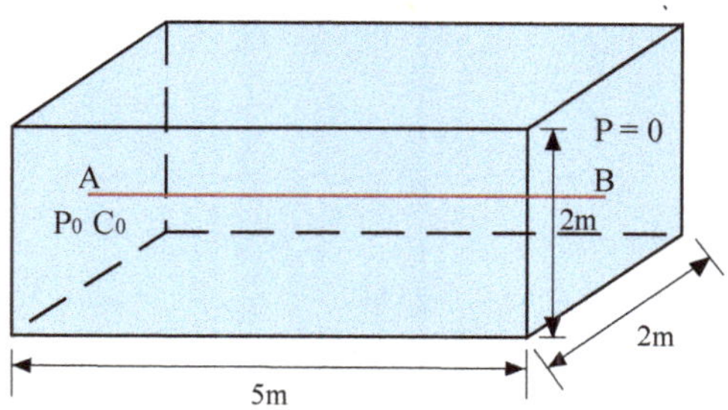

Figure 4. 3D porous rock medium.

In the current simulation, grout is injected into a cuboid domain under a constant pressure and constant concentration at the inlet boundary, while the pressure at the outlet boundary is zero. The parameters used in this case study are listed in Table 2. Figure 5 compares the variations in the pore pressure, particle concentration, rock porosity and permeability under two conditions, namely, (i) filtration and (ii) non-filtration, for the purpose of studying the influence of filtration on the grout penetration process. Under the first condition, the grout penetration parameters and rock characteristic parameters are extracted at time steps of $t = 800$ s, $t = 1600$ s, $t = 2400$ s, $t = 3200$ s and $t = 4000$ s, whereas, under the second condition, they are extracted at a time step of $t = 4000$ s.

Table 2. Parameters for the sensitivity analysis.

Parameter	Symbol	Unit	Value
Injection grout pressure	P	kPa	200
Injection grout concentration	C	kg/m^3	1000
Water viscosity	μ_w	$Pa \cdot s$	0.001
Water density	ρ_w	kg/m^3	1000
Rock density	ρ_s	kg/m^3	2940
Initial matrix porosity	ϕ_0^m	–	0.24
Intrinsic matrix permeability	k_0^m	m^2	1.25×10^{-11}
Hydrodynamic dispersion tensor of	D_m	m^2/s	0
Constant deposition coefficient	λ	$1/s$	6×10^{-5}
Constant negative scalar	b	–	−120
Gravitational acceleration constant	g	m/s^2	9.8

When ignoring the filtration effects, the porosity and permeability of the rock matrix are constant, and the pore pressure and particle concentration decrease linearly. However, in reality, the filtration makes these parameters vary nonlinearly. When considering the filtration effects, Figure 5a shows that the pore pressure decreases quickly with increases in the penetration length and injection time, indicating that it is necessary to increase the injection pressure with time to ensure the designed grout penetration size. A remarkable decrease in the cement concentration is observed, and the position at which this decrease occurs moves away from the injection position with time, as shown in Figure 5b. Furthermore, the concentration of cement particles also decreases quickly with time, which means that, under the same injection concentration, the grout particles move more slowly with time. Figure 5c,d show that the porosity and permeability decrease with time and increase with the distance from the injection position, indicating that the variation rate of the permeability is larger than that of the porosity.

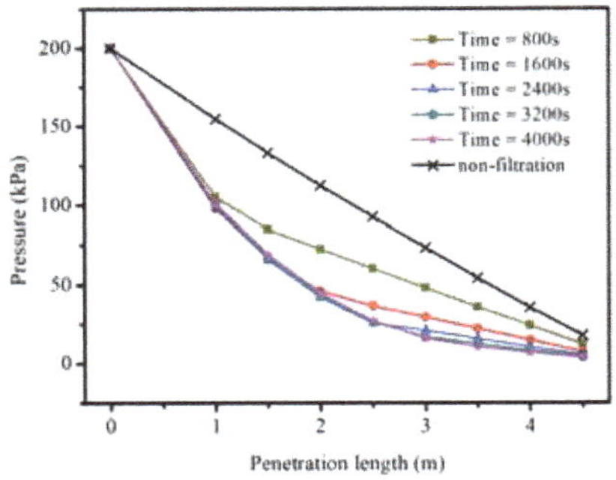

(a) Pressure

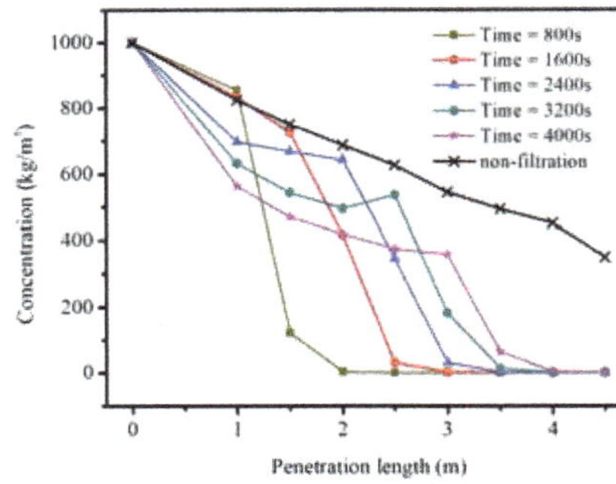

(b) Concentration

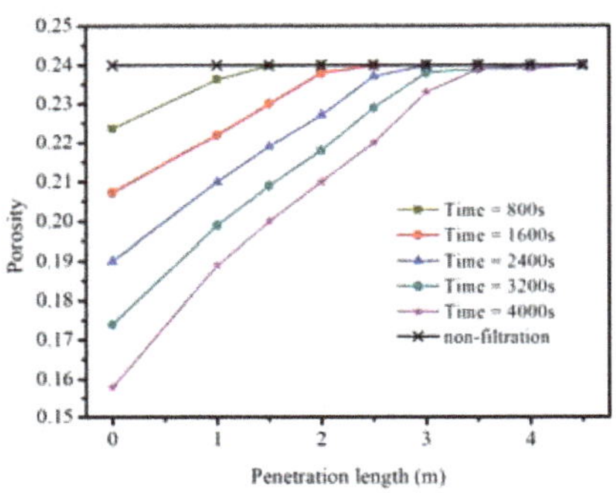

(c) Porosity

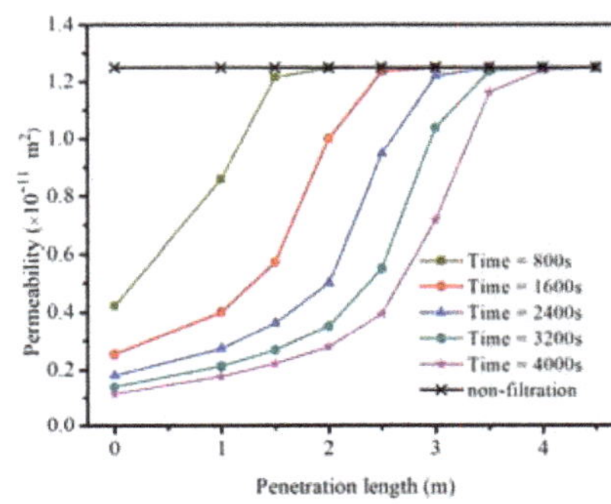

(d) Permeability

Figure 5. Variations in the pore pressure, concentration, porosity and permeability with the penetration length.

4.2.1. Effects of the Grout Injection Pressure

To analyze the influence of the grout injection pressure on the filtration, the injection pressure is varied from 30 kPa to 60 kPa, 100 kPa, 150 kPa and 200 kPa with a total injection time of 4000 s. As shown in Figure 6a, the pore pressure decreases quickly with increasing injection pressure. Figure 6b shows that the concentration decreases with the injection pressure. Although the injection pressures are different, the porosity and permeability of the rock matrix at the inlet boundary are the same, as shown in Figure 6c,d. This indicates that the injection pressure cannot influence the minimum porosity. The porosity and permeability decrease with increasing injection pressure at the same position, which explains the fact that the increase in injection pressure will increase the effect of pore sealing. Thus, in the same position, it is more difficult to inject the grout into the rock under the same constant pressure.

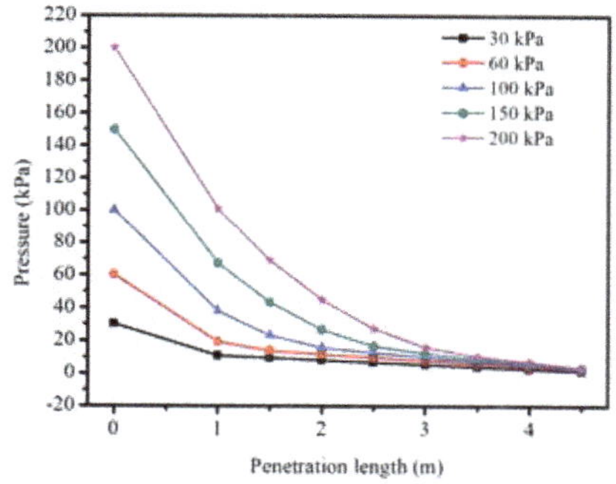

(a) Pressure

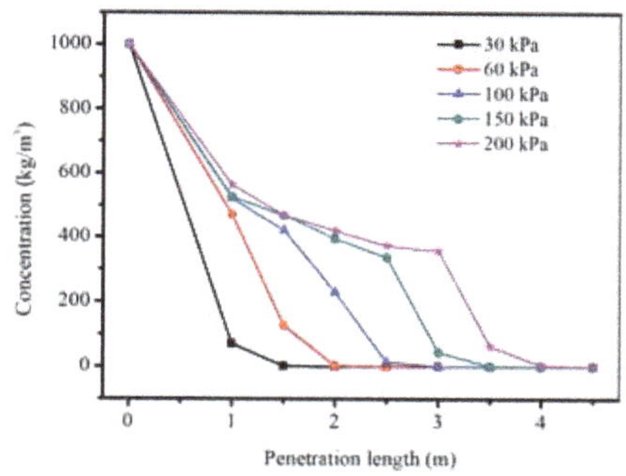

(b) Concentration

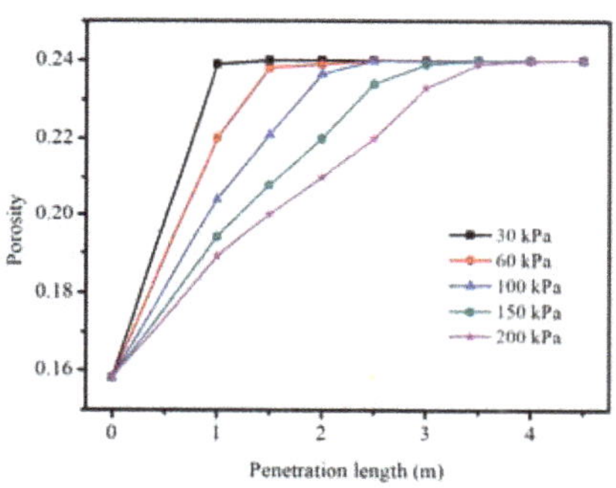

(c) Porosity

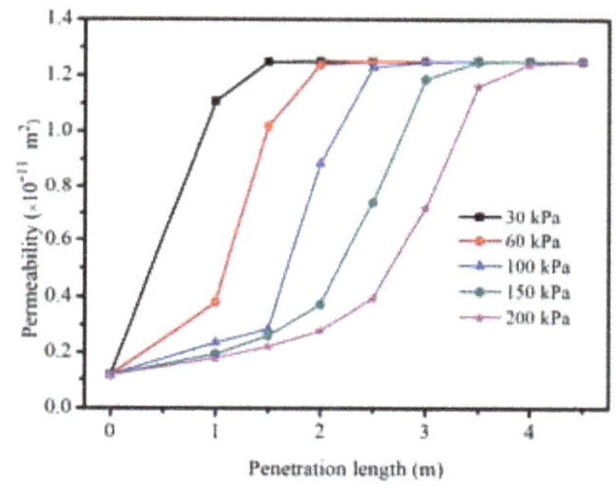

(d) Permeability

Figure 6. Influences of the injection pressure on the pore pressure, concentration, porosity and permeability.

4.2.2. Effects of the Water–Cement Ratio

Figure 7 shows the influence of the water–cement ratio on the filtration. In the simulation, the water–cement ratio is varied from 0.5 to 0.8, 1.0, 2.0 and 4.0. The simulation time is 4000 s. The change in the water–cement ratio is defined as the variation in the initial concentration of injected cement grout particles in the current model. As shown in Figure 7a, the pore pressure decreases with decreasing water–cement ratio. The variation in the pore pressure is almost linear when the grout is relatively dilute. There is a notable breaking point in Figure 7b when the water–cement ratio is lower, expressing that the penetration length of the grout will decrease with decreasing water–cement ratio. The grout with lower water–cement ratio shows more severe filtration effects, which are also indicated by the results of porosities and permeabilities (see Figure 7c,d)

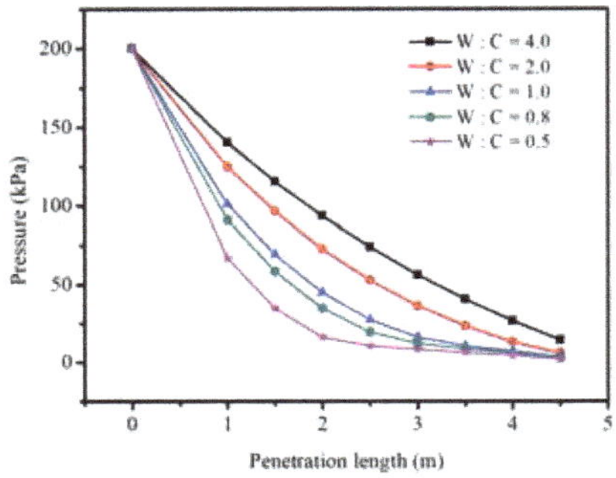

(a) Pressure

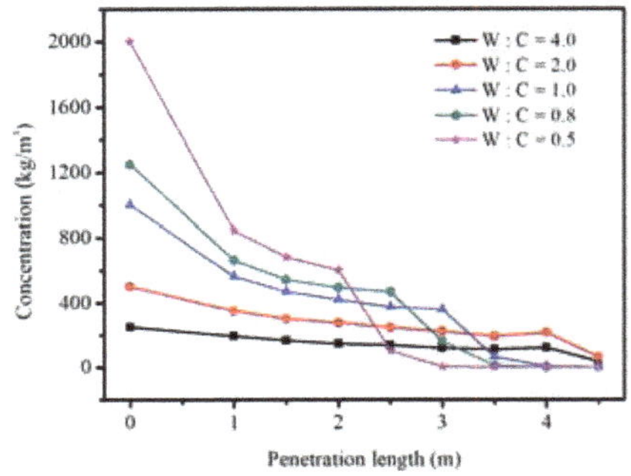

(b) Concentration

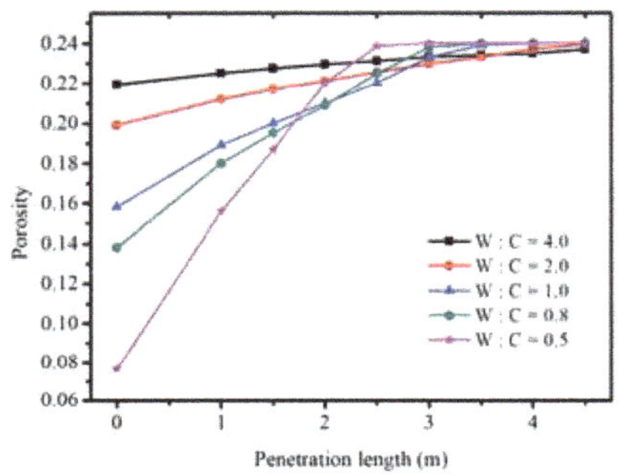

(c) Porosity

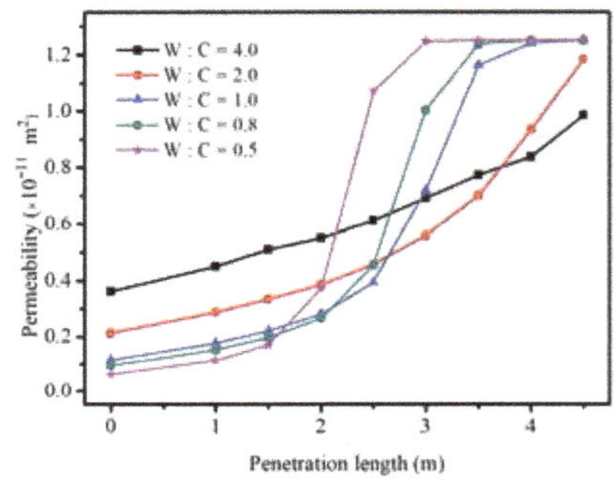

(d) Permeability

Figure 7. Influences of the water–cement ratio on the pore pressure, concentration, porosity and permeability.

4.2.3. Effects of the Grout Injection Rate

In this simulation model, the grout is injected into the rock at a constant rate. The injection rate is chosen as 0.0001, 0.0005, 0.001 , 0.005 and 0.01 m/s. When the injection rate is constant, although filtration has occurred, the pore pressure decreases almost linearly with the penetration length (Figure 8a). In addition, the pressure drops quickly with increasing injection rate. Furthermore, a small variation in the rate produces a significantly large variation in the pressure. Figure 8b shows that the concentration decreases with decreasing injection velocity, and Figure 8c,d show that the porosity and permeability increases with decreasing injection velocity. These observations show that the cement particles can move a longer distance and form a plug over a larger region when the grout velocity is higher. In addition, when the injection rate is low, the filtration needs substantially more time to occur. Thus, increasing the injection velocity can increase the influence of the filtration.

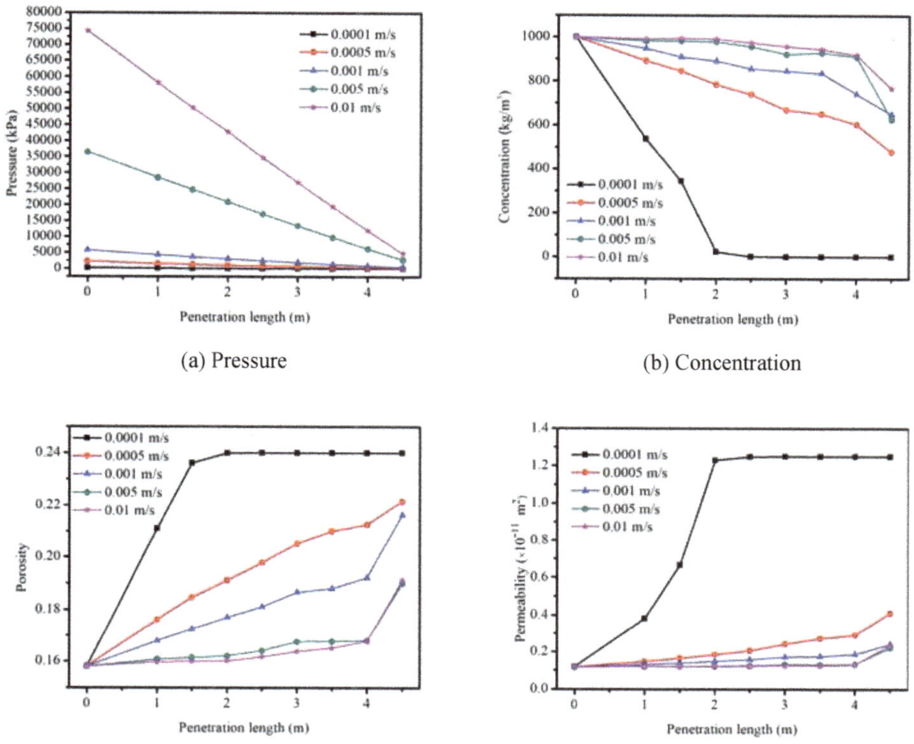

(a) Pressure

(b) Concentration

(c) Porosity

(d) Permeability

Figure 8. Influences of the injection velocity on the pore pressure, concentration, porosity and permeability.

5. Filtration in 3D Fractured Rock

The physical and mechanical properties of the fractures are important geologic features, therein influencing the penetration and filtration of the grout. In this section, the physical parameters of the fracture are varied by changing the inclination angle (α), the dip direction (β) and the aperture. The effects of the physical properties of a fracture on the grout penetration considering filtration are tested and analyzed. Furthermore, the fracture network is generated, and the grout filtration process in a porous medium with a random distribution of fractures is simulated. When the cement particle concentration of grout is lower, the grout becomes relatively dilute and can be regarded as water. Thus, in the current model, the grout penetration range is described as the zone in which the concentration is greater than 450 kg/m^3.

5.1. Effects of the Physical Properties of Fractures

In this simulation, the grout is injected into the rock with a constant injection pressure, and the parameters are as listed in Table 2. The simulation time is set as 4000 s.

The results of grout flow in a single fracture with different inclination angles are given in Figure 9. The fracture extends through the rock. The fracture inclination angle (α) varies from 0° to 15°, 30°, 45°, 60° and 90°. The dip direction of the fracture is 0°, and the aperture is 1 mm. As shown in Figure 9, the existence of a fracture has considerable influences on the filtration process. The grout flows mainly along the fracture, and it flows quickly when approaching the fracture because the conductance of the fracture is much larger than that of the matrix. The inclination angle of the fracture changes the region of grout penetration in the domain, and the distribution of cement particles in the voids changes. In contrast, the filtration effect is inconspicuous in the fracture, as the porosity and permeability of the fracture do not change with the cement concentration or time.

Figure 10 shows the grout penetration distance in the porous medium under two scenarios: (i) considering the filtration effects and (ii) ignoring the filtration effects. Comparing Figure 10a with Figure 10b, the filtration has an obvious and remarkable influence on the grout penetration length. Filtration will stop the grout from flowing, resulting in a maximum grout penetration distance. Figure 11 shows the grout penetration in a porous medium with a single fracture in different dip directions. The dip direction of the fracture is defined to be 30°, 45°, 60° and 75°. The inclination angle is 90°, and the aperture of the fracture is 1 mm. The trend characterizing the grout flow is similar to that in the fractures at different inclination angles. A fracture with a different dip direction will redirect the flow of grout.

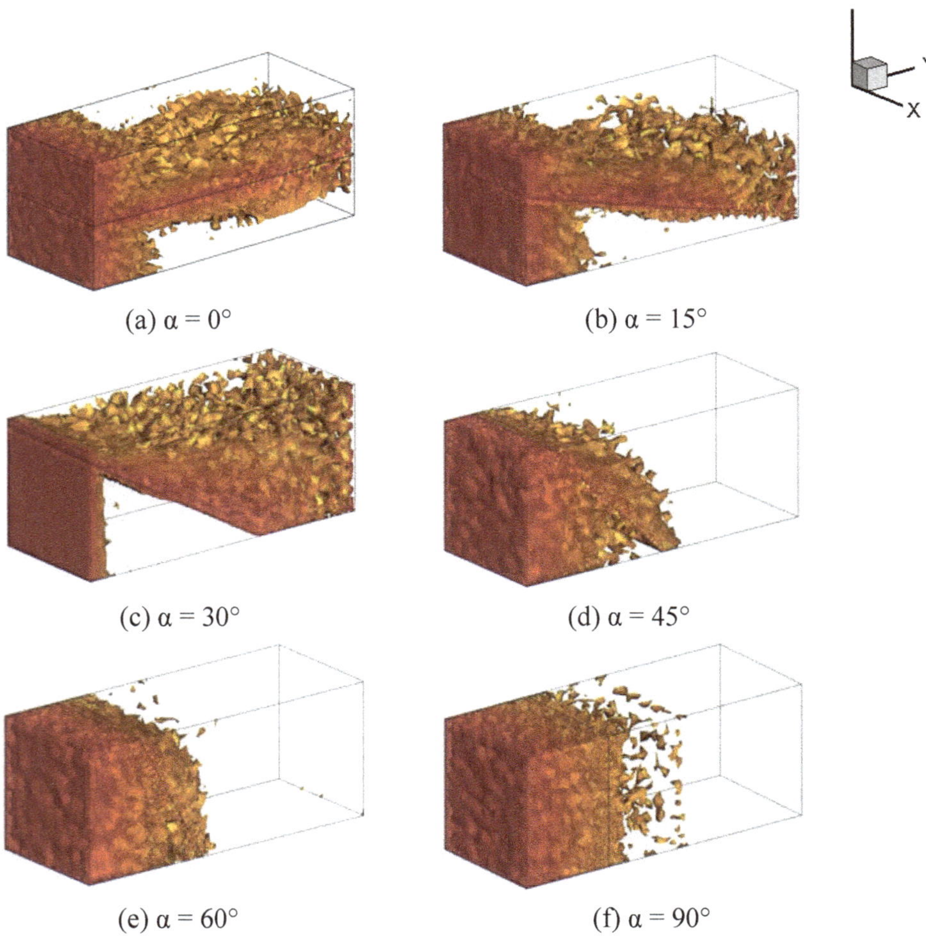

Figure 9. Grout penetration with a single fracture at different inclination angles in the porous medium considering the filtration process.

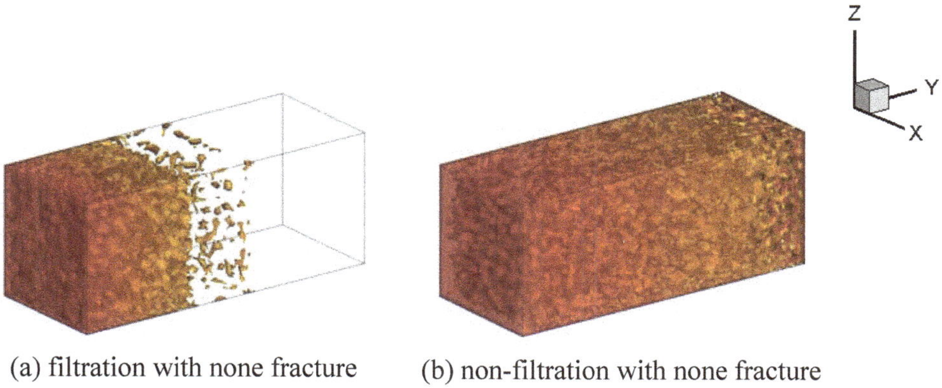

Figure 10. Grout penetration in a porous medium.

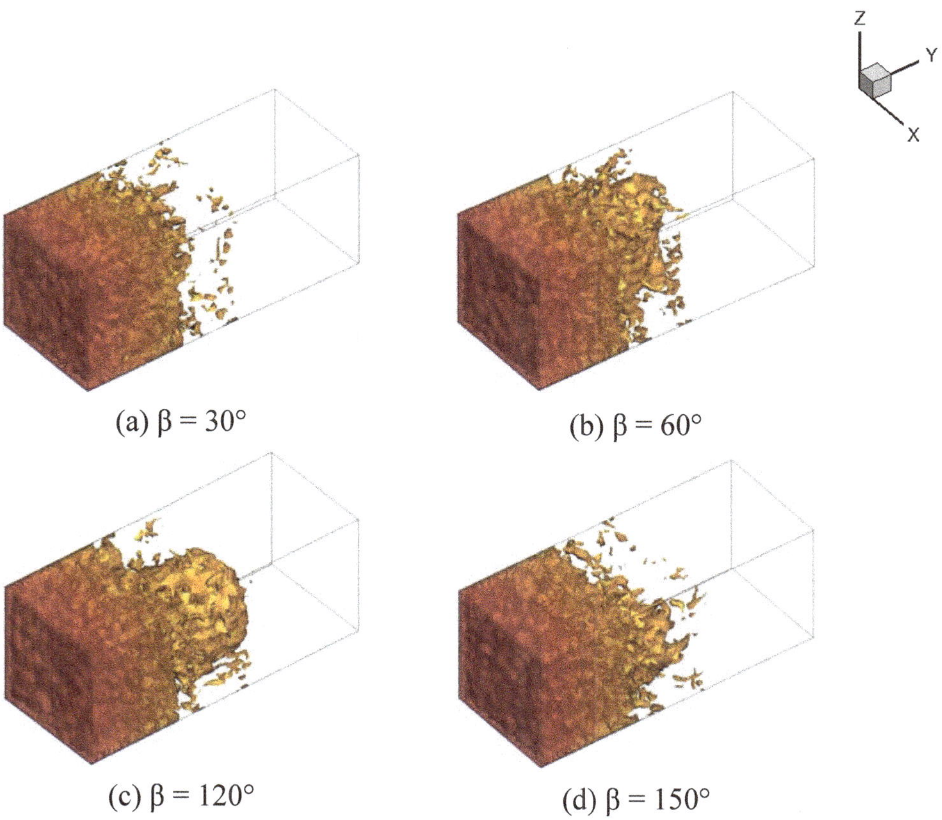

Figure 11. Grout penetration with a single fracture with different dip directions in the porous medium considering the filtration process.

The permeability of the fracture is related to the aperture of the fracture. Figure 12 shows the influence of the fracture aperture on the grout penetration. In this simulation, the inclination angle and dip direction of the fracture are 90° and 60°, respectively. The aperture of the fracture is varied from 0.01 mm to 0.1 mm, 1 mm and 10 mm. In the cases where the permeability of the fracture is substantially larger than that of the rock matrix, the grout flows quickly along the fracture and completely fills the fracture. The matrix pores along the fracture wall become completely clogged by the cement particles. The fracture with a large aperture acts as a path for grout flow, whereas the larger aperture cannot influence the final grout penetration region, as shown in Figure 12a,b because, although the grout penetration into a fracture with a larger aperture can spread quickly, the filtration process in the matrix porosity stops the grout from further penetrating. When the aperture of the fracture is small, the permeability of the fracture is almost the same or less than that of the rock matrix. The velocity of the grout penetration into the fracture is equal to that into the rock matrix; therefore, the fracture cannot redirect the flow of the grout (see Figure 12c,d). Hence, a fracture with a smaller aperture has a minimal effect on the grout filtration process.

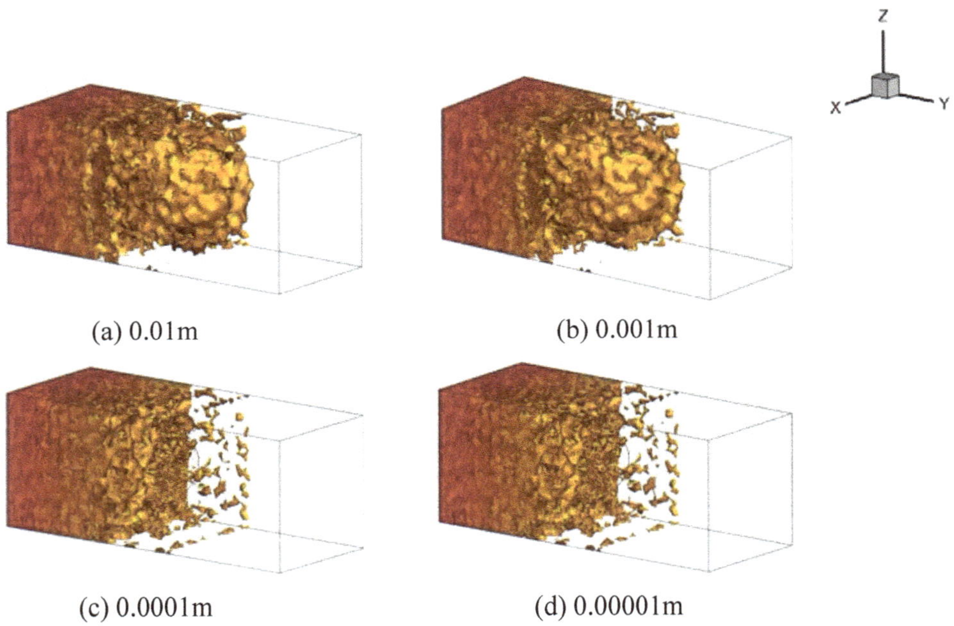

Figure 12. Grout penetration with a single fracture under different apertures in the porous medium considering the filtration process.

5.2. *Effects of the Fracture Network*

The size, dip angle, shape and distribution of fractures in a rock mass are very complicated and diverse. In the current simulation model, a fracture network using a cube model with a size of $5 \times 5 \times 5$ m is generated for exploring the grout penetration in a complicated fractured andporous rock. The aperture of the fracture is set as 0.1 mm. As shown in Figure 13a, two sets of fractures (with ten in each group) are generated randomly. The center coordinate of the fractures follows a normal distribution, and the widths and lengths of the rectangular fractures follow logarithmic distributions. The Fisher coefficient is 22. The randomly distributed parameters for the fracture geometries in this numerical model are listed in Table 3. All fractures are meshed with triangles (shown in Figure 13b), and the rock matrix is meshed with a tetrahedron (shown in Figure 13c) based on the conforming mesh techniques proposed by Wang [71].

Table 3. Parameters adopted for the generation of the fracture models.

Group	Fracture Number	Mean Length (m)		St dev	Dip Angle (Degree)	Dip Direction (Degree)
1	10	x-axis	1.2	0.5	16	40
		y-axis	1	0.5		
2	10	x-axis	1.2	0.5	40	250
		y-axis	1	0.5		

Note: St dev denotes the standard deviation.

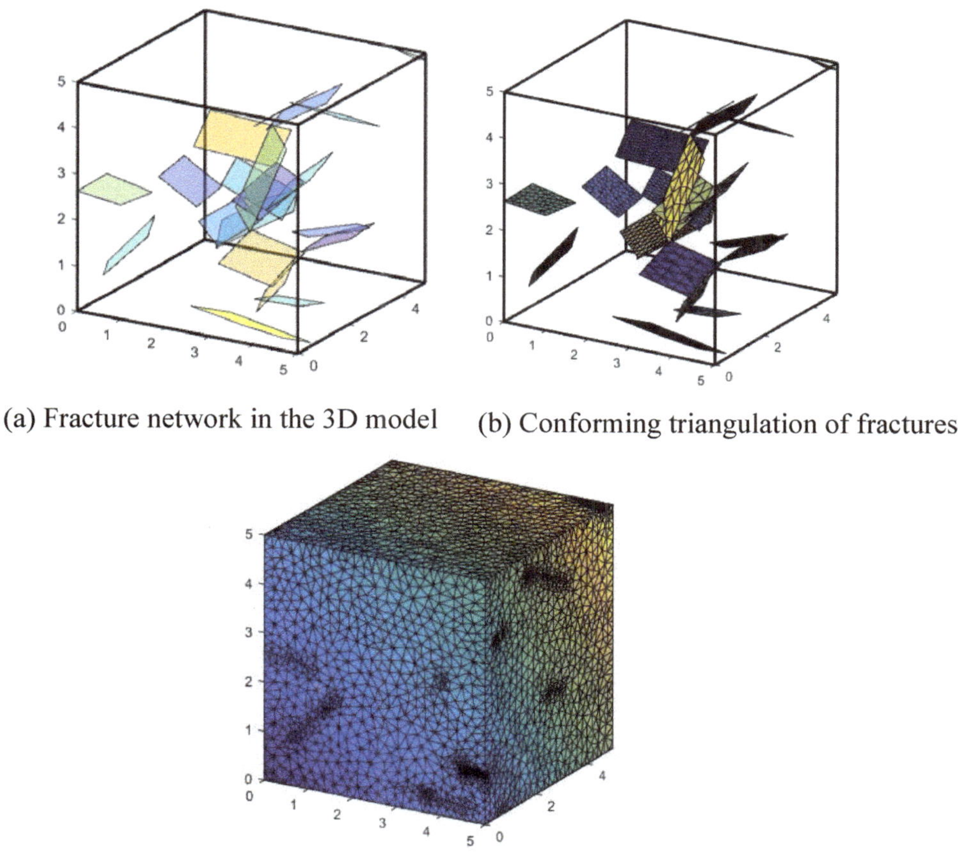

(a) Fracture network in the 3D model (b) Conforming triangulation of fractures

(c) Tetrahedrization of the domain

Figure 13. Distribution of fractures in the porous, fractured rock mass.

The grout is injected into the rock mass with an injection borehole located in the center of the model. The grout flow is modeled and analyzed under conditions with and without fracture networks at a constant injection pressure. The boundary conditions for the top and bottom faces are $\vec{n} \cdot \nabla P = 0$ and $\vec{n} \cdot \nabla C = 0$. The outlet boundaries are considered open boundaries. The pressure is defined as zero, and the concentration gradients are also defined as zero in the normal directions of these boundaries. The constant injection pressure (50 kPa) and the constant cement concentration (1000 $\mathrm{kg/m^3}$) are set in the injection borehole.

Figures 14 and 15 show the grout flow processes influenced by filtration at four time steps. As shown in Figure 14, in the case without fractures, the grout penetration into the matrix is uniform near the injection borehole and forms a spherical grout area because the distribution of the porosity is homogeneous. The grout flow spreads outward from the borehole with time. However, the grout flow almost stops after 2000 s, as it is affected by filtration. When fractures exist in the domain, the range of the grout penetration and the grout direction change significantly, as shown in Figure 15. In the case with fractures, the grout initially penetrates along the fractures upon reaching them, and the grout spread speed is higher in the fractures than in the rock matrix. The grout fills the entire fracture and then flows along the direction of the fracture walls. The grout flow in the matrix stops when the calculation time exceeds 2000 s, while it continues to penetrate into nearby fractures (see Figure 15c,d). The fracture is less influenced by filtration. Thus, a larger pressure or more injection boreholes are needed for the grout to reach the expected point in such a domain.

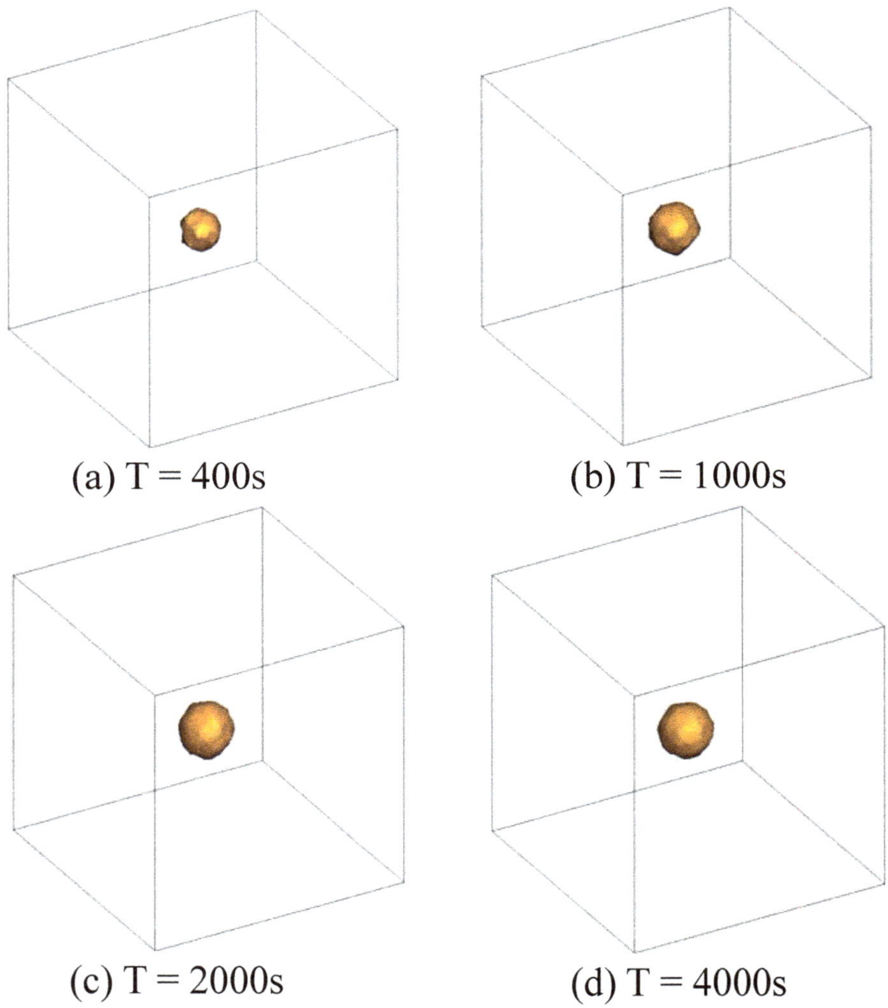

Figure 14. Grout penetration processes at different time steps with a constant injection pressure (P = 50 kPa) in a homogeneous, porous rock considering filtration.

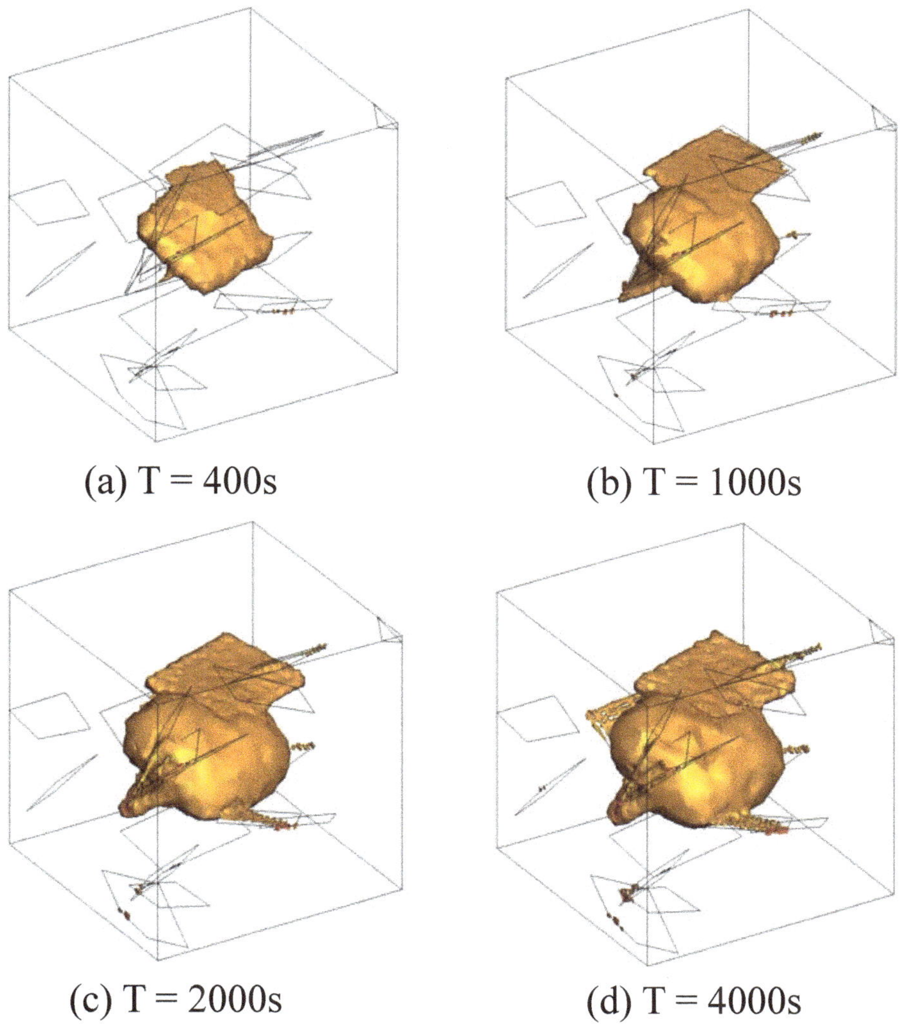

Figure 15. Grout penetration processes at different time steps with a constant injection pressure (P = 50 kPa) in a fractured and porous rock considering filtration.

6. Conclusions

In this study, we propose a numerical 3D UPM to simulate grout flow in fractured and porous rocks considering the effects of filtration. A two-phase model is used for the grout flow. A hyperbolic function is used to determine the relationship between the porosity and permeability that is consistent with experimental results [19]. The proposed model is validated by comparing the numerically obtained results with the experimental results. Then, sensitivity analyses in consideration of the injection pressure, the water–cement ratio of the grout and the injection rate are conducted, thereby demonstrating the reliability of the proposed numerical method. Finally, the influences of the inclination angle, the dip direction, the aperture of the fracture and the complex fracture network on the filtration process are analyzed.

The simulation results indicate that filtration will stop the grout from penetrating farther into the porous and fractured medium. A high grouting pressure, a high water–cement ratio and a high grout injection rate can each increase the retention of cement particles while prolonging the penetration process. Furthermore, the existence of fractures in the domain has considerable influences on the grout flow. The inclination angle and dip direction of the fracture can redirect the flow of the grout and the grout penetration region can change because the permeability of the fracture is much larger than that of the rock matrix; thus, the grout flows quickly along the fracture and completely fills the fracture. When the grout flows into a randomly distributed fracture network, the fractures play a prominent role and influence the flow path of the grout. Although the grout stops penetrating into the rock matrix, it continues spreading into the fractures near it and extends along the fracture edges.

Author Contributions: Conceptualization, G.M. and Z.S.; Methodology, Z.S., X.Y. and Y.Z.; Software, Z.S. and X.Y.; Validation, Z.S. and X.Y.; Formal Analysis, Z.S., X.Y. and W.H.; Investigation, Z.S.; Resources, G.M.; Data Curation, Z.S. and X.Y.; Writing—Original Draft Preparation, X.Y., Z.S. and W.H.; Writing—Review and Editing, Z.S., W.H. and Y.Z.; Visualization, X.Y.; Supervision, G.M.; Project Administration, G.M.; Funding Acquisition, G.M.

Funding: The authors gratefully acknowledge the financial support offered by NSFC 51809069 and Hebei key research and development program 18216110D.

Conflicts of Interest: The authors declare no conflict of interest.

Abbreviations

The following abbreviations are used in this manuscript:

UPM Unified Pipe network Method

References

1. Arenzana, L.; Krizek, R.; Pepper, S. Injection of dilute microfine cement suspensions into fine sands. In Proceedings of the 12th International Conference on Soil Mechanics and Foundation Engineering, Rio de Janeiro, Brazil, 13–18 August 1989.
2. Eriksson, M. Grouting field experiment at the äspö hard rock laboratory. *Tunnel. Undergr. Space Technol.* **2002**, *17*, 287–293. [CrossRef]
3. Lees, E.L.D.; Naudts, A. New developments in permeation grouting: Design and evaluation. *Concr. Int.* **2000**, *22*, 47–52.
4. Nikbakhtan, B.; Osanloo, M. Effect of grout pressure and grout flow on soil physical and mechanical properties in jet grouting operations. *Int. J. Rock Mech. Min. Sci.* **2009**, *46*, 498–505. [CrossRef]
5. Ding, W.; Duan, C.; Zhu, Y.; Zhao, T.; Huang, D.; Li, P. The behavior of synchronous grouting in a quasi-rectangular shield tunnel based on a large visualized model test. *Tunnel. Undergr. Space Technol.* **2019**, *83*, 409–424. [CrossRef]
6. Axelsson, M.; Gustafson, G. The penetracone, a new robust field measurement device for determining the penetrability of cementitious grouts. *Tunnel. Undergr. Space Technol.* **2010**, *25*, 1–8. [CrossRef]
7. Zebovitz, S.; Krizek, R.; Atmatzidis, D. Injection of fine sands with very fine cement grout. *J. Geotechn. Eng.* **1989**, *115*, 1717–1733. [CrossRef]
8. McDowell-Boyer, L.M.; Hunt, J.R.; Sitar, N. Particle transport through porous media. *Water Resour. Res.* **1986**, *22*, 1901–1921. [CrossRef]
9. Sharma, M.; Yortsos, Y. Transport of particulate suspensions in porous media: model formulation. *AIChE J.* **1987**, *33*, 1636–1643. [CrossRef]
10. Kim, Y.S.; Whittle, A.J. Filtration in a porous granular medium: 1. Simulation of pore-scale particle deposition and clogging. *Transp. Porous Med.* **2006**, *65*, 53–87. [CrossRef]
11. Axelsson, M.; Gustafson, G.; Fransson, A. Stop mechanism for cementitious grouts at different water-to-cement ratios. *Tunnel. Undergr. Space Technol.* **2009**, *24*, 390–397. [CrossRef]
12. Hwang, H.; Yoon, J.; Rugg, D.; El Mohtar, C. Hydraulic conductivity of bentonite grouted sand. In *Geo-Frontiers 2011: Advances in Geotechnical Engineering*; American Society of Civil Engineers: Reston, VA, USA, 2011; pp. 1372–1381.
13. Eriksson, M.; Stille, H.; Andersson, J. Numerical calculations for prediction of grout spread with account for filtration and varying aperture. *Tunnel. Undergr. Space Technol.* **2000**, *15*, 353–364. [CrossRef]

14. Abichou, T.; Benson, C.H.; Edil, T.B. Micro-structure and hydraulic conductivity of simulated sand-bentonite mixtures. *Clays Clay Min.* **2002**, *50*, 537–545. [CrossRef]
15. Dupla, J.-C.; Canou, J.; Gouvenot, D. An advanced experimental set-up for studying a monodirectional grout injection process. *Proc. Inst. Civ. Eng.-Ground Improv.* **2004**, *8*, 91–99. [CrossRef]
16. Draganović, A.; Stille, H. Filtration and penetrability of cement-based grout: Study performed with a short slot. *Tunnel. Undergr. Space Technol.* **2011**, *26*, 548–559. [CrossRef]
17. Eriksson, M.; Stille, H. A Method for Measuring and Evaluating the Penetrability of Grouts. In Proceedings of the Third International Conference on Grouting and Ground Treatment, New Orleans, LA, USA, 10–12 February 2003; pp. 1326–1337.
18. Bouchelaghem, F.; Vulliet, L.; Leroy, D.; Laloui, L.; Descoeudres, F. Mathematical and numerical filtration-advection-dispersion model of miscible grout propagation in saturated porous media. *Int. J. Numer. Anal. Methods Geomech.* **2001**, *25*, 1195–1227. [CrossRef]
19. Saada, Z.; Canou, J.; Dormieux, L.; Dupla, J.; Maghous, S. Modelling of cement suspension flow in granular porous media. *Int. J. Numer. Anal. Methods Geomech.* **2005**, *29*, 691–711. [CrossRef]
20. Yoneda, S.; Okabayashi, S.; Baba, O.; Tamura, M.; Mori, A. Permeating properties of ultra-fine cement grout. In *Grouting and Deep Mixing*; Balkema: Rotterdam, The Netherlands, 1996; pp. 107–113.
21. Choo, C.-U.; Tien, C. Analysis of the transient behavior of deep-bed filtration. *J. Colloid Interface Sci.* **1995**, *169*, 13–33. [CrossRef]
22. Herzig, J.; Leclerc, D.; Goff, P.L. Flow of suspensions through porous media-application to deep filtration. *Ind. Eng. Chem.* **1970**, *62*, 8–35. [CrossRef]
23. Chupin, O.; Saiyouri, N.; Hicher, P.-Y. Modeling of a semi-real injection test in sand. *Comput. Geotech.* **2009**, *36*, 1039–1048. [CrossRef]
24. Rege, S.; Fogler, H. *A Network Model for Deep Bed Filtration of Solids And Emulsions*; Tech. Rep.; American Institute of Chemical Engineers: New York, NY, USA, 1988.
25. Tarafdar, S.; Dey, A.; Gupta, B.S. A multiple state stochastic model for deep-bed filtration. *Chem. Eng. Technol. Ind. Chem.-Plant Equip.-Process Eng.-Biotechnol.* **1992**, *15*, 44–50. [CrossRef]
26. Ghidaglia, C.; de Arcangelis, L.; Hinch, J.; Guazzelli, E.Transition in particle capture in deep bed filtration. *Phys. Rev. E* **1996**, *53*, R3028. [CrossRef]
27. Nikiforov, A.; Nikanshin, D. Simulation of transfer of solid particles by a filtration flow. *J. Eng. Phys. Thermophys.* **1998**, *71*, 933–938. [CrossRef]
28. Lee, J.; Koplik, J. Network model for deep bed filtration. *Phys. Fluids* **2001**, *13*, 1076–1086. [CrossRef]
29. Liu, R.; Li, B.; Jiang, Y.; Jing, H.; Yu, L. Relationship between equivalent permeability and fractal dimension of dual-porosity media subjected to fluid–rock reaction under triaxial stresses. *Fractals* **2018**, *26*, 1850072. [CrossRef]
30. Liu, R.; Li, B.; Yu, L.; Jiang, Y.; Jing, H. A discrete-fracture-network fault model revealing permeability and aperture evolutions of a fault after earthquakes. *Int. J. Rock Mech. Min. Sci.* **2018**, *107*, 19–24. [CrossRef]
31. Maghous, S.; Saada, Z.; Dormieux, L.; Canou, J.; Dupla, J. A model for in situ grouting with account for particle filtration. *Comput. Geotech.* **2007**, *34*, 164–174. [CrossRef]
32. Bouchelaghem, F. Multi-scale modelling of the permeability evolution of fine sands during cement suspension grouting with filtration. *Comput. Geotech.* **2009**, *36*, 1058–1071. [CrossRef]
33. Eriksson, M. Prediction of Grout Spread and Sealing Effect. Ph.D. Thesis, KTH Byggvetenskap, Stockholm, Switzerland, 2002.
34. Areias, P.M.; Song, J.; Belytschko, T. Analysis of fracture in thin shells by overlapping paired elements. *Comput. Methods Appl. Mech. Eng.* **2006**, *195*, 5343–5360. [CrossRef]
35. Moës, N.; Belytschko, T. Extended finite element method for cohesive crack growth. *Eng. Fract. Mech.* **2002**, *69*, 813–833. [CrossRef]
36. Wu, J.-Y.; Li, F.-B. An improved stable xfem (is-xfem) with a novel enrichment function for the computational modeling of cohesive cracks. *Comput. Methods Appl. Mech. Eng.* **2015**, *295*, 77–107. [CrossRef]
37. Saloustros, S.; Pela, L.; Cervera, M.; Roca, P. Finite element modelling of internal and multiple localized cracks. *Comput. Mech.* **2017**, *59*, 299–316. [CrossRef]
38. Saloustros, S.; Cervera, M.; Pela, L. Challenges, tools and applications of tracking algorithms in the numerical modelling of cracks in concrete and masonry structures. *Arch. Comput. Methods Eng.* **2018**, 1–45. [CrossRef]

39. Cervera, M.; Pela, L.; Clemente, R.; Roca, P. A crack-tracking technique for localized damage in quasi-brittle materials. *Eng. Fract. Mech.* **2010**, *77*, 2431–2450. [CrossRef]
40. Zhang, Y.; Lackner, R.; Zeiml, M.; Mang, H.A. Strong discontinuity embedded approach with standard sos formulation: Element formulation, energy-based crack-tracking strategy, and validations. *Comput. Methods Appl. Mech. Eng.* **2015**, *287*, 335–366. [CrossRef]
41. Zhang, Y.; Zhuang, X. Cracking elements: A self-propagating strong discontinuity embedded approach for quasi-brittle fracture. *Finite Elem. Anal. Des.* **2018**, *144*, 84–100. [CrossRef]
42. Nikolić, M.; Ibrahimbegovic, A.; Miscevic, P. Brittle and ductile failure of rocks: Embedded discontinuity approach for representing mode i and mode ii failure mechanisms. *Int. J. Numer. Methods Eng.* **2015**, *102*, 1507–1526. [CrossRef]
43. Cervera, M.; Barbat, G.; Chiumenti, M. Finite element modeling of quasi-brittle cracks in 2d and 3d with enhanced strain accuracy. *Comput. Mech.* **2017**, *60*, 767–796. [CrossRef]
44. Cervera, M.; Chiumenti, M.; Codina, R. Mixed stabilized finite element methods in nonlinear solid mechanics: Part I: Formulation. *Comput. Methods Appl. Mech. Eng.* **2010**, *199*, 2559–2570. [CrossRef]
45. Cervera, M.; Chiumenti, M.; Codina, R. Mixed stabilized finite element methods in nonlinear solid mechanics: Part II: Strain localization. *Comput. Methods Appl. Mech. Eng.* **2010**, *199*, 2571–2589. [CrossRef]
46. Cervera, M.; Chiumenti, M.; Benedetti, L.; Codina, R. Mixed stabilized finite element methods in nonlinear solid mechanics. Part III: Compressible and incompressible plasticity. *Comput. Methods Appl. Mech. Eng.* **2015**, *285*, 752–775. [CrossRef]
47. Wu, J.-Y.; Nguyen, V.-P. A length scale insensitive phase-field damage model for brittle fracture. *J. Mech. Phys. Solids* **2018**. [CrossRef]
48. Wu, J.-Y. Robust numerical implementation of non-standard phase-field damage models for failure in solids article. *Comput. Methods Appl. Mech. Eng.* **2018**. [CrossRef]
49. Wu, J.; McAuliffe, C.; Waisman, H.; Deodatis, G. Stochastic analysis of polymer composites rupture at large deformations modeled by a phase field method. *Comput. Methods Appl. Mech. Eng.* **2016**, *312*, 596–634. [CrossRef]
50. Nikolić, M.; Karavelić, E.; Ibrahimbegovic, A.; Miščević, P. Lattice element models and their peculiarities. *Arch. Comput. Methods Eng.* **2018**, *25*, 753–784. [CrossRef]
51. Grassl, P. A lattice approach to model flow in cracked concrete. *Cem. Concr. Compos.* **2009**, *31*, 454–460. [CrossRef]
52. Nikolić, M.; Do, X.N.; Ibrahimbegovic, A.; Nikolić, Ž. Crack propagation in dynamics by embedded strong discontinuity approach: Enhanced solid versus discrete lattice model. *Comput. Methods Appl. Mech. Eng.* **2018**, *340*, 480–499. [CrossRef]
53. Ren, F.; Ma, G.; Fu, G.; Zhang, K. Investigation of the permeability anisotropy of 2d fractured rock masses. *Eng. Geol.* **2015**, *196*, 171–182. [CrossRef]
54. Ren, F.; Ma, G.; Wang, Y.; Fan, L. Pipe network model for unconfined seepage analysis in fractured rock masses. *Int. J. Rock Mech. Min. Sci.* **2016**, *88*, 183–196. [CrossRef]
55. Sun, Z.; Yan, X.; Liu, R.; Xu, Z.; Li, S.; Zhang, Y. Transient analysis of grout penetration with time-dependent viscosity inside 3d fractured rock mass by unified pipe-network method. *Water* **2018**, *10*, 1122. [CrossRef]
56. Bouchelaghem, F.; Vulliet, L.; Leroy, D.; Laloui, L.; Descoeudres, F. Real-scale miscible grout injection experiment and performance of advection–dispersion–filtration model. *Int. J. Numer. Anal. Methods Geomech.* **2001**, *25*, 1149–1173. [CrossRef]
57. Bear, J.; Bachmat, Y. *Introduction to Modeling of Transport Phenomena in Porous Media*; Kluwer Academic Publishers: Dordrecht, The Netherlands, 1990.
58. Kim, Y.S.; Whittle, A.J. Filtration in a porous granular medium: 2. Application of bubble model to 1-d column experiments. *Transp. Porous Media* **2006**, *65*, 309–335. [CrossRef]
59. Kim, Y.S.; Whittle, A.J. Particle network model for simulating the filtration of a microfine cement grout in sand. *J. Geotech. Geoenviron. Eng.* **2009**, *135*, 224–236. [CrossRef]
60. Yoon, J.; Mohtar, C.S.E. A filtration model for evaluating maximum penetration distance of bentonite grout through granular soils. *Comput. Geotech.* **2015**, *65*, 291–301. [CrossRef]
61. Bai, R.; Tien, C. Particle detachment in deep bed filtration. *J. Colloid Interface Sci.* **1997**, *186*, 307–317. [CrossRef] [PubMed]
62. Probstein, R.F. *Physicochemical Hydrodynamics: An Introduction*; John Wiley & Sons: New York, NY, USA, 2005.

63. Zhang, Y.; Pichler, C.; Yuan, Y.; Zeiml, M.; Lackner, R. Micromechanics-based multifield framework for early-age concrete. *Eng. Struct.* , **2013**, *47*, 16–24. [CrossRef]
64. Zhang, Y.; Zeiml, M.; Pichler, C.; Lackner, R. Model-based risk assessment of concrete spalling in tunnel linings under fire loading. *Eng. Struct.* **2014**, *77*, 207–215. [CrossRef]
65. Zhang, Y.; Zeiml, M.; Maier, M.; Yuan, Y.; Lackner, R. Fast assessing spalling risk of tunnel linings under rabt fire: From a coupled thermo-hydro-chemo-mechanical model towards an estimation method. *Eng. Struct.* **2017**, *142*, 1–19. [CrossRef]
66. Babuška, I. Error-bounds for finite element method. *Numer. Math.* **1971**, *16*, 322–333. [CrossRef]
67. White, J.A.; Borja, R.I. Stabilized low-order finite element for coupled solid-deformation/fluid-diffusion and their application to fault zone transients. *Comput. Methods Appl. Mech. Eng.* **2008**, *197*, 4353–4366. [CrossRef]
68. Ren, F.; Ma, G.; Wang, Y.; Li, T.; Zhu, H. Unified pipe network method for simulation of water flow in fractured porous rock. *J. Hydrol.* **2017**, *547*, 80–96. [CrossRef]
69. Ren, F.; Ma, G.; Wang, Y.; Fan, L.; Zhu, H. Two-phase flow pipe network method for simulation of CO_2 sequestration in fractured saline aquifers. *Int. J. Rock Mech. Min. Sci.* **2017**, *98*, 39–53. [CrossRef]
70. Chen, Y.; Ma, G.; Li, T.; Wang, Y.; Ren, F. Simulation of wormhole propagation in fractured carbonate rocks with unified pipe-network method. *Comput. Geotech.* **2018**, *98*, 58–68. [CrossRef]
71. Wang, Y.; Ma, G.; Ren, F.; Li, T. A constrained delaunay discretization method for adaptively meshing highly discontinuous geological media. *Comput. Geosci.* **2017**, *109*, 134–148. [CrossRef]

processes

Article

Numerical Investigation of Hydraulic Fracture Propagation Based on Cohesive Zone Model in Naturally Fractured Formations

Jianxiong Li [1], Shiming Dong [1,*], Wen Hua [1,*], Xiaolong Li [2] and Xin Pan [1]

1 Key Laboratory Deep Underground Science and Engineering, Ministry of Education, College of Architecture and Environment, Sichuan University, Chengdu 610065, China; jianxiongml@126.com (J.L.); pan-px@foxmail.com (X.P.)

2 Sinopec Petroleum Exploration and Production Research Institute, Beijing 100083, China; lixl2018.syky@sinopec.com

* Correspondence: smdong@scu.edu.cn (S.D.); wenhua@scu.edu.cn (W.H.); Tel.: +86-288-541-6486 (S.D. & W.H.)

Received: 11 December 2018; Accepted: 2 January 2019; Published: 8 January 2019

Abstract: Complex propagation patterns of hydraulic fractures often play important roles in naturally fractured formations due to complex mechanisms. Therefore, understanding propagation patterns and the geometry of fractures is essential for hydraulic fracturing design. In this work, a seepage–stress–damage coupled model based on the finite pore pressure cohesive zone (PPCZ) method was developed to investigate hydraulic fracture propagation behavior in a naturally fractured reservoir. Compared with the traditional finite element method, the coupled model with global insertion cohesive elements realizes arbitrary propagation of fluid-driven fractures. Numerical simulations of multiple-cluster hydraulic fracturing were carried out to investigate the sensitivities of a multitude of parameters. The results reveal that stress interference from multiple-clusters is responsible for serious suppression and diversion of the fracture network. A lower stress difference benefits the fracture network and helps open natural fractures. By comparing the mechanism of fluid injection, the maximal fracture network can be achieved with various injection rates and viscosities at different fracturing stages. Cluster parameters, including the number of clusters and their spacing, were optimal, satisfying the requirement of creating a large fracture network. These results offer new insights into the propagation pattern of fluid driven fractures and should act as a guide for multiple-cluster hydraulic fracturing, which can help increase the hydraulic fracture volume in naturally fractured reservoirs.

Keywords: hydraulic fractures; PPCZ; multitude parameters; propagation pattern; stress interference; naturally fracture

1. Introduction

Hydraulic fracturing is an evolving technology that has been used to increase oil recovery in naturally fractured reservoirs [1]. The widely held assumption that hydraulic fractures in these reservoirs are ideal, simple, straight, bi-wing, planar features is not accurate because of stress shadows and the existence of pre-existing natural fractures (NFs) [2]. Multiple-cluster hydraulic fracturing of horizontal wells is a commonly used method in ultra-low permeability and massive natural fracture reservoirs. This technique aims to widen hydraulic fractures (HFs) and generate a large amount of non-planar hydraulic fracture networks (HFNs) [3]. Thus, an accurate mechanism of NF–HF interaction needs to be provided, which ultimately determines the stimulated reservoir volume (SRV) and production.

Various complex fracturing models of naturally fractured formations have been developed to describe the complex interaction mechanism between NFs and HFs, such as the boundary element method (BEM) [1–3], discrete element method (DEM) [4–8], and the finite element method (FEM) based on the cohesive zone method (CZM). Such models intend to overcome the limitations of conventional hydraulic fracturing models, in which HFs are simulated to be planar and bi-wings.

Olson et al. [3,9] proposed a numerical code based on a pseudo-3D displacement discontinuity solution, which provided a preliminary analysis of a complex fracture network with a constant pressure boundary condition. This method, however, could not offer a solution for flow patterns within the fracture network. Zhang et al. [1] developed a BEM program coupling rock deformation and a fluid flow hydraulic fracturing model for a fracture network, which considered multi-fracture propagation and interactions between HFs and NFs. Kan et al. [10] made an improvement to describe the complexities of a HFN, in which the natural fracture could be modeled with many connected linear elements with uniform openings and shearing displacement discontinuities. The disadvantages of the three aforementioned models, including BEM or the displacement discontinuity method, are as follows: they do not achieve an efficient and accurate fracture propagation and do not address the limitations in the mesh shape. Further, discretization for the body boundaries only may lead to serious computation problems and make it challenging to find an iteration solution for unknown boundaries.

The discrete element method (DEM) has been used in modeling hydraulic fracturing networks, in which the generation of HFs can be achieved by a flow network consisting of bond particles or deformable blocks. Hamidi et al. [4] presented a three-dimensional distinct element code (3DEC) to simulate the initiation and propagation of hydraulic fractures induced by hydro-mechanics in a typical fractured formation. Nevertheless, the discrete element method could not be used to model fracture initiation in an intact rock matrix. Therefore, a set of joints were introduced in this method for simulating this process. Zhang et al. [11] used a mixed finite element and discrete element method to develop a model for predicting the initiation and propagation of hydraulic fractures in tight gas reservoirs. As a consequence, a newly complex fracturing model based on a 3D discrete element method was introduced into the naturally fractured formation with multi-layers to simulate the propagation and interaction between hydraulic fracture networks [6,8]. However, the complex fracture network model based on DEM needs a small time increment, which makes it difficult for computation [12].

A representative numerical method for modeling hydraulic fractures is the finite element method based on the CZM, which considers the effects of the fracture tip process zone and softening. Owing to the advantages of the CZM, a finite model based on the CZM was used to simulate full coupling between the incompressible viscous fluid and rock matrix, which was demonstrated to be relatively accurate in hydraulic fracturing under plain strain conditions [13–16]. For the strong non-linear coupling in the hydraulic fracturing process, empirical methods and the linear elastic method, which ignore the effects of the plastic zone and softening, still occupy an important position for design and optimization of hydraulic fracture schemes. However, the CZM model, in the framework of non- linear elastic fracture mechanics provided by Haddad et al. [17], can easily solve this problem. Guo et al. [18,19] used a new model based on the CZM coupled with the seepage-damage field for simulating the interaction between HFs and NFs; multi-layers in the formation were also investigated using this model. The results revealed that the model is highly accurate in predicting fracture initiation and propagation. Moreover, the intersections between HFs and NFs were also in strong agreement with the experimental exercise and analytical solution. Li et al. [12] presented a new pore pressure cohesive element that applied Coulomb's frictional contact model to detail the fracture contact behavior for modeling HFN. In general, the finite element method based on the CZM has significant advantages that make it suitable for modeling the hydraulic fracture process, including fracture tip evolution and a softening effect. Further, the numerical model developed by the CZM is computed using an implicit integration with fewer calculation problems than the dis-continuum method, which uses an explicit integration scheme with a small time increment.

In this study, a fully coupled seepage–stress model based on ABAQUS is developed to simulate the propagation of HFs in naturally fractured reservoirs. Two types of elements—the plain strain element coupling pore fluid stress, and pore pressure cohesive element—are applied to model the rock matrix and fractures, respectively. The zero-thickness pore pressure cohesive zone (PPCZ) elements are incorporated into the finite element model using a python code written by the authors, in which two neighboring quadrangle elements are inserted into one PPCZ element. Different constructions of multiple-cluster hydraulic fracturing are developed to investigate complex propagation pattern between HFs and NFs. Furthermore, various parameters including stress difference, cluster spacing, injection fluid viscosity, and numbers of clusters are investigated for a typical field case. The effect of different parameters on the fracture network are captured with different stress states. The optimal parameters of the fracturing process are determined, and the propagation pattern including NFs and HFs is predicted.

2. Governing Equations

Owing to a modification of the CZM, the latest efforts to model hydraulic fracturing based on ABAQUS have introduced the pore pressure node in cohesive elements since ABAQUS 6.12 [14–17,20,21]. Further, the presence of joint pore pressure nodes among intersection pore pressure cohesive elements makes it possible to model the intersection of fractures. In this model, gradual damage occurs at the intersection of pore pressure nodes, which leads to arbitrary propagation during the fracturing process. Moreover, hydraulic fracturing coupling in situ stress and pore pressure can be solved by the geo-mechanics model with geostatic and soil analyses in standard analysis. Therefore, the results of fully-coupled pore pressure stress can be achieved easily by including the linear and nonlinear material effects and fluid leak-off between cohesive layers and bulk rocks.

2.1. Fluid Flow within the CZM

The fracturing fluid within the CZM model is hypothesized to be an incompressible Newtown fluid governed by the lubrication equation [22]. As shown in Figure 1, both normal and tangential flow can be simulated with the cohesive elements. The shale matrix is considered to be comprised of pore pressure materials, in which fluid loss into the matrix is also taken into account to approximate the field application. Therefore, the global volume balance equation for tangential flow that was formulated by Poiseuille was applied:

$$q = -\frac{w^3}{12\mu}\nabla p_f \tag{1}$$

where q is the fluid volume crossing though the cohesive gap, w is the fracture width, μ is the viscosity of the fracturing fluid, and ∇p_f is the gradient of pressure along the cohesive zone. This equation assumes that the injection rate is proportional to both the pressure gradient and the cube of the fracture width and inversely proportional to the fracturing fluid viscosity.

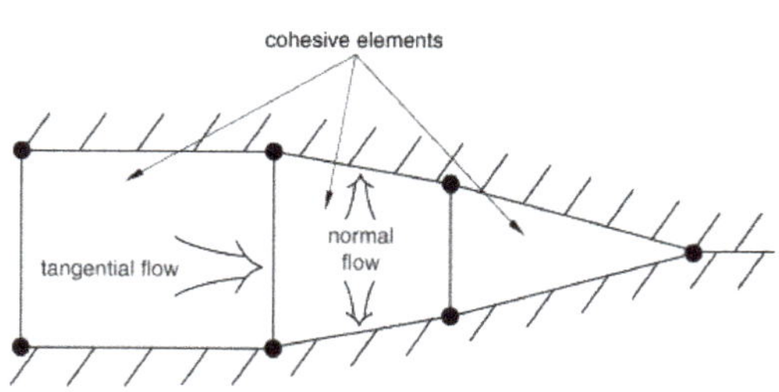

Figure 1. Fracturing flow pattern within the cohesive element.

The normal flow represents the permeation of fracturing fluid by proposing two permeable cohesive layers, in which leak-off coefficients for the two cohesive layers are taken into consideration, as shown in Figure 2. The normal flow is formulated as

$$\begin{aligned} q_t &= c_t(p_i - p_t) \\ q_b &= c_b(p_i - p_b) \end{aligned} \tag{2}$$

where q_t and q_b are the volumes of permeation for the up and side surfaces, respectively; c_t and c_b are the leak-off coefficients for the top and bottom cohesive layers, respectively; p_t and p_b are the pore pressures on the top and bottom surfaces, respectively; and p_i is the pore pressure in the middle of the cohesive element gap.

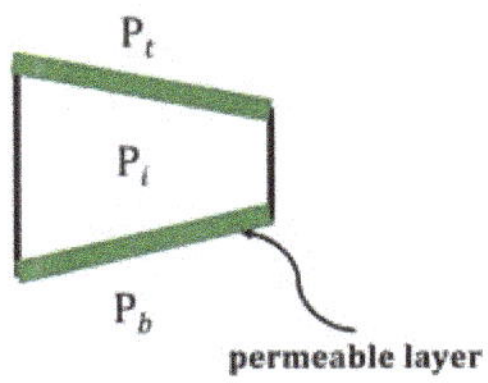

Figure 2. Normal fracturing flow for permeable cohesive surfaces.

For a specific time, the injection fluid flows into the fracture when the cohesive layer fails completely. Injection fluid flows into the fracture, with the tangential flow along the fracture direction and normal flow leak-off into the surrounding formations. The continuity equation of the fracture fluid can be expressed as

$$\frac{\partial w}{\partial t} + \nabla q + (q_t + q_b) = \mathrm{Q(t)}\delta(\mathrm{x,\ y}) \tag{3}$$

where q is the tangential flow mass and Q is the total volume of injection fluid.

2.2. Damage Initiation and Evolution of Cohesive Elements

In this study fluid flow in the fracture network, seepage in a porous media, and the elastic deformation of a rock matrix are fully coupled in the cohesive zone model, unlike the bulk-continuum materials which describe continuum behaviors with stress and strain. The cohesive method uses the traction and separation law for the interface. In the fully coupled model, Biot's [23] effective stress principle is adopted to describe the effective stress due to the rock skeleton and pore pressure. The principle of effective stress is formulated below:

$$\sigma_{ij} = \sigma'_{ij} + \alpha p \delta_{ij} \tag{4}$$

where σ_{ij} is the total stress, σ'_{ij} is the effective stress, α is the Biot constant, p is the pore pressure, and δ_{ij} is the tensor of Kronecker.

The traction–separation damage can be divided into three stages: Initial loading before damage, damage initiation, and damage evolution. The corresponding traction–separation stress vector and stiffness matrix are introduced to find a solution, which can be expressed as:

$$\vec{t} = \begin{Bmatrix} t_s \\ t_n \\ t_t \end{Bmatrix} = \begin{bmatrix} K_{nn} & K_{ns} & K_{nt} \\ K_{ns} & K_{ss} & K_{st} \\ K_{nt} & K_{st} & K_{tt} \end{bmatrix} \begin{Bmatrix} \varepsilon_n \\ \varepsilon_s \\ \varepsilon_t \end{Bmatrix} = \left(\frac{1}{T^{coh}}\right) \vec{K}\vec{\delta} \tag{5}$$

where $\vec{t}$ is the traction stress vector on the cohesive element layer, $\vec{\delta}$ represents the interface separation, T^{coh} is the cohesive element initial gap that is used to calculate the element stiffness, K is the element stiffness, and ε is the strain vector.

In this model, typical zero thickness cohesive elements are embedded in the pore pressure element for modeling an arbitrary propagation of fractures. For the fracture initiation, we used the quadratic nominal stress law to combine the mixed model failure. Damage initiates when the quadratic interaction function reaches a value of one, which involves one normal and two shear nominal stresses. The initiating behavior is shown in Figure 3 and summarized in the function

$$\left(\frac{\langle t_n \rangle}{t_n^0}\right)^2 + \left(\frac{t_s}{t_s^0}\right)^2 + \left(\frac{t_t}{t_t^0}\right)^2 = 1 \tag{6}$$

The metric for damage is a stiffness degradation index, D, which represents the entire damage caused by all stress components. In particular, no damage occurs (D = 0) at the beginning of the simulation, and the cohesive element also satisfies the condition under pure compressive stress. In addition, full degradation of the cohesive element occurs at all integration points when D = 1, which means a zero load carrying capacity for these elements.

$$t_n = \begin{cases} (1-D)T_n, & T_n \geq 0 \\ T_n, & T_n < 0 \end{cases} \tag{7}$$

$$t_s = (1-D)T_s, \tag{8}$$

$$t_t = (1-D)T_t, \tag{9}$$

where T_n, T_s and T_t are the normal and shear stress components predicted by the elastic traction–separation behavior for current separation without damage.

Nevertheless, the behavior of D depends on the nature of material softening, which could be linear and exponential. The index is a function of the effective relative displacement, δ_m, that accounts for combining the effects of δ_n, δ_s and δ_t:

$$\delta_m = \sqrt{\langle \delta_n \rangle^2 + {\delta_s}^2 + {\delta_t}^2} \tag{10}$$

For this particular model, linear softening behavior is represented as

$$D = \frac{\delta_m^f(\delta_m^{\max} - \delta_m^0)}{\delta_m^{\max}(\delta_m^f - \delta_m^0)} \tag{11}$$

where δ_m is the relative effective displacement attained over the course of loading. The superscripts max, f and 0 represent the max relative displacement, displacement when fractures are completely damaged, and the displacement at damage initiation, respectively.

Once fracture initiation occurs, damage can be evaluated using fracture energy theory. The Benzeggagh–Kenane criterion is introduced to model the fracture process. Figure 4 shows the schematic of the mix-mode Benzeggagh–Kenane (BK) criterion during fracture evolution. Moreover, the corresponding formulas of the mechanic criteria can be expressed as

$$G_n^c + (G_s^c - G_n^c)\left(\frac{G_s + G_t}{G_n + G_s + G_t}\right)^{\eta} = G^c \tag{12}$$

where G represents the energy release rate. Superscript c is the critical energy release rate. The subscripts n, s, and t denote the directions of the normal stress and the two shear stresses, and η is a constant for a corresponding material property. The symbol $\langle\ \rangle$ is the Macaulay bracket.

For a negative t, the value in the Macaulay bracket would be zero, which means that the cohesive element has an incompressible property.

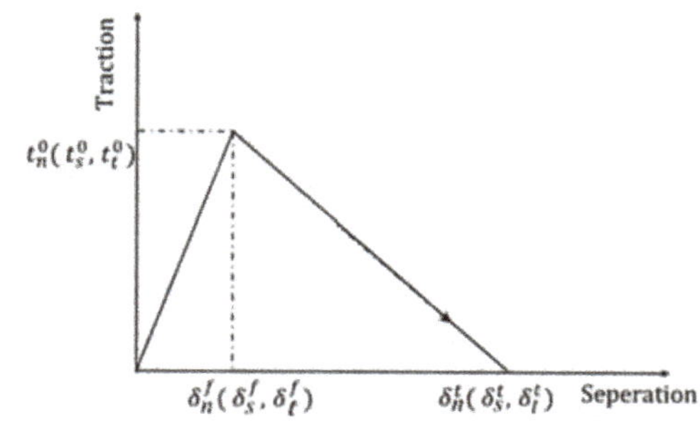

Figure 3. Quadratic nominal stress traction and separation law.

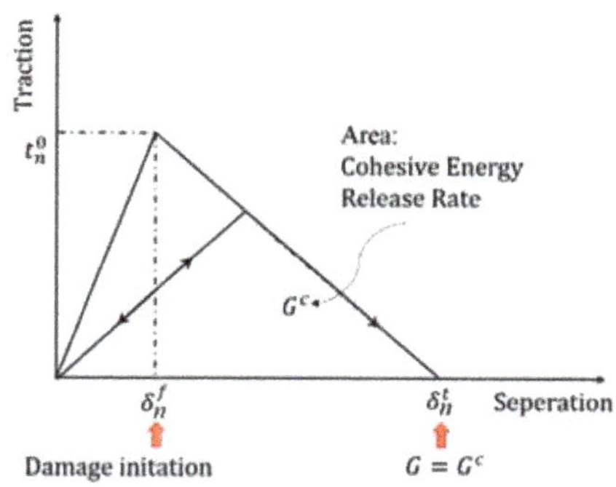

Figure 4. Cohesive linear-softening for the Benzeggagh–Kenane (BK) criterion.

2.3. Equilibrium Equation

In the simulation, the formation is assumed to be a porous medium corresponding to the field rock matrix. Only a single phase of fluid is saturated into the solid skeleton and pores. The total stress within the formation consists of two parts: The total effective stress $\overline{\sigma}$ is determined by the skeleton and the pore pressure p_w. According to the principle of virtual work, the finite element method equilibrium equation can be formulated as

$$\int (\overline{\sigma} - p_w I)\delta\varepsilon dV = \int_S t \cdot \delta v dS + \int_V f \cdot \delta v dV \tag{13}$$

where $\overline{\sigma}$ and p_w are the Boit effective stress and pore pressure, respectively; $\delta\varepsilon$ and δv are the virtual strain rate and virtual velocity, respectively; t and f are the surface displacement per unit area and body force per unit volume, repectively; and I is the unit matrix.

Fluid flow is required in a porous media according to the continuum equation. The change in total fluid mass is equal to the fluid mass crossing the cohesive surface per unit time. Moreover, fluid flow in a porous medium follows Darcy's law. The continuum equation for fluid leak-off can be derived as follows:

$$\frac{1}{J}\frac{\partial}{\partial t}(J\rho_w n_w) + \frac{\partial}{\partial x} \cdot (\rho_w n_w v_w) = 0 \tag{14}$$

where J is the volume change ration; ρ_w, n_w and v_w are the mass density of liquid, porosity of medium and average velocity of liquid relative to the solid phase, respectively; and x is the space vector.

3. Results and Discussion

The two-dimensional symmetrical models were used to model the stratum for multiple-cluster hydraulic fracturing, which extended for 200 m in the direction of horizontal minimal stress (X direction), and for 100 m in the direction of maximal stress (Y direction). There were three types of perforations (two, three and four clusters) with single stage perforations. Each configuration had a distance spacing of 30 m, 40 m, 50 m and 60 m. The natural fractures were designed for an approaching angle of 60° with random lengths of 4–6 m, as shown in Figure 5. However, hydraulic fracturing is a complex nonlinear problem that is coupled with several factors; thus, the viscosity control parameter of a cohesive element was introduced to guarantee convergence. In this work, the viscosity parameter was restricted to 0.01, and its value was verified by the 2016 ABAQUS user's manual. To prevent the mesh size from having an effect, the normal global mesh size was set as 1 m, which can result in fine propagation of fluid driven fracture in the FEM model. Therefore, the optimum numbers of this model are 51,005 elements, containing 17,085 pore pressure plain strain elements (CPE4P) and 33,920 pore pressure cohesive elements (COH2D4P) in this simulation. The CPE4P and COH2D4P elements both contained the pore pressure nodes, which can simulate the pore pressure diffusion in hydraulic fracturing problem. In this model, the typical FEM meshes of CPE4P and COH2D4P were adopted to simulate rock matrix and fluid driven fractures, respectively. Furthermore, parts of the symmetrical models with different approaching angles were set to simulate the interaction for a single HF and NF individually, which extend for 20 × 30 m. The input data used in this simulation are summarized in Table 1, which comes from the shale rock performed by Guo et al. [24]. All models were solved by the ABAQUS standard solver, which can precisely solve the multi-factors coupling problems in hydraulic fracturing. The stress components of the numerical results were described by effective stress; meanwhile, the pore pressure boundaries and initial pore pressure, which adopt hydrostatic pressure, had a value of zero. For that, a wide range of relative stress difference can be attained.

Table 1. Input variations of the numerical simulations.

Categories	Variables	Values
Rock properties	Young's modulus E (GPa)	17.2
	Poisson's ratio v	0.175
	Permeability k (mD)	1
	Porosity (%)	3.65
Cohesive zone properties	Normal nominal stress	1.4 (natural fracture)
	t_n (MPa)	6 (hydraulic fracture)
	First shear nominal stress	8 (natural fracture)
	t_s (MPa)	12 (hydraulic fracture)
	Second shear nominal stress	8 (natural fracture)
	t_t (MPa)	12 (hydraulic fracture)
	Normal fracture energy	300 (natural fracture)
	G_n (J/m^2)	2000 (hydraulic fracture)
	First shear fracture energy	1500 (natural fracture)
	G_s (J/m^2)	3000 (hydraulic fracture)
	Second shear fracture energy	1500 (natural fracture)
	G_t (J/m^2)	3000 (hydraulic fracture)
Pumping parameters	Fluid viscosity μ (mPa·s)	1/50/100
	Injection rate (m^3/s)	0.0001/0.0005/0.001
Initial conditions	Pore pressure (MPa)	51.2
In situ stress	Horizontal stress (MPa)	51.2–66.2

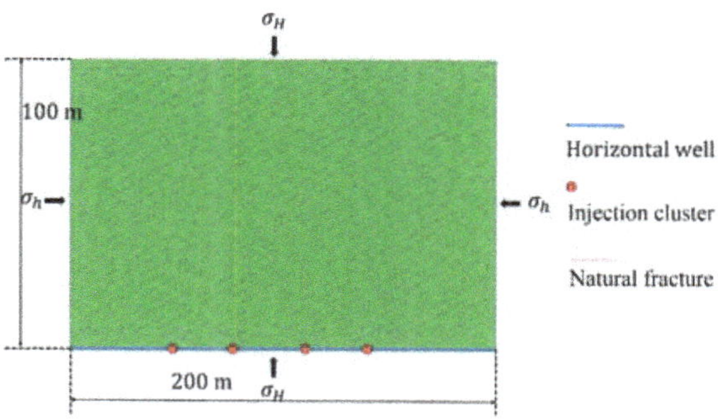

Figure 5. Multiple-cluster model with a 60° approach angle for natural fractures (NFs).

3.1. Interference between a Single Natural Fracture and Hydraulic Fractures

We investigated the influences of the NF, horizontal stress difference, and the approaching angle on fracture propagation under single-cluster hydraulic fracturing. To show compatibility between the numerical and experimental results, various parameters similar to those used in previous experiments were chosen for the numerical simulations, such as the stress difference, NF mechanic properties and intersection angle [25–27]. Even though the scales of the parameter values in the numerical models were different from those of the laboratory fracturing experiments, considering the model dimensions, the numerical simulation results demonstrated in Figure 6 show strong agreement with the experimental results detailed in the literature. The results reveal that the approaching angle of a natural fracture and the stress difference have a strong effect on the morphology of fracture propagation. As shown in Figure 6, when $\nabla\sigma < 2$ MPa and Q = 0.0005 m^3/s, all the HFs propagate along the NF with two fracture directions. However, as $\nabla\sigma$ increases, the HF can open the NF, but only in the direction of the large intersection angle. When the stress difference is large enough, all the HFs will cross the NF to form a long planar fracture, which demonstrates that stress difference is the key factor when HFs intersect with the NF. The HFs tend to cross the NF at any approaching angle when the stress difference is high. Furthermore, a HF is more likely to open and propagate along the NF when the stress difference is low. The smaller the approaching angle of the NF, the higher the critical difference stress for HF crossing will be. The lower the approaching angle of the NF, the easier it is to open two sides of the NF. The numerical simulation results are consistent with experimental results that have been published [25,28,29], as shown in Figure 7. For better visualization, we connected the critical opening pressure at different approaching angles, which is presented as a red line and acts as a comparison standard of our numerical model. It can be seen that HF is more likely to cross the NF above the red critical line; however, the HF is inclined to open the NF below the critical line.

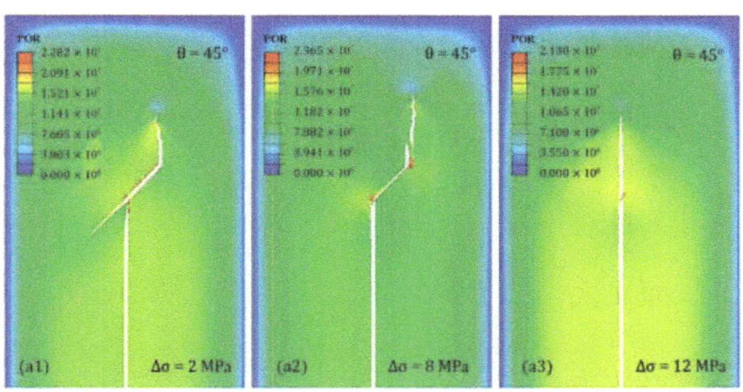

Figure 6. *Cont.*

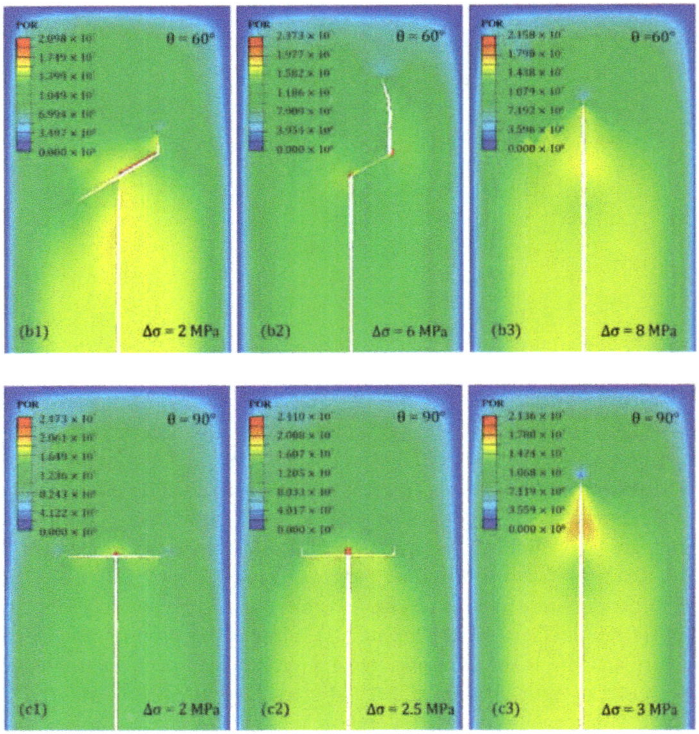

Figure 6. Effect of approaching angle and stress difference on hydraulic fracture propagation. (**a1–a3**) Approaching angle = 45°: stress difference = 2, 8, and 12 MPa, respectively; (**b1–b3**) approaching angle = 60°: stress difference = 2, 6, and 8 MPa, respectively; (**c1–c3**) approaching angle = 90°: stress difference = 2, 2.5, and 3 MPa, respectively.

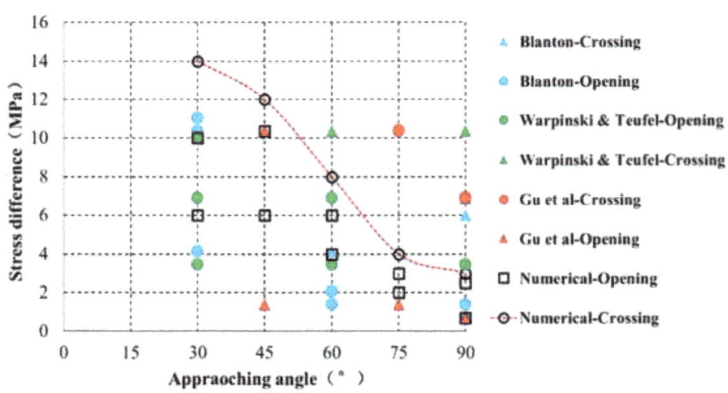

Figure 7. Comparison between numerical simulation and experimental results with different angles of approach and horizontal differential stresses.

3.2. Horizontal Stress Difference

Figure 6 shows that an increase in horizontal stress difference tends to increase the probability of NFs crossing under lower values of the approaching angle. However, the stress shadow resulting from multi-fractures can change this phenomenon in multiple-cluster hydraulic fracturing. Figure 8 shows HF geometries under different stress anisotropies $\nabla\sigma$ in a pre-existing NF system. We assigned the variables as θ = 60°, Q = 0.0005 m^3/s, and μ = 1 mPa·s. All models in this section adopted a two-clusters configuration with a spacing of 40 m. The results show that HFs tend to cross the NFs and form a less complex fracture network as the horizontal stress difference increases, which agrees well with the findings of Zou et al. [7]. In the isotropic case of $\nabla\sigma$ = 2 MPa, the HFs that initiate from the two perforation clusters mainly open and propagate along the NFs, resulting in a highly complex fracture network near the horizontal wellbore. However, the geometry of HFs change with the propagation process owing to stress interference. As shown in Figure 8a, two HFs propagate in opposite directions, deviating from each other, which leads to more opportunities for the right HFs to open NFs and for the left HFs to cross the NFs. When the stress difference increases, the complexity of the fracture network reduces, and fractures propagate in a more straight manner within the same injection process [11]. In addition, during a propagation process, HF growth can be suppressed because of back stress induced by other propagating HFs (Figure 8b,c). Under a horizontal stress difference of 8 MPa (Figure 8d), almost all NFs are crossed by HFs instead of deflecting, which causes less complex fracture networks.

With stronger stress anisotropy, HFs tend to pass through NFs and form smaller curvatures in the propagation path [30], which results in narrower, longer, and less complex fracture networks. Furthermore, the influence of back stress will be more significant, which aggravates the diversion behavior of HFs and changes the crossing phenomenon between HFs and NFs. Therefore, it is beneficial to form more complex fracture networks at lower fracture differences ($\nabla\sigma$ < 8 MPa). As shown in Figure 9, the total fracture network length decreases as the stress difference increases. The total fracture length reduced by 38.26 m when stress difference increased from 2 to 8 MPa, but only reduced by 0.8 m as the stress difference increased from 6 to 8 MPa. This demonstrates that fluid driven fractures can propagate widely and are connected to more natural fractures at lower stress differences, which is useful for connecting the flow channel.

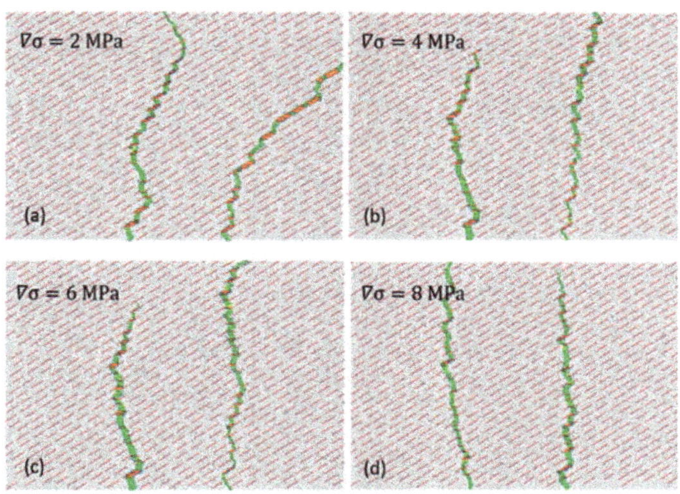

Figure 8. Effect of stress difference on hydraulic fracture propagation. (**a**) Stress difference = 2 MPa; (**b**) stress difference = 4 MPa; (**c**) stress difference = 6 MPa; (**d**) stress difference = 8 MPa.

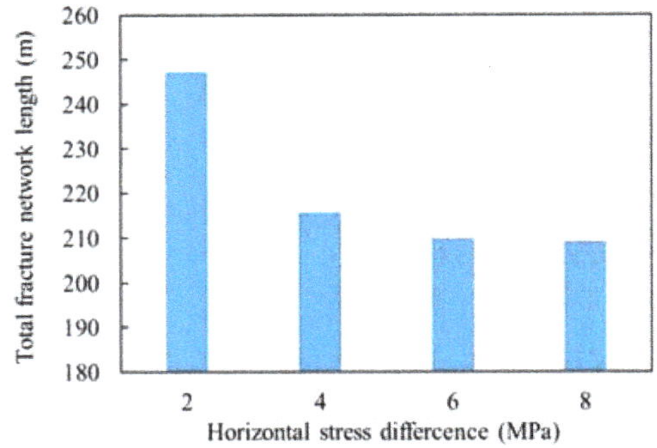

Figure 9. Fracture network lengths at different horizontal stress differences.

3.3. Spacing of Perforation Clusters

When the cluster spacing is small, the mechanical interaction between adjacent fractures is obvious, and a less complex fracture network can be formed [30]. Figure 10 shows the propagation path of multiple fractures in a natural fracture system with a cluster spacing of 30, 40, 50 and 60 m. The parameters $Q = 0.0005\ \text{m}^3/\text{s}$, $\mu = 1$ mPa·s, $\nabla\sigma = 3$ MPa were set for the simulation. It can be seen that different cluster spacing has a significant effect on fracture morphology. As the cluster spacing increases, the influence of the stress shadow among adjacent fractures is alleviated, which results in more opening phenomena among HFs and NFs. Given $S = 30$ m in Figure 10a, the multiple HFs initiate from the wellbore and then intersect with the NFs. With the fracturing process, the stress shadow becomes more severe and obvious suppression can be observed in one of the NFs. Further, the suppressed fractures propagate, becoming wider and shorter. However, unsuppressed fractures extended rapidly, resulting in a longer and narrower fracture network. When the cluster spacing increases (Figure 10b–d), the HFs propagate quickly from the wellbore and experience less suppression compared with the case of $S = 30$ m. The interference of each cluster decreases as cluster spacing increases. Meanwhile, the orientation of the crack tip is also alleviated under this condition, especially when S is larger than 60 m. For the purpose of maximizing the fracture network, deeper propagation should be provided to connect more NFs. Therefore, production can be improved by increasing the perforation cluster spacing within a relative range.

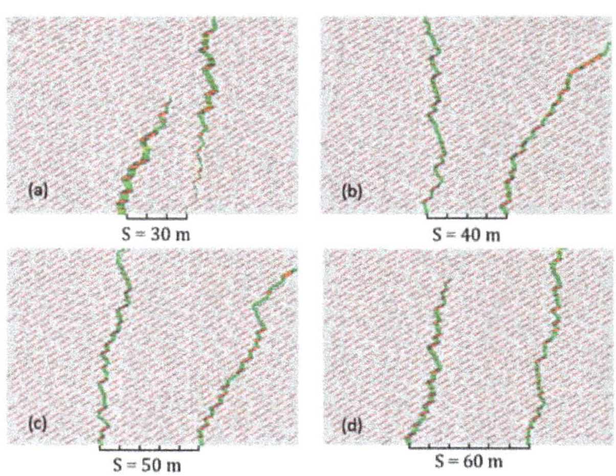

Figure 10. Effect of cluster spacing on hydraulic fracture propagation. (**a**) Cluster spacing = 30 m; (**b**) cluster spacing = 40 m; (**c**) cluster spacing = 50 m; (**d**) cluster spacing = 60 m.

Reducing the cluster spacing is beneficial to improving the complexity of the fracture network under the low stress difference condition [31]. However, numerical simulation in this work gave contrary results: Propagation suppression of multiple fractures occurred with a low stress difference, which caused shorter fracture propagation. Moreover, owing to serious stress interference at lower cluster spacing, a greater amount of crossing can be observed between HFs and NFs, which can reduce the simulated reservoir volume [32]. As shown in Figure 11, the total fracture network length increases when the cluster spacing changes from 30 to 40 m; however, when the spacing is greater than 40 m, the total length does not change significantly. As a result, short perforation cluster spacing should be avoided to increase the complexity of the fracture network. In this case, the optimization of the cluster spacing is recommended to vary from 40 to 60 m, which is suggested by considering not only the influence of stress shadowing but also the controlling area for improving production.

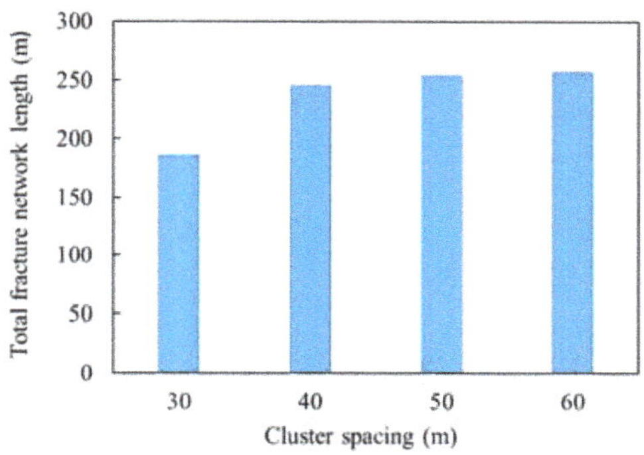

Figure 11. Fracture network lengths at different cluster spacing distances.

3.4. Injection Rate of Fracturing Fluid

The influence of the fracturing fluid on the geometric shape of HFs can be expressed by the injection rate and fluid viscosity. In this section, the influence of the injection rate is discussed. It is concluded that by increasing the injection rate, the fluid injection pressure increases [4]. Owing to a high propagation pressure, HFs have more chance to pass through NFs and tend to form lager planar hydraulic fractures [33]. A pre-existing fracture system with random lengths and an approaching angle of 60° was introduced to investigate the influence of injection rate on the fracture network. The stress difference, fluid viscosity and cluster spacing remained constant at 3 MPa, 1 mPa·s, and 40 m, respectively. As shown in Figure 12, three types of flow rates were used: $Q = 0.0001$ m^3/s, $Q = 0.0005$ m^3/s and $Q = 0.001$ m^3/s. It can be seen that the fracturing of two clusters initiated at the wellbore and reactive NFs at an earlier stage of injection. With the fracturing procedure, the fractures divert in opposite directions at lower injection rates, as shown in Figure 12a,b. However, the morphologies of the two-cluster HFs change less when the fluid rate increases from 0.0001 to 0.0005 m^3/s. When the injection rate reaches 0.001 m^3/s, it has a significant influence on the complexity of the fracture network. As shown in Figure 12c, HFs propagate in a straighter manner perpendicular to the horizontal minimum principle stress, which results in more crossing phenomena and a less complex fracture network.

A high injection rate during the entire process may reduce the complexity of the fracture network due to the high net pressure that will be formed; as a result, the probability of fracture crossing would increase. The total length of the fracture network reduced considerably when the injection rate increased from 0.0005 to 0.001 m^3/s, which resulted in a reduction of flow channel and make the release of gas and oil difficult, as shown in Figure 13. However, a high injection rate can create larger far-field propagation of the fracture network [34], which is mainly due to reduced connection in the near wellbore and deeper propagation of the fracture network. Therefore, various injection rates should be adopted to maximize the complexity of the fracture network. The large fluid pressure within a fracture due to high injection rates can be used to form straighter fracture networks, and deeper propagation at the early stage of fracturing. A low injection rate can open more NFs and creates more diversions in the wider control area in the later period of fracturing. By controlling the injection rates, a large filed of fracture network can be realized.

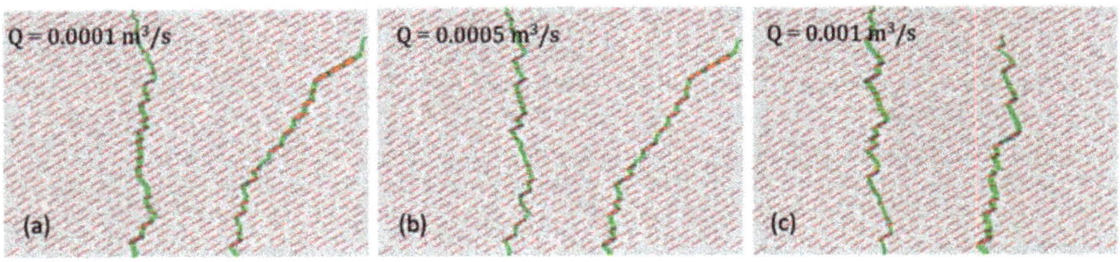

Figure 12. Effect of fluid rate on hydraulic fracture propagation. (**a**) Injection rate = 0.0001 m^3/s; (**b**) injection rate = 0.0005 m^3/s; (**c**) injection rate = 0.001 m^3/s.

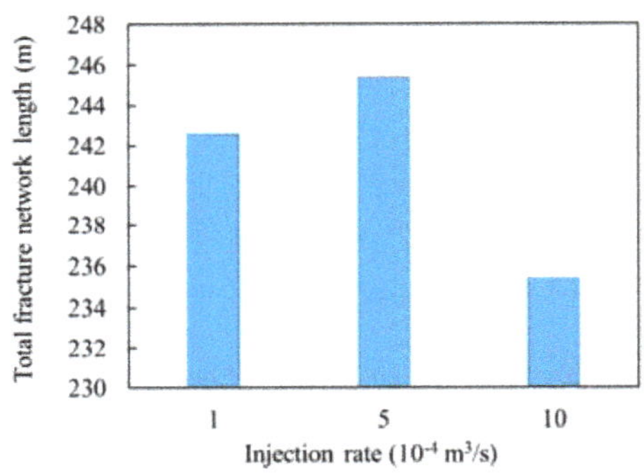

Figure 13. Fracture network lengths at different injection rates.

3.5. Viscosity of the Pumping Fluid

During the fracturing of naturally fractured reservoirs, it was found that the morphology and complexity of the fracture are closely related to the fracture net pressure, which is mainly determined by the fracturing fluid viscosity and injection rate. In this section, three types of numerical models with different viscosities were simulated to investigate the influence of the fracturing viscosity. The results in Figure 14 reveal that fracturing viscosity can affect the morphology and complexity of fractures, especially if it has a high value. In this simulation the stress difference, flow rate, and cluster spacing remained constant at 3 MPa, 0.0005 m^3/s and 40 m, respectively. Figure 14a shows that HFs tend to open more NFs and propagate along the NFs at a low viscosity condition of μ = 1 mPa·s. Meanwhile, the preference for diverting to the opposite direction for each cluster fracture is enhanced with low viscosity fracturing fluid. However, Beguelsdijk et al. [35] found that a fracturing fluid with a high viscosity could induce separate propagation on an optimal fracture surface. For this reason, more opening of NFs may increase the complexity of the fracture network, which will increase the possibility of separate propagation. In addition, more straight fractures will be formed along the direction of maximum horizontal stress, which will result in the formation of a simple fracture network, as shown in Figure 14b,c. The morphology and complexity of fractures will change obviously when the viscosity increases from 1 to 50 mPa·s; however, few transformations can be obtained when the viscosity varies from 50 to 100 mPa·s. Hence, it can be concluded that lower range viscosities have a greater effect on fracture morphology than higher range viscosities.

For the purpose of maximizing the stimulated reservoir volume (SRV), deeper propagation in the far-field region and less tortuosity near the wellbore should be generated for a complex fracture network. A high viscosity during the early stage of fracturing can promote deeper propagation of HFs and reduce the tortuosity near the wellbore. Later, a lower viscosity can activate more NFs and enhance the propagation with the approaching angle, thus forming a more complex fracture network. Figure 15 shows that a lower injection viscosity can lead to the formation of a large fracture network: when the viscosity is 1 mPa·s, the total fracture network length is 254.4 m. Therefore, higher production of oil and gas can be achieved by a lower injection fracturing treatment. More simulations should be conducted to further investigate whether viscosity induce can separate propagation with a random pre-existing NF system.

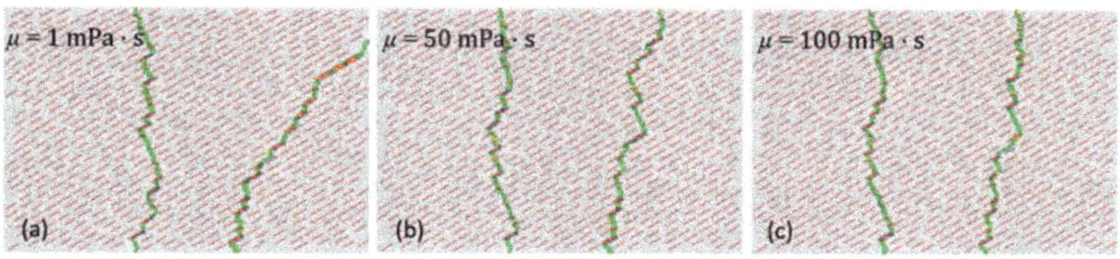

Figure 14. Effect of viscosity on hydraulic fracture propagation. (**a**) Fluid viscosity = 1 mPa·s; (**b**) fluid viscosity = 50 mPa·s; (**c**) fluid viscosity = 100 mPa·s.

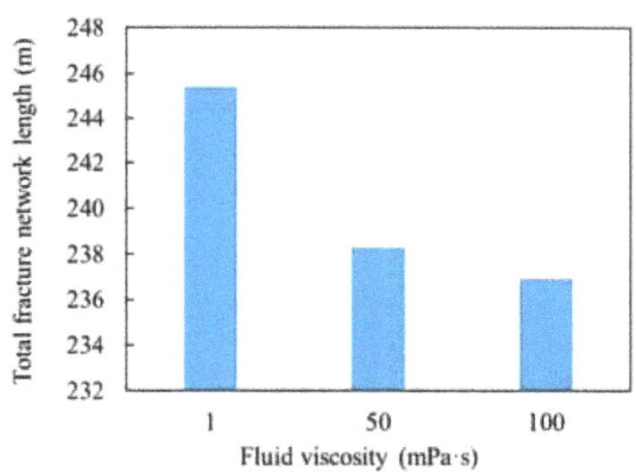

Figure 15. Fracture network lengths achieved with different injection fluid viscosities.

3.6. Number of Perforation Clusters

Different hydraulic fracturing can result in different HF morphologies because of stress interference. For multiple-cluster hydraulic fracturing, the number of perforation clusters has a significant influence on fracture propagation; this can be seen from the six pictures in Figure 16. A fracture's geometric shape is significantly affected by the numbers of clusters at different fracturing times. When the stress difference, fluid rate, cluster spacing and viscosity are fixed at Q = 0.0005 m^3/s, μ = 1 mPa·s, S = 40 m and $\nabla\sigma$ = 3 MPa, the fracture propagation morphologies are comparable in these cases. In this section, three types of perforation configurations (two, three and four clusters) were modeled to investigate the effect of perforation application. The results reveal that the interference of the stress shadow plays a major role in fracture propagation, as it may make the propagation undesirable for production. The resulting mechanical action alters the value and direction of in situ stress, which results in unexpected suppression of the fracture's propagation. As shown in the two-clusters perforation configuration (Figure 16(a1,a2)), back stress causes the left fracture to be suppressed, exhibiting shorter and wider networks compared with the right fracture. The middle fracture seems to be significantly influenced, forming shorter and wider HFs in Figure 16(b1,b2), compared with the fractures at either side. For the four-cluster configuration (Figure 16(c1,c2)), one fracture can break though and propagate into the deep reservoirs rapidly, while the residuary fractures tend to be suppressed more or less. Further, it can be observed from Figure 16 that the fracturing time will affect the morphology of the HFs to differing degrees. All of the configurations show that there is less influence on the morphology of HFs in the early stage of fracturing (i.e., T = 500 s), which is probably due to less spreading of net pressure in the fracture network.

An increase in the perforation number will increase the interaction between HFs and NFs, and the propagation path will be easier to be maintain. Meanwhile, the interference area of the stress shadow will increase with an increase in fracturing time, and the propagation path will be easily altered. As shown in Figure 17, the volume of the fracture network increases rapidly as the number of clusters increases. For instance, the total fracture length increases to 315 m when the number of clusters is set to four. Although larger numbers of cluster can increase the volume of the fracture network, strong interference from multiple clusters can result in considerable suppression of the lateral clusters. In that situation, a shorter and wider facture network will be generated, which goes against fluid release from the formation to the fracture channel. Therefore, lower cluster spacing with multiple clusters should be avoided, as it will increase the serious interference due to stress shadow, which results in a suppressed fracture network.

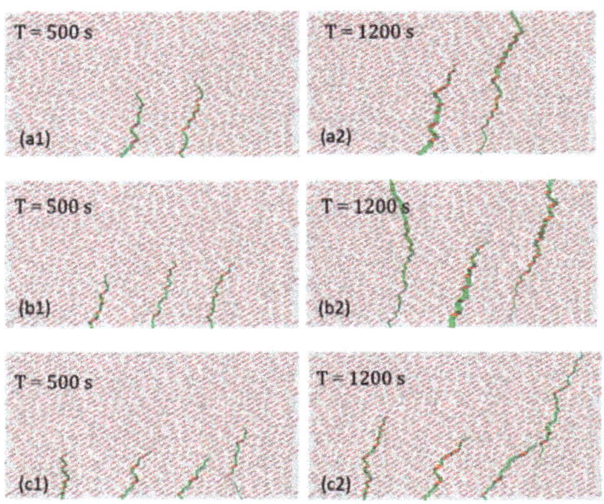

Figure 16. Effect of cluster numbers on hydraulic fracture propagation. (**a1**,**a2**) Two clusters fracturing scenario at fracturing time T = 500, 1200 s; (**b1**,**b2**) three clusters fracturing scenario at fracturing time T = 500, 1200 s; (**c1**,**c2**) four clusters fracturing scenario at fracturing time T = 500, 1200 s.

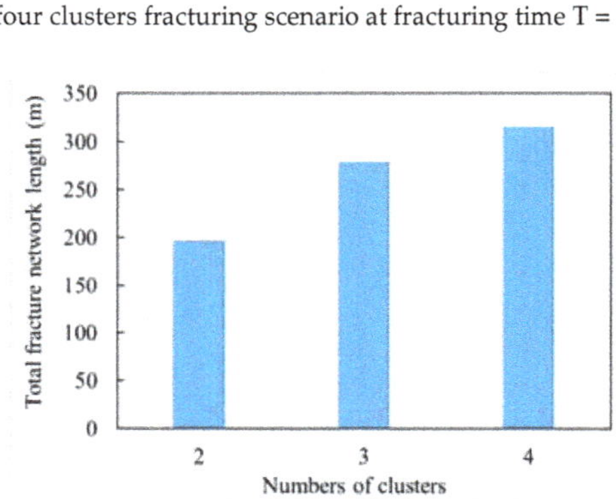

Figure 17. Fracture network lengths corresponding to different numbers of clusters.

4. Conclusions

Using a 2D FEM model based on the PPCZ method, we performed a series of simulation models coupling seepage and stress to investigate the mechanism of fracture network propagation in naturally fractured reservoirs. At the same time, a python code written by the authors was used to incorporate pore pressure elements for modeling the fractures and the joints of the crossed fractures. Several geological and engineering factors influencing the fracture propagation pattern were applied to conduct sensitive analyses. The main propagation rules can be concluded as:

(1) The horizontal differential stress dominates main propagation patterns of the hydraulic fracture network. The morphology of a hydraulic fracture may change from being randomly diverted to more straight as the horizontal differential stress increases. The fluid driven fracture nearly crosses NFs to extend simple transverse fractures when the differential stress exceeds 8 MPa at an approaching angle of 60°, according to our FEM models.
(2) The viscosity and injection rate of the fracturing fluid affect the development of the fracture network at different fracturing times. The fracture network area can reach its maximum with various injection rates and viscosities during different fracturing stages. Using a higher injection rate and fluid viscosity appears to raise the net pressure, which will increase the capacity of a HF to cross the NFs near the wellbore and facilitate a deeper propagation into the formations.
(3) Using a short cluster spacing at 30 m will result in unexpected consequences where one or more fractures will be suppressed to form short and simple transverse fractures. Nevertheless, the phenomenon of suppression could weaken when the cluster spacing distance increases from 40 to 60 m.
(4) Cluster spacing should be taken into consideration as the increase of cluster numbers causes enhanced stress interference. Due to tremendous suppression and merging of fractures under multiple stress interference, increasing the cluster number is meaningless with lower cluster spacing.

Author Contributions: J.L. carried out fracture propagation in hydraulic fracturing and completed the numerical simulation; S.D. designed the study, helped draft the manuscript, and checked the manuscript; W.H. verified the simulation and experiment parameters; X.L. participated in the design of the study and drafted the manuscript; and X.P. participated in the numerical model design and calculation.

Funding: This work was supported by the Sichuan Science and Technology Program (Project No.2018JY0024), and National Natural Science Foundation of China (Project No.11872042).

Conflicts of Interest: The authors declare no conflict of interest.

References

1. Zhang, X.; Jeffrey, R.G. Role of overpressurized fluid and fluid-driven fractures in forming fracture networks. *J. Geochem. Explor.* **2014**, *144*, 194–207. [CrossRef]
2. Zhang, X.; Jeffrey, R.G.; Wu, B. Mechanics of edge crack growth under transient pressure and temperature conditions. *Int. J. Solids Struct.* **2015**, *69–70*, 11–22. [CrossRef]
3. Olson, J.E. Multi-fracture propagation modeling: Applications to hydraulic fracturing in shales and tight gas sands. In Proceedings of the 42nd U.S. Rock Mechanics Symposium: ARMA, San Francisco, CA, USA, 29 June–2 July 2008.
4. Hamidi, F.; Mortazavi, A. A new three dimensional approach to numerically model hydraulic fracturing process. *J. Pet. Sci. Eng.* **2014**, *124*, 451–467. [CrossRef]
5. Yan, C.; Zheng, H. A two-dimensional coupled hydro-mechanical finite-discrete model considering porous media flow for simulating hydraulic fracturing. *Int. J. Rock Mech. Min. Sci.* **2016**, *88*, 115–128. [CrossRef]
6. Zuo, Y.; Ma, X.; Zhang, S.; Zhou, T.; Li, H. Numerical Investigation into the Influence of Bedding Plane on Hydraulic Fracture Network Propagation in Shale Formations. *Rock Mech. Rock Eng.* **2016**, *49*, 3597–3614. [CrossRef]

7. Zou, Y.; Zhang, S.; Ma, X.; Zhou, T.; Zeng, B. Numerical investigation of hydraulic fracture network propagation in naturally fractured shale formations. *J. Struct. Geol.* **2016**, *84*, 1–13. [CrossRef]
8. Fatahi, H.; Hossain, M.M.; Sarmadivaleh, M. Numerical and experimental investigation of the interaction of natural and propagated hydraulic fracture. *J. Nat. Gas Sci. Eng.* **2017**, *37*, 409–424. [CrossRef]
9. Olson, J.E.; Taleghani, A.D. Modeling simultaneous growth of multiple hydraulic fractures and their interaction with natural fractures. In Proceedings of the SPE Hydraulic Fracturing Technology Conference, The Woodlands, TX, USA, 19–21 January 2009.
10. Kan, W.; Olson, J.E. Numerical investigation of complex hydraulic fracture development in naturally fractured reservoirs. *Int. J. Health Geogr.* **2015**, *6*, 1–15.
11. Zhang, S.-C.; Lei, X.; Zhou, Y.-S.; Xu, G.-Q. Numerical simulation of hydraulic fracture propagation in tight oil reservoirs by volumetric fracturing. *Pet. Sci.* **2015**, *12*, 674–682. [CrossRef]
12. Li, Y.; Deng, J.G.; Liu, W.; Feng, Y. Modeling hydraulic fracture propagation using cohesive zone model equipped with frictional contact capability. *Comput. Geotech.* **2017**, *91*, 58–70. [CrossRef]
13. Sarris, E.; Papanastasiou, P. The influence of the cohesive process zone in hydraulic fracturing modelling. *Int. J. Fract.* **2011**, *167*, 33–45. [CrossRef]
14. Carrier, B.; Granet, S. Numerical modeling of hydraulic fracture problem in permeable medium using cohesive zone model. *Eng. Fract. Mech.* **2012**, *79*, 312–328. [CrossRef]
15. Chen, Z. Finite element modelling of viscosity-dominated hydraulic fractures. *J. Pet. Sci. Eng.* **2012**, *88–89*, 136–144. [CrossRef]
16. Shen, X.; Cullick, S. *Numerical Modeling of Fracture Complexity with Application to Production Stimulation*; Elsevier: Amsterdam, The Netherlands, 2012; p. A6.
17. Haddad, M.; Sepehrnoori, K. Cohesive Fracture Analysis to Model Multiple-Stage Fracturing in Quasibrittle Shale Formations. In Proceedings of the Simulia Community Conference, Providence, RI, USA, 19–22 May 2014.
18. Guo, J.; Zhao, X.; Zhu, H.; Zhang, X.; Pan, R. Numerical simulation of interaction of hydraulic fracture and natural fracture based on the cohesive zone finite element method. *J. Nat. Gas Sci. Eng.* **2015**, *25*, 180–188. [CrossRef]
19. Guo, J.; Luo, B.; Lu, C.; Lai, J.; Ren, J. Numerical investigation of hydraulic fracture propagation in a layered reservoir using the cohesive zone method. *Eng. Fract. Mech.* **2017**, *186*, 195–207. [CrossRef]
20. Yao, Y. Linear Elastic and Cohesive Fracture Analysis to Model Hydraulic Fracture in Brittle and Ductile Rocks. *Rock Mech. Rock Eng.* **2012**, *45*, 375–387. [CrossRef]
21. Shin, D.H.; Sharma, M.M. *Factors Controlling the Simultaneous Propagation of Multiple Competing Fractures in a Horizontal Well*; Pearson/Prentice Hall: Upper Saddle River, NJ, USA, 2014; pp. 816–819.
22. Batchelor, G.K. *An Introduction to Fluid Dynamics*; Cambridge University Press: Cambridge, UK, 1967; pp. 1–29.
23. Biot, M.A. General Theory of Three-Dimensional Consolidation. *J. Appl. Phys.* **1941**, *12*, 155–164. [CrossRef]
24. Guo, T.; Zhang, S.; Qu, Z.; Zhou, T.; Xiao, Y.; Gao, J. Experimental study of hydraulic fracturing for shale by stimulated reservoir volume. *Fuel* **2014**, *128*, 373–380. [CrossRef]
25. Blanton, T.L. Propagation of Hydraulically and Dynamically Induced Fractures in Naturally Fractured Reservoirs. In Proceedings of the SPE Unconventional Gas Technology Symposium, Louisville, KY, USA, 18–21 May 1986.
26. Zhou, J.; Jin, Y.; Chen, M. Experimental investigation of hydraulic fracturing in random naturally fractured blocks. *Int. J. Rock Mech. Min. Sci.* **2010**, *47*, 1193–1199. [CrossRef]
27. Yushi, Z.; Shicheng, Z.; Tong, Z.; Xiang, Z.; Tiankui, G. Experimental Investigation into Hydraulic Fracture Network Propagation in Gas Shales Using CT Scanning Technology. *Rock Mech. Rock Eng.* **2015**, *49*, 33–45. [CrossRef]
28. Warpinski, N.R.; Teufel, L.W. Influence of geologic discontinuities on hydraulic fracture propagation. *J. Pet. Technol.* **1984**, *39*, 209–220. [CrossRef]
29. Gu, H.; Weng, X.; Lund, J.B.; Mack, M.G.; Ganguly, U.; Suarez-Rivera, R. Hydraulic Fracture Crossing Natural Fracture at NonOrthogonal Angles, A Criterion, Its Validation and Applications. In *SPE Production & Operations*; Society of Petroleum Engineers: Houston, TX, USA, 2011.
30. Zeng, Q.; Yao, J. Numerical simulation of fracture network generation in naturally fractured reservoirs. *J. Nat. Gas Sci. Eng.* **2016**, *30*, 430–443. [CrossRef]

31. Manchanda, R.; Roussel, N.P.; Sharma, M.M. Factors influencing fracture trajectories and fracturing pressure data in a horizontal completion. In Proceedings of the 46th U.S. Rock Mechanics/Geomechanics Symposium, Chicago, IL, USA, 24–27 June 2012.
32. Bunger, A.P.; Jeffrey, R.G.; Kear, J.; Zhang, X.; Morgan, M. Experimental investigation of the interaction among closely spaced hydraulic fractures. In Proceedings of the 45th U.S. Rock Mechanics/Geomechanics Symposium, San Francisco, CA, USA, 26–29 June 2011.
33. Guo, T.; Zhang, S.; Zou, Y.; Xiao, B. Numerical simulation of hydraulic fracture propagation in shale gas reservoir. *J. Nat. Gas Sci. Eng.* **2015**, *26*, 847–856. [CrossRef]
34. Liu, C.; Shi, F.; Zhang, Y.; Zhang, Y.; Deng, D.; Wang, X.; Liu, H.; Wu, H. High injection rate stimulation for improving the fracture complexity in tight-oil sandstone reservoirs. *J. Nat. Gas Sci. Eng.* **2017**, *42*, 133–141. [CrossRef]
35. Beugelsdijk, L.J.L.; de Pater, C.J.; Sato, K. Experimental hydraulic fracture propagation in a multi-fractured medium. In Proceedings of the SPE Asia Pacific Conference on Integrated Modelling for Asset Management, Yokohama, Japan, 25–26 April 2000.

processes

MDPI

Article

Characterization of Pores and Fractures in Soft Coal from the No. 5 Soft Coalbed in the Chenghe Mining Area

Pan Wei [1,2], Yunpei Liang [1,2], Song Zhao [3], Shoujian Peng [1,2,*], Xuelong Li [1,2,*] and Ran Meng [4]

1 State Key Laboratory of Coal Mine Disaster Dynamics and Control, Chongqing University, Chongqing 400044, China; weipan21@126.com (P.W.); liangyunpei@126.com (Y.L.)
2 College of Resources and Environmental Science, Chongqing University, Chongqing 400044, China
3 Development Department, Tarim Oilfield Company, China National Petroleum Corporation, Korla 841000, China; zhaosong002@126.com
4 College of Safety Science and Engineering, Xi'an University of Science and Technology, Xi'an 710054, China; mcr615@126.com
* Correspondence: sjpeng@cqu.edu.cn (S.P.); lilxcumt@126.com (X.L.); Tel.: +86-186-8087-1822 (S.P.); +86-182-2309-3975 (X.L.)

Received: 29 November 2018; Accepted: 25 December 2018; Published: 31 December 2018

Abstract: The characteristics of the pore structure and gas migration in soft coalbeds are the premise of evaluating gas discharge in soft coalbeds. To explore the pore structure characteristics of soft coal masses, the No. 5 soft coalbed in the eastern zone of Chenghe Mining Area, was investigated and compared with the No. 5 hard coalbed in the western zone. By using a mercury intrusion method, low-temperature liquid nitrogen adsorption, and scanning electron microscopy (SEM), the pore structure characteristics of the No. 5 coalbed were explored. Moreover, based on fractal theory, the pore structure of coal was characterized. The results showed the pores in soft coal mainly appeared as small pores and micropores in which the small pores accounted for nearly half of the total pore volume. Mesopores and macropores were also distributed throughout the soft coal. The mercury-injection and mercury-ejection curves of soft coal showed significant hysteresis loops, implying that pores in coal samples were mainly open while the mercury-injection curve of hard coal was consistent with its mercury-ejection curve, showing no hysteresis loop while having an even segment, which indicated that closed pores occupied the majority of the pore volume in the coal samples. The curves of low-temperature nitrogen adsorption of soft coal all follow an IV-class isotherm. Moreover, the fractal dimensions of soft coal are respectively larger than the fractal dimensions of hard coal. It can be seen that the characterization of pores and fractures of the soft coal was different from the hard coal in the western distinct of the old mining area. The gas prevention and control measures of soft coal should be formulated according to local conditions.

Keywords: coalbed methane (CBM); soft coal masses; pore structure; fractal pore characteristics

1. Introduction

With increasing industrial development, the demand for fossil-fuel energy has constantly increased, and especially that for oil and coal [1]. Since the implementation of reform and opening-up policies, China's economy has grown rapidly, causing China's demand for energy to rise; however, China's coal resource reserve ranks third in the world at a proportion of 11.6%, so the coal-based energy consumption structure of China will not change within the short-term. Therefore, coal resources show a significant promotional effect on the development of China's economy, exerting strategic influences thereon [2,3].

To satisfy the demand for coal and realize high-yield, rapid coal mining, many coal mine accidents occur in China [4,5]. Coal mine accidents which result in more than 10 deaths are generally gas accidents, which occur at the highest frequency and cause the largest damage [6,7]. Besides, coal generates a large amount of greenhouse gas (CO_2) during its near-constant consumption, consequently leading to increasingly severe climate warming. It also causes problems such as the re-distribution of global precipitation, the melting of glaciers and frozen soil, and sea-level rise. Moreover, airborne haze not only threatens the natural ecological balance but also threatens human survival [8–10]. Currently, cities, towns, and villages in China have begun burning gas instead of coal and dismantling numerous coal-burning boilers. The reasons are attributed to two aspects: on the one hand, coal combustion generates greenhouse gases; on the other, gas problem appears in coalbeds during mining [11]. Gas in coalbed is both a source of disasters and energy (as coalbed methane (CBM)). CBM is an associated resource of coal, and as a non-conventional natural gas, is an efficient clean energy source [12,13].

The greenhouse effect of methane is more than 20 times that of CO_2 [14]. For gas prevention in coalbed, gas extraction plays an important role, however, the permeability of coalbed in China is poor and gas extraction efficiency in coalbed is low. Moreover, the gas concentration in the pipeline for gas extraction is low and the technology and equipment available for efficiently utilizing this low-concentration gas remain immature, which results in a low utilization ratio of gas. Additionally, existing equipment and operating costs are high. Therefore, in many mines, the measures necessary for gas utilization are not taken and also the extracted gas is directly emitted to atmosphere, so gas drainage rather than gas extraction and coal mining is implemented [15]. As an efficient clean energy source, CBM has good prospects in terms of being developed to a large-scale and utilized [12]. The efficient extraction of gas in coalbed can reduce the occurrence of gas disasters in mines to ensure safety. To extract gas at a high concentration exerts a positive influence on changes to China's energy structure and reduced greenhouse effect contribution [15].

Coalbed is also called a double-porosity medium owing to its containing tiny pore structures with multiple morphologies [16]. The pore and fracture network in a coal mass becomes the main channel for gas migration in coalbed, therefore, the higher the degree of development of the pore and fracture network in a coal mass is, the higher the gas permeability is, and the higher the gas extraction effect is [17–19]. However, the permeability of coalbed in China is generally poor. Therefore, to improve the connection of pores and fractures in coalbed, it is necessary to increase the permeability of coalbed by utilizing hydraulic fracturing, hydraulic slotting, directional drilling, or deep-hole blasting [20–23]. In particular, the majority of soft coal shows dynamic phenomena or outburst risk owing to it exhibiting various characteristics such as poor permeability, low strength, rapid gas desorption velocity, and high gas content. The thickness of soft coalbed changes and the majority thereof is unstable. The thickness of soft coalbed varies from several centimeters to the height of the whole coalbed and soft coalbed in different areas show a great discrepancy. Thus, investigating the characteristics of pore and fracture structure is important [11].

Pore structure in coal refers to the size, shape, development, and interactive relationship in coalbed. The basic parameters for characterizing pore structure in coal include: pore size, specific pore volume, specific surface area (SSA), porosity, and median pore size [24–27]. Numerous scholars have explored the porosity characteristics of coalbed by utilizing different methods [28–49]. The methods for quantitatively characterizing the porosity characteristics of coalbed mainly include liquid injection methods, non-fluid injection methods, and image analysis methods. The fluid injection methods mainly involve mercury intrusion porosimetry, CO_2 adsorption, liquid nitrogen adsorption, and nuclear magnetic resonance (NMR) [28,31]. Non-fluid injection methods mainly include: small angle X-ray scattering (SAXS), small angle neutron scattering (SANS), focused ion beam-scanning electron microscope (FIB-SEM), and micron-resolution computed tomography (μ-CT) [34,41–43]. The image analysis methods mainly use scanning electron microscopy (SEM), atomic force microscope

(AFM), transmission electron microscope (TEM), and field emission scanning electron microscopy (FESEM) [46–49].

Due to being restricted by experimental method limitations, fluid injection methods are only able to test open pores in coal masses while failing to characterize closed pores: however, fluid injection methods are widely used in coal masses. Generally, when investigating the characteristics of pores and fractures in a coalbed, the pore and fracture structures thereof are quantitatively characterized by combining CO_2 (or low-temperature N_2) adsorption methods with mercury intrusion porosimetry. On this basis, the specific area of pores, pore volume, and pore size distribution characteristics are discussed. Additionally, the fractal characteristics of pores and fractures with different sizes are investigated based on different mathematical models [28–32]. The mercury intrusion experiment is also widely applied, based on which the relationship between pore volume and pressure can be established [33].

In recent years, multiple non-destructive and efficient non-fluid injection technologies have been applied to characterize pores and fractures in coalbed. SAXS/SANS technology is applicable to fragile samples whose core is hard to drill. By using this method, coal samples are not damaged during measurement. Compared with CO_2/N_2 adsorption methods, SAXS/SANS technology can be used for testing under different temperatures and pressures and the acquired pore information is not restricted by the interactive effect between fluids and surfaces, shielding effects, and pore connectivity. It is feasible to analyze information about closed pores in coalbed and various parameters including porosity, pore size distribution, and SSA [34–36]. FIB-SEM and μ-CT technologies can reveal the three-dimensional (3D) continuous change and anisotropic characteristics of pores and fractures [37–41]. FIB-SEM technology is widely used to quantify nano-sized pores in coal of different ranks [42–45].

Image analysis technology can be adopted to observe the pores in coalbed by using micro-observation technology under a microscope to further acquire images and qualify various characteristics such as size, shape, connectivity, and surface morphology of pores. The results are intuitively visible and qualitative and quantitative analyses can be combined with statistical analysis: data on pores obtained by using SEM image analysis technology has become a research hot-spot [46–49]. By using TEM and AFM technologies, one not only can explore the molecular structure and surface characteristics of coal but also can characterize the pore structure [47].

As we can see, different methods for characterizing pore and fracture structures have their own advantages and disadvantages as well as the pore diameter measurement ranges. Investigating the basic characteristics of pore and fracture structures in soft coal from different regions is important when efficiently developing CBM, improving the gas extraction efficiency in coalbed, controlling gas accidents in mines, and reducing the greenhouse effect caused by coalbed gas. The No. 5 coalbed in the western zone of Chenghe Mining Area, Shaanxi Province, China contains hard coal. Heyang, Shanyang, and Xizhuo Coal Mines are located in the new district in the eastern part of the mining area. During the construction of the three new mines, the property of No. 5 coalbed exposed was quite different from that of No. 5 coalbed in the west area. The No. 5 coalbed in the east area is softer and higher gas content than No. 5 coalbed in the west area. On this basis, the characterization of pores and fractures of the No. 5 soft coal newly exposed in the east area were investigated by applying mercury intrusion porosimetry, low-temperature liquid nitrogen adsorption, and SEM: the pore structure of coal was also characterized by use of fractal theory, which can be used for the future gas prevention and control foundation of the new mining area, as well as studying the No. 5 hard coal in the west area. In this way, the pore structure of soft coal in Chenghe Mining Area was investigated and compared with that of the No. 5 hard coalbed of Dongjiahe Coal Mine in the western part of the mining area which are same coalbed and belonged to the same mining area. This provides a theoretical basis for investigating the gas adsorption, desorption, and gas extraction of the No. 5 coalbed in the eastern zone of the new mining area.

2. Materials and methods

2.1. Selection of Coal Samples

The No. 5 main mineable coalbed in Shanyang (SY), Heyang (HY), and Xizhuo (XZ) Coal Mines and the No. 5 main mineable coalbed of Dongjiahe (DJH) Coal Mine were used for the experiments. Moreover, the coal samples were all taken from the newly exposed surface of coalbed. Tables 1 and 2 list the parameters pertaining to the gas in coal samples and the coal-rock analysis result of coal samples, respectively.

HY, SY, and XZ show a low Protodyakonov coefficient and a large initial diffusion velocity (Table 1). Additionally, HY, SY, and XZ are softer, more friable, and cohesive, compared with DJH, and showing unclear macroscopic bedding, and their surfaces appear as clastic, flake-shaped structures, exhibiting typical characteristics of soft coal.

Table 1. Coal sample basic parameters.

Sample	Protodyakonov Coefficient	Initial Speed of Methane Diffusion (mL/s)	Porosity (%)	Adsorption Constant		Industrial Analysis		
				a (cm^3/g)	b (Mpa^{-1})	A_{ad} (%)	M_{ad} (%)	V_{daf} (%)
HY	0.93	22.27	11.19	22.66	1.34	33.28	0.54	18.35
SY	0.87	20.15	11.63	17.37	1.51	34.58	0.66	23.17
XZ	0.73	26.87	10.86	18.76	1.27	30.98	1.71	21.10
DJH	2.21	5.05	8.91	14.12	1.31	11.25	0.57	15.85

Note: HY, SY, XZ, and DJH refer to the coal samples taken from Heyang, Shanyang, Xizhuo, and Dongjiahe Coal Mines, respectively. Soft coal includes HY, SY, and XZ while DJH represents a hard coal sample.

Table 2. Coal components.

Sample	Vitrinite Reflectance (%)	Macerals			Mineral Content (%)
		Vitrinite Content (%)	Inertinite Content (%)	Chitinous Content (%)	
HY	1.57	69.2	30.0	0	0.8
SY	1.76	73.8	18.7	0	7.5
XZ	1.86	65.1	33.6	0.1	1.2
DJH	1.78	40.4	46.6	0	12.8

It can be seen from Table 2 that SY coal samples showed 73.8% vitrinite content while DJH exhibited the lowest vitrinite content, implying that the higher the gelatinization, the higher the vitrinite content. The inertinite content was from 18.7% to 46.6%, which was proportional to the fusinization in the later period of coal formation, and there was little-to-no exinite present.

2.2. Experimental Equipment and Test Methods

2.2.1. Mercury Intrusion Porosimetry

The MK-AutoPore IV9510 mercury injection apparatus (Micromeritics Instrument Corp, Norcross, GA, USA) (Figure 1) was used, with a measurable pore size range of 0.003 to 1000 μm, which can be used with two low-pressure (LP) pump stations and a high-pressure (HP) pump station or four LP pump stations and two HP pump stations. The apparatus, at a pressure of 33,000 or 60,000 psi, allowed volume precision for mercury injection or ejection of better than ±0.1 μL. To eliminate the influences of mineral impurities in coal samples and fractures caused by human factors on the measurement result as far as possible, four types of experimental coal samples were crushed into particles at about 2 mm diameter.

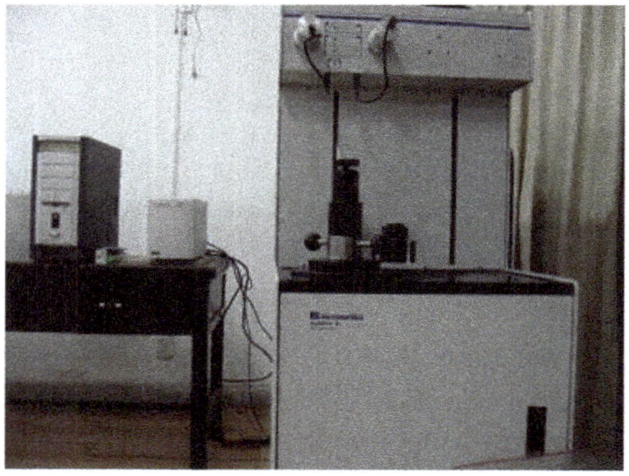

Figure 1. MK-AutoPore IV9510 mercury injection apparatus.

2.2.2. Low-Temperature Nitrogen Adsorption Method

The ASAP2020M (Micromeritics Instrument Corp, Norcross, GA, USA) automatic analyzer for SSA and micropores (Figure 2) was employed. It can be used to analyze various parameters including single- and multi-point BET (Brunauer, Emmett, and Teller) SSA, Langmuir SSA, pore size distribution, and pore volume according to a static gauging method for isothermal physical adsorption. It is a classical method for measuring the samples whose SSA is greater than 1 m^2/g. Before preparing samples, it is necessary to clean the sample cell using ethyl alcohol or acetone and then dry it. Three grams of experimental coal samples (four types of each) were separately weighed and stored. Moreover, the sample cells and tube plugs were labeled.

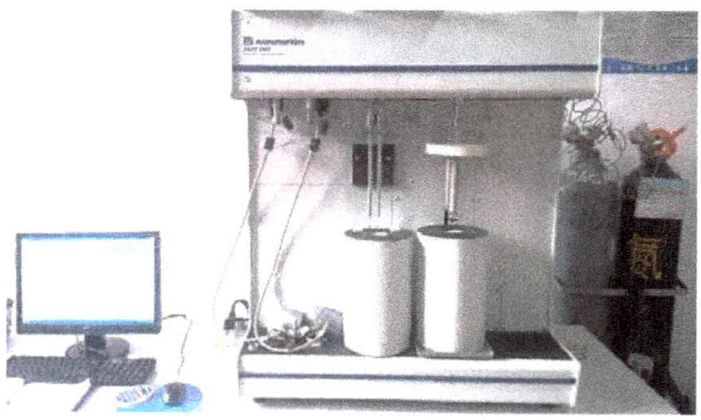

Figure 2. ASAP 2020M automatic analyzer for SSA (specific surface area) and micropores.

2.2.3. SEM Observation

The applied ΣIGMA SEM (Carl Zeiss AG, Oberkochen, Baden-Württemberg, Germany) (Figure 3) was characterized by a digital control system whose acceleration voltage was in the range 0.1 to 30 kV, and it can be magnified by 12,000 to 500,000 times. Equipped with a zoom focusing system with high resolution and high precision, it can be used to analyze microstructures of powdery, blocky, and thin-film samples.

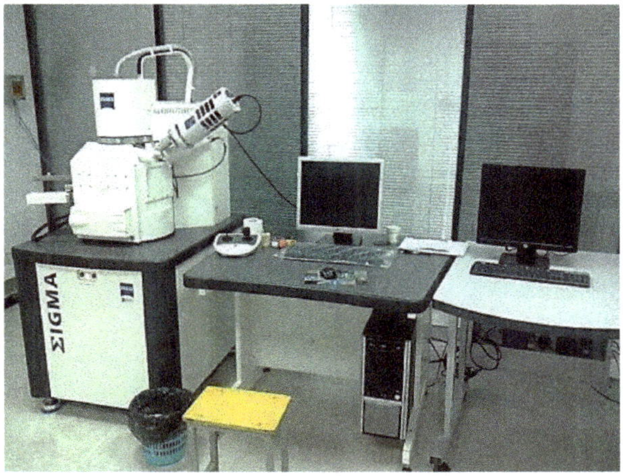

Figure 3. The ΣIGMA SEM.

In the experiment, 1 to 2 cm^3 small-blocky coal samples were taken from coal blocks. Moreover, relatively even natural fracture surfaces were chosen for observation, which were cleaned by gas absorption. Afterwards, the observed surface was plated to form a conducting layer. The loosened fragile soft coal should be treated gently at all times.

3. Results and Discussion

3.1. Characteristics of Pore Structure

3.1.1. Measuring Pore Structure by Mercury Intrusion Porosimetry

The results of mercury intrusion experiments are listed in Table 3: the displacement pressures of soft coal were all larger than those of hard coal. The displacement pressure can be used to measure the permeability of coalbed: the lower the displacement pressure, the better the permeability of a coalbed. The permeability of soft coal was lower than that of hard coal. The mean radius of throat of soft coal was 201.85 nm while the DJH hard rock showed the largest mean radius of throat of 341.29 nm. The mean mercury saturation of soft coal was 78.33% while DJH hard coal exhibited the largest mercury saturation (92.97%). The mean mercury-ejection ratio of soft coal was 57.78% while the mercury-ejection ratio of DJH hard rock was the largest, reaching 97.73%.

Table 3. Results of mercury intrusion experiments.

Sample	Displacement Pressures (MPa)	Mean Radius of Throat (nm)	Mercury Saturation (%)	Mercury Ejection Ratio (%)	Proportions of Pore Volume (%)			
					Macropore	Mesopore	Small Pore	Micropore
HY	0.04	215.51	84.52	65.37	4.53	10.77	44.98	39.72
SY	0.06	176.73	70.48	48.76	6.16	26.48	45.44	21.92
XZ	0.05	213.30	79.99	59.22	7.89	13.03	43.59	35.49
DJH	0.03	341.29	92.97	97.73	1.85	41.74	22.55	33.86

The mean proportions of macropores, mesopores, small pores, and micropores in soft coal were 6.19%, 16.67%, 44.67%, and 32.38%, while those in DJH hard coal were 1.85%, 41.74%, 22.55%, and 33.86%, respectively. It can be seen that small pores and micropores are dominant in soft coal in which small pores accounted for nearly half of the total pore volume, and mesopores and macropores were also evenly distributed. The pores in hard coal mainly appeared as mesopores, followed by micropores, while small pores and macropores accounted for a minor proportion overall. Under the effect of geostress, the gas in the coal mass was mainly adsorbed in micropores and small pores and flowed through mesopores and macropores: however, under the effect of geostress, the pore volume of the coal mass decreased and therefore the porosity of the coal was reduced, thus resulting in a reduction in permeability, which impaired the seepage and migration of gas in coalbed. After the pressure-relief process, the desorbed gas can be more constantly diffused in soft coal than in hard coal.

According to the initial data of mercury intrusion experiment, the mercury-injection and mercury-ejection curves of the four types of coal samples were plotted (Figure 4). The curves were divided into two types: Figure 4a–c are classified as the first type while Figure 4d is belonged to the other.

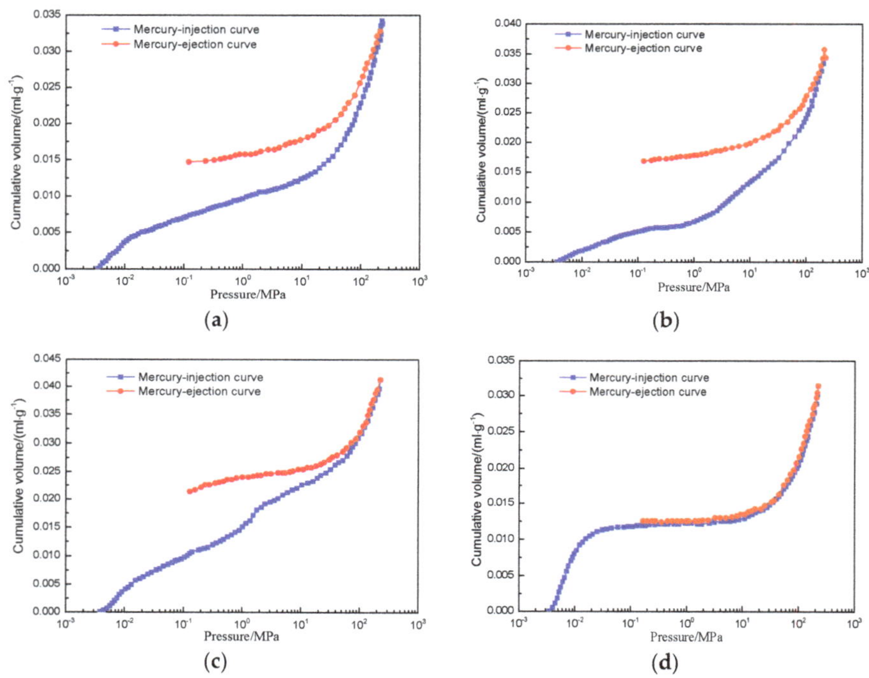

Figure 4. Mercury intrusion curve. (**a**) HY (Heyang); (**b**) SY (Shanyang); (**c**) XZ (Xizhuo); (**d**) DJH (Dongjiahe).

The mercury-intrusion loops shown in Figure 4a–c showed significant hysteresis loops and there was no platform segment in the curves. The injected and ejected amounts of mercury differed, implying that pores in coal samples mainly appeared as open pores. The mercury-injection curves were similar to the mercury-ejection curves, which indicated that certain amounts of semi-closed pores were present in the coal. The absence of a platform segment in the curves implied that the matrix pores in coal exhibited large differences, with non-uniform structures therein. The tectonic stress augmented the connectivity between mesopores and macropores in soft coal, thus improving the capability of pores for storing gas. However, the pores in soft coal were mainly semi-closed and therefore much gas was likely to be released once the pressure on the coalbed was relieved. The curve in Figure 4d showed an even segment and the mercury-injection curve was consistent with the mercury-ejection curve, without a hysteresis loop, which implied that pores in coal masses were mostly closed. The presence of the platform segment in the curve indicated that the matrix pores in coal masses exhibited little difference, with a uniform structure therein. The primary pores in coal masses were compressed or appeared as long, thin gaps. The deformation led to the reduction in volume of coal masses and therefore gas pressures increased. The non-connected microfractures, abundant with gas, comprised a closed system, and once surrounding stress decreased, particularly during mining, a stress concentration zone was formed at the front of working face. In this case, the accumulated gas in coal masses flowed to the working face space.

3.1.2. Measuring Pore Structure by Low-Temperature Nitrogen Adsorption

After analyzing the trend in the adsorption isotherm, the characteristic parameters of pore surfaces in coal masses can be qualified. Moreover, the interaction between the surface of coal masses and gas was analyzed. The low-temperature nitrogen adsorption and desorption curves of coal samples are shown in Figure 5.

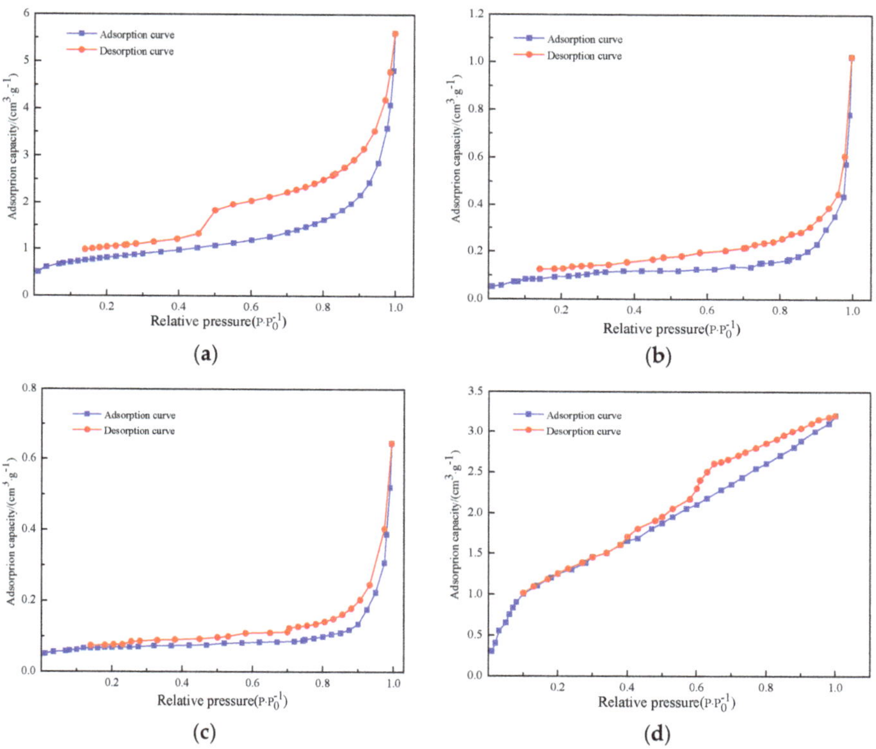

Figure 5. Low-temperature nitrogen adsorption and desorption isotherms. (**a**) HY; (**b**) SY; (**c**) XZ; (**d**) DJH.

The low-temperature nitrogen adsorption and desorption isotherms of soft coal in Figure 5a–c all belong to IV-type isotherms (based on International Union of Pure and Applied Chemistry (IUPAC)), showing different sizes of adsorption hysteresis loops. Shapes of the hysteresis loops can reflect the characteristic of the pore structure. As shown in Figure 5b,c, the adsorption hysteresis loops in the curves of the coal samples were small, indicating that impermeable pores were found in coal samples, and the adsorption curves were bulged, which indicated that there was significant interaction between coal masses and nitrogen and therefore many micropores could be found in the coal. When the relative pressure exceeded 0.8, the adsorption–desorption curves suddenly increased, reflecting the poor permeability of the coal.

The curves in Figure 5a showed a large adsorption hysteresis loop and a significant inflection point, reflecting the presence of flask-shaped pores. It can be seen from the Kelvin formula [50,51] that the corresponding pore size was about 4 nm under a relative pressure of 0.5, which was larger than the diameter of the gas molecules. The presence of flask-shaped pores caused the adsorbed gas to be difficult to desorb. The adsorption curves were not significantly convex, which meant that the interaction between coal masses and nitrogen was weak. When the relative pressure exceeded 0.8, the adsorption–desorption curves suddenly increased. The curves in Figure 5d showed a small adsorption hysteresis loop, and a significant inflection point, reflecting the presence of flask-shaped pores. However, the relative pressure region corresponding to the hysteresis loop was large. After the inflection point of the hysteresis loop, the condensed liquid in the open hole evaporated completely, and then transited to a closed pore at one end. That is why the adsorption curve coincided with the desorption curve in the low relative pressure region. For curves in Figure 5a–d, under a relative pressure of zero, the adsorption capacities of soft and hard coal for nitrogen were both positive because nitrogen was preferentially adsorbed in the micropores in coal masses while soft coal contained more micropores. The superposed effect of adsorption force fields in micropores led to an increased adsorption potential of soft coal. As the relative pressure changed from 0.4 to 1, both soft and hard coal showed different sizes of hysteresis loop.

When adsorption tests were carried out on solid with capillary pores, the pore with corresponding Kelvin radius would be capillary condensation with the increase of relative pressure on the basis of the theory of adsorption and condensation [52,53]. If the pressure was increased and then decompressed, the adsorbent would gradually desorb and evaporate. The relative pressure of condensation and evaporation of the same pore may be same or different due to the different specific shape of the capillary. If the relative pressure of condensation and evaporation was the same, the adsorption curve overlapped the desorption curve. On the contrary, if the two relative pressures were different, the two branches of the adsorption isotherm would separate to form a loop that we called it hysteresis loop [53].

In the desorption process, the relative pressure gradually decreased at the beginning, and the hysteresis loop was generated at the neck of flask-shaped pores due to the difference in shape of the gas-liquid interface during the condensation of the open pores and evaporation. However, due to the existence of pores with different diameters, when the relative pressure drops to a certain value, the corresponding larger pores begin to evaporate, resulting in a decrease in the adsorption amount and a gradual decrease in the desorption line. When the relative pressure dropped to the value corresponding to the inflection point, it meant that the condensed liquid of the smallest open pore was about to evaporate, and while the pressure continued to drop slightly, the liquid in it would gush out immediately. Therefore, there was a sharp drop in the inflection point on the hysteresis loop.

By using a low-temperature nitrogen adsorption method, the experimental data on SSA and pore volume of coal samples can be acquired (Table 4). The value of BET SSA was different from that of Langmuir SSA (Table 4). The reason for this was that the Langmuir model assumes that the adsorption of coal for gas is single-layer adsorption-based while the BET model supposes that the adsorption of coal for gas is multi-layer adsorption-based.

Table 4. Characteristics of pore distribution in coal masses. SSA: specific surface area; BJH: Barrett–Joyner–Halenda.

Sample	BET SSA (m^2/g)	Langmuir SSA (m^2/g)	BJH Pore Volume (m^2/g)	Mean Pore Size (nm)
HY	10.4317	13.2485	0.006214	8.2244
SY	10.4228	12.1175	0.006117	8.9146
XZ	14.3245	16.2412	0.007415	6.2112
DJH	8.2412	10.3258	0.004157	13.5496

The BET SSA of soft coal samples ranged from 10.4228 to 14.3245 m^2/g, with a mean of 11.7263 m^2/g, and that of hard coal samples was 8.2412 m^2/g. The Langmuir SSA of soft coal was between 12.1175 and 16.2412 m^2/g, with a mean of 13.8691 m^2/g, and that of hard coal was 10.3258 m^2/g. Moreover, the Barrett–Joyner–Halenda (BJH) pore volume of soft coal ranged from 0.006117 to 0.006214 cm^3/g (0.006582 cm^3/g on average) while that of hard coal was 0.004157 cm^3/g. The mean pore size of soft coal ranged from 6.2112 to 8.9146 nm (with a mean of 7.7834 nm) while that of hard coal was 13.5496 nm. It can be seen that the BET SSA, Langmuir SSA, and BJH pore volume of soft coal were all larger than those of hard coal while only the mean pore size of the former was lower than that of the latter. The SSA and pore volume of soft coal were larger than those of hard coal while the mean pore size was lower than that of hard coal. This reflected the fact that soft coal is capable of adsorbing a larger volume of gas and there were more gas adsorption sites on its internal surfaces per unit pore volume in soft coal than in hard coal.

According to the mercury intrusion and low-temperature nitrogen adsorption experiments, it can be seen that the lower the pore size, the larger the surface area. The number of small pores in coal was larger than that of mesopores, which was larger than the number of macropores. Therefore, the surface area of pores depends on micropores. It is supposed that pores were connected and therefore the pores in coal were distributed in a form similar to a tree in which macropores represent the trunk while mesopores and micropores refer to branches. Macropores determine the volume and branches determine the pore area so that the tree-shaped pore distribution is more luxuriant in soft coal masses.

3.1.3. Observing Pore Structure by Using SEM

The SEM image of HY coal samples obtained is shown in Figure 6a. The coal particles of HY coal samples generally appeared as irregular polyhedra, with sharp edges, reflecting the high brittleness of the coal samples. The pores showed tearing and irregular zigzag fractures, with different lengths, and pores were developed, showing different sizes, which not only led to the reduction of coal strength but also resulted in the increase of surface area thereof, reflecting the fact that coal masses showed a certain gas-storing capacity.

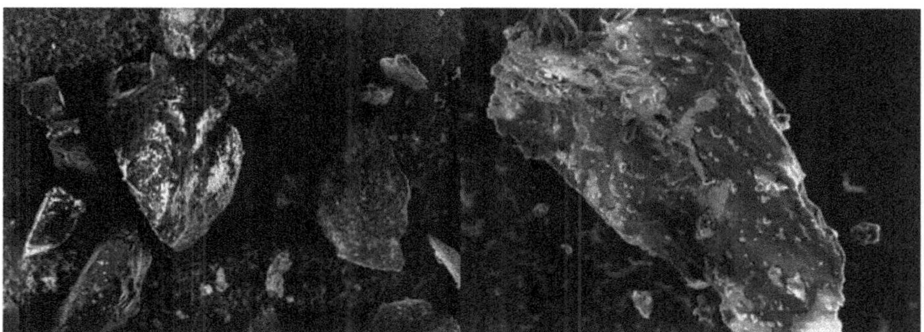

(a)

Figure 6. *Cont.*

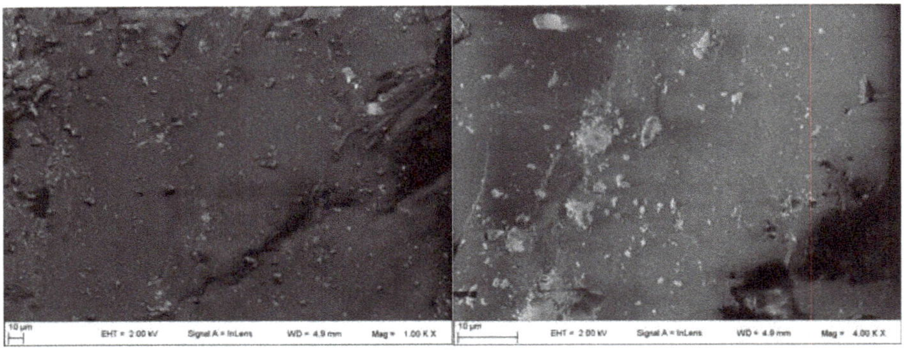

(**b**)

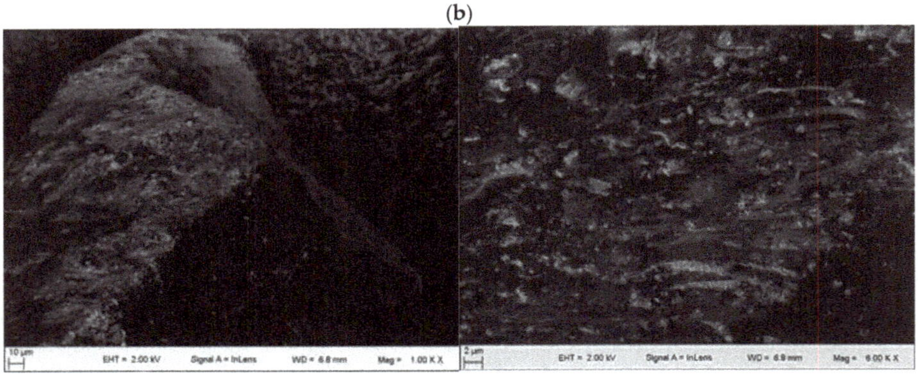

(**c**)

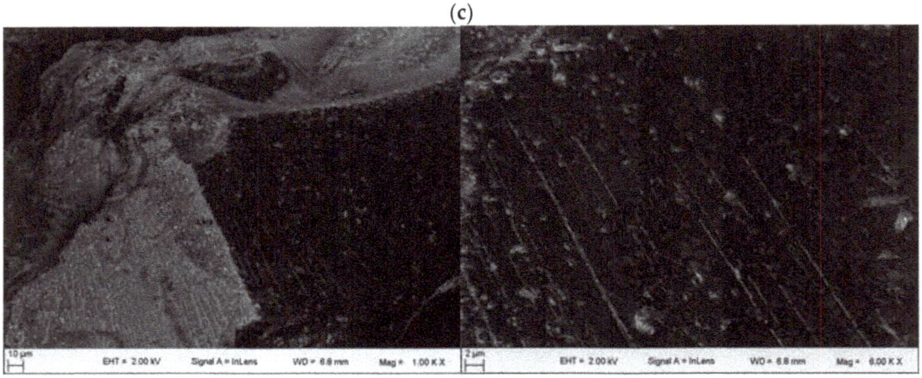

(**d**)

Figure 6. SEM images of the coal samples. (**a**) HY; (**b**) SY; (**c**) XZ; (**d**) DJH.

As shown in Figure 6b, the surface of SY coal samples showed a low roughness and a favorable evenness: a fracture shown in the image cut lengthways through the surface of the coal mass. Small particles were stacked on the surface of large particles and further pits were found due to the exfoliation of small particles. Moreover, some clay substances were distributed on the surface of coal masses and various inorganic components were mixed in coal mass, causing a certain influence on the strength of the coal mass.

XZ presenting a rough fracture surface contained a great number of developed and staggered fractures (Figure 6c). Coal masses were made of countless approximately spherical fine ellipsoids or granules. Small fractures were developed around the macro-fractures and the two kinds of fractures were jointly developed, appearing as irregular mesh- and branch-shaped fractures. Holes were found on the fracture surface on which a large number of stacked strip-shaped milk-white clay substances were distributed.

DJH, with favorable surface evenness, exhibited a high mass density, on which some mineral components were distributed. Moreover, regularly distributed cracks were found on the surface of the coal samples (Figure 6d).

It can be seen from SEM images of coal masses that the microstructures of coal masses showed the following characteristics: micropores were densely distributed where coal masses were fractured; tectonism strengthened the connectivity of pores on the surface of coal masses and the compression direction of compressed micropores was the same as the exfoliation direction of the coal-rock mass, therefore, observing the occurrence of micropores in a coal-rock mass can also reflect the tectonic deformation of soft coal.

3.2. Fractal Characteristics of Pore Structures

3.2.1. Mercury Intrusion Method

It is supposed that there is a cube of side-length R: the cube is divided into m small isometric cubes. A rule is adopted to remove a part of these small cubes and the number of the remaining small cubes is N_{b1}. After undergoing k such operations, the side length of the remaining cubes is $a_k = R/m^k$ and the total number is calculated using Equation (1) [54,55].

$$N_{bk} = \left(\frac{R}{a_k}\right)^{D_b} = \frac{R^{D_b}}{a_k^{D_b}} = \frac{C}{a_k^{D_b}} = Ca_k^{-D_b} \tag{1}$$

where, D_b refers to fractal dimension of pore volume, and $D_b = \log(N_{b1})/\log(m)$.

According to the aforementioned formula, the relationship between the pore volume V_k of coal and a_k can be derived as in Equation (2).

$$\frac{dV_k}{da_k} \infty a_k^{2-D_b} \tag{2}$$

When measuring the pores in coal by mercury intrusion method, the applied pressure $p(r)$ and r satisfy the Washburn equation and the total pore volume under a constant pressure is equal to the volume of mercury injected into the pores (Equation (3)).

$$log[dVp(r)/dp(r)] \infty (4 - D_b) logr \infty (D_b - 4) logp(r) \tag{3}$$

where, $p(r)$ and r denote the external pressure (MPa) and the diameter (nm) of pores in coal samples, respectively.

After plotting the best-fit line $log[dVp(r)/dp(r)]$ and $logp(r)$, slope K can be attained and therefore Equation (4) is acquired.

$$D_b = K + 4 \tag{4}$$

According to experimental data obtained by mercury intrusion porosimetry, the fractal dimensions of the four types of coal samples can be obtained (Table 5).

Table 5. Calculated fractal dimensions of pores.

Sample	Pore Diameter Range (nm)	K	D_b
HY	74.16~100,000	1.0679	2.9321
SY	89.21~100,000	1.1083	2.8917
XZ	77.69~100,000	1.0826	2.9174
DJH	93.45~100,000	1.1298	2.7303

The adsorption of coal for gas is conducted on the surface of pores. The fractal dimension is used for measuring the irregularity and roughness of the surface of pores. The fractal dimension of coal is highly correlated with the complex pore structures and heterogeneous surface area of coal. As shown in Table 5, the fractal dimensions (D_b) of the four types of experimental coal samples differed little: HY coal samples exhibited the largest D_b (2.9321) while D_b for the DJH was the lowest, that is, D_b of soft coal is larger than that of hard coal. This reflects the fact that the surface of pores in soft coal is more irregular and rough compared with that in hard coal.

3.2.2. Low-Temperature Nitrogen Adsorption

According to the experimental data obtained by low-temperature nitrogen adsorption, the fractal dimension of pores on the surface of coal masses can be calculated. Based on the parameters related to the adsorption capacity, Equation (5) is established.

$$ln(Q) = (D-3)ln(ln\frac{p_0}{p}) + C \tag{5}$$

where, Q and p refer to the adsorption capacity (cm^3/g) under equilibrium pressure P and the partial pressure (MPa) of nitrogen, respectively. Moreover, p_0, D and C denote the saturated vapor pressure of gas adsorption, fractal dimension, and a constant, respectively.

In terms of SSA of coal masses, Equation (6) gives

$$ln(A) = (D-2)ln(ln\frac{p_0}{p}) + C \tag{6}$$

where, A refers to the SSA (m^2/g) of the surface of coal masses.

Data for $ln\ (ln\ P_0/P)$ are plotted by separately using $ln\ Q$ and $ln\ A$ to conduct linear fitting. According to the slope of the fitted line, the fractal dimension D can be obtained (Table 6).

Table 6. Calculated fractal dimensions of pore surfaces.

Sample	p/p_0 Range	Fractal Dimensions of Pore Surfaces Calculated by Equation (5)	Fractal Dimensions of Pore Surfaces Calculated by Equation (6)
HY	0.15~0.96	2.34	2.23
SY	0.44~0.99	2.26	2.18
XZ	0.41~0.99	2.31	2.12
DJH	0.44~0.99	1.84	1.81

As shown in Table 6, the calculation result obtained using Equation (5) was slightly different from that using Equation (6), which was mainly caused by the presence of micropores. The difference in size of micropores caused the change in adsorption potential of micropores in coal masses for nitrogen. By analyzing the data in Table 6, the fractal dimensions of pore surfaces in soft coal were between 2.26 and 2.34, with a mean some 1.25 times that of hard coal. The larger the fractal dimension, the larger the surface roughness of coal masses. This indicated that there were more pores per unit volume of soft coal and thereby more sites for adsorbing gas per unit pore space.

3.2.3. Images Taken by SEM

By utilizing the box-counting dimension method [56–58], the fractal characteristics of the pore structure of soft coal masses were analyzed. When observing the surface characteristics of coal samples by using SEM, the inhomogeneity of energy on the surface of coal samples and structural differences on the pore surface led to a significant difference in intensities of secondary reflection electron beams from the SEM. It is supposed that each disparate point corresponds to a geometrical characteristic of the structure. According to fractal theory, within each length of ε cm, the number $N_c(L)$ of geometrical characteristics within the length of L is expressed by Equation (7)

$$N_c(L) \infty a L^{2-D} \tag{7}$$

where, a, D, and L denote the proportionality coefficient, fractal dimension, and the area (cm^2) of the fractal plot, respectively.

According to the box-counting dimension calculation principle, the relationship between the box dimension and the total surface area can be plotted by using "Tri Prism surface area" algorithm of fractal analysis software Fractal Fox 2.0 [56,57]. By fitting the curve, the fractal dimensions of SSA distribution of pores in different coal samples were attained.

By conducting binarization the image, each pixel point in the image is set to be black (0) or white (1) to acquire a series of two-dimensional matrices consisting of 0 s and 1 s, that is, the data file of a binary image. Block division was carried out on the data file, so that the line number of the blocks was equal to the column number (marked as k). The number of blocks was recorded as N_k ($k = 1, 2, 4, \ldots, 2^i$). By using the size of the pixel point as the side length to conduct block division, the box counts were recorded as $N_1, N_2, N_4, \ldots, N_{2i}$. The dimension ε of the pixel point was equal to the number of pixel points in a line of the image and each line and column of the blocks both consisted of k pixel points, with the side length of the blocks being $\varepsilon = k\varepsilon$. By conducting linear fitting on the datum point ($ln\varepsilon_k$, lnN_k), the negative of the slope of the fitted curve represents the box dimension D.

In a similar way, by using "Box counting dimension" algorithm in the Fractal Fox 2.0 software [57,58], the relationship between box dimension and the box number can be acquired. By fitting the curve, the fractal dimension of the pore distribution in these coal samples could be calculated.

According to fractal theory, a larger fractal dimension value indicates higher roughness of a pore surface, a larger number of pores per unit volume, and more developed pores. By analyzing the data in Table 7, the fractal dimension of the pore distribution in SY coal samples was 1.1696 and that of the SSA of pores in HY coal samples was 2.6783, reflecting the fact that the pores in HY and SY coal masses were well developed. Therefore, these coal samples contained more gas adsorption sites per unit pore space, thus augmenting the adsorption capacity for gas. The resulting SEM image only approximately showed the one-dimensional space. If 3D spatial factors were taken into account, the pore volume would have been larger in such coal masses.

Table 7. Calculated fractal dimensions of pores.

Sample	Amplification	Fractal Dimension of Pore Distribution	Correlation Coefficient	Fractal Dimension of Pore SSA	Correlation Coefficient
HY	4.00KX	1.0743	0.9558	2.6783	0.9673
SY	4.00KX	1.1696	0.9545	2.6421	0.9649
XZ	6.00KX	1.1248	0.9512	2.2646	0.9704
DJH	6.00KX	1.0054	0.9578	2.1642	0.9621

3.3. Prospect of Future Work

The eastern part of Chenghe Mining area is a new mining area. The No.5 coalbed exposed is all soft coal, with a relatively large the gas content in the coalbed. However, the western part is the old mining area, and the No. 5 coal seam is hard coal. Moreover, the No. 5 soft coalbed in the eastern part and the No. 5 hard coalbed in the western part are the same coalbed. Under such a background condition, the pore structure characteristics of No. 5 coal seam in Chenghe Mining area are studied in this paper. Furthermore, based on fractal theory, the pore structure of coal was characterized. However, because there was no gas emission in other mines in the west area, only gas emission occurred locally in Dongjiahe Coal Mine, while a large amount of gas emission occurred in the No. 5 soft coal of the coal mines in the east area. Therefore, in order to contrast with No. 5 soft coal in the east area, only the hard coal was sampled in Dongjiahe Coal Mine, and no other coal mines in the western area were sampled the hard coal. Hence, we should study No. 5 hard coal of other coal mines in the western part in the future research plan. At the same time, other methods can be used to study the pore structure characteristics more comprehensively and systematically.

This provides a theoretical basis for investigating the gas adsorption, desorption, and gas extraction of the No. 5 coalbed in the eastern zone of the new mining area. So, based on the study of pore structure, we will have a further study on the adsorption and desorption rules of both soft and hard coal in Chenghe Mining area in order to study the different adsorption and desorption rules between them. Moreover, the layout of drainage boreholes under different drilling radius, different negative pressure of extraction and different layout modes will be studied on the basis of the study of adsorption and desorption law, which will lay a foundation for the prediction of gas emission and the formulation of coalbed gas prevention and control measures in the working face, so as to achieve the safe co-mining of coal and gas.

4. Conclusions

The structural characteristics of pores and fractures in soft coal masses were explored and compared with those of hard coal from the same mining area. The following conclusions are drawn:

(1) The displacement pressure of soft coal was larger than that of hard coal, implying that the permeability of soft coal masses was lower than that of hard coal, with a poor permeability. However, the mean radii of the pore-throats of soft coal were all lower than those of hard coal. Pores in soft coal mainly appeared as small pores and micropores in which small pores accounted for half of the total pore volume, and mesopores and macropores were also evenly distributed. In hard coal, mesopores were widely distributed, followed by micropores, and small pores and macropores accounted for a minority thereof.

(2) The changing trends seen in mercury-injection and mercury-ejection curves of soft coal were the same, showing a significant hysteresis loop, which implied that pores in soft coal samples mainly appeared as open pores. Moreover, the matrix pores in coal showed significant differences, with non-uniform structures appearing therein. During the formation of soft coal, the tectonic stress strengthened the connectivity of mesopores and macropores, thus improving the capacity of pores for storing gas: however, owing to pores in soft coal mainly appearing as semi-closed pores, it was likely to release much gas once the pressure on the coalbed was relieved. Therefore, when mining a soft coalbed, it is necessary to take suitable gas-control measures. The mercury-injection and mercury-ejection curves of hard coal showed no hysteresis loop while having an even segment, which indicated that closed pores accounted for the majority of the pores in these hard coal samples.

(3) The low-temperature nitrogen adsorption curves of soft coal all followed a IV-type isotherm, which showed adsorption hysteresis loops with different sizes under relative pressures of 0.4 to 1. This indicated that soft coal contained more micropores while the superimposed effect of adsorption force fields in micropores strengthened the adsorption potential of soft coal. The BET SSA, Langmuir SSA, and BJH pore volume of soft coal obtained through use of the nitrogen adsorption method were all larger than those of hard coal while only the mean pore size was lower than that of hard coal. This reflected the fact that the soft coal presented a larger gas adsorption capacity and there were more gas adsorption sites on internal surfaces per unit pore space therein.

(4) The fractal dimensions of soft coal calculated by mercury intrusion porosimetry showed subtle differences to those in hard coal while all were larger than those in the latter. The fractal dimensions of the surface area of pores in soft coal, calculated using a low-temperature nitrogen adsorption method, were all larger than those in hard coal. The fractal dimensions of pore distribution and SSA of soft coal attained through SEM were also all larger than those of hard coal. A large fractal dimension of the pore surface of a coal mass meant that the surface of coal masses was rough. In addition, the pores and fractures per unit pore space were better developed and therefore there was favorable

connectivity between pores and fractures, thus providing many adsorption sites for gas to enhance the adsorption capacity.

Author Contributions: Each author has contributed to the present paper. P.W. and R.M. conceived and designed the experiments; P.W., X.L., and S.Z. performed the experiments and analyzed the experimental data. P.W. drafted the manuscript. S.P., Y.L., and X.L. revised the manuscript.

Funding: This work was supported by National Key R & D Program of China (2018YFC0808303), the Basic and Frontier Research Projects of Chongqing (cstc2016jcyjA0117, cstc2018jcyjAX0626) and the Fundamental Research Funds for the Central Universities (2017CDJQJ248825).

Conflicts of Interest: The authors declare no conflict of interest.

References

1. Asif, M.; Muneer, T. Energy supply, its demand and security issues for developed and emerging economies. *Renew. Sustain. Energy Rev.* **2007**, *11*, 1388–1413. [CrossRef]
2. Zhang, Y.; Shao, W.; Zhang, M.; Li, H.J.; Yin, S.J.; Xu, Y.J. Analysis 320 coal mine accidents using structural equation modeling with unsafe conditions of the rules and regulations as exogenous variables. *Accid. Anal. Prev.* **2016**, *92*, 189–201. [CrossRef] [PubMed]
3. Geng, F.; Saleh, J.H. Challenging the emerging narrative: Critical examination of coalmining safety in China, and recommendations for tackling mining hazards. *Saf. Sci.* **2015**, *75*, 36–48. [CrossRef]
4. Chen, H.; Qi, H.; Long, R.; Zhang, M. Research on 10-year tendency of China coal mine accidents and the characteristics of human factors. *Saf. Sci.* **2012**, *50*, 745–750. [CrossRef]
5. Li, X.L.; Li, Z.H.; Wang, E.Y.; Liang, Y.P.; Li, B.L.; Chen, P.; Liu, Y.J. Pattern recognition of mine microseismic (MS) and blasting events based on wave fractal features. *Fractals* **2018**, *26*, 1850029. [CrossRef]
6. Wang, L.; Cheng, Y.P.; Liu, H.Y. An analysis of fatal gas accidents in Chinese coal mines. *Saf. Sci.* **2014**, *62*, 107–113. [CrossRef]
7. Yin, W.; Fu, G.; Yang, C.; Jiang, Z.G.; Zhu, K.; Gao, Y. Fatal gas explosion accidents on Chinese coal mines and the characteristics of unsafe behaviors: 2000–2014. *Saf. Sci.* **2017**, *92*, 173–179. [CrossRef]
8. Nanda, S.; Reddy, S.N.; Mitra, S.K.; Kozinski, J.A. The progressive routes for carbon capture and sequestration. *Energy Sci. Eng.* **2016**, *4*, 99–122. [CrossRef]
9. Yu, C.H.; Huang, C.H.; Tan, C.S. A review of CO_2 capture by absorption and adsorption. *Aerosol Air Qual. Res.* **2012**, *12*, 745–769. [CrossRef]
10. Jokar, S.M.; Rahimpour, M.R.; Shariati, A.; Iulianelli, A.; Bagnato, G.; Vita, A. Pure hydrogen production in membrane reactor with mixed reforming reaction by utilizing waste gas: A case study. *Processes* **2016**, *4*, 33. [CrossRef]
11. Cheng, Y.P.; Wang, L.; Zhang, X.L. Environmental impact of coal mine methane emissions and responding strategies in China. *Int. J. Greenh. Gas Control* **2011**, *5*, 157–166. [CrossRef]
12. Hu, Q.T.; Liang, Y.P.; Wang, H.; Zou, Q.L.; Sun, H.T. Intelligent and integrated techniques for coalbed methane (CBM) recovery and reduction of greenhouse gas emission. *Environ. Sci. Pollut. Res.* **2017**, *24*, 17651–17668.
13. Karacan, C.Ö.; Ruiz, F.A.; Cotè, M.; Phippsc, S. Coal mine methane: A review of capture and utilization practices with benefits to mining safety and to greenhouse gas reduction. *Int. J. Coal Geol.* **2011**, *86*, 121–156. [CrossRef]
14. Ju, Y.; Sun, Y.; Sa, Z.; Pan, J.N.; Wang, J.L.; Hou, Q.L.; Li, Q.G.; Yan, Z.F.; Liu, J. A new approach to estimate fugitive methane emissions from coal mining in China. *Sci. Total. Environ.* **2016**, *543*, 514–523. [CrossRef] [PubMed]
15. Zhou, F.; Xia, T.; Wang, X.; Zhang, Y.F.; Sun, Y.N.; Liu, J.S. Recent developments in coal mine methane extraction and utilization in China: A review. *J. Nat. Gas Sci. Eng.* **2016**, *31*, 437–458. [CrossRef]
16. Zhao, J.L.; Xu, H.; Tang, D.; Mathews, J.P.; Li, S.; Tao, S. A comparative evaluation of coal specific surface area by CO2 and N2 adsorption and its influence on CH4 adsorption capacity at different pore sizes. *Fuel* **2016**, *183*, 420–431. [CrossRef]
17. Zhao, J.L.; Xu, H.; Tang, D.Z.; Mathews, J.P.; Li, S.; Tao, S. Coalbed porosity and fracture heterogeneity of macrolithotypes in the Hancheng Block, eastern margin, Ordos Basin, China. *Int. J. Coal Geol.* **2016**, *159*, 18–29. [CrossRef]

18. Liu, T.; Lin, B.; Zou, Q.; Zhu, C.j. Microscopic mechanism for enhanced coal bed methane recovery and outburst elimination by hydraulic slotting: A case study in Yangliu mine, China. *Greenh. Gases Sci. Technol.* **2016**, *6*, 597–614. [CrossRef]
19. Liu, T.; Lin, B.; Yang, W. Impact of matrix–fracture interactions on coal permeability: Model development and analysis. *Fuel* **2017**, *207*, 522–532. [CrossRef]
20. Li, Q.G.; Lin, B.Q.; Zhai, C. The effect of pulse frequency on the fracture extension during hydraulic fracturing. *J. Nat. Gas Sci. Eng.* **2014**, *21*, 296–303. [CrossRef]
21. Li, Q.G.; Lin, B.Q.; Zhai, C. A new technique for preventing and controlling coal and gas outburst hazard with pulse hydraulic fracturing: A case study in Yuwu coal mine, China. *Nat. Hazards* **2015**, *75*, 2931–2946. [CrossRef]
22. Zou, Q.L.; Lin, B.Q.; Liu, T.; Zhou, Y.; Zhang, Z.; Yan, F.Z. Variation of methane adsorption property of coal after the treatment of hydraulic slotting and methane pre-drainage: A case study. *J. Nat. Gas Sci. Eng.* **2014**, *20*, 396–406. [CrossRef]
23. Zou, Q.; Lin, B.; Zheng, C.; Zheng, C.S.; Hao, Z.Y.; Zhai, C.; Liu, T.; Liang, J.Y.; Yan, F.Z.; Yang, W.; et al. Novel integrated techniques of drilling–slotting–separation-sealing for enhanced coal bed methane recovery in underground coal mines. *J. Nat. Gas Sci. Eng.* **2015**, *26*, 960–973. [CrossRef]
24. Radlinski, A.P.; Mastalerz, M.; Hinde, A.L.; Hainbuchner, M.; Rauch, H.; Baron, M.; Lin, J.S.; Fan, l.; Thiyagarajan, P. Application of SAXS and SANS in evaluation of porosity, pore size distribution and surface area of coal. *Int. J. Coal Geol.* **2004**, *59*, 245–271. [CrossRef]
25. Gryglewicz, G.; Machnikowski, J.; Lorenc-Grabowska, E.; Lota, G.; Frackowiak, E. Effect of pore size distribution of coal-based activated carbons on double layer capacitance. *Electrochim. Acta* **2005**, *50*, 1197–1206. [CrossRef]
26. Ross, D.J.K.; Bustin, R.M. The importance of shale composition and pore structure upon gas storage potential of shale gas reservoirs. *Mar. Pet. Geol.* **2009**, *26*, 916–927. [CrossRef]
27. Gan, H.; Nandi, S.P.; Walker, P.L. Nature of the porosity in American coals. *Fuel* **1972**, *51*, 272–277. [CrossRef]
28. Cai, Y.D.; Liu, D.M.; Pan, Z.J.; Pan, Z.J.; Yao, Y.B.; Li, J.Q.; Qiu, Y.K. Pore structure and its impact on CH_4 adsorption capacity and flow capability of bituminous and subbituminous coals from Northeast China. *Fuel* **2013**, *103*, 258–268. [CrossRef]
29. Clarkson, C.R.; Solano, N.; Bustin, R.M.; Bustin, A.M.M.; Chalmers, G.R.L.; He, L.; Melnichenko, B.; Radliński, A.P.; Blach, T.P. Pore structure characterization of North American shale gas reservoirs using USANS/SANS, gas adsorption, and mercury intrusion. *Fuel* **2013**, *103*, 606–616. [CrossRef]
30. Yao, Y.B.; Liu, D.M.; Tang, D.Z.; Tang, S.H.; Huang, W.H. Fractal characterization of adsorption-pores of coals from North China: An investigation on CH4 adsorption capacity of coals. *Int. J. Coal Geol.* **2008**, *73*, 27–42. [CrossRef]
31. Fu, H.J.; Tang, D.Z.; Xu, H.; Xu, T.; Chen, B.L.; Hu, P.; Yin, Z.Y.; Wu, P.; He, G.J. Geological characteristics and CBM Exploration potential evaluation: A case study in the middle of the southern Junggar Basin, NW China. *J. Nat. Gas Sci. Eng.* **2016**, *30*, 557–570. [CrossRef]
32. Hou, S.H.; Wang, X.M.; Wang, X.J.; Yuan, Y.D.; Pan, S.D.; Wang, X.M. Pore structure characterization of low volatile bituminous coals with different particle size and tectonic deformation using low pressure gas adsorption. *Int. J. Coal Geol.* **2017**, *183*, 1–13. [CrossRef]
33. Kenvin, J.; Jagiello, J.; Mitchell, S.; Pérez-Ramírez, J. Unified method for the total pore volume and pore size distribution of hierarchical zeolites from argon adsorption and mercury intrusion. *Langmuir* **2015**, *31*, 1242–1247. [CrossRef]
34. Melnichenko, Y.B.; Radlinski, A.P.; Mastalerz, M.; Cheng, G.; Rupp, J. Characterization of the CO2 fluid adsorption in coal as a function of pressure using neutron scattering techniques (SANS and USANS). *Int. J. Coal Geol.* **2009**, *77*, 69–79. [CrossRef]
35. Okolo, G.N.; Everson, R.C.; Neomagus, H.W.J.P.; Roberts, M.J.; Sakurovs, R. Comparing the porosity and surface areas of coal as measured by gas adsorption, mercury intrusion and SAXS techniques. *Fuel* **2015**, *141*, 293–304. [CrossRef]
36. Zhao, Y.X.; Sun, Y.F.; Liu, S.M.; Wang, K.; Jiang, Y.D. Pore structure characterization of coal by NMR cryoporometry. *Fuel* **2017**, *190*, 359–369. [CrossRef]

37. Karpyn, Z.T.; Alajmi, A.; Radaelli, F.; Halleck, P.M.; Grader, A.S. X-ray CT and hydraulic evidence for a relationship between fracture conductivity and adjacent matrix porosity. *Eng. Geol.* **2009**, *103*, 139–145. [CrossRef]
38. Watanabe, N.; Ishibashi, T.; Hirano, N.; Ohsaki, Y.; Tsuchiya, Y.; Tamagawa, T.; Okabe, H.; Tsuchiya, N. Precise 3D numerical modeling of fracture flow coupled with X-ray computed tomography for reservoir core samples. *SPE J.* **2011**, *16*, 683–691. [CrossRef]
39. Golab, A.; Ward, C.R.; Permana, A.; Lennox, P.; Botha, P. High-resolution three-dimensional imaging of coal using microfocus X-ray computed tomography, with special reference to modes of mineral occurrence. *Int. J. Coal Geol.* **2013**, *113*, 97–108. [CrossRef]
40. Jing, Y.J.; Armstrong, R.T.; Ramandi, H.L.; Mostaghimi, P. Coal cleat reconstruction using micro-computed tomography imaging. *Fuel* **2016**, *181*, 286–299. [CrossRef]
41. Jing, Y.; Armstrong, R.T.; Mostaghimi, P. Digital coal: Generation of fractured cores with microscale features. *Fuel* **2017**, *207*, 93–101. [CrossRef]
42. Liu, S.; Sang, S.; Wang, G.; Ma, J.S.; Wang, X.; Wang, W.F.; Du, Y.; Wang, T. FIB-SEM and X-ray CT characterization of interconnected pores in high-rank coal formed from regional metamorphism. *J. Pet. Sci. Eng.* **2017**, *148*, 21–31. [CrossRef]
43. Zhou, S.D.; Liu, D.M.; Cai, Y.D.; Yao, Y.B.; Li, B. 3D characterization and quantitative evaluation of pore-fracture networks of two Chinese coals using FIB-SEM tomography. *Int. J. Coal Geol.* **2017**, *174*, 41–54. [CrossRef]
44. Li, Z.T.; Liu, D.M.; Cai, Y.D.; Shi, Y.L. Investigation of methane diffusion in low-rank coals by a multiporous diffusion model. *J. Nat. Gas Sci. Eng.* **2016**, *33*, 97–107. [CrossRef]
45. Li, Z.T.; Liu, D.M.; Cai, Y.D.; Ranjith, P.G.; Yao, Y.B. Multi-scale quantitative characterization of 3-D pore-fracture networks in bituminous and anthracite coals using FIB-SEM tomography and X-ray μ-CT. *Fuel* **2017**, *209*, 43–53. [CrossRef]
46. Louw, E.B.; Mitchell, G.D.; Wang, J.; Winans, R.E.; Mathews, J.P. Constitution of drop-tube-generated coal chars from vitrinite-and inertinite-rich South African coals. *Energy Fuels* **2015**, *30*, 112–120. [CrossRef]
47. Zhou, H.; Wilkes, G.L. Comparison of lamellar thickness and its distribution determined from d.s.c. SAXS, TEM and AFM for high-density polyethylene films having a stacked lamellar morphology. *Polymer* **1997**, *38*, 5735–5747. [CrossRef]
48. Agbabiaka, A.; Wiltfong, M.; Park, C. Small angle X-ray scattering technique for the particle size distribution of nonporous nanoparticles. *J. Nanoparticles* **2013**, *2013*, 11. [CrossRef]
49. Lee, S.; Fischer, T.B.; Stokes, M.R.; Klingler, R.J.; Ilavsky, J.; McCarty, D.K.; Wigand, M.; Derkowski, A.; Winans, R.E. Dehydration effect on the pore size, porosity, and fractal parameters of shale rocks: Ultrasmall-angle X-ray scattering study. *Energy Fuels* **2014**, *28*, 6772–6779. [CrossRef]
50. Thomson, W. 4. On the equilibrium of vapour at a curved surface of liquid. *Proc. R. Soc. Edinb.* **1872**, *7*, 63–68. [CrossRef]
51. Chen, P.; Tang, X.Y. The research on the adsorption of nitrogen in low temperature and micro-pore properties in coal. *J. China Coal Soc.* **2001**, *26*, 552–556.
52. Yan, J.M.; Zhang, P.Y.; Gao, J.C. *Adsorption and Cohesion*; Science Press: Beijing, China, 1986; pp. 113–137.
53. De Boer, J.H. *The Structure and Properties of Porous Materials*; Butterworths: London, UK, 1958; Volume 389, pp. 68–94.
54. Li, Y.H.; Gao, Q.L.; Rudolph, V. Compressibility and fractal dimension of fine coal particles in relation to pore structure characterisation using mercury porosimetry. *Part. Part. Syst. Charact.* **1999**, *16*, 25–31. [CrossRef]
55. Zhu, J.F.; Liu, J.Z.; Yang, Y.M.; Cheng, J.; Zhou, J.H.; Cen, K.F. Fractal characteristics of pore structures in 13 coal specimens: Relationship among fractal dimension, pore structure parameter, and slurry ability of coal. *Fuel Process. Technol.* **2016**, *149*, 256–267. [CrossRef]
56. Ai, T.; Zhang, R.; Zhou, H.W.; Pei, J.L. Box-counting methods to directly estimate the fractal dimension of a rock surface. *Appl. Surf. Sci.* **2014**, *314*, 610–621. [CrossRef]

57. Li, J.; Du, Q.; Sun, C. An improved box-counting method for image fractal dimension estimation. *Pattern Recognit.* **2009**, *42*, 2460–2469. [CrossRef]
58. Sarkar, N.; Chaudhuri, B.B. An efficient differential box-counting approach to compute fractal dimension of image. *IEEE Trans. Syst. Man Cybern.* **1994**, *24*, 115–120. [CrossRef]

processes

MDPI

Article

2D Plane Strain Consolidation Process of Unsaturated Soil with Vertical Impeded Drainage Boundaries

Minghua Huang [1,2,*] **and Dun Li** [1,2]

[1] Key Laboratory of Building Safety and Energy Efficiency of the Ministry of Education, Hunan University, Changsha 410082, China; lidunhnu@163.com

[2] Institute of Geotechnical Engineering, Hunan University, Changsha 410082, China

* Correspondence: huangminghua@hnu.edu.cn; Tel.: +86-152-7315-5772

Received: 26 November 2018; Accepted: 19 December 2018; Published: 21 December 2018

Abstract: The consolidation process of soil stratum is a common issue in geotechnical engineering. In this paper, the two-dimensional (2D) plane strain consolidation process of unsaturated soil was studied by incorporating vertical impeded drainage boundaries. The eigenfunction expansion and Laplace transform techniques were adopted to transform the partial differential equations for both the air and water phases into two ordinary equations, which can be easily solved. Then, the semi-analytical solutions for the excess pore-pressures and the soil layer settlement were derived in the Laplace domain. The final results in the time domain could be computed by performing the numerical inversion of Laplace transform. Furthermore, two comparisons were presented to verify the accuracy of the proposed semi-analytical solutions. It was found that the semi-analytical solution agreed well with the finite difference solution and the previous analytical solution from the literature. Finally, the 2D plane strain consolidation process of unsaturated soil under different drainage efficiencies of the vertical boundaries was illustrated, and the influences of the air-water permeability ratio, the anisotropic permeability ratio and the spacing-depth ratio were investigated.

Keywords: consolidation process; unsaturated soil; semi-analytical solution; impeded drainage boundary; excess pore-pressures

1. Introduction

The consolidation process of soil stratum is a common issue in geotechnical engineering and it has captured a great deal of attention in the geotechnical community. In the mid-1920s, Terzaghi [1] put forward the classical one-dimensional (1D) consolidation process for saturated soil, after which some researchers derived the solutions to 1D consolidation process for saturated soil subjected to different loading and boundary conditions [2–5]. With the problem raised in practical engineering, much attention has been attracted to the consolidation process of unsaturated soil since the 1960s. Scott [6] discussed the consolidation process of unsaturated soil with occluded air bubbles, and Barden [7,8] presented a study of the compacted unsaturated soil. Furthermore, Fredlund and Hasan [9] proposed a 1D consolidation theory by driving two partial differential equations (PDEs) to describe the dissipation processes of excess pore-air and -water pressures. Dakshanamurthy and Fredlund [10] extended this theory to two-dimensional (2D) plane strain consolidation process by referring the concept of 2D heat diffusion. These theories have been widely accepted in geotechnical engineering and have inspired numerous research projects on the consolidation process of unsaturated soil [11–15].

In order to ensure that analytical solutions were available, most of the previous studies treated the top and bottom drainage boundaries of the soil layer as fully drained or undrained [16]. However, these boundaries are generally partially drained (i.e., impeded drainage) in most practical consolidation processes. For example, a sand cushion covered on the top surface of the soil layer, which is commonly used to facilitate the drainage, will become partially drained when its drainage capacity is not effective.

The underlying layer will exhibit impeded drainage characteristic when its void ratio is relatively small. In order to describe the partially drained effect, Gray [17] defined an impeded drainage boundary using the third type boundary, which can also realize the fully drained and undrained boundaries (i.e., the first and second type boundaries) by changing the values of the boundary parameters. Base on this boundary, a number of solutions have been proposed to solve the 1D consolidation process of saturated soil [16–20] and that of unsaturated soil [21–23]. These solutions help to understand the impeding effect (i.e., drainage efficiency) of the covered sand cushion and underlying layer on the 1D consolidation process. However, in the study of the 2D plane strain consolidation processes, particularly for unsaturated soil, there appears no solution incorporating the impeded drainage boundaries. The objective of this paper was to develop a general solution to the 2D plane strain consolidation process of unsaturated soil under such impeded drainage boundaries, that is, the top surface and bottom base of the soil layer were considered to be partially drained.

Based on the consolidation equations proposed by Dakshanamurthy and Fredlund [10], this paper attempted to derive semi-analytical solutions to the 2D plane strain consolidation process of unsaturated soil by incorporating the vertical impeded drainage boundaries. To obtain the final solutions, the eigenfunction expansion and Laplace transform techniques were used to transform PDEs of both air and water phases into two ordinary equations. Then, the solutions for the excess pore-pressures and the soil layer settlement were derived in the Laplace domain, and the final results can be computed by performing the numerical inversion of Laplace transform. Furthermore, two comparisons against the finite difference solution and the previous analytical solution were made to verify the accuracy of the proposed solution. Finally, some computations were conducted to study the 2D plane strain consolidation process of unsaturated soil incorporating vertical impeded drainage boundaries.

2. Mathematical Model

2.1. Governing Equations

In geotechnical practice, vertical drain assisted with preloading is one of the most commonly used techniques to improve the soft soil. The vertical drain can facilitate the consolidation process by providing a shorter drainage path in the horizontal direction. Figure 1 illustrates a referential consolidation system of a homogeneous unsaturated soil layer with two installed vertical drains, in which the vertical boundaries are presumed to be impeded drainage to the air and water phases. This system extends the problem considered by Ho et al. [14,15].

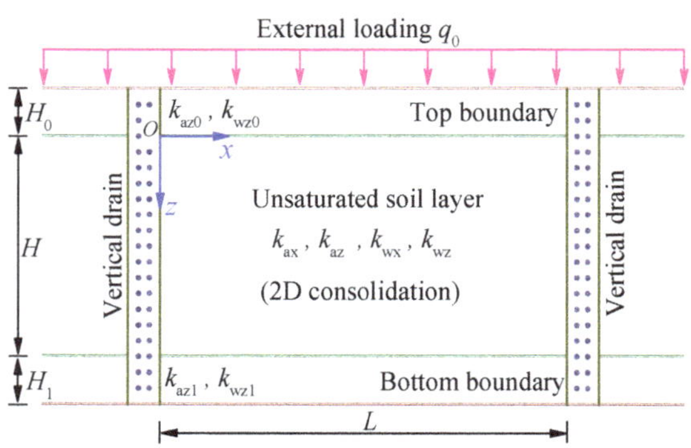

Figure 1. Two-dimensional (2D) consolidation system of unsaturated soil with vertical impeded drainage boundaries.

The thickness of the soil layer was H, and the internal spacing of vertical drains was L. In the soil layer, the coefficients of permeability with respect to the air and water in the x-direction were k_{ax}

and k_{wx}, and those in z-direction were k_{az} and k_{wz}, respectively. On and under the soil layer were two impedance layers, and their thicknesses were H_0 and H_1, respectively. These two impedance layers were considered to be permeable in z-direction only. In the upper impedance layer, the coefficients of permeability with respect to the air and water were k_{a0} and k_{w0}, respectively; those in the down impedance layer were k_{a1} and k_{w1}, respectively. The external load q_0 was assumed to be constant and uniformly distributed on the top of the upper impedance layer.

For a constant external load instantaneously applied, the governing equations for 2D plane strain consolidation process of unsaturated soil can be expressed as follows [10,14,15]:

$$\frac{\partial u_a}{\partial t} + C_a \frac{\partial u_w}{\partial t} + c_{vx}^a \frac{\partial^2 u_a}{\partial x^2} + c_{vz}^a \frac{\partial^2 u_a}{\partial z^2} = 0 \tag{1a}$$

$$\frac{\partial u_w}{\partial t} + C_w \frac{\partial u_a}{\partial t} + c_{vx}^w \frac{\partial^2 u_w}{\partial x^2} + c_{vz}^w \frac{\partial^2 u_w}{\partial z^2} = 0 \tag{1b}$$

where u_a and u_w are the excess pore-air and -water pressures, respectively (kPa); C_a and C_w are the interactive constants associated with the air and water phases, respectively; c_{vx}^a and c_{vx}^w are the consolidation coefficients with respect to the air and water phases in x-direction, respectively (m^2/s); and c_{vz}^a and c_{vz}^w are the consolidation coefficients with respect to the air and water phases in z-direction, respectively (m^2/s). These consolidation parameters can be expressed as follows:

$$C_a = \left[\left(\frac{2m_1^a}{m_2^a} - 1 \right) - \frac{n(1-S_r)}{m_2^a (u_a^0 + u_{atm})} \right]^{-1} \tag{2a}$$

$$c_{vx}^a = \frac{k_{ax} R \Theta}{gM} \left[m_2^a \left(u_a^0 + u_{atm} \right) \left(\frac{2m_1^a}{m_2^a} - 1 \right) - n(1-S_r) \right]^{-1} \tag{2b}$$

$$c_{vz}^a = \frac{k_{az} R \Theta}{gM} \left[m_2^a \left(u_a^0 + u_{atm} \right) \left(\frac{2m_1^a}{m_2^a} - 1 \right) - n(1-S_r) \right]^{-1} \tag{2c}$$

$$C_w = \frac{2m_1^w}{m_2^w} - 1 \tag{2d}$$

$$c_{vx}^w = \frac{k_{wx}}{m_2^w \gamma_w} \tag{2e}$$

$$c_{vz}^w = \frac{k_{wz}}{m_2^w \gamma_w} \tag{2f}$$

where m_1^a and m_1^w are the coefficients of air and water volume change with respect to the change of the net normal stress, respectively (kPa^{-1}); m_2^a and m_2^w are the coefficients of air and water volume change with respect to the change of suction, respectively (kPa^{-1}); n is the soil porosity; S_r is the degree of saturation; u_{atm} is the atmospheric pressure (kPa); Θ is the absolute temperature (K); R is the universal air constant (8.314 $J \cdot mol^{-1} \cdot K^{-1}$); g is the gravitational acceleration (9.8 m/s^2); M is the molecular mass of air phase (0.029 kg/mol); and γ_w is the unite weight of water (9.8 kN/m^3).

2.2. Initial and Boundary Conditions

It was assumed that immediately after the application of the external load, the initial excess pore-air and -water pressures were distributed uniformly throughout the entire stratum. That is:

$$u_a(x,z,0) = u_a^0, \quad u_w(x,z,0) = u_w^0 \tag{3}$$

where u_a^0 and u_w^0 are the values of the initial excess pore-air and -water pressures, respectively.

The horizontal (i.e., left and right) boundaries, which were defined by two vertical drains, were all considered to permeable to the air and water phases. That is:

$$u_a(0,z,t) = u_a(L,z,t) = u_w(0,z,t) = u_w(L,z,t) = 0 \quad (4)$$

The vertical (i.e., top and bottom) boundaries, which were separately defined by sand cushion and underlying layer, were considered to be partially permeable to the air and water phases. That is:

$$\frac{\partial u_a(x,0,t)}{\partial z} - \frac{R_{ta}}{H}u_a(x,0,t) = 0, \quad \frac{\partial u_w(x,0,t)}{\partial z} - \frac{R_{tw}}{H}u_w(x,0,t) = 0 \quad (5a)$$

$$\frac{\partial u_a(x,H,t)}{\partial z} + \frac{R_{ba}}{H}u_a(x,H,t) = 0, \quad \frac{\partial u_w(x,H,t)}{\partial z} + \frac{R_{bw}}{H}u_w(x,H,t) = 0 \quad (5b)$$

where $R_{ta} = k_{az0}H/k_{az}H_0$ and $R_{tw} = k_{wz0}H/k_{wz}H_0$, and they are two dimensionless characteristic factors describing the drainage efficiency of the top surface to the air and water phases, respectively; $R_{ba} = k_{az1}H/k_{az}H_1$ and $R_{bw} = k_{wz1}H/k_{wz}H_1$, and they are two dimensionless characteristic factors describing the drainage efficiency of the bottom base to the air and water phases, respectively. When it is assumed that $R_{ta} = R_{tw} = \infty$ and $R_{ba} = R_{bw} = \infty$, the vertical boundaries are double drainage (i.e., top-base drainage [14,15]). For this scenario, both the top and bottom surfaces were permeable to the air and water phases. When it is assumed that $R_{ta} = R_{tw} = \infty$ and $R_{ba} = R_{bw} = 0$, the vertical boundaries are single drainage (i.e., top drainage [14,15]). For this scenario, the top surface was permeable to the air and water phase, whereas the bottom base was impermeable to the air and water phases.

3. Solution Formulation and Verification

In general, unsaturated soil is complex in natural property and engineering behavior due to the lack of homogeneity, the complicate texture, the interaction of multiphase and the nonlinear feature. In order to handle the mathematical derivation easily, some essential assumptions were made along with those used in the 2D plane strain consolidation theory of unsaturated soil [10,14,15].

The prominent assumptions are listed as follows:

(1) The soil stratum was considered to be homogeneous;
(2) The soil particle and pore water were incompressible;
(3) The flows of the air and water phases were continuous and independent, and simultaneously, they only took place in two directions (i.e., the x- and z-directions);
(4) During the consolidation process, the deformation occurring was small and the coefficients of the volume change for the soil stratum remained constant;
(5) The consolidation parameters with respect to the air phase (i.e., C_a, c_{vx}^a and c_{vz}^a) and those with respect to the water phase (i.e., C_w, c_{vx}^w and c_{vz}^w) remained constant during the consolidation process; and
(6) The effect of air diffusing through water, air dissolving in water, the movement of water vapor and temperature change were ignored.

The above assumptions were not completely accurate for all cases. The consolidation parameters may change because of the variations in the soil properties, such as the permeability coefficients, degree of saturation and porosity, and moreover, the modulus for the soil and the water phase were nonlinear. However, in order to obtain the solution to the consolidation equations of unsaturated soil more easily, it may be acceptable to assume that these soil properties remain constant during the transient process for a small stress increment. Moreover, these constant properties have been commonly adopted to study the consolidation process of unsaturated soil in some existing analytical and semi-analytical methods proposed by Lo et al. [11,14,15], Zhou et al. [12], Shan et al. [13] and Wang et al. [21–23].

3.1. Solution Formulation

Equations (1a) and (1b) can be rewritten as follows:

$$\frac{\partial u_a}{\partial t} = \frac{1}{1 - C_a C_w}\left[C_a c_{vz}^w \left(\kappa_w \frac{\partial^2 u_w}{\partial x^2} + \frac{\partial^2 u_w}{\partial z^2}\right) - c_{vz}^a \left(\kappa_a \frac{\partial^2 u_a}{\partial x^2} + \frac{\partial^2 u_a}{\partial z^2}\right)\right] \tag{6a}$$

$$\frac{\partial u_w}{\partial t} = \frac{1}{1 - C_a C_w}\left[C_w c_{vz}^a \left(\kappa_a \frac{\partial^2 u_a}{\partial x^2} + \frac{\partial^2 u_a}{\partial z^2}\right) - c_{vz}^w \left(\kappa_w \frac{\partial^2 u_w}{\partial x^2} + \frac{\partial^2 u_w}{\partial z^2}\right)\right] \tag{6b}$$

where $\kappa_a = k_{ax}/k_{az}$ and $\kappa_w = k_{wx}/k_{wz}$, and they are two anisotropic permeability ratios of the air and water phases, respectively. Through defining two alternative variables ϕ_1 and ϕ_2, Equations (6a) and (6b) can be transformed into:

$$\frac{\partial \phi_1}{\partial t} = Q_1\left(\frac{\partial^2 \phi_1}{\partial z^2} + \rho_{11}\frac{\partial^2 \phi_1}{\partial x^2} + \rho_{21}\frac{\partial^2 \phi_2}{\partial x^2}\right) \tag{7a}$$

$$\frac{\partial \phi_2}{\partial t} = Q_2\left(\frac{\partial^2 \phi_2}{\partial z^2} + \rho_{12}\frac{\partial^2 \phi_1}{\partial x^2} + \rho_{22}\frac{\partial^2 \phi_2}{\partial x^2}\right) \tag{7b}$$

where:

$$\phi_1 = u_a + c_{21}u_w,\ \phi_2 = c_{12}u_a + u_w \tag{8}$$

$$\rho_{11} = \frac{\kappa_w c_{12}c_{21} - \kappa_a}{c_{12}c_{21} - 1},\ \rho_{21} = \frac{\kappa_a c_{21} - \kappa_w c_{21}}{c_{12}c_{21} - 1},\ \rho_{12} = \frac{\kappa_w c_{12} - \kappa_a c_{12}}{c_{12}c_{21} - 1},\ \rho_{22} = \frac{\kappa_a c_{12}c_{21} - \kappa_w}{c_{12}c_{21} - 1} \tag{9}$$

The details of derivation process and meanings of c_{21} and c_{12} can be found in Appendix A. It is noteworthy that Equations (7a) and (7b) will become two uncoupled PDEs when $\kappa_a = \kappa_w$, and their solutions can be obtained more easily than coupled ones. However, for the sake of generality the solutions are derived below for the coupled case of Equations (7a) and (7b).

Substituting Equation (4) into Equation (8) gives:

$$\phi_1(0,z,t) = \phi_1(L,z,t) = \phi_2(0,z,t) = \phi_2(L,z,t) = 0 \tag{10}$$

These boundary conditions above are homogeneous; thus, the general solutions of Equations (7a) and (7b) can be expressed as follows:

$$\phi_1(x,z,t) = \sum_{i=1}^{\infty} \varphi_1^i(z,t)\cdot\sin(\mu_i x) \tag{11a}$$

$$\phi_2(x,z,t) = \sum_{i=1}^{\infty} \varphi_2^i(z,t)\cdot\sin(\mu_i x) \tag{11b}$$

where φ_1^i and φ_2^i are the generalized Fourier coefficients of ϕ_1 and ϕ_2 with respect to x, respectively; and $\mu_i = i\pi/L$.

Substituting Equations (11a) and (11a) into Equations (7a) and (7b) result in:

$$\frac{\partial \varphi_1^i}{\partial t} = Q_1\left(\frac{\partial^2 \varphi_1^i}{\partial z^2} - \rho_{11}\mu_i^2\varphi_1^i - \rho_{21}\mu_i^2\varphi_2^i\right) \tag{12a}$$

$$\frac{\partial \varphi_2^i}{\partial t} = Q_2\left(\frac{\partial^2 \varphi_2^i}{\partial z^2} - \rho_{12}\mu_i^2\varphi_1^i - \rho_{22}\mu_i^2\varphi_2^i\right) \tag{12b}$$

Implementing the Laplace transforms on Equations (12a) and (12b), and rearranging gives:

$$Q_1\frac{d^2\widetilde{\varphi}_1^i}{dz^2} = \left(s + Q_1\rho_{11}\mu_i^2\right)\widetilde{\varphi}_1^i + Q_1\rho_{21}\mu_i^2\widetilde{\varphi}_2^i - \varphi_1^{i0} \tag{13a}$$

$$Q_2\frac{\partial^2\widetilde{\varphi}_2^i}{\partial z^2} = Q_2\rho_{12}\mu_i^2\widetilde{\varphi}_1^i + \left(s + Q_2\rho_{22}\mu_i^2\right)\widetilde{\varphi}_2^i - \varphi_2^{i0} \tag{13b}$$

where:

$$\varphi_1^{i0} = \vartheta_i\left(u_a^0 + c_{21}u_w^0\right),\ \varphi_2^{i0} = \vartheta_i\left(c_{12}u_a^0 + u_w^0\right) \tag{14}$$

and $\vartheta_i = 2\left[1-(-1)^i\right]/(\mu_i L)$.

The solutions of Equations (13a) and (13b) subjected to the impeded drainage boundaries are:

$$\widetilde{\varphi}_1^i = \chi_1 e^{\lambda_1 z} + \chi_2 e^{-\lambda_1 z} + \psi_2\left(\chi_3 e^{\lambda_2 z} + \chi_4 e^{-\lambda_2 z}\right) + \chi_5 \tag{15a}$$

$$\widetilde{\varphi}_2^i = \psi_1\left(\chi_1 e^{\lambda_1 z} + \chi_2 e^{-\lambda_1 z}\right) + \chi_3 e^{\lambda_2 z} + \chi_4 e^{-\lambda_2 z} + \chi_6 \tag{15b}$$

The derivation details and meanings of χ_1~χ_6, λ_1, λ_2, ψ_1 and ψ_2 can be found in Appendix B.

Implementing the Laplace transform on Equation (A22) gives:

$$\widetilde{u}_a = \sum_{i=1}^{\infty}\left(\frac{\widetilde{\varphi}_1^i - c_{21}\widetilde{\varphi}_2^i}{1-c_{12}c_{21}}\right)\sin(\mu_i x),\ \widetilde{u}_w = \sum_{i=1}^{\infty}\left(\frac{\widetilde{\varphi}_2^i - c_{12}\widetilde{\varphi}_1^i}{1-c_{12}c_{21}}\right)\sin(\mu_i x) \tag{16}$$

By substituting Equations (15a) and (15b) into Equation (16), the solutions for the excess pore-air and -water pressures can be obtained in the Laplace domain. The results are:

$$\widetilde{u}_a = \sum_{i=1}^{\infty}\frac{\sin(\mu_i x)}{1-c_{12}c_{21}}\left[\alpha_1\left(\chi_1 e^{\lambda_1 z} + \chi_2 e^{-\lambda_1 z}\right) + \alpha_2\left(\chi_3 e^{\lambda_2 z} + \chi_4 e^{-\lambda_2 z}\right) + \alpha_5\right] \tag{17a}$$

$$\widetilde{u}_w = \sum_{i=1}^{\infty}\frac{\sin(\mu_i x)}{1-c_{12}c_{21}}\left[\alpha_3\left(\chi_1 e^{\lambda_1 z} + \chi_2 e^{-\lambda_1 z}\right) + \alpha_4\left(\chi_3 e^{\lambda_2 z} + \chi_4 e^{-\lambda_2 z}\right) + \alpha_6\right] \tag{17b}$$

where α_1~α_6 are presented in Equation (A26).

According to the method of two stress-state variables, the volumetric strain of unsaturated soil induced by a constant external load instantaneously applied during the 2D plane strain consolidation process is [14]:

$$\varepsilon_v(x,z,t) = (m_2^s - 2m_1^s)\left[u_a(x,z,t) - u_a^0\right] - m_2^s\left[u_w(x,z,t) - u_w^0\right] \tag{18}$$

where ε_v is the volumetric strain; $m_1^s = m_1^w + m_1^a$, is the coefficient of the soil volume change with respect to the change in net normal stress; $m_2^s = m_2^w + m_2^a$ is the coefficient of the soil volume change with respect to the change in suction.

Applying the Laplace transform to Equation (18) gives:

$$\widetilde{\varepsilon}_v(x,z,s) = (m_2^s - 2m_1^s)\left[\widetilde{u}_a(x,z,s) - u_a^0/s\right] - m_2^s\left[\widetilde{u}_w(x,z,s) - u_w^0/s\right] \tag{19}$$

Integrating Equation (19) along both the x- and z-directions, we can obtain the average settlement of the soil layer in the Laplace domain (denoted as $\widetilde{W}$) as follows:

$$\begin{aligned}\widetilde{W} = & \int_0^H\int_0^L \widetilde{\varepsilon}_v(x,z,s)dxdz \\ & = \sum_{i=1}^{\infty}\frac{1-(-1)^i}{(1-c_{12}c_{21})\mu_i}\{\left[(\alpha_1-\alpha_3)m_2^s - 2\alpha_1 m_1^s\right](\chi_1 e^{\lambda_1 H} - \chi_2 e^{-\lambda_1 H} - \chi_1 + \chi_2)/\lambda_1 \\ & +\left[(\alpha_2-\alpha_4)m_2^s - 2\alpha_2 m_1^s\right](\chi_3 e^{\lambda_2 H} - \chi_4 e^{-\lambda_2 H} - \chi_3 + \chi_4)/\lambda_2 - 2\alpha_5 m_1^s H \\ & +(\alpha_5-\alpha_6)m_2^s H\} - \left[(m_2^s - 2m_1^s)u_a^0 - m_2^s u_w^0\right]HL/s\end{aligned} \tag{20}$$

3.2. Solution Verification

Using Crump's method (see Appendix C) [24] to implement the numerical inversion of Laplace transform (NILT) on Equations (17a), (17b) and (20), the semi-analytical solutions for the excess pore-air and -water pressures and the soil layer settlement were obtained in the time domain. In this part, two comparisons against the previous analytical solution [14] and the finite difference solution were performed to verify the accuracy of the semi-analytical solution. The details of finite difference solution to Equations (1a) and (1b) with vertical impeded drainage boundaries are presented in Appendix D. The soil properties adopted in these comparisons were assumed as follows [14]: L = 2 m, H = 4 m, n = 0.50, S_r = 80%, $k_{wz} = 10^{-10}$ m/s, $k_{az} = 10^{-7}$~10^{-12} m/s, $m_1^s = -2.5 \times 10^{-4}$ kPa^{-1}, $m_2^s = 0.4\ m_1^s$, $m_1^w = -0.5 \times 10^{-4}$ kPa^{-1}, $m_2^w = 4\ m_1^w$, u_a^0 = 20 kPa, u_w^0 = 40 kPa, u_{atm} = 100 kPa and Θ = 293 K.

As stated, when the values of $R_{ta} = R_{tw} = R_t$ and $R_{ba} = R_{bw} = R_b$ approach infinity, the vertical impeded drainage boundaries can be degenerated into those of the top-base drainage system; when $R_t = \infty$ and $R_t = 0$, the vertical impeded drainage boundaries can be simplified into those of the top drainage system. Compared with the results from Ho et al. [14], the variations of the excess pore-air and -water pressures were computed using the semi-analytical solution. Because of limited space, only the isotropic permeability condition for the air and water phases (i.e., $\kappa_a = \kappa_w = 1$) was considered. For the graphic presentation, the point of investigation was located at x = 1 m and z = 4 m. Figure 2 shows the variations of the excess pore-air and -water pressures under the top-base drainage system, and those under the top drainage system are illustrated in Figure 3. In these two figures, the semi-analytical solution and the previous analytical solution were individually abbreviated to SAS and PAS, respectively. It can be observed that the results of the semi-analytical solution with $R_b = R_t = \infty$ were the same as those of previous analytical solution for the top-base drainage system, while the results of the semi-analytical solution with $R_b = 0$ and $R_t = \infty$ were identical to those of the previous analytical solution for the top drainage system.

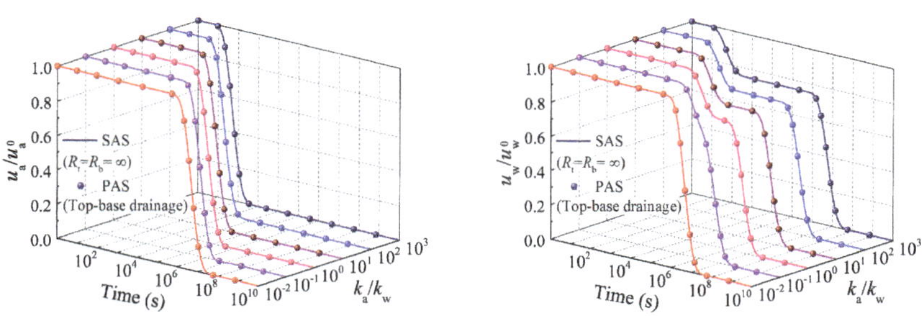

(a) (b)

Figure 2. Excess pore-pressures under top-base drainage boundary calculated from semi-analytical solution (SAS) and previous analytical solution (PAS): (**a**) air phase and (**b**) water phase.

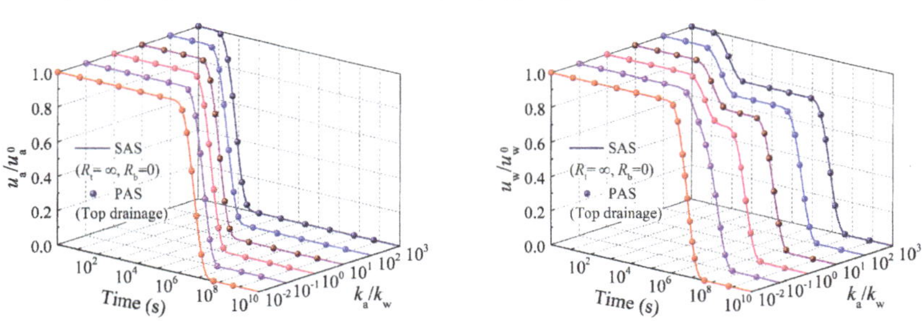

(a) (b)

Figure 3. Excess pore-pressures under top drainage boundary calculated from semi-analytical solution (SAS) and previous analytical solution (PAS): (**a**) air phase and (**b**) water phase.

In order to verify the accuracy of the semi-analytical solution under vertical impeded drainage boundary, another comparison against the finite difference solution was performed. The excess pore-air and -water pressures under the vertical boundaries with $R_b = R_t = 10$ were computed using both the semi-analytical solution and the finite difference solution. For the graphic presentation, the point of investigation also was located at x = 1 m and z = 4 m. Figure 4 illustrates the variations of the excess pore-air and -water pressures, where finite difference solution is abbreviated to finite difference solution (FDS). It can be found that the excess pore-air and -water pressures calculated from the semi-analytical solution were in line with those of the finite difference solution.

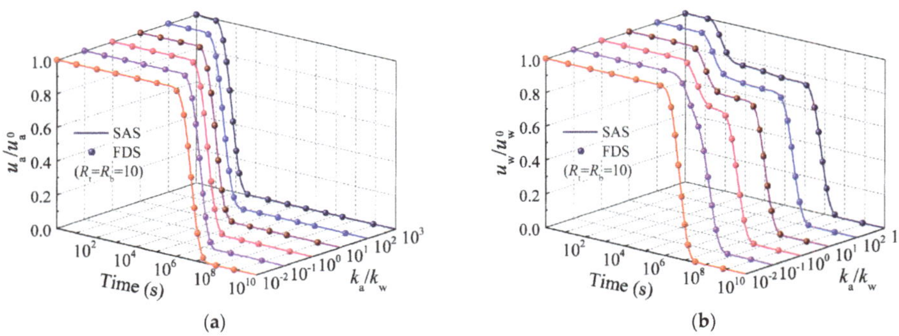

Figure 4. Excess pore-pressures under impeded drainage boundary calculated from semi-analytical solution (SAS) and finite difference solution (FDS): (**a**) air phase and (**b**) water phase.

Therefore, these comparisons showed that the proposed semi-analytical solution incorporating the vertical impeded drainage boundaries was correct and it is also feasible for the 2D plane strain consolidation processes of unsaturated soil with top-base drainage and top drainage boundaries.

4. 2D Consolidation Process of Unsaturated Soil with Vertical Impeded Drainage Boundaries

Based on the proposed semi-analytical solution above, some computations were performed to illustrate the 2D plane strain consolidation process of unsaturated soil with the vertical impeded drainage boundaries. The variations of the excess pore-air and -water pressures and the average settlement of the soil layer were presented, and the influences of the drainage efficiency of vertical boundary, the air-water permeability ratio, the anisotropic permeability ratio and the spacing-depth ratio were investigated. Because of limited space, the drainage efficiencies of top surface and bottom base for air phase were only considered to be the same as those for water phases, that is, $R_{ta} = R_{tw} = R_t$ and $R_{ba} = R_{bw} = R_b$. The parameters of the soil property were assumed to be the same as those used in the previous section.

4.1. Consolidation Process with Different Drainage Efficiencies

The vertical drainage boundaries can be classified into symmetrical and asymmetrical ones according to the drainage efficiencies of top and bottom boundaries (i.e., R_t and R_b). When $R_t = R_b$, the vertical drainage boundaries are symmetrical; they are asymmetrical when $R_t \neq R_b$. For these two types of vertical boundaries, the excess-pore pressures and the soil layer settlement were computed using the proposed semi-analytical solutions considering different drainage efficiencies. For the graphic presentation, the point of investigation on the excess-pore pressures was located at x = 1 m and z = 2.5 m. Figures 5 and 6, respectively, show the variations of the excess pore-pressures and the soil layer settlement under the symmetrical impeded drainage boundary, in which R_t = R_b vary from 0.2 to 100.

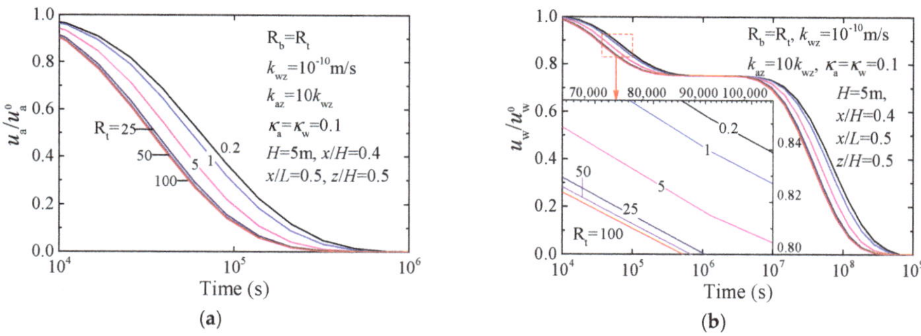

Figure 5. Variations of excess pore-pressures under symmetrical vertical impeded drainage boundary with different drainage efficiencies: (**a**) air phase and (**b**) water phase.

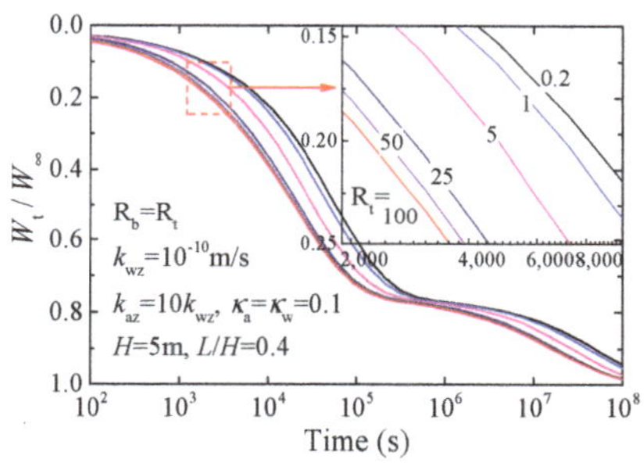

Figure 6. Variations of soil layer settlement under symmetrical vertical impeded drainage boundary with different drainage efficiencies.

It can be observed that the drainage efficiency has a significant influence on the dissipation processes of the excess pore-pressures and the development of the soil layer settlement. Larger $R_t = R_b$ generally resulted in a greater dissipation rate of the excess pore-pressures and a larger settlement rate of the soil layer. Figure 5a graphically indicates that the normalized curves of the excess pore-air pressures gradually shifted to the left with the increase of $R_t = R_b$. Figure 5b illustrates the dissipation pattern of water phase and it exhibits the typical double inverse S curves, in which the influence of the drainage efficiency can be separated into two stages: the first one emerged during the dissipation process of the excess pore-air water, and the second one was caused by the dissipation process of the excess pore-air water. Between these two stages, the plateau period emerged on the dissipation curve of the excess pore-air water. Similarly, the influence of the drainage efficiency on the soil layer settlement can also be divided into two stages based on the typical double inverse S curves, as shown in Figure 6. In addition, it can be also found that the differences of the excess pore-pressures and the soil layer settlement induced by different drainage efficiencies decreased with the increase of $R_t = R_b$. When $R_t = R_b$ were close to 100, the differences of the excess pore-pressures and the soil layer settlement were negligible. This indicates that the corresponding boundary can be considered to be permeable to air and water phases.

Figures 7 and 8, respectively, illustrate the variations of the excess pore-pressures and the soil layer settlement under the asymmetrical vertical impeded drainage boundary, in which R_b = 0 or ∞ and R_t = 1 or 25. Being the same as those in Figures 5 and 6, a bigger R_t delivered a larger dissipation rate of the excess pore-pressure, and further resulted in a larger settlement rate. The dissipation patterns of the excess pore-air pressure are typical inverse S curves, whereas those of the excess pore-water pressure and the soil layer settlement are typical double inverse S ones with two obvious affected stages. Compared with the results under the symmetrical vertical impeded drainage boundary, Figure 7a,b indicate that the excess pore-pressures under the vertical boundaries of $R_b = 0$ dissipated more slowly and consequently resulted in a smaller settlement rate (see Figure 8a). In contrast, Figure 7c,d show that the excess pore-pressures under the vertical boundaries of $R_b = \infty$ dissipated faster and consequently induced a larger settlement rate (see Figure 8b). This phenomenon happened because of the different bottom boundaries. The bottom boundary with $R_b = 0$ was impermeable and it might impede the drainage of the air and water phases; the bottom boundary with $R_b = \infty$ was permeable and could accelerate the drainage process of the air and water phases.

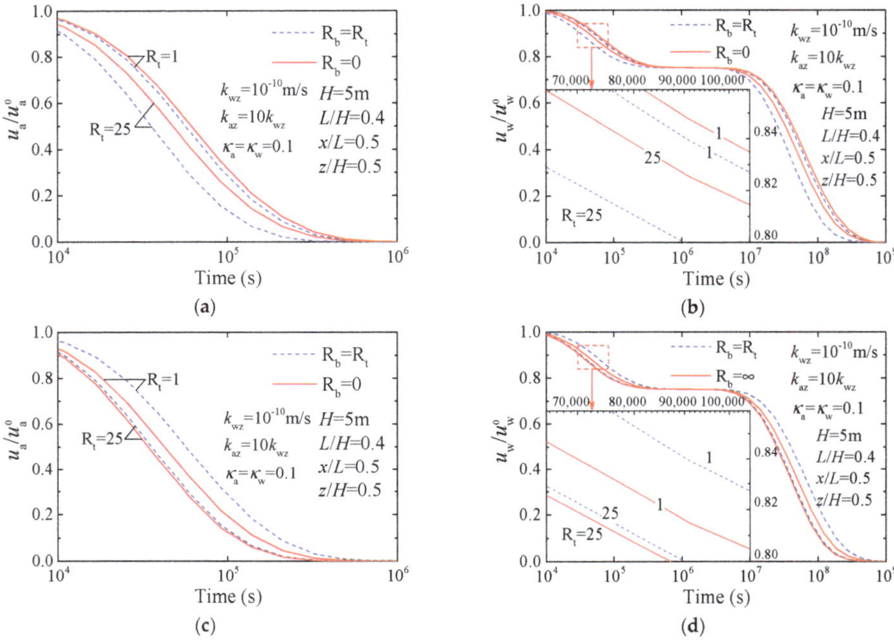

Figure 7. Variations of excess pore-pressures under asymmetrical vertical impeded drainage boundary with different drainage efficiencies: (**a**,**b**) $R_b = 0$ and (**c**,**d**) $R_b = \infty$.

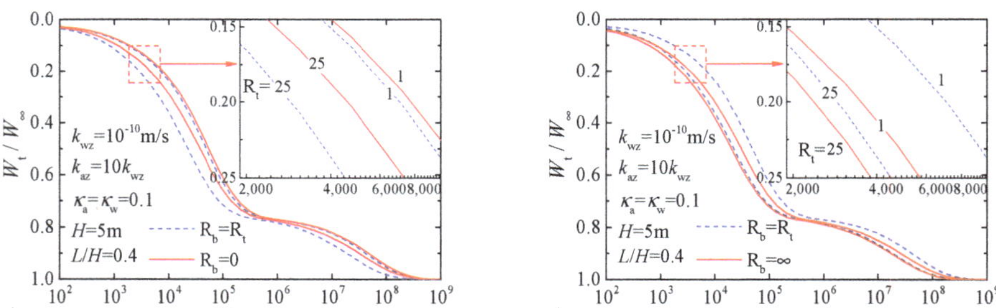

Figure 8. Variations of soil layer settlement under asymmetrical vertical impeded drainage boundary with different drainage efficiencies: (**a**) $R_b = 0$ and (**b**) $R_b = \infty$.

4.2. Consolidation Process with Different Air-Water Permeability Ratios

The air-water permeability ratio was defined as the ratio between the vertical permeability coefficient of air phase and that of water phase, i.e., k_{az}/k_{wz}. Moreover, the anisotropic permeability conditions of these phases were assumed to be the same, i.e., $\kappa_a = \kappa_w$. Figure 9 shows the changes of the excess pore-pressures under different values of k_{az}/k_{wz}, in which the vertical impeded drainage boundaries are given by $R_b = R_t = 5$ and $R_b = R_t = \infty$, respectively. The corresponding variations of the soil layer settlement are shown in Figure 10. It can be observed that the air-water permeability ratio had a great influence on the dissipation processes of the excess pore-pressures and the development of the soil layer settlement. Larger k_{az}/k_{wz} delivered faster dissipation of the excess pore-air pressure at the later stage, and resulted in faster dissipation of the excess pore-water pressure at the intermediate stage. After that, the excess pore-air pressure was fully dissipated, the plateau period emerged on the normalized curve of the excess pore-water pressure, and then, different curves tended to gradually convergence into a single one (see Figure 9b). A similar characteristic can be found in the normalized curves of the soil layer settlement (see Figure 10). Moreover, the bigger k_{az}/k_{wz} can obviously prolong the plateau periods above. In comparison with the results obtained when $R_b = R_t = \infty$, the excess pore-pressures and soil layer settlement obtained when $R_b = R_t= 5$ were smaller; that is, it needed more time to complete the consolidation process. This phenomenon happened as a result of the impeding effects of the top and bottom drainage boundaries given by $R_b = R_t= 5$, and it was identical to the results under 1D consolidation with the impeded drainage boundaries [21].

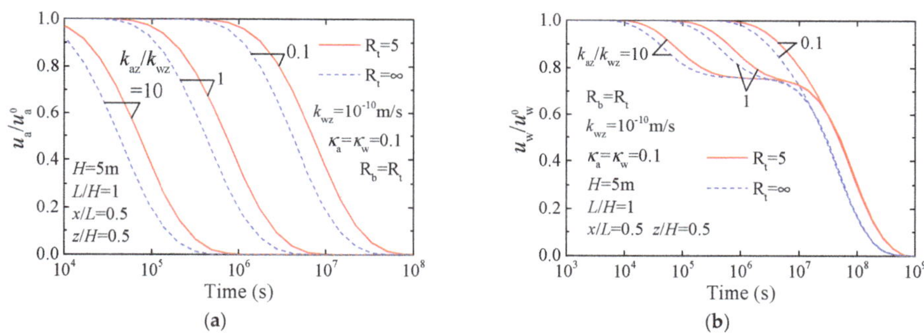

Figure 9. Variations of excess pore-pressures under vertical impeded drainage boundary with different air-water permeability ratios: (a) air phase and (b) water phase.

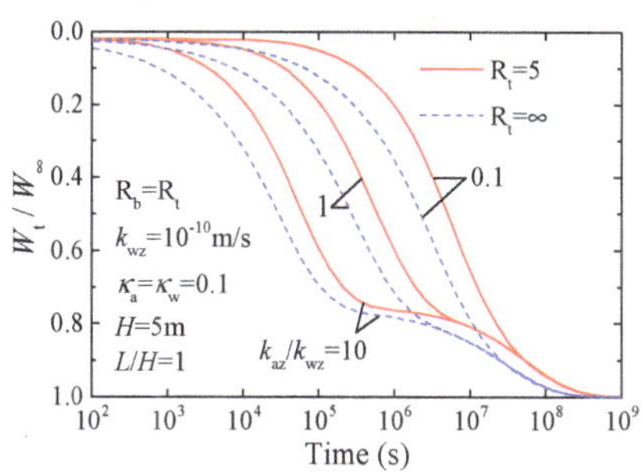

Figure 10. Variations of soil layer settlement under vertical impeded drainage boundary with different air-water permeability ratios.

4.3. Consolidation Process with Different Anisotropic Permeability Ratios

In this part, for simplicity, the anisotropic permeability ratio of air phase was only considered to be the same as that of water phase, i.e., $\kappa_a = \kappa_w$. The vertical impeded drainage boundaries are given by $R_b = R_t = 5$ and $R_b = R_t = \infty$. Figures 11 and 12 illustrate the variations of the excess pore-pressures and the soil layer settlement under different values of $\kappa_a = \kappa_w$, respectively. It was obvious that the anisotropic permeability ratio had a great effect on the dissipation processes of the excess pore-pressures and the development of the soil layer settlement. The dissipation with a larger $\kappa_a = \kappa_w$ tended to progress relatively faster than that with a smaller one. The reason is that the vertical drains installed in the soil layer provide a shorter drainage path in the horizontal direction, and they allow the excess pore-pressures dissipate faster when $\kappa_a = \kappa_w$ is larger. Being the same as the results in Figures 5 and 6, the normalized curves of the excess pore-air pressures were a group of typical inverse S ones with single influence stage while those of the excess pore-water pressures and the soil layer settlements were typical double inverse S curves with two obvious influence stages (see Figures 11 and 12). These curves shifted to the left with an increase of $\kappa_a = \kappa_w$. Being the same as those observed in Figures 9 and 10, under a given value of $\kappa_a = \kappa_w$, the vertical boundary condition of $R_b = R_t = 5$ induced slower dissipations of the excess pore-pressures and slower development of the soil layer settlement than the condition of $R_b = R_t = \infty$. For these two boundary conditions, it can be observed that the larger $\kappa_a = \kappa_w$ was, the smaller the difference between these results was. That is, the drainage efficiencies of the vertical impeded boundaries affected the 2D consolidation behavior more slightly when $\kappa_a = \kappa_w$ was larger (e.g., 4). The reason is that for the scenario of larger $\kappa_a = \kappa_w$, most of the excess pore-pressures were dissipated horizontally, and the vertical drainage effect was small and could even be neglected.

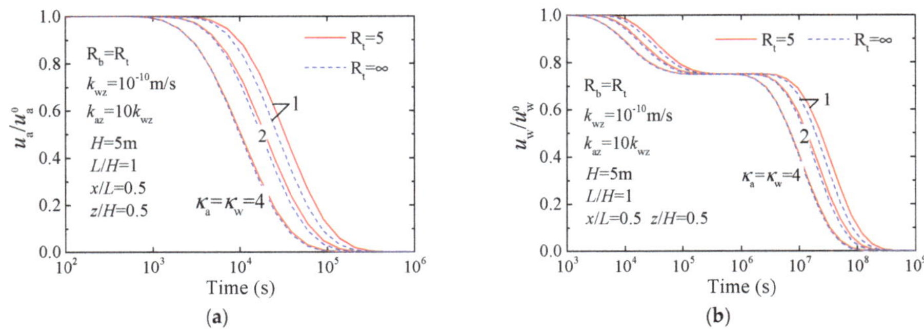

Figure 11. Variations of excess pore-pressures under vertical impeded drainage boundary with different anisotropic permeability ratios: (**a**) air phase and (**b**) water phase.

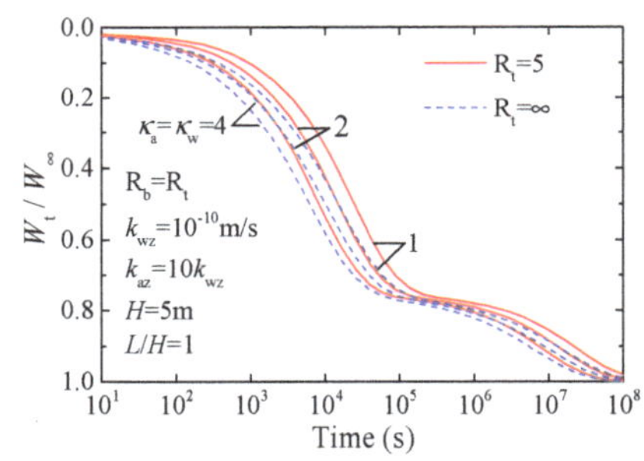

Figure 12. Variations of soil layer settlement under vertical impeded drainage boundary with different anisotropic permeability ratios.

4.4. Consolidation Process with Different Spacing-Depth Ratios

For simplicity, only the isotropic permeability condition for air and water phases was considered in this part, and the vertical impeded drainage boundaries were assumed to be symmetrical. Figures 13 and 14, respectively, illustrate the changes in the excess pore-pressures and soil layer settlement under different spacing-depth ratios (i.e., L/H), in which the vertical impeded drainage boundaries are given by $R_b = R_t = 5$ and $R_b = R_t = \infty$. It can be observed that a smaller L/H delivered a faster dissipation of the excess-pore pressures due to a shorter horizontal drainage path. As a result, the soil layer settlement develops more quickly. Being the same as the results in Figures 5 and 6, there was a single influence stage on the dissipation pattern of the excess pore-air pressure while there were two obvious influence stages on the dissipation of the excess pore-water pressure and on the development curve of the soil layer settlement (see Figures 13 and 14). Under a given L/H, the excess pore-pressures under the vertical boundary condition of $R_b = R_t = 5$ dissipated more slowly that those under the vertical boundary condition of $R_b = R_t = \infty$. The phenomenon was similar to that observed in Figures 9 and 10. For these two boundary conditions, it was obvious that the difference between their results decreased with the increase of L/H. That is, the drainage efficiencies of vertical impeded boundaries had fewer effects on the 2D consolidation behavior when the horizontal drainage path was shorter (e.g., L/H = 0.5). The reason is that most of the excess pore-pressures were dissipated horizontally for the scenario of shorter horizontal drainage path.

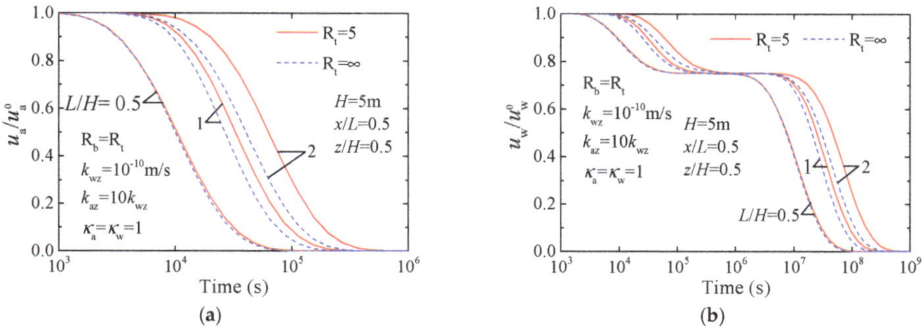

Figure 13. Variations of excess pore-pressures under vertical impeded drainage boundary with different spacing-depth ratios: (**a**) air phase and (**b**) water phase.

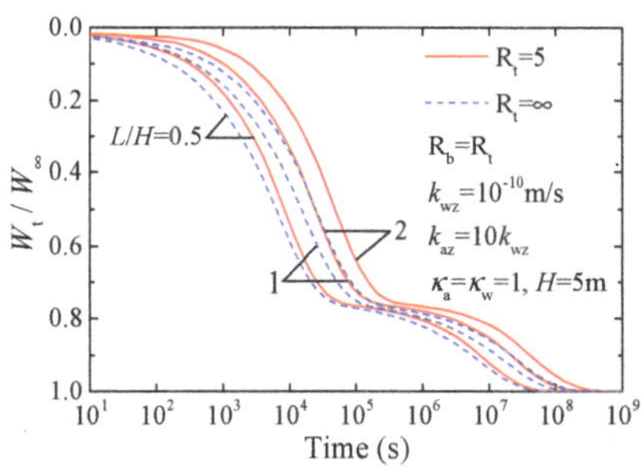

Figure 14. Variations of soil layer settlement under vertical impeded drainage boundary with different spacing-depth ratios.

5. Conclusions

In this paper, a semi-analytical solution to 2D plane strain consolidation process of unsaturated soil incorporating vertical impeded drainage boundaries was obtained using eigenfunction expansion and Laplace transform techniques. By performing the inversion of Laplace transform, some computations were performed to study the consolidation processes under different drainage efficiencies of the vertical boundaries, air-water permeability ratios, anisotropic permeability ratios and spacing-depth ratios. As a result of the analysis the following conclusions can be drawn.

(a) The present semi-analytical solution was validated to be consistent with the finite difference solution and the previous analytical solution in the literature, and it is more general in the drainage boundary compared with the existing analytical solution.
(b) The drainage efficiency of the vertical boundary had a great influence on the dissipation processes of the excess pore-pressures and the development of the soil layer settlement. The larger drainage efficiency delivered the faster dissipation of the excess pore-pressures and the faster development

of the soil layer settlement. When the drainage efficiency was large enough (i.e., $R_t \geq 100$ or $R_b \geq 100$), the corresponding boundary can be considered to be fully drained.

(c) The consolidation process with a larger air-water permeability ratio tended to progress more quickly than that with a smaller one. Either a larger anisotropic permeability ratio (i.e., larger horizontal permeability coefficient) or a smaller spacing-depth ratio delivered faster dissipation of the excess pore pressures and faster development of the soil layer settlement. In comparison with the results under top-base drainage boundary, the impeded drainage boundary with limited drainage efficiency slowed down the consolidation process significantly.

Author Contributions: M.H. proposed the idea, conceptualization, and mathematical derivation as well as helping to write and organize the paper; and D.L. technically supported the content, verified the accuracy and writing-original draft preparation.

Funding: This research was partially supported by the National Key Research and Development Plan of China (Grand No. 2016YFC0800200), the National Natural Science Foundation of China (Grant Nos. 51508180 and 51578231) and the China Postdoctoral Science Foundation (Grand No. 2015M570677).

Conflicts of Interest: The authors have no conflict of interest to declare.

Appendix A : Transformation of Equations (6a) and (6b)

There has been a common alternative variable method that has been used to rewrite the 1D consolidation equations of unsaturated soil [12,21–23]. In this appendix, it was adopted to transform the governing equations of 2D plane strain consolidation process. Multiplying Equations (6a) and (6b) by the arbitrary constants c_1 and c_2, respectively, and then adding them together results in:

$$\frac{\partial(c_1 u_a + c_2 u_w)}{\partial t} = (c_2 W_a + c_1 A_a)\left(\kappa_a \frac{\partial^2 u_a}{\partial x^2} + \frac{\partial^2 u_a}{\partial z^2}\right) + (c_1 A_w + c_2 W_w)\left(\kappa_w \frac{\partial^2 u_w}{\partial x^2} + \frac{\partial^2 u_w}{\partial z^2}\right) \tag{A1}$$

where:

$$A_a = -\frac{c_{vz}^a}{1 - C_a C_w},\ A_w = \frac{C_a c_{vz}^w}{1 - C_a C_w},\ W_a = \frac{C_w c_{vz}^a}{1 - C_a C_w},\ W_w = -\frac{c_{vz}^w}{1 - C_a C_w} \tag{A2}$$

Introducing a constant Q that satisfies the following relationships:

$$Qc_1 = c_2 W_a + c_1 A_a \tag{A3}$$

$$Qc_2 = c_1 A_w + c_2 W_w \tag{A4}$$

In order to make these relationships above hold true, the constant Q must satisfy:

$$(Q - A_a)(Q - W_w) = A_w W_a \tag{A5}$$

Equation (A5) is a quadratic equation of Q, and its two roots Q_1 and Q_2 are obtained as follows:

$$Q_{1,2} = \frac{1}{2}\left[A_a + W_w \pm \sqrt{(A_a - W_w)^2 + 4A_w W_a}\right] \tag{A6}$$

When $Q = Q_1$, the solutions of Equations (A3) and (A4) are denoted as c_{11} and c_{21}; when $Q = Q_2$, the solutions are denoted as c_{12} and c_{22}. Moreover, without loss of generality, it is possible to assume that $c_{11} = c_{22} = 1$, and consequently c_{21} and c_{12}, can be obtained as follows:

$$c_{21} = \frac{A_w}{Q_1 - W_w},\ c_{12} = \frac{W_a}{Q_2 - A_a} \tag{A7}$$

Then, by defining two alternative variables ϕ_1 and ϕ_2 as presented in Equation (8), Equation (A1) can be transformed into Equations (7a) and (7b), respectively.

Appendix B : Derivation of Solution to Equations (13a) and (13b)

Equations (13a) and (13b) can be rewritten as the following linear second-order system:

$$\mathbf{Q}\frac{d^2\widetilde{\boldsymbol{\varphi}}}{dz^2} = A\widetilde{\boldsymbol{\varphi}} - \boldsymbol{\varphi}^0 \quad \text{(A8)}$$

where:

$$\mathbf{Q} = \begin{bmatrix} Q_1 & 0 \\ 0 & Q_2 \end{bmatrix}, \mathbf{A} = \begin{bmatrix} s+\rho_{11}Q_1\mu_i^2 & \rho_{21}Q_1\mu_i^2 \\ \rho_{12}Q_2\mu_i^2 & s+\rho_{22}Q_2\mu_i^2 \end{bmatrix}, \widetilde{\boldsymbol{\varphi}} = \begin{Bmatrix} \widetilde{\varphi}_1^i \\ \widetilde{\varphi}_2^i \end{Bmatrix}, \boldsymbol{\varphi}^0 = \begin{Bmatrix} \varphi_1^{i0} \\ \varphi_2^{i0} \end{Bmatrix} \quad \text{(A9)}$$

Let $\widetilde{\boldsymbol{\varphi}}^* = \widetilde{\boldsymbol{\varphi}} - \mathbf{A}^{-1}\boldsymbol{\varphi}^0$, then, Equation (A8) can be simplified into a homogeneous system:

$$\frac{d^2\widetilde{\boldsymbol{\varphi}}^*}{dz^2} = \mathbf{Q}^{-1}\mathbf{A}\widetilde{\boldsymbol{\varphi}}^* \quad \text{(A10)}$$

Based on the eigenvalue method for the homogeneous system [25,26], the general non-constant solutions of Equation (A10) can be constructed using the following vector functions:

$$\widetilde{\boldsymbol{\varphi}}^* = \boldsymbol{\psi}e^{\lambda z} \quad \text{(A11)}$$

where λ is a scalar and $\boldsymbol{\psi}$ is a constant vector.

Substituting Equation (A11) into Equation (A10) results in the following eigenvalue problem:

$$\left(\mathbf{Q}^{-1}\mathbf{A} - \lambda^2\mathbf{I}\right)\boldsymbol{\psi} = 0 \quad \text{(A12)}$$

where $\mathbf{I}$ is a unit matrix, and λ^2 is an eigenvalue of $\mathbf{Q}^{-1}\mathbf{A}$ with corresponding eigenvector $\boldsymbol{\psi}$.

The characteristic equation with respect to $\mathbf{Q}^{-1}\mathbf{A}$ is:

$$\det\left(\mathbf{Q}^{-1}\mathbf{A} - \lambda^2\mathbf{I}\right) = \begin{vmatrix} s/Q_1+\rho_{11}\mu_i^2-\lambda^2 & \rho_{21}\mu_i^2 \\ \rho_{12}\mu_i^2 & s/Q_2+\rho_{22}\mu_i^2-\lambda^2 \end{vmatrix} = 0 \quad \text{(A13)}$$

where $\det(\cdot)$ denotes the determinant of a matrix.

Solving Equation (A13) for the eigenvalue λ^2, results in:

$$\begin{matrix} \lambda_1^2 \\ \lambda_2^2 \end{matrix} = \frac{(Q_1+Q_2)s + (\rho_{11}+\rho_{22})Q_1Q_2\mu_i^2 \pm \sqrt{\eta}}{2Q_1Q_2} \quad \text{(A14)}$$

where $\eta = \left[(Q_2-Q_1)s + (\rho_{11}-\rho_{22})Q_1Q_2\mu_i^2\right]^2 + 4\rho_{21}\rho_{12}Q_1^2Q_2^2\mu_i^4$.

It should be noted that, using Crump's method to implement inverse Laplace transform on a given function, the value of s can be freely chosen according to the poles of the given function and the requirement of the relative errors. Therefore, the value of s can be reasonably chosen to hold that $\eta \neq 0$ (i.e., $\lambda_1^2 \neq \lambda_2^2$). Thus, the eigenvectors of Equation (A12) can be obtained as follows:

$$\boldsymbol{\psi}_1 = \begin{Bmatrix} 1 & \psi_1 \end{Bmatrix}^{\mathrm{T}}, \boldsymbol{\psi}_2 = \begin{Bmatrix} \psi_2 & 1 \end{Bmatrix}^{\mathrm{T}} \quad \text{(A15)}$$

where $\boldsymbol{\psi}_1$ and $\boldsymbol{\psi}_2$ two linear-independent vectors; the superscript T represents the transposition of vector; the parameters ψ_1 and ψ_2 are:

$$\psi_1 = -\frac{\rho_{12}Q_2\mu_i^2}{s+\rho_{22}Q_2\mu_i^2-\lambda_1^2Q_2}, \psi_2 = -\frac{\rho_{21}Q_1\mu_i^2}{s+\rho_{11}Q_1\mu_i^2-\lambda_2^2Q_1} \quad \text{(A16)}$$

Using Equations (A14) and (A15), the general solution of the homogenous system (A10) can be expressed as:

$$\widetilde{\boldsymbol{\varphi}}^{*}=\left\{\begin{array}{c}\chi_1 e^{\lambda_1 z}+\chi_2 e^{-\lambda_1 z}+\psi_2(\chi_3 e^{\lambda_2 z}+\chi_4 e^{-\lambda_2 z})\\ \psi_1(\chi_1 e^{\lambda_1 z}+\chi_2 e^{-\lambda_1 z})+\chi_3 e^{\lambda_2 z}+\chi_4 e^{-\lambda_2 z}\end{array}\right\} \tag{A17}$$

where χ_1, χ_2, χ_3 and χ_4 are arbitrary functions of s, and can be determined from the top and bottom boundary conditions.

Then, the general solution of the original system (A8) (i.e., Equations (13a) and (13b)) can be obtained as follows:

$$\widetilde{\boldsymbol{\varphi}}=\left\{\begin{array}{c}\chi_1 e^{\lambda_1 z}+\chi_2 e^{-\lambda_1 z}+\psi_2(\chi_3 e^{\lambda_2 z}+\chi_4 e^{-\lambda_2 z})+\chi_5\\ \psi_1(\chi_1 e^{\lambda_1 z}+\chi_2 e^{-\lambda_1 z})+\chi_3 e^{\lambda_2 z}+\chi_4 e^{-\lambda_2 z}+\chi_6\end{array}\right\} \tag{A18}$$

where:

$$\chi_5=\frac{(s+Q_2\rho_{22}\mu_i^2)\varphi_1^{i0}-Q_1\rho_{21}\mu_i^2\varphi_2^{i0}}{(s+Q_1\rho_{11}\mu_i^2)(s+Q_2\rho_{22}\mu_i^2)-Q_1Q_2\rho_{12}\rho_{21}\mu_i^4} \tag{A19}$$

$$\chi_6=\frac{(s+Q_1\rho_{11}\mu_i^2)\varphi_2^{i0}-Q_2\rho_{12}\mu_i^2\varphi_1^{i0}}{(s+Q_1\rho_{11}\mu_i^2)(s+Q_2\rho_{22}\mu_i^2)-Q_1Q_2\rho_{12}\rho_{21}\mu_i^4} \tag{A20}$$

Taking the derivation with respect to Equation (A18) gives:

$$\frac{d\widetilde{\boldsymbol{\varphi}}}{dz}=\left\{\begin{array}{c}\lambda_1(\chi_1 e^{\lambda_1 z}-\chi_2 e^{-\lambda_1 z})+\psi_2\lambda_2(\chi_3 e^{\lambda_2 z}-\chi_4 e^{-\lambda_2 z})\\ \psi_1\lambda_1(\chi_1 e^{\lambda_1 z}-\chi_2 e^{-\lambda_1 z})+\lambda_2(\chi_3 e^{\lambda_2 z}-\chi_4 e^{-\lambda_2 z})\end{array}\right\} \tag{A21}$$

Substituting Equations (11a) and (11b) into Equation (8), and solving it for u_{a} and u_{w} gives:

$$u_{\mathrm{a}}=\sum_{i=1}^{\infty}\left(\frac{\varphi_1^i-c_{21}\varphi_2^i}{1-c_{12}c_{21}}\right)\sin(\mu_i x),\ u_{\mathrm{w}}=\sum_{i=1}^{\infty}\left(\frac{\varphi_2^i-c_{12}\varphi_1^i}{1-c_{12}c_{21}}\right)\sin(\mu_i x) \tag{A22}$$

By substituting Equation (A22) into Equations (5a) and (5b), the top and bottom boundary conditions can be expressed as follows:

$$\left(\frac{\partial}{\partial z}-\frac{R_{\mathrm{ta}}}{H}\right)\left[\varphi_1^i(0,t)-c_{21}\varphi_2^i(0,t)\right]=0,\ \left(\frac{\partial}{\partial z}-\frac{R_{\mathrm{tw}}}{H}\right)\left[\varphi_2^i(0,t)-c_{12}\varphi_1^i(0,t)\right]=0 \tag{A23}$$

$$\left(\frac{\partial}{\partial z}+\frac{R_{\mathrm{ba}}}{H}\right)\left[\varphi_1^i(H,t)-c_{21}\varphi_2^i(H,t)\right]=0,\ \left(\frac{\partial}{\partial z}+\frac{R_{\mathrm{bw}}}{H}\right)\left[\varphi_2^i(H,t)-c_{12}\varphi_1^i(H,t)\right]=0 \tag{A24}$$

Implementing a Laplace transform on Equations (A23) and (A24), and further substituting these transform results into Equations (A18) and (A21) leads to:

$$\mathbf{D}\boldsymbol{\chi}=\mathbf{R} \tag{A25}$$

where:

$$\mathbf{D}=\begin{bmatrix}\alpha_1(\lambda_1 H-R_{\mathrm{ta}}) & -\alpha_1(\lambda_1 H+R_{\mathrm{ta}}) & \alpha_2(\lambda_2 H-R_{\mathrm{ta}}) & -\alpha_2(\lambda_2 H+R_{\mathrm{ta}})\\ \alpha_3(\lambda_1 H-R_{\mathrm{tw}}) & -\alpha_3(\lambda_1 H+R_{\mathrm{tw}}) & \alpha_4(\lambda_2 H-R_{\mathrm{tw}}) & -\alpha_4(\lambda_2 H+R_{\mathrm{tw}})\\ \alpha_1(\lambda_1 H+R_{\mathrm{ba}})e^{\lambda_1 H} & -\alpha_1(\lambda_1 H-R_{\mathrm{ba}})e^{-\lambda_1 H} & \alpha_2(\lambda_2 H+R_{\mathrm{ba}})e^{\lambda_2 H} & -\alpha_2(\lambda_2 H-R_{\mathrm{ba}})e^{-\lambda_2 H}\\ \alpha_3(\lambda_2 H+R_{\mathrm{bw}})e^{\lambda_1 H} & -\alpha_3(\lambda_1 H-R_{\mathrm{bw}})e^{-\lambda_1 H} & \alpha_4(\lambda_2 H+R_{\mathrm{bw}})e^{\lambda_2 H} & -\alpha_4(\lambda_2 H-R_{\mathrm{bw}})e^{-\lambda_2 H}\end{bmatrix}$$
$$\mathbf{R}=\left\{\begin{array}{cccc}\alpha_5 R_{\mathrm{ta}} & \alpha_6 R_{\mathrm{tw}} & -\alpha_5 R_{\mathrm{ba}} & -\alpha_6 R_{\mathrm{bw}}\end{array}\right\}^T,\ \boldsymbol{\chi}=\left\{\begin{array}{cccc}\chi_1, & \chi_2 & \chi_3 & \chi_4\end{array}\right\}^{\mathrm{T}},\ \alpha_1=1-c_{21}\psi_1,$$
$$\alpha_2=\psi_2-c_{21},\ \alpha_3=\psi_1-c_{12},\ \alpha_4=1-c_{12}\psi_2,\ \alpha_5=\chi_5-c_{21}\chi_6 \text{ and } \alpha_6=\chi_6-c_{12}\chi_5 \tag{A26}$$

Solving Equation (A25) for $\boldsymbol{\chi}$ gives:

$$\boldsymbol{\chi}=\mathbf{D}^{-1}\mathbf{R} \tag{A27}$$

For the case of $R_{ta} = R_{tw} = R_t$ and $R_{ba} = R_{bw} = R_b$, i.e., the effects of impeded drainage on air and water phases at top and bottom boundaries can be considered to be the same, $\chi_\ell(\ell = 1\sim4)$ can be obtained as follows:

$$\begin{aligned}
\chi_1 &= (\chi_5 - \chi_6\psi_2)\left[(\lambda_1 H - R_b)R_t e^{-\lambda_1 H} + (\lambda_1 H + R_t)R_b\right]/\chi_7 \\
\chi_2 &= (\chi_5 - \chi_6\psi_2)\left[(\lambda_1 H + R_b)R_t e^{\lambda_1 H} + (\lambda_1 H - R_t)R_b\right]/\chi_7 \\
\chi_3 &= (\chi_6 - \chi_5\psi_1)\left[(\lambda_2 H - R_b)R_t e^{-\lambda_2 H} + (\lambda_2 H + R_t)R_b\right]/\chi_8 \\
\chi_4 &= (\chi_6 - \chi_5\psi_1)\left[(\lambda_2 H + R_b)R_t e^{\lambda_2 H} + (\lambda_2 H - R_t)R_b\right]/\chi_8 \\
\chi_7 &= 2(\psi_1\psi_2 - 1)\left[(\lambda_1^2 H^2 + R_b R_t)\sinh(\lambda_1 H) + \lambda_1 H(R_b + R_t)\cosh(\lambda_1 H)\right] \\
\chi_8 &= 2(\psi_1\psi_2 - 1)\left[(\lambda_2^2 H^2 + R_b R_t)\sinh(\lambda_2 H) + \lambda_2 H(R_b + R_t)\cosh(\lambda_2 H)\right]
\end{aligned} \quad \text{(A28)}$$

Appendix C : Inverse Laplace Transform by Crump's Method

The Laplace transform of a real function $f(t)$, $t \geq 0$, is defined as:

$$F(s) = \int_0^\infty e^{-st} f(t)dt \quad \text{(A29)}$$

Throughout, it should be assumed that $f(t)$ is piecewise continuous and of exponential order β (i.e., $|f(t)| \leq \mathcal{M}e^{\beta t}$). In this case, the transform function $F(s)$ is defined for $\mathrm{Re}(s) > \beta$.

Starting with $F(s)$, Crump's method estimates the values of the inverse transform $f(t)$ based on the summation in Durbin's Fourier series approximation [27]:

$$f(t) \approx \frac{e^{bt}}{T}\left\{\frac{1}{2}F(b) + \sum_{k=1}^{\infty} Re\left[F\left(b + \frac{jk\pi}{T}\right)e^{\frac{jk\pi}{T}t}\right]\right\} \quad \text{(A30)}$$

where $j = \sqrt{-1}$; b and T are two real parameters. The series in Equation (A29) generally converges very slowly. In order to evaluate this series efficiently, Crump used the epsilon-algorithm to speed the convergence. The epsilon-algorithm is given as [24]:

$$\varepsilon_{\ell+1}^{(m)} = \varepsilon_{\ell-1}^{(m+1)} + \frac{1}{\varepsilon_\ell^{(m+1)} - \varepsilon_\ell^{(m)}}, \text{ for } \ell, m = 1, 2, \cdots, 2N+1 \quad \text{(A31)}$$

where $\varepsilon_0^{(m)} = 0$ and $\varepsilon_1^{(m)}$ is the mth partial sum of the series in Equation (A30). Then, the sequence of $\varepsilon_1^{(1)}, \varepsilon_3^{(1)}, \varepsilon_5^{(1)}, \ldots, \varepsilon_{2N+1}^{(1)}$ is a sequence of successive approximations to the sum of the series that will often better approximate the sum than the untransformed one.

The values of the parameters are chosen as [24,28]: (1) $T = 0.8 \times \max(0.01, t_{\max})$, where $t_{\max}$ is the largest t-value that desires the inverse transform; (2) $b = \beta_b - \ln(0.1 \times E_r)/(2T)$, where β_b should be specified equal to, or slightly larger than β; E_r is the required relative error in the values of the inverse transform; thus, it must be in the range [0, 1).

Appendix D : Finite difference solution to Equations (1a) and (1b)

The Forward time and central space (FTCS) difference scheme [29,30], as shown in Figure A1, was adopted to develop the finite difference solution to Equations (1a) and (1b). Using this scheme, Equations (1a) and (1b) can be expressed as follows:

$$\frac{u_a^{i,j,\ell+1} - u_a^{i,j,\ell}}{\Delta t} + C_a \frac{u_w^{i,j,\ell+1} - u_w^{i,j,\ell}}{\Delta t} + \frac{c_{vx}^a}{\Delta x^2} f_x^a + \frac{c_{vz}^a}{\Delta z^2} f_z^a = 0 \quad \text{(A32)}$$

$$\frac{u_w^{i,j,\ell+1} - u_w^{i,j,\ell}}{\Delta t} + C_w \frac{u_a^{i,j,\ell+1} - u_a^{i,j,\ell}}{\Delta t} + \frac{c_{vx}^w}{\Delta x^2} f_x^w + \frac{c_{vz}^w}{\Delta z^2} f_z^w = 0 \quad \text{(A33)}$$

where:

$$f_{\mathrm{x}}^{\mathrm{a}} = u_{\mathrm{a}}^{i-1,j,\ell} - 2u_{\mathrm{a}}^{i,j,\ell} + u_{\mathrm{a}}^{i+1,j,\ell},\ f_{\mathrm{z}}^{\mathrm{a}} = u_{\mathrm{a}}^{i,j-1,\ell} - 2u_{\mathrm{a}}^{i,j,\ell} + u_{\mathrm{a}}^{i,j+1,\ell},$$
$$f_{\mathrm{x}}^{\mathrm{w}} = u_{\mathrm{w}}^{i-1,j,\ell} - 2u_{\mathrm{w}}^{i,j,\ell} + u_{\mathrm{w}}^{i+1,j,\ell},\ f_{\mathrm{z}}^{\mathrm{w}} = u_{\mathrm{w}}^{i,j-1,\ell} - 2u_{\mathrm{w}}^{i,j,\ell} + u_{\mathrm{w}}^{i,j+1,\ell}. \quad \text{(A34)}$$

Equations (A32) and (A33) can be rearranged as follows:

$$u_{\mathrm{a}}^{i,j,\ell+1} = u_{\mathrm{a}}^{i,j,\ell} + \frac{1}{1-C_{\mathrm{a}}C_{\mathrm{w}}}\left(C_{\mathrm{a}}\beta_{\mathrm{x}}^{\mathrm{w}}f_{\mathrm{x}}^{\mathrm{w}} + C_{\mathrm{a}}\beta_{\mathrm{z}}^{\mathrm{w}}f_{\mathrm{z}}^{\mathrm{w}} - \beta_{\mathrm{x}}^{\mathrm{a}}f_{\mathrm{x}}^{\mathrm{a}} - \beta_{\mathrm{z}}^{\mathrm{a}}f_{\mathrm{z}}^{\mathrm{a}}\right) \quad \text{(A35)}$$

$$u_{\mathrm{w}}^{i,j,\ell+1} = u_{\mathrm{w}}^{i,j,\ell} + \frac{1}{1-C_{\mathrm{a}}C_{\mathrm{w}}}\left(C_{\mathrm{w}}\beta_{\mathrm{x}}^{\mathrm{a}}f_{\mathrm{x}}^{\mathrm{a}} + C_{\mathrm{w}}\beta_{\mathrm{z}}^{\mathrm{a}}f_{\mathrm{z}}^{\mathrm{a}} - \beta_{\mathrm{x}}^{\mathrm{w}}f_{\mathrm{x}}^{\mathrm{w}} - \beta_{\mathrm{z}}^{\mathrm{w}}f_{\mathrm{z}}^{\mathrm{w}}\right) \quad \text{(A36)}$$

where:

$$\beta_{\mathrm{x}}^{\mathrm{a}} = c_{\mathrm{vx}}^{\mathrm{a}}\Delta t/(\Delta x)^2,\ \beta_{\mathrm{z}}^{\mathrm{a}} = c_{\mathrm{vz}}^{\mathrm{a}}\Delta t/(\Delta z)^2,\ \beta_{\mathrm{x}}^{\mathrm{w}} = c_{\mathrm{vx}}^{\mathrm{w}}\Delta t/(\Delta x)^2,\ \beta_{\mathrm{z}}^{\mathrm{w}} = c_{\mathrm{vz}}^{\mathrm{w}}\Delta t/(\Delta z)^2 \quad \text{(A37)}$$

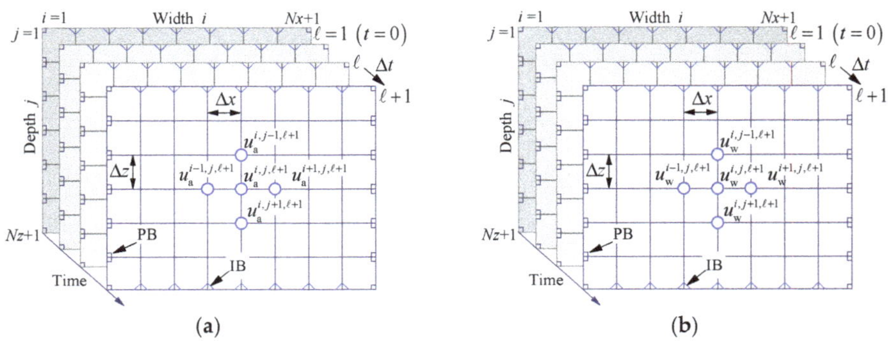

Figure A1. Forward time and central space (FTCS) difference scheme for the governing equations: (**a**) air phase and (**b**) water phase.

The initial and boundary conditions can be obtained as follows:

(i) The initial excess pore-air and pore-water pressures ($i = 1 \sim Nx+1,\ j = 1 \sim Nz+1$):

$$u_{\mathrm{a}}^{i,j,1} = u_{\mathrm{a}}^{0},\ u_{\mathrm{w}}^{i,j,1} = u_{\mathrm{w}}^{0} \quad \text{(A38)}$$

(ii) The permeable drainage boundaries (PB) at both side surfaces ($j = 1 \sim Nz+1$):

$$u_{\mathrm{a}}^{1,j,\ell} = u_{\mathrm{a}}^{Nx+1,j,\ell} = u_{\mathrm{w}}^{1,j,\ell} = u_{\mathrm{w}}^{Nx+1,j,\ell} = 0 \quad \text{(A39)}$$

(iii) The impeded drainage boundary (IB) at the top surface ($i = 2 \sim Nx$):

$$u_{\mathrm{a}}^{i,1,\ell+1} = u_{\mathrm{a}}^{i,1,\ell} + \frac{1}{1-C_{\mathrm{a}}C_{\mathrm{w}}}\left(C_{\mathrm{a}}\beta_{\mathrm{x}}^{\mathrm{w}}f_{\mathrm{x},\,j=1}^{\mathrm{w}} + C_{\mathrm{a}}\beta_{\mathrm{z}}^{\mathrm{w}}f_{\mathrm{zt}}^{\mathrm{w}} - \beta_{\mathrm{x}}^{\mathrm{a}}f_{\mathrm{x},j=1}^{\mathrm{a}} - \beta_{\mathrm{z}}^{\mathrm{a}}f_{\mathrm{zt}}^{\mathrm{a}}\right) \quad \text{(A40)}$$

$$u_{\mathrm{w}}^{i,1,\ell+1} = u_{\mathrm{w}}^{i,1,\ell} + \frac{1}{1-C_{\mathrm{a}}C_{\mathrm{w}}}\left(C_{\mathrm{w}}\beta_{\mathrm{x}}^{\mathrm{a}}f_{\mathrm{x},j=1}^{\mathrm{a}} + C_{\mathrm{w}}\beta_{\mathrm{z}}^{\mathrm{a}}f_{\mathrm{zt}}^{\mathrm{a}} - \beta_{\mathrm{x}}^{\mathrm{w}}f_{\mathrm{x},j=1}^{\mathrm{w}} - \beta_{\mathrm{z}}^{\mathrm{w}}f_{\mathrm{zt}}^{\mathrm{w}}\right) \quad \text{(A41)}$$

where $f_{\mathrm{zt}}^{\mathrm{a}} = 2u_{\mathrm{a}}^{i,2,\ell} + 2(\Delta z R_{\mathrm{ta}}/H - 1)u_{\mathrm{a}}^{i,1,\ell}$, $f_{\mathrm{zt}}^{\mathrm{w}} = 2u_{\mathrm{w}}^{i,2,\ell} + 2(\Delta z R_{\mathrm{tw}}/H - 1)u_{\mathrm{w}}^{i,1,\ell}$.

(iv) The impeded drainage boundary (IB) at the bottom base ($i = 2 \sim Nx$):

$$u_{\mathrm{a}}^{i,Nz+1,\ell+1} = u_{\mathrm{a}}^{i,Nz+1,\ell} + \frac{1}{1-C_{\mathrm{a}}C_{\mathrm{w}}}\left(C_{\mathrm{a}}\beta_{\mathrm{x}}^{\mathrm{w}}f_{\mathrm{x},j=Nz+1}^{\mathrm{a}} + C_{\mathrm{a}}\beta_{\mathrm{z}}^{\mathrm{w}}f_{\mathrm{zb}}^{\mathrm{w}} - \beta_{\mathrm{x}}^{\mathrm{a}}f_{\mathrm{x},j=Nz+1}^{\mathrm{a}} - \beta_{\mathrm{z}}^{\mathrm{a}}f_{\mathrm{zb}}^{\mathrm{a}}\right) \quad \text{(A42)}$$

$$u_{\rm w}^{i,Nz+1,\ell+1} = u_{\rm w}^{i,Nz+1,\ell} + \frac{1}{1-C_{\rm a}C_{\rm w}}\left(C_{\rm w}\beta_{\rm x}^{\rm a} f_{{\rm x},j=Nz+1}^{\rm a} + C_{\rm w}\beta_{\rm z}^{\rm a} f_{\rm zb}^{\rm a} - \beta_{\rm x}^{\rm w} f_{{\rm x},j=Nz+1}^{\rm a} - \beta_{\rm z}^{\rm w} f_{\rm zb}^{\rm w}\right) \quad \text{(A43)}$$

where $f_{\rm zb}^{\rm a} = 2u_{\rm a}^{i,Nz,\ell} + 2(\Delta z R_{\rm ba}/H - 1)u_{\rm a}^{i,Nz+1,\ell}$, $f_{\rm zb}^{\rm w} = 2u_{\rm w}^{i,Nz,\ell} + 2(\Delta z R_{\rm bw}/H - 1)u_{\rm w}^{i,Nz+1,\ell}$.

Based on Equations (A32)–(A43), a computer program can be developed to provide the numerical results of the excess pore-air and pore-water pressures under vertical impeded drainage boundaries. In order to make the computer scheme converge, the time interval and space internal should be reasonably chosen and controlled. For the domain as shown in Figure A1, the mesh sizes should satisfy the following stability conditions [29,30]:

$$\beta_{\rm x}^{\rm a} + \beta_{\rm z}^{\rm a} \leq 0.5,\ \beta_{\rm x}^{\rm w} + \beta_{\rm z}^{\rm w} \leq 0.5 \quad \text{(A44)}$$

References

1. Terzaghi, K. *Theoretical Soil Mechanics*; Wiley & Sons: New York, NY, USA, 1943.
2. Schiffman, R.L. Consolidation of soil under time-dependent loading and varying permeability. In *Highway Research Board Proceeding, Proceedings of the Thirty-Seventh Annual Meeting of the Highway Research Board, Washington, DC, USA, 6–10 January 1958*; Highway Research Board: Washington, DC, USA; Volume 37, pp. 584–617.
3. Baligh, M.M.; Levadoux, J.N. Consolidation theory of cyclic loading. *J. Geotech. Eng. Div.* **1978**, *104*, 415–431.
4. Xie, K.H.; Xie, X.Y.; Li, B.H. Analytical theory for one-dimensional consolidation of clayey soils exhibiting rheological characteristics under time-dependent loading. *Int. J. Numer. Anal. Met. Geomech.* **2008**, *32*, 1833–1855. [CrossRef]
5. Liu, J.C.; Griffiths, D.V. A general solution for 1D consolidation induced by depth- and time-dependent changes in stress. *Géotechnique* **2015**, *65*, 66–72. [CrossRef]
6. Scott, R.F. *Principles of Soil Mechanics*; Addison-Wesley Pub. Co.: Boston, MA, USA, 1963.
7. Barden, L. Consolidation of compacted and unsaturated clays. *Géotechnique* **1965**, *15*, 267–286. [CrossRef]
8. Barden, L. Consolidation of clays compacted dry and wet of optimum water content. *Géotechnique* **1974**, *24*, 605–625. [CrossRef]
9. Fredlund, D.G.; Hasan, J.U. One-dimensional consolidation theory: Unsaturated soils. *Can. Geotech. J.* **1979**, *16*, 521–531. [CrossRef]
10. Dakshanamurthy, V.; Fredlund, D.G. Moisture and air flow in an unsaturated soil. In Proceedings of the Fourth International Conference on Expansive Soils, Denver, CO, USA, 16–18 June 1980; Volume 1, pp. 514–532.
11. Ho, L.; Fatahi, B.; Khabbaz, H. Analytical solution for one-dimensional consolidation of unsaturated soils using eigenfunction expansion method. *Int. J. Numer. Anal. Met. Geomech.* **2014**, *38*, 1058–1077. [CrossRef]
12. Zhou, W.H.; Zhao, L.S.; Li, X.B. A simple analytical solution to one-dimensional consolidation for unsaturated soils. *Int. J. Numer. Anal. Met. Geomech.* **2014**, *38*, 794–810. [CrossRef]
13. Shan, Z.D.; Ling, D.S.; Ding, H.J. Exact solutions for one-dimensional consolidation of single-layer unsaturated soil. *Int. J. Numer. Anal. Met. Geomech.* **2012**, *36*, 708–722. [CrossRef]
14. Ho, L.; Fatahi, B.; Khabbaz, H. A closed form analytical solution for two-dimensional plane strain consolidation of unsaturated soil stratum. *Int. J. Numer. Anal. Met. Geomech.* **2015**, *39*, 1665–1692. [CrossRef]
15. Ho, L.; Fatahi, B. Analytical solution for the two-dimensional plane strain consolidation of an unsaturated soil stratum subjected to time-dependent loading. *Comput. Geotech.* **2015**, *67*, 1–16. [CrossRef]
16. Xie, K.H.; Xie, X.Y.; Gao, X. Theory of one dimensional consolidation of two-layered soil with partially drained boundaries. *Comput. Geotech.* **1999**, *24*, 265–278. [CrossRef]
17. Gray, H. Simultaneous consolidation of contiguous layers of unlike compressible soils. *Trans. ASCE* **1945**, *110*, 1327–1356.
18. Mesri, G. One-dimensional consolidation of a clay layer with impeded drainage boundaries. *Water Resour. Res.* **1973**, *9*, 1090–1093. [CrossRef]
19. Schiffman, R.L.; Stein, J.R. One-dimensional consolidation of layered system. *J. Soil. Mech. Found. Div. ASCE* **1970**, *96*, 1090–1093.

20. Cai, Y.Q.; Liang, X.; Wu, S.M. One-dimensional consolidation of layered soils with impeded boundaries under time-dependent loadings. *Appl. Math. Mech. Engl.* **2004**, *25*, 937–944.
21. Wang, L.; Sun, D.A.; Xu, Y.F. Semi-analytical solutions to one-dimensional consolidation for unsaturated soils with semi-permeable drainage boundary. *Appl. Math Mech. Engl.* **2017**, *38*, 1439–1458. [CrossRef]
22. Wang, L.; Sun, D.A.; Li, L.Z.; Li, P.C.; Xu, Y.F. Semi-analytical solutions to one-dimensional consolidation for unsaturated soils with symmetric semi-permeable drainage boundary. *Comput. Geotech.* **2017**, *89*, 71–80. [CrossRef]
23. Wang, L.; Sun, D.A.; Qin, A.F. Semi-analytical solution to one-dimensional consolidation for unsaturated soils with semi-permeable drainage boundary under time-dependent loading. *Int. J. Numer. Anal. Met. Geomech.* **2017**, *41*, 1636–1655. [CrossRef]
24. Crump, K.S. Numerical inversion of Laplace transform using a Fourier series approximation. *JACM* **1976**, *23*, 89–96. [CrossRef]
25. Goode, S.W. *Differential Equations and Linear Algebra*, 2nd ed.; Prentice-Hall Inc.: Upper New Jersey River, NJ, USA, 2000.
26. Zwillinger, D. *Handbook of Differential Equations*, 3rd ed.; Academic Press: San Diego, CA, USA, 1997.
27. Durbin, F. Numerical inversion of Laplace transforms: An efficient improvement to Dubner and Abate's method. *Comput. J.* **1974**, *17*, 371–376. [CrossRef]
28. Wang, L.; Sun, D.A.; Li, P.C.; Xie, Y. Semi-analytical solution for one-dimensional consolidation of fractional derivative viscoelastic saturated soils. *Comput. Geotech.* **2017**, *83*, 30–39. [CrossRef]
29. Morton, K.W.; Mayers, D. *Numerical Solution of Partial Differential Equations: An Introduction*, 2nd ed.; Cambridge University Press: Cambridge, UK, 2005.
30. Croft, D.R.; Lilley, D.G. *Heat Transfer Calculations Using Finite Difference Equations*; Applied Science Publishers LTD: London, UK, 1977.

processes

MDPI

Article

Investigation of Deep Mine Shaft Stability in Alternating Hard and Soft Rock Strata Using Three-Dimensional Numerical Modeling

Xiaoming Sun [1,2,*], Gan Li [1,2], Chengwei Zhao [1,2], Yangyang Liu [1,2] and Chengyu Miao [1,2]

1 State Key Laboratory for Geomechanics & Deep Underground Engineering, China University of Mining & Technology, Beijing 100083, China; ligan303@126.com (G.L.); zcw13683071036@126.com (C.Z.); yyang_liu@126.com (Y.L.); mcyiaohengu@126.com (C.M.)

2 School of Mechanics and Civil Engineering, China University of Mining & Technology, Beijing 100083, China

* Correspondence: 108830@cumtb.edu.cn or sunxiaoming@cumtb.edu.cn; Tel.: +86-159-1095-7882

Received: 8 October 2018; Accepted: 4 December 2018; Published: 20 December 2018

Abstract: The problem of shaft instability has always been a major difficulty in deep mining practices. The shaft fracture has a high probability of being located near the aquifers and the soft–hard rock contact zone. This paper describes the deformation and stress characteristics of surrounding rock and the shaft lining under the interactive geological conditions under soft and hard rock strata in Anju coal mine, Shandong Province, China. Using the Method of Geological Strength Index (GSI) and considering the rock-softening characteristics of water, the parameters of rock mass are calibrated. By means of the 3DEC-trigon method, the variation characteristics of surrounding rock and the shaft lining are simulated. After shaft excavation, under the condition of no support, shear failure and tensile failure occur in shallow surrounding rock shafts, and a pressure relief zone is formed. Shear failure is the main destruction mode in deep surrounding rock. Because of the different strengths of the surrounding rock, the deformation of the surrounding rock is significantly different. After the surrounding rock is softened by water absorption, the difference is magnified. The maximum shear stress and plastic zone appear near the interface between soft and hard rock. Under the condition of shaft lining support, uneven deformation of surrounding rock surely leads to nonlinear variation of pressure on the shaft lining. Under the action of an inhomogeneous pressure field, partial shear failure occurs in the shaft lining, and the shear failure area increases after the surrounding rock is softened by water. Because of the nonlinear deformation of the shaft lining, it is easy to produce stress concentration and bending moment near the interface between hard and soft strata. The control methods of advance grouting and pressure relief excavation are proposed to improve the stability of the shaft, and a good effect is gained.

Keywords: mine shaft; alternate strata; surrounding rock; shaft lining; relief excavation

---

1. Introduction

As the coal resource has been exhausted in shallow mining areas, deep mining has become increasingly common, and the depth of the mine shaft has also increased greatly. Deep geotechnical engineering needs to face special geological conditions such as high geostress, high geotemperature and high karst water pressure, which make the engineering more difficult and dangerous [1]. The shaft is an important passageway connecting ground and underground space, as a guttural project of roadway engineering.

However, the phenomenon of shaft fracture has occurred frequently and caused serious damage and great loss to coal mine enterprises, and it has restricted deep mine mining. Through a large number of investigations on shaft rupture in the Huang Huai area, Wang J et al. [2] opined that the location of shaft fractures is mostly located near the aquifers and the contact zone of soft rock interlaced with hard rock. There are mainly five hypotheses on the mechanism of shaft failure of coal mines [3–5]: shaft failure by new tectonic movement, the construction of the shaft, groundwater seepage and land subsidence, temperature stress induced under variable temperature, and the vertical additional stress induced by mining and dewatering. However, the mechanism of the failure of a deep shaft in the soft–hard rock interface is still unclear and less studied.

Some experts have studied the physical and mechanical properties of alternate rock stratum. A. Yassaghi et al. [6] pointed out that the contact zone rock mass properties were significantly reduced, especially when wet. Through long-term detailed observations of the contact zone in the tunnel, it was proved that the tunnel-wall convergence in the contact zone was 3% higher than that in the normalized tunnel. Wenkai Feng et al. [7] found that deformation failure occurred many times during the Mounigou tunnel construction, and in the soft–hard rock contact zone the deformation failure was most obvious. The numerical simulation research (FLAC-3D) shows that there are obvious stress concentrations and differentiations near the interface of the soft-and-hard rock stratum, and that the surrounding rock deformation failure belongs to the type of soft rock plastic-squeezing failure.

The main and auxiliary shaft linings of the Baodian coal mine and Donghuantuo coal mine were destroyed during construction, and the destroyed location was near the interface between quaternary stratum and bedrock. The location of shaft lining failure in the Qianjiaying coal mine is near the stratum interface between sandstone and clay layers. The Chen Silou main shaft, Xinglong main shaft, Yangcun main/auxiliary shaft, Taiping main/ventilation shaft, Tongting main shaft and Pan Yidong main shaft failures occurred near the interface of strata. As shown in Figure 1, we think that because of the different mechanical parameters of the soft and hard rock, the deformation and failure characteristics of the rock are different under the effects of the ground stress and water. As a result, the pressure on the shaft lining is also different, which may cause a stress concentration phenomenon, thus causing the damage of the shaft lining.

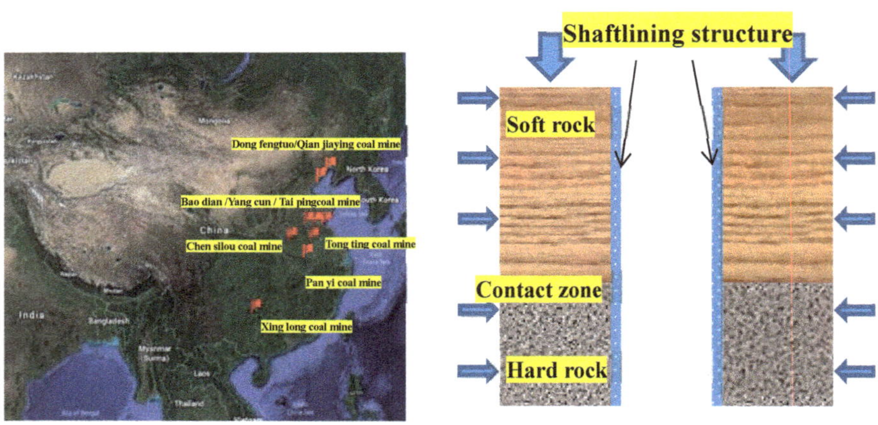

Figure 1. Schematic image of stress in shaft.

It is hard to determine the stress and deformation failure characteristics of shaft in the soft–hard rock contact zone, depending on theoretical analysis only. Therefore, the mainly used method in our study is numerical simulation. At present, in shaft research, the applied methods on numerical analysis include the distinct element method and the finite element method [8,9]. Comparing with the results of the finite element method, the accuracy of displacement and stress given by the discrete element method of the paper is more precise and can explicitly simulate the fracture and failure process of an underground excavation [10,11]. The 3 Dimension Distinct Element Code (3DEC) is one of the most commonly used DEM software programs, and precisely reflects the stress and strain characteristics in 3D space. In the 3DEC simulation, to overcome the defects of only allowing fracture development along predefined discontinuities, the 3DEC-trigon method approach is used to research shaft stability [12–14].

In this thesis, 3DEC-trigon method is applied to simulate the mine shaft stability in soft–hard rock based on a case study at AN JU mine shaft. In the process of reasonable parameter selection, X-ray diffraction can be used to detect mineral content in rock samples. With the uniaxial compression testing system, the rock strength parameters can be obtained with different moisture content levels. According to the geologic description (appearance and structure of a rock mass), the method of Geological Strength Index (GSI) is adopted to evaluate the rock mass integrity and strength. In the numerical scheme, the trigon blocks are generated and bonded through contacts with predefined dimensions. The parameters of polygons and contacts in the trigon program are calibrated by the rock mass properties obtained through GSI. Based upon laboratory tests and previous research findings, a 3D calculated model of a shaft is established with the 3DEC-trigon method. The aim of this study is to understand the progressive failure mechanism of the mine shaft in the soft–hard rock stratum. The property of rock wetting that affects the shaft stability is also considered.

2. Engineering Properties of Deep Shaft-Surrounding Rock

2.1. The Engineering Status

The AN JU mine is in Jining City in China's Shandong Province (Figure 2a). The depth of the auxiliary shaft is 1008 m and its diameter is 6 m. The surrounding rock of the shaft is mainly composed of mudstone and sandstone ranging from 900–950 m deep, and the detailed strata histogram is illustrated in Figure 3. The strata in the area can be divided into Quaternary strata, Jurassic stratum, Permian Shanxi Formation, Lower Shihezi Formation, Carboniferous–Permian Taiyuan Formation, Carboniferous Benxi Formation and Ordovician. Coal-bearing strata are Shanxi and Taiyuan Formations. There are many faults and structural fissures in this area. The larger faults include five north–south faults, 28 NE–NE faults and seven NW faults. Aquifers in this area mainly include a Quaternary aquifer, Jurassic sandstone aquifer, and deep Ordovician aquifer. In this paper, the target location is located in the Jurassic aquifer. The average thickness of the aquifer is about 690.5 m. It is mainly medium-fine sandstone and belongs to the fractured confined aquifers. According to borehole data, four parts in the Jurassic aquifer are divided, as shown in the Table 1. The C40 cast-in-situ concrete construction technique is applied in this region and the thickness of concrete is 600 mm (Figure 2b).

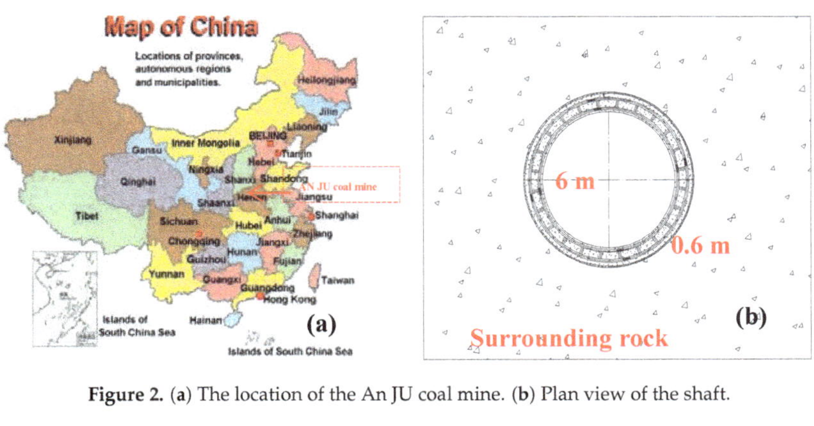

Figure 2. (a) The location of the An JU coal mine. (b) Plan view of the shaft.

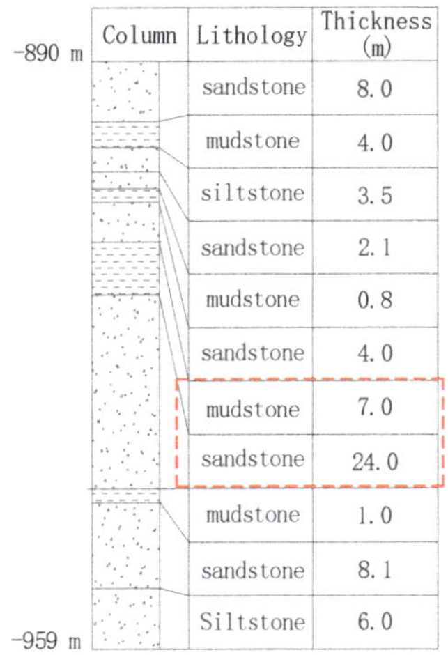

Figure 3. Lithological descriptions of the shaft.

Table 1. Hydrogeologic nature of rock mass.

Aquifers	Shaft Depth	Lithology	Joint Distribution
I	355–505 m	Jurassic sandstone	Fracture density and openness lower
II	639–759 m	Sandstone	Dense vertical joints
III	736–871 m	Sandstone	High degree of openness, more filler
IV	850–1005 m	Sandstone/mudstone	High degree of openness, strong connectivity

2.2. Rock Mass Properties

The shaft stability shows a rather pronounced variation with rock mechanical properties. Mineral component and moisture content of rocks have a dominant influence on rock mechanical properties [14]. Water is one of the important factors affecting rock strength. The interaction between water and rock includes mechanical action, physical action and chemical action. The mechanical action includes the influence of hydrostatic pressure and hydrodynamic pressure. Physical action mainly refers to the process of water softening, mudding, lubrication, drying and wetting, and freezing and thawing of rock, which change the physical and mechanical properties of rock and degrade the inherent mechanical properties of rock. Chemical actions mainly include dissolution, ion exchange, hydration and hydrolysis. The influence factors of water absorption characteristics of rocks include the influence of clay minerals, and the void structure and microstructure of clay minerals. In order to provide the parameter basis for the numerical simulation, X-ray diffraction and uniaxial compression tests are carried out for onsite rock specimens to understand the mineral components and mechanical properties of surrounding rock in the AN JU deep shaft. The X-ray diffraction (TTR III Multifunctional X-ray Diffraction Instrument, Rigaku, Tokyo City of Zhaodao, Japan) of rock samples is analyzed. The diffraction spectrums are shown in Figure 4. The analysis results are as follows: The clay mineral content of the mudstone is 50.3%, the largest proportion of clay minerals is montmorillonite; the clay mineral content of the sandstone is 10.8%, the largest proportion of clay minerals is kaolinite. In order to understand the effect of water on the mechanical properties of rock, the EHF-EG200KN-type full digital hydraulic servo test system (2000 kN Rock Uniaxial and Triaxial Test System, Changchun Testing Machine Co., Ltd, Beijing, China) is used for uniaxial test of natural-saturated-state rock samples (Figure 5).

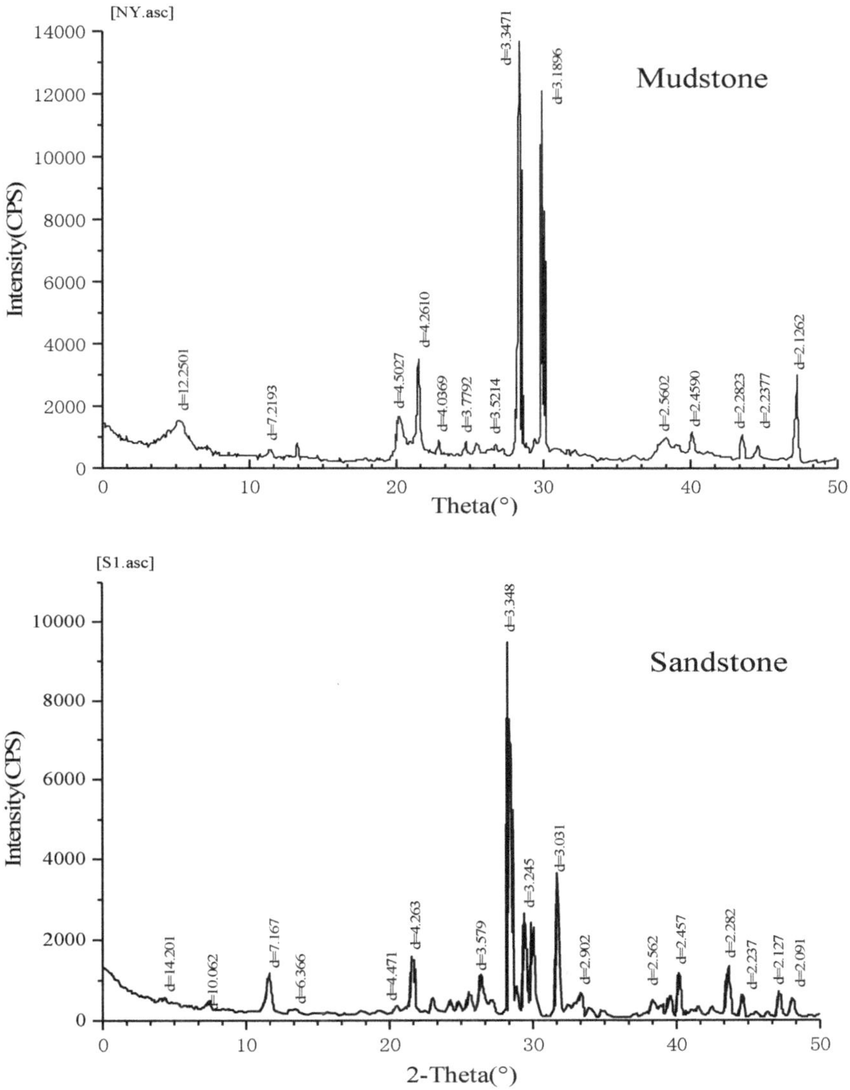

Figure 4. X-ray diffraction analysis of mudstone and sandstone specimens.

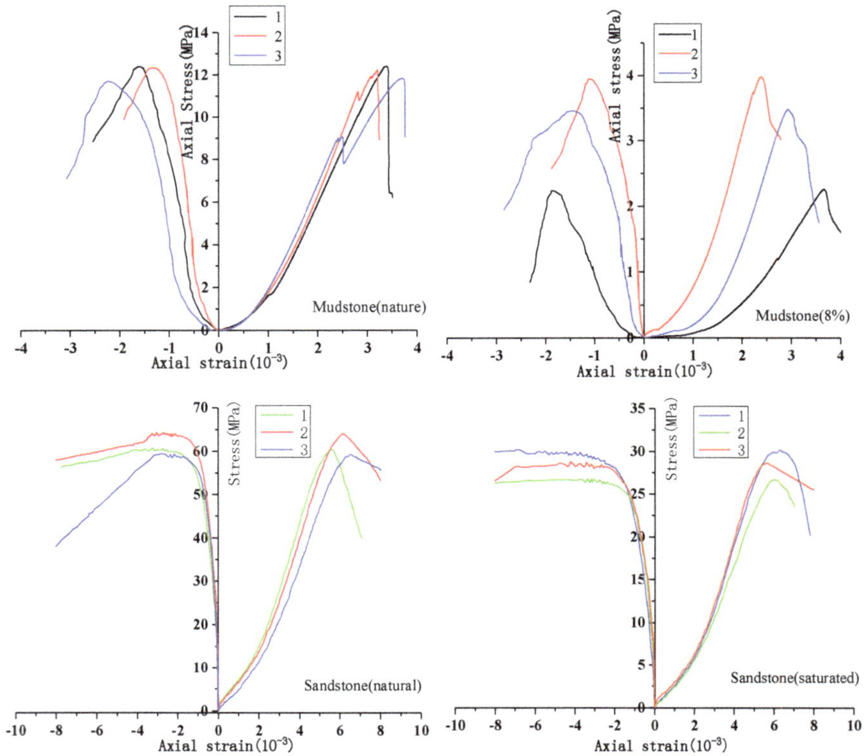

Figure 5. The whole stress–strain curve of rock.

The rock specimen properties are listed in Table 2. The average uniaxial compressive strength of mudstone under natural conditions is 12.12 MPa, and that of sandstone is 61.13 MPa. After saturated water absorption, sandstone strength is 28.52 MPa, and mudstone strength is 3.25 MPa when water absorption is 8%.

Table 2. The table of uniaxial compression experiment calculation.

Lithology	σ_c/MPa	σ_c/MPa (Average)	E/GPa	E/GPa (Average)	u
Mudstone (8%)	2.25 3.96 3.47	3.25	0.8 0.86 0.86	0.84	0.34
Mudstone (natural)	12.38 12.19 11.80	12.12	2.99 2.94 2.89	2.94	0.29
Sandstone (saturated)	28.720 26.734 30.101	28.51	7.72 7.91 7.10	7.5	0.27
Sandstone (natural)	64.149 60.505 59.350	61.33	12.22 12.74 12.82	12.59	0.21

During the stability of jointed rock mass evaluation, the joint characteristics, weathering degree and roughness of rock mass should be considered. Generally speaking, the strength of jointed rock mass is less than that of intact rock mass [15]. At present, the commonly used methods for the engineering classification of rock mass include RMR (Rock Mass Rating), RQD (Rock Quality Designation), RMI (Rock Mass Index) and GSI [14]. The deformation modulus of rock mass is calculated by using the GSI (Geological Strength Index) on the basis of the study of the rock mass strength criteria for many years, such as from E. Hoek and so on. E. Hoek built the correlation model between rock mass modulus and the deformation modulus of the intact rock, and proved the rationality of the complex stability index [16]. The strength of the rock mass is calculated with the E. Hoek (2002) method [17].

$$\frac{E_{rm}}{E_i} = \left(0.02 + \frac{1 - D/2}{1 + e^{(60 + 15D - GSI)/11}}\right) \tag{1}$$

$$\sigma_{cmass} = \sigma_{ci}\frac{(m_b + 4s - a(m_b - 8s))}{2(1 + a)(2 + a)}\left(\frac{m_b}{4 + s}\right)^{as - 1} \tag{2}$$

$$m_b = m_i \exp\left(\frac{GSI - 100}{28 - 14D}\right) \tag{3}$$

$$s = \exp\left(\frac{GSI - 100}{9 - 3D}\right) \tag{4}$$

$$a = \frac{1}{2} + \frac{1}{6}(e^{-\frac{GSI}{15}} - e^{-\frac{20}{3}}) \tag{5}$$

The calculation methods of rock mass properties (E_{rm}: rock mass modulus; σ_{cmass}: uniaxial compressive strength of rock mass) are based on the Geological Strength Index (GSI), the deformation modulus of the intact rock (E_i), the uniaxial compressive strength of the intact rock (σ_{ci}) and the rock mass damage factor (D = 0.5). Rock mass properties are listed and described in Table 3.

Table 3. Intact rock properties and rock mass properties.

Lithology	Intact Rock		GSI	Constant				Rock Mass	
	σ_{ci} (MPa)	E_i (GPa)		m_i	m_b	s	a	σ_{cmass} (MPa)	E_{rm} (GPa)
Mudstone (8%)	3.25	0.84	37	5	0.249	0.000225	0.514	0.84	0.055
Mudstone (natural)	12.12	2.94	37	5	0.249	0.000225	0.514	3.14	0.189
Sandstone (saturated)	28.51	7.59	40	19	1.091	0.000335	0.511	7.39	0.58
Sandstone (natural)	61.33	12.59	40	19	1.091	0.000335	0.511	15.89	0.96

2.3. Mechanics Model of Shaft Lining Stability through Soft–Hard Rock Contact Zone

As shown in Figure 6, a simplified structural mechanics model for shaft lining forces is established. P1 represents the vertical stress of rock strata, and increases with the increase of burial depth, while P2 represents the vertical stress of the shaft lining, including the vertical component of self-weight stress and additional stress of the shaft lining. Due to the different mechanical properties of rock strata, lateral deformation produced by vertical loads is different. Under the action of vertical stress, the surrounding soft rock of the shaft would produce large lateral deformation, and the shaft lining would endure heavy load (q1). However, the lateral deformation of hard rock under vertical stress is smaller, and the lateral load on the shaft lining is also smaller (q2). As shown in Figure 6, the shaft lining produces irregular deformation near the soft–hard rock contact zone, and the stress concentration is produced under the pressure P2, which could lead to damage of the shaft lining.

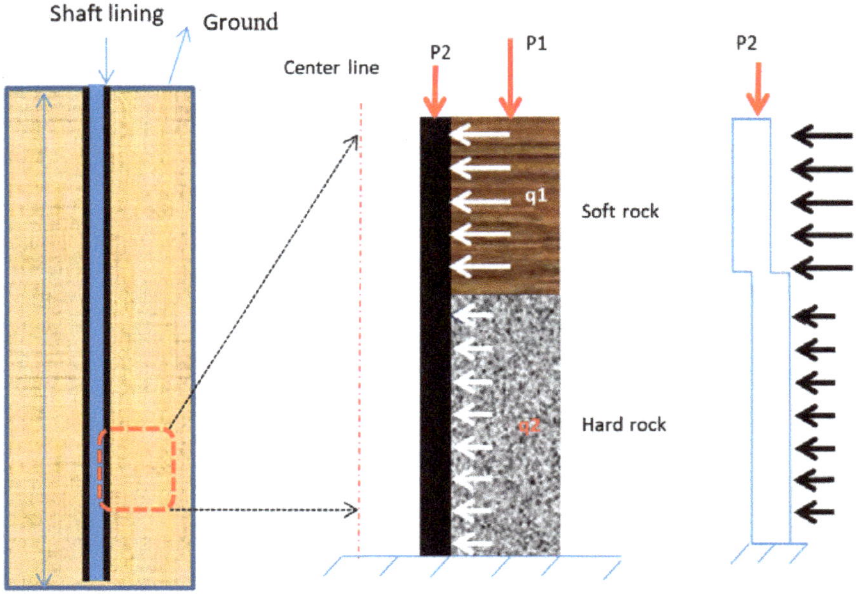

Figure 6. Simplified structural mechanics model for shaft lining forces.

3. Numerical Model Establishment

3.1. The Determination of Parameters in 3DEC-Trigon

The behavior of the rock mass is controlled by parameters that can be divided into the following two groups: (1) block model parameters—density, bulk modulus (K), shear modulus (G), cohesion (C), internal friction angle (φ), tensile strength (σ_t); and (2) contact model parameters—normal stiffness (K_n), shear stiffness (K_s), cohesion (C^j), internal friction angle (φ^j), and tensile strength ($\sigma_t{}^j$). The edge length of the irregular blocks ($\Delta z_{\min}$) is set to 0.1 m in the surrounding rock area of the shaft. According to the rock mechanics parameters obtained from laboratory and calculation (Tables 1 and 2), K, G, K_n and K_s can be calculated as follows [18]:

$$K = \frac{E}{3(1 - 2\mu)} \tag{6}$$

$$G = \frac{E}{2(1 + \mu)} \tag{7}$$

$$K_n = 10\left[\frac{K + \frac{3}{4}G}{\Delta z_{\min}}\right] \tag{8}$$

$$K_s = 0.4K_n \tag{9}$$

Other parameters required in 3DEC were obtained from a series of simulated rock mass compression tests. If the rock mass is small—for example, it contains only one structural plane—then the mechanical properties of the rock mass in a certain direction will be significantly controlled by this structural plane. If the rock mass size continues to increase, including more structural planes, the mechanical properties of rock mass in this direction will depend on the combination of structural planes. When the rock mass size increases to a certain extent, the combination of structural planes in this direction tends to be fixed, and the influence on the mechanical properties of rock mass tends to be stable. The smallest rock mass size with a combination of structural planes that is stable is called the REV (Representative Elementary Volume) [19]. According to numerical experiments, we select REV = 4 × 4 × 8 m, when $\Delta z_{\min} = 0.4$. According to a series of numerical simulation experiments, the numerical simulation parameters are determined as shown in Table 4. The uniaxial experimental curve of rock mass is shown in Figure 7.

Table 4. Parameters for the rock mass.

Lithology	Block Properties						Contact Properties				
	Density (kg/m^3)	K (GPa)	G (GPa)	C^b (MPa)	φ^b	$\sigma^b{}_t$ (MPa)	K^n (GPa)	K^s (GPa)	C^j (MPa)	φ^j	$\sigma^j{}_t$ (MPa)
Mudstone (8%)	2320	0.08	0.02	0.6	15	0.16	8.38	3.28	0.5	15	0.09
Mudstone (natural)	1900	0.15	0.073	0.9	21	0.4	20.50	8.21	2.94	21	0.69
Sandstone (saturated)	2730	0.42	0.230	1.6	21	0.5	59.17	23.67	3.15	25	3.07
Sandstone (natural)	2550	0.55	0.400	2.2	32	1.0	85.21	34	6.63	32	6.32
Concrete (C40)	2500	0.53	0.42	2.1	31	1.2	87.23	33	6.75	31	6.43

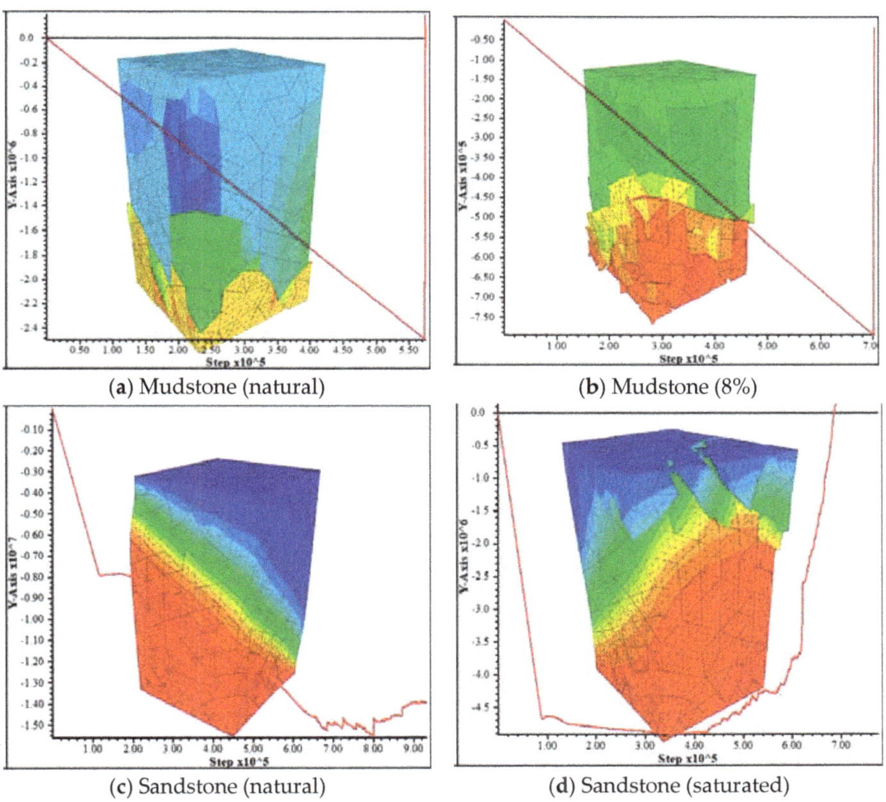

(**a**) Mudstone (natural) (**b**) Mudstone (8%)

(**c**) Sandstone (natural) (**d**) Sandstone (saturated)

Figure 7. Simulated uniaxial compressive test for the rock mass.

3.2. *Numerical Mode and Simulation Schemes of Shaft*

As shown in the Figure 8, a numerical model based on 3DEC numerical simulation software is established. The size of the model is 50 (x width) × 50 (y width) × 15 (z height) m, containing 135,210 blocks. The mudstone (7 m) and sandstone (7 m) strata traversed by the shaft in the mine were chosen to be research objects. The shaft is about 6 m in diameter, and the shaft lining thickness is about 0.6 m. According to the results of the relative crustal stress test, vertical stress mainly is compression stress in depth, the vertical stress is measured to 26 MPa, the maximum horizontal stress is 23 MPa along the x-axis, the minimum horizontal stress is 18 MPa, and the simulated depth is 905 m. The upper boundary and horizontal boundary adopt the stress boundary condition, and the bottom boundary condition is fixed z-direction displacement. The surrounding rock is divided by joints with a maximum length of 0.4 m near the shaft, and joints with a length of about 1 m are used in other parts.

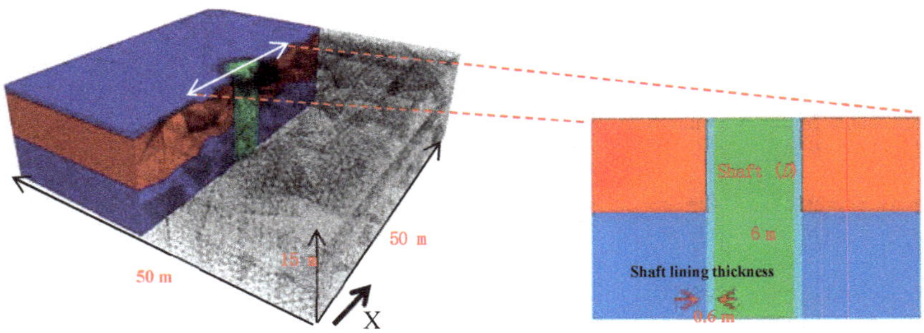

Figure 8. Numerical model for the surrounding rock of the shaft.

Through the removal of blocks in the shaft, excavation is simulated, and the water softening of rock mass is simulated by the weakening of joints and block parameters within the range of 1.5 m surrounding the shaft lining. In order to understand the deformation and stress characteristics of surrounding rock and shaft lining at the soft–hard rock contact zone, the following numerical simulation schemes are adopted in the mining area: (1) deformation and stress characteristics of surrounding rock in the natural state of surrounding rock, without shaft lining support; (2) deformation and stress characteristics of surrounding rock and shaft lining in the natural state of surrounding rock with shaft lining support; (3) deformation and stress characteristics of surrounding rock with rock water absorption softening.

4. Simulation Result Analysis

4.1. *Displacement Analysis*

Figure 9 shows the displacement curves of the monitoring points after shaft excavation. The results showed that all monitoring curves increased at the beginning of a period of time, then they showed a steady trend gradually. The deformation of surrounding rock under natural conditions is shown in Figures 9b and 10a. Maximum deformation of mudstone is 1.2 m, and that of sandstone is 0.2 m. Figures 9c and 10b show the deformation characteristics of surrounding rock after water absorption and softening. Compared with rock in natural conditions, the deformation of mudstone increased by 0.4 m and sandstone by 0.2 m after softening. Figures 9d and 10c,d respectively indicate the displacement characteristics of concrete shaft lining under natural conditions and water softening. The maximum displacement of the shaft lining is parallel to the direction of maximum principal stress. The displacement of the shaft lining in the mudstone formation (33 mm) is larger than that in sandstone stratum (28 mm). This feature is more obvious after water absorption and softening of rock, where the deformation is 34 mm and 58 mm, respectively. Because of the nonuniform deformation of the shaft lining, there will be larger bending moments at the junction between the hard rock and the weak rock.

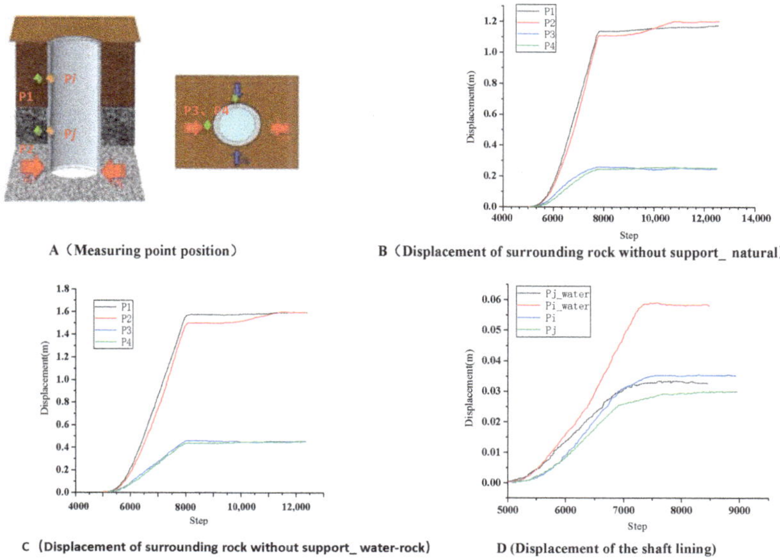

Figure 9. Displacement curves of the monitoring points after shaft excavation.

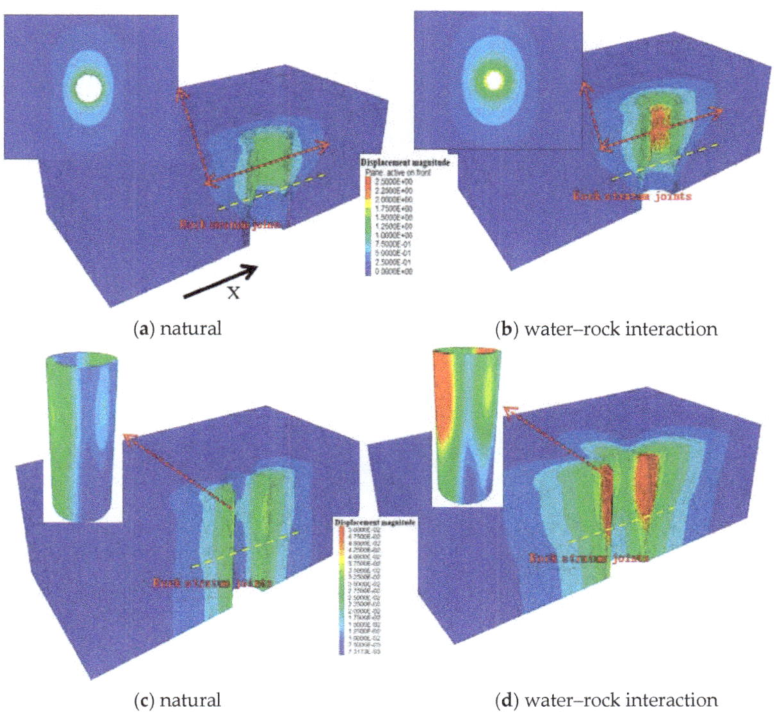

Figure 10. Simulated displacement map of the surrounding rock under unsupported conditions and under concrete supporting (C40) conditions.

4.2. Stress Analysis

Simulated distribution characteristics of max shear stress under unsupported and concrete supporting conditions are presented in Figure 11a–d, and water softening of surrounding rock is also considered. It can be seen in Figure 11a,b that under unsupported conditions, the pressure relief zone appears obviously in the surrounding rock of the shaft, and the maximum shear stress appears near the interface between the hard and soft rock strata. The formation of a pressure relief zone is mainly caused by plastic failure of surrounding rock. Figure 11b demonstrates that the maximum shear stress distribution of surrounding rock after water absorption softening is obviously higher than that of the natural state, and the influence of upper soft rock on hard rock is greater. The scope of the pressure relief zone also expanded obviously after water absorption softening. Figure 11c,d illustrates the max shear stress characteristics of the concrete shaft lining, which is in the natural state, and water-absorbing soft rock condition of surrounding rock, respectively. The result indicated that the shear stress of the shaft lining in the soft rock section is much greater than that in the hard rock stratum. This kind of imbalance is more prominent after water absorption and softening of the surrounding rock.

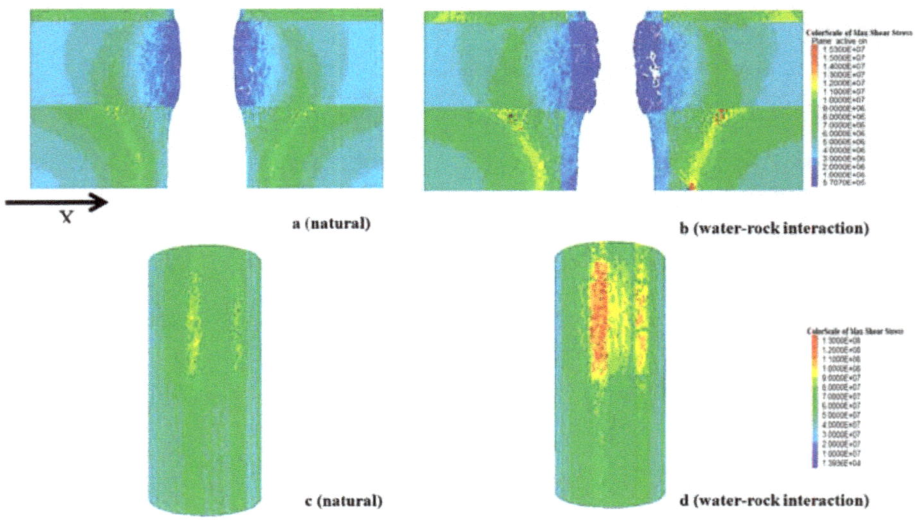

Figure 11. Simulated distribution characteristics of max shear stress under unsupported and concrete supporting conditions.

4.3. Plastic Zone Analysis

This paper assumes that the surrounding rock of the shaft is a homogeneous isotropic medium. Firstly, the stress distribution characteristics of surrounding rock before and after excavation are determined by elasticity theory. Then, according to the H-B strength criterion, the yield failure of the surrounding rock under this stress field can be judged. The H-B strength criterion is shown in Formula (10) [20],

$$\sigma_1 = \sigma_3 + \sigma_c \left(m_b \frac{\sigma_3}{\sigma_c} + s \right)^a \tag{10}$$

where σ_1—maximum principal stress, σ_3—minimum principal stress, σ_c—uniaxial compressive strength, m_a, s, a—the parameters reflecting the characteristics of rock mass. At $\sigma_3 + \sigma_c \left(m_b \frac{\sigma_3}{\sigma_c} + s \right)^a - \sigma_1 = 0$, the surrounding rock is at the critical point of elasticity and plasticity.

According to the theory of elasticity, the stress value of a point in the surrounding rock of the shaft is obtained by the polar coordinate method. The analysis diagram of the surrounding rock of the shaft is shown in Figure 12.

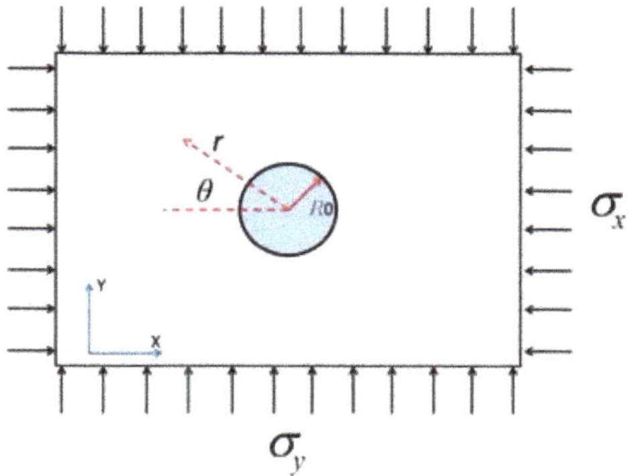

Figure 12. The analysis diagram of the surrounding rock of the shaft.

$$\begin{bmatrix} \sigma_1 = \frac{\sigma_r + \sigma_\theta}{2} + \sqrt{\left(\frac{\sigma_r - \sigma_\theta}{2}\right)^2 + (\tau_{r\theta})^2} \\ \sigma_3 = \frac{\sigma_r + \sigma_\theta}{2} - \sqrt{\left(\frac{\sigma_r - \sigma_\theta}{2}\right)^2 + (\tau_{r\theta})^2} \end{bmatrix} \tag{11}$$

$$\begin{cases} \sigma_r = \frac{\sigma_y}{2}\left[(1+\lambda)(1-\frac{R_0^2}{r^2}) + (1-\lambda)(1-4\frac{R_0^2}{r^2}+3\frac{R_0^4}{r^4})\cos 2\theta\right] \\ \sigma_\theta = \frac{\sigma_y}{2}\left[(1+\lambda)(1+\frac{R_0^2}{r^2}) - (1-\lambda)(1+3\frac{R_0^4}{r^4})\cos 2\theta\right] \\ \tau_{r\theta} = \frac{\sigma_y}{2}\left[(1-\lambda)(1+2\frac{R_0^2}{r^2}-3\frac{R_0^4}{r^4})\sin 2\theta\right] \end{cases} \tag{12}$$

According to Formulas (10)–(12), where σ_y, σ_x is the horizontal in-situ stress, λ is the ratio of horizontal in-situ stress which is <1, and R_0 is the radius of roadway, the range of the plastic zone is related to the uniaxial compressive strength and rock integrity of rocks when the size of the shaft and in-situ stress are determined.

The damage modes of the shaft-surrounding rock and concrete shaft lining are mainly composed of tensile failure and shear failure. Figure 13a,b shows the failure mode of the surrounding rock of the shaft without support. The shear failure and tensile failure coexist in the 5 m range of surrounding rock. The deep surrounding rock parts are mainly shear failure. It is especially pointed out that Figure 12b takes into account the water absorption softening of surrounding rock. In the horizontal direction, the plastic zone of the surrounding rock is elliptical, and the long axis is perpendicular to the maximum principal stress direction. Under the influence of incompatible deformation of surrounding rock, the soft rock stratum produces lateral additional stress on the hard rock stratum, which makes the plastic zone expand nonlinearly at the interface between the soft and hard rock strata. After the water softening of the surrounding rock, the plastic area of the surrounding rock increases due to tensile failure. Figure 12c,d shows that the failure of the concrete shaft lining is mainly shear failure. After water absorption softening, the plastic zone of the shaft lining increases obviously.

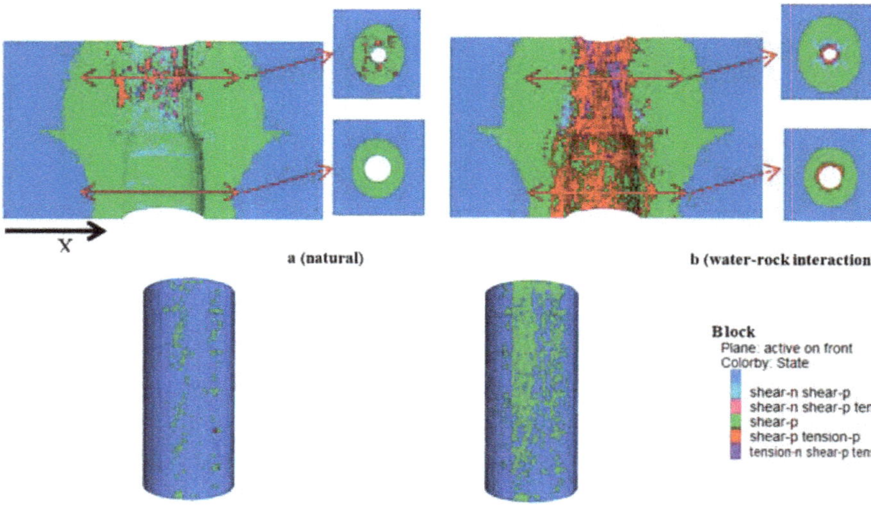

Figure 13. Distribution of plastic zone.

4.4. Mechanism Analysis of the Shaft Deformation in Soft–Hard Rock Strata

The deformation and failure mechanisms of the shaft and shaft lining in alternate hard and soft rock strata are as follows:

(1) Under high stress, rock masses with different strengths have different deformation ratios. The target location of this study is deep at 900 m, and the vertical stress and horizontal stress are relatively high. After the shaft is excavated, the shallow rock mass around the shaft is destroyed under the action of shear stress and tensile stress. With the failure of rock mass, a certain range of loosened zones occur, and the stress is propagated from the surface to the interior, eventually forming a plastic zone dominated by shear failure. The destruction of rock is accompanied by the large deformation of surrounding rock, and the smaller the rock strength, the greater the deformation of surrounding rock. In addition, the surrounding rock contains a large number of clay minerals, so it will soften and expand when exposed to water.

(2) Additional stress in soft rock stratum. The soft rock stratum will produce larger shear stress than the hard rock stratum near the interface, leading to a large plastic zone near the rock interface. The numerical simulation shows that the water–rock interaction aggravates the damaging process of the rock mass.

(3) Stress concentrations due to squeezing action of surrounding rock. The uncoordinated deformation of strata will inevitably lead to asymmetric deformation of the shaft lining. The deformation of the shaft lining in the hard rock stratum decreases suddenly, which will cause the bending moment to increase and the stress concentration to increase abruptly. The deformation of the shaft lining parallel to the maximum principal stress is obviously higher than that in other directions.

5. Case Study

5.1. Control Principle of the Surrounding Rock

Numerical results show that the uncoordinated deformation of surrounding rock and the water absorption softening of rock have a significant effect on shaft stability. Therefore, advance grouting should be adopted in the water-bearing section of the shaft. On the one hand, an artificial water-isolating layer is established between the fractured rock mass and the aquifer to reduce the water absorption and softening phenomenon of surrounding rock, while on the other hand, the strength of the rock mass is enhanced. In order to reduce the pressure of the mine shaft lining, the stress of the surrounding rock should be released to utilize the utmost self-bearing capacity of the rock mass. Under these control measures of surrounding rock, the deformation of the Anju coal mine shaft is in the safe range.

5.2. Construction Technique of Deep Shaft Advance Grouting

Urea–formaldehyde resin and oxalic acid are used as grouting materials. The grouting material manufacturer is Wenchang Petrochemical Co. Ltd. of Zibo City, Shandong Province, China. The product model is U20 urea–formaldehyde resin. The grouting parameters are as follows: the range of grouting concentration is 10–50%; initial viscosity: 1.7–15 cP (20°); bond strength: 1.7–3.7 MPa; sand consolidation strength: 1.2–9.5 MPa; setting time: 1.08–35.33 min; gelling properties: 3.17–47.5 min. As shown in Figure 14, pre-grouting pipes are uniformly arranged around the shaft and the center, the inclination of grouting pipes is 80°, and the number of grouting pipes is determined according to the water gushing. In order to achieve an ideal grouting effect, technical parameters should be adjusted according to borehole detection, stability of surrounding rock, shaft water gushing and grouting observations in the actual construction.

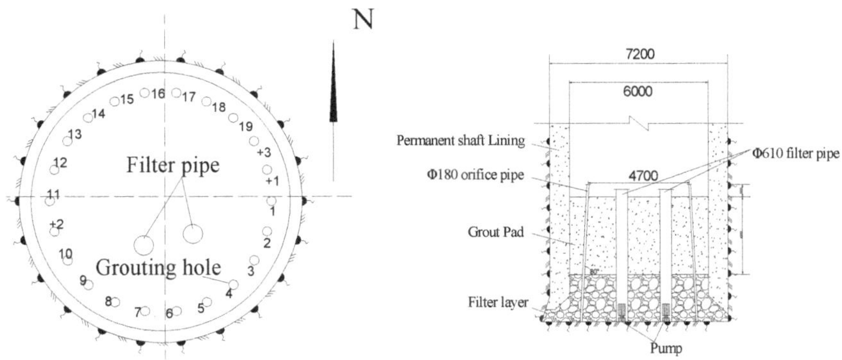

Figure 14. Layout of shaft pre-grouting construction.

5.3. Pressure Unloading of Surrounding Rock

The shaft passing through the deep soft rock stratum is divided into three steps, which can fully release the surrounding rock pressure and deformation of the shaft. Firstly, a small section with radius 3000 mm is used to drive 1.4 m. The process is controlled at about 16 h. The second step is to use a full section heading of 1.4 m. Finally, the construction of the concrete shaft lining is conducted by use of concrete (C40), and the thickness is 0.6 m.

6. Conclusions

This paper presents a numerical simulation study on the deformation mechanism of surrounding rock, and the stress characteristics of the shaft lining, in alternating hard and soft rock strata. With the project background of the Anju coal mine auxiliary shaft (900 m section) in Shandong Province, the research mainly focuses on the surrounding rock strength, water's rock-softening characteristics, deformation and failure rules of the shaft-surrounding rock, and stress characteristics of the shaft lining.

The test results show that the uniaxial compressive strength of mudstone decreases by 3.72 times at 8% water content, and that of sandstone decreases by 2.15 times at the saturation state. Using the GSI evaluation index, the parameters of numerical simulation under different rock mechanics conditions were determined. Deformation and failure characteristics of surrounding rock under the natural state and water absorption softening state were studied by using the 3DEC-trigon method.

It is considered that after the excavation, the shallow surrounding rock will be destroyed and the stress relaxation zone will be formed under the condition of no support. The failure mode of shallow surrounding rock comprises the coexistence of shear failure and tensile failure. The failure mode of deep surrounding rock is mainly shear failure. The deformation of surrounding rock is closely related to the strength of rock strata, and there is obvious nonlinear change in the vertical direction. This phenomenon is more obvious after water absorption softening of rock strata. Soft rock strata will produce larger lateral additional stress on hard rock strata near the interface of strata. It shows that the surrounding rock pressure will cause the shearing failure of the shaft lining. After the water softening effect on the surrounding rock, the plastic area increases obviously in the shaft lining. In the area where the shaft lining displacement changes abruptly, significant bending moments and stress concentration will be generated. In addition, the swelling of clay minerals in the surrounding rock will be a threat to the shaft.

In order to reduce the threat of water–rock coupling to the shaft lining and improve rock mass strength, a pre-grouting control system was adopted near the aquifer section. The method of surrounding rock pressure relief is adopted to improve the bearing capability of surrounding rocks, reduce the uneven deformation of the shaft lining and improve the stability of the shaft. The case study can provide reference for the failure process and control technology of the shaft under deep alternating hard and soft rock strata.

Author Contributions: Conceptualization, X.S. and G.L.; Methodology, X.S.; Software, G.L.; Validation, C.Z., Y.L.; Formal Analysis, C.M.; Investigation, G.L.; Resources, X.S.; Data Curation, G.L.; Writing-Original Draft Preparation, X.S.; Writing-Review & Editing, G.L.; Visualization, C.Z.; Supervision, X.S.; Project Administration, X.S.; Funding Acquisition, X.S.

Funding: This work was supported by the National Key Research and Development Plan of China (Grant No. 2016YFC0600901), the National Natural Science Foundation of China (Grant No. 51874311), the Special Fund of Basic Research and Operating of China University of Mining & Technology, Beijing (Grant No. 2009QL03), and the Yueqi Outstanding Scholar Award Program of China University of Mining and Technology, Beijing.

Conflicts of Interest: The authors declare no conflicts of interest.

References

1. He, M.C.; Xie, H.P.; Peng, S.P.; Jiang, Y.D. Study on rock mechanics in deep mining engineering. *Chin. J. Rock Mech. Eng.* **2005**, *24*, 2803–2813.
2. Wang, J.; Luo, N.; Bai, Z. On the relation between interlayer glide caused by coal extraction and the shaft rupture occurring in coal mines in huanghuai area. *Chin. J. Rock Mech. Eng.* **2003**, *22*, 1072–1077.
3. Zhang, W.Q.; Lu, Y.-H.; Gong, H.Y.; Meng, X.J.; Guan, Y.Z. Causes analysis of shaft wall damage inYan-Teng Mining District and its prevention and contyolling method. *Rock Soil Mech.* **2004**, *25*, 1977–1980.
4. Jing, L.; Liu, F.; Gao, Q.; Yang, R. Rupture stress of shaft wall in mine due to ground subsidence. *Chin. J. Rock Mech. Eng.* **2004**, *23*, 3274–3280.
5. Liu, J.L.; Chen, L.W.; Wang, J.L. Characteristic analysis of temperature stresses of shaft wall. *Rock Soil Mech.* **2011**, *32*, 2386–2390.
6. Yassaghi, A.; Salari-Rad, H. Squeezing rock conditions at an igneous contact zone in the Taloun tunnels, Tehran-Shomal freeway, Iran: A case study. *Int. J. Rock Mech. Min. Sci.* **2005**, *42*, 95–108. [CrossRef]
7. Feng, W.; Huang, R.; Li, T. Deformation analysis of a soft–hard rock contact zone surrounding a tunnel. *Tunn. Undergr. Space Technol. Inc. Trenchless Technol. Res.* **2012**, *32*, 190–197. [CrossRef]
8. Wang, L.; Cheng, Y.P.; Ge, C.G.; Chen, J.X.; Li, W.; Zhou, H.X.; Wang, H.F. Safety technologies for the excavation of coal and gas outburst-prone coal seams in deep shafts. *Int. J. Rock Mech. Min. Sci.* **2013**, *57*, 24–33. [CrossRef]
9. Walton, G.; Kim, E.; Sinha, S.; Sturgis, G.; Berberick, D. Investigation of shaft stability and anisotropic deformation in a deep shaft in Idaho, United States. *Int. J. Rock Mech. Min. Sci.* **2018**, *105*, 160–171. [CrossRef]

10. Bai, Q.S.; Tu, S.H.; Zhang, C.; Zhu, D. Discrete element modeling of progressive failure in a wide coal roadway from water-rich roofs. *Int. J. Coal Geol.* **2016**, *167*, 215–229. [CrossRef]
11. Gao, F.; Stead, D. Discrete element modelling of cutter roof failure in coal mine roadways. *Int. J. Coal Geol.* **2013**, *116–117*, 158–171. [CrossRef]
12. Chen, W.; Konietzky, H.; Abbas, S.M. Numerical simulation of time-independent and -dependent fracturing in sandstone. In Proceedings of the ACM/IEEE International Conference on Software Engineering, Florence, Italy, 16–24 May 2015; pp. 129–138.
13. Gao, F.Q.; Stead, D. The application of a modified Voronoi logic to brittle fracture modelling at the laboratory and field scale. *Int. J. Rock Mech. Min. Sci.* **2014**, *68*, 1–14. [CrossRef]
14. Yang, S.Q.; Chen, M.; Jing, H.W.; Chen, K.F.; Meng, B. A case study on large deformation failure mechanism of deep soft rock roadway in Xin'An coal mine, China. *Eng. Geol.* **2017**, *217*, 89–101. [CrossRef]
15. Lu, S.; Xu, M. Determination and application of modulus of deformation of rock masses based on GSI system. *Chin. J. Rock Mech. Eng.* **2009**, *28*, 2736–2742.
16. Hoek, E.; Diederichs, M.S. Empirical estimation of rock mass modulus. *Int. J. Rock Mech. Min. Sci.* **2006**, *43*, 203–215. [CrossRef]
17. Hoek, E.; Carranza-Torres, C.; Curkum, B. Hoek-Brown failure criterion—2002 Edition. In Proceedings of the Fifth North American Rock Mechanics Symposium, Toronto, ON, Canada, 7–10 July 2002; Volume 1, pp. 267–273.
18. Itasca Consulting Group, Inc. *3DEC User Manual*; Itasca Consulting Group, Inc.: Minneapolis, MN, USA, 2016.
19. Kulatilake, P.H.S.W.; Panda, B.B. Effect of Block Size and Joint Geometry on Jointed Rock Hydraulics and REV. *J. Eng. Mech.* **2000**, *126*, 850–858. [CrossRef]
20. Wu, L.; Adoko, A.C.; Li, B. An Illustration of Determining Quantitatively the Rock Mass Quality Parameters of the Hoek–Brown Failure Criterion. *Rock Mech. Rock Eng.* **2018**, *51*, 1063–1076. [CrossRef]

processes

MDPI

Article

Experimental Study on Compression Deformation and Permeability Characteristics of Grading Broken Gangue under Stress

Yu Zhang [1], Wei Zhou [2,*], Ming Li [1,*] and Zhanqing Chen [1]

1 State Key Laboratory for Geomechanics and Deep Underground Engineering, China University of Mining and Technology, Xuzhou 221116, China; zhangyucumt1981@126.com (Y.Z.); chenzhanqing@vip.163.com (Z.C.)

2 College of Mining Engineering, China University of Mining and Technology, Xuzhou 221116, China

* Correspondence: loutian1982@126.com (W.Z.); mingl@cumt.edu.cn (M.L.); Tel.: +86-138-1531-4985 (W.Z.); +86-135-8547-7939 (M.L.)

Received: 16 October 2018; Accepted: 4 December 2018; Published: 9 December 2018

Abstract: As the important raw material for backfill mining, broken gangue's deformation and permeability characteristics directly affect the deformation of the overlying strata above the filling space. In this paper, through lateral compression and pressed seepage tests, the deformation and permeability characteristics of broken gangue as a function of the stress level and grading features were studied. This research indicates that the stress of broken gangue increases exponentially with an increase in strain, and the compression modulus and compression rate present a positive correlation. The samples with discontinuous grading are more difficult to compress than the continuous grading samples, and the discontinuous grading samples are tighter in accordance with the increase in compression rate. At the same time, the change range of the seepage velocity and permeability of the broken gangue decreases. Positive correction between the grading index of the broken gangue and the effect of reducing the permeability of samples is more obvious under axial compression, and less axial stress is needed to achieve the same permeability level for discontinuous grading. This paper can provide an important test basis for the design of grading parameters and the prediction of filling effects of broken gangue on backfill mining.

Keywords: permeability characteristics; grading broken gangue; compressive stress; compression deformation

1. Introduction

Many high fragmentation and porous particle structures from broken rock mass are produced during geological movement, such as fault fractured zones [1–3] and the fillings of collapse columns [4–6]. In mine construction, large amounts of broken rock mass are required, such as for gangue filling mining backfill [7,8], tailing dam construction [9], and reinforcement technology for rock roadways [10,11]. Broken rock mass still has a certain residual bearing capacity, and its deformation and seepage behaviours have a great influence on the internal fluid and gas migration. Therefore, it is of significant importance to study the deformation and permeability of broken rock mass for safety and green mining.

The deformation characteristics of broken rock mass are basic mechanical properties that greatly impact the research of controlling surface subsidence by filling mining [12,13], building tunnels and roadways passing through a fault fracture zone [14], and fixing foundations and subgrades [15]. Dong et al. (2015) studied the influence of mass density, moisture content and confining pressure on the deformation of broken rock mass [16]. The results of this research not only gave the characteristics of stress-strain relationship that show hardening at first and then softening, but also indicated that the properties of broken sandstone under low confining pressure are characterized by shear contraction and that the high confining pressure will result in shear dilation. In addition, broken gangue has a high sensitivity to the rate of water content. Yin et al. (2012) analysed the plastic deformation and stability of rock mass under dynamic loads, and studied the reasonable critical failure strain which showed that the critical stress increased in proportion to confining pressure [17]. In other words, the critical stress and confining pressure have a positive correlation.

The permeability of broken rock mass has great significance to the safe and efficient production of mines, especially to water-conductive fault zone [18], water prevention and control of the surrounding rock of collapse columns [19,20], and in situ solution mining and migration of mining pollutants [21]. Many scholars have made great achievements in studying the penetrating characteristics of broken rock mass by testing. Chai et al. (2002) provided the mechanism of water bursting caused by deformation and failure according to the nonlinear seepage characteristics of broken rock mass [22]. Using a pressure permeability test, Min et al. (2004) found that the permeating properties of broken rock mass change greatly with the displacement of the load [23]. Through these experiments, the relations of the pore pressure gradient and the seepage velocity of the broken rock mass were obtained. Finally, the results indicated the existence of nonlinear seepage characteristics. However, due to the limitation of experimental equipment and technology, there have been few studies on the relationship between the deformation under loads and permeability of broken rock mass. Some scholars regarded the change in porosity as a deformable characteristic of broken rock mass and established the relationship between the porosity and permeability [24–26]. The basic reason for the deformation of broken rock mass is the change in the stress state, but the relationship between permeability and bearing stress changes is unclear. In addition, the particle size is the basic property of broken rock mass, this property has a great impact on stocking states, pore structure, and deformation under loads. The above broken rock mass are the ideal single particle size or a simple proportion of mixed sample. However, some relevant studies had shown that the mixed particles with a certain size grading characteristic can be piled up into a tight state. Therefore, the grading features have a certain effect on the deformation and permeability [27,28].

The research object selected in this paper is broken gangue, a by-product of coal mining. The deformation and permeability characteristics of broken gangue were analysed through experimental results by confined compression and pressure seepage tests, and evaluate the influence of stress and grading on compressibility and permeability of broken rock mass. The results can provide an important basis for the technology of gangue filling mining.

2. Test System and Sample Preparation

2.1. Test Samples

In this experiment, coal gangue (Zhangji Colliery, Huainan, China) is used as the test target. Coal gangue is the waste produced by coal mining and processing, which is the raw material for backfill mining. The bulk gangue is initially broken into small pieces in the crusher (Zhengzhou Hengxing Heavy Equipment Co., Ltd., Zhengzhou, China); then, the broken gangue particles with different particle sizes are obtained by using a crusher and grading sifter (Gaofu Machinery Co., Ltd., Xinxiang, China). The apertures of the sifter are 5, 10, 15, 20, and 25 mm, and by using the sifter, a single particle size between the ranges of 0–5, 5–10, 10–15, 15–20, and 20–25 mm were obtained, respectively, as shown in Figure 1.

Figure 1. Different particle sizes of gangue and a grading sifter.

The samples used in the experiment were obtained by mixing the single-grain gangue particles in certain proportions. According to the particle interference theory of G. A. Wegmouth [29], in order to achieve the densest state after mixing particles of different sizes, the pores between large-sized particles should be filled by smaller-sized particles, and the particle size of the filler particles cannot be larger than the gap between the large particles. The present grading method designs are the continuous grading method and the discontinuous grading method; the difference between the methods is that the particle sizes determined by the continuous grading method are consecutive, whereas some particles would be lost by the other method.

According to the theory of Talbot Grading [30], the proportion of particle diameter gangue in continuous grading samples could be described by Equation (1):

$$P_i = \left(\frac{d_i}{d_{\max}}\right)^n \tag{1}$$

where P_i is the proportion of the particle size of d_i, $d_{\max}$ is the largest diameter particle in the mixed sample (mm), and n is the power exponent of Talbot.

The n of the Talbot power exponent can be equal to 0.1, 0.3, 0.5, 0.7, and 0.9, and the maximum particle size $d_{\max}$ is 25 mm. According to Equation (1), the proportion of particles d_i can be calculated for different continuous grading samples. Since some particle sizes are lost in the samples by discontinuous grading, discontinuous grading samples can be obtained by rejecting the first-order particle size by continuous grading. The maximum and minimum sizes are required to retain in mixed-grade samples, which are framework grain and pore-filled particles, respectively. Samples with the sizes of 5–10, 10–15, and 15–20 mm are removed in the continuous grading, which correspond to the index m with 0.3, 0.5 and 0.7 mm, respectively. The removed particles will be supplemented by smaller samples with original proportions. The particle sizes of continuous-grade and discontinuous-grade samples is shown in Table 1. Where n is continuous grading index, m is discontinuous grading index, d_i is the particle size, and P_i is the proportion of the particle size of d_i.

Table 1. Proportion of particles in different grading index samples.

Grading Index		P_i(%)				
		d_1 (0–5 mm)	d_2 (5–10 mm)	d_3 (10–15 mm)	d_4 (15–20 mm)	d_5 (20–25 mm)
n	0.1	85.13	6.11	3.78	2.77	2.21
	0.3	61.70	14.26	9.83	7.73	6.48
	0.5	44.72	18.52	14.21	11.98	10.56
	0.7	32.41	20.24	17.28	15.60	14.46
	0.9	23.49	20.35	19.31	18.66	18.19
m	0.3	75.97	0	9.83	7.73	6.48
	0.5	54.77	22.69	0	11.98	10.56
	0.7	39.64	24.76	21.14	0	14.46

From Table 1, it can be found that the distribution of the particle size of the broken gangue samples with different Talbot grading indices has great differences. The small particle proportion decreases gradually with the increase in m and n, while that for the large particles is opposite. The particle proportion less than 5 mm decreases with the increase in m and n; however, the size larger than 5 mm exhibits an opposite variation trend. For example, when n increases from 0.1 to 0.9, the proportion of particle size 0–5 mm is reduced by 61.64%, while the proportion for the particle sizes of 5–10, 10–15, 15–20, and 20–25 mm are increased by 14.24, 15.53, 15.89, and 15.98%, respectively. At the same time, it can be found that the proportion of small particles with discontinuous grading is much larger than that of continuous grading, but both have the same proportion of large particles in the test. For example, when the values of m and n are equal and are taken as 0.3, 0.5, and 0.7 mm, the particles of 0–5 mm in discontinuous grading are higher than that of continuous grading by 14.27%, 10.05% and 16.15%, respectively. The particle proportion of different grading samples directly reflects the size of the broken gangue and determines the differences of pore structure. According to the particle size grading in Table 1, five groups of continuous grading samples and three groups of discontinuous grading ones were obtained after combining the filtered gangue particles of different sizes.

2.2. Experimental Setup

The test system consists of a DDL600 electronic universal testing machine (Changchun Mechanical Science Research Institute Co., Ltd., Changchun, China), osmosis device, seepage circuit, hydraulic pump, and hydraulic accessories, as shown in Figure 2. Among them, the seepage circuit consists of a hydraulic pump, pressure sensor, reversing valve, globe valve, flow sensor. The osmotic device is composed of a cylinder, piston, bottom plate, rims, permeable plate. Additionally, the axial force of the DDL600 electronic universal testing machine ranges from 0 to 600 kN, with the measurement accuracy of ±0.5%. The axial displacement measurement ranges from 0 to 80 mm with the accuracy of ±0.5%; the inner diameter of permeation cylinder is 100 mm, with the capacity of 1.57×10^6 mm^3.

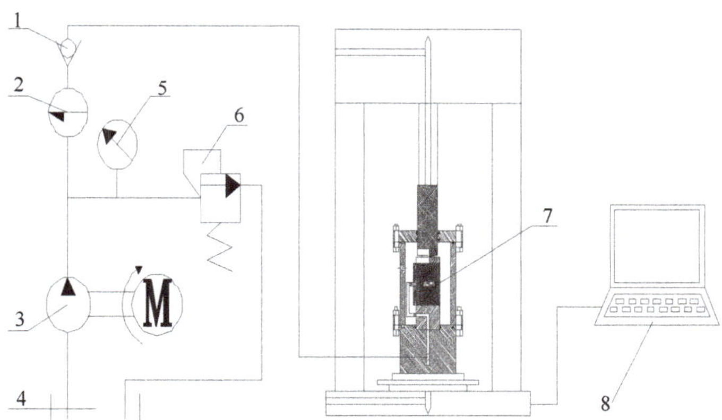

Figure 2. Osmosis system: (1) check valve; (2) flow meter; (3) hydraulic pump; (4) penetrant box; (5) pressure gauge; (6) relief valve; (7) osmosis device; and (8) computer.

Before the experiment, the stability, reliability, and sealing of the entire test system were tested. The osmosis device in the test system is shown in Figure 3. Samples are filled into the osmosis device, and then they are placed on the DDL600 electronic universal testing machine with the upper end contact to the check valve set to 1, and the lower exposure connected to the collecting pail. Before the test, the sample must be saturated in order to fill the piping of the system with osmotic fluid. During the test, the permeate flow rate was obtained by weighing the amount of water infiltrated in the collection tank, and the axial deformation was directly obtained by the test machine data acquisition system.

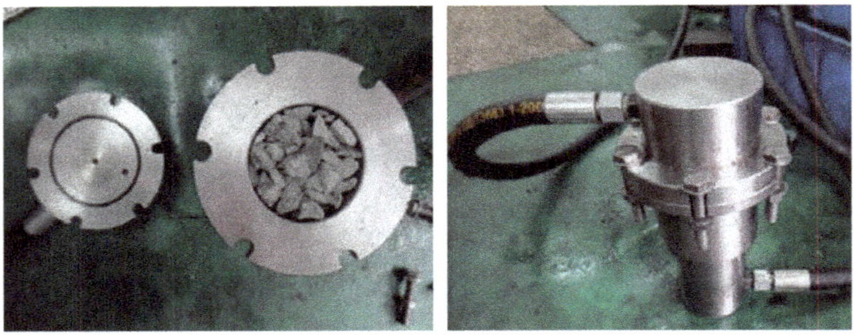

Figure 3. Filling method of the sample in the osmosis device.

3. Parameter Calculation

3.1. Characteristic Parameters of Compression Deformation

The compression modulus is often used to measure the constrictive strength of broken rock under the condition of lateral restraint, and it describes the change rate of stress relative to the strain, which can be expressed as follows [2]:

$$E_t = \frac{d\sigma_1}{d\varepsilon_1} \tag{2}$$

The differential approximate form of Equation (2) [2] is as follows:

$$E_t = \frac{\Delta\sigma_1}{\Delta\varepsilon_1} \tag{3}$$

The compression of broken rock described by the compression ratio is [2]:

$$\gamma = \frac{\Delta h}{H} \tag{4}$$

where Δh represents the height change of samples at a certain stress level, and H is the initial height of the sample. The compressive rate of broken rock is equal to the value of strain ε in the condition of lateral restraint.

3.2. Permeability Parameter

The porosity of the samples can be calculated by the following Equation (5) [31]:

$$\phi = 1 - \frac{m}{\rho_0 A(h_0 - h_i)} \tag{5}$$

where m is the quality of broken rock, (kg); ρ_0 is the density of sandstone, (kg/m^3); A is the cross-sectional area of broken rock, (m^2); h_0 is the initial height of broken sandstone, (m); and h_i is the height of the sample at the loaded time of t_i for test machine, (m).

The seepage of porous media of broken rock particles is consistent with the relation of Forchheimer [32]:

$$-\frac{\partial p}{\partial h} = \frac{\mu v}{k} + \rho\beta v^2 \tag{6}$$

where p is the osmotic pressure, (MPa); h is the height of the porous medium, (m); ρ is the water density, 1000 kg/m^3; μ is the water viscosity, (Pa·s); v represents the seepage velocity, (m/s); k is the permeability, (m^2); and β is the non-Darcy flow factor, (m^{-1}).

By testing, the time series V_i of cumulative volume of exudate can be obtained, in (t_i,t_{i+1}) time period, the average flow is Q_i, and v_i is the average seepage velocity in (t_i,t_{i+1}) time period:

$$v_i = \frac{Q_i}{A} = \frac{4Q_i}{\pi d^2} \tag{7}$$

where d is the diameter of outlet pipe section.

Since the outlet of penetration connects to the atmosphere, the pressure gradient J at the moment of i is [33]:

$$J = \frac{\Delta P}{h_i} = \frac{P_1 - P_2}{h_i} = \frac{P_1}{h_i} \tag{8}$$

where $\Delta P = P_1 - P_2$ represents the osmotic pressure difference among the samples, and P_1 and P_2 are the pressure in the osmotic inlet and outlet, i.e., $P_2 = 0$. h_i is the height of the sample at the moment of i.

According to Equation (6), the seepage velocity at the moment of i and i + 1 is v_i and v_{i+1} respectively, and J is regarded as the pressure gradient. According to the relationship of $v - J$ in i and i + 1, the permeability k_i and non-Darcy flow β-factor at time i would be obtained [33]:

$$\begin{cases} J = \frac{\mu v_i}{k} + \rho\beta {v_i}^2 \\ J = \frac{\mu v_{i+1}}{k} + \rho\beta {v_{i+1}}^2 \end{cases} \tag{9}$$

$$\begin{cases} k_i = \frac{\mu v_i v_{i+1}}{J(v_i + v_{i+1})} \\ \beta_i = -\frac{J}{\rho v_i v_{i+1}} \end{cases} \tag{10}$$

4. The Evolution Laws of Deformation and Permeability of Broken Gangue

4.1. The Characteristics of Compression Deformation of Broken Gangue

The samples are given index n values of 0.1, 0.3, 0.5, 0.7, and 0.9 by continuous grading and index m values of 0.3, 0.5 and 0.7 by discontinuous grading with confined compression. The displacement loading method with a loading rate of 0.05 mm/s is applied to all the samples. The relations between stress and strain ($\sigma - \varepsilon$), as well as the relations between the compression modulus and compression ratio ($E_t - \gamma$), of continuous and discontinuous grading samples, are shown in Figure 4. From Figure 4, it can be concluded that the relationships of stress-strain and compression ratio-modulus can all be well fitted by exponential functions under the two different conditions.

As seen from Figure 4a,b, the stress-strain relationship of different continuous-grade samples is approximately the same. With the increase in axial strain, the stress increases exponentially, and the compression modulus is compatible with the increase in the compression ratio. This result means that the greater the compression rate is, the harder it is for the gangue to be compressed. Compared to the stress-strain relationship, it can be observed that, with the decrease in the Talbot exponent, the stress and compressive modulus increase quickly. The smaller the sample's Talbot index is, the more difficult it is to compress. As seen from Figure 4a, with the smaller Talbot index, smaller particles with higher content are filled closely, so the sample is more difficult to compress. Similarly, the relationships of stress-strain and compressive modulus-ratio of the discontinuous grading samples shown in Figure 4c,d are similar to those of the continuous-grade samples, with an increase in stress and strain according to the exponential function. The smaller the discontinuous grading index is, the faster the compression modulus increases with increasing compression rate, i.e., the samples are harder to compress. In addition, there is a great difference in the compression deformation of samples by continuous and discontinuous grading. Taking the samples with index of 0.5 by discontinuous and continuous grading as the examples, when the strain is 0.1, the stress of these samples are 3.35 and 2.76 MPa, respectively, and the compressive modulus are 1.12 and 0.96 MPa; when the strain is 0.25, the stress of the samples is 43.6 and 35.9 MPa, and the compression modulus are 11.8 and 9.6 MPa.

It is indicated that samples with discontinuous grading are more difficult to compress than those with continuous grading, and with an increasing compression rate, the discontinuous grading samples are denser, because the particles are also denser in arrangement.

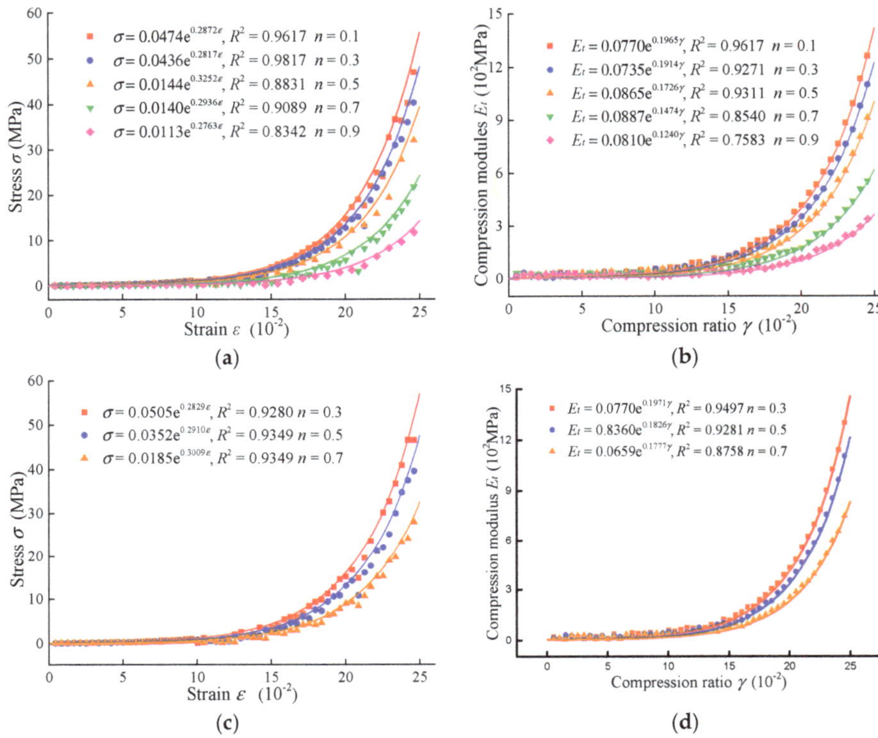

Figure 4. The compression properties of broken gangue: (**a**) relationship of stress-strain by continuous grading; (**b**) relationship of compression modulus-ratio by continuous grading; (**c**) relationship of stress-strain by discontinuous grading; and (**d**) the relationship of the compression modulus-ratio by discontinuous grading.

4.2. The Evolution Characteristics of Permeability of Grading Broken Gangue

To study the permeability of broken rock in the whole process of compression, the permeability parameters of grading broken gangue were tested. At the same time, the change law of exudate is recorded with the change in the compression rate. To reduce the influence of osmotic pressure on the compression process, the osmotic pressure was fixed at 2 MPa. The change law of the total flow rate and seepage velocity of exudate with the compression rate under different grading conditions are shown in Figure 5a,b. In addition, the permeability and non-Darcy flow β-factor variation trends with the compression rate are shown in Figure 5c,d, respectively (note: Figure 5b can be obtained by Equation (7), and Figure 5c,d can be obtained by Equation (10)).

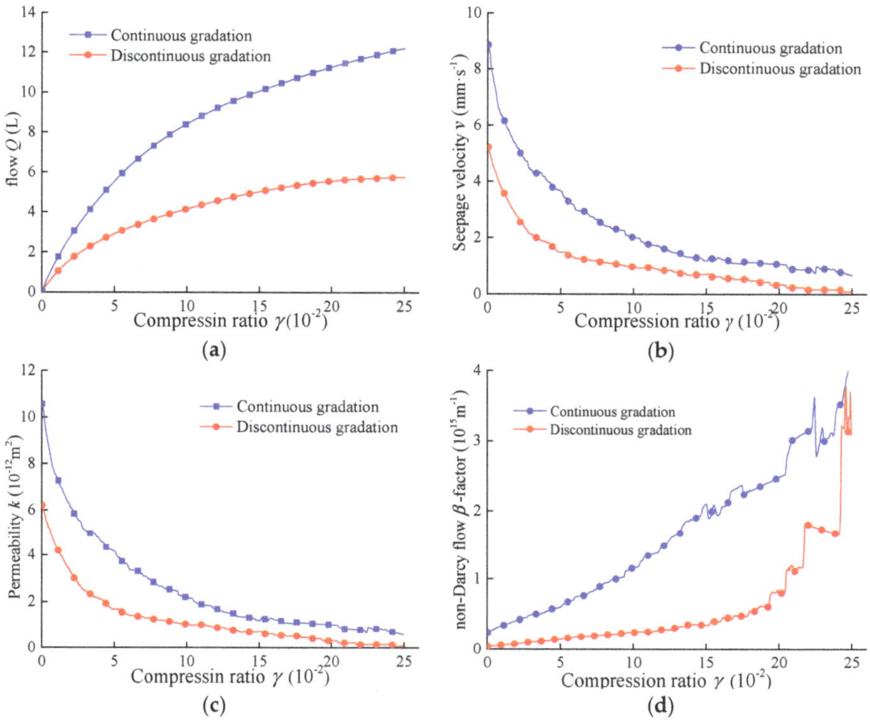

Figure 5. Evolution law of seepage characteristic of broken gangue under compression condition: (**a**) evolution law of cumulative flow; (**b**) evolution law of seepage velocity; (**c**) evolution law of permeability; and (**d**) evolution law of β-factor for non-Darcy flow.

As seen from Figure 5, the evolution law of permeability of broken gangue by discontinuous and continuous grading under axial loading conditions is basically the same. The cumulative flow and non-Darcy flow β-factor show a nonlinear increasing trend with increasing compression rate, but with an increasing rate contrary to the above situation. With increasing compression rate, the seepage velocity and permeability are characterized by decreasing nonlinearly. In addition, there are obvious differences in the evolution law of permeability for discontinuous and continuous grading. Compared with continuous-grade broken gangue, the increased rate of cumulative flow of the discontinuous-grade samples is smaller under the same compression rate, and the cumulative flow is also smaller. With an increasing compression rate, the seepage velocity and permeability of discontinuous and continuous grading of broken gangue gradually decrease, with a nonlinear decreasing trend for the decreasing rate; and the seepage velocity and permeability of the discontinuous grading sample are smaller at initial values, that the decrease rate is faster and that the final value is smaller. With the increasing compression rate, the non-Darcy flow β-factor shows a linear increasing trend by continuous grading, which is the opposite of the discontinuous grading samples. However, the final non-Darcy flow β-factor of these samples was similar.

From Figure 4, broken gangue was denser and less compressible with a high compression rate. Therefore, the variation range of cumulative flow, seepage velocity, and permeability all decrease with the increasing compression rate while, for a higher compression rate of broken gangue, the pore is smaller and fluid resistance is greater, which contributes heavily to nonlinear seepage; as a result, the non-Darcy flow β-factor is larger.

As seen from Figure 5, the smaller particle size of broken gangue under discontinuous grading occupies a higher proportion of space; the pores between large particles are easily filled; and mixed samples are denser, so the seepage velocity and permeability are small in the initial state. The pores are filled densely with small particles. Thus, the samples are more difficult to compress, and the decreasing speed of permeability becomes steadily smaller under a high compression rate.

4.3. *Influence of Stress on the Permeability of Grading Broken Rock*

The above study shows that compressibility and permeability of broken gangue depends on the compression state, while the compressive state of the broken gangue is related to axial stress. To study the relationship between axial stress and permeability of broken rock, the axial stress of 5 MPa is used to compress grading broken gangue first. Then, after fixing the load displacement and testing permeability parameter of broken gangue under the present conditions by a steady-state method, the permeability tests under 10, 15, 20, 25, and 30 MPa are completed. Finally, the permeability tests of broken gangue samples were completed in the same way with continuous and discontinuous grading.

In Figure 6, the permeability of broken gangue under different compressive stress is displayed. It can be seen from the graph that the permeability of broken gangue shows a negative exponential function, decreasing with the increase in axial compressive stress. The permeability decreases rapidly when the compressive stress is less than 15 MPa; when the compressive stress is more than 15 MPa, the permeability decreases gently. This result shows that axial compression has significance in reducing the permeability of broken gangue when the compressive stress is less than 15 MPa; however, when the compressive stress is greater than 15 MPa, the effect is not obvious.

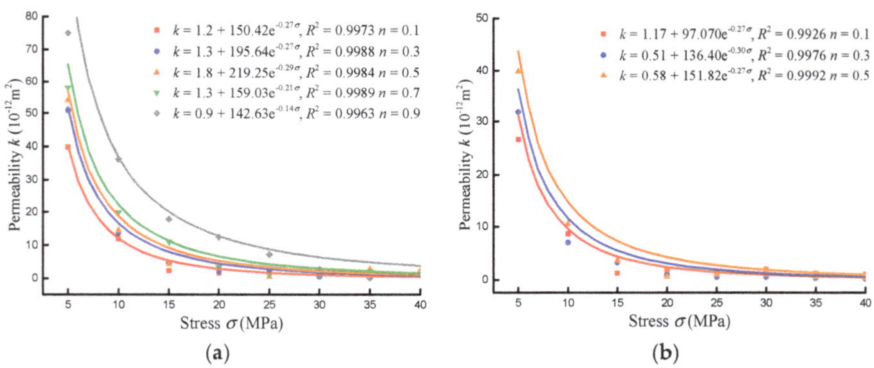

Figure 6. Permeability of broken gangue with different compressive stress: (**a**) continuous grading; and (**b**) discontinuous grading.

Second, under the same stress, the larger the grading index, the higher the permeability of broken gangue is, and the more significant the axial compression is to reduce the permeability of the sample. For example, for continuous grading samples, when the stress increases from 5 MPa to 40 MPa, the permeability of $n = 0.9$ decreases from 76 to 7 μm^2; for $n = 0.1$, the permeability decreases from 39 to 4 μm^2; and the change amplitudes is 69 and 35 μm^2, respectively.

In addition, the axial stress of discontinuous grading gangue must be smaller in order to achieve the same permeability level. For $n = 0.5$, axial stress of continuous grading should be set to 10 MPa, and that for discontinuous grading should be set at 8 MPa to make the permeability of samples reduce to 20 μm^2; in continuous grading, the data should be 15 MPa, while with discontinuous grading requiring 12 MPa to make the permeability of samples reduce to 10 μm^2.

It can be found that, for grading broken gangue, with a higher proportion of small particles, the particles are easier to adjust by sliding and rolling due to axial stress in order to achieve a tighter state. However, in the process of compression, samples with many large particle sizes must overcome the dislocation of large particles. Therefore, it is more difficult to compress them.

These conclusions are quite consistent with previous research by Harianto Rahardjo [34]. In the mixed gangue particle samples, the pores generated by accumulation of large particles are usually occupied by small particles, which result in the higher compressibility of discontinuous grading samples. Therefore, considering the permeability-mechanical response, appropriate backfill material properties need to be considered in the application of gangue filling mining project. After the coal has been mined, the backfill material plays the supporting role of the roof rock mass, which is a critical concern for controlling surface subsidence [12].

5. Conclusions

The broken rock mass is widely distributed in underground engineering, and it is very important to the stability of these engineering, especially, the deformation characteristic and permeability could be the pivotal factor in the migration of underground water. The deformation and permeability of broken gangue as a function of the stress level and grading features were studied through confined compression and pressed seepage tests. This research indicates that with the increase in axial strain, the stress of broken gangue increases with an exponential function, and the compression modulus is increased with increasing compression ratio, which indicates that the greater compression rate of broken gangue is, the harder it is to be compressed. Samples by discontinuous grading are harder to compress than those of continuous grading, and the former is denser with the increase in compression rate. Discontinuous grading has a larger number of small particle components and a denser grain arrangement based on both tests. Therefore, samples by discontinuous grading are more difficult to compress. With increasing compression rate, the change range of seepage velocity and permeability of broken gangue decreases, with a higher compression rate of broken gangue and smaller pores, the fluid has higher resistance. All these factors lead to stronger nonlinear characteristics of seepage, the larger the grading index of broken gangue is, the more remarkable effect of axial compression on reducing the permeability of the samples. To achieve the same permeability level, compared to that of continuous grading, the axial stress by the discontinuous grading sample is smaller. Finally, through the research and discussion in this paper, the mixed gangue particle samples with smaller grading index are more suitable for backfill material, and the performance of discontinuous grading samples is better in resistance of compression and penetration. Therefore, the broken gangue with discontinuous grading and small grading index is more suitable as filling material, meanwhile fault fractured zones and the collapse columns with the same characteristics are less prone to water inrush accidents.

Author Contributions: Y.Z. contributed to the experimental design and sourcing of the laboratory equipment and raw materials; W.Z. carried out the permeability analysis of grading broken gangue; M.L. contributed to the experimental operation, data collection, and writing of the manuscript; and Z.C. also contributed in the experimental design and correction of the manuscript.

Funding: This paper was support by the National Key Research and Development Plan (2016YFC0501103), National Natural Science Foundation of China (51574222, 51704281).

Acknowledgments: The authors gratefully acknowledge Qiang Li, who has given us many useful ideas and suggestions.

Conflicts of Interest: The authors declare no conflict of interest.

References

1. Manouchehrian, A.; Cai, M. Numerical modeling of rockburst near fault zones in deep tunnels. *Tunn. Undergr. Space Technol.* **2018**, *80*, 164–180. [CrossRef]
2. Yin, Q.; Jing, H.W.; Su, H.J.; Zhao, H.H. Experimental study on mechanical properties and anchorage performances of rock mass in the fault fracture zone. *Int. J. Geomech.* **2018**, *18*, 04018067. [CrossRef]
3. Zheng, Y.L.; Zhang, Q.B.; Zhao, J. Challenges and opportunities of using tunnel boring machines in mining. *Tunn. Undergr. Space Technol.* **2016**, *57*, 287–299. [CrossRef]
4. Sparks, R.S.J.; Wilson, L.; Hulme, G. Theoretical modeling of the generation, movement, and emplacement of pyroclastic flows by column collapse. *J. Geophys. Res. Solid Earth* **1978**, *83*, 1727–1739. [CrossRef]

5. Sparks, R.S.J. A model for the formation of ignimbrite by gravitational column collapse. *J. Geol. Soc. Lond.* **1976**, *132*, 441–451. [CrossRef]
6. Utili, S.; Zhao, T.; Houlsby, G.T. 3D DEM investigation of granular column collapse: Evaluation of debris motion and its destructive power. *Eng. Geol.* **2015**, *186*, 3–16. [CrossRef]
7. Huang, Y.; Li, J.; Song, T.; Kong, G.; Li, M. Analysis on Filling Ratio and Shield Supporting Pressure for Overburden Movement Control in Coal Mining with Compacted Backfilling. *Energies* **2016**, *10*, 31. [CrossRef]
8. Ma, C.Q.; Li, H.; Zhang, P. Subsidence prediction method of solid backfilling mining with different filling ratios under thick unconsolidated layers. *Arab. J. Geosci.* **2017**, *10*, 511. [CrossRef]
9. Zhang, L.T. Summary on the dam-break of tailing pond. *J. Hydraul. Eng. ASCE* **2013**, *44*, 594–600.
10. Wang, F.T.; Zhang, C.; Wei, S.F.; Zhang, X.G.; Guo, S.H. Whole section anchor–grouting reinforcement technology and its application in underground roadways with loose and fractured surrounding rock. *Tunn. Undergr. Space Technol.* **2016**, *51*, 133–143.
11. Wu, X.Z.; Jiang, Y.J.; Guan, Z.C.; Wang, G. Estimating the support effect of energy-absorbing rock bolts based on the mechanical work transfer ability. *Int. J. Rock Mech. Min. Sci.* **2018**, *103*, 168–178. [CrossRef]
12. Cao, W.H.; Wang, X.F.; Li, P.; Zhang, D.S.; Sun, C.D.; Qin, D.D. Wide Strip Backfill Mining for Surface Subsidence Control and Its Application in Critical Mining Conditions of a Coal Mine. *Sustainability* **2018**, *10*, 700. [CrossRef]
13. Li, J.; Huang, Y.; Qiao, M.; Chen, Z.; Song, T.; Kong, G.; Gao, H.; Guo, L. Effects of Water Soaked Height on the Deformation and Crushing Characteristics of Loose Gangue Backfill Material in Solid Backfill Coal Mining. *Processes* **2018**, *6*, 64. [CrossRef]
14. Cai, W.; Dou, L.M.; Li, Z.L.; He, J.; He, H.; Ding, Y.L. Mechanical Initiation and Propagation Mechanism of a Thrust Fault: A Case Study of the Yima Section of the Xiashi-Yima Thrust (North Side of the Eastern Qinling Orogen, China). *Rock Mech. Rock Eng.* **2015**, *48*, 1–19. [CrossRef]
15. Nakashima, S.; Kishida, K.; Adachi, T.; Izu, Y. Experimental Study on the Stability of Dam Foundation in Consideration of the Effect of the Concrete Plug Treatments. *Fish. Sci.* **2000**, *66*, 915–923.
16. Dong, H.; Hu, Z.R.; Fu, H.L.; Chen, C.; Chen, X.W. Analysis of deformation characteristics of eluvial gravel soil under mass density and confining pressure. *J. Cent. South Univ.* **2015**, *46*, 3879–3887. [CrossRef]
17. Yin, Z.Q.; Li, X.B.; Jin, J.F.; He, X.-Q.; Du, K. Failure characteristics of high stress rock induced by impact disturbance under confining pressure unloading. *Trans. Nonferrous Met. Soc. China* **2012**, *22*, 175–184. [CrossRef]
18. Scheingross, J.S.; Minchew, B.M.; Mackey, B.H.; Simons, M.; Lamb, M.P.; Hensley, S. Fault-zone controls on the spatial distribution of slow-moving landslides. *Geol. Soc. Am. Bull.* **2013**, *125*, 473–489. [CrossRef]
19. Li, H.; Bai, H.B.; Wu, J.J.; Ma, Z.; Ma, K.; Wu, G.; Du, Y.; He, S. A Cascade Disaster Caused by Geological and Coupled Hydro-Mechanical Factors—Water Inrush Mechanism from Karst Collapse Column under Confining Pressure. *Energies* **2017**, *10*, 1938. [CrossRef]
20. Yin, Q.; Jing, H.W.; Ma, G.W.; Su, H.; Liu, R. Investigating the roles of included angle and loading condition on the critical hydraulic gradient of real rock fracture networks. *Rock Mech. Rock Eng.* **2018**, *51*, 3167–3177. [CrossRef]
21. Mudd, G.M. Critical review of acid in situ leach uranium mining: 1. USA and Australia. *Environ. Geol.* **2001**, *41*, 390–403. [CrossRef]
22. Chai, J.R. Analysis of nonlinear seepage through fracture network in rock mass. *J. Hydrodyn.* **2002**, *17*, 217–221.
23. Min, K.B.; Rutqvist, J.; Tsang, C.F.; Jing, L. Stress-dependent permeability of fractured rock masses: A numerical study. *Int. J. Rock Mech. Min. Sci.* **2004**, *41*, 1191–1210. [CrossRef]
24. Noiriel, C.; Gouze, P.; Bernard, D. Investigation of porosity and permeability effects from microstructure changes during limestone dissolution. *Geophys. Res. Lett.* **2004**, *31*, 1183–1186. [CrossRef]
25. Liu, M.; Shabaninejad, M.; Mostaghimi, P. Predictions of permeability, surface area and average dissolution rate during reactive transport in multi-mineral rocks. *J. Pet. Sci. Eng.* **2018**, *170*, 130–138. [CrossRef]
26. Moosavi, S.A.; Goshtasbi, K.; Kazemzadeh, E.; Bakhtiari, H.A.; Esfahani, M.R.; Vali, J. Relationship between porosity and permeability with stress using pore volume compressibility characteristic of reservoir rocks. *Arab. J. Geosci.* **2014**, *7*, 231–239. [CrossRef]
27. Tang, Z.; Dong, X.; Yang, Y.; Ma, L. Research on the relationship between grain composition and repose angle of coal gangue in Dongkuang mine, Heshan city, Guangxi, China. *J. Earth Sci.* **2014**, *25*, 309–314. [CrossRef]

28. Hunter, R.P.; Bowman, E.T. Visualisation of seepage induced suffusion and suffosion within internally erodible granular media. *Geotechnique* **2017**, *68*, 918–930. [CrossRef]
29. Bush, R.T. A theory of particle interference based upon the uncertainty principle. *Lettere al Nuovo Cimento* **1983**, *36*, 241–244. [CrossRef]
30. Miao, X.X.; Li, S.C.; Liu, W.Q. Experimental Study of Seepage Properties of Broken Sandstone under Different Porosities. *Transp. Porous Media* **2011**, *86*, 805–814. [CrossRef]
31. Zhou, J.Q.; Hu, S.H.; Fang, S.; Chen, Y.F.; Zhou, C.B. Nonlinear flow behavior at low Reynolds numbers through rough-walled fractures subjected to normal compressive loading. *Int. J. Rock Mech. Min. Sci.* **2015**, *80*, 202–218. [CrossRef]
32. Meraj, M.A.; Shehzad, S.A.; Hayat, T.; Abbasi, F.M.; Alsaedi, A. Darcy-Forchheimer flow of variable conductivity Jeffrey liquid with Cattaneo-Christov heat flux theory. *Appl. Math. Mech.* **2017**, *38*, 557–566. [CrossRef]
33. Yin, Q.; Ma, G.W.; Jing, H.W.; Wang, H.; Su, H.; Wang, Y.; Liu, R. Hydraulic properties of 3D rough-walled fractures during shearing: An experimental study. *J. Hydrol.* **2017**, *555*, 169–184. [CrossRef]
34. Rahardjo, H.; Satyanaga, A.; D'Amore, G.A.R.; Leong, E.-C. Soil–water characteristic curves of gap-graded soils. *Eng. Geol.* **2012**, *125*, 102–107. [CrossRef]

processes

MDPI

Article

Effects of Cyclic Wetting-Drying Conditions on Elastic Modulus and Compressive Strength of Sandstone and Mudstone

Shiyuan Huang [1,2,*], Junjie Wang [1,2,*], Zhenfeng Qiu [1,2] and Kai Kang [3]

1 Key Laboratory of Hydraulic and Waterway Engineering of the Ministry of Education, Chongqing Jiaotong University, Chongqing 400074, China; qiuzhenfeng3012@126.com
2 National Engineering Research Center for Inland Waterway Regulation, Chongqing Jiaotong University, Chongqing 400074, China
3 Department of Engineering and Ecological Geology, Faculty of Geology, Lomonosov Moscow State University, Moscow 119991, Russia; kevinkang8@mail.ru
* Correspondence: cqjtdxhsy@163.com (S.H.); wangjunjiehhu@163.com (J.W.); Tel.: +86-155-2320-5027 (S.H.); +86-189-9648-0863 (J.W.)

Received: 29 October 2018; Accepted: 16 November 2018; Published: 22 November 2018

Abstract: The influence of water on the mechanical properties of rock is vital for determining the rock stability when subjected to changes of water conditions. In this paper, a series of uniaxial compression tests were conducted to investigate effects of cyclic wetting and drying on the mechanical properties of sandstone and mudstone collected from Chongqing city, China. The results showed that both elastic modulus and uniaxial compressive strength of sandstone and mudstone were reduced by wetting and drying cycles, and that the degradation rate of the two mechanic parameters of mudstone was always larger than sandstone. The parameters, including water adsorption, degradation degree of elastic modulus, degradation degree of uniaxial compressive strength, increase with the increase of the wetting-drying cycles (N). The relationship between these three parameters and the value of $N + 1$ could be well fitted by logarithmic curves. The average degradation degree was also used to describe the degradation of per time wetting-drying cycles. It is found that the average degradation degree of elastic modulus and uniaxial compressive strength decrease with the increase of wetting-drying cycles. Moreover, the relationships between the mechanical properties and the porosity are presented, which can be fitted by linear curves. In the cyclic wetting-drying process, the elastic modulus and the uniaxial compressive strength decreased with the porosity increasing, and the degradation rates of sandstone mechanic parameters were higher than those of mudstone.

Keywords: cyclic wetting-drying; deterioration; sandstone; mudstone; elastic modulus; uniaxial compressive strength

1. Introduction

Sandstone and mudstone are two common sedimentary rocks. It is well known that sandstone has been the most common construction material through human history all over the world due to its high strength, while mudstone is characterized by low permeability and high hydrophily, as well as easy softening and decay [1]. In Chongqing city of China, the strata composed of interbedded red sandstone and mudstone are widely distributed. Because they are difficult to divide [2], the mixture of sandstone and mudstone particles is often used as a main fill material in many hydraulic and civil engineering works in Chongqing city [3]. While this material is filled in or along a large reservoir bank, it could be subjected to cyclic wetting and drying caused by cyclic lowering and rising of water level, which may affect the mechanical behavior of the mixture, and then reduces stability of filled structures [4–6]. The mechanical properties of the sandstone and mudstone particles mixture mainly depend on its parent rocks (sandstone and mudstone). In order to use the mixture of sandstone and mudstone particles reasonably in the civil engineering, the effect of cyclic wetting and drying on the mechanic properties of the parent rocks need to be studied first.

In recent years, the water-interaction has been viewed as a crucial factor which influences the stability of geotechnical engineering constructions and it's becoming an important research hotspot. The effect of cyclic wetting and drying on the physical and mechanical properties of geotechnical materials such as rock and soil have been studied by many researchers, including bulk density, porosity, weight loss [7], water absorption [7], wave velocity [8], elastic modulus [9–11], fracture toughness [12], tensile strength [13,14], static uniaxial or triaxial compressive strength [15–17], dynamic uniaxial compressive strength [18], shear strength [9,10,19], granulometric change [20–23], etc. The results indicated that the degradation with different degrees of the physical and mechanic properties of rock materials occurred after cyclic wetting and drying.

Table 1 shows the recent experimental research by different scholars of the effects of cyclic wetting-drying on the mechanical parameters of various rocks. The mechanical parameters deteriorate to different degrees under the cyclic wetting-drying conditions. Hua et al. [12] found an obvious reduction in static tensile strength after cyclic wetting-drying treatment. Zhou et al. [18] found an obvious decrease in dynamic tensile strength of sandstone under cyclic wetting-drying treatment. Yao et al. and Deng et al. [9–11] found an evident reduction in the elastic modulus (*EM*) and the uniaxial compressive strength (*UCS*) of sandstone after cyclic wetting-drying treatment. However, some scholars [15,17] did not find an obvious decrease in the mechanical parameters of different sandstones due to cyclic wetting-drying treatment, but they indicated that the possible small strength reduction could be covered up by the inter-sample variability inherent in the strength measurements. Moreover, Lin et al. [13] observed that the *UCS* of sandstone with 11% chlorite reduced by 20% after 60 wetting-drying cycles, while for sandstone with 1% chlorite, the reduction was not obvious. Their study found that chlorite is dissolved and leached out more easily and that the porosity of sandstone increases, which leads to a strength reduction, with the chlorite content of the sandstone as the critical factor.

Table 1. Effects of cyclic wetting-drying on the mechanical properties of various sandstone by different researchers.

Researches	Specimen Type	Specimen Size (mm)	Wetting and Drying Methods	N_{max}	Test Method and Testing Condition	Mechanical Properties
Hale et al. (2003) [15]	Six kinds of sandstones from northeastern Ohio and western Pennsylvania	Φ54 × 108	Wetting method: submerged in water for 24 h Drying method: oven-dried at 110 °C for 24 h	50	Uniaxial compression tests under dry condition	*UCS*: no obvious decreasing trend
Lin et al. (2005) [13]	Tertiary sandstones from northern Taiwan	Φ55 × 125	Wetting method: submerged in water for 24 h Drying method: not specified	60	Uniaxial compression tests under dry condition	*UCS* (sandstone with 11% chlorite): decrease obviously, *UCS* (sandstone with 1% chlorite): no obvious decreasing trend
Yao et al. (2010) [9]	Red sandstone from China	Φ50 × 100	Wetting method: forced saturation for 8 h, then immersed in water for 24 h Drying method: oven-dried at 105 °C for 24 h, then cooled to room temperature	8	Uniaxial compression tests under wet condition	*UCS*: decrease obviously, *EM*: decrease obviously
Deng et al. (2012) [10,11]	Sandstone from Three Gorges Reservoir Region of China	Φ50 × 100	Wetting method: submerged in water from Yangtze river for 30 days Drying method: air-dried at room temperature for 5 days	6	Uniaxial compression tests under wet condition	*UCS*: decrease obviously, *EM*: decrease obviously

Table 1. *Cont.*

Researches	Specimen Type	Specimen Size (mm)	Wetting and Drying Methods	N_{max}	Test Method and Testing Condition	Mechanical Properties
Hua et al. (2015) [12]	Sandstone from Ziyang of China	Φ75 × 25	Wetting method: submerged in water for 48 h Drying method: oven-dried at 105 °C for 24 h	7	Cracked Brazilian disk splitting tests under wet condition	*TS*: decrease obviously, *FT*: decrease obviously,
Khanlari et al. (2015) [17]	Red sandstones from central Iran	Φ54 × 108	Wetting method: submerged in water for 24 h Drying method: oven-dried at 110 °C for 24 h	40	Uniaxial compression tests under dry condition	*UCS*: no obvious decreasing trend
Zhao et al. (2016) [14]	Sandstone with a low clay mineral content from Chongqing of China	Φ50 × 25	Wetting method: submerged in water for 18 h Drying method: air-dried at room temperature for 72 h	15	Brazilian tensile tests under dry condition	*TS*: no obvious decreasing trend
Zhou et al. (2017) [18]	Fine-grained sandstone from Kunming of China	Φ50 × 25	Wetting method: submerged purified water for 48 h at 25 °C Drying method: air-dried at 25 °C for more than 100 h	50	Dynamic compression tests under dry condition	*DEM*: decrease obviously *DUCS*: decrease obviously

UCS represents uniaxial compressive strength, *EM* represents elastic modulus, *TS* represents tensile strength, *FT* represents fracture toughness, *DUCS* represents dynamic uniaxial compressive strength, *DEM* represents dynamic tensile strength.

These differences mentioned above indicate that sandstone mineralogy may decide its sensitivity to cyclic wetting-drying treatments. However, the investigations have been mainly concentrated on sandstone, and the report on the effect of cyclic wetting and drying on the mechanic properties of mudstone is scarce. In this present study, the effect of cyclic wetting-drying conditions on the physical and mechanic properties of sandstone and mudstone were investigated by uniaxial compression tests.

2. Test Material and Specimen Preparation

The test materials in this study came from Chongqing Jiaotong University, Chongqing city, China. The selected sandstone and mudstone were formed in Mesozoic, and the colors were fresh gray and aubergine, respectively. According to ISRM [24], specimen processing was conducted as follows: Initially, the sandstone and mudstone cylinders were obtained from the rock mass by drilling cores; later, the cylinders were made into pieces of size 50 mm in diameter and 100 mm in thickness (Figure 1). The initial water content (w), natural density (ρ_n) and dry density (ρ_d) of the sandstone specimens are 1.41%, 2.46 g/cm^3, and 2.43 g/cm^3, respectively. The w, ρ_n and ρ_d of the mudstone specimens are 1.75%, 2.42 g/cm^3, and 2.37 g/cm^3, respectively. Before testing, the specimens were stored in dry conditions with a temperature of 20 °C.

Figure 1. Typical sandstone and mudstone specimens.

3. Testing Scheme and Testing Procedure

Wetting-drying cycle process, as an important factor affecting the mechanical properties of rocks, is focused on in this study. A uniaxial compression test in the laboratory was carried out, it is used to observe the effects of cyclic wetting-drying on rocks, which is called as wetting-drying uniaxial compression test (WDUCT).

In the WDUCT, the selected specimens were submerged in water for 24 h firstly, then these specimens were then dried in an electro-thermostatic blast oven at 110 °C for 12 h. Later, they were cooled to temperature of 20 °C for 12 h. The whole process is regarded as a single wetting-drying cycle (see Figure 2).

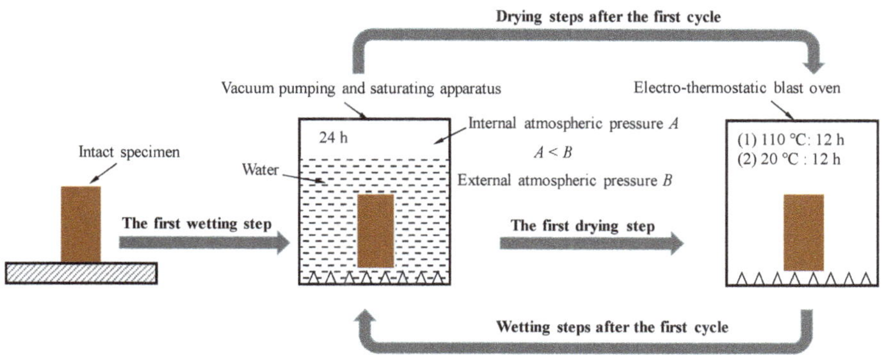

Figure 2. Schematic experimental process of the cyclic wetting-drying treatment.

Before the test, both sandstone and mudstone specimens were subjected to 1, 5, 10, 15, 20 and 40 wetting and drying cycles, and 3 specimens were subjected to each of the above treatments. Meanwhile, 3 untreated sandstone specimens and 3 untreated mudstone specimens were also tested to determine the initial mechanical properties of the specimens without wetting-drying cycles.

In addition, because of the material composition, mudstone is sensitive to moisture [25,26]. When a mudstone specimen immerged into water, it will be swelled, and as time goes on, it will be disintegrated. In the process of the wetting and drying, mudstone specimens will be disturbanced unavoidably, which will fail to complete a whole uniaxial compression test. Therefore, a protection device for mudstone was designed (shown in Figure 3), it was made up of a fixation frame, two porous stones and a rubber spring. This device could ensure the specimen was completely immersed in water without particles flaking, and the garter spring will not limit the deformation of the specimen in the wetting and drying process.

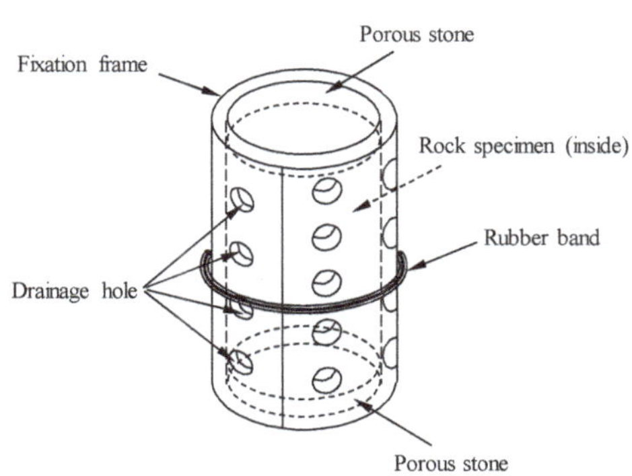

Figure 3. A protection device for mudstone specimen.

However, as the tests went on, it was found that after 13 wetting and drying cycles, the selected mudstone specimens were crumbed seriously which can not be tested further, so the values of wetting and drying cycles (N) of mudstone specimens were corrected as 1, 5, 10, 12. The testing schemes are summarized in Table 2. In total, 21 sandstone specimens and 15 mudstone specimens were tested in wetting and drying cycles. Then, uniaxial compressive tests were performed under drying condition for the wetting-drying specimens. These tests were conducted in the Fundamental Mechanical Laboratory of Chongqing Jiaotong University, where the RMT-150C rock mechanic uniaxial compression testing system(Institute of Rock and Soil Mechanics, Chinses Academy of Sciences, Wuhan, China.) was used for loading (Figure 4).

Table 2. Testing scheme.

Test Material	Number of Wetting-Drying Cycles (N)
Sandstone	1, 5, 10, 15, 20, 40
Mudstone	1, 5, 10, 12

Figure 4. RMT-150C Rock mechanics testing system.

4. Results and Analyses

4.1. *Effects of Cyclic Wetting-Drying on Water Adsorption*

Before uniaxial compressive tests, the water absorption of each specimen could be calculated according to the weight variation at different wetting-drying cycles:

$$w = \frac{m_w - m_d}{m_d} \times 100\% \tag{1}$$

where w (%) is the water adsorption of the specimen, m_w is the mass of the saturated specimen, m_d is the mass of the dried specimen, respectively.

Through Equation (1), the water adsorption of sandstone and mudstone could be obtained. The water adsorption of sandstone was respectively 2.21%, 2.75%, 2.63%, 3.02%, 3.02%, 3.32% and 3.37%, corresponding to the value of N is 0, 1, 5, 10, 15, 20 and 40. And the water adsorption of sandstone was respectively 1.98%, 2.63%, 3.39%, 3.67% and 3.87%, corresponding to the value of N is 0, 1, 5, 10 and 12. With the value of N increasing, the water adsorption of sandstone and mudstone increased gradually.

Figure 5 shows the effects of the cyclic wetting and drying on the water absorption of sandstone and mudstone. The water adsorption increased along a logarithmic curve with the increment of the value of $N + 1$. The variation of water adsorption with the increment of the value of $N + 1$ could be fitted by a logarithmic curve as follows:

$$\text{Sandstone } w_s(N) = 0.287\ln(N+1) + 2.312 \; R^2 = 0.87 \tag{2}$$

$$\text{Mudstone } w_m(N) = 0.706\ln(N+1) + 2.057 \; R^2 = 0.99 \tag{3}$$

where $w_s(N)$, $w_m(N)$ is the water adsorption of sandstone and mudstone after N times wetting and drying, %; N is the number of cyclic wetting and drying, and the range of N for sandstone is from 0 to 40, for mudstone is from 0 to 12; R^2 is a coefficient of determination, which indicates the mudstone is more sensitive to the water.

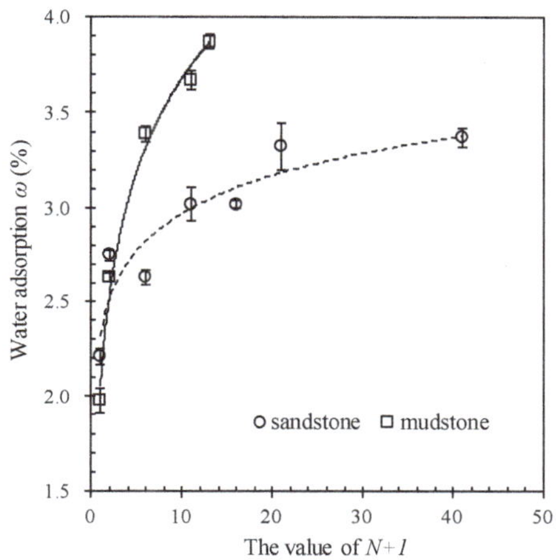

Figure 5. Relationship between the water adsorption and the value of $N + 1$.

Meanwhile, the porosity of the specimens after different wetting-drying cycles were measured by water saturation method [27]. According to the weight and volume of the specimens which were measured in the wetting-drying process, the porosity was calculated from the equation as follows:

$$n = \frac{V_v}{V} \times 100\% = \frac{m_w - m_d}{m_w - m_{sub}} \times 100\% \tag{4}$$

where n is the porosity of the rock specimen, V is the volume of the rock specimen, V_v is the volume of pores, m_w is the mass of the saturated specimen, m_d is the mass of the dried specimen and m_{sub} is the mass of the saturated specimen submerged in water.

Figure 6 shows the relationship between porosity and water absorption of these specimens after different wetting-drying cycles, which indicates that the water absorption and the porosity have a good linear relationship (Equations (5) and (6)) in both sandstone and mudstone specimens.

$$\text{Sandstone } w_s = 0.4034n - 0.0012 \; R^2 = 0.99 \tag{5}$$

$$\text{Mudstone } w_m = 0.6052n - 2.0041 \; R^2 = 0.93 \tag{6}$$

where w is the water adsorption of the rock specimen, %; n is the porosity of the rock specimen.

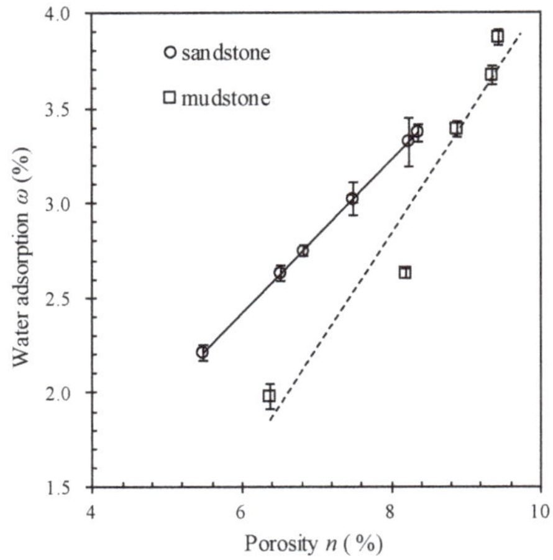

Figure 6. Relationship between the water adsorption and porosity of specimens.

As is known, the water storage capacity depends on the porosity of the rock specimens, and water absorption is the external manifestation of the size and distribution of cracks in the rock. The increase of water absorption indicates that the cracks in the rock samples increase and extend.

During the wetting and drying processes, both the water adsorption and porosity of these specimens increased, in which that of mudstone changed greater. FU [28] investigated the relationship between the water adsorption of moderate weathered sandstone specimens and the value of $N + 1$, shown in Equation (7).

$$w_s(N) = 0.3226\ln(N + 1) + 3.2344 \; R^2 = 0.99 \quad (7)$$

It is found that the change rate of water adsorption of moderate weathered sandstone is lower than the slightly weathered sandstone in this work, which indicated that the water adsorption and the porosity were related to the weathering degree of rock materials.

4.2. Stress-Strain Curves

The deformation and failure process of a rock generally undergoes five stages: (1) The micro-cracks compaction stage: The rock is compacted in the initial loading and the stress-strain curve is supposed to be concave, this can be due to the micro-cracks inside the rock closing under the external force; (2) The elastic deformation stage: The stress-strain curve shows an approximate linear relationship up to the yield strength; (3) The yield stage: The micro crack developed and accumulated until the rock specimen is completely failed, while the strength reaches the peak strength. (4) The strain softening stage: after the rock reaches the peak strength, the stress decreases with the strain increasing rapidly, the macroscopic fracture surface is formed with the development of the cracks. (5) The plastic flow stage: After the rock fractures, it still has bearing capacity, which is named residual strength.

The stress-strain curves of the selected specimens after cyclic wetting and drying from WDUCT are shown in Figures 7 and 8. The letter N represents the number of cyclic wetting and drying.

Due to the limitation of the test apparatus, these curves here did not always follow the above 5 stages, and it was difficult to obtain a complete plastic flow stage. For the sandstone specimens, with increment of N value, the micro-cracks compaction stage (stage 1) becomes wider and the slope of the stress-strain curve in the elastic deformation stage (stage 2) was reduced significantly. This is mainly because of the internal micro-crack growth and expansion during wetting-drying cycles. When the number of wetting and drying cycles reaches 15, it's clear that the plastic property of sandstone was obviously enhanced, the yield platform was obvious and the yield stage (stage 3) became longer. For the mudstone specimens, the effect of wetting-drying cycles on the stress-strain curves was not obvious. However, it could be found that the peak strength of both sandstone and mudstone after cyclic wetting and drying was less than that of intact specimens, the peak strength of both sandstone and mudstone decreased with the increase of the value of N.

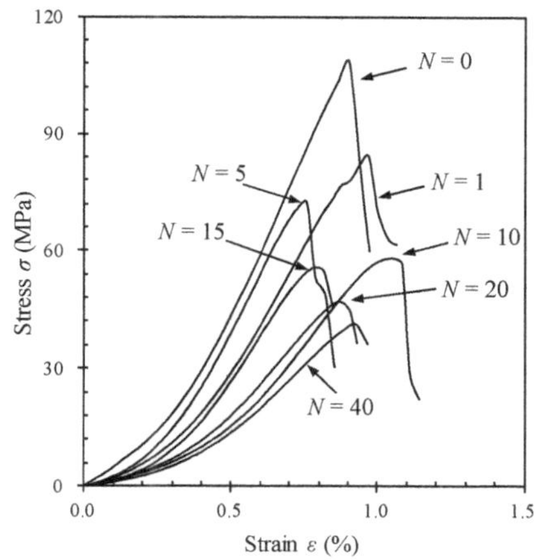

Figure 7. Stress-strain curves of sandstone specimens after wetting-drying cycles.

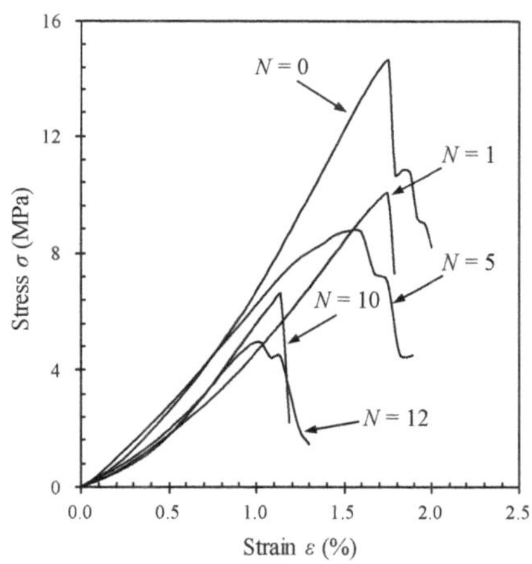

Figure 8. Stress-strain curves of mudstone specimens after wetting-drying cycles.

Based on the results presented in Figure 9, the fracture patterns of these specimens showed obvious brittle fractures for specimens under less wetting-drying cycles, the fracture surfaces distributed vertically. However, with the increase of N value, the sandstone specimens and mudstone specimens became more softened respectively, the angles of the fracture surfaces changed, with obvious sloping fractures, showing shear failure mode. Thus, the cyclic wetting and drying process had a significant influence on the failure features of sandstone that change from brittle pattern to ductile pattern.

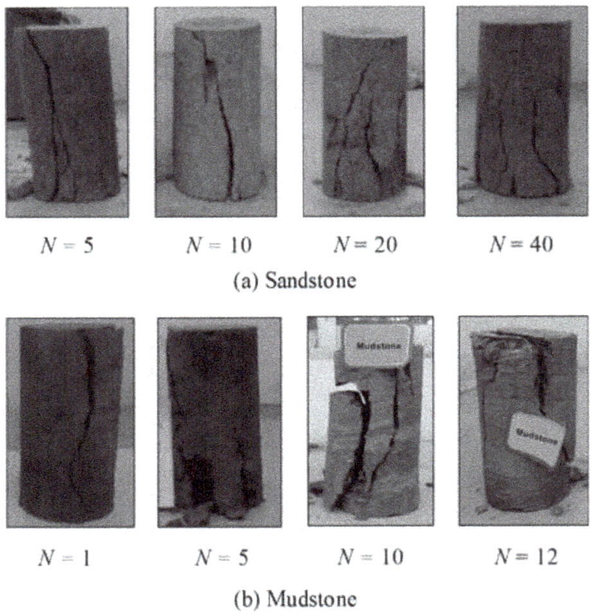

Figure 9. Failure forms of the specimens.

4.3. Effects of Cyclic Wetting-Drying on Uniaxial Compressive Strength

From Figures 7 and 8, it is found that the *UCS* decreases with the increase of the value of *N*, especially for the first cycle.

The damage of any material due to cyclic wetting and drying could be expressed as a value of "Degradation Degree (*D*)", *D* indicates the change in the material strength. The mechanical properties based on *D* have been used in the past to measure the changes in rock strength due to cyclic wetting and drying conditions, freeze-thaw conditions, and thermal treatment [29]. Here this method was used, in which Degradation Degree was calculated as:

$$D_{UCS} = (1 - UCS_N/UCS_0) \times 100\% \tag{8}$$

where D_{UCS} is the total degradation degree of *UCS*, UCS_N is the compressive strength of specimens after *N* wetting and drying cycles, and UCS_0 is the uniaxial compressive strength without wetting and drying. According to the test results, the *UCS* decreased with the increase in the value of *N*. When the value of *N* was respectively 1, 5, 10, 15, 20 and 40, the *UCS* of sandstone was reduced by 50.00%, 23.75%, 34.53%, 43.88%, 48.43% and 56.45%, 60.46%. When the value of *N* was respectively 1, 5, 10 and 12, the *UCS* of mudstone was reduced by 30.63%, 36.87%, 52.55% and 61.29%. At the beginning of testing, the *UCS* degenerated significantly, with the increase in the number of wetting-drying cycles, the water-rock interaction weakened gradually, and the decrease rates of *UCS* of sandstone and mudstone tended to slow down.

Figure 10 shows the degradation degree of *UCS* of sandstone and mudstone calculated by Equation (7). The evolution of the D_{UCS} indicated that sandstone and mudstone had undergone continuous but different amount of degradation after cyclic wetting and drying. The D_{UCS} of mudstone is larger than that of sandstone. For example, when the value of *N* reached 10, the D_{UCS} of sandstone and mudstone were 43.88%, 52.55%, respectively. In addition, the relationship between the D_{UCS} of sandstone and mudstone and the value of *N* + 1 could be fitted by a logarithmic curve as follows:

$$\text{Sandstone } D_{UCS} = 15.548 \ln(N + 1) + 6.198 \; R^2 = 0.96 \tag{9}$$

$$\text{Mudstone } D_{UCS} = 20.381 \ln(N + 1) + 5.909 \; R^2 = 0.91 \tag{10}$$

The differential coefficients of the above equations can describe the degradation rates of the *UCS*, which can be expressed as follows:

$$D' = r/(N + 1) \tag{11}$$

where the coefficient r represents the degradation rate. It was clear that the degradation rate of mudstone (r = 20.381) was higher than that of sandstone (r = 15.548).

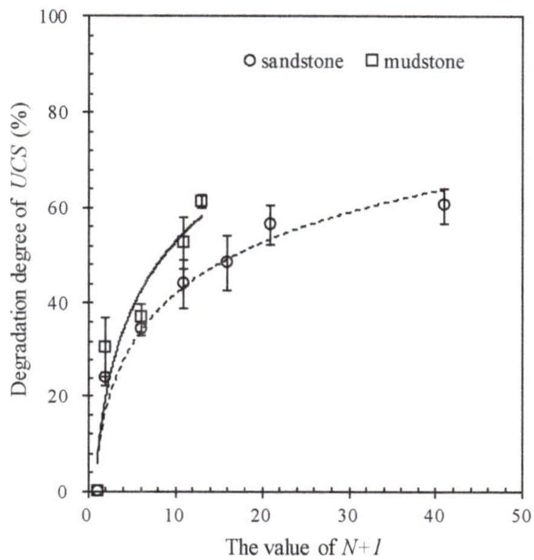

Figure 10. Relationship between the degradation degree of the *UCS* and the value of N + 1.

4.4. Effects of Cyclic Wetting-Drying on the Elastic Modulus

From Figures 7 and 8, it can be seen that with the value of N increasing, the *EM* (elastic modulus) decreased gradually. At the first wetting and drying cycle, the *EM* of sandstone and mudstone degenerated rapidly. With the increase in the value of N, the rate of *EM* decrease tended to slow down. The degradation degree of *EM* is defined by the following formula:

$$D_{EM} = (1 - EM_N/EM_0) \times 100\% \tag{12}$$

where D_{EM} is the total degradation degree of *EM*, EM_N is the elastic modulus after N wetting and drying cycles, and EM_0 is the elastic modulus of the intact specimen without wetting and drying action.

According to the test results, the *EM* decreased with the increase of the value of N. When the value of N was respectively 1, 5, 10, 15, 20 and 40, the *EM* of sandstone was reduced by 14.55%, 15.68%, 45.92%, 35.38%, 48.92% and 55.22%. When the value of N was respectively 1, 5, 10 and 12, the *EM* of mudstone was reduced by 44.74%, 44.74%, 51.13% and 51.13%.

Figure 11 shows the degradation degree of *EM* of sandstone and mudstone calculated by Equation (13). The change of the D_{EM} indicated that sandstone and mudstone had undergone continuous but different amount of degradation after cyclic wetting and drying. The relationship of the D_{UCS} of sandstone and mudstone and "N + 1" could be fitted by a logarithmic curve as follows:

$$\text{Sandstone } D_{EM} = 14.872 \ln(N + 1) + 6.198 \; R^2 = 0.89 \tag{13}$$

$$\text{Mudstone } D_{EM} = 16.102 \ln(N + 1) + 14.362 \; R^2 = 0.68 \tag{14}$$

It was clear that the degradation rate of mudstone (r = 16.102) was larger than that of sandstone (r = 14.548). Obviously, the effects of cyclic wetting and drying on the *UCS* and *EM* of mudstone were more intense than that of sandstone. Gautam et al. [30] indicated that rocks with higher clay contents slake more rapidly and extensively than those with lower clay contents. As is known, mudstone is a fine-grained sedimentary rock, and its original constituents were clays or muds, while most sandstone is composed of quartz or feldspar.

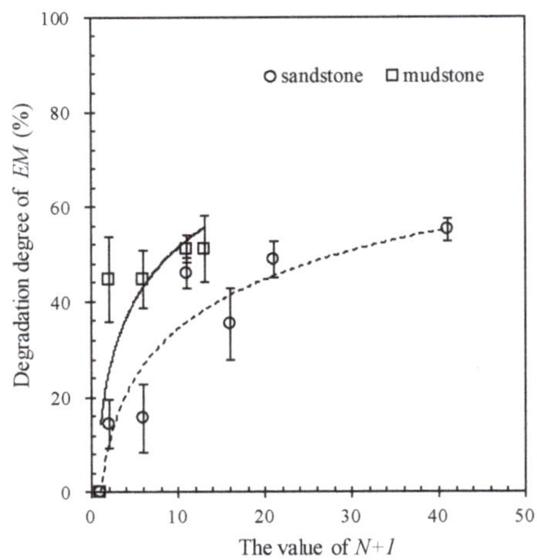

Figure 11. Relationship between the degradation degree of the *EM* and the value of $N + 1$.

4.5. Average Degradation Degree

From Figures 10 and 11, the average degradation degree (*AD*), which represents the degradation degree of per wetting and drying cycle, can be easily calculated, shown in Figures 12 and 13. While N value increased from 1 to 40, the D_{UCS} value of sandstone reduced from 23.75% to 1.51%, the D_{EM} value of sandstone reduced from 14.55% to 1.38%. While the N value increased from 1 to 12, the D_{UCS} value of mudstone reduced from 30.63% to 5.11%, the D_{EM} value of mudstone reduced from 44.74% to 4.26%. The relationships between the average degradation degree (degree/time) of the two properties and the N value can be expressed by a power function as shown in Equation (15) to Equation (18).

$$\text{Sandstone } AD_{UCS} = 23.512\ N - 0.732\ R^2 = 0.99 \quad (15)$$

$$\text{Sandstone } AD_{EM} = 12.779\ N - 0.591\ R^2 = 0.89 \quad (16)$$

$$\text{Mudstone } AD_{UCS} = 43.778\ N - 0.945\ R^2 = 0.99 \quad (17)$$

$$\text{Mudstone } AD_{EM} = 28.833\ N - 0.740\ R^2 = 0.98 \quad (18)$$

For the mudstone specimens, the maximum value of N is 12 in WDUCT, which is much less than that of sandstone specimens. In order to make an overall comparison of AD_{UCS} and AD_{EM} in sandstone and mudstone, several curves are plotted by Equation (15) to Equation (18), which are shown in Figures 14 and 15. From these figures, the range of N value is the same, and it is clear that AD_{UCS} and AD_{EM} of sandstone and mudstone all decrease with the increment of N value. As shown in Figure 14, it can be seen that at the beginning of wetting and drying cycles, the difference between the AD_{UCS} value of sandstone and the AD_{UCS} value of mudstone was large, which showed the AD_{UCS} value of mudstone was larger than the AD_{UCS} value of mudstone, with the value of N increasing, the two values tended to converge. While from Figure 15, the difference between the AD_{EM} value of sandstone and AD_{EM} value of mudstone is basically the same with the value of N increasing, which shows that the AD_{EM} value of sandstone is always larger than that of mudstone.

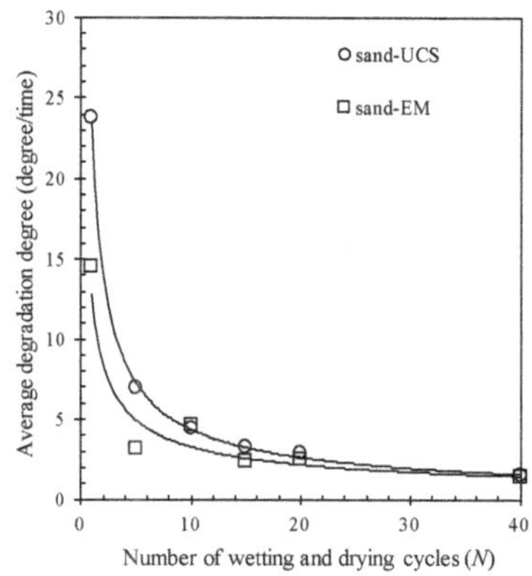

Figure 12. Average degradation degree of *UCS* and *EM* of sandstone after wetting-drying cycles (*UCS* represents uniaxial compressive strength, *EM* represents elastic modulus).

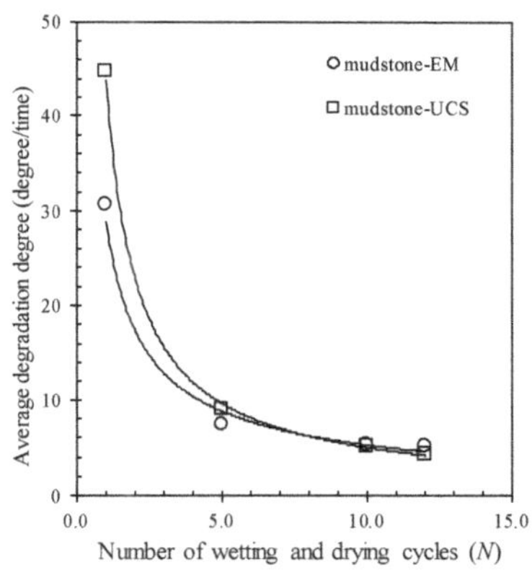

Figure 13. Average degradation degree of *UCS* and *EM* of mudstone after wetting-drying cycles.

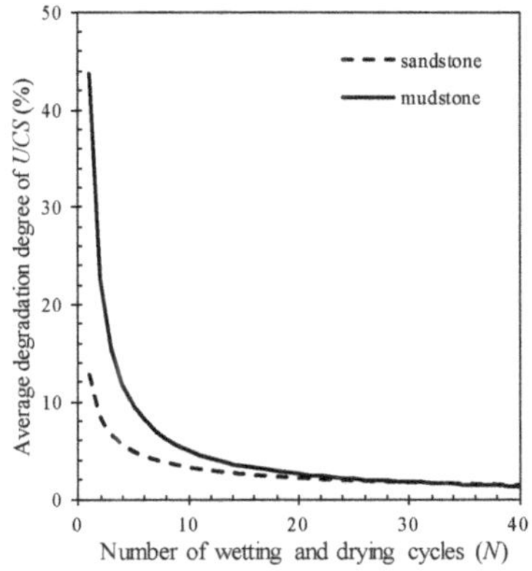

Figure 14. Fitted curves of average degradation degree of *UCS* and wetting and drying cycles.

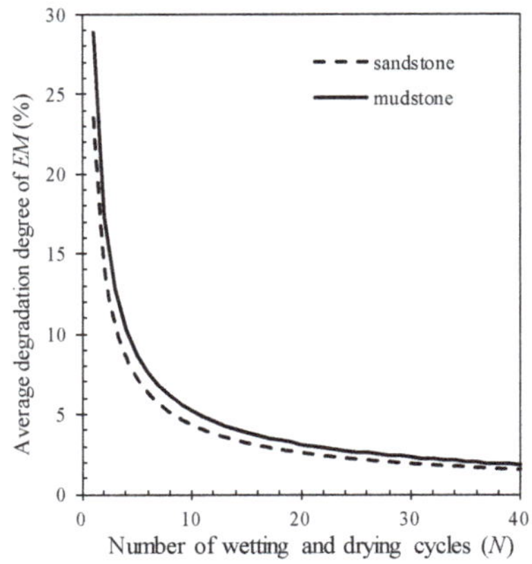

Figure 15. Fitted curves of average degradation degree of *EM* and wetting and drying cycles.

5. Discussion

As mentioned above, the *UCS* and the *EM* of the selected specimens could be calculated by the value of *N*, but it was hard to know the specific value of *N* in the practical engineering. As is known, the water absorption and porosity represent the volume of void and cracks in rock specimens, therefore they can be used to define the damage induced by wetting and drying cycles. Also, because of the good linear relationship between the two parameters as mentioned above, the water absorption and the porosity could be considered equivalent. Figure 16 further shows the relationship between the *UCS* and the porosity of selected specimens, and Figure 17 shows the relationship between the *EM* and the porosity of selected specimens, both of them can be fitted by linear curves as follows:

$$UCS = \mathrm{a}n + \mathrm{b} \quad (19)$$

$$EM = \mathrm{c}n + \mathrm{d} \quad (20)$$

where n is the porosity of the specimen measured after N times wetting and drying cycles, a, b, c and d are the coefficients of these equations.

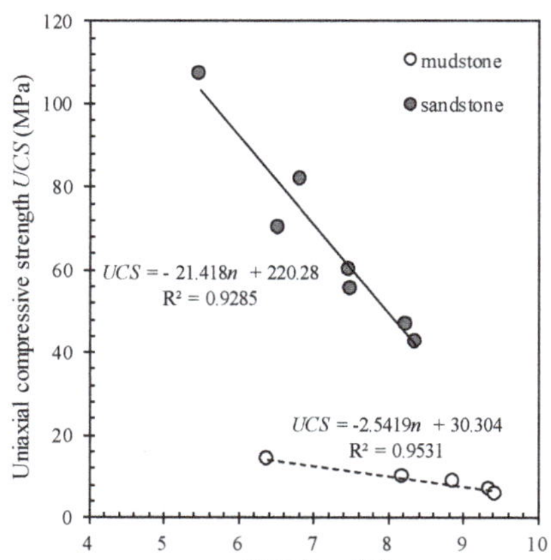

Figure 16. Relationship between the *UCS* and the porosity.

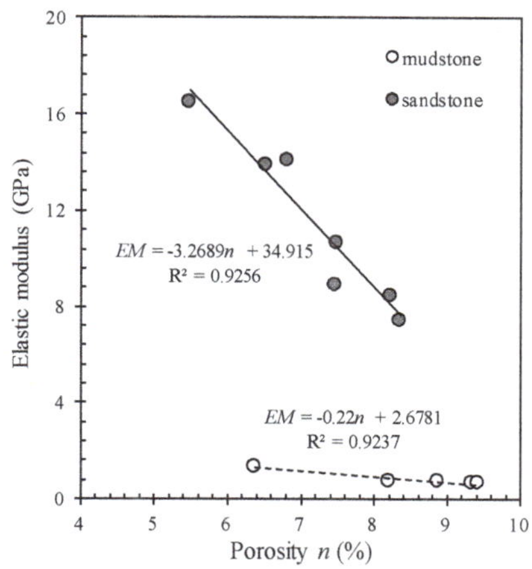

Figure 17. Relationship between the *EM* and the porosity.

From Figures 16 and 17, it is clear that both of the *EM* and the *UCS* of sandstone and mudstone were decreasing along linear curves with the porosity increasing. With the porosity value increasing, the *EM* and the *UCS* of sandstone decreased more rapidly than those of mudstone, which indicated that the deterioration in the mechanical properties of mudstone does not depend on the difference of water content during wetting and drying processes.

As is known, sandstone is a kind of sedimentary rock, composed of quartz and/or feldspar. While mudstone is a fine-grained sedimentary rock of which original constituents is clay. Therefore, it can be assumed that as the porosity value increases, the mechanical properties of a rock which has more clay decrease more rapidly.

6. Conclusions

In order to investigate the effects of cyclic wetting-drying conditions on elastic modulus and compressive strength of sandstone and mudstone, a series of uniaxial compression tests in laboratory were carried out. Based on analyses of experimental data, the following conclusions can be drawn:

(1) Wetting-drying cycles have a remarkable influence on physical and mechanical properties of sandstone and mudstone. The elastic modulus (*EM*) and uniaxial compressive strength (*UCS*) of sandstone and mudstone are reduced by wetting and drying cycles, while the water adsorption and porosity increase.

(2) With the increment of the number of wetting-drying cycles (N), the water adsorption, the degradation degree of *EM* and the degradation degree of *UCS* increase, and the relationship between these three parameters and the value of $N + 1$ could be fitted by logarithmic curves.

(3) While using the average degradation degree (*AD*) to describe the degradation degree of every wetting-drying cycle, it is found that as the value of N increased, the *AD* of *EM* and *UCS* decreased.

(4) The relationships between the mechanical properties and the porosity can be fitted by linear curves. In the cyclic wetting-drying process, the *EM* and the *UCS* decrease with the porosity increasing, and the decay rates of sandstone are rapider than those of mudstone.

Author Contributions: S.H. conceived the experiments, S.H. and J.W. designed the experiments and wrote the initial draft of the manuscript. S.H., J.W. and Z.Q. analyzed the data and wrote the final manuscript. K.K. was working on revising the content of the manuscript. S.H., J.W. and K.K. made revisions of the manuscript.

Funding: This work was funded by the Graduate Education Innovation Fund of Chongqing, China (Grant nos. CYB17126), the Open Fund of Key Laboratory of Hydraulic and Waterway Engineering of the Ministry of

Education of China (Grant nos. SLK2018B07) and the National Natural Science Foundation of China (Grant no. 51479012, U1865103).

Conflicts of Interest: The authors declare no conflict of interest.

References

1. Hammon, W.S.; Zeng, X.; Corbeanu, R.M.; Mcmechan, G.A. Estimation of the spatial distribution of fluid permeability from surface and tomographic gpr data and core, with a 2-d example from the ferron sandstone, utah. *Geophysics* **2002**, *67*, 1505–1515. [CrossRef]
2. Chongqing Geological and Mineral Resource Exploration and Development Corporation (CGMREDC). *Chongqing Geological Map (Scale 1:500,000)*; Chongqing Yangtze River Map Printing Factory: Chongqing, China, 2013. (In Chinese)
3. Wang, J.; Qiu, Z.; Deng, W. Shear Strength of a Crushed Sandstone-Mudstone Particle Mixture. *Mar. Georesour. Geotechnol.* **2014**, *1*, 33–37. [CrossRef]
4. Jia, G.W.; Zhan, T.L.T.; Chen, Y.M.; Fredlund, D.G. Performance of a large-scale slope model subjected to rising and lowering water levels. *Eng. Geol.* **2009**, *106*, 92–103. [CrossRef]
5. Gao, Y.; Zhu, D.; Leig, H.; Qin, H. Stability analysis of three-dimensional slopes under water drawdown conditions. *Can. Geotech. J.* **2014**, *51*, 1355–1364. [CrossRef]
6. Luo, F.; Zhang, G. Progressive failure behavior of cohesive soil slopes under water drawdown conditions. *Environ. Earth Sci.* **2016**, *75*, 973. [CrossRef]
7. Sumner, P.D.; Loubser, M.J. Experimental sandstone weathering using different wetting and drying moisture amplitudes. *Earth Surf. Process. Landf.* **2010**, *33*, 985–990. [CrossRef]
8. Kassab, M.A.; Weller, A. Study on p-wave and s-wave velocity in dry and wet sandstones of tushka region, egypt. *Egypt. J. Pet.* **2015**, *24*, 1–11. [CrossRef]
9. Yao, H.; Zhang, Z.; Zhu, C.; Shi, Y.; Yuan, L. Experimental study of mechanical properties of sandstone under cyclic drying and wetting. *Rock Soil Mech.* **2010**, *31*, 3704–3708, 3714. (In Chinese)
10. Deng, H.; Li, J.; Zhu, M.; Wang, K.; Wang, L.; Deng, C. Experimental research on strength deterioration rules of sandstone under "saturation-air dry" circulation function. *Rock Soil Mech.* **2012**, *33*, 3306–3312. (In Chinese)
11. Deng, H.; Zhou, M.; Li, J.; Sun, X.; Huang, Y. Creep degradation mechanism by water-rock interaction in the red-layer soft rock. *Arab. J. Geosci.* **2016**, *9*, 601. [CrossRef]
12. Hua, W.; Dong, S.; Li, Y.; Xu, J.; Wang, Q. The influence of cyclic wetting and drying on the fracture toughness of sandstone. *Int. J. Rock Mech. Min. Sci.* **2015**, *78*, 331–335. [CrossRef]
13. Lin, M.L.; Jeng, F.S.; Tsai, L.S.; Huang, T.H. Wetting weakening of tertiary sandstones—Microscopic mechanism. *Environ. Geol.* **2005**, *48*, 265–275. [CrossRef]
14. Zhao, Z.; Yang, J.; Zhang, D.; Peng, H. Effects of wetting and cyclic wetting-drying on tensile strength of sandstone with a low clay mineral content. *Rock Mech. Rock Eng.* **2016**, *50*, 485–491. [CrossRef]
15. Hale, P.A. A laboratory investigation of the effects of cyclic heating and cooling, wetting and drying, and freezing and thawing on the compressive strength of selected sandstones. *Environ. Eng. Geosci.* **2003**, *9*, 117–130. [CrossRef]
16. Özbek, A. Investigation of the effects of wetting-drying and freezing–thawing cycles on some physical and mechanical properties of selected ignimbrites. *Bull. Eng. Geol. Environ.* **2014**, *73*, 595–609. [CrossRef]
17. Khanlari, G.; Abdilor, Y. Influence of wet–dry, freeze–thaw, and heat–cool cycles on the physical and mechanical properties of upper red sandstones in central iran. *Bull. Eng. Geol. Environ.* **2015**, *74*, 1287–1300. [CrossRef]
18. Zhou, Z.; Cai, X.; Chen, L.; Cao, W.; Zhao, Y.; Xiong, C. Influence of cyclic wetting and drying on physical and dynamic compressive properties of sandstone. *Eng. Geol.* **2017**, *220*, 1–12. [CrossRef]
19. Zhang, B.; Zhang, J.; Sun, G. Deformation and shear strength of rockfill materials composed of soft siltstones subjected to stress, cyclical drying/wetting and temperature variations. *Eng. Geol.* **2015**, *190*, 87–97. [CrossRef]
20. Esposito, L.; Esposito, A.W.; Pasculli, A.; Sciarra, N. Particular features of the physical and mechanical characteristics of certain Phlegraean pyroclastic soils. *Catena* **2013**, *104*, 186–194. [CrossRef]

21. Pasculli, A.; Sciarra, N.; Esposito, L.; Esposito, A.W. Effects of wetting and drying cycles on mechanical properties of pyroclastic soils. *Catena* **2017**, *156*, 113–123. [CrossRef]
22. Bonn, D.; Eggers, J.; Indekeu, J.; Meunier, J.; Rolley, E. Wetting and spreading. *Rev. Modern Phys.* **2009**, *81*, 739–805. [CrossRef]
23. Cascini, L.; Sorbino, G.; Cuomo, S.; Ferlisi, S. Seasonal effects of rainfall on the shallow pyroclastic deposits of the Campania region (southern Italy). *Landslides* **2014**, *11*, 779–792. [CrossRef]
24. Brown, E.T. *Rock Characterization Testing and Monitoring: ISRM Suggested Methods*; Pergamon Press: Oxford, UK, 1981.
25. Tan, Y.; Mo-Zhen, H.; Zhou, W.; Zuo, Q.; Wang, H.; Bo, Y. Effects of drying-wetting cycle and loading on compressive property of mudstone. *Rock Soil Mech.* **2016**, *37*, 2165–2171.
26. Zhang, D.; Chen, A.; Xiong, D.; Liu, G. Effect of moisture and temperature conditions on the decay rate of a purple mudstone in southwestern china. *Geomorphology* **2013**, *182*, 125–132. [CrossRef]
27. Chen, T.C.; Yeung, M.R.; Mori, N. Effect of water saturation on deterioration of welded tuff due to freeze-thaw action. *Cold Reg. Sci. Technol.* **2004**, *38*, 127–136. [CrossRef]
28. Fu, Y. Study on Water-Rock Interaction with the Cyclic Drying-Wetting Effect on Rock. Doctor's Thesis, Chongqing University, Chongqing, China, 2010.
29. Talukdar, M.; Roy, D.G.; Singh, T.N. Correlating mode-I fracture toughness and mechanical properties of heat-treated crystalline rocks. *J. Rock Mech. Geotech. Eng.* **2018**, *10*, 91–101. [CrossRef]
30. Gautam, T.P.; Shakoor, A. Comparing the slaking of clay-bearing rocks under laboratory conditions to slaking under natural climatic conditions. *Rock Mech. Rock Eng.* **2016**, *49*, 19–31. [CrossRef]

processes

Article

Feasibility of Using Gangue and Fly Ash as Filling Slurry Materials

Hao Wang [1,*], Jian Jiao [1], Yumin Wang [1] and Weisheng Du [2]

1 College of Resources & Safety Engineering, China University of Mining & Technology, Beijing, D11 Xueyuan Road, Haidian District, Beijing 100083, China; jiaojian@student.cumtb.edu.cn (J.J.); Emmie.WYM@outlook.com (Y.W.)

2 State Key Laboratory of Hydroscience and Engineering, Tsinghua University, Haidian District, Beijing 100084, China; duweisheng1225@126.com

* Correspondence: TBP1600101006@student.cumtb.edu.cn; Tel.: +86-188-1139-7619

Received: 14 October 2018; Accepted: 19 November 2018; Published: 22 November 2018

Abstract: When used as filling aggregates, the physical and mechanical properties of gangue and fly ash are crucial for the design of filling and transportation systems. The mineral composition of gangue and fly ash affects the filling body's strength, and the fluidity of the slurry affects the mining engineering process. In this study, gangue and fly ash samples were characterized by X-ray diffraction to ensure their suitability as filling materials. Tests were carried out with an Intelligent Torque Rheometer, and the optimal ratio of the slurry's components was determined. After fitting the data using ORIGIN software, the following curves were obtained: (1) rheological parameters versus slurry mass fraction; (2) rheological parameters versus component ratio. On the basis of the curves, the recommended ratio of the components in the paste-like slurry was determined. We found that the mass concentration of the slurry must be kept strictly below 78%, and the optimal slurry composition includes coal gangue, fly ash, and a gelling agent at a weight ratio of 8:3:1. In order to illustrate the effectiveness of the filling technology, surface sink curves from different filling techniques in two closely situated positions of the Xinyang Coal Mine were compared. The results showed that the use of a paste-like slurry can effectively reduce surface subsidence. Furthermore, it will lead to cost-effective engineering designs for the sustainable development of coal mines.

Keywords: green mining; paste-like slurry; XRD; intelligent torque rheometer; on-site monitoring

1. Introduction

Gangue and fly ash are common industrial waste products, which are occupying an increasingly large area of land and polluting the environment. The coal mining method of paste-like slurry filling exemplifies the application and characteristics of green mining from the perspective of both resource recycling and waste reuse. In 2017, China's coal production was about 3.5×10^9 tons, of which 2.5×10^9 tons came from underground mining [1]. Coal gangue is a kind of waste rock generated during coal mining and processing. It is collected from roofs, floors, mezzanines, and meteorites during roadway tunneling, and washed gangue is generated during coal washing. According to incomplete statistics, about 5×10^9 tons of waste gangue has been deposited on the surface, and over 1600 large-scale waste dumps occupy a landmass of approximately 1.5×10^7 hm^2 [2,3]. The annual increase in the mining subsidence area is about 5×10^4 hm^2; the area related to coal resource exploitation is about 0.2 hm^2 of subsidence per 10^4 tons of coal yield. Sulfur, phosphorus, carbon, and dozens of components in waste coal (wasted coal refers to the coal mixed in the gangue or attached to the surface of the gangue) combined with spontaneous combustion cause various degrees of pollution in the atmosphere, water, soil, and so forth. In addition, if rainwater infiltrates a spontaneously combusting waste dump, the rapid expansion of heated air will cause an explosion [4]. Increased fuel consumption has led to an increasingly prominent contradiction between the concern for resource exhaustion and environmental destruction and the interest in economic development [5,6].

There are several methods for filling a goaf: solid or direct backfilling, hydraulic filling, cemented filling, and so on [7–9]. Every filling method has its disadvantages. As for the gangue filling method, it is difficult to monitor the filling body's strength and stability, while, for the cement filling process, the performance of the cementing material is strict [10]; in addition, with the paste filling method it is easy to block the pipe because of the high concentration of the filling material [11], whereas the high water filling technology is uneconomical [12]. In the 2000s, paste-like backfilling methods were developed in coal mines. The slurry was made of gangue and fly ash at concentrations that could be higher than 75%, and a gelling agent was added at a concentration of about 3–5% in order to rapidly consolidate the body and achieve high strength. The high mass concentration of the slurry resulted in an appearance similar to toothpaste. The paste-like filling technology has many remarkable advantages, such as extensive resource recovery, low cost, minimum environmental pollution, simple preparation technology, efficient transportation, and a filling body with high strength [13,14]. Applied to coal mining under buildings, railways, and water bodies, the technology effectively controls the surrounding rock reformation and subsidence, raises the resource recovery ratio, and realizes its goal without waste and harm [15].

A paste-like slurry is a mixture of gangue particles, fly ash, cement, and water. Gangue particles and fly ash are aggregates in the paste-like slurry. The widely present floc structures in water at a certain mass concentration maintain fluid homogeneity [16], and the coarse gangue particles are deposited slowly and freely in the homogeneous slurry. With the increase in mass concentration, the coarse particles are increasingly enveloped and become components of the slurry. When the slurry mass concentration exceeds a certain critical threshold, the entire fluid is converted into a homogeneous slurry [17]. The fluid is termed a heterogeneous–homogeneous composite fluid, which is a typical kind of Bingham fluid with good fluidity. The resistance of the composite flow is much lower than the resistance of a non-homogeneous flow in the same conditions [14]. Generally speaking, the effective and successful filling of the goaf with the slurry should guarantee a strength of the filling body that is sufficient to support roofs and to effectively transport material [16–18]. The physical composition and form of the material have a significant effect on the strength of the filling body [15,19]. X-ray diffraction (XRD) is usually used to identify the mineralogical phases of natural and anthropogenic materials. In this study, gangue and fly ash sample characterizations were performed by XRD first, and then a torque rheometer was used to determine the ratio of the slurry components. On the basis of a large number of experiments, the following curves were obtained: (1) rheological parameters versus slurry mass concentration and (2) rheological parameters versus composition ratio. On the basis of the trends of the curves, the optimum concentration of the slurry components and the optimum group distribution ratio were determined. An ill-proportioned ratio will inevitably affect the rheological properties of the slurry, thereby causing filling accidents, and the filling cost will rise. Limited by experimental equipment, the ratio of the paste-like slurry components has always been based on experience or a large number of similar tests [20–22]. This is a waste of materials and work. The new equipment described here not only avoids these drawbacks but also improves the accuracy of the results.

There have been many studies on surface sink caused by mining [23–26]. Previous research has shown that filling materials could meet the requirements to support the roof. Research has further shown the feasibility of using gangue and fly ash as filling materials and the resulting fluidity of the slurry, as well as the high efficiency of coal production and its environmental sustainability.

2. X-Ray Diffraction Experiment

2.1. Sample Preparation

As the main parts of the filling slurry, the basic physical and mechanical properties of gangue and fly ash are crucial for the design of filling and transportation systems. The mineral composition of washed gangue and fly ash affects the filling body's strength, and the fluidity of the slurry affects the planning process.

Gangue samples were collected from the Xinyang Coal Mine in Gaoyang Town, Xiaoyi City, Shanxi Province, and fly ash was collected from a thermal power station in Wutong Town, which is 20 miles away from Gaoyang Town.

2.2. X-Ray Diffraction of Samples

The equipment used was a D/Max-3B X-ray diffractometer manufactured by Rigaku Corporation, Tokyo, Japan. Powder X-ray diffraction profiles were measured using the diffractometer (with Cu-Kα radiation) equipped with a curved graphite monochromator. The slit system was set as follows: DS: 1°; RS: 1°; SS: 0.15 mm; RSM: 0.6°. The voltage and electric current of the X-ray generator were 35 kV and 30 mA. The speed of continuous scanning in the qualitative analysis was 3°/min, and the sampling interval was 0.02°. The speed of continuous scanning in the quantitative analysis was 0.25°/min, and the sampling interval was 0.01°.

On the basis of the K-value method of the Chinese standard (GB5225-86), the composition of washed gangue was determined and is shown in Table 1, while the X-ray diffraction spectrum is shown in Figure 1. The composition of fly ash is shown in Table 2, and the X-ray diffraction spectrum is shown in Figure 2.

Table 1. Chemical composition of washed gangue.

Sample	Na_2O	MgO	Al_2O_3	SiO_2	K_2O	CaO	Fe_2O_3	P	S
Washed Gangue (%)	1.2	1.2	24.3	53.1	1.3	5.4	1.9	0.03	0.8

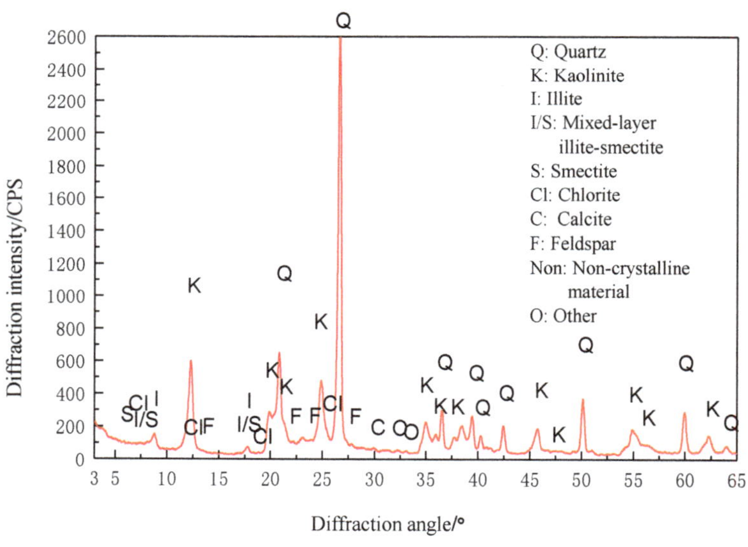

Figure 1. X-ray diffraction spectrum of washed gangue.

Table 2. Chemical composition of fly ash.

Sample	Na_2O	MgO	Al_2O_3	SiO_2	K_2O	CaO	Fe_2O_3	P	S
Fly Ash (%)	1.5	1.2	25.5	48.8	0.6	4.2	4.7	0.07	1.0

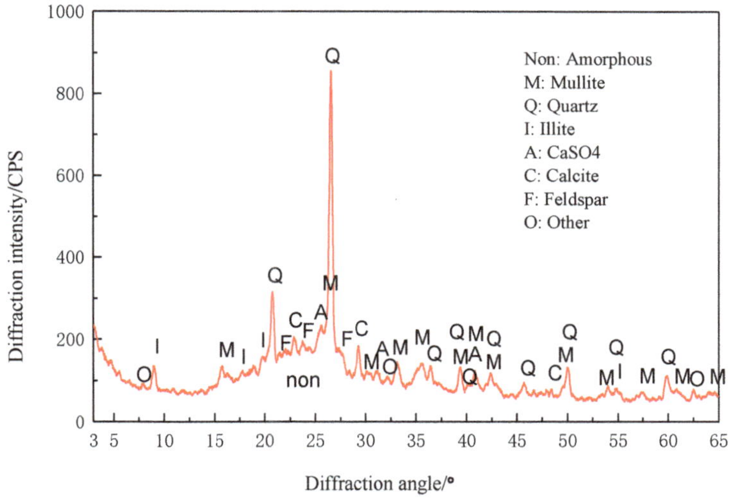

Figure 2. X-ray diffraction spectrum of fly ash.

Comparative analyses with the standard powder diffraction data of various substances, provided by the Powder Diffraction Federation International Data Center (JCPDS-ICDD), were performed. The analyses show that the main components of the washed gangue were quartz, kaolinite, some illite, mixed-layer illite–smectite, non-crystalline materials, and a small quantity of other minerals, some of which may be considered coal or another substance. The main components of fly ash were non-crystalline; it contained slightly more quartz than gangue, some minerals, such as mixed-layer illite–smectite, and illite, and small amounts of calcite, $CaSO_4$, and feldspar. The non-crystalline material in the sample can be produced by heating the clay at a moderate temperature. The results showed that SiO_2 was the main component of gangue and fly ash, an attribute that would impart high strength to the filling body.

3. Development of the Intelligent Torque Rheometer and Rheological Experiments of the Paste-like Slurry

3.1. Development of the Intelligent Torque Rheometer

The paste-like slurry was made of coarse particles, fine particles, and water. The coarse particle was coal gangue, crushed to a size smaller than 20 mm in situ. The fine particle was fly ash, and the gelling agent was Portland cement.

Because the maximum size of the coarse particles in the paste-like slurry was about 20 mm, the slurry could not be measured by normal rheometers. On the basis of rheometer principles and a torque sensor, an intelligent torque rheometer was developed. Figure 3 presents the intelligent torque rheometer, and Figure 4 shows a schematic of the Intelligent Torque Rheometer.

The most important part of the intelligent torque rheometer is a high-sensitivity iron piece leaf which can measure small deformations. When the iron piece leaf rotates at a speed of ω, the slurry will be driven to rotate in different layers. The slurry's initial resistance causes the iron leaf to deform. The shape variables (γ) occurring at different radii (r) are different. The angular velocity of the position at $r + dr$ is $\omega + d\omega$.

$$d\gamma = \frac{(r+dr)(\omega+d\omega)dt - (r+dr)\omega dt}{dr} = \frac{rd\omega - drd\omega}{dr} \approx r\frac{d\omega}{dr}dt \quad (1)$$

$$\frac{d\gamma}{dt} = r\frac{d\omega}{dr} \quad (2)$$

With the initial shear stress of the slurry expressed as τ_0, Equation (3) can be established:

$$\tau = \tau_0 + \eta\frac{d\gamma}{dt} = \tau_0 + \eta r\frac{d\omega}{dr} \tag{3}$$

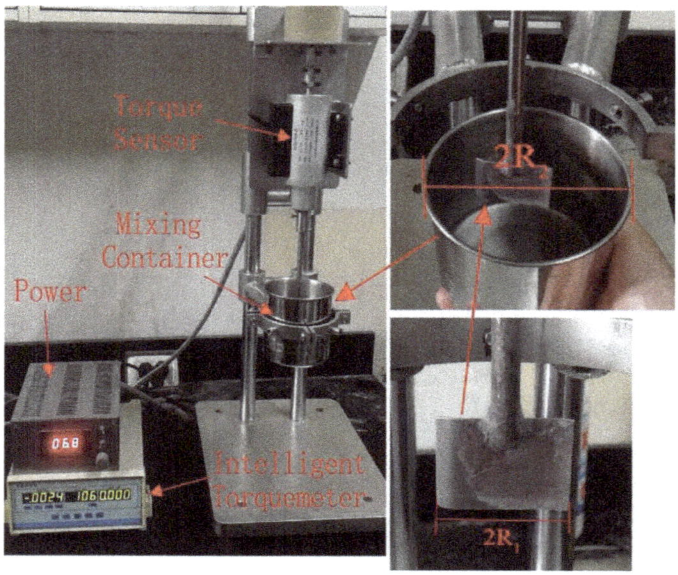

Figure 3. Intelligent Torque Rheometer.

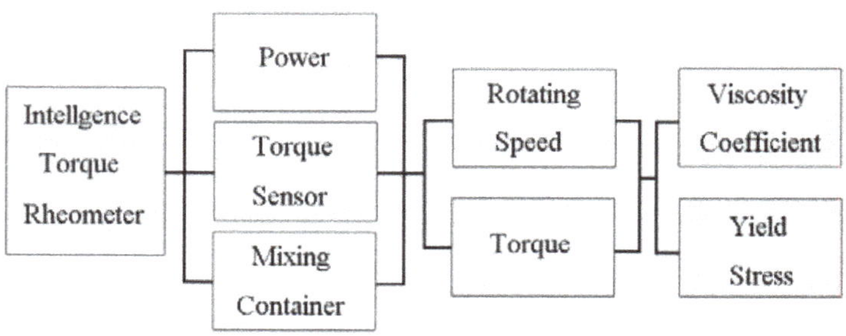

Figure 4. Schematic of the Intelligent Torque Rheometer.

The moment M is:

$$M = \tau \cdot 2\pi rhr = 2\pi r^2 h(\tau_0 + \eta r\frac{d\omega}{dr}) \tag{4}$$

$$d\omega = (\frac{M}{2\pi h\eta r^3} - \frac{\tau_0}{\eta r})dr \tag{5}$$

Then, the formula's integral is obtained:

$$\int_0^\alpha d\omega = \int_{R_1}^{R_2}(\frac{M}{2\pi h\eta r^3} - \frac{\tau_0}{\eta r})dr \tag{6}$$

Finally, the rheological parameters of the slurry can be calculated by Equation (7),

$$\omega = \frac{M}{4\pi h\eta}\left(\frac{1}{R_1^2} - \frac{1}{R_2^2}\right) - \frac{\tau_0}{\eta} ln\frac{R_2}{R_1} \tag{7}$$

$$M = \frac{4\pi h\eta}{\frac{1}{R_1^2} - \frac{1}{R_2^2}} \cdot \omega + \tau_0 \cdot \frac{4\pi h}{\frac{1}{R_1^2} - \frac{1}{R_2^2}} \cdot ln\frac{R_2}{R_1} \tag{8}$$

where R_1 is the radius of the high-sensitivity iron piece leaf; R_2 is the radius of the mixing container; h is the height of the high-sensitivity iron piece leaf; M is the torque; η is the viscosity coefficient; τ_0 is the initial shear stress; ω is the rotating speed.

3.2. Calibration for the Intelligent Torque Rheometer

In order to obtain the true torque, the Anton Paar RheolabQC rheometer (Anton Paar RheolabQC rheometer, Graz, Austria) was used as a calibration instrument for accurately testing the rheological properties of Newtonian and non-Newtonian fluids and calibrate the intelligent torque rheometer. Figure 5 shows the Anton Paar RheolabQC rheometer.

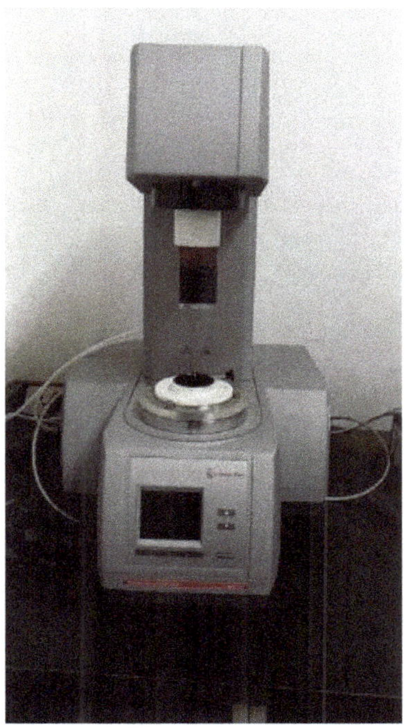

Figure 5. Anton Paar RheolabQC Rheometer.

The measured torque, named $M1$, and the actual torque, named M, can be represented by Equation (9):

$$M = kM_1 + a \tag{9}$$

where M is the actual torque; M_1 is the measured torque.

The torques and rotation speeds of slurries with different concentrations were measured using the RheolabQC rheometer and the intelligent torque rheometer and were then compared. Equation (9) was used in the process. Finally, the values of k and a were obtained. The test indicators of the RheolabQC rheometer are shown in Table 3.

Table 3. Viscometer test indicators.

Power	Measuring Range (η)	Shear Stress (τ)	Shear Rate (Ds)	Rotating Speed (n)
220 ± 22 V 50 ± 0.5 Hz	1 to 10^9 mPa·s	0.5 to 10^4 Pa	10^{-2} to 6500	0.01~1200 r/min

Since the calibration instrument is mainly suitable for measuring a non-Newtonian fluid slurry, the experimental material was a mixture of tailings and fly ash. The slurry parametric test results are shown in Table 4.

Table 4. Slurry parametric test.

Mass Concentration	Viscosity Coefficient/Pa·s	Initial Shear Stress/Pa	The Relation of M–ω
54%	0.040	1.626	$M = 1.76 \times 10^{-5}\omega + 1.66 \times 10^{-4}$
56%	0.042	3.179	$M = 1.89 \times 10^{-5}\omega + 3.21 \times 10^{-4}$
58%	0.045	5.961	$M = 2.09 \times 10^{-5}\omega + 5.88 \times 10^{-4}$
60%	0.074	8.847	$M = 3.31 \times 10^{-5}\omega + 8.74 \times 10^{-4}$

The rheological properties corresponding to different concentrations showed that the initial shear stress increased as the slurry concentration increased. When the slurry concentration was 54–58%, the plastic viscosity of the slurry changed gently. When the concentration was higher than 60%, the plastic viscosity of the slurry increased drastically. The rheological properties curve of the slurry was fitted by ORIGIN software (Northampton, MA, USA) and showed obvious characteristics of a Bingham fluid. Using ORIGIN, the measured torques were fitted with the actual torque, and the fitted regression line is shown in Figure 6.

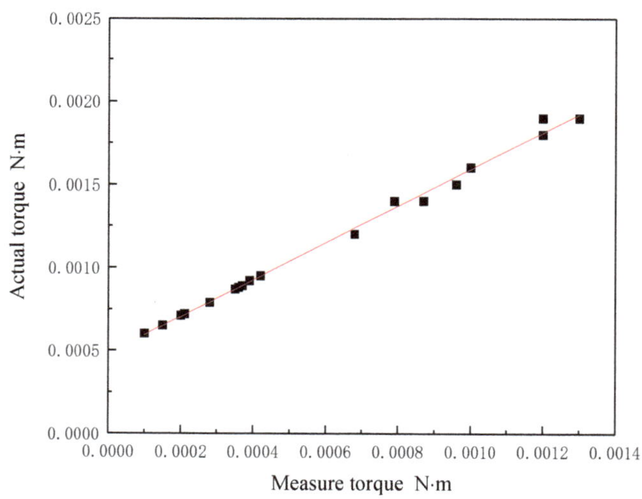

Figure 6. Torque comprehensive regression curve.

The formula relating the measured torque to the actual torque is shown in Equation (10):

$$M = 1.107M_1 + 4.825 \times 10^{-4} \tag{10}$$

According to Equation (8), the parameters of the Intelligent Torque Rheometer are: R_1 = 2 cm, R_2 = 5 cm, h = 3 cm. The relation between torque and rotating speed can be determined by Equation (11):

$$M = 1.787 \times 10^{-4}\eta \cdot \omega + 1.638 \times 10^{-4}\tau_0 \tag{11}$$

Table 5 shows the test values under different concentrations.

Table 5. Test values under different concentrations.

Mass Concentration	Rotating Speed r/min	Measured Torque N·m	Actual Torque N·m
54%	220	0.00015	0.00065
	260	0.00021	0.00072
	330	0.00035	0.00087
	370	0.00037	0.00089
	400	0.00042	0.00095
56%	240	0.0001	0.0006
	270	0.0002	0.00071
	340	0.00028	0.00079
	380	0.00036	0.00088
	420	0.00039	0.00092
58%	260	0.00087	0.0014
	300	0.00096	0.0015
	330	0.001	0.0016
	410	0.0012	0.0018
	460	0.0012	0.0019
60%	220	0.00068	0.0012
	260	0.00079	0.0014
	320	0.001	0.0016
	340	0.0012	0.0018
	380	0.0013	0.0019

In order to reduce measurement errors with the intelligent torque rheometer, the following requirements should be met:

(1) Avoiding high-speed rotations during the process;
(2) Basing the weight of the slurry on the width of the mixing head leaf, the immersion depth, and the outer cylinder;
(3) Using the average of multiple torque measurements to reduce the experimental error.

3.3. Rheological Properties of the Paste-Like Slurry

The rheological parameters of the filling slurry can be used to evaluate the pipe resistance of the slurry, that is, the head loss. There are many filling processes for coal mines, and the materials used are also different. They directly affect the solidification performance of the filling body. Therefore, the content of each material needs to be within a specific interval.

The experiments were carried out as described below, and the experimental data were fitted by ORIGIN.

(1) Five different mass concentrations were chosen: 74%, 75%, 76%, 78%, and 79%. Table 6 presents the ratios of the slurry composition at different mass concentrations.

Table 6. Ratios of the slurry composition at different mass concentrations.

Mass Concentration	Gangue (g)	Fly Ash (g)	Portland Cement (g)	Water (g)
74%	296	113.8	34.2	156
75%	300	115.4	34.6	150
76%	304	116.9	35.1	144
78%	312	120	36	132
79%	316	121.5	36.5	126

(2) With a mass concentration of 76% and a ratio of cement to fly ash of 1:3, the ratios of fine particles to coarse particles were varied: 1:1, 1:1.5, 1:2, 1:2.5, 1:3. Table 7 presents the ratio of the slurry composition at different ratios of fine particles to coarse particles.

Table 7. Ratio of the slurry composition at different ratios of fine particles to coarse particles.

Ratio of Fine Particles to Coarse Particles	Gangue (g)	Fly Ash (g)	Portland Cement (g)	Water (g)
1:1	228	171	57	144
1:1.5	273.6	136.8	45.6	144
1:2	304	114	38	144
1:2.5	326	97.5	32.5	144
1:3	342	85.5	28.5	144

(3) With a mass concentration of 76% and the content of the coarse particles constant, the ratios of cement to fly ash content were varied: 1:3, 1:3.5, 1:4, 1:4.5, 1:5. Table 8 presents the ratio of the slurry composition at different ratios of fly ash to cement in the slurry.

Table 8. Ratio of the slurry composition at different ratios of fly ash to cement in the slurry.

Ratio of Fine Particles Versus Coarse Particles	Gangue (g)	Fly Ash (g)	Portland Cement (g)	Water (g)
1:3	296	114.2	45.8	144
1:3.5	296	120	40	144
1:4	296	124.4	35.6	144
1:4.5	296	128	32	144
1:5	296	131	29	144

Figure 7 shows that higher mass concentrations increased the torque at the same rotating speed. The relationship between torque and rotating speed could be determined from the fitted curve in Figure 7. According to formula (11), the rheological parameters could also be obtained. The results are in Table 9.

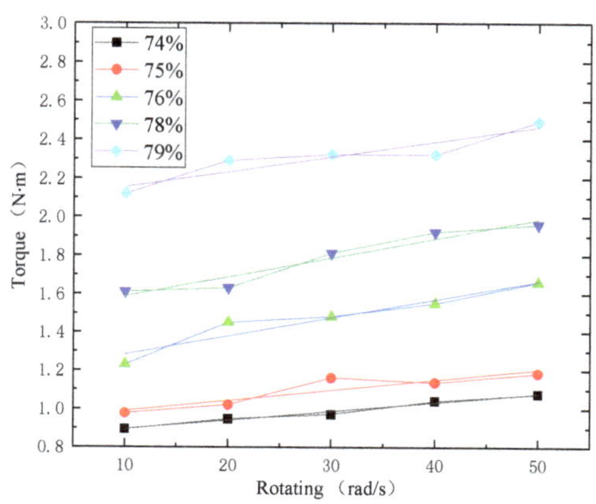

Figure 7. Relation between rotating speed and torque at different concentrations.

Table 9. Relation between rotating speed and torque and the rheological parameters at different concentrations of the slurry.

Concentrations	Relation Between Rotating Speed and Torque	Viscosity Coefficient (Pa·s)	Initial Shear Stress (Pa)
74%	$M = 4.33 \times 10^{-4}\,\omega + 8.51 \times 10^{-3}$	2.42	51.89
75%	$M = 4.67 \times 10^{-4}\,\omega + 9.19 \times 10^{-3}$	2.61	56.06
76%	$M = 4.85 \times 10^{-4}\,\omega + 1.25 \times 10^{-2}$	2.71	76.12
78%	$M = 5.07 \times 10^{-4}\,\omega + 1.66 \times 10^{-2}$	2.83	101
79%	$M = 5.33 \times 10^{-4}\,\omega + 2.17 \times 10^{-2}$	2.98	132.55

Figure 7 shows the relation between rotating speed and torque at different mass concentrations. Table 9 shows the relations between rotating speed and torque that arose from different mass concentrations of the slurry, along with the resulting rheological parameters.

Figure 8 shows the relation between the viscosity coefficient and the slurry mass concentration; Figure 9 shows the relation between the initial shear stress and the slurry mass concentration.

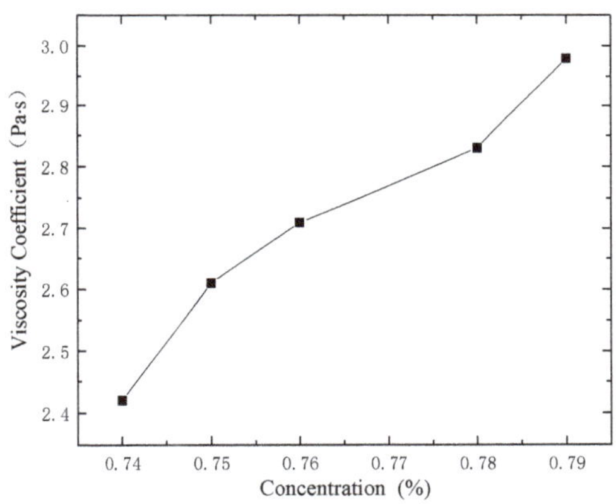

Figure 8. Relation between viscosity coefficient and slurry mass concentrations.

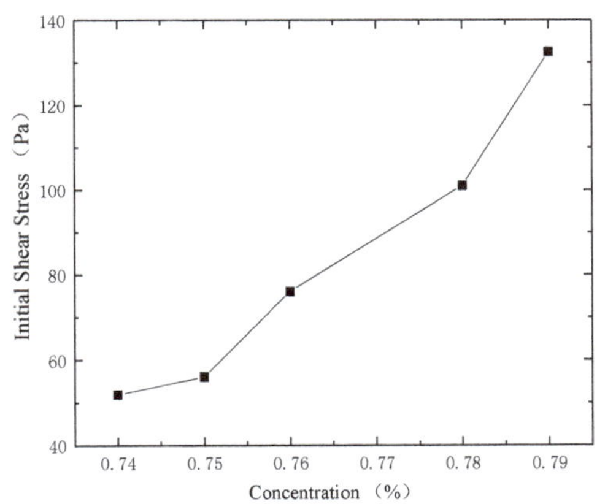

Figure 9. Relation between initial shear stress and slurry mass concentrations.

The viscosity coefficient increased with the increase in the mass concentration, but the changing trend was different at different segments of the concentration curve. The viscosity coefficient increased sharply in the 74–75% segment, while it increased gently in the 75–78% segment. The viscosity coefficient growth trend in the 78–79% segment was the most obvious. Considering the safety of the filling system and the transmission efficiency, the mass concentration should be controlled below 78%.

The initial yield stress increased with the increase of the slurry mass concentration in the 74–79% segment, and the change trend varied with the concentration. The increase in the 74–75% segment was not obvious. Then, the increase rate was more obvious in the 75–76% segment than in the adjacent intervals. The most obvious increase trend was in the 78–79% segment. Considering the safety of the filling system and the transmission efficiency, the mass concentration should be controlled below 78%.

Figure 10 shows the relation between rotating speed and torque at different ratios of fine particles to coarse particles. It shows that the torque increased with the increasing ratio of fine particles to coarse particles at the same rotating speed. The slope of the fitted curves is significantly larger than the slope of the curves in Figure 7.

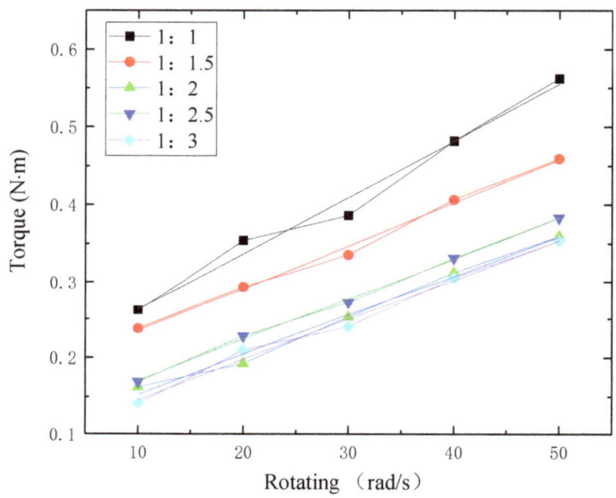

Figure 10. Relation between rotating speed and torque at different ratios of fine particles to coarse particles.

On the basis of Figure 10, Table 10 shows the relations between rotating speed and torque that arose from different ratios of fine particles to coarse particles, along with the resulting rheological parameters of the slurry.

Table 10. Relation between rotating speed and torque at different ratios of fine particles to coarse particles of the slurry and the resulting rheological parameters.

Ratio of Fine Particles to Coarse Particles	Relation Between Rotating Speed and Torque	Viscosity Coefficient (Pa·s)	Initial Shear Stress (Pa)
1:1	$M = 6.60 \times 10^{-4}\ \omega + 2.08 \times 10^{-3}$	3.67	126.9
1:1.5	$M = 5.66 \times 10^{-4}\ \omega + 1.80 \times 10^{-3}$	3.16	109.5
1:2	$M = 5.02 \times 10^{-4}\ \omega + 1.12 \times 10^{-2}$	2.96	68.53
1:2.5	$M = 5.16 \times 10^{-4}\ \omega + 1.23 \times 10^{-2}$	2.88	74.68
1:3	$M = 4.82 \times 10^{-4}\ \omega + 1.12 \times 10^{-2}$	2.69	68.52

Figure 11 shows the relation between the viscosity coefficient and different ratios of fine particles to coarse particles; Figure 12 shows the relation between the initial shear stress and different ratios of fine particles to coarse particles.

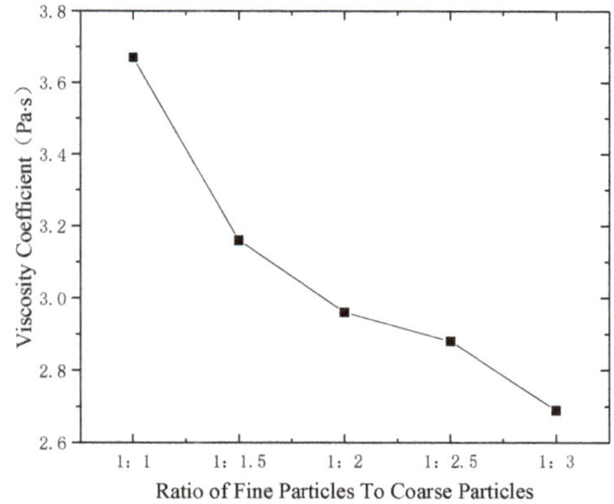

Figure 11. Relation between viscosity coefficient and different ratios of fine particles to coarse particles.

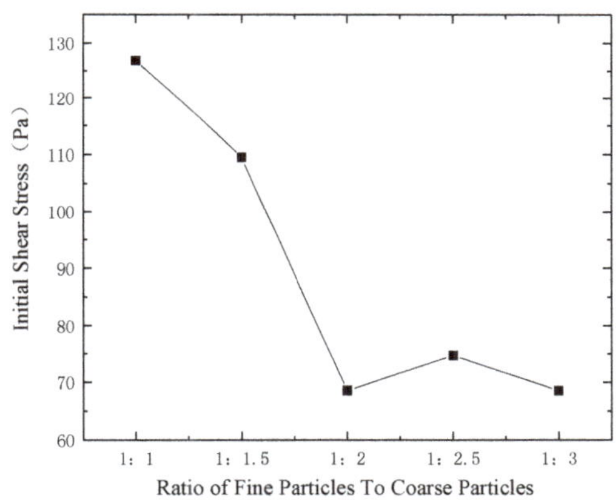

Figure 12. Relation between initial shear stress and different ratios of fine particles to coarse particles.

The curve shows that the viscosity coefficient decreased with the increase in the ratio of fine particles to coarse particles. In particular, when the ratio increased from 1:1 to 1:1.5, the viscosity coefficient decreased significantly. The decreasing trend eased slightly when the ratio increased from 1:1.5 to 1:2, but the slope was still steeper than in the 1:2 to 1:2.5 segment. The decreasing trend from 1:2.5 to 1:3 was more obvious than in the 1:2 to 1:2.5 segment. The curve shows that the fine particle concentration was the key factor driving the viscosity coefficient change. Small changes in fine particle content will cause the viscosity coefficient to change remarkably.

The curve shows that the initial yield stress tended to decrease first and then fluctuated with the increase in the ratio of fine particles to coarse particles. When the content ratio changed from 1:1 to 1:2, the initial yield stress decreased notably; however, when the content ratio increased from 1:2 to 1:2.5, the yield stress became larger. Then, the initial yield stress decreased again when the ratio changed from 1:2.5 to 1:3. It can be seen from the trend of the rheological parameters that the best ratio of fine particles to coarse particles is 1:2.

Figure 13 shows the relation between rotating speed and torque at different ratios of cement to fly ash in the slurry. It shows that the torque increased with the ratio of cement to fly ash in the slurry at the same rotating speed. The slopes of the fitted curves are roughly equal to the slopes of the curves in Figure 7.

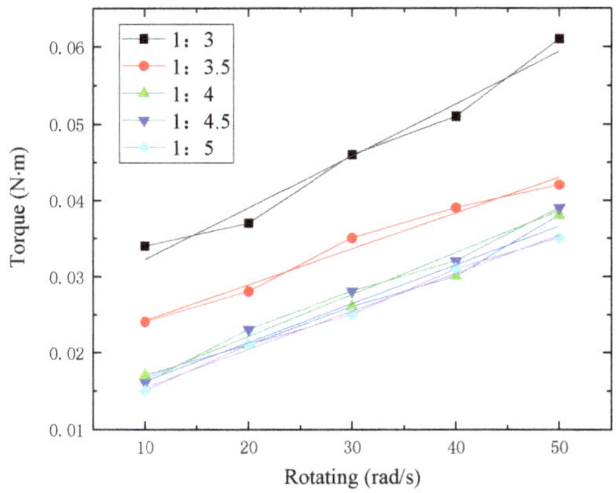

Figure 13. Relation between rotating speed and torque at different ratios of cement to fly ash in the slurry.

Table 11 shows the relations between rotating speed and torque at different ratios of fine particles to coarse particles and the resulting rheological parameters of the slurry, based on Figure 13.

Table 11. Relation between rotating speed and torque at different ratios of cement to fly ash and the resulting rheological parameters of the slurry.

Ratio of Cement versus Fly Ash	Relation Between Rotating and Torque	Viscosity Coefficient (Pa·s)	Initial Shear Stress (Pa)
1:3	$M = 4.99 \times 10^{-4}\,\omega + 9.00 \times 10^{-3}$	2.79	54.88
1:3.5	$M = 4.67 \times 10^{-4}\,\omega + 9.99 \times 10^{-3}$	2.61	60.9
1:4	$M = 5.02 \times 10^{-4}\,\omega + 1.11 \times 10^{-2}$	2.81	67.56
1:4.5	$M = 5.08 \times 10^{-4}\,\omega + 1.21 \times 10^{-2}$	2.84	72.98
1:5	$M = 5.39 \times 10^{-4}\,\omega + 1.59 \times 10^{-2}$	2.99	109.75

Figure 14 shows the relation between the viscosity coefficient and different ratios of cement to fly ash; Figure 15 shows the relation between the initial shear stress and different ratios of cement to fly ash.

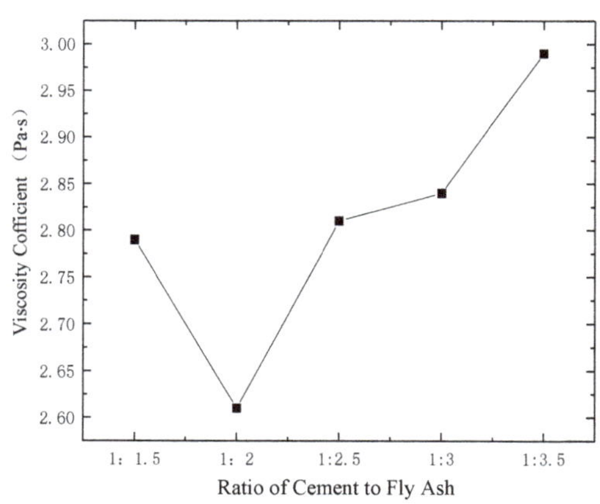

Figure 14. Relation between viscosity coefficient and different ratios of cement to fly ash.

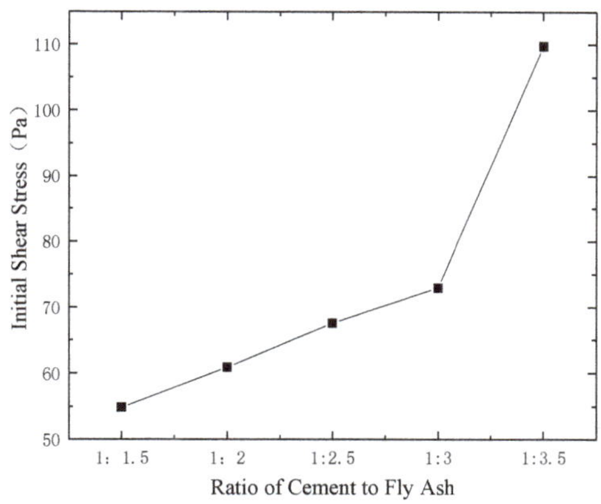

Figure 15. Relation between initial shear stress and different ratios of cement to fly ash.

The curve shows that the viscosity coefficient tended to decrease first and then increased with the ratio of cement to fly ash. When the content ratio changed from 1:1.5 to 1:2, the initial yield stress decreased; however, when the content ratio increased from 1:2 to 1:3.5, the viscosity coefficient became large. The change in the segment from 1:2 to 1:2.5 was the most obvious of all. The segment from 1:2.5 to 1:3 changed more moderately than the segment from 1:3 to 1:3.5.

The initial yield stress increased with the cement and fly ash content ratio. However, the increasing tendency for the ratio change from 1:3 to 1:3.5 was more obvious than for the change in ratio from 1:1.5 to 1:3. Considering that the cost of cement is higher than that of fly ash, the cement content should be reduced as much as possible, i.e., with minimal sacrifice of the filling body's strength. Taking into account the slurry's flowability and the filling cost, the best ratio of cement to fly ash content is 1:3. In summary, the optimum ratio of Portland cement to fly ash to gangue is 1:3:8.

4. Evaluation of the Effectiveness of Slurry Filling

The paste-like slurry described in this paper was utilized in the Xinyang Coal Mine. The single-strike longwall mining method was used in this mine, with a mining height of 2.2 m. The average dip angle of the coal seam was 6°, and the average strike length of the working face was 569.7 m. The production capacity was 4.7 million t/a. The mining process alternated with the filling process. The mining face layout and filling face layout are shown in Figure 16.

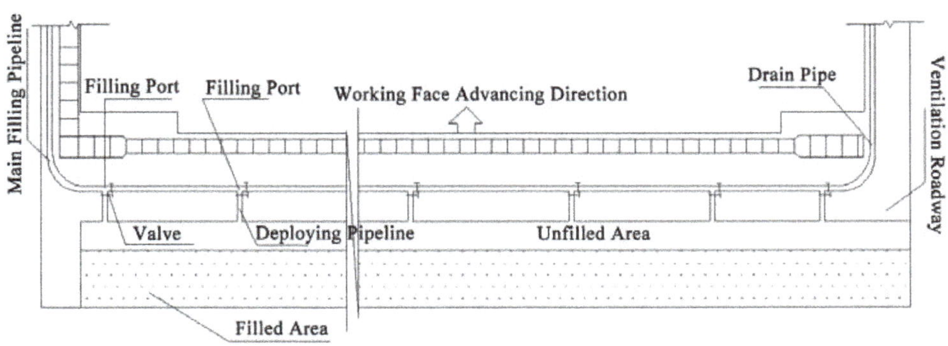

Figure 16. Layout diagram of the filling face.

According to the "Safety Regulations for Coal Mines (2016)" [27], coal pillars are left under the coal mine industrial square facilities to protect these constructions from subsidence and damage.

The filling process of Xinyang Coal Mine was divided into two stages: The first stage consisted of filling by using high-concentration cementation filling at the front of the industrial square facilities. The use of a water-reducing admixture can retain the fluidity of the slurry with less water content, which is a unique feature of this filling method. This slurry was made of cement, fly ash, gangue, and a water-reducing admixture. The ratio of cement/fly ash/gangue/water-reducing admixture was 1:2:5:0.1. The mass concentration of the slurry was 74%. During the second stage, because of the presence of a large number of buildings and agricultural facilities on the surface, a smaller surface sink was allowed. The paste-like slurry filling method was used at this stage. The surface sink data was obtained by a monitoring station on the surface. The layout of the surface monitoring points is shown in Figure 17.

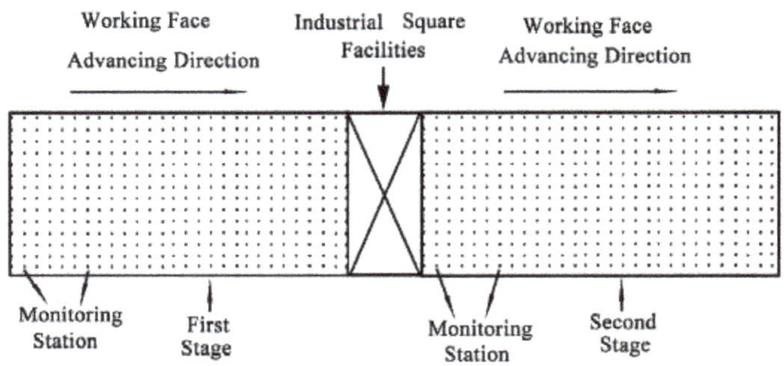

Figure 17. Layout diagram of the monitoring stations.

Table 12 presents the brick–concrete structure damage grades from the standard of the China Coal Association. Table 13 presents the surface deformation after using the high-concentration filling slurry. Table 14 presents the surface deformation after filling with the paste-like slurry. According to the "Building, Water Body, Railway and Main Mine Road Coal Pillar Retention and Coal Mining Regulations" [28] enacted by the China Coal Association, the surface deformation of the high-concentration cement filling process is graded as Class II, and the surface deformation of the paste-like slurry filling process is graded as Class I, meeting the requirements for surface building protection.

Table 12. China Coal Association's brick–concrete structure damage grade system.

Damage Level	Surface Deformation			Damage Level
	Horizontal Deformation ε (mm/m)	Curvature k (10^{-3}/m)	Tilt Angle i (mm/m)	
I	≤2.0	≤0.2	≤3.0	Negligible Damage or Slight Damage
II	≤4.0	≤0.4	≤6.0	Mild Damage
III	≤6.0	≤0.6	≤10.0	Moderate Damage
IV	>6.0	>0.6	>10.0	Severe Damage

Table 13. Surface deformation after filling with the high-concentration filling slurry.

Surface Deformation	Vertical Displacement (mm)	Horizontal Displacement (mm)	Tilt Angle (mm/m)	Curvature (mm/m^2)	Horizontal Deformation (mm/m)
Maximum Deformation	189	87/−87	3.60/−3.60	0.06/−0.10	2.03/−2.95

Table 14. Surface deformation after filling with the paste-like slurry.

Surface Deformation	Vertical Displacement (mm)	Horizontal Displacement (mm)	Tilt Angle (mm/m)	Curvature (mm/m^2)	Horizontal Deformation (mm/m)
Maximum Deformation	82	34/−34	1.13/−1.13	0.02/−0.04	0.57/−1.26

The surface sink curves of the two methods are shown in Figure 18.

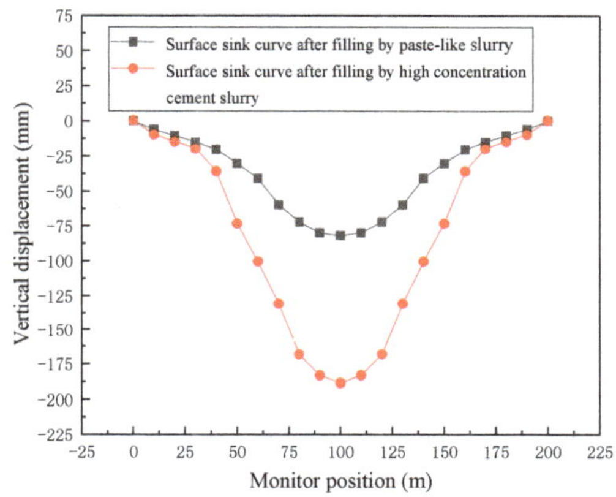

Figure 18. Surface sink curves of two filling slurries at the same filling rate.

Both surface sink-fitted curves were drawn using data collected from monitoring after six months. Figure 18 shows that: (1) the two curves are similar to a parabola and both high-concentration cementation filling and paste-like filling could effectively control surface sinking; (2) the area near the center of the goaf had the maximum surface sink; in the first stage, the variation of the radius, within a radius of 25 m, is not very obvious, then, when the radius changed from 25 m to 75 m, the sinking amount is sharply reduced from 180 mm to 30 mm; in the second stage, the sedimentation amount reduced from 75 mm to 25 mm; both curves have obvious changes; (3) when the radius is larger than 75 m, the two curves tend to be gentle again, up the boundary of the goaf; (4) the amount of surface sink when filling with the paste-like slurry was about 45% of the maximum amount of the surface sink when filling with the high-concentration cement slurry, indicating that the paste-like filling can better control surface subsidence.

5. Conclusions

The use of coal gangue and fly ash as filling materials is of significance to the sustainable development of mining. Our experimental results show that gangue and fly ash are suitable as filling materials. For practical use, the mass concentration of the slurry and the ratio of the components need to be controlled strictly.

XRD was used to analyze the gangue and fly ash. Their composition indicates that both have high strength characteristics and can be used as filling materials to support roof stability. The intelligent torque rheometer provides a new way to obtain the rheological parameters of a slurry that contains coarse particles. The equipment can also be used for other slurries containing large particles, such as a coarse coal slurry transported by pipeline. The optimized slurry ratio is determined from the viewpoint of rheology. Determining the rheological parameters according to the curves obtained from the measurements is a convenient method that produces accurate results.

The technology discussed in this paper is suitable for mines that employ paste-like backfilling. It may be useful to consider the strength of the optimized slurry after its consolidation with different component ratios, but that requires another research method. It may be that a lower filling rate can also control the sinking of the land surface, but this needs to be addressed by future studies.

Author Contributions: H.W. conceived and designed the experiments; H.W. and J.J. performed the experiments; H.W. and W.D. analyzed the data; Y.W. contributed analysis tools; H.W. wrote the paper.

Funding: This research received no external funding.

Conflicts of Interest: The authors declare no conflict of interest.

References

1. China National Coal Association. Available online: http://www.coalchina.org.cn/page/tjsj.htm (accessed on 20 November 2018).
2. Miao, X.X.; Qian, M.G. Research on green mining of coal resources in China: Current status and future prospects. *J. Min. Saf. Eng.* **2009**, *26*, 1–14. (In Chinese) [CrossRef]
3. Qian, M.G. On sustainable coal mining in China. *J. Chin. Coal Soc.* **2010**, *35*, 529–534. (In Chinese)
4. Li, J.; Huang, Y.; Qiao, M.; Chen, Z.; Song, T.; Kong, G.; Gao, H.; Guo, L. Effects of Water Soaked Height on the Deformation and Crushing Characteristics of Loose Gangue Backfill Material in Solid Backfill Coal Mining. *Processes* **2018**, *6*, 64. [CrossRef]
5. Liang, Y.; Liang, H.; Zhu, S. Mercury emission from spontaneously ignited coal gangue hill in Wuda Coalfield, Inner Mongolia, China. *Fuel* **2016**, *182*, 525–530. [CrossRef]
6. Wang, S.; Wang, X. Potentially useful elements (Al, Fe, Ga, Ge, U) in coal gangue: a case study in Weibei coal mining area, Shaanxi Province, northwestern China. *Environ. Sci. Pollut. Res. Int.* **2018**, *25*, 11893–11904. [CrossRef] [PubMed]
7. Miao, X.X. Progress of fully mechanized mining with solid backfilling technology. *J. Chin. Coal Soc.* **2012**, *37*, 1247–1255. [CrossRef]
8. Dickout, M.H. Filling effect on mining and properties of backfill materials. In Proceedings of the Jubilee Symposium on Mine Filling, North West Queensland Branch, Australia, August 1973; pp. 6–12.
9. Feng, G.M. Research on the Superhigh-water Packing Material and Filling Mining Technology and Their Application. Ph.D. Thesis, China University of Mining and Technology, Xuzhou, China, 2009.
10. Sun, H.; Liu, Q.; Liu, W. Commercial scale trial of a new consolidated backfill technology of full-range tailings with high water content. *Nonferrous Met.* **1993**, *45*, 7–13.
11. Zhao, C.; Hu, N. Development and application of cementing filling material. *Gold* **2008**, *29*, 25–29. [CrossRef]
12. Wang, X.; Xiao, W.; Wang, X.; Xiao, Z. Study on rheological properties of full tailing paste filling slurry of Jinchuan mine. *Min. Metall. Eng.* **2002**, *22*, 13–16. [CrossRef]
13. Soete, J.; Claes, S.; Claes, H.; Janssens, N.; Cnudde, V.; Huysmans, M.; Swennen, R. Lattice Boltzmann Simulations of Fluid Flow in Continental Carbonate Reservoir Rocks and in Upscaled Rock Models Generated with Multiple-Point Geostatistics. *Geofluids.* **2017**, *2017*. [CrossRef]
14. Lin, T. Research on the Flow Properties of the Gangue Paste-Like Slurry. Ph.D. Thesis, China University of Mining and Technology, Beijing, China, 2016.
15. Fall, M.; Pokharel, M. Coupled effects of sulphate and temperature on the strength development of cemented tailings backfill: Portland cement-paste backfill. *Cem. Concr. Compos.* **2010**, *32*, 819–828. [CrossRef]
16. Huang, Y.C. *Mine Filling Theory and Technology*; Metallurgical Industry Press: Beijing, China, 2014; pp. 59–60. ISBN 978-7-5024-6509-4.
17. Fei, X. *Slurry and Granular Materials Conveying Hydraulics*; Tsinghua University Press: Beijing, China, 1994; p. 17. ISBN 978-7-3020-1384-6.
18. Huang, Y.; Sun, H. Experimental Study on the Rheological Characteristic of Paste-like Slurry with Tailings as Aggregate. *Metal Mine* **2003**, *324*, 8–10. [CrossRef]
19. Fall, M.; Benzaazous, M.; Ouellet, S. Experiment characterization of the influence of tailings fineness and density on the quality of cemented paste backfill. *J. Mater. Sci.* **2005**, *18*, 41–44. [CrossRef]
20. Qiu, Z.; Cao, T.; Li, Y.; Wang, J.; Chen, Y. Rheological Behavior and Modeling of a Crushed Sandstone-Mudstone Particle Mixture. *Processes* **2018**, *6*, 192. [CrossRef]

21. Fan, H.; Guo, Z.; Huang, W. Experiment of Mechanics Properties of Gangue Cementing Stowing Body. *Coal Min. Technol.* **2015**, *20*, 83–86. [CrossRef]
22. Wang, Y.; Zhu, W.; Xie, G. Comparative analysis on the structure and properties of two paste-like cemented materials. *Nonferrous Met. (Min. Sec.)* **2017**, *69*, 71–75. [CrossRef]
23. Wang, J.; Li, Y.; Zhou, X.; Wu, L.; Wei, S. Ground movement caused by mining under thick alluvium. *J. Chin. Coal Soc.* **1997**, *22*, 18–21. [CrossRef]
24. Guo, G.; Wang, Y.; Ma, Z. A new method for ground subsidence contral in coal mining. *J. Chin. Univ. Min. Technol.* **2004**, *33*, 150–153. [CrossRef]
25. Xu, J.; You, Q.; Zhu, W. Theoretical study of strip filling to control mining subsidence. *J. Chin. Coal Soc.* **2007**, *32*, 119–122. [CrossRef]
26. Zhao, J.; Liu, C.; Li, J.; Wang, W. Three-Dimensional geological modeling and surface damage in gully area due to shallow coal seam mining. *J. Min. Saf. Eng.* **2018**, *35*, 969–977.
27. State Administration of Work Safety. Safety Regulations for Coal Mines. Available online: http://www.chinasafety.gov.cn/ (accessed on 20 November 2018).
28. National Coal Mine Safety Administration. Building, Water Body, Railway and Main Mine Road Coal Pillar Retention and Coal Mining Regulations. Available online: http://www.chinacoal-safety.gov.cn/ (accessed on 20 November 2018).

processes

MDPI

Article

The Effect of Collision Angle on the Collision and Adhesion Behavior of Coal Particles and Bubbles

Qiming Zhuo, Wenli Liu *, Hongxiang Xu *, Xiaopeng Sun, He Zhang and Wei Liu

School of Chemical and Environmental Engineering, China University of Mining and Technology (Beijing), Beijing 100083, China; zhuoqiming92@126.com (Q.Z.); xiaopeng201210@126.com (X.S.); Zhe9127@163.com (H.Z.); 13699151899@163.com (W.L.)

* Correspondence: liuwenli08@163.com (W.L.); xuhongxiang001@cumtb.edu.cn (H.X.)

Received: 8 October 2018; Accepted: 31 October 2018; Published: 5 November 2018

Abstract: The collision and adhesion behavior of particles and bubbles is the key to flotation. Many scholars have investigated the collision and adhesion law of regularly shaped and homogeneous particles (glass beads, glass fiber), but the particles in flotation cells are irregular and heterogeneous. Therefore, it is necessary to take actual coal samples as the research object. First, based on previous research, a particle–bubble collision and adhesion behavior measurement device was set up to study free falling coal particles with different surface properties colliding and adhering to a bubble (db = 1.0 mm). Then bituminous coal from Inner Mongolia was taken as the test object, and the collision and adhesion process of a large amount of coal particles was traced. The entire process is photographically recorded by a camera and analyzed frame by frame through a self-designed software. Finally, the relationship between collision angle and initial settlement position (initial), particle velocity (process), and adhesion efficiency (result) was studied by taking the collision angle as the cut-in point. It was found that both the distribution range of the initial settlement position and the particle central distribution interval are expanding outward with the increase of collision angle. The resistance layer has an important influence on the velocity of particles. The collision angle had an effect on adhesion efficiency and the adhesion efficiency of low-density particles was higher than that of high-density particles.

Keywords: flotation; coal particle; collision angle; initial settlement position; particle velocity; adhesion efficiency

1. Introduction

Flotation is a kind of technology that takes bubbles as the carrier to recover useful minerals according to the difference between particle surface properties. It has the advantages of large throughput, highly precise separation, and simple operation, so it has been widely used in the separation of fine minerals such as coal, copper, and zinc ore [1–3]. The interaction between particles and bubbles will affect the flotation results. At present, it is generally believed that the process of particle–bubble interaction can be divided into three sub processes [4]: (1) particles collide with bubbles, (2) particles adhere to bubbles and form a stable combination, and (3) unstable adhesion particles detach from bubbles.

When a particle collides with a bubble and squeezes the free water between them, there is a thin liquid film between them [5]. As the particle keeps approaching the bubble, the thin liquid film gradually thins until it reaches the critical liquid film thickness. With the rupture of the thin liquid film, the particle adheres to the bubble and the extension of three-phase contact makes the adhesion more firm. Adhesion is influenced by the particle's surface properties, the state of the fluid, and the chemical environment of the solution.

In recent years, many scholars have done experiments on the movement of particles on the surface of bubbles. The research method used is mainly particle sedimentation, that is, an injector is used to create a bubble on top of the needle in the water tank. The particles settle under the action of gravity. The collision and adhesion process of particles is observed by a high-speed camera [6–8].

At present, materials with homogeneous surface properties (glass beads, glass fiber) are usually used as test objects. It is found that the adhesion behavior is affected by the hydrophobicity of the particles [9–11]. Hydrophobic glass beads will jump in the bubbles, while this phenomenon is not observed if the test material is hydrophilic glass beads. Hydrophilic glass beads only slide along the upper hemisphere of the bubble and will detach if they cross the bubble equator, but hydrophobic glass beads will continue to slide along the bubble surface and eventually adhere to the bottom of the bubble after crossing the equator [12,13]. In addition to hydrophobicity, the shape of the particle also has a great influence on adhesion [14]. The irregular convex surface of particles can speed up the liquid film drainage and puncture the liquid film. The induction time of polygonal particles is much lower than that of spherical particles [15]. It was found that the collision angle also affects adhesion through studying the motion of hydrophobic glass fiber. Studies have shown that when the collision angle was less than 30°, the glass fiber adhered to the bubble along the long axis and adhered stably. When the collision angle was greater than 30°, the glass fiber adhered to the bubble along the short axis and was susceptible to the fluid environment [16].

In summary, most of the tested objects have been materials with regular shape and homogeneous surface properties (glass beads, glass fiber). Actual minerals have not been studied in depth because of their complex surface properties and irregular shape. Research work in other fields have shown that material heterogeneity has an important influence on many processes of particles, such as the strength and deformation behavior and micro-cracking process of minerals. Therefore, the influence of particle heterogeneity should be considered in the research of flotation [17–19]. In this paper, actual coal samples are taken as the research object and the relationship between collision angle and initial settling position, particle velocity, and adhesion efficiency is studied from the point of view of particle group.

2. Materials and Methods

2.1. Experimental Device

A three-dimensional (3D) diagram of the test device is shown in Figure 1. The main part of the device is composed of a funnel micromoving device, a bubble micromoving device, a bubble-generating device, a feeding funnel, a water tank, a camera, and a light-emitting diode (LED) array light source.

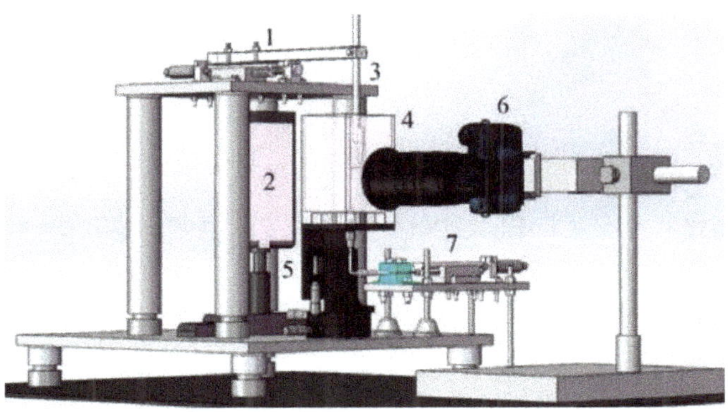

Figure 1. Three-dimensional (3D) diagram of experimental apparatus: 1: funnel micromoving device; 2: light-emitting diode (LED) array light source; 3: feeding funnel; 4: water tank; 5: bubble micromoving device; 6: camera; 7: bubble-generating device.

The characteristics of the device are as follows: (1) The funnel position is adjusted by the funnel micromoving device to ensure that the particles settle in the depth of focus, and the adjustment accuracy is 0.01 mm. (2) The bubble micromoving device can move bubbles in the direction of XYZ, and the moving precision is 0.01 mm. (3) The bubble diameter can be adjusted to 0.5–2.0 mm. (4) The material of the water tank is PMMA (Polymethyl methacrylate) and the light transmittance rate is 92%. The water tank size is 80 mm × 90 mm × 150 mm. (5) The LED array light source, used to provide illumination, consists of 8 × 12 LED beads. The luminous flux is 1250 lm and the color-rendering index is 95.

2.2. Experimental Methods

The conversion relationship between the pixel and the actual distance is first established in the experiment. The method is as follows: (1) Generate a bubble with the bubble-generation device. (2) Move the bubble through the bubble micromoving device. Take 1 photo per 0.50 mm and move 5 times. (3) Calculate the actual distance of 1 pixel according to the actual moving distance of the bubble and the pixel movement of the bubbles in the image. The bubble size can be adjusted according to the conversion relationship, and the bubble diameter is adjusted to 1 mm in this experiment.

The relative position of the feeding funnel and the bubble is shown in Figure 2. The outlet of the funnel is 4 mm from the top of the bubble and the particle enters the water tank through the feeding funnel. No external force is applied in the feeding process, and particles only settle under gravity. Each experimental video was about 5 h, and the final count of effective particles was about 1200.

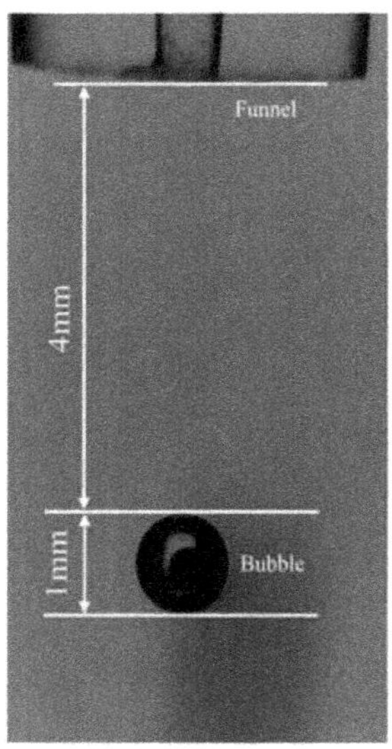

Figure 2. Diagram of relative position between feeding funnel and bubble.

2.3. Video Processing Software.

The method of video processing is to combine artificial watching video and image processing software (Image-Pro-Plus 6.0) in the early stage. This method will consume a lot of manpower and cannot get enough parameters of particles. Besides, there are differences between different people's results for the same video. To solve this problem, video processing software was programmed on the MATLAB platform to realize fully automated processing for the experimental video.

The software contains two modules. The first module is the ruler processing module, which is to get the actual distance corresponding to the pixels in the video by precisely moving the bubbles. The second module is the video processing module, which is to track all particles in the video. Then the parameters of particles and bubbles are analyzed. These parameters include the particles trajectories, the particles velocity, the collision point between particles and bubbles, the collision angle, the particles area, the particles circumference, the particles equivalent diameter and the particles clarity [20].

2.3.1. Principle of Software

The principle of video processing software is to extract each frame of the experimental video for processing. Each frame is processed by image interception and grayscale transformation. The grayscale image is transformed into a binary image by the OTSU algorithm. The OTSU algorithm is an efficient adaptive threshold algorithm. Although the gray-scale average algorithm is simple in calculation, the segmentation precision is lower for low-contrast images. The maximum entropy algorithm involves logarithmic operation and the operation speed is low. Because the object in this experiment is the experiment video which involves a large number of images and the computation is huge so it is not suitable for the maximum entropy algorithm. After the binarization process finished, the target area is processed by filling, denoising, and segmentation, and the locations of particles are obtained by background subtraction. Particle tracking was determined by comparing the shape, characteristic length, equivalent circle diameter, and position before and after the two frames to determine whether the particles in the two frames belonged to the same particle. In the process of video processing, the particle position is represented by centroid coordinates and the particle velocity is calculated by dividing the moving distance of the particles in the two frames by the time. The particle size is expressed in equivalent circle diameter, which is equal to the diameter of the projection area of the particle. Figure 3 shows a diagram of the software.

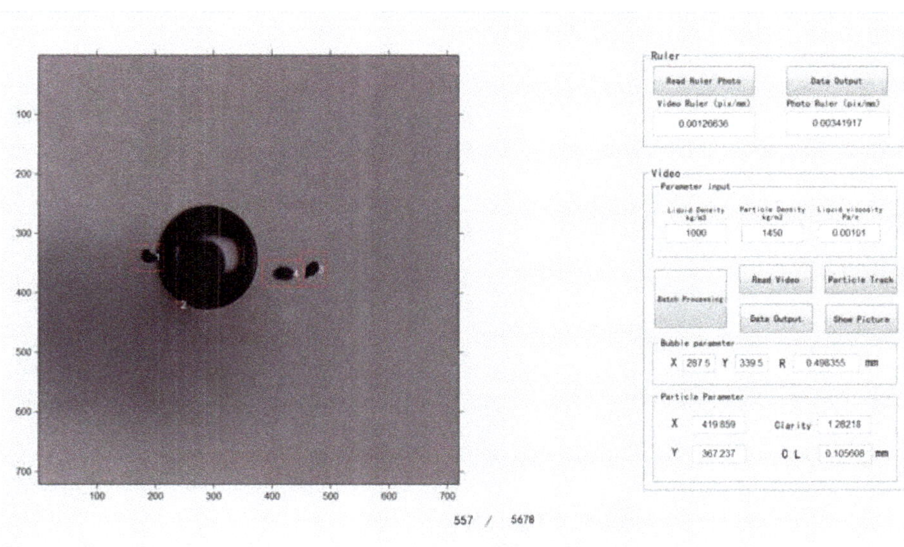

Figure 3. Diagram of software.

2.3.2. Collision Angle

The collision angle is the angle between the connection of the collision point to the bubble center and the vertical direction. Because the actual shape of coal particles is irregular, it is difficult to get the collision angle by using the particle centroid coordinate. To accurately represent the relative position of particle and bubble, the center of the bubble is defined as the coordinate origin, the horizontal direction is the x-axis, and the vertical direction is the z-axis.

A schematic of the collision angle solving method is shown in Figure 4. The first step is to find the edge pixel coordinates of particle and bubble to calculate the distance between the particle's edge pixels (the solid point in Figure 4) to the bubble's edge pixels and find the minimum value. Then the coordinates corresponding to the minimum distance values of the first time less than 1 pixel are obtained, which is the collision point. This method is verified by the velocity of the particle. When the particle collides with the bubble, the velocity will drop to the minimum. The angle between the centroid point of the particle velocity minimum to the central point of the bubble and the z-axis is compared with the collision angle. It was found that the difference between the two angles was $\pm 3°$.

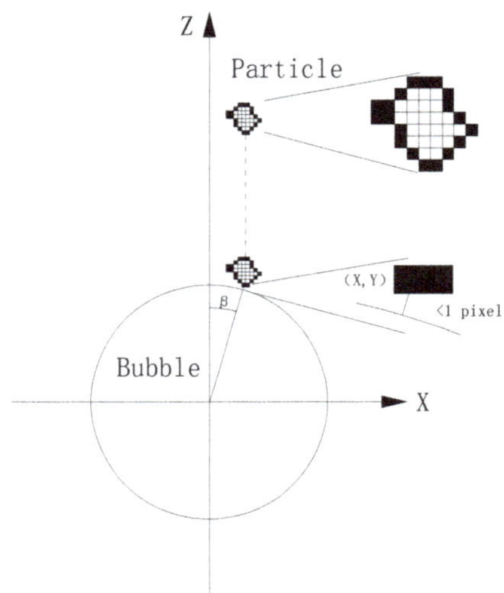

Figure 4. Schematic of collision angle solving method.

2.4. Materials

The test coal samples were bituminous coal from Gongwusu ore district, which is located in the northwest edge of Ordos basin and belong to Zhuozishan coalfield. The coal-bearing stratum in Zhuozishan coalfields are mainly composed of delta facies, tidal-flat facies and meandering river facies. The main coal-bearing stratum in the coalfield are located in the Permo-Carboniferous Taiyuan formation and Shanxi formation. The coalfield contains 18 coal seams which are sequentially numbered No. 1–No. 18 from top to bottom. The main coal seam is No. 9 coal seam of Shanxi formation and No. 16 coal seam of Taiyuan formation and the coal-bearing sedimentary sequence from Gongwusu mine is shown in Figure 5 [21,22].

In this experiment, the coal sample was taken from No. 16 coal seam and the coal sample was pre-treated before the experiment. First, the coal samples were wet-screened to obtain samples of particle size 0.15–0.10 mm. Subsequently, float-and-sink tests were carried out to obtain coal samples with density of -1.3 g/cm^3, 1.4–1.5 g/cm^3, and +1.7 g/cm^3. The air dry basis ash (A_{ad}) of the 3 samples was 3.62%, 18.12%, and 42.96% and the static contact angles were 100.5°, 87.1°, and 65.6°, respectively.

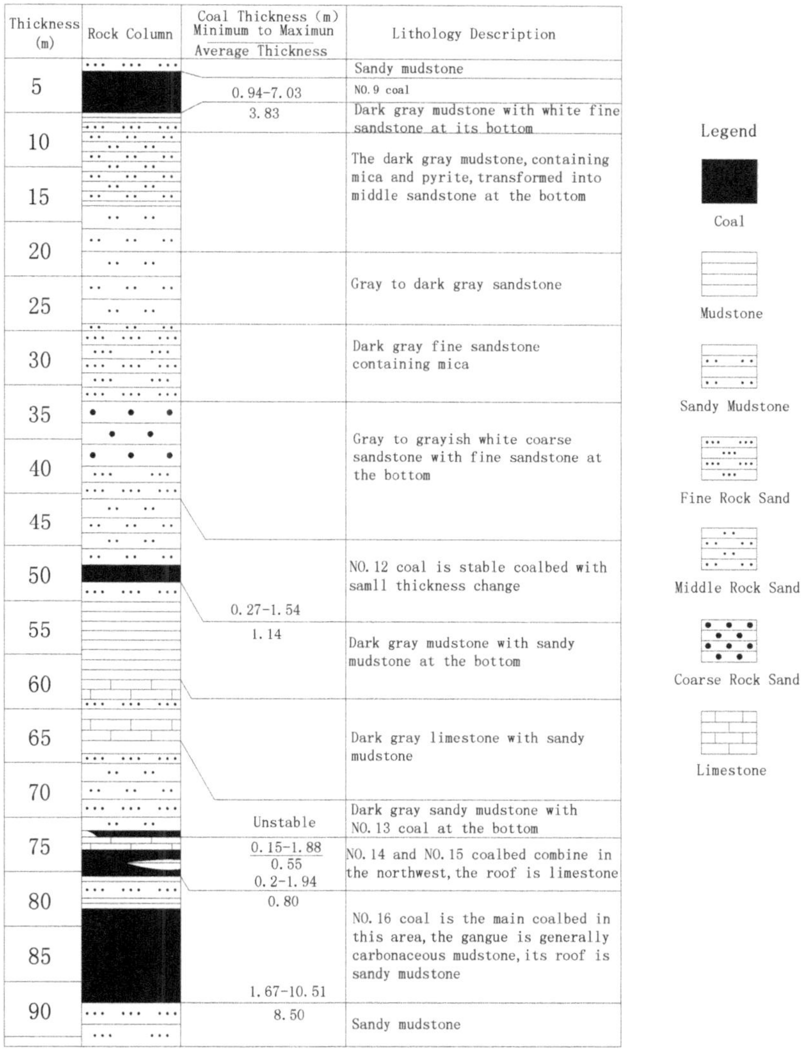

Figure 5. The coal-bearing sedimentary sequence from Gongwusu mine.

The experimental liquid environment was deionized water with conductivity of 0.25 μs/cm. We added 600 mL deionized water to the water tank, ensuring that the liquid surface completely submerged the funnel inlet. In each test, 5 mg coal particles and 200 mL deionized water were mixed evenly with a magnetic stirrer. The video resolution was 1280 × 720, the ISO was 100, the FPS was 50, the value of the aperture was 5.6, the shutter speed was 1/500, and the test environment temperature was 25 ± 1 °C.

3. Results and Discussion

3.1. Collision Angle versus Initial Settlement Position

Due to the influence of water flow disturbance, fine particles coming out from the funnel outlet will have a radial oscillation. To ensure the verticality of particle trajectory, the initial settlement position was set at 2 mm from the origin of the coordinates.

During the statistical process, it was found that the particle number distributions in different collision angle ranges were quite different. Taking the coal samples with density 1.4–1.5 g/cm^3 as an example, Figure 6 shows the ratio of number of particles in each collision angle to total number of particles.

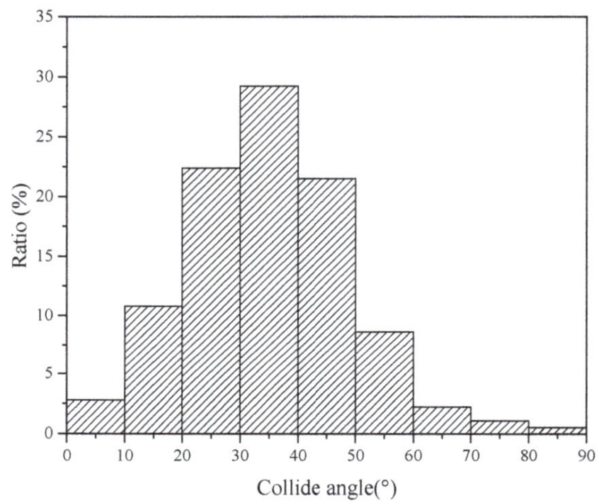

Figure 6. Ratio of particles in each collision angle.

As shown in Figure 6, the number of particles in the middle collision angle range had more than two sides. At a collision angle of 20–50°, the number of particles accounted for 73.10% of the total particles. When the collision angle was less than 10° or greater than 60°, the number of particles was very small; when the collision angle was greater than 60°, the ratio was only 3.88%.

Particle flow around the bubble in the process of approaching was analyzed, resulting in the trajectory of particles deviating from the z-axis. If the initial settlement position was far from the z-axis, it was difficult for particles to collide with the bubble due to the influence of the flow around the bubble, resulting in a smaller number of particles at a collision angle greater than 60°.

To study the relationship between the initial settlement position and the collision angle, first, the distance between the initial settlement position and the z-axis of the particles in the range of the collision angles was calculated with an interval of 10°. Then the particle ratio in each interval was calculated with an interval of 0.05 mm.

Figure 7 shows the distribution of the initial settlement position corresponding to each collision angle. To visually reflect the ratio of particles in the initial settlement position, a change in color is used to indicate a change in the ratio of particle numbers. When the collision angle was greater than 60°, there were few particles that could collide with bubbles, and they had no statistical meaning.

It can be seen from Figure 7 that the initial settlement position of the particles shows two change rules as collision angle increased. First, the initial settlement position range continued to expand outward. Second, the central distribution interval of the particles in the initial settlement position also expanded outward.

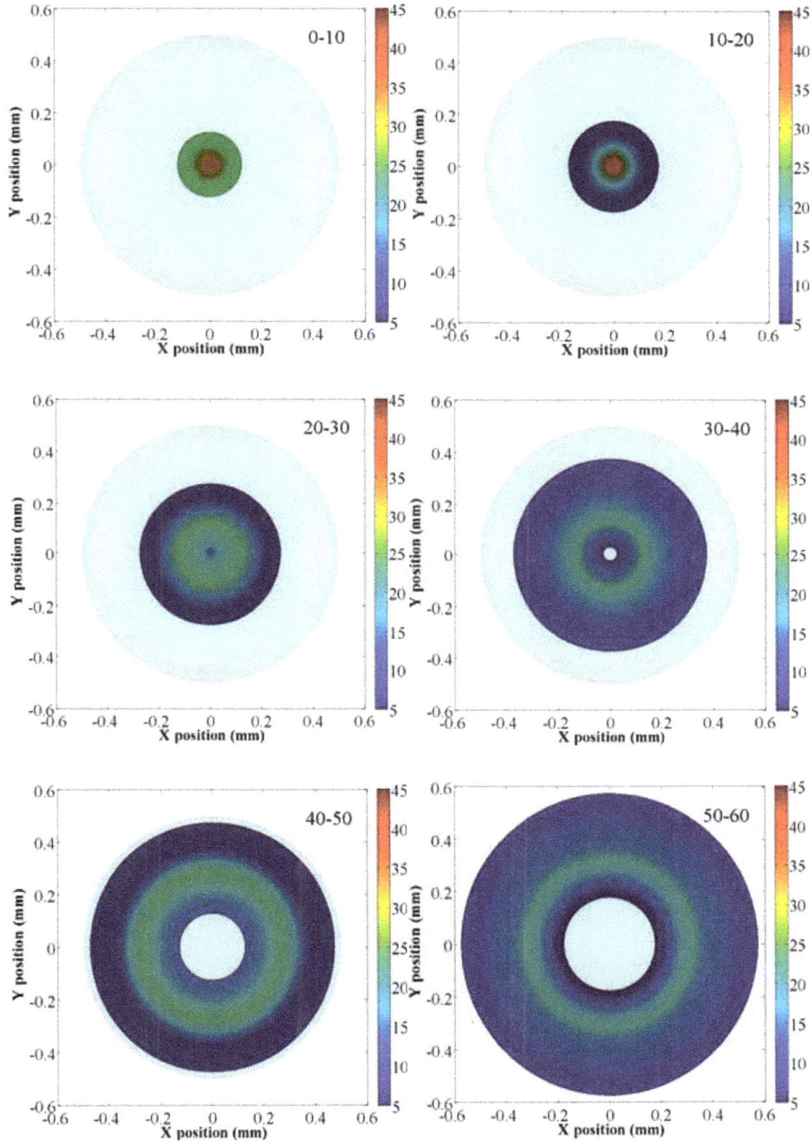

Figure 7. Distribution of initial settlement position of each collision angle.

When the collision angle increased from 0–10° to 20–30°, the initial settlement position extended from 0.15 mm to 0.30 mm from the z-axis. When the collision angle increased to 30–40°, a hollow interval appeared. That means that no particle collided with the bubble in the interval near the z-axis. As the collision angle increased, the area of the hollow interval expanded continuously. When the collision angle increased to 50–60°, the interval between 0–0.20 mm was hollow. At collision angles of 0–20°, the particle central distribution interval was 0–0.10 mm, and the ratio of particles in this interval was over 70%. With the increase of collision angle, the central distribution interval of particles expanded outward. When the collision angle increased to 20–30° and 30–40°, the main distribution intervals of particles were 0.05–0.25 mm and 0.10–0.30 mm, respectively. The ratio of particles in these two intervals was 79.95% and 70.78%. When the collision angle increased to 40–50°, the particles were mainly distributed in the interval of 0.20–0.40 mm. The ratio of particles in the interval reached 78.07%.

3.2. Collision Angle versus Particle Velocity

During the sedimentation process, the velocity of particles is affected by the relative position of particles and bubbles. Figure 8 shows the actual trajectory of six particles; the density of these particles was between 1.4 and 1.5 g/cm^3, and these particles were obtained by float-and-sink experiment. The collision angles were 5.86°, 15.03°, 21.33°, 36.99°, 41.95°, and 50.91°, in ascending order.

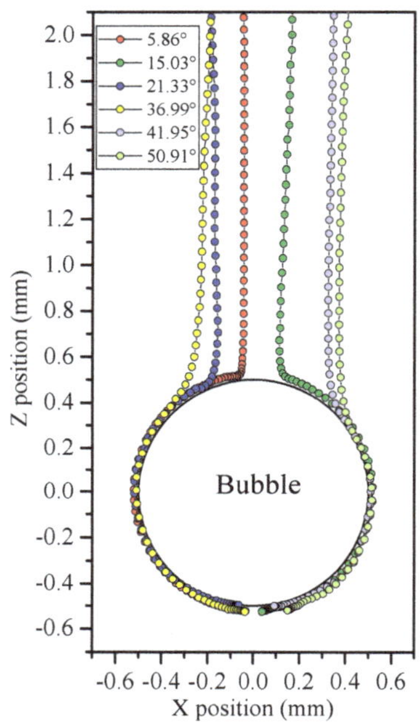

Figure 8. Trajectory of coal particles.

Figure 9 shows a diagram of particle velocity corresponding to the particles in Figure 8. The particle velocity was very stable when the particle was far from the bubble, and the velocity was the terminal velocity. When the distance between particles and bubbles decreased to a certain distance, the velocity of particles decreased sharply and achieved the minimum at the collision point.

As the particles began to slide on the surface of the bubble, the particle velocity gradually increased. When the particles slid to the equator of the bubble, the particle velocity increased to the maximum. At this time, the particle velocity was similar to the terminal velocity. As the particles crossed the bubble equator, the particle velocity began to decrease, eventually reducing to 0, and finally the particle adhered to the bottom of the bubble.

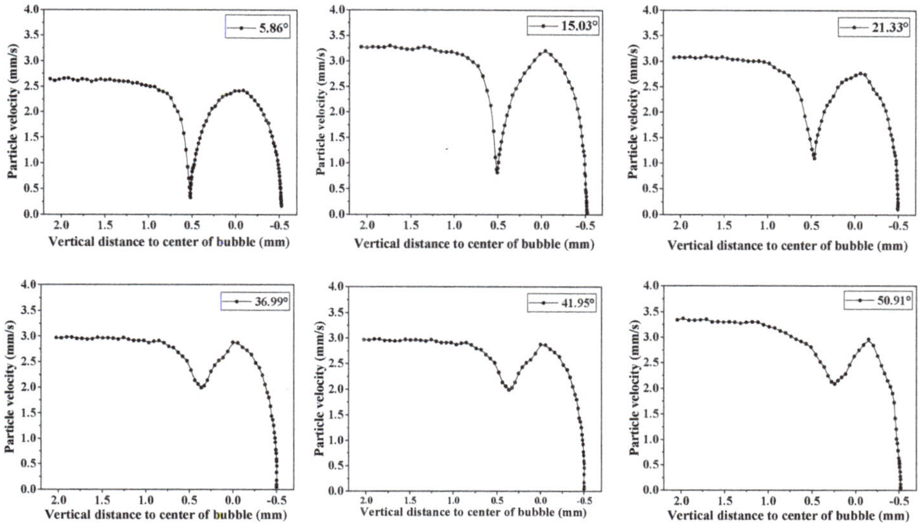

Figure 9. Diagram of particle velocity.

As shown in Figure 9, the particle velocity began to decrease before collision, and this process is related to the collision angle. The idea of resistance layer was introduced when exploring the effect of bubbles on particle velocity. Once the particle enters the resistance layer, the velocity decreases until it collides with the bubble. To quantify the range of resistance layer, the position where the particle velocity decreases by 10% is defined as the boundary of the resistance layer. Figure 10 shows the resistance layer boundary of the three coal samples. Figure 11 shows the distance from the resistance layer boundary to the bubble center at different collision angles.

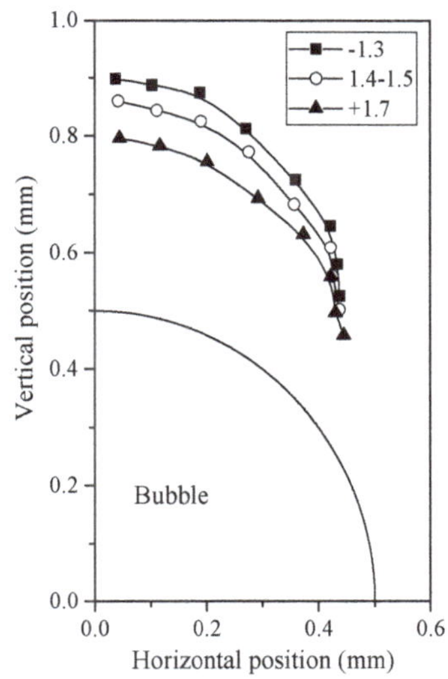

Figure 10. Boundary of coal sample resistance layer.

It is known from Figure 11 that the thickness of the resistance layer around the bubble is not constant, but is related to the collision angle and the properties of particles. The thickness of the resistance layer gradually thins as the collision angle increases. When the collision angle increases to

50–60°, the thickness of the resistance layer is reduced to 0.24 mm. As the particle density increases, the thickness of the resistance layer is compressed. When the particle density increases from -1.3 g/cm^3 to $+1.7$ g/cm^3, the resistance layer thickness of the three coal samples is 0.39 mm, 0.36 mm, and 0.29 mm at a collision angle of 0–20°, and the resistance layer thickness of low-density particles at each collision angle is larger than that of high-density particles.

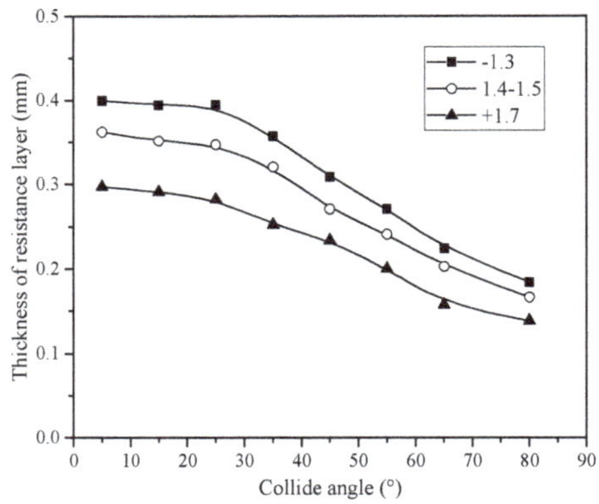

Figure 11. Boundary of coal sample resistance layer to center of bubble.

In addition, when the particle collides with the bubble at different collision angles, the particle velocity changes with the collision angle. The terminal velocity of the six particles in Figure 8 is around 3 ± 0.5 mm/s. The particle velocity at the collision point increases with the increase of collision angle. Figure 12 shows the relationship between particle velocity at the collision point and collision angle. The relationship between the decreased ratio of particle velocity at the collision point and the collision angle is shown in Figure 13.

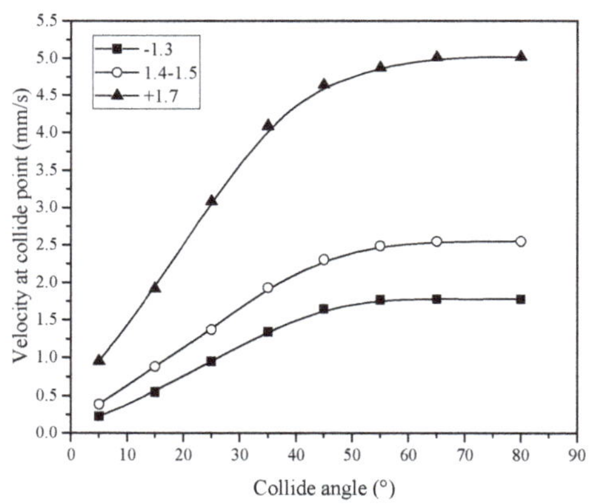

Figure 12. Particle velocity at collision point.

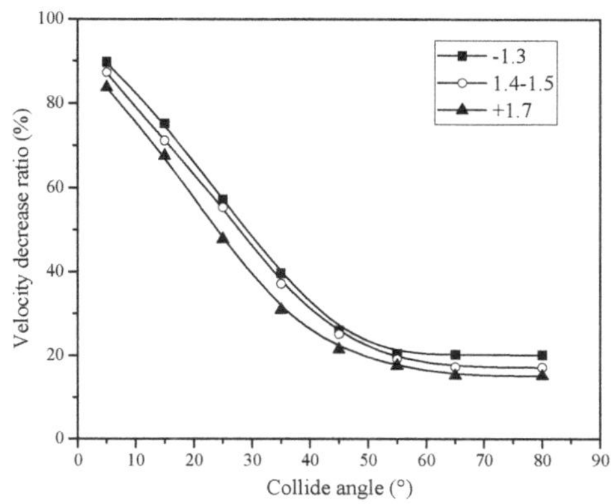

Figure 13. Reduction ratio of particle velocity at collision point.

When the collision angle is less than 50°, the particle velocity at the collision point increases approximately linearly with the increase of collision angle. When the collision angle is greater than 50°, the particle velocity at the collision point tends to be stable and the decreased particle velocity ratio accounts for about 20% of the terminal velocity. The reduction of particle velocity at the collision point is related to the particle density, and the velocity reduction of low-density particles is larger than that of high-density particles, but the difference is not obvious. When the collision angle is 0–10°, the decreased velocity ratio of three coal samples is 89.78%, 87.39%, and 83.84%, respectively. When the collision angle increases to 50–60°, the decreased velocity ratio is 20.54%, 19.20%, and 17.20%.

3.3. Collision Angle versus Adhesion Efficiency

The particle will slide along the surface of the bubble after colliding with the bubble. If the particle–bubble interaction force can puncture the thin liquid film during the sliding period, a three-phase contact will be formed, and the particle will adhere to the bubble. Conversely, the particle will slide from the surface of the bubble.

Adhesion efficiency is the ratio of the number of particles adhering to the bubble at a certain position to the total number of particles at that position. The adhesion efficiency of the three coal samples was fitted by logistic model, and the adjusted R-square is above 0.95. This means that the fitting equation can reflect the change of adhesion efficiency with the collision angle (Figure 14). Taking the 1.4–1.5 g/cm^3 coal samples as an example to analyze, when the collision angle is less than 20°, the adhesion efficiency is over 90%; as the collision angle increases, the adhesion efficiency decreases rapidly. When the collision angle increases to 40–50°, the adhesion efficiency drops to 37.82%. At this collision angle, it is hard for the particle to adhere to the bubble. When the collision angle is greater than 60°, the adhesion efficiency is almost 0.

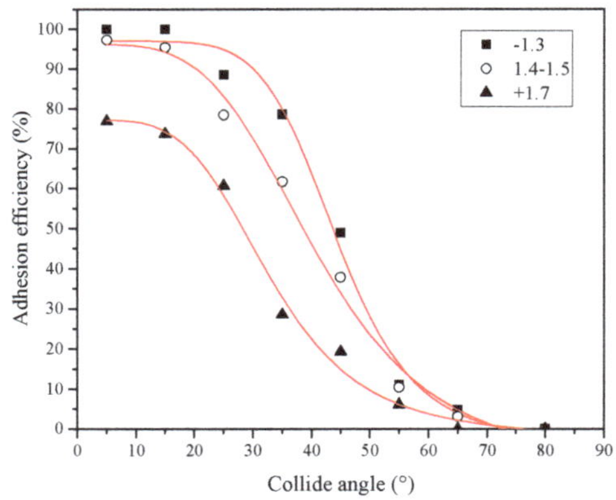

Figure 14. Adhesion efficiency of coal samples.

Analysis shows that the differences of the particles in the same coal sample are not huge. According to the experimental results of Section 3.2, if the particle collides with the bubble at a small collision angle, the particle velocity at the collision point is very small, which means there is a big loss of the particle's kinetic energy. This kinetic energy is used to drain the thin liquid film. As the collision angle increases, the particle velocity at the collision point increases gradually, which means the particle's ability to puncture the thin liquid film decreases, so the adhesion efficiency decreases with the increase of collision angle. The adhesion efficiency of the three coal samples decreases with the increase of coal sample density under the same collision angle.

To quantify adhesion efficiency, the area enclosed by the adhesion efficiency curve is calculated by the Gauss–Legendre numerical integration method, which is used to indicate the level of the coal sample's adhesion efficiency. This value is defined as the static water adhesion angle. The static water adhesion angles of the three coal samples were 38.35°, 33.62°, and 22.70°. This means that the static water adhesion angle decreases with the increase of coal sample density.

Analysis shows that adhesion behavior is affected by the hydrophobicity of the particle. Nguyen used AFM (Atomic Force Microscope, Veeco, Santa Barbara, CA, USA) to determine the interaction force between hydrophobic glass beads and bubbles, and found that there was a strong attraction force between them. This force will break the thin liquid film and make the glass beads adhere to the bubbles, but the interaction force between hydrophilic glass beads and bubbles is expressed as repulsion [23]. Xie [24] confirmed the results by measuring the interaction force between sphalerite and bubbles.

The static contact angles of the three coal samples were 100.5°, 87.1°, and 65.6°. This means that the hydrophobicity of particles decreased with the increase of particle density. As the particle density increases, the proportion of hydrophilic part in the coal increases. The probability of contacting the bubble increases, thereby causing the adhesion efficiency of the coal sample to decrease.

4. Conclusions

The study of the relative motion between particles and bubbles is very important for the understanding of flotation mechanism. A device was used to study the collision and attachment processes between particles and bubbles. A large number of coal particles were tracked and the relationship between the collision angle and the particle initial settling position, the particle velocity and the adhesion efficiency was studied.

1. The number of particles in different collision angles varies greatly. The particles are mainly distributed in the range of 20–50°. With the increase of collision angle, both the initial settlement range and the central distribution interval of the particles expand outward gradually.

2. Particle velocity is affected by the distance between the particles and the bubbles. After the particles enter the resistance layer, the velocity decreases immediately. The thickness of the resistance layer keeps constant and then slowly thins as the collision angle increases. The thickness of the resistance layer of low-density particles is larger than that of high-density particles at all positions. With the increase of collision angle, the particle velocity at the collision point decreases. The proportion is reduced.

3. Adhesion efficiency decreases with the increase of collision angle, and the adhesion efficiency of low-density particles is higher than that of high-density particles.

Author Contributions: All authors contributed equally to this study. Q.Z., W.L., and H.X. conceived of and designed the study. Q.Z. developed the experimental apparatus. Q.Z. and W.L. developed the software. Q.Z., H.Z., and P.X. performed the experiments. Q.Z., W.L., and H.X. wrote and modified the paper.

Funding: This research was funded by the National Natural Science Foundation of China (NO. 51604280).

Conflicts of Interest: The authors declare no conflict of interest.

References

1. Mouat, J. The development of the flotation process: Technological change and the genesis of modern mining, 1898–1911. *Aust. Econ. Hist. Rev.* **1996**, *36*, 3–31. [CrossRef]
2. Ejtemaei, M.; Gharabaghi, M.; Irannajad, M. A review of zinc oxide mineral beneficiation using flotation method. *Adv. Colloid Interface Sci.* **2014**, *206*, 68–78. [CrossRef] [PubMed]
3. Aghazadeh, S.; Mousavinezhad, S.K.; Gharabaghi, M. Chemical and colloidal aspects of collector less flotation behavior of sulfide and non-sulfide minerals. *Adv. Colloid Interface Sci.* **2015**, *225*, 203–217. [CrossRef] [PubMed]
4. Dai, Z.; Fornasiero, D.; Ralston, J. Particle-bubble collision models—A review. *Adv. Colloid Interface Sci.* **2000**, *85*, 231–256. [CrossRef]
5. Mao, M.; Zhang, J.; Yoon, R.H. Is There a Thin Film of Air at the Interface between Water and Smooth Hydrophobic Solids? *Langmuir* **2004**, *20*, 1843–1849. [CrossRef]
6. Albijanic, B.; Ozdemir, O.; Nguyen, A.V. A review of induction and attachment times of wetting thin films between air bubbles and particles and its relevance in the separation of particles by flotation. *Adv. Colloid Interface Sci.* **2010**, *159*, 1–21. [CrossRef] [PubMed]
7. Verrelli, D.I.; Koh, P.T.L.; Nguyen, A.V. Particle–bubble interaction and attachment in flotation. *Chem. Eng. Sci.* **2011**, *66*, 5910–5921. [CrossRef]
8. Wang, W.; Zhou, Z.; Nandakumar, K. Attachment of individual particles to a stationary air bubble in model systems. *Int. J. Miner. Process.* **2003**, *68*, 47–69. [CrossRef]
9. Xing, Y.; Gui, X.; Pan, L. Recent experimental advances for understanding bubble-particle attachment in flotation. *Adv. Colloid Interface Sci.* **2017**, *246*, 105–132. [CrossRef] [PubMed]
10. Chi, M.P.; Nguyen, A.V.; Miller, J.D. Investigations of bubble–particle interactions. *Int. J. Miner. Process.* **2003**, *72*, 239–254.
11. Verrelli, D.I.; Koh, P.T.L.; Bruckard, W.J. Variations in the induction period for particle–bubble attachment. *Miner. Eng.* **2012**, *36–38*, 219–230. [CrossRef]
12. Wang, W.; Zhou, Z.; Nandakumar, K. Effect of surface mobility on the particle sliding along a bubble or a solid sphere. *J. Colloid Interface Sci.* **2003**, *259*, 81–88. [CrossRef]
13. Nguyen, A.V.; Evans, G.M. Movement of fine particles on an air bubble surface studied using high-speed video microscopy. *J. Colloid Interface Sci.* **2004**, *273*, 271–277. [CrossRef] [PubMed]
14. Verrelli, D.I.; Bruckard, W.J.; Koh, P.T.L. Particle shape effects in flotation. Part 1: Microscale experimental observations. *Miner. Eng.* **2014**, *58*, 80–89. [CrossRef]
15. Hassas, B.V.; Caliskan, H.; Guven, O. Effect of roughness and shape factor on flotation characteristics of glass beads. *Colloids Surf. A Physicochem. Eng. Asp.* **2016**, *492*, 88–99. [CrossRef]
16. Lecrivain, G.; Petrucci, G.; Rudolph, M. Attachment of solid elongated particles on the surface of a stationary gas bubble. *Int. J. Multiphase Flow* **2015**, *71*, 83–93. [CrossRef]
17. Yoni, I.; Simon, E. Impact of grain size and rock composition on simulated rock weathering. *Earth Surf. Dyn.* **2018**, *6*, 319–327. [CrossRef]

18. Nezhad, M.M.; Fisher, Q.J.; Gironacci, E.; Rezania, M. Experimental study and numerical modeling of fracture propagation in shale rocks during brazilian disk test. *Rock Mech. Rock Eng.* **2018**, *51*, 1755–1775. [CrossRef]
19. Jun, P.; Louis, N.Y.W.; Cee, I. T. Influence of grain size heterogeneity on strength and microcracking behavior of crystalline rocks. *J. Geophys. Res.* **2017**, *122*, 1054–1073. [CrossRef]
20. Zhuo, Q.; Liu, W.; Liu, W.; Kai, P. Experimental study on the attachment behavior of coal particles and bubbles. *J. China Coal Soc.* **2018**, *43*, 2029–2035. [CrossRef]
21. Kang, J. *Distribution of Elements and Enrichment Mechanism of Mineral Matter in the Wuhai C-P Coals*; China University of Mining & Technology: Beijing, China, 2015; pp. 17–23.
22. Shifeng, D.; Deyi, R.; Yuegang, T.; Longyi, S. Distribution, isotopic variation and origin of sulfur in coals in the Wuda coalfield, Inner Mongolia, China. *Int. J. Coal Geol.* **2002**, *51*, 237–250. [CrossRef]
23. Nguyen, A.V.; Evans, G.M.; Nalaskowski, J. Hydrodynamic interaction between an air bubble and a particle: Atomic force microscopy measurements. *Exp. Therm. Fluid Sci.* **2004**, *28*, 387–394. [CrossRef]
24. Xie, L.; Shi, C.; Wang, J. Probing the interaction between air bubble and sphalerite mineral surface using atomic force microscope. *Langmuir* **2015**, *31*, 24–38. [CrossRef] [PubMed]

processes

MDPI

Article

Experimental Study on Feasibility of Enhanced Gas Recovery through CO_2 Flooding in Tight Sandstone Gas Reservoirs

Fengjiao Wang [1], Yikun Liu [1,*], Chaoyang Hu [1,*], Yongping Wang [2], Anqi Shen [1] and Shuang Liang [1]

1 Key Laboratory for Enhanced Oil & Gas Recovery of the Ministry of Education, College of Petroleum Engineering, Northeast Petroleum University, Daqing 163318, China; wangfengjiao8699@126.com (F.W.); anqi1986@163.com (A.S.); liangshuang21@126.com (S.L.)

2 CNOOC China Limited, Tianjin Branch, Tianjin 300459, China; wangyp50@cnooc.com.cn

* Correspondence: shenliujili@163.com (Y.L.); huchaoyang@nepu.edu.cn (C.H.)

Received: 15 October 2018; Accepted: 31 October 2018; Published: 2 November 2018

Abstract: The development of natural gas in tight sandstone gas reservoirs via CH_4-CO_2 replacement is promising for its advantages in enhanced gas recovery (EGR) and CO_2 geologic sequestration. However, the degree of recovery and the influencing factors of CO_2 flooding for enhanced gas recovery as well as the CO_2 geological rate are not yet clear. In this study, the tight sandstone gas reservoir characteristics and the fluid properties of the Sulige Gasfield were chosen as the research platform. Tight sandstone gas long-core displacement experiments were performed to investigate (1) the extent to which CO_2 injection enhanced gas recovery (CO_2-EGR) and (2) the ability to achieve CO_2 geological storage. Through modification of the injection rate, the water content of the core, and the formation dip angle, comparative studies were also carried out. The experimental results demonstrated that the gas recovery from CO_2 flooding increased by 18.36% when compared to the depletion development method. At a lower injection rate, the diffusion of CO_2 was dominant and the main seepage resistance was the viscous force, which resulted in an earlier CO_2 breakthrough. The dissolution of CO_2 in water postponed the breakthrough of CO_2 while it was also favorable for improving the gas recovery and CO_2 geological storage. However, the effects of these two factors were insignificant. A greater influence was observed from the presence of a dip angle in tight sandstone gas reservoirs. The effect of CO_2 gravity separation and its higher viscosity were more conducive to stable displacement. Therefore, an additional gas recovery of 5% to 8% was obtained. Furthermore, the CO_2 geological storage exceeded 60%. As a consequence, CO_2-EGR was found to be feasible for a tight sandstone gas reservoir while also achieving the purpose of effective CO_2 geological storage especially for a reservoir with a dip angle.

Keywords: CO_2 flooding; supercritical CO_2; CO_2 geological storage; tight sandstone gas reservoirs; enhanced gas recovery

---

1. Introduction

"Gas flooding" typically using CO_2, N_2, or air has become one of the leading enhanced oil recovery (EOR) technologies for residual oil development in conventional reservoirs [1–4]. Unfortunately, "gas flooding" for natural gas reservoirs is currently only in the research and development stage. Tight gas reservoirs are one of the most important areas of unconventional natural gas exploration and development in the world and they have rich resource reserves [5]. However, a tight gas reservoir is characterized by poor reservoir properties, strong heterogeneity, and complicated pore-throat structures. The main traditional method for gas recovery is depletion development, but the recovery is only approximately 35% [6,7]. In order to increase recovery from tight gas reservoirs, a new method of enhanced gas recovery (EGR) is urgently required. The phase state of CO_2 is easily transformed into the supercritical state [8–10] when the temperature exceeds the critical temperature (31.26 °C) and the pressure exceeds the critical pressure (7.29 MPa). Due to tight/shale gas reservoirs generally having great depths, it is easy for CO_2 to reach the supercritical state if it is injected into these reservoirs. Theoretically, supercritical CO_2 effectively displaces the natural gas and improves gas recovery in tight/shale gas reservoirs due to its higher density, higher viscosity, and lower diffusion rate.

There have been a substantial number of detailed investigations on gas adsorption characteristics in recent years, which aim to understand the mechanism of CH_4 displacement by CO_2 in coal reservoirs. Littke [11] studied the adsorption and desorption abilities of CO_2 and CH_4 under various temperature and pressure conditions. The adsorption capacity of CO_2 was higher when compared to CH_4. Liang [12] experimentally investigated the displacement mechanism underlying the driving out of coal-bed methane by gaseous CO_2 and discovered that the permeability of CO_2 was beyond two orders of magnitude higher when compared to CH_4. This result was explained by the differences in the physical properties of the two gases, which are combined with the competitive adsorption effect. Zeng [13] theoretically established an internally consistent adsorption-strain-permeability model to describe the adsorption capacity of coal reservoirs to CH_4 and CO_2 and the displacement process of CH_4 by CO_2. The results indicated that the adsorption capacity of CO_2 was two to five times that of CH_4. In general, CO_2 is chosen for injection into tight sandstone gas reservoirs to achieve EGR based on the following aspects. First, CH_4 in tight gas reservoirs primarily exists in an adsorbed state and CO_2 has a stronger adsorption capacity than CH_4 under the same conditions [14,15]. Second, since the mixing speed of CO_2 and CH_4 is slower when compared to pressure recovery, the injection of CO_2 can increase the formation pressure and displacement pressure gradient. Consequently, the flow velocity increases, which effectively gathers and drives the flow of CH_4 in the reservoirs [16]. Moreover, the maintenance of pressure can also provide pressure support to prevent the formation of subsidence and water invasion [17]. In addition, injecting CO_2 into tight sandstone gas reservoirs not only can achieve the purpose of EGR but also realize CO_2 geological storage, which is of great significance for mitigating the global greenhouse effect [18,19]. For the previously mentioned reasons and the fact that CO_2 flooding can increase coal seam recovery, CO_2 flooding is feasible for EGR from tight sandstone gas reservoirs, theoretically. According to reports, only three pilot projects have existed globally [20,21]. CO_2 storage pilot experiments were carried out in three gas fields including Beihai K12-B in Holland [22], Budafa in Hungary [23], and Algeria [10], but all were mainly concerned with achieving CO_2 geological storage. CO_2-EGR is widely utilized in medium-permeability and high-permeability gas reservoirs both domestically and overseas while experimental studies on CO_2-EGR in tight sandstone gas reservoirs have rarely been reported.

In this work, the reservoir characteristics and fluid properties of the Sulige Gasfield were chosen as the research platform while displacement experiments using combined natural long-cores were performed to investigate the variables affecting CO_2-EGR. The extent of CO_2-EGR and CO_2 geological storage was measured by using various injection rates, dry and aqueous cores, and formation dip angles. In this scenario, the feasibility of CO_2-EGR in tight sandstone gas reservoirs using the displacement mechanism is examined and the geological storage of CO_2 is also discussed and explained.

2. Experimental Section

2.1. Cores

During experimentation, natural tight cores were obtained from the Sulige Gasfield of Ordos Basin. In order to simulate the actual geological conditions of tight gas sandstone reservoirs, long-core displacement experiments were carried out. Since a natural tight long-core with a length of 1 m was obviously unrealistic for the experiments, a series of natural small-sized cores were combined to form a long-core, which was used to perform the displacement experiments of natural gas with CO_2. The long-core consisted of 32 natural tight short-cores that were 3.0×10^{-4} m in length and 2.5×10^{-4} m in diameter. The total long-core was approximately 1 m in length. Even though this combination of natural short-cores might result in the end effect, the placement of a piece of filter paper between every two short-cores could effectively ameliorate it [24]. Using weighted averages, the average permeability and porosity of the long-core were 0.325×10^{-3} μm^2 and 9.36%, respectively. In addition, a piece of tight sandstone core was used to perform the capillary pressure tests. The diameter of the core was 2.54×10^{-2} m and the length was 5×10^{-2} m. The core permeability and porosity were 0.307×10^{-3} μm^2 and 9.41%, respectively.

2.2. Fluid

The gas for the experiments was CH_4 of high purity with the purity exceeding 99.95%. The CO_2 purity for the experiments exceeded 99.9%. The simulated formation water for the experiments was prepared according to the ion content and salinity of the actual formation water in the Sulige Gasfield, which is presented in Table 1.

Table 1. Mineral composition of the water sample in experiments.

K^+	Na^+	Ca^{2+}	Mg^{2+}	Sr^{2+}	Cl^-	$SO_4{}^{2-}$	$CO_3{}^{2-}$	$HCO_3{}^-$	OH^-	Total Salinity
(mg/L)										
1541	15,221	12,344	2539	650	59,774	744	none	257	none	102,000

2.3. Experimental Setup

The experimental setup for the long-core displacement experiments was produced in France and it was mainly composed of an injection system, a displacement system, and a production system. A total of six pressure measurement points were set up from the injection end to the outlet end and the pressure variation throughout the entire process was observed by a monitoring system. The experimental installation is presented in Figure 1.

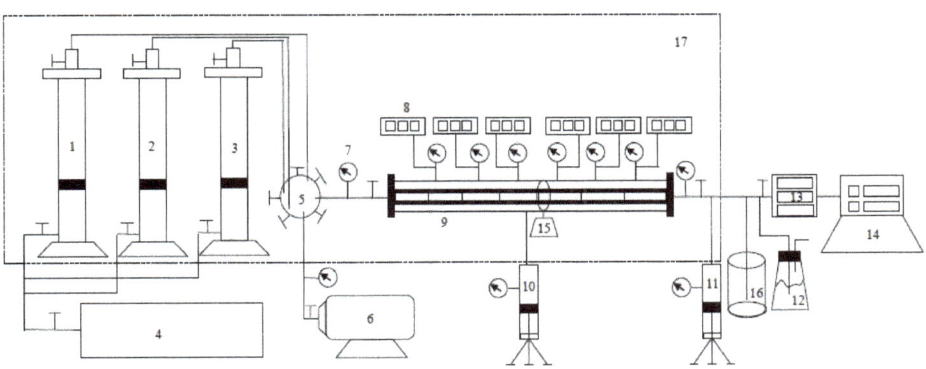

Figure 1. Schematic diagram of the experimental setup (1. Intermediate container (high purity CH_4), 2. Intermediate container (simulated water), 3. Intermediate container (high purity CO_2), 4. Pump, 5. Sluice with five gates, 6. Vacuum pump, 7. Pressure gage, 8. Electronic pressure gauge, 9. The long-core, 10. Confining pressure pump, 11. Back pressure pump, 12. Conical flask, 13. Electronic pressure gauge, 14. Chromatographic analyzer, 15. Angle regulator of core, 16. Beaker, 17. Thermostat box).

In addition, a high-pressure semipermeable plate instrument was utilized for capillary pressure testing.

2.4. Experimental Method

The experiments were conducted under the conditions of 110 °C and 30 MPa, which match the conditions in the actual tight gas reservoir of the Sulige Gasfield.

2.4.1. Measurement of Capillary Pressure

The effect of the injection rate on the results of CO_2 flooding CH_4 can be interpreted by comparing the viscous pressure drop generated by the flow and the capillary pressure. Therefore, we conducted an experiment using the semipermeable plate method to measure the capillary force [25] and then we measured the correlation of the capillary force with water saturation. First, the core was saturated with water. Then the displacement pressure difference between the two ends was established by using vacuum extraction. This pressure difference is balanced with a certain capillary force and different displacement pressure differences correspond to different capillary forces. Under a displacement pressure difference, the non-wet phase fluid (gas) displaces the wet phase fluid (the formation water) in the core. Therefore, the saturation of the wet phase (water) decreases as the displacement pressure difference increases. According to the displacement pressure balanced with the capillary pressure and the corresponding wet phase (water) saturation in the core during the displacement process, the correlation of the capillary force with the wet phase (water) saturation can be obtained. According to the research of Zou et al. [26,27], for a tight sandstone gas core, a hydrophilic semipermeable plate with a threshold pressure of 3 MPa should be selected. Thus, after the wet phase fluid saturates the semipermeable plate, the wet phase fluid can only pass through the semipermeable plate as a result of the capillary pressure until the displacement pressure is not less than the threshold pressure of the semipermeable plate. It is worth noting that the equilibration time should exceed 72 h for each pressure point.

2.4.2. Measurement of CH_4 Recovery and CO_2 Storage Efficiency through CO_2 Flooding

Generally, higher water saturation always exists in tight sandstone gas reservoirs in China. As a consequence, irreducible water saturation under conditions of confining pressure and a displacement pressure drop should be established in the experiment first. The simulated formation water was injected into the long-core and CH_4 was used to displace the water until the simulated formation water did not flow out of the long-core outlet.

The experiments were performed according to the following procedures.

(1) The short tight sandstone cores were added to the long-core holder in order, which was placed into the thermostat box. The displacement flow path was connected, according to the experimental device diagram presented in Figure 1.
(2) The long-core was vacuum-pumped. The thermostat box was set to 110 °C and the long-core confining pressure was set to 30 MPa.
(3) The irreducible water saturation status of the combined tight long-core was obtained under conditions of confining pressure and a displacement pressure drop. The simulated formation water was injected into the closed long-core holding system from intermediate container 2. Consequently, high-purity CH_4 gas was injected into the system from intermediate container 1. In this process, the simulated formation water was displaced by CH_4 until it did not flow out of the long-core outlet, which indicated that this procedure was over.
(4) Pure CH_4 was continuously injected into the closed long-core holding system with a constant pressure of 8 MPa. The tight sandstone long-core was fully saturated with CH_4 and the outlet valve was closed during the entire saturation process. The CH_4 saturation process was considered to be completed when the inlet pressure was stable for more than 12 h.
(5) The intermediate container 3, which was filled with high-purity CO_2, was connected to the displacement device system. The pressures of the long-core inlet and outlet were 12 MPa and 8 MPa, respectively. This meant that the displacement differential pressure was 4 MPa. Each time

that a 0.1 pore volume (PV) of CO_2 was injected into the long-core, the pressure at each pressure point was recorded. Furthermore, the gas production at the outlet was also recorded and the gas contents of CH_4 and CO_2 were analyzed with the chromatography analyzer. The characteristics of CO_2 migration and breakthrough of the front edge were monitored in real time. Additionally, the displacement efficiency of CO_2 flooding CH_4 was calculated. When the gas content of CO_2 at the outlet exceeded 98%, the experiments were over.

(6) Subsequent to each group of experiments, the tight sandstone long-core was vacuum-pumped and steps (2)–(5) were repeated.

3. Results and Discussion

3.1. CH_4 Recovery through CO_2 Flooding and CO_2 Storage Efficiency under Constant Pressure Displacement

Under a confining pressure of 30 MPa, the irreducible water saturation of the combined tight long-core was 41.05% in this experiment. According to the previously mentioned experimental procedures, CO_2 was injected to displace CH_4 at a constant pressure of 4 MPa until the gas content of CO_2 at the outlet exceeded 98%. The CO_2 mole fraction collected at the outlet and the CO_2 recovery calculated are shown in Figure 2a,b. Under the experimental temperature and pressure conditions, CO_2 was in the supercritical state. The results demonstrated that the displacement front of supercritical CO_2 would be miscible with CH_4, which requires an extended duration of this process.

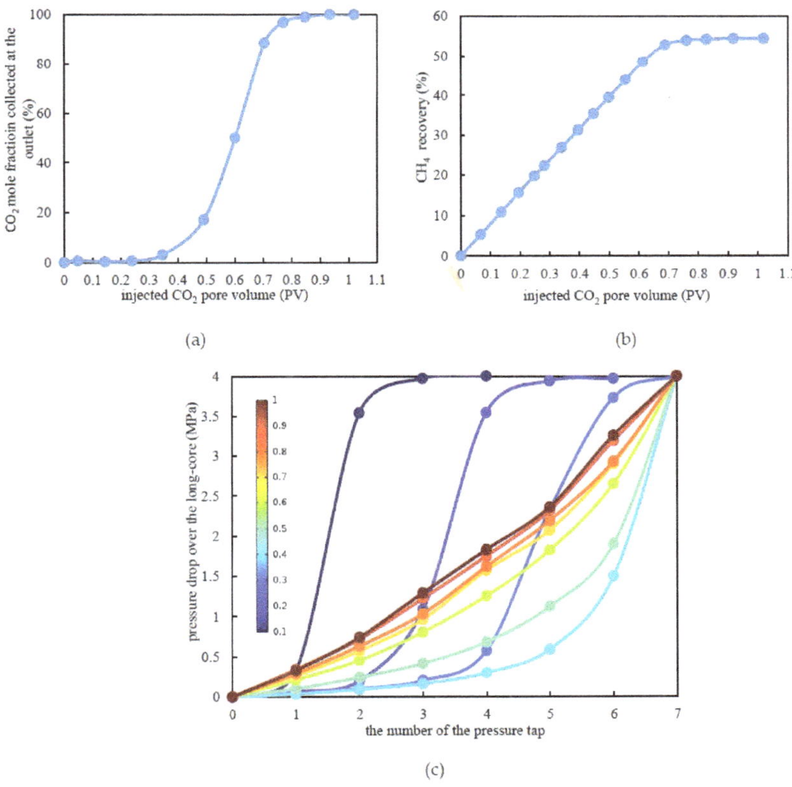

Figure 2. Experimental results under constant differential pressure displacement. (**a**) CO_2 mole fraction at the outlet versus the injected CO_2 pore volume. (**b**) CH_4 recovery by CO_2 flooding. (**c**) Pressure drop over a tight long-core versus the pore volume during CO_2 flooding (Note: the colored lines represent the various injected CO_2 pore volumes).

According to Figure 2a, as the pore volume (PV) of the injected CO_2 increased to 0.4 PV, a current of CO_2 was detected at the outlet, which indicated that displacement front breakthrough had occurred. Following this, the CO_2 mole fraction collected at the outlet increased almost linearly with the pore volume of the injected CO_2 until it reached 0.7 PV. When it reached approximately 0.8 PV, almost no CH_4 was detected at the outlet, which indicates that the CH_4 recovery had reached its maximum. It can be clearly observed from Figure 2c that the pressure drop was mainly diminished at the displacement leading edge. Before breakthrough of the leading edge, the higher the pore volume of the injected CO_2, the further the pressure propagation, and, for the same pressure test point, the pressure drop decreased. As the injected CO_2 reached 0.4 PV, breakthrough occurred at the displacement leading edge. Subsequently, with the increase in the pore volume of injected CO_2, the distribution of the pressure drop was more uniform over the whole long-core. With the formation of continuous flow channels, the pressure drop tended to be stable. When the injected CO_2 pore volume was 1.0 PV, the pressure drop curve was close to linear.

As seen in Figure 2b, the CH_4 recovery from CO_2 flooding was approximately 53.36%, which is an 18.36% increase when compared to the depletion development method for tight sandstone gas reservoirs. Therefore, it can be concluded that it is feasible to achieve EGR in tight sandstone gas reservoirs through CO_2 flooding.

3.2. CH_4 Recovery through CO_2 Flooding and CO_2 Storage Efficiency under Various Injection Rates

In order to determine the effect of the injection rate on EGR and CO_2 storage efficiency, four groups of experiments were conducted using injection rates of 0.2, 0.4, 0.6, and 0.8 mL/min, respectively. The experimental results are presented in Figure 3.

The results demonstrated that, with the injection rate conditions of 0.2, 0.4, 0.6, and 0.8 mL/min, displacement front breakthrough occurred when the injected CO_2 pore volume was 0.34, 0.42, 0.47, and 0.60 PV, respectively (Figure 3a). The lower the displacement velocity is, the earlier the occurrence of displacement front breakthrough is. This result occurred because the diffusion of CO_2 dominated under a lower injection rate. Moreover, the viscous pressure drop, which was generated by the flow and the capillary pressure between the gas and the water, constituted the percolation resistance. At a lower injection rate, because the water in the tiny capillaries could barely be driven and the long-core was already in a state of near-irreducible water saturation, the capillary pressure had little effect on the gas flow. CO_2 mainly flowed in the channels formed by larger-sized pores and throats. The main seepage resistance was the viscous force between the gas and the water attached to the pore surface while the viscous force was relatively lower at a lower injection rate. This was also a reason for gas breakthrough occurring sooner. Under a higher injection rate, in order to overcome higher seepage resistance, the injection pressure was apparently higher, which caused displacement of the additional water in the tiny capillaries. Consequently, the water saturation further decreased and the capillary pressure increased (Figure 4). When the water saturation decreased from 41.5% to 40.75%, the capillary force rapidly increased from 2.09 to 3.52 MPa. That means that the water saturation only decreased by less than 1.8% but caused an increase in a capillary force up to 168%. In this case, the capillary pressure and the viscous force increased simultaneously, but the seepage resistance was too high, which requires more time for the breakthrough to occur.

Figure 3b shows the changing degrees of CH_4 recovery with varying injection rates. There were large differences in the degree of CH_4 recovery for different time frames of CO_2 breakthrough. At CO_2 leading edge breakthrough, the CH_4 recovery was 43.24%, 46.10%, 50.46%, and 56.25%, which corresponds to the CO_2 injection rates of 0.2, 0.4, 0.6, and 0.8 mL/min. The difference between the maximum and the minimum degree of recovery was 13.01%. However, at the end of displacement, CH_4 recovery corresponding to the four injection rates above was 56.39%, 57.32%, 58.00%, and 59.80%. The CH_4 recovery difference was only 3.4% over the entire displacement range. This phenomenon demonstrated that the miscible behavior between supercritical CO_2 and CH_4 mainly occurred at the leading edge of the displacement and it required more time to occur under lower displacement velocity conditions. Therefore, the miscible band range was also relatively higher. In addition, a continuous flow channel gradually formed following breakthrough of the leading edge and, subsequently, no additional water outflow occurred at the outlet. This meant that the capillary pressure was essentially constant and the viscous force was slightly decreased. In general, slight changes existed in the swept volume of CO_2 flooding at various injection rates. Consequently, the injection rates had slight differences during EGR by CO_2 flooding.

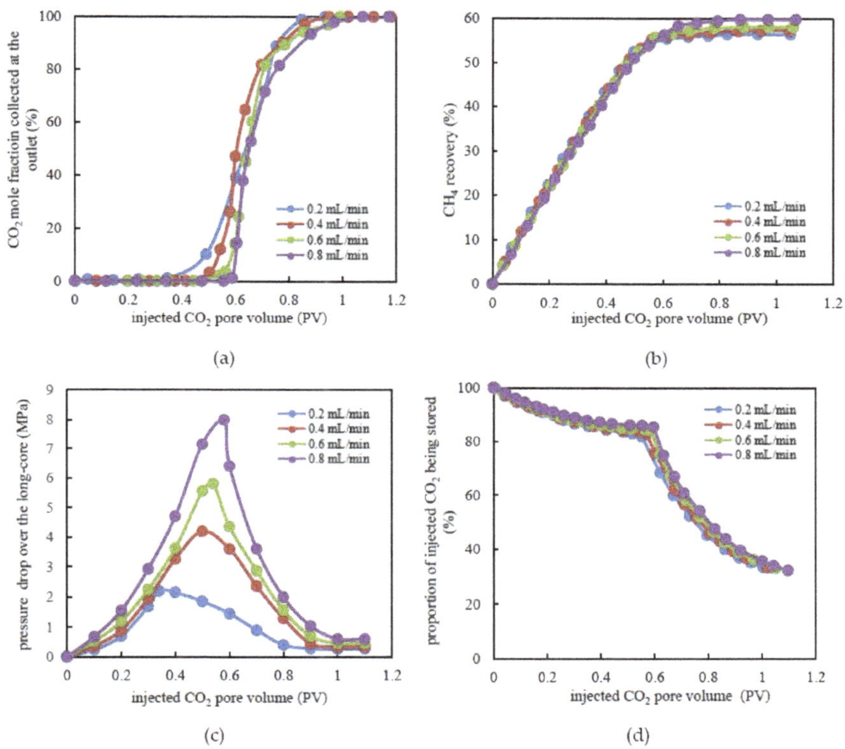

Figure 3. Experimental results under various injection rates. (**a**) CO_2 mole fraction at the outlet versus injected CO_2 pore volume. (**b**) CH_4 recovery by CO_2 flooding. (**c**) Pressure drop over tight long-core versus the pore volume during CO_2 flooding. (**d**) The proportion of injected CO_2 being stored versus an injected CO_2 pore volume.

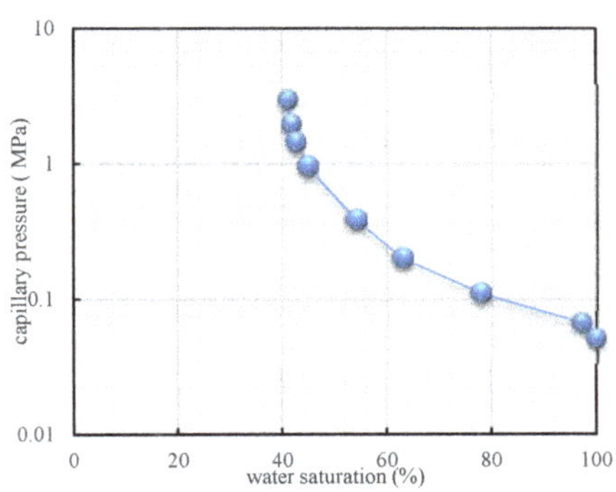

Figure 4. Curve of capillary pressure versus water saturation.

During CO_2 flooding, the increase in the pressure drop accelerated with the injected CO_2 pore volume prior to the displacement front breakthrough (Figure 3c). The pressure drop over the long-core attained its maximum value prior to the moment of the breakthrough. At the different breakthrough points, the pressure drop over the long-core was 2.12, 4.07, 5.89, and 8.16 MPa. Subsequently, the pressure drop decreased gradually and the speed decreased more quickly for higher injection rates of displacement under the same injected CO_2 pore volume conditions. At a lower injection rate, breakthrough of the leading edge mainly depended on the diffusion of CO_2 molecules. Consequently, the pressure drop became slower after breakthrough. As a flowing single gas gradually becomes a mixture of gases, the pressure drop was mainly concentrated in the latter half of the long-core. The closer to the outlet the pressure drop concentration is, the higher the resistance to seepage is.

The proportion of CO_2 being stored is defined as the ratio of the amount of CO_2 storage to the total amount of CO_2 injected. In the initial stage, most of the injected CO_2 remained in the long-core and the proportion of CO_2 being stored was higher (Figure 3d). From the injection of CO_2 to breakthrough of the leading edge, the proportion of the stored CO_2 decreased from 100% to 84.85%, 85.07%, 86.76%, and 87.64%, which corresponds to CO_2 injection rates of 0.2, 0.4, 0.6, and 0.8 mL/min, respectively. Following CO_2 breakthrough, the proportion of stored CO_2 decreased rapidly and it was almost unaffected by the injection rate. The final proportion of CO_2 being stored was 32.50%. This occurred due to the fact that both CO_2 and CH_4 had sufficient time to achieve sufficient miscibility with a lower injection rate. Therefore, additional CO_2 was stored subsequently to the leading edge breakthrough. To a certain extent, a higher CO_2 injection rate could improve gas recovery. However, it would also lead to larger pressure loss in a short time, which is not conducive to sustainable, stable production. As a consequence, the actual production system for developing tight sandstone gas reservoirs should take into account the joint determination of EGR and the economic payback period.

3.3. *CH_4 Recovery through CO_2 Flooding and CO_2 Storage Efficiency under Dry and Aqueous Cores*

In general, higher water saturation is common in the tight sandstone gas reservoirs of China [5]. In one group of experiments, the tight long-core was saturated with high-purity and dry CH_4. In the other group, an aqueous long-core with an irreducible water saturation of 41.5% existed under the confining pressure of 30 MPa. The injection rate was 0.2 mL/min.

The results demonstrated that the displacement front broke through when the injected CO_2 pore volume was 0.4 PV for the dry long-core while it was up to 0.5 PV for the aqueous core (Figure 5a). For the former, when the injected CO_2 was 0.76 PV, the CO_2 mole fraction collected at the outlet was 100% and no more CH_4 was displaced. For the latter, this situation did not occur until the injected CO_2 was 0.95 PV. Due to the presence of water, a portion of injected CO_2 initially dissolved and formed unstable carbonic acid in the water. In the same area, when the carbonic acid was saturated in the aqueous solution, CO_2 began to displace CH_4. Consequently, leading edge breakthrough occurred slightly later in time for the aqueous long-core with CO_2 flooding. When the CO_2 leading edge broke through, the CH_4 recovery was 48.78% and 52.48%, which corresponds to the dry long-core and the aqueous core. However, at the end of displacement, the corresponding CH_4 recovery was 54.68% and 56.23%, respectively (Figure 5b). In previous studies [28], it was shown that, due to the strong adsorption capacity of the tight sandstone rock surface, a tiny fraction of pores and throats might be blocked by water. Consequently, CH_4 exists in relatively larger-sized pores and throats, so it is more easily displaced. The amount of gas in this part was quite limited. Therefore, CO_2-EGR for the aqueous long-core was slightly enhanced when compared to the dry long-core.

As shown in Figure 5c, there was less variation in the range of pressure drop values for the aqueous long-core when compared to the dry long-core. Due to a portion of injected CO_2 being dissolved in the formation water, under the same injected CO_2 pore volume, the pressure drop was relatively lower when compared to the dry long-core. The higher the pressure, the higher the proportion of dissolved CO_2 in the formation water. At the point of CO_2 leading edge breakthrough, the pressure drop over the dry long-core and the aqueous long-core was 2.85 and 2.32 MPa. Following breakthrough of the leading edge, for the aqueous long-core, the seepage resistance was lower when compared to the dry long-core. Therefore, the pressure drop decreased more slowly.

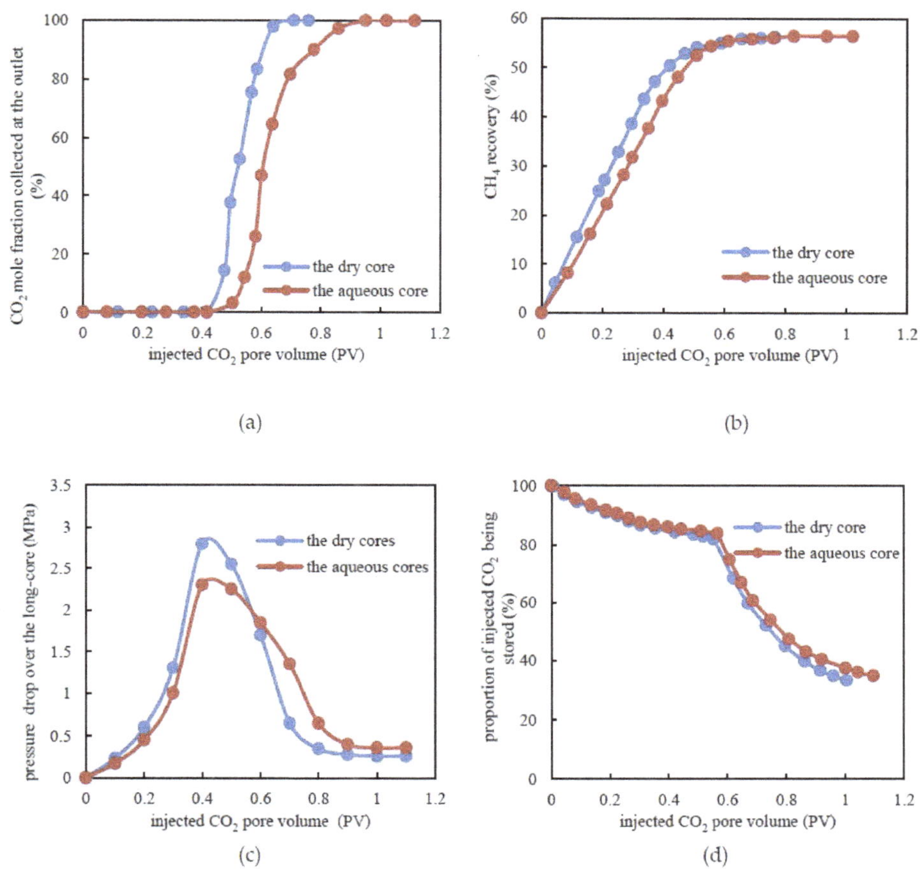

Figure 5. Experimental results under tight dry and aqueous cores. (**a**) CO_2 mole fraction at the outlet versus injected CO_2 pore volume. (**b**) CH_4 recovery by CO_2 flooding. (**c**) Pressure drop over tight long-core versus the pore volume during CO_2 flooding. (**d**) The proportion of injected CO_2 being stored versus an injected CO_2 pore volume.

From the injection of CO_2 to the breakthrough of the leading edge, the proportion of the stored CO_2 decreased from 100% to 84.03% and 84.80%, which corresponds to the dry long-core and the aqueous core, respectively (Figure 5d). Following breakthrough of CO_2, the proportion of CO_2 being stored decreased rapidly for the two situations mentioned above. At the end of displacement, the proportion of CO_2 being stored was 33.07% and 34.92%. In summary, due to the dissolution of CO_2 in the formation water, the total CO_2 being stored in the aqueous tight long-core was about 5% higher than that of the dry long-core during the whole displacement process. For the aqueous long-core, CO_2 was mainly stored by two mechanisms. One was to displace CH_4 and occupy its former position and the other way was to dissolve in the formation water. Accordingly, a slightly improved geological storage effect existed for CO_2 in tight sandstone reservoirs with formation water. This indicates that CO_2 flooding is more suitable for an aqueous tight sandstone gas reservoir (such as in China) since it results in a higher degree of enhanced gas recovery and improved CO_2 storage. In addition, with a stable injection rate, since the CO_2-EGR method can more or less mitigate the hindrance of water relative to the gas flow, controlling the water saturation could be regarded as a major factor rather than a primary factor in the development process of actual tight sandstone gas reservoirs.

3.4. CH_4 Recovery through CO_2 Flooding and CO_2 Storage Efficiency under Various Formation Dip Angle Conditions

Since the density of supercritical CO_2 was higher when compared to CH_4, the injected CO_2 tended to deposit at the reservoir bottom. Therefore, three groups of experiments were carried out to study the formation angle effect on EGR through CO_2 flooding. In the displacement experiments, the injection end of the long-core was at a lower position and the corresponding outlet was at a higher position with formation dip angles of 3° and 5°, respectively, relative to the actual formation conditions. In addition, a group of experiments was designed for comparison with the long-core placed horizontally. The injection rate was 0.2 mL/min.

The results demonstrated that the displacement front breakthrough occurred when the injected CO_2 pore volume was 0.37, 0.51, and 0.60 PV, which corresponds to the horizontally placed long-core and dip angles of 3° and 5°, respectively (Figure 6a). The higher the dip angle, the longer the miscible band and the later CO_2 breakthrough occurred. Since the stage before CH_4 breakthrough is an important period for CO_2 storage, such experimental results indicated that the CO_2 storage rate could be increased by at least around 1.5 times at this stage for a tight sandstone gas reservoir even with a small dip angle. When the CO_2 leading edge broke through, the CO_2 recovery was 40.49%, 53.03%, and 59.06%, which corresponds to the horizontally placed long-core and dip angles of 3° and 5°, respectively (Figure 6b). In addition, at the end of displacement, the corresponding CH_4 recovery was 54.49%, 59.65%, and 62.95%, respectively. Due to the higher density of supercritical CO_2, it was easier for CH_4 to move from the bottom to the higher part. Moreover, the gravitational differentiation between CO_2 and CH_4 limited the mixing of gases in the vertical direction, which was beneficial to the displacement. Furthermore, as the viscosity of supercritical CO_2 was higher when compared to CH_4, a favorable mobility ratio could improve the displacement stability. An additional 5% to 8% recovery of CH_4 could be achieved when compared to the horizontally placed long-core due to the dip angles (Figure 6b).

In terms of a pressure drop over the long-core, a relatively higher seepage resistance existed in the long-core with a dip angle at the early stage, which leads to a higher pressure drop for the same injected CO_2 pore volume (Figure 6c). At the point of CO_2 leading edge breakthrough, the pressure drop over the long-core was 2.40, 5.15, and 7.9 MPa, which corresponds to the horizontally placed long-core and dip angles of 3° and 5°, respectively. Subsequent to leading edge breakthrough, the pressure difference decreased faster in the slanted long-core. For the long-cores with a dip angle, the gravity increased the flow pressure at the injection end while the CH_4 at the bottom was displaced to a higher position by CO_2. The higher the dip angle, the higher the pressure drop over the long-core and the larger the stored volume of CO_2. The geological rate of CO_2 refers to the ratio of the amount of CO_2 storage to the pore volume of the core. Consequently, CO_2 could occupy the additional space that previously belonged to CH_4. Furthermore, the storage rate of injected CO_2 would be higher under the same conditions (Figure 6d). From the injection of CO_2 to breakthrough of the leading edge, the stored rate of CO_2 gradually reached 40.49%, 52.68%, and 56.29%, which corresponds to the horizontally placed long-core and dip angles of 3° and 5°, respectively. At the end of the displacement, the corresponding storage rate of CO_2 was eventually 54.88%, 59.65%, and 62.90%. Thus, the geological rate of CO_2 increased by 4.77% and 8.02% for the long-cores with dip angles of 3° and 5°. Therefore, the CH_4 recovery and CO_2 storage could be improved CO_2 flooding in tight gas sandstone reservoirs with a dip angle.

In addition, before CO_2 breakthrough, the pressure drop over the long-core with a dip angle of 5° was approximately four times that without a dip angle under the same injection rate and it stored a large amount of displacement energy. This suggested that the initial bottom-hole pressure should be appropriately higher in tight sandstone gas reservoirs with a dip angle. The purpose is to reduce the rapid coning entry of CH_4 and CO_2 after the gas leading edge breakthrough. Then the bottom-hole pressure should be slowed as the development progresses in order to effectively extend the effective period of CO_2-EGR and improve the storage rate of CO_2.

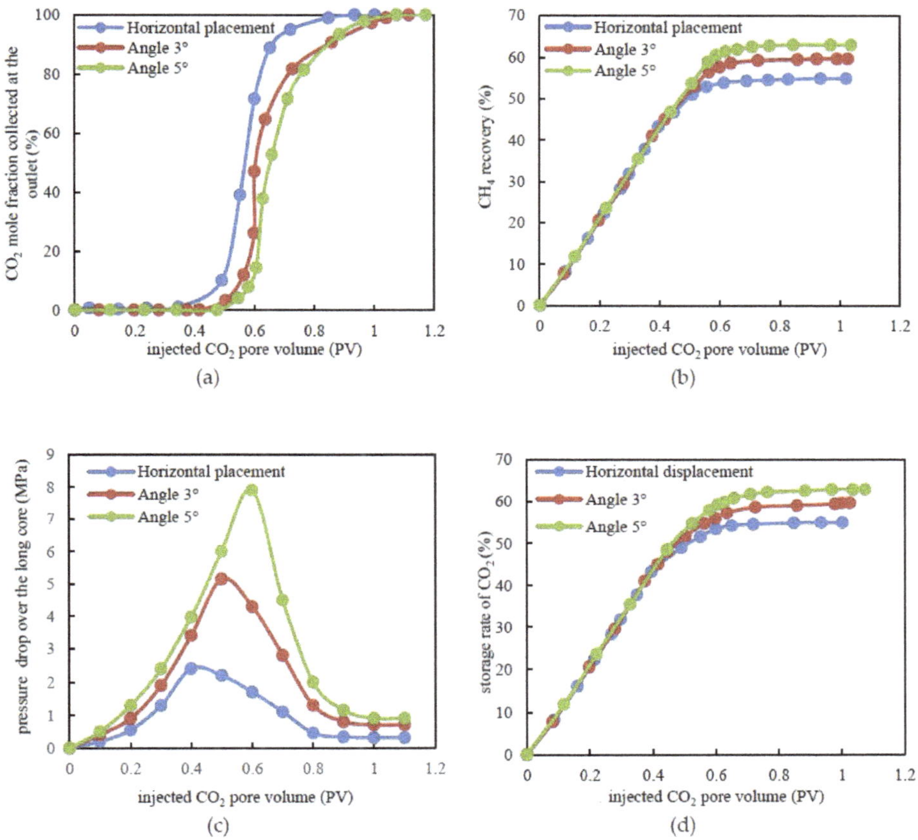

Figure 6. Experimental results under various formation dip angle conditions. (**a**) CO_2 mole fraction at the outlet versus injected CO_2 pore-volume. (**b**) CH_4 recovery by CO_2 flooding. (**c**) Pressure drop over tight long-core versus the pore-volume during CO_2 flooding. (**d**) The storage rate of CO_2 versus injected CO_2 pore-volume.

4. Conclusions

In this work, the extent of CO_2-EGR and CO_2 geological storage in a tight sandstone gas reservoir were investigated by carrying out a series of long-core displacement experiments with the purpose of analyzing the feasibility of CO_2-EGR. The effects of the injection rate, the water content of the core, and the ability to achieve CO_2 geological storage are discussed in detail and their implications for CO_2-EGR are summarized. It is found that the CH_4 recovery through CO_2 flooding is approximately 53.36%, which is an 18.36% increase when compared to the depletion development method for tight sandstone gas reservoirs. Additionally, the CO_2 geological rate is around 60%. The CH_4 recovery and CO_2 geological rate are related to the time of CO_2 leading edge breakthrough and the pressure drop distribution along the long-core. First, a higher injection rate improves the CH_4 recovery and CO_2 geological rate to some extent. However, a higher CO_2 injection rate leads to greater pressure loss. Therefore, a reasonable developing strategy for tight sandstone gas reservoirs should consider the combined effect of EGR and economic payback period. Second, for the aqueous tight long-core due to the dissolution of CO_2 in the formation water, the CH_4 recovery and the CO_2 geological rate are increased by 1.5% and 5%, respectively. The water phase reduces the pressure loss and extends the time to the gas leading edge breakthrough. Therefore, when a CO_2-EGR method is utilized to develop a tight sandstone gas reservoir, controlling water saturation could be regarded as a major factor rather than a primary factor. Third, compared with the horizontally placed long-core, the CH_4 recovery increases by an additional 5% to 8% for an inclined long-core with dip angles of 3° and 5°. In addition, the geological rate of CO_2 increases by 4.77% and 8.02%. Additionally, based on the pressure drop over the long-core before CO_2 breakthrough, the initial bottom-hole pressure should be appropriately higher in tight sandstone gas reservoirs with a dip angle. Then the bottom-hole pressure should be slowed as the development progresses in order to extend the effective period of CO_2-EGR and improve the geological rate of CO_2. This research is of great significance for guiding a successful CO_2-EGR process for the actual development of tight sandstone gas reservoirs.

Author Contributions: F.W. and Y.L. put forward the idea of the experiments in this paper; C.H. and Y.W. designed a series of experiments scheme and wrote the paper; A.S. and S.L. contributed to the results analysis and post-processing. All authors reviewed the manuscript.

Funding: This work was supported by the Natural Science Fund of China project (51804076) and the Natural Science for Youth Foundation of Heilongjiang Province (Grant No. QC2018047).

Conflicts of Interest: The authors declare no conflict of interest.

References

1. Peter, G.H.; Bath, O.B. Status report on miscible/immiscible gas flooding. *J. Pet. Sci. Eng.* **1989**, *2*, 103–117. [CrossRef]
2. Santos, L.; Marcondes, F.; Sepehrnoori, K. A 3D compositional miscible gas flooding simulator with dispersion using Element-based Finite-Volume method. *J. Pet. Sci. Eng.* **2013**, *112*, 61–68. [CrossRef]
3. Chen, M.; Wang, H.; Liu, Y.; Ma, L.; Wu, D.; Wang, S. Corrosion behavior study of oil casing steel on alternate injection air and foam liquid in air-foam flooding for enhance oil recovery. *J. Pet. Sci. Eng.* **2017**, *165*, 970–977. [CrossRef]
4. Hosseini, S.; Alfi, M.; Nicot, J.; Nunez, L. Analysis of CO_2 storage mechanisms at a CO_2-EOR site, Cranfield, Mississippi. Greenhouse. *Gases Sci. Eng.* **2018**, *27*, 218–229. [CrossRef]
5. Lai, F.; Li, Z.; Zhang, W.; Dong, H.; Kong, F.; Jiang, Z. Investigation of Pore Characteristics and Irreducible Water Saturation of Tight Reservoir Using Experimental and Theoretical Methods. *Energy Fuels* **2018**, *32*, 1–18. [CrossRef]
6. Wang, H.; Ma, F.; Tong, X.; Liu, Z.; Zhang, X.; Wu, Z.; Li, D.; Wang, B.; Xie, Y.; Yang, L. Assessment of global unconventional oil and gas resources. *Pet. Explor. Dev.* **2016**, *43*, 850–862. [CrossRef]
7. Guo, S.; Lyu, X.; Zhang, Y. Relationship between tight sandstone reservoir formation and hydrocarbon charging: A case study of a Jurassic reservoir in the eastern Kuqa Depression, Tarim Basin, NW China. *J. Nat. Gas Eng.* **2018**, *52*, 304–316. [CrossRef]
8. Wang, F.; Liu, Y.; Hu, C.; Shen, A.; Liang, S.; Cai, B. A Simplified Physical Model Construction Method and Gas-Water Micro Scale Flow Simulation in Tight Sandstone Gas Reservoirs. *Energies* **2018**, *11*, 1559. [CrossRef]
9. Gomez, A.; Briot, P.; Raynal, L.; Broutin, P.; Gimenez, M.; Soazic, M.; Cessat, P.; Saysset, S. ACACIA Project—Development of a Post-Combustion CO_2 Capture Process. Case of the DMXTM Process. *Oil Gas Sci. Technol.* **2014**, *69*, 1121–1129. [CrossRef]
10. Shi, Y.; Jia, Y.; Pan, W.; Huang, L.; Yan, J.; Zheng, R. Potential evaluation on CO_2-EGR in tight and low-permeability reservoirs. *Nat. Gas Ind. B* **2017**, *37*, 62–69. [CrossRef]
11. Littke, R.; Krooss, B.; Merkel, A.; Gensterblum, Y. Gas saturation and CO_2 enhancement potential of coalbed methane reservoirs as a function of depth. *AAPG Bull.* **2014**, *98*, 395–420. [CrossRef]
12. Liang, W.; Wu, D.; Zhao, Y. Experimental study of coalbeds methane replacement by carbon dioxide. *Chin. J. Rock Mech. Eng.* **2010**, *29*, 665–673. (In Chinese)

13. Zeng, Q.; Wang, Z.; Liu, L.; Ye, J. Modeling CH_4 Displacement by CO_2 in Deformed Coalbeds during Enhanced Coalbed Methane Recovery. *Energy Fuels* **2018**, *32*, 1942–1955. [CrossRef]
14. Dang, Y.; Zhao, L.; Lu, X.; Xu, J.; Sang, P.; Guo, S.; Zhu, H.; Guo, W. Molecular simulation of CO_2/CH_4 adsorption in brown coal: Effect of oxygen-, nitrogen-, and sulfur-containing functional groups. *Appl. Surf. Sci.* **2017**, *423*, 33–42. [CrossRef]
15. Murray, R.; Derek, B.; Ferus, I.; Cui, X.; Cory, T.; Mike, C. A Laboratory Study of CO_2 Interactions within Shale and Tight Sand Cores-Duvernay, Montney and Wolfcamp Formations. In Proceedings of the SPE Canana Unconventional Resources Conference, Calgary, AB, Canada, 13–14 March 2018. [CrossRef]
16. Xu, C.; Cai, J.; Yu, Y.; Yan, K.; Li, X. Effect of pressure on methane recovery from natural gas hydrates by methane-carbon dioxide replacement. *Appl. Energy* **2018**, *217*, 527–536. [CrossRef]
17. Stanfield, P. Use of low- and High-IFT fluid Systems in experimental and numerical modeling of Systems that mimic CO_2 storage in deep saline Formations. *J. Pet. Sci. Eng.* **2015**, *129*, 97–109. [CrossRef]
18. Darcis, M.; Class, H.; Flemisch, B.; Helmig, R. Sequential Model Coupling for Feasibility Studies of CO_2 Storage in Deep Saline Aquifers Couplage séquentiel des modèles numériques appliqué aux études de faisabilité du stockage de CO_2 en aquifère salin profond. *Oil Gas Sci. Technol.* **2011**, *66*, 93–103. [CrossRef]
19. Tian, W.; Li, A.; Ren, X.; Josephine, Y. The threshold pressure gradient effect in the tight sandstone gas reservoirs with high water saturation. *Fuel* **2018**, *226*, 221–229. [CrossRef]
20. Kubus, P. CCS and CO_2-storage possibilities in Hungary. In Proceedings of the SPE International Conference on CO_2 Capture, Storage, and Utilization, New Orleans, LA, USA, 10–12 November 2010. [CrossRef]
21. Birkedal, K.; Hauge, L.; Graue, A.; Ersland, G. Transport Mechanisms for CO_2-CH_4 Exchange and Safe CO_2 Storage in Hydrate-Bearing Sandstone. *Energies* **2015**, *8*, 4073–4095. [CrossRef]
22. Vandeweijer, V.; Meer, B.; Hofstee, C. Monitoring the CO_2, injection site: K12-B. *Energy Procedia* **2011**, *4*, 5471–5478. [CrossRef]
23. Zhang, Y.; Wang, Y.; Xue, F.; Wang, Y.; Ren, B.; Zhang, L.; Ren, S. CO_2 foam flooding for improved oil recovery: Reservoir simulation models and influencing factors. *J. Pet. Sci. Eng.* **2015**, *133*, 838–850. [CrossRef]
24. Fakhr, M.; Rashidi, A.; Giti, A.; Akbarnejad, M.; Masoud, Z. Parametric Study for the Growth of Single-walled Carbon Nanotubes over Co-Mo/Mgo in Fluidized Bed Reactor by CCVD Method. *J. Pet. Sci. Eng.* **2010**, *4*, 28–34. [CrossRef]
25. Bottero, S.; Hassanizadeh, S.; Kleingeld, P. From Local Measurements to an Upscaled Capillary Pressure–Saturation Curve. *Transp. Porous Media* **2011**, *88*, 271–291. [CrossRef]
26. Zou, C.; Zhu, R.; Wu, S.; Yang, Z.; Tao, S.; Yuan, X.; Hou, L.; Yang, H.; Xu, C.; Li, D.; et al. Type, characteristics, genesis and prospects of conventional and unconventional hydrocarbon accumulation: Taking tight gas in China as an instance. *Acta Pet. Sin.* **2012**, *33*, 173–187. [CrossRef]
27. Zhang, C.; Zhang, C.; Zhang, Z.; Qin, R.; Yu, J. Comparative experimental study of the core irreducible water saturation of tight gas reservoir. *Nat. Gas Geosci.* **2016**, *27*, 352–358. [CrossRef]
28. Hu, H.; Cheng, Y. Modeling by computational fluid dynamics simulation of pipeline corrosion in CO_2-containing oil-water two phase flow. *J. Pet. Sci. Eng.* **2016**, *146*, 134–141. [CrossRef]

processes

MDPI

Article

The Impact of Oriented Perforations on Fracture Propagation and Complexity in Hydraulic Fracturing

Liyuan Liu [1,2], Lianchong Li [1,*], Derek Elsworth [2], Sheng Zhi [2] and Yongjun Yu [1]

[1] Center for Rock Instability and Seismicity Research, Northeastern University, Shenyang 110819, China; liyuan426@163.com (L.L.); yuyongjun5658@163.com (Y.Y.)

[2] Department of Energy and Mineral Engineering, EMS Energy Institute and G3 Center, Pennsylvania State University, University Park, State College, PA 16802, USA; elsworth@psu.edu (D.E.); suz140@psu.edu (S.Z.)

* Correspondence: lilianchong@mail.neu.edu.cn; Tel.: +86-138-9888-6235

Received: 30 September 2018; Accepted: 26 October 2018; Published: 1 November 2018

Abstract: To better understand the interaction between hydraulic fracture and oriented perforation, a fully coupled finite element method (FEM)-based hydraulic-geomechanical fracture model accommodating gas sorption and damage has been developed. Damage conforms to a maximum stress criterion in tension and to Mohr–Coulomb limits in shear with heterogeneity represented by a Weibull distribution. Fracturing fluid flow, rock deformation and damage, and fracture propagation are collectively represented to study the complexity of hydraulic fracture initiation with perforations present in the near-wellbore region. The model is rigorously validated against experimental observations replicating failure stresses and styles during uniaxial compression and then hydraulic fracturing. The influences of perforation angle, in situ stress state, initial pore pressure, and properties of the fracturing fluid are fully explored. The numerical results show good agreement with experimental observations and the main features of the hydraulic fracturing process in heterogeneous rock are successfully captured. A larger perforation azimuth (angle) from the direction of the maximum principal stress induces a relatively larger curvature of the fracture during hydraulic fracture reorientation. Hydraulic fractures do not always initiate at the oriented perforations and the fractures induced in hydraulic fracturing are not always even and regular. Hydraulic fractures would initiate both around the wellbore and the oriented perforations when the perforation angle is >75°. For the liquid-based hydraulic fracturing, the critical perforation angle increases from 70° to 80°, with an increase in liquid viscosity from 10^{-3} Pa·s to 1 Pa·s. While for the gas fracturing, the critical perforation angle remains 62° to 63°. This study is of great significance in further understanding the near-wellbore impacts on hydraulic fracture propagation and complexity.

Keywords: hydraulic fracturing; gas fracturing; oriented perforation; fracture propagation; damage mechanics; fluid viscosity

1. Introduction

Hydraulic fracturing is commonly used to improve productivity from hydrocarbon reservoirs and has been a key technique in accessing unconventional shale reservoirs. As such, it is recognized as one of the most effective stimulation techniques for enhanced recovery from reservoirs [1,2]. Hydraulic fracturing is used not only to create macro-scale fractures but is also complicit in connecting and reactivating pre-existing fractures at all scales to create transport pathways within reservoirs. The transport linkage between preexisting flaws (i.e., perforations, joints, and natural fractures) are often critical to hydrocarbon production from reservoirs. These natural fractures or other flaws can induce complex fracture geometries and flow pathways resulting from both tensile and shear failure [3–5]. Implicit in this is understanding fracture initiation and propagation in the near-wellbore region as of great significance for hydraulic fracture treatment and injectivity test interpretation [6–8].

Since most wells are cased and require to be perforated to access the reservoir, the majority of hydraulic fracturing treatments are conducted through perforations. Perforations play a vital role in the origin of complex fracture geometries [6,9–12]. In fractured reservoirs, perforations are often the sole hydraulic connection between the fractured reservoir and the wellbore. Hydraulic fractures initiated from preexisting perforations will reorient themselves normal to the direction of the minimum far-field stress as they propagate out from the wellbore [2,4,13–16]. Near the wellbore, multiple sub-parallel fractures may be generated due to fracture initiation processes such as perforations and natural flaws located at the open-hole wellbore wall. Hydraulic fractures commonly initiate either from perforations located at the cased wellbore or from preexisting natural fractures intersecting an uncased wellbore [17]. However, when the perforation angle and the horizontal stress difference are large, dual fractures may be induced along perforations and the maximum stress direction. This means that the fractures formed by hydraulic fracturing from orientated perforations may be very complex. With regard to moving boundary problems in hydraulic fracturing, many model reduction techniques have been proposed to achieve the desired accuracy and computational efficiency [18–20]. Recently, several efforts have been made to determine an optimal perforation condition utilizing the integrated reduced-order model [21] and 3D fracture-propagation model [22]. Near-wellbore effects contain the fracture curvature, pressure enhancement in perforation hole and fracture branching, which are crucial issues regarding stimulated reservoir volume thus production. Therefore, in this paper, phenomena on the hydraulic fracture curving and branching near wellbore are focused. Furthermore, a finite element method (FEM)-based damage model is proposed to simulate the complete evolution of hydraulic fractures initiation and propagation. The impacts of oriented perforations on the near-wellbore fracture complexity are examined in terms of fracture trajectory, horizontal differential stress, fracturing fluid viscosity, and wellbore pressure.

Mechanical and transport properties of the reservoir may be significantly altered during stimulation [23–28]. In consideration of mentioned above, the impact of perforations and pre-existing features in reservoirs on the evolution of fluid-driven fractures, their interaction with pre-existing features and the resulting complexity of the evolving fracture network are investigated. This response is explored through a novel fully coupled model of rock damage and fluid flow for both slightly compressible and compressible fluids. Damage conforms to a maximum stress criterion in tension and to Mohr–Coulomb limits in shear with heterogeneity represented by a Weibull distribution. Fracturing fluid flow, rock deformation and damage, and fracture propagation are collectively represented to study the complexity of hydraulic fracture initiation with perforations present in the near-wellbore region. Specifically, the main objective of this study is thus to investigate the effects of preexisting perforation angle, horizontal differential stress, fracturing fluid, and initial pore pressure on fracture propagation and complexity. Research findings are applied to optimize perforation design in order to minimize high treatment pressures and to control complexity in the near-wellbore fracture region.

2. Conceptual Model

The near-wellbore fractures initiated from two symmetrically oriented perforations considered in this study are shown in Figure 1. The 2D hydraulic fracture model considers a cross-section through a vertical well with its axis aligned in the direction of maximum or intermediate principal stress. These two perforations are represented by two preexisting fractures (2D perforations rather than tunnels), which serve as initiation sites for the propagation of longitudinal hydraulic fractures. On the condition that the perforated interval is long, or the well is not cemented and cased, these hydraulic fractures and their propagation within the transverse plane are assumed to propagate under plane-strain conditions [17,29]. After initiation, and beyond the influence of the wellbore, the hydraulic fractures will reorient themselves from the perforation to the orientation of the minimum principal stress. However, when the perforation angles are large, more than two hydraulic fractures may initiate in a homogeneous rock instead of just a single zigzag shape fracture, as described in Figure 1.

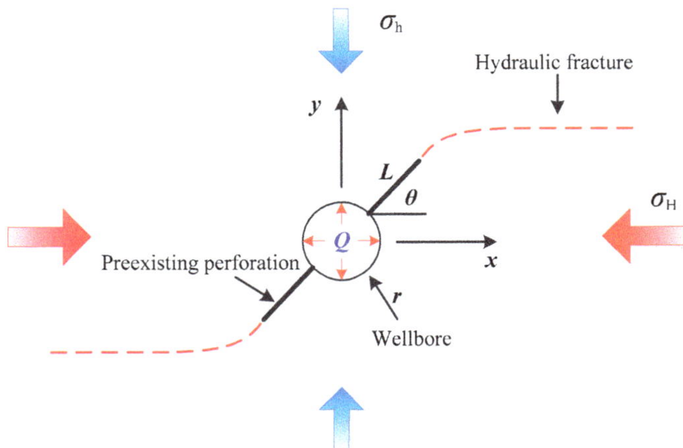

Figure 1. Hydraulic fractures initiate from two symmetrically disposed preexisting perforations.

The in situ stress field is defined in terms of the in-plane maximum horizontal stress σ_H and minimum horizontal stress σ_h. The preexisting perforation length L is assumed to be very small relative to the wellbore radius r—therefore the impact of preexisting perforation length on hydraulic fracture path can be neglected. The perforation is defined by its azimuth θ with respect to the orientation of maximum horizontal stress σ_H (x-axis). If the perforation tunnel is oriented in the direction of the maximum principal stress ($\theta = 0$), the hydraulic fractures would propagate directly in this (x-axis) direction. When the tunnels are oblique to this direction ($\theta \neq 0$) the hydraulic fractures must ultimately reorient themselves parallel to the direction of the maximum horizontal stress when they exit the stress shadows of the near-wellbore region and of the perforation. Beyond such stress-shadows, the hydraulic fractures tend to propagate in the plane of least resistance perpendicular to the minimum horizontal stress.

3. Formulation of the Conceptual Model

During hydraulic fracturing, fluids will penetrate into the porous medium surrounding both the borehole and the perforation tunnel. Thus, the evolution of the fracture(s) then results from this coupled phenomenon of fluid flow, solid deformation, and damage. The governing equations for this fully coupled hydrological-mechanical damage behavior were developed based on conservation of mass, momentum, and energy. The effects of rock damage on the evolving mechanical and transport parameters are also considered.

3.1. Governing Equations for Mechanical Response

The poroelastic constitutive relation for rock deformation is given by Reference [30]:

$$\varepsilon_{ij} = \frac{1}{2G}\sigma_{ij} - \left(\frac{1}{6G} - \frac{1}{9K}\right)\sigma_{kk}\delta_{ij} + \frac{\alpha}{3K}p\delta_{ij} + \frac{1}{3}\varepsilon_s\delta_{ij} \tag{1}$$

where G is the equivalent shear modulus (Pa), K is the bulk modulus (Pa), p is the fluid pressure (Pa), δ_{ij} is the Kronecker delta, α is the Biot coefficient, and ε_s is the adsorption strain (m/m).

The gas adsorption strain may be represented by the Langmuir isotherm, as follows [31]:

$$\varepsilon_s = \frac{\varepsilon_L p}{p + P_L} \tag{2}$$

where ε_L is the Langmuir strain constant (m/m) and P_L is the Langmuir pressure constant (Pa).

From this, a modified Navier-type constitutive deformation [32], in terms of displacement u_i (m) and fluid pressure p (Pa), can be expressed as:

$$Gu_{i,kk} + \frac{G}{1-2v}u_{k,ki} - \alpha p_{,i} - K\varepsilon_{s,i} + f_{,i} = \rho_s \frac{\partial^2 u_i}{\partial t^2} \tag{3}$$

where the effects of fluid pressure and injection rate on the rock deformation are defined as Equation (3).

3.2. Governing Equations for Fluid Flow

For the purpose to study the effect of different fracturing fluids on fracture complexity, fluid flow for both slightly compressible (liquids) and compressible (gases) fluids are presented, respectively.

3.2.1. Slightly Compressible Fluids

In general, the equation of state is defined with regard to the fluid compressibility, c_f (Pa^{-1}), as:

$$c_f = \frac{1}{\rho}\frac{\partial \rho}{\partial p} \tag{4}$$

Assuming that the fluid compressibility c_f is constant over a prescribed range of pressure, after integration, Equation (4) can be written as:

$$\rho = \rho_0 e^{c_f(p-p_0)} \tag{5}$$

where ρ_0 is the density (kg·m^{-3}) at the reference pressure p_0 (Pa). According to Taylor series expansion, this may be expressed as:

$$\rho = \rho_0\left[1 + c_f(p-p_0) + \frac{1}{2}c_f^2(p-p_0)^2 + \cdots\right] \tag{6}$$

with an approximate result given by:

$$\rho \approx \rho_0\left[1 + c_f(p-p_0)\right] \tag{7}$$

Similarly, the rock compressibility may be defined as:

$$c_r = \frac{1}{\phi}\frac{\partial \phi}{\partial p} \tag{8}$$

and again, after integration, this may be written as:

$$\phi = \phi_0 e^{c_r(p-p_0)} \tag{9}$$

where ϕ_0 is the porosity at p_0. Similarly, this may be approximated as:

$$\phi \approx \phi_0[1 + c_r(p-p_0)]. \tag{10}$$

Neglecting the effect of gravity, which will be small in relation to fluid pressure gradients, fluid flow may be defined in terms of Darcy's law as:

$$\mathbf{q} = -\frac{k}{\mu_f}\nabla p \tag{11}$$

where k is the equivalent permeability of the rock (m^2) and μ_f is the fluid viscosity (Pa·s).

According to the mass conservation equation, the governing equation for fluid transport in a porous medium is defined as:

$$\frac{\partial m}{\partial t} + \nabla \cdot (\rho \mathbf{q}) = Q_s \tag{12}$$

where m is fluid mass, ρ is fluid density (kg·m^{-3}), $\mathbf{q}$ is the Darcy velocity vector and Q_s is the source or sink of the compressible fluid defined in terms of mass rate (kg·m^{-3}·s^{-1}).

Supposing that the rock is fluid-saturated, the fluid content per unit volume can be expressed as $m = \rho\phi$, where ϕ is the porosity. For slightly compressible fluids the partial derivative of m with respect to time can be written as:

$$\frac{\partial m}{\partial t} = \phi \frac{\partial \rho}{\partial t} + \rho \frac{\partial \phi}{\partial t} \tag{13}$$

The total compressibility may be defined as:

$$c_t = c_f + \frac{\phi_0}{\phi} c_r \tag{14}$$

Combining Equations (4)–(14) yields the governing equation for slightly compressible fluid flow in the fracture network as:

$$\phi \rho c_t \frac{\partial p}{\partial t} + \nabla \cdot \left(-\frac{\rho k}{\mu_f} \nabla p \right) = Q_s \tag{15}$$

enabling solution for evolving flux or pressure boundary conditions and the evolution of permeability with damage.

3.2.2. Compressible Fluids

For compressible fluid (gas) the compressibility c_g is both much larger than for a slightly compressible fluid (liquid) and variable with pressure. In such a case, the pressure-volume-temperature (PVT) relation for a non-ideal or real gas can be written as:

$$\rho = \frac{pM}{ZRT} \tag{16}$$

where M is the molecular weight (kg/mol), Z is the gas compressibility factor (Pa^{-1}), R is the universal gas constant (J·mol^{-1}·K^{-1}), and T is the temperature (K). Assuming that the gas compressibility and viscosity are constant, then the governing relation is defined as:

$$\phi \frac{M}{ZRT} \frac{\partial p}{\partial t} + \nabla \cdot \left(-\frac{\rho k}{\mu_g} \nabla p \right) = Q_s \tag{17}$$

accommodating the pressure dependent compressibility of the fluid.

3.3. Governing Equations Accommodating Rock Heterogeneity and Damage Evolution

To characterize heterogeneity at the mesoscopic-level, mechanical parameters such as strength, Young's modulus, and Poisson's ratio may be assigned according to the Weibull distribution. This distribution defines the parameter according to a probability density function as [33]:

$$f(u) = \frac{m}{u_0} \left(\frac{u}{u_0} \right)^{m-1} e^{-(\frac{u}{u_0})^m} \tag{18}$$

where u is the parameter of interest (such as strength, Young's modulus, or Poisson's ratio); the scale parameter u_0 is related to the average of the element parameter, and m is the shape parameter of the Weibull distribution function. m is defined as the degree of rock homogeneity and called the homogeneity index. The variations of $f(u)$ with respect to m are shown in Figure 2. Obviously, a higher m value represents a more homogeneous rock.

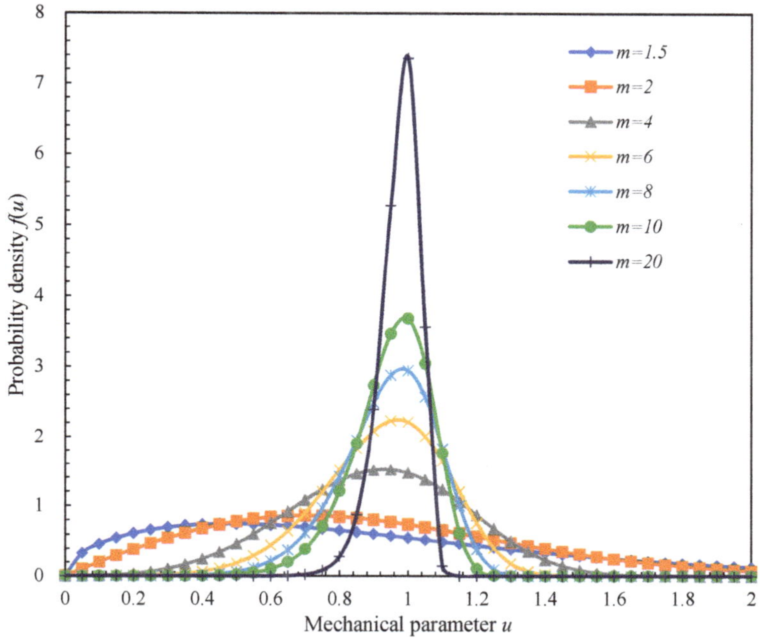

Figure 2. Distributions of rock properties for different homogeneous indices (Mechanical parameter u_0 is 1).

As illustrated in Figure 3, rock damage in tension or shear is initiated when its state of stress (positive for compression) satisfies the maximum tensile stress criterion or the Mohr–Coulomb criterion, respectively as [27,33]:

$$F_1 = -\sigma_3 - f_{t0} = 0,\ F_2 = \sigma_1 - \sigma_3 \frac{1 + \sin\theta}{1 - \sin\theta} - f_{c0} = 0 \quad (19)$$

where f_{t0} and f_{c0} are uniaxial tensile and compressive strength (Pa), respectively, σ_1 and σ_3 are first and third principal stresses (Pa), respectively, θ is the internal frictional angle (°), and F_1 and F_2 are two damage threshold functions (Pa).

Based on the theory of elastic damage, the elastic modulus of the damaged rock is expressed as [33]:

$$E = (1 - D)E_0 \quad (20)$$

where D represents the damage variable, and E and E_0 are the Young's modulus (Pa) of the damaged and the undamaged element, respectively. In the present study, the element, as well as its damage, is assumed isotropic, so the E, E_0 and D parameters are all scalars. Under any stress and initial conditions, the tensile stress criterion is applied preferentially. In other words, the maximum tensile stress criterion is first applied to judge whether the elements are first damaged in tension or not. Only elements that survive this test will be checked for damage in shear using the Mohr–Coulomb criterion.

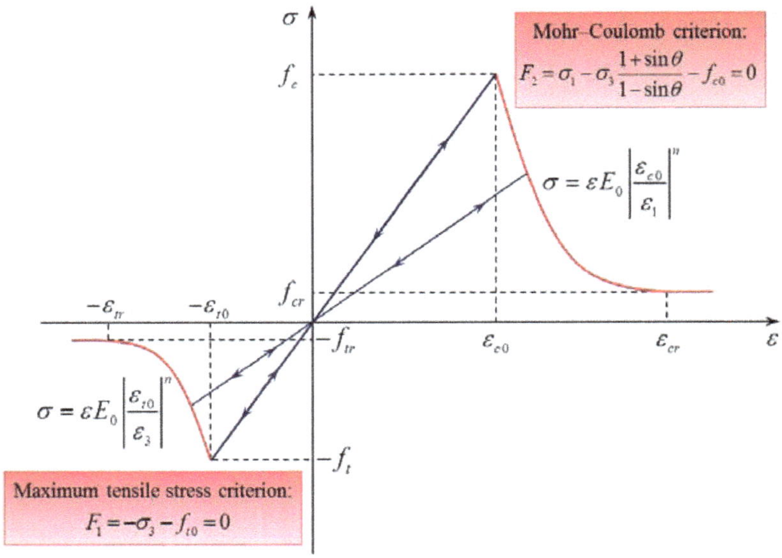

Figure 3. The damage constitutive law for rock under uniaxial stress condition.

In terms of the damage constitutive law shown in Figure 3, the damage variable for the rock can be calculated as [33]:

$$D = \begin{cases} 0 & F_1 < 0 \quad and \quad F_2 < 0 \\ 1 - \left|\frac{\varepsilon_{t0}}{\varepsilon_3}\right|^n & F_1 = 0 \quad and \quad dF_1 > 0 \\ 1 - \left|\frac{\varepsilon_{c0}}{\varepsilon_1}\right|^n & F_2 = 0 \quad and \quad dF_2 > 0 \end{cases} \tag{21}$$

where ε_1 and ε_3 are the major and minor principal strains (m/m), respectively. ε_{t0} and ε_{c0} are the maximum tensile and maximum compressive principal strains (m/m) when tensile and shear damage occurs, respectively, and n is a constitutive coefficient specified as 2.0. In Equation (19), when $F_1 < 0$ and $F_2 < 0$ the applied stress is insufficient to satisfy the maximum tensile stress criterion and the Mohr–Coulomb failure criterion, respectively. $F_1 = 0$ and $\mathrm{d}F_1 > 0$ implies rock damage in the tensile mode when the stress state satisfies the maximum tensile stress criterion and the rock is still under load. $F_2 = 0$ and $\mathrm{d}F_2 > 0$ implies rock damage in the shear mode when the stress state satisfies the Mohr–Coulomb failure criterion and the rock remains loaded.

The damage variables of Equation (21) remain in the range from 0 to 1.0 regardless of the form or magnitude of damage. However, negative and positive damage magnitudes are respectively adapted for damage in tension and in shear, merely to allow visualization of the two damage modes in post-processed figures. In this regard, tensile damage is represented as negative numbers ($-1 \leq D < 0$), while shear damage is represented as positive numbers ($0 < D \leq 1$). In the respect of the irreversibility of damage, the damage variable may only increment monotonically from zero during loading ($dF_1 > 0$ or $dF_2 > 0$) and remain unchanged during unloading ($dF_1 < 0$ or $dF_2 < 0$). In this respect, the damage, defined by Equation (20), reduces the elastic modulus E and the shear modulus G of the rock, via to Equations (20) and (21).

4. Model Validation against Experimental Observations

In this section, the hydraulic fracturing geomechanical model proposed in this study is validated against observed mechanisms of uniaxial compression and hydraulic fracturing [11].

4.1. Comparisons of Breakdown Pressure and Fracture Geometry

The relationship between axial stress and axial strain during modeled rupture of a uniaxial specimen is shown in Figure 4 relative to the mechanical parameters recovered from experiment [11]. Apparent from Table 1 and Figure 4 is that the experimental data adequately replicate the anticipated stress-strain behavior. Additionally, it should be noted that the simulation of uniaxial compression is conducted to provide basic mechanical properties to inform the numerical simulation of hydraulic fracturing conducted in the same material. A visual comparison between experimental results [11] and the numerical simulations is shown in Figure 5. As can be seen from Figure 5, the fracture pattern recovered from the numerical model is consistent with experimental observations. Hydraulic fractures initiate outward along the perforation and as they propagate away from the near-wellbore towards the region of unaltered in-situ stress they reorient themselves perpendicular to the minimum horizontal stress.

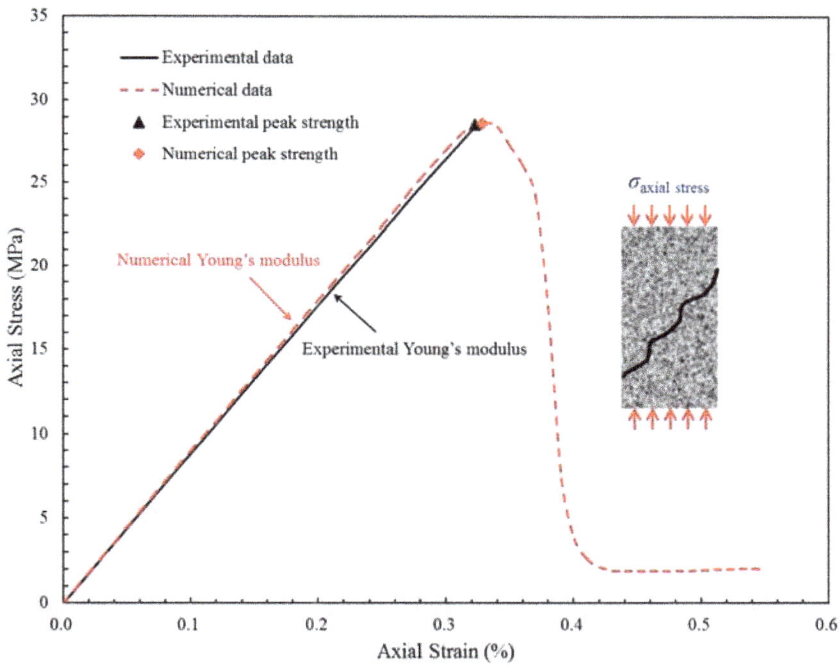

Figure 4. Simulation of uniaxial compression of a sample compared with experimental observations by Chen et al. [11].

Table 1. Comparison of experimental data and numerical data.

Symbol	Experimental Data	Numerical Data	$\Delta E/E$	$\Delta\sigma_c/\sigma_c$	$\Delta\sigma_t/\sigma_t$
Young's modulus, E (GPa)	8.402	8.51	1.29%	-	-
Uniaxial compressive strength, σ_c (MPa)	28.34	28.57	-	0.81%	-
Tensile strength, σ_t (MPa)	2.59	2.56	-	-	−1.16%

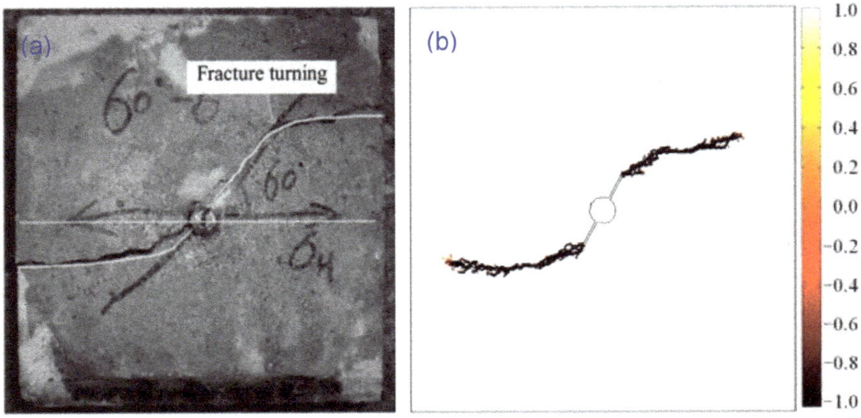

Figure 5. Comparison of model simulation results with experimental results. (**a**) Experimental fracture pattern from Chen et al. [11]. (**b**) Fracture pattern obtained from numerical simulation.

4.2. Geometric Model and Boundary Conditions

The geometry and boundary conditions of the specimen are simplified as shown in Figure 6. The numerical specimen, 300 mm on edge with a 10 mm diameter borehole at the center is subjected to a biaxial horizontal stress. The boundary conditions correspond to a confining pressure σ_H applied on the top boundary and σ_h applied on the right boundary with rollers applied along both the left side and the base. There are no-flux conditions on all boundaries except for the borehole wall into which a constant fluid injection rate is applied. The fluid injection rate in the experiments is 2.1×10^{-9} $m^3 \cdot s^{-1}$. The numerical mechanical properties for the simulation are listed in Table 2.

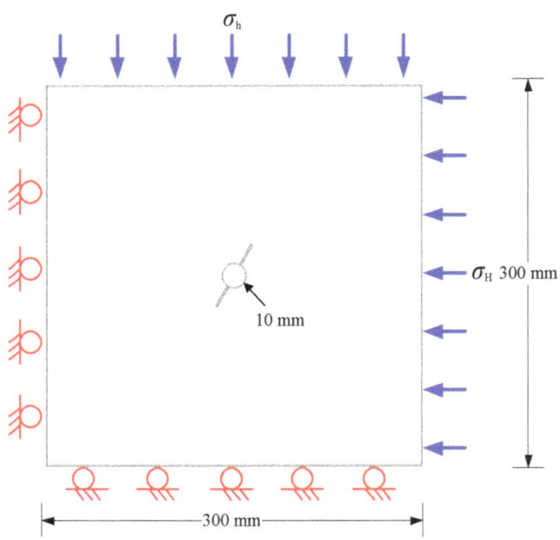

Figure 6. Geometry of the numerical model used to investigate near-wellbore hydraulic fracturing processes.

Table 2. Input parameters for the near-wellbore hydraulic fracturing model.

Symbol	Value	Unit
Homogeneity index, m	10	-
Mean value of the elasticity modulus, E	8.737	GPa
Mean value of uniaxial compressive strength, $\overline{f}_c$	45.53	MPa
Mean value of uniaxial tensile strength, $\overline{f}_t$	5.69	MPa
Poisson ratio, ν	0.23	-
Initial porosity, ϕ_0	1.85	%
Initial permeability, k_0	1.0×10^{-16}	m^2
Initial pore pressure, p_0	0.1	MPa
Viscosity, μ	133	mPa·s

4.3. *Effects of Preexisting Perforation Orientation*

Perforation orientation has a significant effect on the geometry of the hydraulic fracture initiated from the wellbore. The azimuth and length of the preexisting perforation are two important factors that affect hydraulic fracture propagation and pattern. An optimal preexisting perforation should initiate only a single wing fracture with minimum tortuosity at an achievable breakdown pressure [9,34]. Numerical simulations for cases with preexisting perforation angles of 0°, 15°, 30°, 45°, 60°, and 75° are conducted to examine the response. As shown in Figures 7 and 8, a larger perforation angle induces a relatively larger curvature during the reorientation of the hydraulic fractures and takes a long distance for the hydraulic fracture to rotate to the direction completely aligned with the direction of the maximum horizontal stress. A larger perforation angle requires a higher breakdown pressure to initiate the hydraulic fracture. In addition, another observation is that hydraulic fractures are initiated both around the wellbore and preexisting fractures when the perforation angle is 75°. Additionally, hydraulic fractures do not always initiate from preexisting perforations. The fractures formed by hydraulic fracturing with oriented perforations have a complex morphology. Therefore, for perforations in field operations, perforations with a high angle are generally not recommended and should be avoided as much as possible [14]. Specifically, there is a significant difference in breakdown pressure between numerical and experimental results, when the perforation angle is 75°. This may be due to experimental inconsistencies or due to heterogeneity in the sample, since it is expected that the breakdown pressure would increase with an increase of the perforation angle [9,14,17].

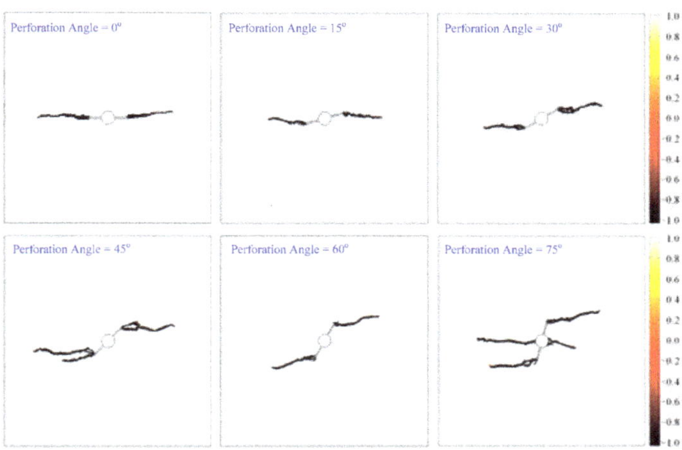

Figure 7. Numerical results of near-wellbore fracture patterns for different preexisting perforation angles.

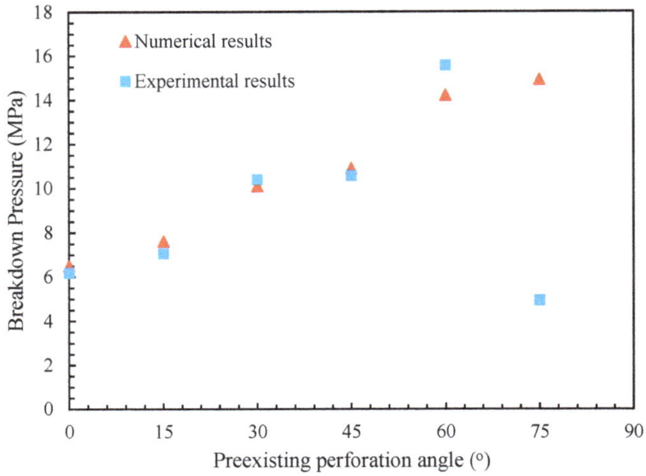

Figure 8. Comparison of experimental results and numerical results for breakdown pressures.

5. Analysis of Near-Wellbore Hydraulic Fracture Complexity

In situ stress, initial pore pressure, and fluid viscosity are important factors that affect the tortuosity and extension pressure of near-wellbore hydraulic fractures. In order to fully understand hydraulic fracture behavior in the field, sensitivity studies are conducted to investigate the effects of in situ horizontal differential stress, initial pore pressure, and fracturing fluid viscosity.

5.1. Effects of Horizontal Differential Stress

In order to investigate the effects of different in situ stresses on hydraulic fracture behavior, stress ratios λ of 1, 1.2, 1.4, 1.6, 1.8, and 2 are selected (stress ratios define the ratio between maximum horizontal stress and minimum horizontal stress). The magnitude of σ_h is held constant at 20 MPa. Stress ratio has a significant effect on both hydraulic facture propagation and reorientation, as shown in Figure 9, where different stress ratios correspond to different fracture morphologies. When the stress ratio is equal to unity (hydrostatic) the hydraulic fractures initiate and propagate along the direction of the oriented perforation. As the stress ratio λ is proportional to the difference between σ_H and σ_h, the larger this difference, the larger the propensity for the hydraulic fractures to reorient themselves to the maximum horizontal stress direction [35]. A larger stress ratio results in both a smaller curvature during reorientation and occurs at a shorter distance for the fractures to reorient their direction to align with the direction of the maximum horizontal stress. Moreover, based on the results presented in Figure 10 and Table 3, it appears that the stress ratio also has a significant influence on the initiation pressure and breakdown pressure during hydraulic fracturing. It is apparent that there are clear reductions both in the initiation pressure and the breakdown pressure with increasing stress ratio.

Table 3. Initiation pressure and breakdown pressure for different stress ratios.

Stress Ratio	Perforation Angle (°)	Initiation Pressure (MPa)	Breakdown Pressure (MPa)
1.0	45	30.3	54.5
1.2	45	29.4	53.7
1.4	45	28.3	52.3
1.6	45	27.9	51.6
1.8	45	26.5	49.5
2.0	45	25.2	46.8

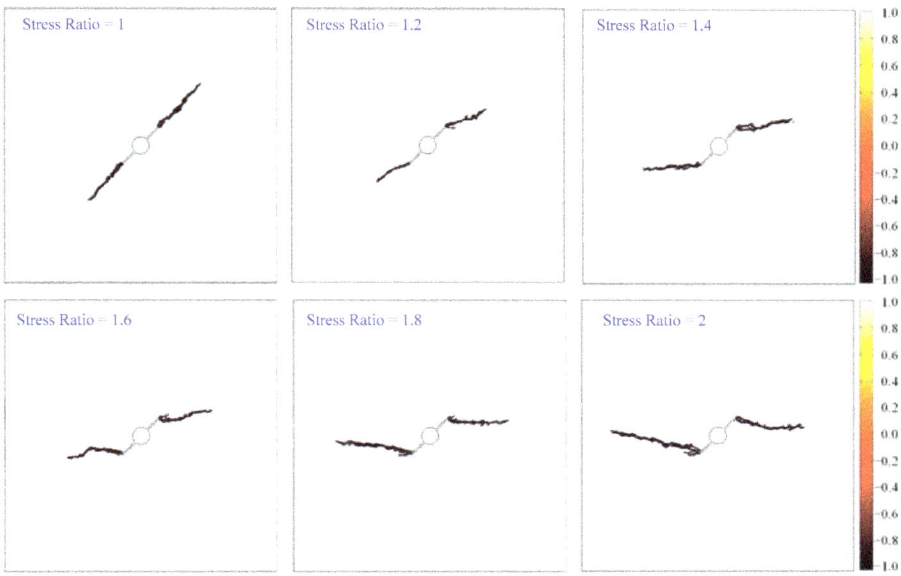

Figure 9. Simulation results of fracture tortuosity/complexity of hydraulic fracturing produced at various stress ratios.

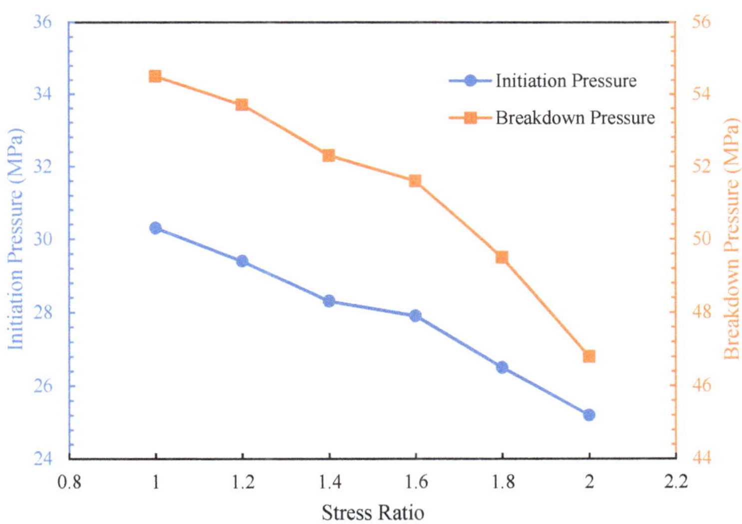

Figure 10. Initiation and breakdown pressures with different stress ratios.

As mentioned above, a larger perforation angle results in a complicated fracture morphology, which hydraulic fractures are initiated both from perforations and wellbore, as shown in Figure 7 (perforation angle equals 75°). In order to avoid to forming such complicated fracture morphology, the critical perforation angle is defined, which is used as the minimum perforation angle to form complex fracture morphology. A series of numerical simulations have been conducted to identify the changes of critical perforation angle with stress ratios and fracturing fluids. The results of critical perforation angle against stress ratio are shown in Figure 11. Under the condition of hydrostatic pressure, the hydraulic fractures initiate and propagate along the direction of the oriented perforation, and thus will not form complex fracture morphology. The larger the stress ratio is, the less the critical perforation angle is, and more easily to form complicated fracture morphology. Such complex fracture morphology can result in difficult transportation of proppants and a significant increase in the required injection pressure for propagating hydraulic fractures, which are an undesirable phenomenon that should be avoided as much as possible in field operations [10,14].

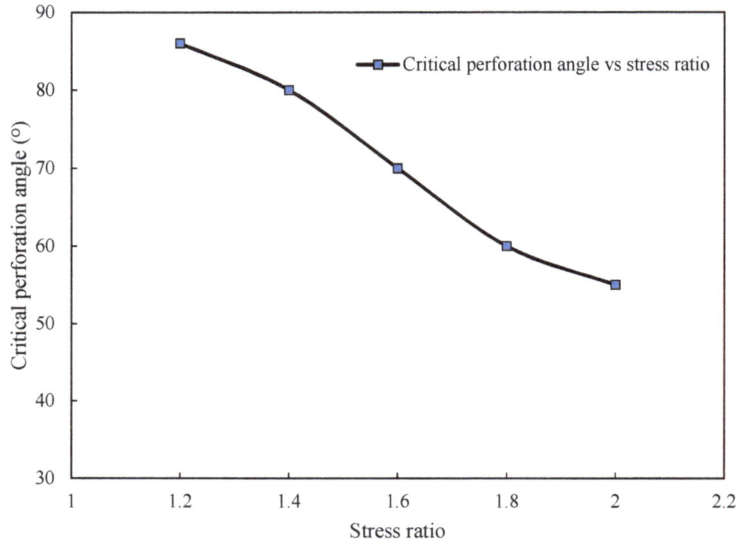

Figure 11. The relationship between critical perforation angle and stress ratio.

5.2. *Effects of Initial Pore Pressure*

With constant total stresses of σ_h (20 MPa) and σ_H (32 MPa), the initial pore pressure is varied from 2 MPa to 16 MPa. The purpose of these numerical simulations is to examine the effects of initial pore pressure on initiation pressure and breakdown pressure. An increase of fluid pore pressure can decrease static friction and thereby facilitate fracture propagation on favorably oriented planes in a deviatoric stress field. As shown in Figures 12 and 13, the resulting differential initiation pressure (P_c-P_0) and breakdown pressure (P_b-P_0) decrease with an increase in the initial pore pressure. This is consistent with the D–C (Detournay–Cheng) criterion [36]. To incorporate an effective stress law into the D–C criterion, the geomechanical model proposed in this study correctly describes the relationship between breakdown pressure and the far-field stress in hydraulic fracturing [37]. The effects of pore pressure on the initiation pressure are illustrated by the initiation equation based on poroelastic theory [38]. The initiation pressure decreases with an increase in the initial pore pressure [39]. There are two classical approaches to define initiation pressure in terms of far-field stresses [38,40]. These represent behavior for both nonpenetrating injection fluids [40], and for penetrating fluids [38], with initiation pressures given by:

$$\begin{cases} P_{H-W} - P_0 = T + 3(\sigma_h - \alpha P_0) - (\sigma_H - \alpha P_0) \\ P_{H-F} - P_0 = \frac{T+3(\sigma_h - \alpha P_0) - (\sigma_H - \alpha P_0)}{2 - \alpha \frac{1-2\nu}{1-\nu}} \end{cases} \tag{22}$$

where $P_{H\text{-}W}$ and $P_{H\text{-}F}$ are the initiation pressures related to Hubbert–Willis [40] and Haimson–Fairhurst [38] equations, respectively, T is the tensile strength of the rock, and σ_h and σ_H are the far-field principal stresses. As $0 \leq \phi \leq \alpha \leq 1$ (ϕ is rock porosity) and $0 \leq \nu \leq 0.5$ (ν is Poisson' ratio of rock) for rock, then obtain $0 \leq \alpha(1-2\nu)/(1-\nu) \leq 1$. Therefore, the initiation pressure for a penetrating fluid is always smaller than (or equal to) that for a nonpenetrating fluid. It can be seen from Figure 12 that the resulting differential initiation pressure (P_c-P_0) during gas fracturing is significantly different from that for liquid fracturing. The obvious effect of gas fracturing is in the reduction of the initiation pressure. This is in good agreement with Equation (22). In addition, as shown in Figures 12 and 13, for water-based fracturing, the resulting P_c-P_0 and P_b-P_0 decrease linearly with an increase in the initial pore pressure. However, for gas fracturing, the resulting P_c-P_0 and P_b-P_0 indicate a nonlinear decrease with an increase in initial pore pressure. The influences of gas penetration complicate the mechanism of the fracturing process. The gas penetration not only alters the pore pressure in the reservoir, but also the gas adsorption-induced damage modifies the mechanical properties of the reservoir rock [27,41,42]. According to the Langmuir adsorption isotherm, as shown in Equation (2), with an increase in pore pressure, the sorption capacity and volumetric strain simultaneously increase with the adsorptive strain potentially resulting in additional rock damage. This adsorption-induced damage is fully coupled to the gas fracturing model proposed in this paper. As a consequence, both the initiation pressure and breakdown pressure of gas fracturing show nonlinear decreases with increasing pore pressure.

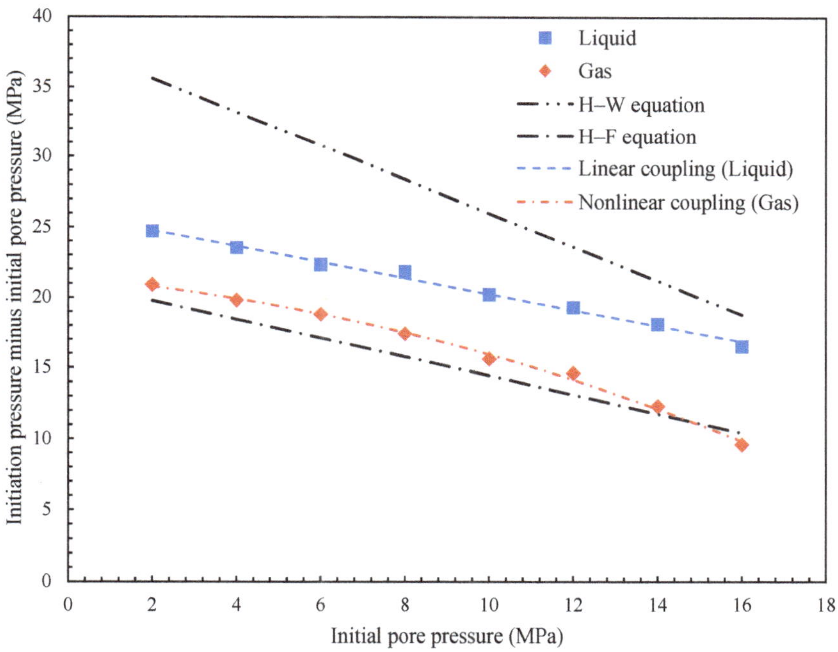

Figure 12. Relationship between initiation pressure and initial pore pressure.

5.3. Effects of Fracturing Fluids

The impacts of both slightly compressible (liquid) and compressible (gas) fluids in fracturing performance were examined to better understand fracture behavior during hydraulic fracturing with liquids or gases. Fluid viscosity is an important variable that influences fracture propagation - the higher the fluid viscosity the slower the local flow rate in the fracture and the higher the fluid pressure under the same injection rate [43,44]. As shown in Figure 14, when the viscosity of the liquid is high the breakdown pressure is large. Conversely, as the gas has a lower viscosity, the breakdown pressure is also smaller. This fact is illustrated in Figure 14, which shows that the relationship between breakdown pressure and fluid viscosity follows a power-law function. In addition, the higher the fracturing fluid viscosity, the more complex the fracture geometry local to the wellbore. This is because a fluid with relatively higher viscosity has a relatively slower flow rate in the near-wellbore region [45]. Figure 15 shows the variations of critical perforation angle with different viscosities of fracturing fluids. For the liquid-based hydraulic fracturing, the critical perforation increases with an increase in liquid viscosity. While for the gas fracturing, the critical perforation angle does not change a lot with the changes of gas viscosity. Although it is essential to create a dominant hydraulic fracture for hydrocarbons recovery, one should carefully design the fracturing fluid to avoid damage to the near-wellbore fracture system.

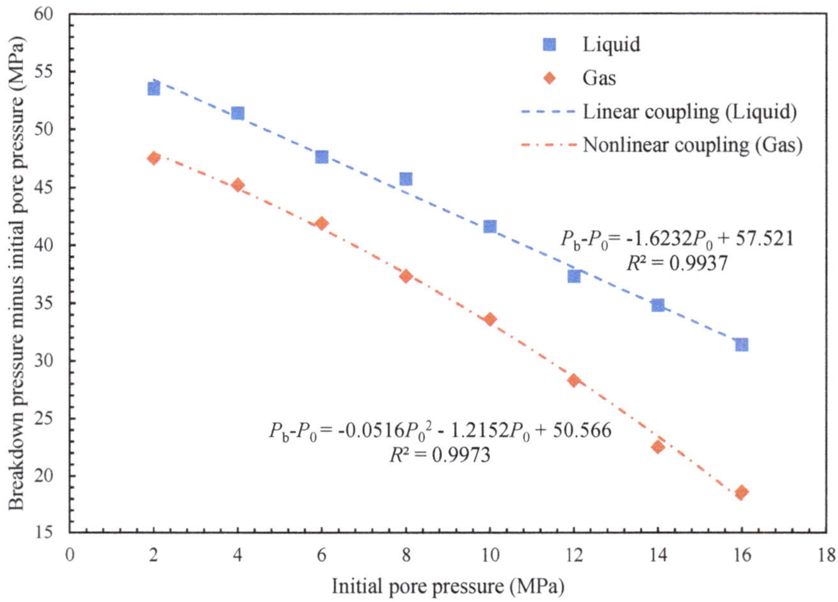

Figure 13. Breakdown pressure as a function of initial pore pressure.

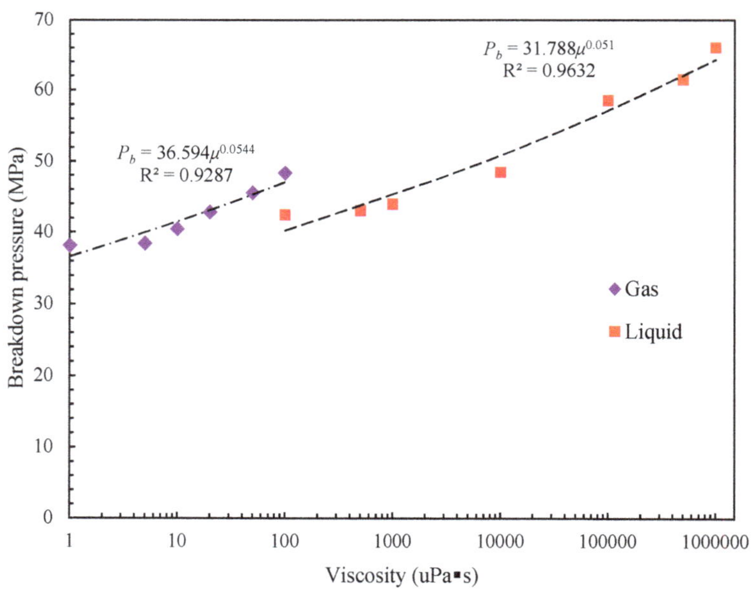

Figure 14. Relationship between breakdown pressure and fracturing fluid viscosity.

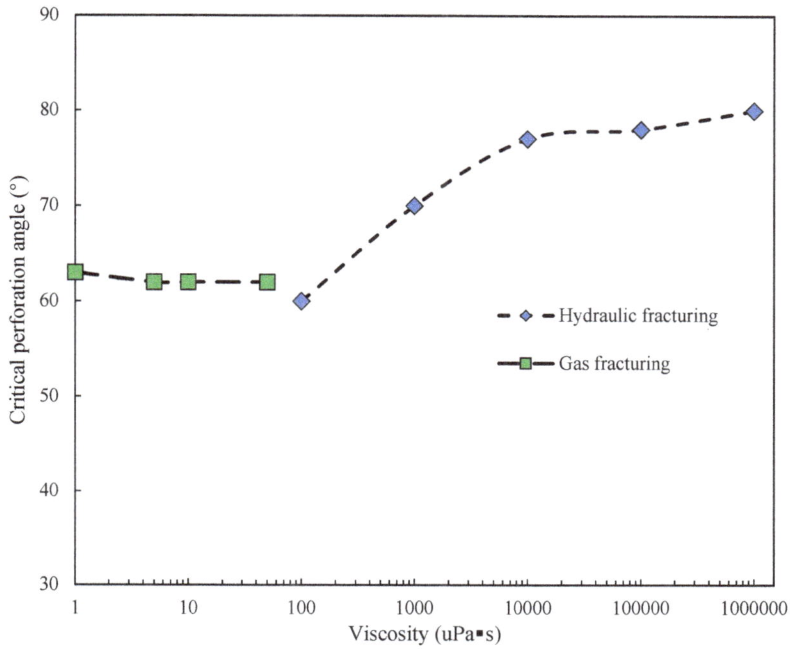

Figure 15. The relationship between critical perforation angle and stress ratio.

6. Conclusions

In this study, a fully-coupled geomechanical hydraulic fracture model is proposed under the rubric of FEM-based damage mechanics. This approach accommodates maximum tensile stress theory with shear failure according to Mohr–Coulomb theory and is applied to predict fracture propagation and complexity in heterogeneous rocks. The physical process involves fully coupling fluid flow in the fracture and pressure diffusion into the surrounding porous rock while simultaneously accommodating rock damage, deformation, and fracture propagation. The form of the near-wellbore fractures initiating from two symmetrically-disposed radial perforations are analyzed. The effects of perforation angle, horizontal differential stress, fracturing fluid, and initial pore pressure on hydraulic fracturing are investigated. The following conclusions are derived from the numerical simulations.

A larger perforation angle induces a relatively larger curvature of the fracture during hydraulic fracture reorientation and requires a longer distance for the hydraulic fracture to rotate to the direction that is completely aligned with the direction of the maximum horizontal stress. Hydraulic fractures do not always initiate at the oriented perforations and the fractures induced during hydraulic fracturing are not always even and regular. Hydraulic fractures are initiated both around the wellbore and the oriented perforations when the perforation angle is >75°. A larger perforation angle may cause a more complex fracture morphology. Therefore, perforations at a high angle with respect to the maximum principal stress are generally not recommended.

Stress ratio has a significant influence on fracture initiation pressure, breakdown pressure, and fracture propagation. Under the condition of hydrostatic pressure ($\lambda = 1$), the hydraulic fractures initiate and propagate along the direction of the oriented perforation, and thus will not form complex fracture morphology. The resulting fracture propagation does not always follow the directions provided by the perforations but rather reorients to the direction of the maximum horizontal stress. A larger stress ratio results in reduced fracture initiation and breakdown pressures. A larger stress ratio has a smaller critical perforation angle, thus, more easily to form complex fracture morphology.

The resulting differential initiation pressure and breakdown pressure decrease with an increase in the initial pore pressure. Particularly, the initiation pressures of the numerical simulations are in good agreement with the solutions of the Hubbert–Willis [40] and Haimson–Fairhurst [38] equations. In addition, since the adsorption-induced strain is fully coupled to the gas fracturing model, the behavior during gas fracturing is significantly different to that during water-based fracturing. The gas penetration not only alters the pore pressure in the reservoir but the gas adsorption-induced damage also modifies the mechanical properties of the reservoir rock.

The fracturing fluid viscosity influences the initiation pressure, breakdown pressure, and critical perforation angle by affecting the fluid flow behavior and the evolution of the poroelastic stress around the wellbore. The fracture initiation pressure and breakdown pressure increase with an increase in the viscosity. For the liquid-based hydraulic fracturing, breakdown pressure increases 55.6%, and the critical perforation angle increases from 70° to 80°, when the viscosity is increased to 1 Pa·s. While for the gas fracturing, breakdown pressure increases 26.8% when the viscosity is increased to 10^{-4} Pa·s. Notably, the critical perforation angle, remaining 62° to 63°, and does not change a lot with the varies of gas viscosity. The critical perforation angle is sensitive to the viscosity of liquid and is insensitive to the viscosity of gas.

The proposed model can be used to determine an optimal perforation condition in various hydraulic fracturing. Research findings are applied to optimize perforation design in order to minimize high treatment pressures and to control complexity in the near-wellbore fracture region. This study is of great significance in further understanding the near-wellbore impacts on hydraulic fracture propagation and complexity.

Author Contributions: L.L. (Liyuan Liu), L.L. (Lianchong Li), S.Z. and Y.Y. conceived and designed the numerical simulations; D.E. contributed analysis methods; L.L. (Liyuan Liu) and D.E. wrote and modified the paper.

Funding: This work was funded by the National Key Research and Development Program of China (Grant No. 2016YFC0801607), National Science Foundation of China (Grant Nos. 51525402, 51374049, and 51761135102), the Fundamental Research Funds for the Central Universities of China (Grant Nos. N170108028 and N170106003), and Science and Technology Major Project of Anhui Province (Grant No. 17030901023). This work also sponsored by China Scholarship Council.

Conflicts of Interest: The authors declare no conflicts of interest.

References

1. Cherny, S.; Chirkov, D.; Lapin, V.; Muranov, A.; Bannikov, D.; Miller, M.; Willberg, D.; Medvedev, O.; Alekseenko, O. Two-dimensional modeling of the near-wellbore fracture tortuosity effect. *Int. J. Rock Mech. Min.* **2009**, *46*, 992–1000. [CrossRef]
2. Abdollahipour, A.; Marji, M.F.; Bafghi, A.Y.; Gholamnejad, J. Simulating the propagation of hydraulic fractures from a circular wellbore using the displacement discontinuity method. *Int. J. Rock Mech. Min.* **2015**, *80*, 281–291. [CrossRef]
3. Palmer, I.; Cameran, J.; Moschovidis, Z.; Ponce, J. Natural fractures influence shear stimulation direction. *Oil Gas J.* **2009**, *107*, 37–43.
4. Zhou, J.; Chen, M.; Jin, Y.; Zhang, G.Q. Analysis of fracture propagation behavior and fracture geometry using a tri-axial fracturing system in naturally fractured reservoirs. *Int. J. Rock Mech. Min.* **2008**, *45*, 1143–1152. [CrossRef]
5. Ye, Z.; Janis, M.; Ghassemi, A.; Riley, S. Laboratory Investigation of fluid flow and permeability evolution through shale fractures. In Proceedings of the Unconventional Resources Technology Conference, Austin, TX, USA, 24–26 July 2017.
6. Hossain, M.M.; Rahman, M.K.; Rahman, S.S. Hydraulic fracture initiation and propagation: Roles of wellbore trajectory, perforation and stress regimes. *J. Pet. Sci. Eng.* **2000**, *27*, 129–149. [CrossRef]
7. Fallahzadeh, S.H.; Rasouli, V.; Sarmadivaleh, M. An investigation of hydraulic fracturing initiation and near-wellbore propagation from perforated boreholes in tight formations. *Rock Mech. Rock Eng.* **2015**, *48*, 573–584. [CrossRef]

8. Feng, Y.; Gray, K.E. Discussion on field injectivity tests during drilling. *Rock Mech. Rock Eng.* **2017**, *50*, 493–498. [CrossRef]
9. Sepehri, J.; Soliman, M.Y.; Morse, S.M. Application of extended finite elementmethod to simulate hydraulic fracture propagation from orientedperforations. In Proceedings of the SPE Hydraulic Fracturing Technology Conference, The Woodlands, TX, USA, 3–5 February 2015.
10. Zhang, G.; Chen, M. Complex fracture shapes in hydraulic fracturing with orientated perforations. *Pet. Explor. Dev.* **2009**, *36*, 103–107. [CrossRef]
11. Chen, M.; Jiang, H.; Zhang, G.Q.; Jin, Y. The experimental investigation of fracture propagation behavior and fracture geometry in hydraulic fracturing through oriented perforations. *Pet. Sci. Technol.* **2010**, *28*, 1297–1306. [CrossRef]
12. Mahrer, K.D. A review and perspective on far-field hydraulic fracture geometry studies. *J. Pet. Sci. Eng.* **1999**, *24*, 13–28. [CrossRef]
13. Haimson, B.C.; Cornet, F.H. ISRM suggested methods for rock stress estimation—Part 3: Hydraulic fracturing (hf) and/or hydraulic testing of pre-existing fractures (HTPF). *Int. J. Rock Mech. Min.* **2003**, *40*, 1011–1020. [CrossRef]
14. Feng, Y.; Gray, K.E. Modeling near-wellbore hydraulic fracture complexity using coupled pore pressure extended finite element method. In Proceedings of the 51st US Rock Mechanics/Geomechanics Symposium, San Francisco, CA, USA, 25–28 June 2017; American Rock Mechanics Association: San Francisco, CA, USA, 2017.
15. Daneshy, A.A. A study of inclined hydraulic fractures. *Soc. Pet. Eng. J.* **1973**, *13*, 61–68. [CrossRef]
16. Feng, Y.; Gray, K.E. Modeling of curving hydraulic fracture propagation from a wellbore in a poroelastic medium. *J. Nat. Gas Sci. Eng.* **2018**, *53*, 83–93. [CrossRef]
17. Zhang, X.; Jeffrey, R.G.; Bunger, A.P.; Thiercelin, M. Initiation and growth of a hydraulic fracture from a circular wellbore. *Int. J. Rock Mech. Min.* **2011**, *48*, 984–995. [CrossRef]
18. Narasingam, A.; Siddhamshetty, P.; Kwon, J.S.I. Temporal clustering for order reduction of nonlinear parabolic PDE systems with time-dependent spatial domains: Application to a hydraulic fracturing process. *Aiche J.* **2017**, *63*, 3818–3831. [CrossRef]
19. Sidhu, H.S.; Narasingam, A.; Siddhamshetty, P.; Kwon, J.S.I. Model order reduction of nonlinear parabolic PDE systems with moving boundaries using sparse proper orthogonal decomposition: Application to hydraulic fracturing. *Comput. Chem. Eng.* **2018**, *112*, 92–100. [CrossRef]
20. Gu, Q.Y.; Hoo, K.A. Model-based closed-loop control of the hydraulic fracturing process. *Ind. Eng. Chem. Res.* **2015**, *54*, 1585–1594. [CrossRef]
21. Siddhamshetty, P.; Wu, K.; Kwon, S.I. Optimization of simultaneously propagating multiple fractures in hydraulic fracturing to achieve uniform growth using data-based model reduction. *Chem. Eng. Res. Des.* **2018**, *136*, 675–686. [CrossRef]
22. Wu, K.; Olson, J.E. Mechanisms of simultaneous hydraulic-fracture propagation from multiple perforation clusters in horizontal wells. *SPE J.* **2016**, *21*, 1000–1008. [CrossRef]
23. Rutqvist, J.; Jeanne, P.; Dobson, P.F.; Garcia, J.; Hartline, C.; Hutchings, L.; Singh, A.; Vasco, D.W.; Walters, M. The northwest geysers EGS demonstration project, California—Part 2: Modeling and interpretation. *Geothermics* **2016**, *63*, 120–138. [CrossRef]
24. Wang, H.Y. Numerical modeling of non-planar hydraulic fracture propagation in brittle and ductile rocks using xfem with cohesive zone method. *J. Pet. Sci. Eng.* **2015**, *135*, 127–140. [CrossRef]
25. Schmitt, D.R.; Zoback, M.D. Diminished pore pressure in low-porosity crystalline rock under tensional failure: Apparent strengthening by dilatancy. *J. Geophys. Res. Solid Earth* **1992**, *97*, 273–288. [CrossRef]
26. Zhu, W.C.; Liu, J.; Tang, C.A.; Zhao, X.D.; Brady, B.H. Simulation of progressive fracturing processes around underground excavations under biaxial compression. *Tunn. Undergr. Space Technol.* **2005**, *20*, 231–247. [CrossRef]
27. Liu, L.Y.; Zhu, W.C.; Wei, C.H.; Elsworth, D.; Wang, J.H. Microcrack-based geomechanical modeling of rock-gas interaction during supercritical CO_2 fracturing. *J. Pet. Sci. Eng.* **2018**, *164*, 91–102. [CrossRef]
28. Li, Z.C.; Li, L.C.; Li, M.; Zhang, L.Y.; Zhang, Z.L.; Huang, B.; Tang, C.A. A numerical investigation on the effects of rock brittleness on the hydraulic fractures in the shale reservoir. *J. Nat. Gas Sci. Eng.* **2018**, *50*, 22–32. [CrossRef]

29. Zhang, X.; Jeffrey, R.G. The role of friction and secondary flaws on deflection and re-initiation of hydraulic fractures at orthogonal pre-existing fractures. *Geophys. J. Int.* **2006**, *166*, 1454–1465. [CrossRef]
30. Zhang, H.B.; Liu, J.S.; Elsworth, D. How sorption-induced matrix deformation affects gas flow in coal seams: A new Fe model. *Int. J. Rock Mech. Min.* **2008**, *45*, 1226–1236. [CrossRef]
31. Robertson, E.P.; Christiansen, R.L. Modeling laboratory permeability in coal using sorption-induced-strain data. *SPE Reserv. Eval. Eng.* **2007**, *10*, 260–269. [CrossRef]
32. Detournay, E.; Cheng, A.H.D. 5—Fundamentals of poroelasticity. In *Analysis and Design Methods*; Fairhurst, C., Ed.; Pergamon: Oxford, UK, 1993; pp. 113–171.
33. Zhu, W.C.; Tang, C.A. Micromechanical model for simulating the fracture process of rock. *Rock Mech. Rock Eng.* **2004**, *37*, 25–56. [CrossRef]
34. Behrmann, L.A.; Nolte, K.G. Perforating requirements for fracture stimulations. *SPE Drill. Complet.* **1998**, *14*, 228–234. [CrossRef]
35. Li, X.; Wang, J.H.; Elsworth, D. Stress redistribution and fracture propagation during restimulation of gas shale reservoirs. *J. Pet. Sci. Eng.* **2017**, *154*, 150–160. [CrossRef]
36. Detournay, E.; Cheng, A. Influence of pressurization rate on the magnitude of the breakdown pressure. In Proceedings of the 33th US Symposium on Rock Mechanics, Santa Fe, NM, USA, 3–5 June 1992; American Rock Mechanics Association: Santa Fe, NM, USA, 1992.
37. Song, I.; Haimson, B. Effect of pressurization rate and initial pore pressure on the magnitude of hydrofracturing breakdown pressure in Tablerock Sandstone. In Proceedings of the 38th U.S. Rock Mechanics Symposium, Washington, DC, USA, 7–10 July 2001; American Rock Mechanics Association: Washington, DC, USA, 2001.
38. Haimson, B.; Fairhurst, C. Initiation and extension of hydraulic fractures in rocks. *Soc. Pet. Eng. J.* **1967**, *7*, 310–318. [CrossRef]
39. Guo, F.; Morgenstern, N.R.; Scott, J.D. Interpretation of hydraulic fracturing breakdown pressure. *Int. J. Rock Mech. Min. Sci. Geomech. Abstr.* **1993**, *30*, 617–626. [CrossRef]
40. Hubbert, M.K.; Willis, D.G.W. Mechanics of hydraulic fracturing. *Trans. Soc. Pet. Eng. AIME* **1957**, *210*, 153–168.
41. Zhu, W.C.; Liu, L.Y.; Liu, J.S.; Wei, C.H.; Peng, Y. Impact of gas adsorption-induced coal damage on the evolution of coal permeability. *Int. J. Rock Mech. Min.* **2018**, *101*, 89–97. [CrossRef]
42. Liu, L.Y.; Zhu, W.C.; Wei, C.H.; Ma, X.H. Mechanical model and numerical analysis of mechanical property alterations of coal induced by gas adsorption. *Rock Soil Mech.* **2018**, *39*, 1500–1508. [CrossRef]
43. Detournay, E. Propagation regimes of fluid-driven fractures in impermeable rocks. *Int. J. Geomech.* **2004**, *4*, 35–45. [CrossRef]
44. Zhang, X.; Jeffrey, R.G.; Thiercelin, M. Deflection and propagation of fluid-driven fractures at frictional bedding interfaces: A numerical investigation. *J. Struct. Geol.* **2007**, *29*, 396–410. [CrossRef]
45. Wang, J.; Elsworth, D.; Wu, Y.; Liu, J.; Zhu, W.; Liu, Y. The influence of fracturing fluids on fracturing processes: A comparison between water, oil and SC-CO2. *Rock Mech. Rock Eng.* **2017**, *51*, 299–313. [CrossRef]

processes

MDPI

Article

Experimental Development Process of a New Fluid–Solid Coupling Similar-Material Based on the Orthogonal Test

Shiliang Liu [1,*] and Weitao Liu [2,*]

[1] School of Resources and Geosciences, China University of Mining and Technology, Xuzhou 221116, China
[2] School of Mining and Safety Engineering, Shandong University of Science and Technology, Qingdao 266590, China
* Correspondence: lslcumt@cumt.edu.cn (S.L.); wtliu@sdust.edu.cn (W.L.); Tel.: +86-136-4520-7719 (S.L.); +86-532-8605-7172 (W.L.)

Received: 27 September 2018; Accepted: 24 October 2018; Published: 1 November 2018

Abstract: Similar-material, composed of different raw materials with different properties, is similar to the physical and mechanical properties of geotechnical media, in which raw material proportioning is an important means to control the performance of similar-material in physical simulation. On this basis, a new fluid–solid coupling similar-material was developed through proportioning tests, in which similar-material is mixed with river sand, calcium carbonate, talc powder, white cement, vaseline, antiwear hydraulic oil. The optimum proportioning test development process was established. First, the proportioning test scheme was designed based on the orthogonal test. Subsequently, test specimens were produced to obtain parameters such as density, compressive strength, tensile strength, and permeability coefficient. Then, by increasing the ingredients of the proportioning, the evolution law of parameters was obtained by range and variance analysis. Finally, four multiple linear regression equations between the parameters and similar-material ingredients were obtained, and the optimum proportioning of ingredients was further determined for different requirements. The results indicate that the selected raw materials and their proportioning method are feasible, and the results were also verified in a coal mine floor water inrush by physical simulation test. The experimental development process of a fluid–solid coupling similar-material can provide a reference for similar-material under different demand conditions.

Keywords: orthogonal test; fluid–solid coupling theory; similar-material; regression equation; optimum proportioning

1. Introduction

Theoretical derivation, numerical simulation and similar simulation tests are three main research methods in complex geotechnical engineering [1,2]. Similar simulation test, a reproduction of the real physical entities, can reflect the basic physical and mechanical properties of rock and soil medium based on the similarity principle [3–6]. Similar simulation test is characterized by a short cycle, low cost, and results of visual image [7,8]. To carry out similar simulation tests, the similar-material must have properties similar to those of the physical and mechanical properties of the geotechnical medium. Similar-material is composed of different raw materials with different properties, the proportioning of which is an important means to control the performance of similar-material in physical simulation for different requirements. Therefore, selection of raw materials and their proportioning is of great significance for similar simulation experiments in geotechnical engineering.

At present, selection of raw materials and their proportioning has been widely studied, but the focus of the research has been concentrated on single-phase similar-material with general solid phase [9–18], which does not mention hydration characteristics of similar-material. In addition, the effect of temperature and humidity on similar-material and its control measures have been investigated, providing evidence for the reliability of simulation results [19,20]. For the investigation on fluid–solid coupling similar-material, some progress has been achieved [21–27]. For example, Wang et al. [28] adopted sand and gypsum as the aggregate, vaseline and silicone oil as the cementing agent, and selected the reasonable non-hydrophilic material proportioning, configuring the fluid–solid coupling similar-material for water-protection mining. Wang et al. [29] developed new types of similar-material of fault and surrounding rock, i.e., similar-material of fault consists of sand, talc powder, gypsum, bentonite and paraffin liquid, and similar-material of surrounding rock is composed of sand, barite powder, talc powder, white cement and latex, which were applied to simulate the water inrush occurring in the Yonglian tunnel of Jiangxi Province, China. Although the abovementioned fluid–solid coupling similar-materials were applied to the geotechnical engineering, there are still some problems. (1) Similar-material containing gypsum have water absorption, water softening and disintegration problems, and it is difficult to meet the requirements of similar-material deformation and permeability; (2) Similar-material with paraffin can solve the problem of similar-material disintegration in water, but it is characterized with complexity of the configuration process, strict requirements for temperature conditions, and long preparation period. (3) Most of the effects of material ingredients proportioning on similar-material properties have been qualitatively analyzed, but there is a lack of quantitative research and a definite set of methods for determining the proportioning of similar-material for different needs in similar simulation test.

On this basis, according to the characteristics of fluid–solid coupling similar-material, the raw materials of similar-material were first selected, i.e., river sand, calcium carbonate, talc powder, white cement, vaseline, and antiwear hydraulic oil. Second, specimen parameters, such as density, compressive strength, tensile strength, and permeability coefficient were tested. Third, the analysis of similar-material properties was quantitatively carried out by range analysis, variance analysis, and regression method. Eventually, the research results were applied to the physical simulation test of the water inrush from mining coal floor.

2. Materials and Method

2.1. Materials

The fluid–solid coupling similarity theory and the raw materials of similar-material are the foundation of the optimum proportioning of ingredients.

2.1.1. Fluid–Solid Coupling Similarity Theory

The fluid–solid coupling similarity theory is mainly used to determine the similarity between the solid and fluid in the same system. The fluid–solid coupling mathematical model of continuum is used to derive the similar equation between the model and the prototype parameters [30].

Seepage equation:

$$K_x \frac{\partial^2 p}{\partial^2 x} + K_y \frac{\partial^2 p}{\partial^2 y} + K_z \frac{\partial^2 p}{\partial^2 z} = S \frac{\partial p}{\partial t} + \frac{\partial e}{\partial t} + W \tag{1}$$

where K_x, K_y, K_z are the permeability coefficients in x, y, and z directions (cm/s), $K_x = K_y = K_z$; p is the water pressure (MPa); S is the storage coefficient; e is the volume strain; W is the source sink term.

Equilibrium equation:

$$\sigma_{ij,i} + X_j = \rho \frac{\partial^2 u_i}{\partial t^2} \tag{2}$$

where $\sigma_{ij,i}$ is the stress tensor; X_j is the volume force (N/m^3); ρ is the density (g/cm^3); u_i is the displacement (cm).

Effective stress equation:

$$\sigma_{ij} = \overline{\sigma_{ij}} + \alpha \delta p \tag{3}$$

where σ_{ij} is the total stress tensor; $\overline{\sigma_{ij}}$ is the effective stress tensor; α is the effective stress coefficient of Biolt; δ is the Kronker mark; ρ is the density (g/cm^3).

Based on Equations (1)–(3), the similar equations between the model and the prototype parameters are as follows [31].

$$C_G \frac{C_u}{C_l^2} = C_\lambda \frac{C_e}{C_l} = C_G \frac{C_e}{C_l} = C_\gamma = C_\rho \frac{C_u}{C_t^2} \tag{4}$$

where C_G, C_u, C_l, C_λ, C_γ, C_e, C_ρ, and C_t are the similarity ratios of the shear modulus, displacement, model size, Lame constant, bulk density, volumetric strain, density, and time, respectively.

$$C_k = \frac{\sqrt{C_l}}{C_\gamma} \tag{5}$$

where C_l, C_λ, and C_k are the similarity ratios of the model size, Lame constant, and permeability coefficient, respectively.

2.1.2. Similar-Material Components

The fluid–solid coupling similar-material not only meets the requirements of solid deformation and permeability, but also conforms to the characteristics that the material is non-hydrophilic and has no obvious water softening. On this basis, river sand (Xishanlou Company, Qingdao, China), calcium carbonate (Xishanlou Company, Qingdao, China) and talc powder (Xishanlou Company, Qingdao, China)were selected as the aggregate, white cement (Xishanlou Company, Qingdao, China) and vaseline (Xishanlou Company, Qingdao, China)were selected as the cementing agent, and antiwear hydraulic oil (Xishanlou Company, Qingdao, China) was selected as the regulator based on the fluid–solid coupling theory. The raw materials of similar-material are shown in Figure 1.

(a) (b) (c) (d) (e) (f)

Figure 1. The raw materials of similar-material: (**a**) River sand; (**b**) Calcium carbonate; (**c**) Talc powder; (**d**) White cement; (**e**) Vaseline; (**f**) Antiwear hydraulic oil.

(1) Aggregate: river sand, calcium carbonate, and talc powder

The grain diameter of river sand is less than 1 mm; calcium carbonate has an average particle size in the range 1–3 μm; and talc powder is 625 orders. Calcium carbonate and talc powder were cemented into larger aggregates with river sand, increasing the strength and cohesion of the material.

(2) Cementing agent: white cement and vaseline

The white cement with P.O32.5 quality has low price, and is a type of good water cementing material, which can control the whole process by improving the compressive strength, elasticity, and cohesion. Vaseline, light brown block ointment, is insoluble in water and has excellent plasticity and cementation capacity.

(3) Regulator: antiwear hydraulic oil

Antiwear hydraulic oil can reduce material dryness and increase the non-hydrophilic characteristics.

2.2. Methodology

The methodology consists mainly of three parts: the orthogonal test schemes of similar-material proportion, the fabrication of specimens, and the index parameters test of specimens.

2.2.1. Orthogonal Test Schemes of Similar-Material Proportion

The orthogonal experimental design method, the main method of partial factor design, was introduced to study the similar-material proportion. In the experiment, the result is called the index, and the parameters which may affect the test indexes are called factors. The specific conditions for each factor to be compared in the experiment are called levels [32,33].

The orthogonal test design method proposed in this study can be separated into three steps:

Step 1: Determine the factors. Four factors were set up, namely: A, the percentage of river sand in aggregate; B, the mass ratio of calcium carbonate and talc powder; C, the mass ratio of cement and vaseline; D, the percentage of hydraulic oil in the total mass of similar-material.

Step 2: Set levels for each factor. Five levels were set for each factor, as listed in Table 1.

Step 3: Design the orthogonal test design scheme by SPSS software (IBM SPSS Statistics 25.0-2017, IBM China Company Limited, Beijing, China, 2017) [34,35]. The orthogonal test design scheme was four factors and five levels, which can be expressed as L_{25} (5^4). The level values of each factor were set as input in the orthogonal experimental design module of SPSS software, leading to the schemes, as listed in Table 2.

Table 1. Orthogonal test design level.

Level	Factors			
	A (%)	B	C	D (%)
1	75	3:4	3:7	2
2	80	4:4	4:6	3
3	85	5:4	5:5	4
4	90	6:4	6:4	5
5	95	7:4	7:3	6

Table 2. The orthogonal test schemes of similar-material.

Schemes	A (%)	B	C	D (%)
S1	1 (75)	1 (3:4)	1 (3:7)	1 (2)
S2	1	2 (4:4)	2 (4:6)	2 (3)
S3	1	3 (5:4)	3 (5:5)	3 (4)
S4	1	4 (6:4)	4 (6:4)	4 (5)
S5	1	5 (7:4)	5 (7:3)	5 (6)
S6	2 (80)	1	2	3
S7	2	2	3	4
S8	2	3	4	5
S9	2	4	5	1
S10	2	5	1	2
S11	3 (85)	1	3	5
S12	3	2	4	1
S13	3	3	5	2
S14	3	4	1	3
S15	3	5	2	4
S16	4 (90)	1	4	2
S17	4	2	5	3
S18	4	3	1	4
S19	4	4	2	5
S20	4	5	3	1
S21	5 (95)	1	5	4
S22	5	2	1	5
S23	5	3	2	1
S24	5	4	3	2
S25	5	5	4	3

Note: The total mass of the similar-material in each scheme is 2000 g.

2.2.2. Fabricating Specimens

The fabrication process of similar-material specimen is as follows (Figure 2).

Figure 2. The fabrication process of similar-material specimen: (**a**) Weighing; (**b**) Stirring; (**c**) Heating vaseline; (**d**) Adding hydraulic oil; (**e**) Compacting; (**f**) Demolding; (**g**) Partial specimen; (**h**) PVC tube.

(1) Aggregate, cementing agent and regulator were weighed proportionately.
(2) The aggregate and cement were mixed evenly, followed by adding water 0.5 times of cement.
(3) Vaseline was heated to a liquid state and poured into the above mixture.
(4) Antiwear hydraulic oil was added and stirred.
(5) The well-mixed materials were loaded into a mold and compacted. The mold for testing specimen tensile strength is a PVC tube with a height of 25 mm and an inner diameter of 45 mm (Figure 2h).
(6) Demolded and labeled, specimens were maintained for three days at room temperature.

2.2.3. Testing Index Parameters of Specimens

The fluid–solid coupling similar-material needs to meet the requirements of solid deformation, permeability, and hydration characteristics. Therefore, the index parameters of specimen such as hydrophilicity test, tensile strength, compressive strength, and permeability coefficient were tested.

(1) Hydrophilicity test

The water absorption rate can be used to indicate the hydrophilicity of similar-material [36,37]. The greater the water absorption rates of the specimens, the stronger the hydrophilicity. The water absorption rate can be represented by Equation (6).

$$a = \frac{m_{\text{water}}}{m_{\text{dry}}} \times 100\% \tag{6}$$

where a is the water absorption rate (%); m_{water} is the specimen weight after immersion (g); m_{dry} is the specimen weight before immersion (g).

The weight, height, and diameter of the specimen before immersion were measured, and then density (ρ) was calculated, as listed in Table 3. In addition, the degree of disintegration of specimens in water was observed after the specimens were immersed for 1 day, 2 days, and 3 days respectively (Figure 3) [38,39]. With increased immersion time, the water absorption rate increased gradually, and no disintegration phenomenon was observed. The change trend of water absorption rate is shown in Figure 4, indicating that the water absorption rate increased significantly from 1 day to 2 days of immersion time, while water absorption increased slightly from 2 days to 3 days of immersion time, during which the specimen was saturated and the water absorption rate reached a plateau.

Table 3. The index parameters of specimens (*PrePEM*: pre-peak elastic modulus; *PostPEM*: post peak elastic modulus; *BB*: brittle behavior; *RTC*: the ratio of uniaxial tensile strength and uniaxial compressive strength).

Schemes	σ_c (MPa)	*PrePEM* (MPa)	*PostPEM* (MPa)	*BB*	σ_t (MPa)	*RTC*	ρ (g/cm^3)	K (cm/s)
S1	0.268	24.15	9.05	2.67	0.028	1/9.6	1.766	3.15×10^{-5}
S2	0.228	22.39	10.38	2.16	0.021	1/10.9	1.761	2.08×10^{-5}
S3	0.250	23.06	9.00	2.56	0.025	1/10.0	1.884	2.95×10^{-4}
S4	0.278	24.55	9.15	2.68	0.026	1/10.7	1.879	2.85×10^{-6}
S5	0.461	40.31	14.21	2.84	0.048	1/9.6	1.832	1.23×10^{-6}
S6	0.272	24.38	9.33	2.61	0.024	1/11.3	1.778	1.09×10^{-4}
S7	0.400	43.92	15.17	2.90	0.033	1/12.1	1.851	2.58×10^{-6}
S8	0.435	51.20	19.89	2.57	0.042	1/10.4	1.886	8.79×10^{-5}
S9	0.859	103.13	43.00	2.40	0.095	1/9.1	1.892	3.25×10^{-6}
S10	0.472	45.18	16.09	2.81	0.050	1/9.4	1.939	2.06×10^{-6}
S11	0.500	55.23	18.79	2.94	0.045	1/11.1	1.731	8.13×10^{-5}
S12	1.005	98.76	34.58	2.86	0.101	1/10.0	1.837	8.45×10^{-6}
S13	1.103	112.40	42.00	2.68	0.099	1/11.1	1.843	1.25×10^{-7}
S14	0.521	39.87	15.05	2.65	0.046	1/11.3	1.809	5.17×10^{-5}
S15	0.365	40.26	16.35	2.46	0.040	1/9.1	1.792	3.09×10^{-5}
S16	0.910	106.90	40.61	2.63	0.080	1/11.4	1.895	2.00×10^{-7}
S17	1.223	121.20	49.88	2.43	0.102	1/12.0	1.894	1.56×10^{-7}
S18	0.538	40.26	16.35	2.46	0.053	1/10.2	1.728	4.25×10^{-5}
S19	0.502	42.29	17.23	2.45	0.053	1/9.5	1.755	8.09×10^{-6}
S20	0.786	55.23	18.79	2.94	0.070	1/11.2	1.786	1.02×10^{-6}
S21	1.311	135.60	52.40	2.59	0.111	1/11.8	1.736	8.79×10^{-8}
S22	0.531	39.83	15.05	2.65	0.057	1/9.3	1.798	7.59×10^{-6}
S23	0.656	45.92	17.17	2.67	0.058	1/11.3	1.722	9.93×10^{-5}
S24	0.715	66.20	29.20	2.23	0.079	1/9.1	1.725	1.21×10^{-6}
S25	1.116	108.69	43.60	2.49	0.101	1/11.1	1.722	2.29×10^{-7}

Figure 3. Testing water absorption rate.

(2) Testing compressive strength σ_c and brittle behavior (*BB*)

The compressive strength was tested using a SHIMADZU electronic universal testing machine (AG-X250, SHIMADZU China Company Limited, Beijing, China). In order to simulate permeability evolution caused by the fluid–solid coupling material rupture and its effect on the mechanical properties, the *BB* of similar-material was considered, reflecting the ratio of the pre-peak elastic modulus (*PrePEM*) and post peak elastic modulus (*PostPEM*) in the uniaxial compressive strength test. When *PrePEM*/*PostPEM* < 3, the material belongs to brittle failure material [40]. The test results of compressive strength and brittle behavior of the specimens are shown in Table 3, indicating that the compressive strength variation was in the range 0.228–1.311 MPa, indicating that the test specimens met the requirements of brittle materials.

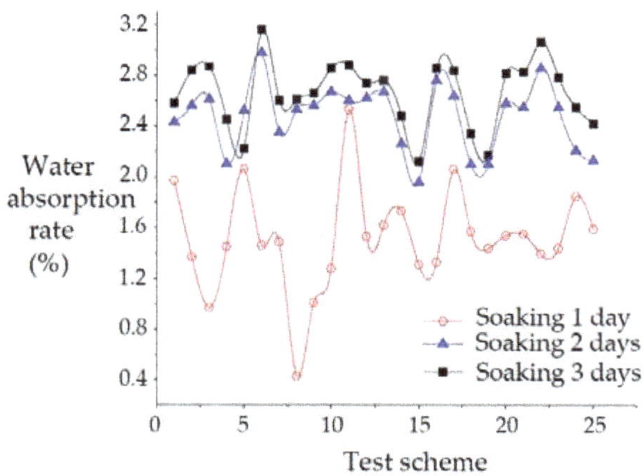

Figure 4. The trend of water absorption rate.

(3) Testing tensile strength σ_t

The tensile strength of the specimen was measured by the Brazil splitting method [41]. Similar-material tensile strength was obtained by Equation (7), as listed in Table 3.

$$\sigma_t = \frac{2P_v}{\pi D L_h} \tag{7}$$

where D is the diameter of the specimen (mm); P_v is the failure load value of the specimen (N); and L_h is the thickness of the specimen (mm).

In Table 3, the tensile strength variation was in the range 0.021–0.111 MPa. The ratio of uniaxial tensile strength and uniaxial compressive strength (*RTC*) of rock is nearly 0.1, which is one of the important differences between the rock material and other materials. In Table 3, *RTC* is in the range 1/11.8–1/9.1, which is close to the rock average *RTC* of 0.1 [42]. Therefore, the mechanical properties of similar-material and rock are similar.

(4) Testing permeability coefficient K

The permeability coefficient K is one of the indexes, reflecting the permeability of similar-material. In the similar simulation test, because the permeability of the material and the water flow are small, the water level is difficult to control; therefore, the variable head test is used in the laboratory [23], which is the main test principle, as expressed by Equation (8).

$$K = \frac{bL}{At} \ln \frac{\Delta h_1}{\Delta h_2} \tag{8}$$

where K is the permeability coefficient (cm/s); b is the glass tube sectional area (cm^2); A is the specimen sectional area (cm^2); L is the specimen height (cm); Δh_1, Δh_2 are initial water head difference and water head difference after t time (cm), respectively. In Table 3, the variation range of the permeability coefficient of similar material specimen is 8.79×10^{-8}–2.95×10^{-4} cm/s.

3. Results and Discussion

Based on the measured values for specimen parameters, the qualitative and quantitative relationships between the specimen index parameters and similar-material proportioning ingredients were obtained. In order to obtain the optimum proportioning of ingredients, using a coal mine floor water inrush by similar simulation test as a case, four multiple linear regression equations were further calculated, providing the optimum proportioning of the ingredients.

3.1. Results

The relationship between the specimen index parameters, i.e., density, compressive strength, tensile strength, and permeability coefficient, and four factors in the orthogonal test schemes, was analyzed quantitatively and qualitatively by the range and variance analyses. The variance analysis is based on the Minitab software (Minitab 17, TechMax Information Technical Co., Ltd., Shanghai, China, 2014) [43].

3.1.1. Density Analysis

The specimen density was analyzed as follows: first, a qualitative analysis was studied by the range analysis; furthermore, a quantitative analysis was carried out to obtain the quantitative relationship between the specimen density and four factors by the variance analysis.

(1) Range analysis of density

Range analysis consists of two stages: calculation and judgment. K_{jm} is the sum of the experimental indexes corresponding to the j factor, m level is in the range analysis. $K_{jm\text{-}a}$ is the average value of K_{jm} and R_j is the range of the j column factor, reflecting the variation range of the test index. The larger the R_j, the greater the effect of the factors on the test indicators, which can determine the primary and secondary factors.

① Calculation

Combining Tables 2 and 3, the corresponding test index and the average value of the first level A_1 of factor A are as follows: K_{A1} = 1.766 + 1.761 + 1.884 + 1.879 + 1.832 = 9.122, $K_{A1\text{-}a}$ = $K_{A1}/5$ = 1.824. Similarly, K_{A2} = 9.346, $K_{A2\text{-}a}$ = $K_{A2}/5$ = 1.869; K_{A3} = 9.013, $K_{A3\text{-}a}$ = $K_{A3}/5$ = 1.803; K_{A4} = 9.058, $K_{A4\text{-}a}$ = $K_{A4}/5$ = 1.812; K_{A5} =8.703, $K_{A5\text{-}a}$ = $K_{A5}/5$ = 1.741.

Then, based on the range R definition, it was concluded that $R_A = K_{A2\text{-}a} - K_{A5\text{-}a}$= 0.128, R_B = 0.047, R_C = 0.082, R_D = 0.036, as shown in Table 4.

② Judgement

In Table 4, it was seen that $R_A > R_C > R_B > R_D$. Therefore, the order of the factors that affects the specimen density is A > C > B > D.

Figure 5 shows the intuitive analysis chart of effective factors on specimen density. Specimen density increases with increasing percentage of river sand in aggregate, because the density of the coarse aggregate (river sand) is smaller than that of fine aggregate (calcium carbonate and talc powder). In addition, density increases with increasing the mass ratio of cement and vaseline.

(2) Variance analysis of density

The abovementioned density values of different levels in four factors were used as the input in the Minitab software. Then, the variance analysis of density was carried out in the variance analysis module of the Minitab software, as shown in Table 5. *Seq SS*, *Adj SS*, and *Adj MS* are the sum of the squares of deviations, adjusted squares sum of deviations, and adjusted squares sum of mean-square error, respectively. *F* value is an index in variance analysis. p value is used to determine the significance of a factor. Generally, $p < 0.05$ is significant, and $p < 0.01$ is very significant. The larger the F value, the smaller the p value, indicating a more reliable result.

Table 4. The range analysis of the parameters of similar-materials in the orthogonal test.

Factors		A	B	C	D	Sum of Test Results
	K_1	9.122	8.906	9.040	9.003	
	K_2	9.346	9.141	8.808	9.163	
	K_3	9.013	9.063	8.977	9.087	
	K_4	9.058	9.060	9.219	8.986	
	K_5	8.703	9.071	9.197	9.002	
ρ	$K_{1\text{-}a}$	1.824	1.781	1.808	1.801	$\sum = 45.241$
	$K_{2\text{-}a}$	1.869	1.828	1.762	1.833	
	$K_{3\text{-}a}$	1.803	1.813	1.795	1.817	
	$K_{4\text{-}a}$	1.812	1.812	1.844	1.797	
	$K_{5\text{-}a}$	1.741	1.814	1.839	1.800	
	R	0.128	0.047	0.082	0.036	
	K_1	1.485	3.261	2.330	3.574	
	K_2	2.438	3.387	2.023	3.428	
	K_3	3.494	2.982	2.651	3.382	
	K_4	3.959	2.875	3.744	2.892	
	K_5	4.329	3.200	4.957	2.429	
σ_c	$K_{1\text{-}a}$	0.297	0.652	0.466	0.715	$\sum = 20.705$
	$K_{2\text{-}a}$	0.488	0.677	0.405	0.686	
	$K_{3\text{-}a}$	0.699	0.596	0.530	0.676	
	$K_{4\text{-}a}$	0.792	0.575	0.749	0.578	
	$K_{5\text{-}a}$	0.866	0.640	1.191	0.686	
	R	0.378	0.102	0.586	0.229	
	K_1	0.148	0.288	0.234	0.352	
	K_2	0.244	0.314	0.196	0.329	
	K_3	0.331	0.277	0.252	0.298	
	K_4	0.358	0.299	0.350	0.263	
	K_5	0.406	0.309	0.455	0.245	
σ_t	$K_{1\text{-}a}$	0.0296	0.0576	0.0468	0.0704	$\sum = 1.487$
	$K_{2\text{-}a}$	0.0488	0.0628	0.0392	0.0658	
	$K_{3\text{-}a}$	0.0662	0.0554	0.0504	0.0596	
	$K_{4\text{-}a}$	0.0716	0.0598	0.07	0.0526	
	$K_{5\text{-}a}$	0.0812	0.0618	0.091	0.049	
	R	0.0516	0.0074	0.0518	0.0214	
	K_1	3.51×10^{-4}	2.22×10^{-4}	1.35×10^{-4}	1.44×10^{-4}	
	K_2	2.05×10^{-4}	3.96×10^{-5}	2.68×10^{-4}	2.44×10^{-5}	
	K_3	1.72×10^{-4}	5.25×10^{-4}	6.81×10^{-4}	7.56×10^{-4}	
	K_4	5.20×10^{-5}	6.71×10^{-5}	9.96×10^{-5}	7.89×10^{-5}	
	K_5	1.08×10^{-5}	3.54×10^{-5}	4.85×10^{-6}	1.86×10^{-4}	
K	$K_{1\text{-}a}$	7.02×10^{-5}	4.44×10^{-5}	2.70×10^{-5}	2.88×10^{-5}	$\sum = 1.19 \times 10^{-3}$
	$K_{2\text{-}a}$	4.10×10^{-5}	7.92×10^{-6}	5.36×10^{-5}	4.88×10^{-6}	
	$K_{3\text{-}a}$	3.44×10^{-5}	1.05×10^{-4}	1.36×10^{-4}	1.51×10^{-4}	
	$K_{4\text{-}a}$	1.04×10^{-5}	1.34×10^{-5}	1.99×10^{-5}	1.58×10^{-5}	
	$K_{5\text{-}a}$	2.16×10^{-6}	7.08×10^{-6}	9.70×10^{-7}	3.72×10^{-5}	
	R	7.00×10^{-5}	9.79×10^{-5}	1.35×10^{-4}	1.46×10^{-4}	

Table 5. The variance analysis of similar-material density (*Seq SS*: the sum of the squares of deviations; *Adj SS*: adjusted squares sum of deviations; *Adj MS*: adjusted squares sum of mean-square error).

Variance Sources	Free Degree	*Seq SS*	*Adj SS*	*Adj MS*	*F*	*p*
A	4	0.042940	0.042940	0.010735	2.69	0.109
B	4	0.005942	0.005942	0.001486	0.37	0.823
C	4	0.022829	0.022829	0.005707	1.43	0.309
D	4	0.004546	0.004546	0.001137	0.28	0.880
Error	8	0.031966	0.031966	0.003996		

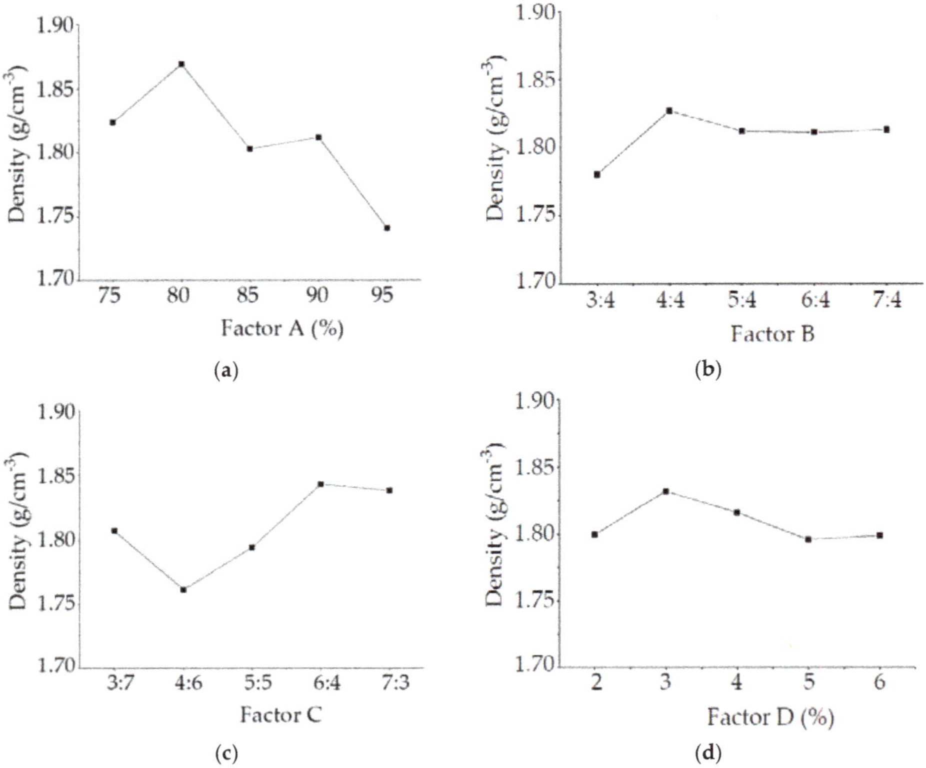

Figure 5. Intuitive analysis chart of density influencing factor: (**a**) Factor A; (**b**) Factor B; (**c**) Factor C; (**d**) Factor D.

In Table 5, p values are $\geq$5%, illustrating that the affecting factors have no significant effect on the specimen density. The factors order affecting the specimen density is A > C > B > D based on p values, and this result is consistent with the range analysis of density.

In the Minitab regression analysis, the indexes of the regression model include the standard deviation of the regression model error (S_e), the percentage of the regression model error in the total error ($R\text{-}Sq$) and the adjusted $R\text{-}Sq$. $R\text{-}Sq$ is used to show that the model is in line with the data, and the larger the value, the better the regression model and the data. The larger the $R\text{-}Sq$ value, the better the fit between the regression model and the data.

Based on the regression analysis module in the Minitab software, the multiple linear regression equation between the density and four factors were obtained [44], as shown by Equation (9).

$$\rho = 1.80964 - 0.06904A - 0.00456B + 0.02976C + 0.00924D \tag{9}$$

where S_e = 0.0632119, $R\text{-}Sq$ = 80.46%, the adjusted $R\text{-}Sq$ = 75.39%, verifying the reliability of the similar-material density regression model.

3.1.2. Compressive Strength Analysis

The compressive strength analysis method was similar to that of the specimen density.

(1) Range analysis of compressive strength

Similarly, $R_C > R_A > R_D > R_B$. Therefore, the order of the factors affecting the specimen density is C > A > D > B. Figure 6 shows the intuitive analysis chart of effective factors of specimen compressive strength. Specimen compressive strength increases when increasing the percentage of river sand in the aggregate and the mass ratio of cement and vaseline.

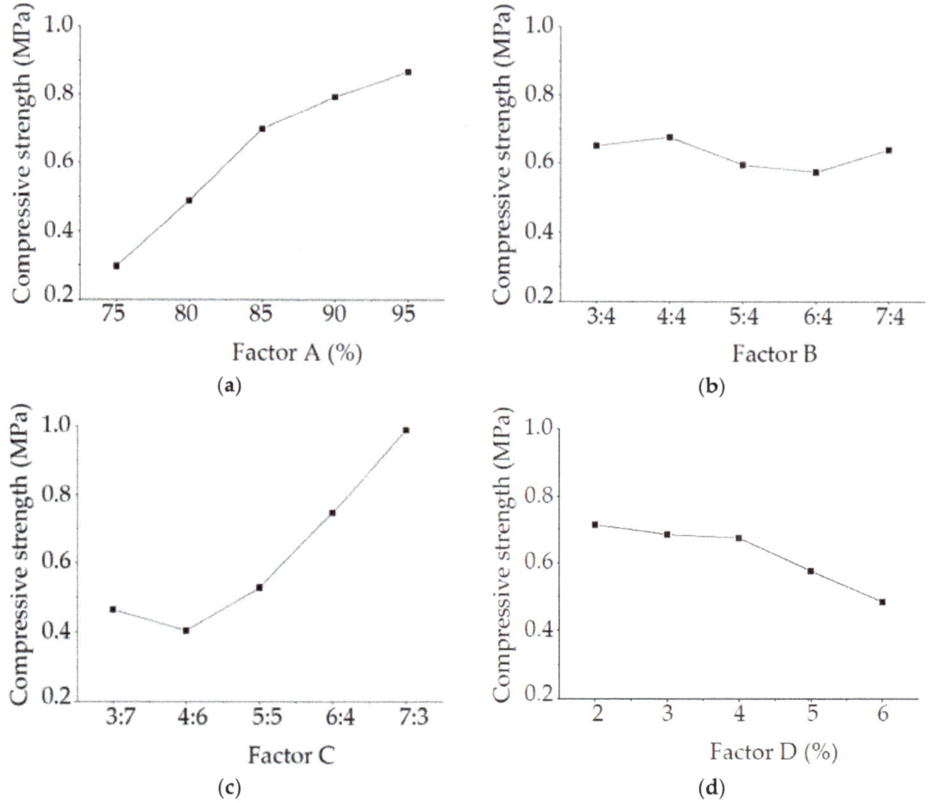

Figure 6. Intuitive analysis chart of compressive strength influencing factor: (**a**) Factor A; (**b**) Factor B; (**c**) Factor C; (**d**) Factor D.

(2) Variance analysis of compressive strength

Equation (10) was obtained based on the Minitab regression analysis.

$$\sigma_c = -2.04562 + 3.2768A - 0.13096B + 0.469376C - 5.75D \quad (10)$$

where S_e = 0.173737, $R\text{-}Sq$ =94.31%, the adjusted $R\text{-}Sq$ =82.94%, indicating the reliability of the similar-material compressive strength regression model.

3.1.3. Tensile Strength Analysis

The tensile strength analysis method was similar to that of the specimen density.

(1) Range analysis of tensile strength

Similarly, $R_C > R_A > R_D > R_B$. Therefore, the order of the factors affecting the specimen density is C > A > D > B. Figure 7 shows the intuitive analysis chart of factors affecting the specimen tensile strength. Specimen tensile strength increases with increasing the percentage of river sand in aggregate and the mass ratio of cement and vaseline.

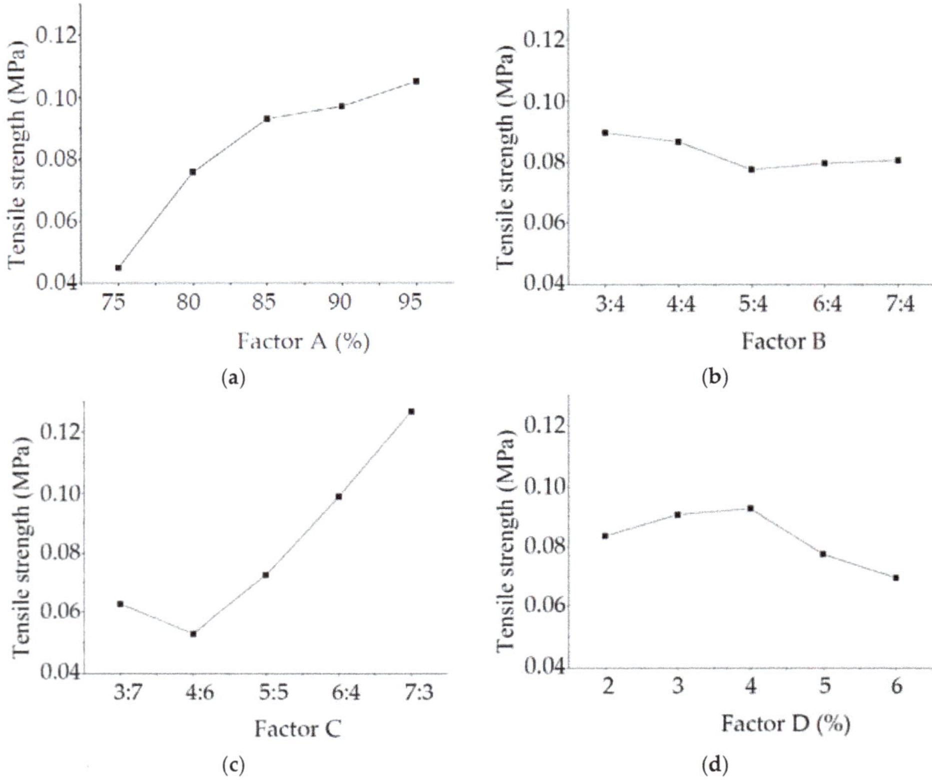

Figure 7. Intuitive analysis chart of tensile strength influencing factor: (**a**) Factor A; (**b**) Factor B; (**c**) Factor C; (**d**) Factor D.

(2) Variance analysis of tensile strength

Equation (11) was obtained based on the Minitab regression analysis.

$$\sigma_t = -0.19218 + 0.2916\text{A} - 0.00784\text{B} + 0.0439676\text{C} - 0.424\text{D} \quad (11)$$

where S_e = 0.0146990, $R\text{-}Sq$ = 95.06%, the adjusted $R\text{-}Sq$ = 85.17%, confirming the reliability of the similar-material tensile strength regression model.

3.1.4. Permeability Coefficient Analysis

The permeability coefficient analysis method was similar to that of the specimen density.

(1) Range analysis of permeability coefficient

The order of on permeability coefficients is as follows: $R_D > R_C > R_B > R_A$. Therefore, the order of the factors affecting the specimen density is D > C > B > A. Figure 8 shows the intuitive analysis chart of influence factors affecting the specimen permeability coefficient. With increasing the percentage of antiwear hydraulic oil in the total mass of similar materials, the mass ratio of cement and vaseline, and the mass ratio of calcium carbonate and talc powder, the permeability coefficient first increases and then decreases.

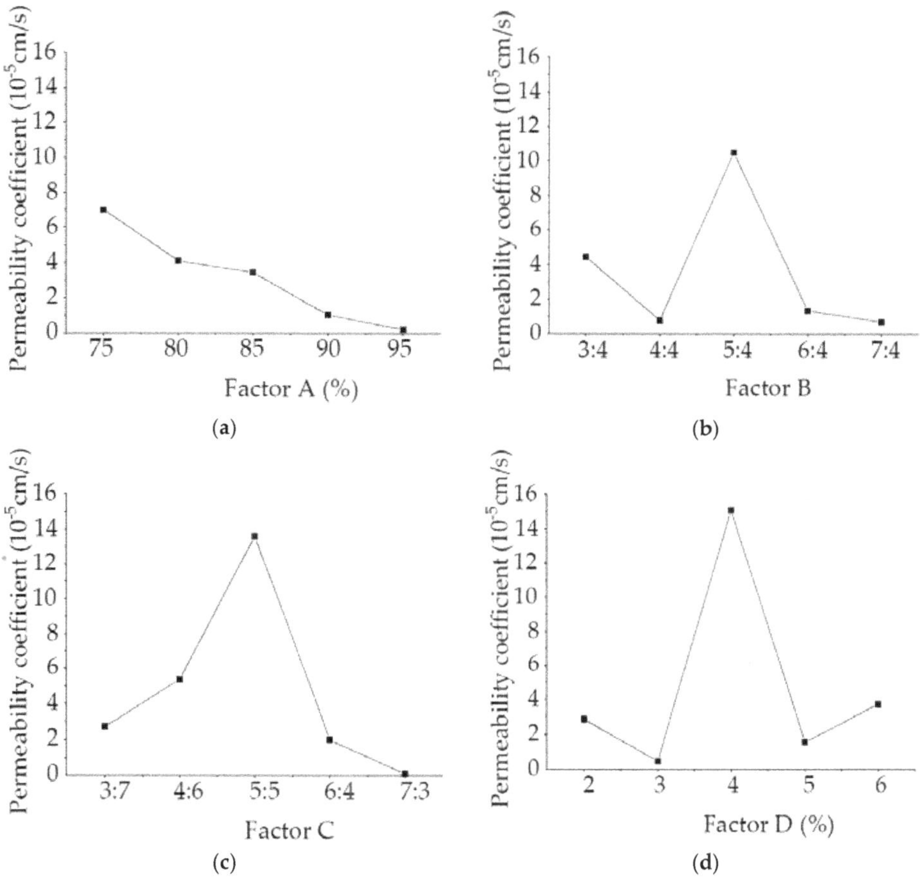

Figure 8. Intuitive analysis chart of permeability coefficient influencing factor: (**a**) Factor A; (**b**) Factor B; (**c**) Factor C; (**d**) Factor D.

(2) Variance analysis of permeability coefficient

Equation (12) was obtained based on the Minitab regression analysis.

$$K = 0.00031441 - 0.0002555\text{A} - 0.00002766\text{B} - 0.00003062\text{C} + 0.00027941\text{D} \quad (12)$$

where S_e = 0.0000412355, *R-Sq* = 86.20%, the adjusted *R-Sq* = 78.60%, indicating the reliability of similar-material permeability coefficient regression model.

3.2. Discussion

The optimum proportion of the fluid–solid coupling similar-material is the key to study geotechnical engineering problems in similar simulation test, obtained by Equations (9)–(12).

Water inrush from the coal floor is a typical fluid–solid coupling problem in geotechnical engineering. In order to explore the mechanism of water inrush from mining coal floor, the floor water inrush of the 15 coal seams in a colliery was taken as an application example. Based on the fluid–solid coupling similarity theory and original parameters of sandstone stratum and mudstone stratum in aquifuge, the test model parameters of aquifuge were calculated, as shown in Table 6. On this basis, the optimum ratio of similar-material in the aquifuge strata was also obtained by Equations (9)–(12), as listed in Table 7.

Table 6. The original rock and model parameter of aquifuge.

Stratum		ρ (g/cm^3)	σ_c (MPa)	σ_t (MPa)	K (cm/s)
Mudstone	Protolith	1.815	54.25	9.65	3.55×10^{-6}
	Model	1.801	0.261	0.048	2.52×10^{-7}
Sandstone	Protolith	1.903	92.30	16.25	1.69×10^{-5}
	Model	1.893	0.458	0.081	1.20×10^{-6}

Table 7. The optimal ratio of similar-material in aquifuge

Stratum	A (%)	B	C	D (%)	Sand: Calcium Carbonate: Talc Powder: White Cement: Vaseline: Antiwear Hydraulic Oil
Mudstone	87.45	1.27	1.19	4.78	12.46:1.00:0.79:0.69:0.89:0.79
Sandstone	75.32	1.43	1.74	3.65	5.19:1.00:0.70:0.51:0.29:0.29

According to the optimum ratio of mudstone and sandstone in aquifuge, six specimens were prepared to verify the requirements of the hydrophilicity, compressive strength, tensile strength, brittleness and permeability coefficient. Figure 9 shows the stress and strain curves of mudstone and sandstone in aquifuge, indicating that the specimen strength immersed for three days was greater than 85% of the specimen without immersion; these results show that this similar-material can meet the practical requirements of aquifuge. Furthermore, similar-material model test was laid using the optimal ratio of in aquifuge, and the results are consistent with the mining practice, verifying that the new fluid–solid coupling similar-material is feasible and effective in solving the geotechnical engineering problem.

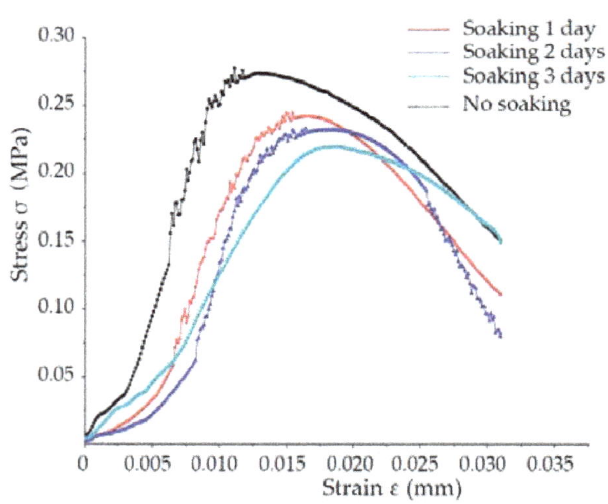

(a)

Figure 9. *Cont.*

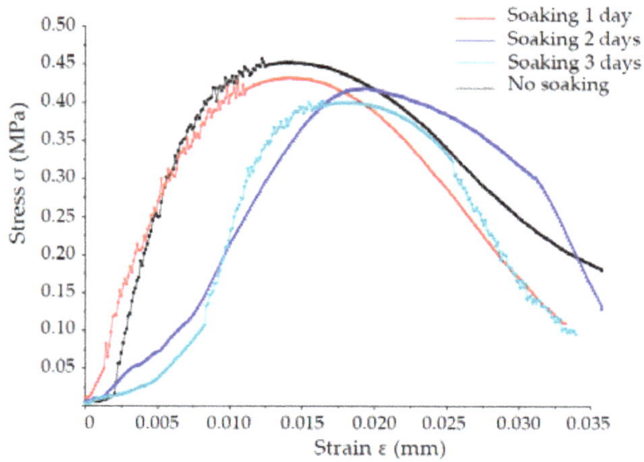

(b)

Figure 9. The stress and strain curves of similar-material: (**a**) Mudstone; (**b**) Sandstone.

4. Conclusions

An experimental development process of a new fluid–solid coupling similar-material based on the orthogonal test was established. First, river sand, calcium carbonate and talc powder, white cement and vaseline, and antiwear hydraulic oil were used as the aggregate, cementing agent, and regulator, respectively, making up the raw materials of similar-material. Second, the proportioning test scheme was designed based on the orthogonal test. Test specimens were produced to obtain parameters, such as density, compressive strength, tensile strength, and permeability coefficient. Third, the qualitative and quantitative relationships between the specimen parameters and similar-material proportioning ingredients were obtained through range and variance analysis. That is to say, the density, compressive strength, and tensile strength increase when increasing the mass ratio of cement and vaseline. With increasing the percentage of hydraulic oil in aggregate, the mass ratio of cement and vaseline, and the mass ratio of calcium carbonate and talc powder, the permeability coefficient first increases and then decreases. Fourth, four multiple linear regression equations between the specimen parameters and similar-material ingredients were obtained, and the optimum proportioning of ingredients was further determined for different requirements. Finally, taking a coal mine floor water inrush by similar simulation test as a case, based on the four multiple linear regression equations, the optimum proportioning of the sand, calcium carbonate, talc powder, white cement, vaseline, and antiwear hydraulic oil of the mudstone and sandstone in the aquifuge was determined as 12.46:1.00:0.79:0.69:0.89:0.79, 5.19:1.00:0.70:0.51:0.29:0.29, respectively, which were successfully applied. The newly developed fluid–solid coupling similar-material can provide a reference for similar-material under different demand conditions.

Author Contributions: All the authors contributed to this paper. S.L. conceived and designed the experiments schemes and wrote the paper. General supervision was provided by W.L. All authors have read and approved the final manuscript.

Funding: The study has been financially supported by the National Natural Science Foundation of China under Grant Nos. 51274135, 51428401, and 51034003, the National High Technology Research and Development Program (863 Program) of China under Grant No. 2015AA016404-4, and the State Key Research and Development Program of China under Grant No. 2017YFC0804108.

Conflicts of Interest: The authors declare no conflict of interest.

References

1. Li, Z.K.; Lu, D.R.; Nakayama, H.; Xi, J.H.; Sun, J.S. Developement and applicaiton of new technology for 3D geomechanical model test of large underground houses. *Chin. J. Rock Mech. Eng.* **2003**, *22*, 1430–1436.
2. Cui, X.M.; Miu, X.X.; Su, D.G.; Ma, W.M. Error analysis in similar material simulation test of the movement of rock strata and surface. *Chin. J. Rock Mech. Eng.* **2002**, *21*, 1827–1830.
3. Chapman, D.N.; Ahn, S.K.; Hunt, D.V.L.; Chan, A.H.C. The Use of Model Tests to Investigate the Ground Displacements Associated with Multiple Tunnel Construction in Soil. *Tunn. Undergr. Space Technol.* **2006**, *21*, 413. [CrossRef]
4. Jeon, S.; Kim, J.; Seo, Y.; Hong, C. Effect of a Fault and Weak Plane on the Stability of a Tunnel in Rock—A Scaled Model Test and Numerical Analysis. *Int. J. Rock Mech. Min.* **2004**, *41*, 658–663. [CrossRef]
5. Fumagalli, E. *Statical and Geomechanical Model*; Springer: New York, NY, USA, 1973; pp. 25–36.
6. Kim, S.H.; Burd, H.J. Model testing of closely spaced tunnels in clay. *Geotechnique* **1998**, *48*, 375–388. [CrossRef]
7. Gu, D.Z. *Similar Materials and Similar Models*; China University of Mining and Technology Press: Xuzhou, China, 1995.
8. Fu, X.M.; Deng, R.G. *Indoor Rock Mechanics Test*; Southwest Jiao Tong University Press: Chengdu, China, 2012; pp. 56–58.
9. Chen, S.J.; Wang, H.L.; Zhang, J.W.; Xing, H.L.; Wang, H.L. Experimental study on low-strength similar-material proportioning and properties for coal mining. *Adv. Mater. Sci. Eng.* **2015**, *2015*, 1–6. [CrossRef]
10. Chen, S.J.; Wang, H.L.; Zhang, J.W.; Xing, H.L.; Wang, H.L. Low-Strength Similar Materials for Backfill Mining: Insight from Experiments on Components and Influence Mechanism. *Geotech. Test. J.* **2015**, *38*, 929–935. [CrossRef]
11. Chen, L.W.; Bai, S.W. Proportioning test study on similar of rock burst tendency of brittle rock mass. *Rock Soil Mech.* **2006**, *27*, 1050–1054.
12. Meguid, M.A.; Saada, O.; Nunes, M.A.; Mattar, J. Physical modeling of tunnels in soft ground: A review. *Tunn. Undergr. Space Technol.* **2008**, *23*, 185–198. [CrossRef]
13. Wu, B.T.; Zhu, H.H.; Xu, W.Q.; Ming, T. Research study of similar material for weak surrounding rock mass of class IV. *Rock Soil Mech.* **2013**, *34*, 109–116.
14. Zhang, Q.Y.; Li, S.C.; Guo, X.H.; Li, Y.; Wang, H.P. Research and development of new typed cementitious geotechnical similar material for iron crystal sand and its application. *Rock Soil Mech.* **2008**, *29*, 2126–2130.
15. Zhang, Q.Y.; Liu, D.J.; Jia, C.; Shen, X.; Liu, J.; Duan, K. Development of geomechanical model similitude material for salt rock oil-gas storage medium. *Rock Soil Mech.* **2009**, *30*, 3581–3586.
16. Zhang, S.T.; Dai, L.C.; Wang, B.; Cao, Y. Experiment study on mixture ratio of similar material for simulation of coal and gas outburst. *Coal Sci. Technol.* **2015**, *43*, 76–81.
17. Liu, J.H.; Li, W.X.; Liu, Y.S.; Liu, B.G. A method for determining the ratio of similar material to simulate porous water-bearing stratum. *Rock Soil Mech.* **2018**, *39*, 657–664. [CrossRef]
18. Zhang, F.; Ma, G.; Tao, Y.Q.; Liu, X.; Feng, D.; Li, R. Proportioning experiment of similar material for coal and rock model test. *Coal Geol. Explor.* **2018**, *46*, 119–124.
19. Zha, J.F.; Li, H.Z.; Guo, G.L.; Wang, J.T. Influence of temperature and humidity on similar material and its control measures. *Environ. Earth Sci.* **2017**, *76*, 740. [CrossRef]
20. Li, H.Z.; Guo, G.L.; Zha, J.F. Study on time-varying characteristics of similar material model strength and the regulation measures. *Environ. Earth Sci.* **2017**, *76*, 518. [CrossRef]
21. Li, L.J.; Qian, M.G.; Yin, Y.Q. Simulation of similar materials for water inrush from coal floor. *Coal Geol. Explor.* **1996**, *25*, 33–36.
22. Sun, W.B.; Zhang, S.C.; Li, Y.Y.; Lu, C. Development application of solid-fluid coupling similar material for floor strata and simulation test of water-inrush in deep mining. *Chin. J. Rock Mech. Eng.* **2015**, *31*, 2665–2670.
23. Sun, W.B.; Zhang, S.C.; Guo, W.J.; Liu, W.T. Physical simulation of high-pressure water inrush through the floor of a deep mine. *Mine Water Environ.* **2017**, *36*, 542–549. [CrossRef]
24. Chen, J.T.; Yin, L.M.; Sun, W.B.; Lu, C.; Zhang, S.C.; Sun, X.Z. Development and application for new solid-fluid coupling similar material of deep floor aquifuge. *Chin. J. Rock Mech. Eng.* **2015**, *34*, 3956–3964.

25. Li, S.C.; Feng, X.D.; Li, S.C.; Li, L.P.; Li, G.Y. Research and development of a new similar material for solid-fluid coupling and its application. *Chin. J. Rock Mech. Eng.* **2010**, *29*, 281–288.
26. Li, S.C.; Zhou, Y.; Li, L.P.; Zhang, J.; Song, S.G. Development and application of a new similar material for underground engineering fluid-solid coupling model test. *Chin. J. Rock Mech. Eng.* **2012**, *31*, 1128–1137.
27. Yu, L.Y.; Jing, H.W.; Xu, B.S.; Wang, Y.C. Solid-fluid coupling analogous material test for subsea tunnel. *J. Cent. South Univ. (Sci. Technol.)* **2015**, *46*, 983–990. [CrossRef]
28. Wang, H. The Development and Application of Solid-Liquid Coupling Similar Stimulation Non-Hydrophilic Material for Water-Protection Mining. Master's Thesis, China University of Mining and Technology, Xuzhou, China, 2015.
29. Wang, K.; Li, S.C.; Zhang, Q.S.; Zhang, X.; Li, L.P. Development and application of new similar materials of surrounding rock for a fluid-solid coupling model test. *Rock Soil Mech.* **2016**, *37*, 2521–2533. [CrossRef]
30. Zhao, B.; Wen, G.C.; Sun, H.T.; Sun, D.L.; Yang, H.M.; Cao, J.; Dai, L.C.; Wang, B. Similarity criteria and coal-like material in coal and gas outburst physical simulation. *Int. J. Coal Sci. Technol.* **2018**, *5*, 167–178. [CrossRef]
31. Hu, Y.Q.; Zhao, Y.S.; Yang, D. Simulation theory & method of 3D solid-liquid coupling. *J. Liaoning Tech. Univ.* **2007**, *26*, 204–206.
32. Wang, W.Z. *Design and Analysis of Experiments*; Higher Education Press: Beijing, China, 2004; pp. 23–25.
33. Gao, R.; Yan, H.; Ju, F.; Mei, X.C.; Wang, X.L. Influential factors and control of water inrush in a coal seam as the main aquifer. *Int. J. Min. Sci. Technol.* **2018**, *28*, 187–193. [CrossRef]
34. Guo, Z.G. *Methods of Social Statistics: Application of SPSS Software*; Renmin University of China Press: Beijing, China, 1999.
35. Li, K.; Li, H.F. Response characteristics analysis of mine water filled structure with ground-tunnel transient electromagnetic method. *J. China Univ. Sci. Technol.* **2018**, *47*, 1113–1122.
36. Wu, R.A.; Wei, Y.J.; Ji, C.L. Analysis of deformation mechanism of slope with soft and hard rock inter-bedded structure in Three Gorges reservoir area: A case study of Xiangcheng elementary school in Yunyang. *J. Catastrophol.* **2018**, *33*, 212–217. [CrossRef]
37. Cao, P.; Wang, H.; Jin, J.; Hao, R.Q.; Fan, W.C. Experimental study of the fracture failure of sandstone containing hole and fissure under seepage water pressure. *J. China Univ. Sci. Technol.* **2018**, *47*, 240–246.
38. Esposito, L.; Esposito, A.W.; Pasculli, A.; Sciarra, N. Particular features of the physical and mechanical characteristics of certain Phlegraean pyroclastic soils. *Catena* **2013**, *104*, 186–194. [CrossRef]
39. Pasculli, A.; Sciarra, N.; Esposito, L.; Esposito, A.W. Effects of wetting and drying cycles on mechanical proprerties of pyroclastic soils. *Catena* **2017**, *156*, 113–123. [CrossRef]
40. Cao, R.H.; Lin, H.; Cao, P. Strength and failure characteristics of brittle jointed rock-like specimens under uniaxial compression: Digital speckle technology and a particle mechanics approach. *Int. J. Min. Sci. Technol.* **2018**, *28*, 669–677. [CrossRef]
41. Peng, S.J.; Chen, C.C.; Xu, J.; Zhang, H.L.; Tang, Y.; Nie, W.; Zhao, K. Loading rate dependency of rock stress-strain curve based on Brazil splitting test. *Chin. J. Rock Mech. Eng.* **2018**, *37*, 3247–3252. [CrossRef]
42. Zhou, H.; Meng, F.Z.; Zhang, C.Q.; Xu, R.C.; Lu, J.J. Quantitative evaluation of rock brittleness based on stress-strain curve. *Chin. J. Rock Mech. Eng.* **2014**, *33*, 1114–1122.
43. Wu, L.Y. *Introduction to MINITAB Software: The Most Practical Statistical Analysis Tutorial*; Higher Education Press: Beijing, China, 2012; pp. 78–81.
44. Feng, H.Y. The influence of pressure and temperature on dispensing quantity in dispensing technology based on MINITAB regression analysis method. *Commod. Qual.* **2017**, *9*, 229–230.

processes

MDPI

Article

High Mixing Efficiency by Modulating Inlet Frequency of Viscoelastic Fluid in Simplified Pore Structure

Meng Zhang [1,†], Yunfeng Cui [2,†], Weihua Cai [1,*], Zhengwei Wu [3], Yongyao Li [1], Fengchen Li [1,4] and Wu Zhang [5,*]

1 School of Energy Science and Engineering, Harbin Institute of Technology, Harbin 150001, China; meng.zhang_china@outlook.com (M.Z.); yy-li@stu.hit.edu.cn (Y.L.); lifch@hit.edu.cn (F.L.)
2 School of Mechanical Engineering, Shanghai Jiao Tong University, Shanghai 200000, China; yunfeng.cui@sjtu.edu.cn
3 Department of Biomedical Engineering and Biotechnology, University of Massachusetts Lowell, Lowell, MA 01854, USA; zhengwei_wu@student.uml.edu
4 Sino-French Institute of Nuclear Engineering and Technology, Sun Yat-Sen University, Zhuhai 519082, China
5 School of Electrical and Electronic Engineering, Nanyang Technological University, Singapore 639798, Singapore
* Correspondence: caiwh@hit.edu.cn (W.C.); zh0002wu@outlook.com (W.Z.)
† These authors contributed equally to this work.

Received: 26 September 2018; Accepted: 16 October 2018; Published: 1 November 2018

Abstract: Fluid mixing plays an essential role in microscale flow systems. Here, we propose an active mixing approach which enhances the mixing of viscoelastic fluid flow in a simplified pore T-junction structure. Mixing is actively controlled by modulating the driving pressure with a sinusoidal signal at the two inlets of the T-junction. The mixing effect is numerically investigated for both Newtonian and viscoelastic fluid flows under different pressure modulation conditions. The result shows that a degree of mixing as high as 0.9 is achieved in viscoelastic fluid flows through the T-junction mixer when the phase difference between the modulated pressures at the two inlets is 180°. This modulation method can also be used in other fluid mixing devices.

Keywords: mixer; viscoelastic fluid; pore structure

1. Introduction

Mixing processes aim to generate a homogenous solution of multiple components in natural and engineering flows [1,2]. Such processes have been widely applied in chemical analysis [3,4], biological analysis [5,6], heat and mass transfer for microfluidic devices [7,8], fluid dynamic analysis in a porous medium [9–12], and numerous relevant small-scale research areas [13–15]. Efficient mixing is often desired for complete interaction within multicomponent systems, microreaction platforms, and mixing and transfers of small volumes at pore-scales [16,17]. Mixing can be accomplished through chaotic flow or a turbulent flow [8,18,19]; however, due to the small-scales, such flows are difficult to achieve in microfluidic systems. The realization of turbulent flow in microchannels is much more complex and challenging than in larger tubes or in open space, where the flow rate is severely limited by the driving pressure. At low flow rates, where flows are often in the laminar, low Reynolds number regime, mixing must rely on molecular diffusion between the different fluid layers, the thicknesses of which are much larger than the characteristic diffusion length. Thus, instead of relying on inertial effects, other strategies to induce mixing, like introducing the elastic effect at microscale laminar flow, may enhance the mixing effect [20].

Many microfluidic designs use fluid mechanisms to overcome the absence of turbulence [8,18,21]. These mechanisms include splitting–recombining [22], twisting [23–25], transversal flows [9], vortices [26,27], and chaotic advection [1,28–30]. These strategies can be categorized into passive and active ways [2,31,32], which depend on the application of an external energy source. They operate by the same inherent principle to generate transversal components of velocity, causing the flow to become unstable, and the thickness of fluid layers to decrease.

Mixing is also a critical process in many subsurface engineering operations, such as oil recovery [29,33]. After a secondary recovery by water flooding of a subsurface hydrocarbon reservoir, the oil and aqueous phases coexist, and additional fluid may be injected to perform enhanced oil recovery (EOR). Fluids injected during EOR include surfactant solutions [34], polymer solutions [35], and aqueous nanoparticle suspensions [36,37], and the mixing of the new aqueous fluid with the previously-injected water is an important goal of the flooding operation. In particular, polymer solutions are always applied as the final chasing fluid following other EOR fluids [38] because the high viscosity of polymer solutions can effectively suppress viscous fingering and thereby stabilize the injection front [11,39,40]. Consequently, how the chasing polymer solution mixes with the preceding EOR fluids is important in determining the overall EOR efficiency. However, in addition to being highly viscous, polymer solutions can also be viscoelastic. Viscoelasticity has been shown to significantly affect flow in porous media [41–44] or even bring about local instabilities [41,45,46]. The instability of viscoelastic fluids in porous media has been widely observed, and presents a promising opportunity for enhanced mixing [47,48], but little is understood about how to actively control these instabilities to achieve this purpose.

In this paper, we modeled a T-junction chamber as a pore element to simulate the mixing effect of unstable viscoelastic fluid flows [49–53]. The instability of the viscoelastic fluid was induced by perturbing the flow through controlled driving pressures at the inlets. This mechanism can be considered an active method for the enhancement of mixing within a microfluidic channel when the flow state is laminar and uniaxial across the pore-like channel. The driving pressure at the two inlets of the mixer was modulated by time-dependent sinusoidal signals. The degree of mixing was compared under constant pressure as well as when the pressure was modulated in one or both inlets. The effects of amplitude, frequency, and phase difference of the driving pressures were compared for viscoelastic fluid flows and Newtonian fluid flows. Finally, the results were experimentally validated using a microfluidic device.

2. Numerical Schemes

2.1. Computational Model of T-Junction Micromixer

The computational model used in this study to simulate the T-junction micromixer is shown in Figure 1. Fluid components of different concentrations flow into two inlets I_1 and I_2 from both sides of the junction and convergence begins at the junction point X. Then, the mixing fluid flows through a narrow converging channel that feeds into an expanded mixing channel where the two fluids will mix with each other. The component concentration at a cross section S in the mixing channel is monitored to determine the mixing efficiency. The length and width of the inlet channels are $L_0 = 20$ and $W_0 = 1$, respectively, while the converging channel has dimensions of $L_1 = 6$ and $W_1 = 0.5$ and the mixing channel has $L_2 = 20$ and $W_2 = 6$. The micromixer structure has a depth of $d = 1$, and the outlet is set at the end of the mixing channel.

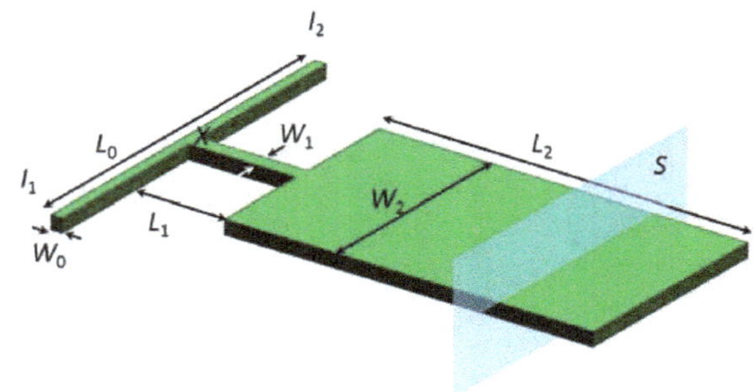

Figure 1. The computational model of T-junction micromixer.

2.2. Governing Equations

Flows in the T-junction micromixer were investigated using a solver based on the open-source software OpenFoam. The velocity, pressure, and concentration fields were numerically simulated based on the fundamental governing equations. The continuous equation and Navier–Stokes (N–S) equation were applied to solve the incompressible Newtonian fluid flow:

$$\nabla\cdot\mathbf{U} = 0 \tag{1}$$

$$\frac{\partial\rho\mathbf{U}}{\partial t} + \rho\mathbf{U}\cdot\nabla\mathbf{U} = -\nabla p + \mu\nabla^2\mathbf{U} \tag{2}$$

where $\mathbf{U}$ represents the velocity vector of flow; p is the pressure; ρ is the density; and μ is the dynamic viscosity of the fluid.

The convection–diffusion equation was applied to solve the component concentrations in the flow:

$$\frac{\partial C}{\partial t} + \mathbf{U}\cdot\nabla C = D\nabla^2 C \tag{3}$$

where C is the component concentration of the fluid and D is the diffusion coefficient of fluid.

For the incompressible viscoelastic fluid flow, the N–S equation is modified as:

$$\frac{\partial\rho\mathbf{U}}{\partial t} + \rho\mathbf{U}\cdot\nabla\mathbf{U} = -\nabla p + \mu\nabla^2\mathbf{U} + \nabla\cdot\boldsymbol{\tau}_p \tag{4}$$

where $\boldsymbol{\tau}_p$ is the elastic stress, which can be derived from the constitutive equation of viscoelastic fluid. The constitutive equation describes the relation between molecule deformation and elastic stress in the fluid:

$$\boldsymbol{\tau}_p = k_B T(f(r)\mathbf{C} - \mathbf{I}) \tag{5}$$

where k_B is the Boltzmann constant; T is the absolute temperature of the fluid; $f(r)$ is the Peterlin function; $\mathbf{C}$ is the conformation tensor of polymer molecules; and $\mathbf{I}$ is the Kronecker symbol for the unit tensor. The constitutive equation is further derived from Equation (5) as:

$$\boldsymbol{\tau}_p = \eta_p(f(r)\mathbf{C} - \mathbf{I})/\lambda \tag{6}$$

where η_p is the solute viscosity and λ is the relaxation time of the solution. The transport equation of the conformation tensor is then expressed as:

$$\frac{\partial\mathbf{C}}{\partial t} + \mathbf{U}\cdot\nabla\mathbf{C} = \mathbf{C}\cdot\nabla\mathbf{U} + (\nabla\mathbf{U})^T\cdot\mathbf{C} - [\mathbf{I} + \alpha(\mathbf{C} - \mathbf{I})]\cdot[f(r)\mathbf{C} - \mathbf{I}]/\lambda \tag{7}$$

where α is a parameter that relates to the anisotropy of drag encountered by flowing polymer segments [54]. Different $f(r)$ and α are selected in different constitutive models. Here, we assumed that the viscoelastic fluid is linearly extended during the flow, so that $f(r) = 1$ and $\alpha = 0$. This satisfies the flow situation of a Boger fluid in the Oldroyd-B constitutive model. Therefore, Equations (6) and (7) are simplified as:

$$\boldsymbol{\tau}_p = \eta_p(\mathbf{C} - \mathbf{I})/\lambda \tag{8}$$

$$\frac{\partial \mathbf{C}}{\partial t} + \mathbf{U}\cdot\nabla\mathbf{C} = \mathbf{C}\cdot\nabla\mathbf{U} + (\nabla\mathbf{U})^T\cdot\mathbf{C} - \mathbf{I}\cdot(\mathbf{C} - \mathbf{I})/\lambda \tag{9}$$

Moreover, in order to gain more universal computational results, the dimensionless governing equations were used based on the following:

$$\nabla^* = H\cdot\nabla,\ \mathbf{U}^* = \frac{\mathbf{U}}{U_i},\ t^* = \frac{tU_i}{H},\ p^* = \frac{p}{\rho U_i{}^2},\ C^* = \frac{C}{C_{\max}} \tag{10}$$

where H is the hydraulic diameter of the channel; U_i is the center velocity of the fluid at the inlet channel; ρ is the fluid density; and C is the component concentration in the fluid. We defined the dimensionless Reynolds number Re and Schmidt number Sc as:

$$Re = \frac{\rho U_i H}{\mu},\ Sc = \frac{\nu}{D} \tag{11}$$

where ν is the kinematic viscosity of the fluid. The dimensionless governing continuous equation and the N–S equation can then be expressed as:

$$\nabla^*\cdot\mathbf{U}^* = 0 \tag{12}$$

$$\frac{\partial \mathbf{U}^*}{\partial t^*} + \mathbf{U}^*\cdot\nabla^*\mathbf{U}^* = -\nabla^* p^* + \frac{1}{Re}\nabla^{2*}\mathbf{U}^*, \tag{13}$$

and the convection–diffusion equation is expressed as:

$$\frac{\partial C^*}{\partial t^*} + \mathbf{U}^*\cdot\nabla^* C^* = \frac{1}{Re\cdot Sc}\nabla^{2*}C^* \tag{14}$$

Meanwhile, for a viscoelastic fluid, we defined the dimensionless parameter Weissenberg number Wi and the viscosity ratio β as:

$$Wi = \frac{\lambda U_i}{D}, \beta = \frac{\eta_s}{\eta_s + \eta_p} \tag{15}$$

where λ is the relaxation time of the viscoelastic fluid; η_p is the solute kinetic viscosity; η_s is the solvent kinetic viscosity. Thus, the dimensionless N–S equation for viscoelastic fluid flow and the conformation tensor transport equation are modified as:

$$\frac{\partial \mathbf{U}^*}{\partial t^*} + \mathbf{U}^*\cdot\nabla^*\mathbf{U}^* = -\nabla^* p^* + \frac{\beta}{Re}\nabla^{2*}\mathbf{U}^* + \frac{1-\beta}{ReWi}\nabla^*\mathbf{C} \tag{16}$$

$$\frac{\partial \mathbf{C}}{\partial t^*} + \mathbf{U}^*\cdot\nabla^*\mathbf{C} = \mathbf{C}\cdot\nabla^*\mathbf{U}^* + (\nabla^*\mathbf{U}^*)^T\cdot\mathbf{C} - \frac{\mathbf{I}\cdot(\mathbf{C} - \mathbf{I})}{Wi} \tag{17}$$

For both Newtonian fluid and viscoelastic fluid cases, the dimensionless density, dimensionless viscosity, and Re were all set at 1.0, while Sc was set at 10^{-8}. In addition, in the viscoelastic fluid case, the dimensionless relaxation time λ was set at 5.0, the dimensionless solution viscosity was set at 0.4, and dimensionless solvent viscosity was set at 0.6.

2.3. Numerical Methods

In this work, the fluid in the micromixer initially had a component concentration of 0. The dimensionless component concentration C_1 was 0 for the fluid entering through inlet I_1, while $C_2 = 1$ for the fluid entering through inlet I_2. The driving pressure is P_1 at inlet I_1 and P_2 at inlet I_2.

The no-slip condition was imposed on all channel surfaces, and a fully developed flow condition was used at the mixer outlet.

The first-order Euler implicit scheme was used for time marching in the unsteady transport equations of Equations (14), (16), and (17), with a small dimensionless time step $\delta t = 10^{-3}$. The convection terms in Equations (14) and (16) were discretized by the QUICK scheme, while the bounded MINMOD scheme was used to discretize the convection terms in Equation (17). Pressure–velocity coupling was handled by the PISO algorithm.

2.4. The Definition of Mixing Efficiency

Given component concentrations $C_1 = 0$ and $C_2 = 1$, a full mixing of the two fluids will result in a concentration of 0.5 when two averaged flow rates are the same. The degree of mixing is calculated by taking the standard deviation of concentration at all meshed points in cross-section S, defined as:

$$A = \sqrt{\sum_{i=1}^{N} \frac{(x_i - E)^2}{N}} \tag{18}$$

where N is the number of meshed points in the cross-section; x_i is the concentration at point i; and E is the full mixing concentration, which is 0.5. Equation (18) is valid if the fluid density at each mesh point is the same. However, the flow rate is higher at the center of the channel and lower at the channel sides due to the no-slip boundary condition and fluid viscosity. Therefore, the deviation needs to be adjusted as:

$$B = \sqrt{\sum_{i=1}^{N} \frac{(x_i - E)^2}{N} \frac{V_i}{V_{mean}}} \tag{19}$$

where V_i is the velocity at the mesh point i and V_{mean} is the average velocity in cross-section S. B ranges from 0 to 0.5, where $B = 0$ means full mixing while $B = 0.5$ indicates no mixing. We then define the degree of mixing as:

$$M = 1 - \frac{\sqrt{\sum_{i=1}^{N} \frac{(x_i - E)^2}{N} \frac{V_i}{V_{mean}}}}{\mu} \tag{20}$$

Therefore, $M = 1$ indicates full mixing, while $M = 0$ when no mixing occurs. We will use this index to present the mixing efficiency in the following sections.

3. Results and Discussion

3.1. Mixing in Condition of Constant Pressure

We first investigated fluid mixing in the T-junction micromixer when both inlets I_1 and I_2 are under the same constant pressure $P_0 = 3500$. Figure 2a,b compare the component concentration distribution in the Newtonian fluid and viscoelastic fluid cases after the fluid has flowed in the micromixer for a long time ($t = 120$). The color map indicates the normalized component concentration (red represents $C = 1$ and blue represents $C = 0$). It can be seen that the two injected fluids are well separated in the converging channel. In the mixing channel, green appears at the center area along the flow direction. The concentration gradually changed between 0 and 1 and reached 0.5 in the center, indicating a high mixing here. Compared to the Newtonian fluid case, the mixing area is larger in the viscoelastic fluid. Since the simulation conditions for two fluids were identical except that the viscoelastic fluid had a nonzero elasticity, the improvement in mixing must originate from elastic stress in the viscoelastic fluid.

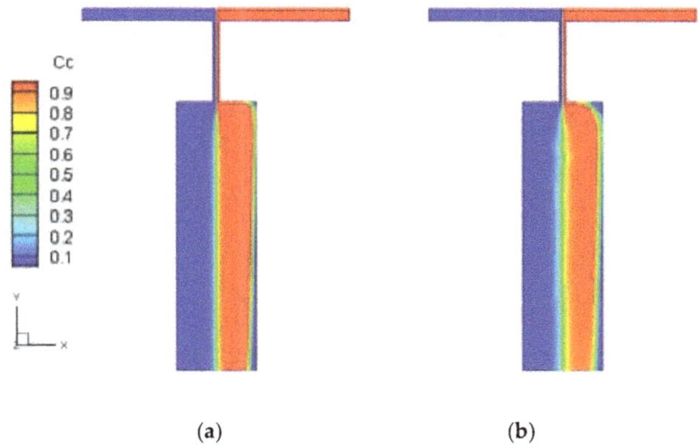

Figure 2. Concentration distribution in the Newtonian fluid case (**a**) and viscoelastic fluid case (**b**) under constant driving pressure at both inlets.

The elastic stress effect was further confirmed by investigating the flow streamlines and fluid angular velocity, as shown in Figure 3. In the Newtonian fluid case, the streamlines bend 90° downward at the convergence point and there is no angular velocity in the flow. By contrast, the streamlines in viscoelastic fluid in the same region first bend up slightly before bending down. A nonzero angular velocity is observed in the narrow channel, indicating convection of two viscoelastic fluids and more significant mixing. The degree of mixing M was analyzed at the cross-section S, which was set at a distance of $L_s = 12$ below the channel expansion point. For both cases, M increases until it becomes stable starting from $t = 80$, but their magnitudes differ: M plateaus at 0.15 for the Newtonian fluid case and 0.25 for the viscoelastic fluid case, as shown in Figure 4. This quantitatively shows that mixing was improved in viscoelastic fluid case.

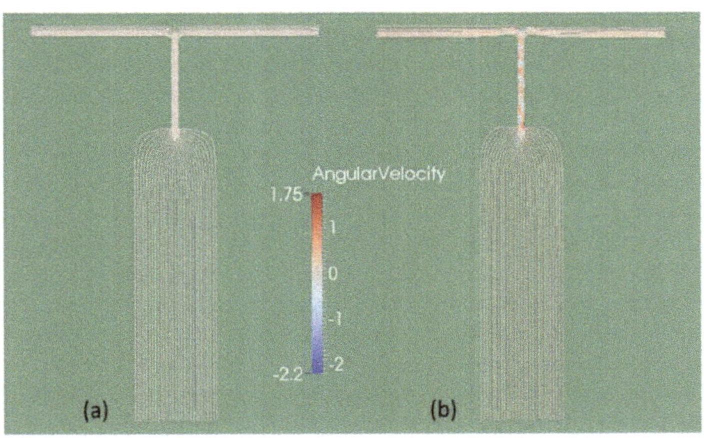

Figure 3. Streamlines and angular velocity in the Newtonian fluid case (**a**) and viscoelastic fluid case (**b**) under constant driving pressure at both inlets.

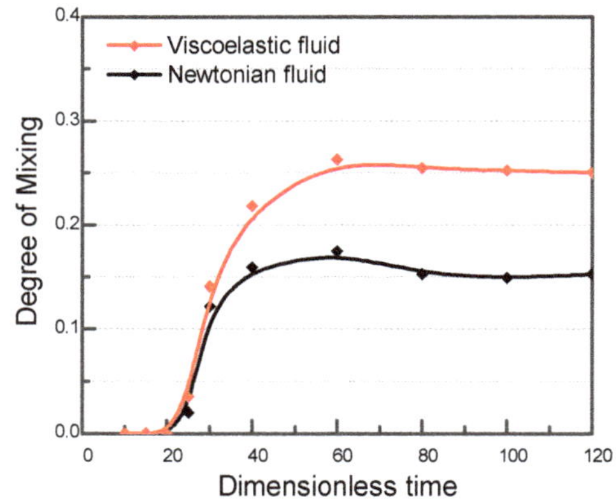

Figure 4. Degree of mixing in the Newtonian fluid case (**black line**) and viscoelastic fluid case (**red line**) under constant driving pressure at both inlets.

3.2. Mixing in Single-Side Pressure Oscillation

We modulated the driving pressure P_1 at inlet I_1 with a sinusoidal factor while keeping the pressure P_2 at inlet I_2 constant:

$$P_1(t) = P_0[1 + A\sin(2\pi f t)] \tag{21a}$$

$$P_2 = P_0 \tag{21b}$$

where A and f are the amplitude and frequency of the modulating signal, respectively. A, f, and P are all dimensionless. A was set at 0.8 to ensure a large pressure modulation while keeping the absolute pressure positive during the driving process. f was set at 1. P_0 was the same as that in the constant pressure case, 3500. The change in the degree of mixing in cross-section S within one period was studied when pressures in Equation (21) were applied to the viscoelastic fluid case. Under the above driving pressure, the concentration distributions at t = 120 in the Newtonian fluid and viscoelastic fluid cases are shown in Figure 5, respectively. The transition region between the blue (low concentration) and red (high concentration) fluids expanded significantly when the pressure was modulated compared to when the pressure was constant, indicating that pressure modulation greatly improves mixing efficiency. Meanwhile, the concentrations in the converging channel were no longer uniformly distributed along the channel due to the pressure changes over time. At times $P_1 < P_2$, higher concentration fluid from inlet I_2 will enter the converging channel. The converse happens when $P_1 > P_2$. In other words, the two fluids will alternatively enter the mixing channel in larger amounts. Due to the large width of the subsequent expanded channel, the small amounts of fluids introduced during one pressure cycle will start to expand and convect with each other. In this way, better mixing can be realized. In addition, we can also see that the larger discontinuous portion of fluid alternated more in the converging channel for the viscoelastic fluid case compared to the Newtonian fluid case, which indicates a higher mixing efficiency in the viscoelastic fluid. This was confirmed by investigating the degree of mixing at the cross-section S for both fluids. Figure 6 compares the mixing of a viscoelastic fluid with a different relaxation time and a Newtonian fluid. Mixing increased with the relaxation time and became stable at t = 80 for all fluids. The Newtonian fluid had the lowest degree of mixing of 0.62. Mixing gradually increased from 0.63 to 0.70 as the relaxation time λ increased from 1 to 10. This is because a higher relaxation time indicates a larger elastic stress, which leads to more convection in the fluid. Compared to the condition of constant pressure, the degree of mixing increased from 0.15 to 0.61 for the Newtonian fluid case under modulated pressure and from 0.25 to 0.68 for a viscoelastic fluid with λ of 5.

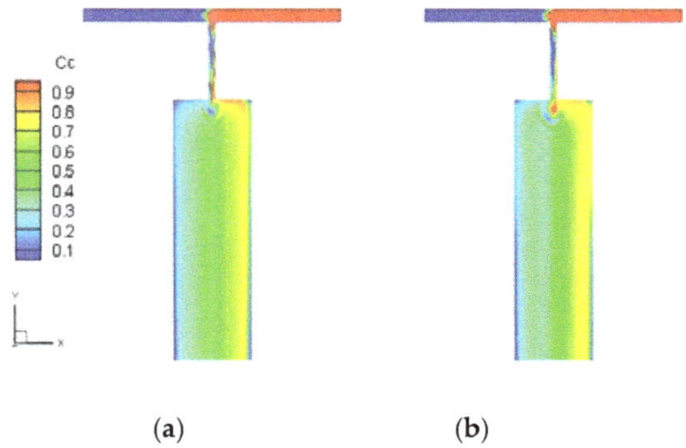

Figure 5. Concentration distribution in the Newtonian fluid case (**a**) and viscoelastic fluid case (**b**) under modulated pressure at the inlet I_1 and constant pressure at the inlet I_2.

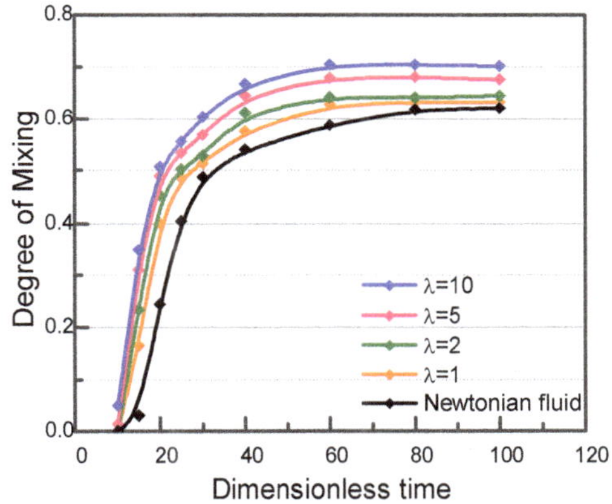

Figure 6. Degree of mixing in the Newtonian fluid case and viscoelastic fluid case with different relaxation time under modulating driving pressure at the inlet I_1 and constant pressure at the inlet I_2.

The flow streamlines and flow angular velocity were also investigated within one period, shown in Figure 7a,b for the Newtonian fluid case and Figure 7c,d for the viscoelastic fluid case. For both fluids, the stream flowed mainly from inlet I_1 at t = 119.2 (Figure 7a,c) and mainly from inlet I_2 at t = 119.7 (Figure 7b,d). The streamlines were smooth, and no angular velocity was observed in the Newtonian fluid case. However, the viscoelastic fluid formed flow vortexes alternating between the right and left sides of convergence points at times t = 119.2 and t = 119.7, respectively. The fluctuation in flow vortexes indicates enhanced convection between two fluids, which in turn leads to better mixing.

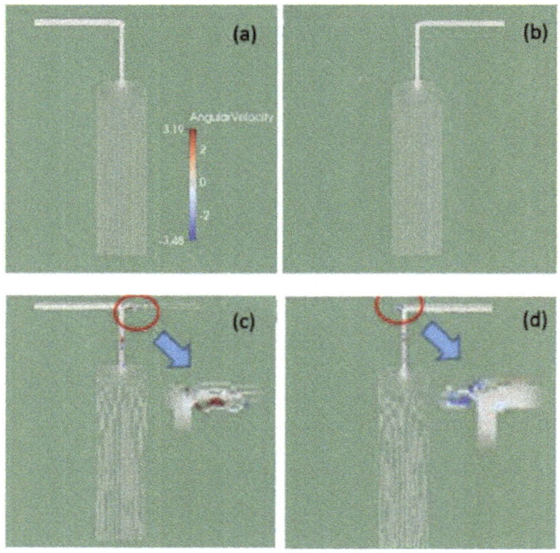

Figure 7. Streamlines and angular velocity in the Newtonian fluid case (**a**,**c**) and viscoelastic fluid case (**b**,**d**) under modulated pressure at the inlet I_1 and constant pressure at the inlet I_2.

3.3. *Mixing in Double-Sided Pressure Oscillation*

Since a single inlet modulation improves mixing, we next investigated the effect of modulating the driving pressure at both inlets. Here, we modulated both driving pressures P_1 and P_2 with sinusoidal signals of the same amplitude A and frequency f, but with independent phase delays φ_1 and φ_2:

$$P_1(\mathrm{t}) = P_0[1 + A\sin(2\pi f\mathrm{t} + \varphi_1)] \tag{22a}$$

$$P_2(\mathrm{t}) = P_0[1 + A\sin(2\pi f\mathrm{t} + \varphi_2)] \tag{22b}$$

and we examined the dependence of mixing on A, f, and $\Delta\varphi = \varphi_1 - \varphi_2$, the phase difference between the two pressures.

Setting A = 3500 and f = 1, the concentration distribution of viscoelastic fluid when $\Delta\varphi$ varies from 0 to π was investigated at t = 120, shown in Figure 8. The high and low concentration fluids are symmetrically distributed on both sides of the T-junction when $\Delta\varphi$ = 0 (Figure 8a) because the driving pressure at both inlets is always the same. Only very small amounts of the two fluids mix with each other along the center of the channel. However, the distribution symmetry is broken as $\Delta\varphi$ starts to increase. A small fluctuation of the fluids in the converging channel was observed when $\Delta\varphi = \pi/6$, and it was magnified at the expansion point when the fluids entered the expanded channel, as shown in Figure 8b. As a consequence, the size of the mixing region starts to increase. As $\Delta\varphi$ is increased to $\pi/3$, the two fluids start to alternate along the converged channel (Figure 8c) and the blue and red flows start to become discontinuous at the expansion point. Furthermore, the green region increases, while narrow streams of the low and high concentration fluids remain observable near the left and right sidewalls, respectively. As $\Delta\varphi$ is further increased to $\pi/2$ (Figure 8d), the green region occupies most of the expansion channel, and isolated spots of unmixed fluid appear at the expansion point, as shown in the insert of Figure 8d. This means that the instantaneous pressure difference between the two inlets was large enough to allow the fluids to enter the converged channel separately. For $\Delta\varphi$ between $2\pi/3$ to π (Figure 8e–g), the green region occupies almost all of the expansion channel, indicating good mixing. The degree of mixing for different $\Delta\varphi$ is plotted in Figure 9. When $\Delta\varphi$ = 0, the mixing degree stabilizes at 0.2 after t = 20. The stabilization time t increases to 40, 60 and 80 for $\Delta\varphi$ of $\pi/6$, $\pi/3$, and $\pi/2$, respectively, and the corresponding maximum degrees of mixing increase to 0.45,

0.63 and 0.73, respectively. The degree of mixing reaches a system-wide maximum value of 0.82 when $\Delta\varphi = \pi$. A phase difference of π maximizes the difference in the amount of fluid entering from each inlet at any instant in time. The fluid volumes alternate during each period, and this alternating fluid pattern expands in the wide mixing channel. As a result, the diffusion area is increased. By contrast, when $\Delta\varphi = 0$, the fluids from the inlets flow side by side along the channel direction and the diffusion area is only along the length of the center line.

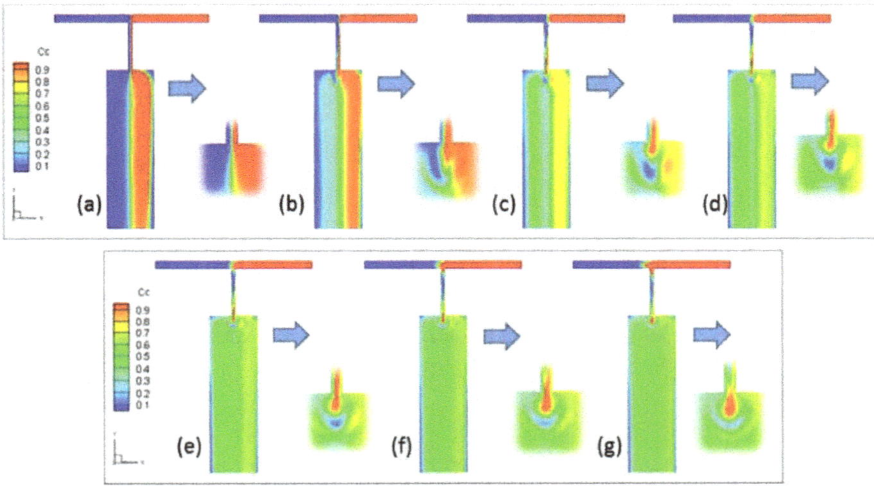

Figure 8. Concentration distribution in viscoelastic fluid case under modulated pressure at both inlets with phase difference (**a**) 0, (**b**) $\pi/6$, (**c**) $\pi/3$, (**d**) $\pi/2$, (**e**) $2\pi/3$, (**f**) $5\pi/6$, and (**g**) π.

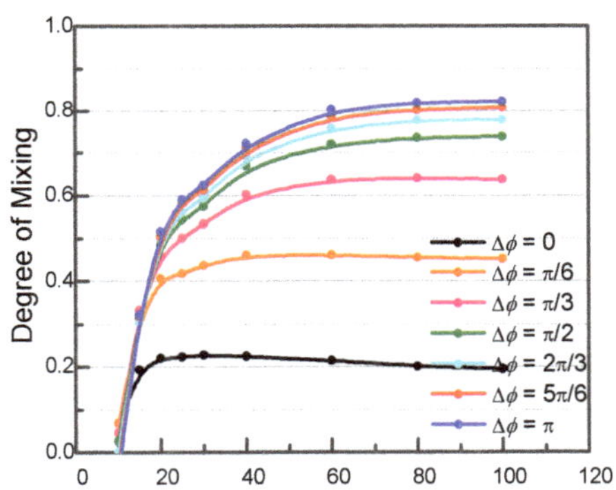

Figure 9. Degree of mixing in viscoelastic fluid case under modulated pressure at both inlets with phase difference from 0 to π.

At a higher frequencies, less fluid will enter the converged channel during each period. As shown in Figure 10a,b, the component mixing maps are compared for the viscoelastic fluid case when $f = 1$ and $f = 1.5$, respectively, with fixed parameters $A = 3500$ and $\Delta\varphi = \pi$. The green region shrinks significantly when f increases from 1 to 1.5. In addition, larger regions of unmixed fluid occupy the left and right sides of the expanded channel. This indicates that a higher modulating frequency undermines the mixing effect. This can be better understood by looking at the concentration distributions in the inlets and in the converging channel. At time $t = 120$, the pressure at inlet I_2 is higher than that at inlet I_1, and more of the red fluid will be injected. At a lower modulating frequency, the difference in the amount of fluid flowing into the converging channel from each side is larger in each cycle; this can be observed in the converging channel and results in a higher mixing efficiency overall. The final degrees of mixing, which increase from 0.61 to 0.90 as the pressure oscillation frequency decreases from $f = 2$ to $f = 0.5$, are shown in Figure 11.

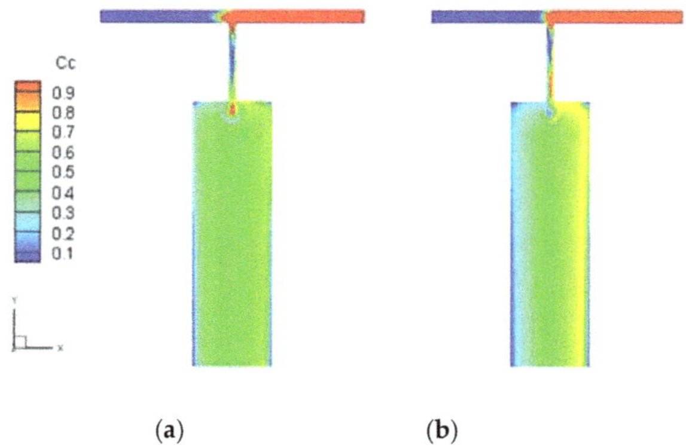

Figure 10. Concentration distribution in viscoelastic fluid case under modulated pressure at both inlets with modulating frequency of $f = 1$ (**a**) and $f = 1.5$ (**b**).

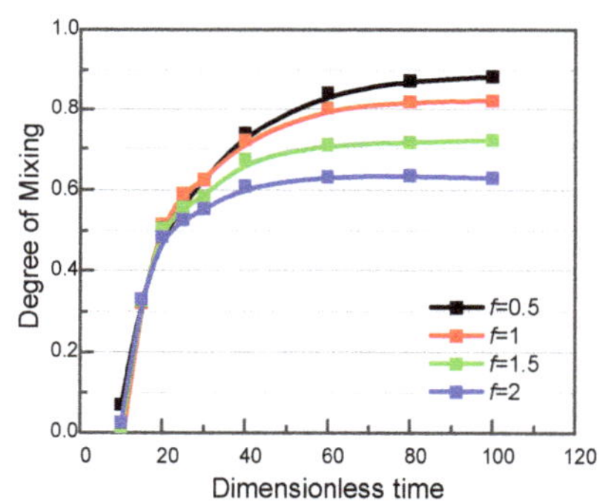

Figure 11. Degree of mixing in viscoelastic fluid case under modulated pressure at both inlets with different modulating frequencies.

In addition, we can also increase the degree of mixing by increasing the alternating volume within one cycle using a higher pressure level at the inlets. To do this, P_0 in Equation (22) is varied between 3500 to 6000 while setting $f = 1$ and $\Delta\varphi = \pi$. By increasing P_0, the final degree of mixing in the viscoelastic fluid case also increases from 0.82 to 0.91, as shown in Figure 12.

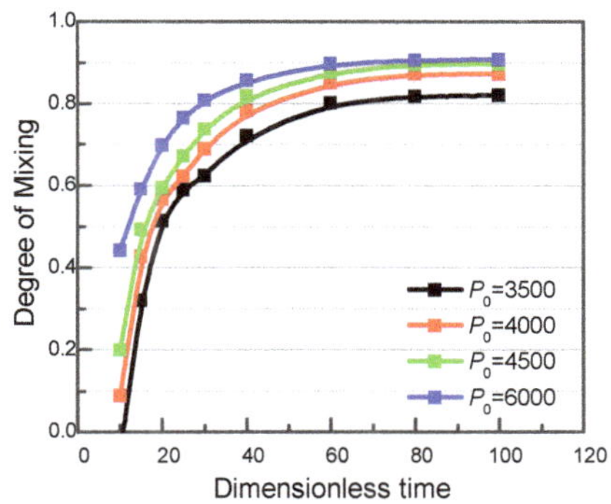

Figure 12. Degree of mixing in viscoelastic fluid case under modulated pressure at both inlets with different amplitudes.

4. Experimental Results

To validate the results of the simulations, the mixing efficiency was experimentally studied here in a PDMS-glass bonded microfluidic chip, as shown in Figure 13a. A mold for the PDMS structure was fabricated from SU8 photoresist on a silicon wafer using a standard photolithography process. The PDMS is a mixture of base and crosslinker with a ratio of 10:1. It was poured on the SU8 mold and cured at 65 °C for 4 h. To complete the chip, the cured PDMS was bonded to a glass slide after oxygen plasma treatment. The size of the microchannel was scaled to 100 μm based on the model in Figure 1. The length and width of the inlet channel were L_0 = 2 mm and W_0 = 100 μm, respectively. The converging channel was L_1 = 600 μm in length and W_1 = 50 μm in width. The mixing channel was L_2 = 2 mm in length and W_2 = 600 μm in width. The entire micromixer structure had a depth of d = 100 μm, and the outlet was directly at the end of the mixing channel.

A Newtonian glycerol solution was injected into the two inlets of the T-junction mixer. The fluid at one inlet contained 10 μg/mL of Rhodamine B, while the other did not, so the mixing degree could be determined by quantifying the Rhodamine B concentration. With the limit of one programmable pump (PHD ULTRA^{TM} 4400), we fixed the flow rate at 500 μL/h on one inlet, and modulated the flow rate on the other side under three different conditions: (i) Constant flow rate of 500 μL/h; (ii) alternating flow rate between 0 and 1000 μL/h at a frequency f = 0.5 Hz and duty cycle of 50%; (iii) alternating flow rate between 0 and 1000 μL/h at a frequency f = 0.1 Hz and duty cycle of 50%. The recorded flow pattern is shown in Figure 13b. The color map represents the Rhodamine B concentration, which is normalized from 0 to 1: Yellow denotes a concentration of 1 and blue, a concentration of 0. For case (i), the flow rates at both inlets were the same constant value and the mixing region was narrow, indicating a small degree of mixing. For case (ii), the mixing region was increased when the flow rate on one side alternated at f = 0.5 Hz. The mixing region was further increased when the frequency decreased to 0.1 Hz in case (iii). According to the normalized concentrations of Rhodamine B, the degree of mixing at different flow distances along the channel is calculated using Equation (20) and plotted in Figure 13c. The origin indicates the point at which the fluid enters the mixing channel. The degree of mixing gradually increases with the flow distance, and stabilizes at 0.20 for case (i), 0.25 for case (ii), and 0.37 for case (iii). This experimentally demonstrates the enhancement in mixing when the flow is modulated on one side compared to having a constant driving flow at both sides, which agrees well with the simulation results. This also confirms that mixing is increased as the alternating frequency decreases.

(a)

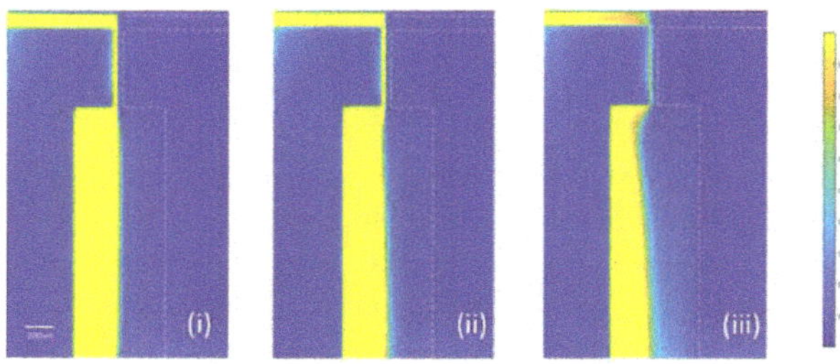

(b)

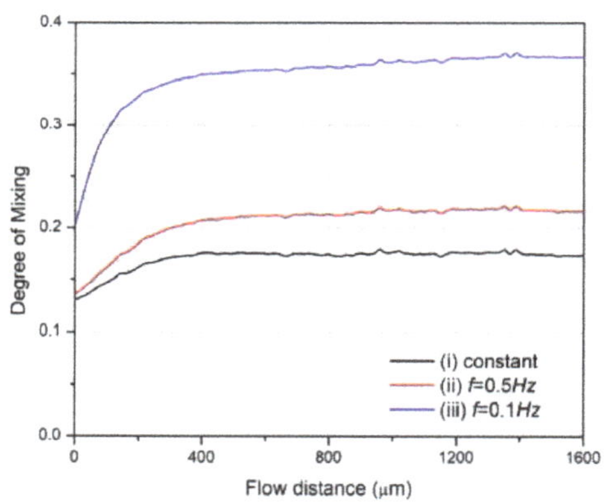

(c)

Figure 13. (**a**) The fabricated microfluidic chip for the T-mixer; (**b**) Recorded flow pattern of the glycerol solution with/without Rhodamine B on left/right side and the flow rate for right side (**i**) remains constant at 500 μL/h; (**ii**) alternates with frequency f = 0.5 Hz; (**iii**) alternates with frequency f = 0.1 Hz. (**c**) The measured degree of mixing for case (**i–iii**).

5. Conclusions

In conclusion: A T-junction micromixer was modeled, and mixing efficiencies were numerically compared for both a Newtonian fluid and a viscoelastic fluid as a function of driving pressure amplitude and modulation frequency. The degree of mixing is higher in a viscoelastic fluid than in a Newtonian fluid due to elastic stress in the viscoelastic fluid. Different modulations of the driving pressures at each inlet of the micromixer were explored. Under constant driving pressures, the degree of mixing was relatively low, 0.15 for the Newtonian fluid case and 0.25 for the viscoelastic fluid case.

When the driving pressures were modulated with a sinusoidal factor at one inlet while being held constant at the other, the degree of mixing increased to 0.62 and 0.67 for the Newtonian fluid and the viscoelastic fluid, respectively. When the driving pressures at both inlets of the micromixer were modulated with a sinusoidal factor, the mixing efficiency could be increased significantly by controlling the phase difference between the modulated pressures. The degree of mixing reached a maximum value of 0.82 using a viscoelastic fluid with a phase difference of π. Mixing enhancement arising from a low-frequency, single-inlet alternating modulation of the flow rate was experimentally demonstrated for a Newtonian glycerol solution. The method of modulating the driving pressure to enhance mixing may have potential applications in chemical engineering and in flow through porous media.

Author Contributions: M.Z. raised the original idea, wrote part of code for simulation software, analyzed the data, carried on the experiment, and wrote the paper. Y.C. simulated the data. W.C. revised the paper and coordinated this work. Z.W. did the experiment. Y.L. simulated the data. F.L. supervised on the simulation and the code for simulation software. W.Z. discussed the original idea, analyzed the simulation and experimental data, composed, and revised the paper.

Funding: This work is supported by this project is supported by the China Scholarship Council (No. 201606120130).

Conflicts of Interest: The authors declare no conflicts of interest.

References

1. Stroock, A.D.; Dertinger, S.K.W.; Ajdari, A.; Mezić, I.; Stone, H.A.; Whitesides, G.M. Chaotic mixer for microchannels. *Science* **2002**, *295*, 647–651. [CrossRef] [PubMed]
2. Lee, C.Y.; Fu, L.M. Recent advances and applications of micromixers. *Sens. Actuators B Chem.* **2018**, *259*, 677–702. [CrossRef]
3. Roberge, D.M.; Ducry, L.; Bieler, N.; Cretton, P.; Zimmermann, B. Microreactor technology: A revolution for the fine chemical and pharmaceutical industries? *Chem. Eng. Technol.* **2005**, *28*, 318–323. [CrossRef]
4. Ehrfeld, W.; Hessel, V.; Löwe, H. *Microreactors—New Technology for Modern Chemistry*; Wiley-VCH: Weinheim, Germany, 2000; p. 288. ISBN 3-527-29590-9.
5. Kefala, I.N.; Papadopoulos, V.E.; Karpou, G.; Kokkoris, G.; Papadakis, G.; Tserepi, A. A labyrinth split and merge micromixer for bioanalytical applications. *Microfluid. Nanofluid.* **2015**, *19*, 1047–1059. [CrossRef]
6. Lang, Q.; Ren, Y.; Hobson, D.; Tao, Y.; Hou, L.; Jia, Y.; Hu, Q.; Liu, J.; Zhao, X.; Jiang, H. In-plane microvortices micromixer-based AC electrothermal for testing drug induced death of tumor cells. *Biomicrofluidics* **2016**, *10*, 64102. [CrossRef] [PubMed]
7. Hessel, V.; Löwe, H.; Schönfeld, F. Micromixers—A review on passive and active mixing principles. *Chem. Eng. Sci.* **2005**, *60*, 2479–2501. [CrossRef]
8. Abed, W.M.; Whalley, R.D.; Dennis, D.J.C.; Poole, R.J. Experimental investigation of the impact of elastic turbulence on heat transfer in a serpentine channel. *J. Non-Newtonian Fluid Mech.* **2016**, *231*, 68–78. [CrossRef]
9. Ye, Y.; Chiogna, G.; Cirpka, O.A.; Grathwohl, P.; Rolle, M. Experimental investigation of transverse mixing in porous media under helical flow conditions. *Phys. Rev. E* **2016**, *94*, 13113. [CrossRef] [PubMed]
10. Ouyang, Y.; Xiang, Y.; Zou, H.; Chu, G.; Chen, J. Flow characteristics and micromixing modeling in a microporous tube-in-tube microchannel reactor by CFD. *Chem. Eng. J.* **2017**, *321*, 533–545. [CrossRef]
11. Li, J.S.; Li, Q.; Cai, W.H.; Li, F.C.; Chen, C.Y. Mixing Efficiency via Alternating Injection in a Heterogeneous Porous Medium. *J. Mech.* **2018**, *34*, 167–176. [CrossRef]
12. Cai, J.; Wei, W.; Hu, X.; Liu, R.; Wang, J. Fractal characterization of dynamic fracture network extension in porous media. *Fractals* **2017**, *25*. [CrossRef]
13. Elvira, K.S.; Solvas, X.C.; Wootton, R.C.R.; Demello, A.J. The past, present and potential for microfluidic reactor technology in chemical synthesis. *Nat. Chem.* **2013**, *5*, 905–915. [CrossRef] [PubMed]
14. Floquet, C.F.A.; Sieben, V.J.; MacKay, B.A.; Mostowfi, F. Determination of boron concentration in oilfield water with a microfluidic ion exchange resin instrument. *Talanta* **2016**, *154*, 304–311. [CrossRef] [PubMed]
15. Liu, R.; Jiang, Y.; Li, B.; Wang, X. A fractal model for characterizing fluid flow in fractured rock masses based on randomly distributed rock fracture networks. *Comput. Geotech.* **2015**, *65*, 45–55. [CrossRef]
16. Knight, J.B.; Vishwanath, A.; Brody, J.P.; Austin, R.H. Hydrodynamic focusing on a silicon chip: Mixing nanoliters in microseconds. *Phys. Rev. Lett.* **1998**, *80*, 3863–3866. [CrossRef]

17. Squires, T.M.; Quake, S.R. Microfluidics: Fluid physics at the nanoliter scale. *Rev. Mod. Phys.* **2005**, *77*, 977–1026. [CrossRef]
18. Xia, H.M.; Wang, Z.P.; Koh, Y.X.; May, K.T. A microfluidic mixer with self-excited 'turbulent' fluid motion for wide viscosity ratio applications. *Lab Chip* **2010**, *10*, 1712–1716. [CrossRef] [PubMed]
19. Lemenand, T.; Della Valle, D.; Habchi, C.; Peerhossaini, H. Micro-mixing measurement by chemical probe in homogeneous and isotropic turbulence. *Chem. Eng. J.* **2017**, *314*, 453–465. [CrossRef]
20. Wu, J.W.; Xia, H.M.; Zhang, Y.Y.; Zhu, P. Microfluidic mixing through oscillatory transverse perturbations. *Mod. Phys. Lett. B* **2018**, *32*. [CrossRef]
21. Stone, H.A.; Stroock, A.D.; Ajdari, A. Engineering flows in small devices: Microfluidics toward a lab-on-a-chip. *Annu. Rev. Fluid Mech.* **2004**, *36*, 381–411. [CrossRef]
22. Schönfeld, F.; Hessel, V.; Hofmann, C. An optimised split-and-recombine micro-mixer with uniform 'chaotic' mixing. *Lab Chip* **2004**, *4*, 65–69. [CrossRef] [PubMed]
23. Sivashankar, S.; Agambayev, S.; Mashraei, Y.; Li, E.Q.; Thoroddsen, S.T.; Salama, K.N. A "twisted" microfluidic mixer suitable for a wide range of flow rate applications. *Biomicrofluidics* **2016**, *10*. [CrossRef] [PubMed]
24. Lee, N.Y.; Yamada, M.; Seki, M. Development of a passive micromixer based on repeated fluid twisting and flattening, and its application to DNA purification. *Anal. Bioanal. Chem.* **2005**, *383*, 776–782. [CrossRef] [PubMed]
25. Jafari, O.; Rahimi, M.; Kakavandi, F.H. Liquid-liquid extraction in twisted micromixers. *Chem. Eng. Process.* **2016**, *101*, 33–40. [CrossRef]
26. Lin, M.X.; Hyun, K.A.; Moon, H.S.; Sim, T.S.; Lee, J.G.; Park, J.C.; Lee, S.S.; Jung, H.I. Continuous labeling of circulating tumor cells with microbeads using a vortex micromixer for highly selective isolation. *Biosens. Bioelectron.* **2013**, *40*, 63–67. [CrossRef] [PubMed]
27. Bensaid, S.; Deorsola, F.A.; Marchisio, D.L.; Russo, N.; Fino, D. Flow field simulation and mixing efficiency assessment of the multi-inlet vortex mixer for molybdenum sulfide nanoparticle precipitation. *Chem Eng. J.* **2014**, *238*, 66–77. [CrossRef]
28. Niu, X.; Lee, Y.K. Efficient spatial-temporal chaotic mixing in microchannels. *J. Micromech. Microeng.* **2003**, *13*, 454–462. [CrossRef]
29. Burghelea, T.; Segre, E.; Bar-Joseph, I.; Groisman, A.; Steinberg, V. Chaotic flow and efficient mixing in a microchannel with a polymer solution. *Phys. Rev. E Stat. Nonlinear Soft Matter Phys.* **2004**, *69*. [CrossRef] [PubMed]
30. Simonnet, C.; Groisman, A. Chaotic mixing in a steady flow in a microchannel. *Phys. Rev. Lett.* **2005**, *94*. [CrossRef] [PubMed]
31. Cai, G.; Xue, L.; Zhang, H.; Lin, J. A review on micromixers. *Micromachines* **2017**, *8*, 274. [CrossRef]
32. Lee, C.Y.; Wang, W.T.; Liu, C.C.; Fu, L.M. Passive mixers in microfluidic systems: A review. *Chem. Eng. J.* **2016**, *288*, 146–160. [CrossRef]
33. Gan, H.Y.; Lam, Y.C.; Nguyen, N.T. Polymer-based device for efficient mixing of viscoelastic fluids. *Appl. Phys. Lett.* **2006**, *88*. [CrossRef]
34. Xu, K.; Liang, T.; Zhu, P.; Qi, P.; Lu, J.; Huh, C.; Balhoff, M. A 2.5-D glass micromodel for investigation of multi-phase flow in porous media. *Lab Chip* **2017**, *17*, 640–646. [CrossRef] [PubMed]
35. Needham, R.B.; Doe, P.H. Polymer Flooding Review. *J. Pet. Technol.* **1987**, *39*, 1503–1507. [CrossRef]
36. Xu, K.; Zhu, P.; Tatiana, C.; Huh, C.; Balhoff, M. A microfluidic investigation of the synergistic effect of nanoparticles and surfactants in macro-emulsion based EOR. In Proceedings of the SPE—DOE Improved Oil Recovery Symposium Proceedings, Tulsa, OK, USA, 11–13 April 2016.
37. Xu, K.; Zhu, P.; Colon, T.; Huh, C.; Balhoff, M. A microfluidic investigation of the synergistic effect of nanoparticles and surfactants in macro-emulsion-based enhanced oil recovery. *SPE J.* **2017**, *22*, 459–469. [CrossRef]
38. Olajire, A.A. Review of ASP EOR (alkaline surfactant polymer enhanced oil recovery) technology in the petroleum industry: Prospects and challenges. *Energy* **2014**, *77*, 963–982. [CrossRef]
39. Jha, B.; Cueto-Felgueroso, L.; Juanes, R. Fluid mixing from viscous fingering. *Phys. Rev. Lett.* **2011**, *106*. [CrossRef] [PubMed]
40. Jha, B.; Cueto-Felgueroso, L.; Juanes, R. Synergetic fluid mixing from viscous fingering and alternating injection. *Phys. Rev. Lett.* **2013**, *111*. [CrossRef] [PubMed]

41. James, D.F.; McLaren, D.R. The laminar flow of dilute polymer solutions through porous media. *J. Fluid Mech.* **1975**, *70*, 733–752. [CrossRef]
42. Peters, E.C.; Petro, M.; Svec, F.; Fréchet, J.M.J. Molded Rigid Polymer Monoliths as Separation Media for Capillary Electrochromatography. 1. Fine Control of Porous Properties and Surface Chemistry. *Anal. Chem.* **1998**, *70*, 2288–2295. [CrossRef] [PubMed]
43. Rodriguez, S.; Romero, C.; Sargenti, M.L.; Müller, A.J.; Sáez, A.E.; Odell, J.A. Flow of polymer solutions through porous media. *J. Non-Newtonian Fluid Mech.* **1993**, *49*, 63–85. [CrossRef]
44. Stavland, A.; Jonsbråten, H.C.; Lohne, A.; Moen, A.; Giske, N.H. Polymer flooding—Flow properties in porous media versus rheological parameters. In Proceedings of the 72nd European Association of Geoscientists and Engineers Conference and Exhibition 2010: A New Spring for Geoscience, Barcelona, Spain, 14–17 June 2010; pp. 3292–3306.
45. Liu, R.; Li, B.; Jiang, Y. Critical hydraulic gradient for nonlinear flow through rock fracture networks: The roles of aperture, surface roughness, and number of intersections. *Adv. Water Resour.* **2016**, *88*, 53–65. [CrossRef]
46. Liu, R.; Li, B.; Jiang, Y. A fractal model based on a new governing equation of fluid flow in fractures for characterizing hydraulic properties of rock fracture networks. *Comput. Geotech.* **2016**, *75*, 57–68. [CrossRef]
47. Li, Z.; Kim, S.J. Pulsatile micromixing using water-head-driven microfluidic oscillators. *Chem. Eng. J.* **2017**, *313*, 1364–1369. [CrossRef]
48. Glasgow, I.; Aubry, N. Enhancement of microfluidic mixing using time pulsing. *Lab Chip* **2003**, *3*, 114–120. [CrossRef] [PubMed]
49. Krupa, K.; Nunes, M.I.; Santos, R.J.; Bourne, J.R. Characterization of micromixing in T-jet mixers. *Chem. Eng. Sci.* **2014**, *111*, 48–55. [CrossRef]
50. Gao, Z.; Han, J.; Bao, Y.; Li, Z. Micromixing efficiency in a T-shaped confined impinging jet reactor. *Chin. J. Chem. Eng.* **2015**, *23*, 350–355. [CrossRef]
51. Liu, Z.; Guo, L.; Huang, T.; Wen, L.; Chen, J. Experimental and CFD studies on the intensified micromixing performance of micro-impinging stream reactors built from commercial T-junctions. *Chem. Eng. Sci.* **2014**, *119*, 124–133. [CrossRef]
52. Oualha, K.; Ben Amar, M.; Michau, A.; Kanaev, A. Observation of cavitation in exocentric T-mixer. *Chem. Eng. J.* **2017**, *321*, 146–150. [CrossRef]
53. Li, B.; Liu, R.; Jiang, Y. Influences of hydraulic gradient, surface roughness, intersecting angle, and scale effect on nonlinear flow behavior at single fracture intersections. *J. Hydrol.* **2016**, *538*, 440–453. [CrossRef]
54. Giesekus, H. A simple constitutive equation for polymer fluids based on the concept of deformation-dependent tensorial mobility. *J. Non-Newtonian Fluid Mech.* **1982**, *11*, 69–109. [CrossRef]

processes

MDPI

Article

Geotechnical Properties of Effluent-Contaminated Cohesive Soils and Their Stabilization Using Industrial By-Products

Muhammad Irfan [1], Yulong Chen [2,*], Muhammad Ali [3], Muhammad Abrar [3], Ahmed Qadri [3] and Osama Bhutta [3]

1 Birudo Engineers, 107/C, PAF Officers Colony, Lahore 54810, Pakistan; irfan@uet.edu.pk
2 Department of Hydraulic Engineering, Tsinghua University, Beijing 100084, China
3 Department of Civil Engineering, University of Engineering and Technology Lahore, Lahore 54890, Pakistan; m.ali.92@outlook.com (M.A.); engr.abrarmughal@gmail.com (M.A.); iahmedwaheed@hotmail.com (A.Q.); osamabhutta123@gmail.com (O.B.)
* Correspondence: chen_yl@tsinghua.edu.cn

Received: 18 September 2018; Accepted: 17 October 2018; Published: 22 October 2018

Abstract: The unchecked and unnoticed disposal of industrial leachates is a common malpractice in developing countries. Untreated effluents from industries drastically deteriorate the soil, altering nearly all of its characteristics. An increase in urbanization has led to construction on these deteriorated lands. In this study, the chemical impact of two industrial effluents, dyeing (acidic) and tannery (basic), is studied on two cohesive soils, i.e., high plastic clay (CH) and low plastic clay (CL). Properties such as liquid limit, plasticity index, specific gravity, maximum dry density, unconfined compressive strength, swell potential, swell pressure, and compression indices decrease with effluent contamination, with the exception of the basic effluent, for which the trend changes after a certain percentage. This study also examines the time variation of properties at different effluent percentages, finding that unconfined compressive strength of both soils increases with time upon dyeing (acidic) contamination and decreases with tannery (basic). The stabilizing effect of two industrial by-products, i.e., marble dust and ground granulated blast furnace slag (GGBFS) have been evaluated. Unlike their proven positive effect on uncontaminated soils, these industrial by-products did not show any significant stabilization effect on leachate-contaminated cohesive soils, thereby emphasizing the need to utilize special remediation measures for effluent treated soils.

Keywords: effluents; soil properties; cohesive soils; contamination; time variation; stabilization

1. Introduction

It has long been recognized that changes in pore fluid chemistry can exert a strong effect on soil strength characteristics [1]. In recent years, the growing rate of soil contamination has prompted a number of studies in which the effects of chemicals on the geotechnical properties of fine-grained soils have been studied [2]. Industrial effluents are often discharged untreated, a practice which contaminates water channels (and ultimately the soil) through seepage.

A substantial volume of effluent waste is generated annually. These effluents consist of toxic heavy metals and highly reactive chemical compounds which, upon disposal, alter the shape and composition of the soil structure following certain chemical reactions. Such reactions cause changes in the soil behavior related to various geotechnical properties. In recent years, a number of studies have been performed to investigate the effect of pore water chemistry on the strength characteristics of soil. Anson and Hawkins [3], Moore and Brunsden [4], and Tiwari et al. [5] have studied the effect of calcium and sodium chloride on the residual shear strength of weathered mudstone, and reported that high concentrations of ions in the pore water can significantly decrease the diffuse double layer of clay particles, a process that would lead to a greater residual shear strength. Ratnaweera and Meegoda [6] have researched the influence of organic fluids on the stress–strain behavior of soil and noted that changes in strength mostly depended on the dielectric constant of the organic fluids. Gratchev et al. [7,8] performed studies on the effect of the pore fluid chemistry on the undrained cyclic behavior of artificial soil mixtures and natural clays and reported that the cyclic shear strength can be greatly influenced by the type of ions and their concentrations in the pore water. Gratchev and Towhata [9] reported that a decrease in pH correlates with increases in compression indices and liquid limit and plastic limit of soils, due to the dissolution of calcium carbonate. Sunil et al. [10,11] dealt with an attempt to study the effect of leachate on the Atterberg limits, shear strength properties and chemical characteristics of lateritic soil. The liquid limit and the plasticity index of the lateritic soils increases with municipal solid waste leachate concentration. The increase in clay content of lateritic soil after interaction with the leachate led to an increase in cohesion parameter and decreases of the friction angle. The pH measurements of lateritic soil contaminated with municipal solid waste (MSW) leachate indicated an increase in pH values. This is also accompanied by a slight increase in the cation exchange capacity of the soil. The change in chemical characteristics of lateritic soil contributed due to the addition of leachate may be detrimental to foundation concrete in real field conditions.

A geotechnical engineer's responsibility has therefore become cumbersome as conventional geotechnical principles and theories fail to account for the behavior of contaminated soils. It has been necessary to modify design methodologies to consider the effects of contaminants on soil properties to ensure accurate design of foundations and underground structures in the era of industrial evolution.

Rapid urbanization in various metropolitan areas of Pakistan has forced developers to construct in places that were previously used as dumping sites for industrial effluents. Geotechnical investigation reveals that the top soil layer in major part of Pakistan is mainly composed of cohesive soils [12]; thus, these soils were selected for this study in an attempt to understand the effects of industrial effluents on the geotechnical properties of contaminated cohesive soils. Furthermore, mitigation of such effects using industrial by-products has been discussed.

2. Materials and Methods

2.1. Soil Samples

Two samples of locally available natural cohesive soils were used in this study. The samples were obtained close to the soil surface and were representative of natural alluvial deposits common in major parts of Punjab province in Pakistan. One of the samples was procured from Nandipur near Gujranwala, Punjab and the other from Defence Housing Authority (DHA), Lahore. Grain size distribution, illustrated in Figure 1, shows the Nandipur and DHA soil samples to be classified as CH (A-6 (10)) and CL (A-7-6 (20)) per the Unified (AASHTO) Soil Classification System. To avoid the possible influence of vegetation, fill material and other organic matter, soil samples were procured from around 1.0 m deep test pits rather than from the surface. X-ray diffraction (XRD) analysis of the soil samples (see Figure 2) revealed the presence of kaolinite in CL soil and illite in CH soil as dominant minerals. The soil samples' physical and chemical properties are summarized in Table 1.

Table 1. Physical and chemical properties of soil samples.

Properties	CH	CL
Classification	High plastic clay	Low plastic clay
pH	7.6	7.3
Liquid limit	54.6%	31.18%
Plastic limit	26.61%	17.35%
Specific gravity	2.71	2.78
Max. dry density	18.12 kN/m^3	19.57 kN/m^3
Optimum moisture content	15.7%	9.5%
Swell potential	5.4%	2.2%
Swell pressure	307.92 kPa	132.58 kPa

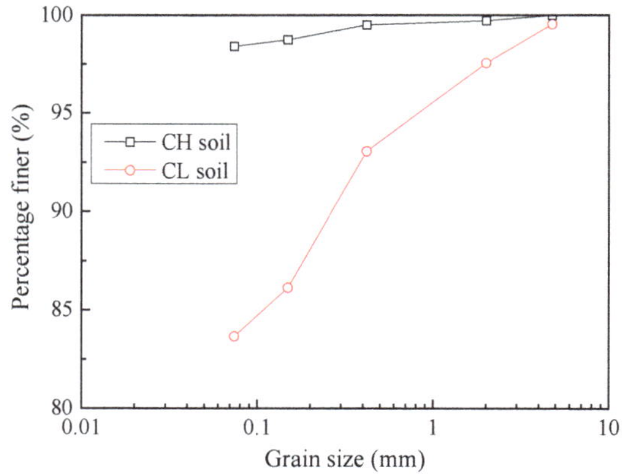

Figure 1. Grain size distribution of Nandipur (CH) and DHA (CL) soil samples.

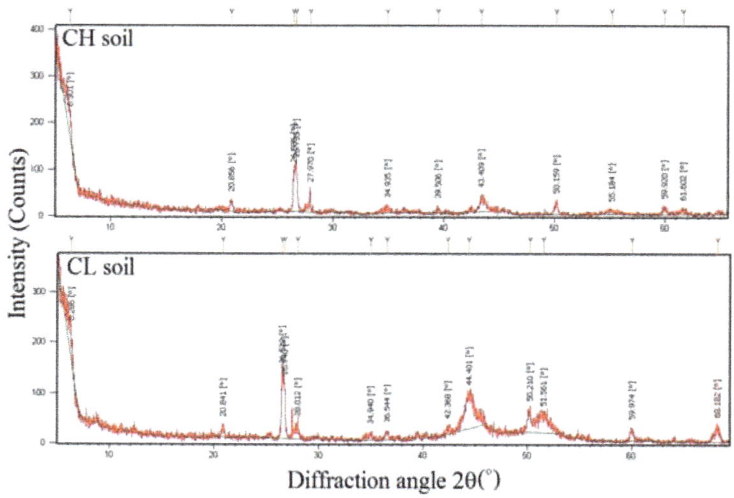

Figure 2. XRD analysis of soil samples.

2.2. Industrial Effluents

Two industrial effluents, representing extreme ends of the pH scale, were selected to examine their effects on the engineering characteristics of cohesive soil. An acidic effluent collected from the outfall of a dyeing industry (see Figure 3) and a basic effluent collected from a tannery (see Figure 4) were used in this research. Fresh samples of industrial effluents were collected from drains of a dyeing industry located in Sheikhupura, Punjab, whereas wastes from a tannery were procured from Sialkot, Punjab. The collected effluent samples were subjected to chemical examination, the summary of these results are presented in Table 2.

Figure 3. Effluent from dyeing industry (Acidic).

Figure 4. Effluent from tannery industry (Basic).

Table 2. Properties of industrial effluents.

Tests Performed	Tannery Effluent	Dyeing Effluent
Color	Dark greenish	Dark reddish
pH	11.5	5.1
Total solids (g/L)	25.43	1.673
Total dissolved solids (mg/L)	19.78	910
Total suspended solids (mg/L)	603	330
Suspended solids (g/L)	6	1
BOD (mg/L)	4445	1980
COD (mg/L)	5000	2100
Chlorides (mg/L)	230	175
Specific gravity	1.012	1.007
Sulphates (mg/L)	281	198
Nitrates (mg/L)	34	28

2.3. Industrial By-Products

Industrial by-products were used to stabilize the effluent-contaminated soils obtained in phase I of this research. Specifically, by-products from the marble industry (marble dust) and steel industry (ground granulated blast furnace slag (GGBFS) were procured from local industries in Lahore; their chemical composition is summarized in Table 3.

Table 3. Properties of industrial by-products.

Material	Constituents	Percentage (%)
Marble dust	SiO_2	6.2
	Fe_2O_3	0.8
	CaO	30.1
	Al_2O_3	4.8
Ground granulated blast furnace slag	CaO	30–50
	SiO_2	28–38
	Al_2O_3	8–24
	MgO	1–18

2.4. Testing Scheme

A systematic methodology was adopted to prepare the leachate contaminated soil specimens. The cohesive soil samples were first oven-dried and pulverized. Industrial effluents were then thoroughly mixed with the soil samples at the indicated percentages of 0%, 5%, 10%, 15%, and 20% by dry weight of soil. The soil-effluent blends were left for 48 h for maturation before initiating laboratory testing. The contaminated soil samples were then air-dried and sieved through 4.75 mm mesh to remove coarse particles as well as large lumps. The dried and sieved soil was stored in airtight containers ready for laboratory testing. Sixteen polluted specimens, with different effluent proportions, were prepared for each soil type. A typical flow-chart depicting the specimen preparation process is shown in Figure 5.

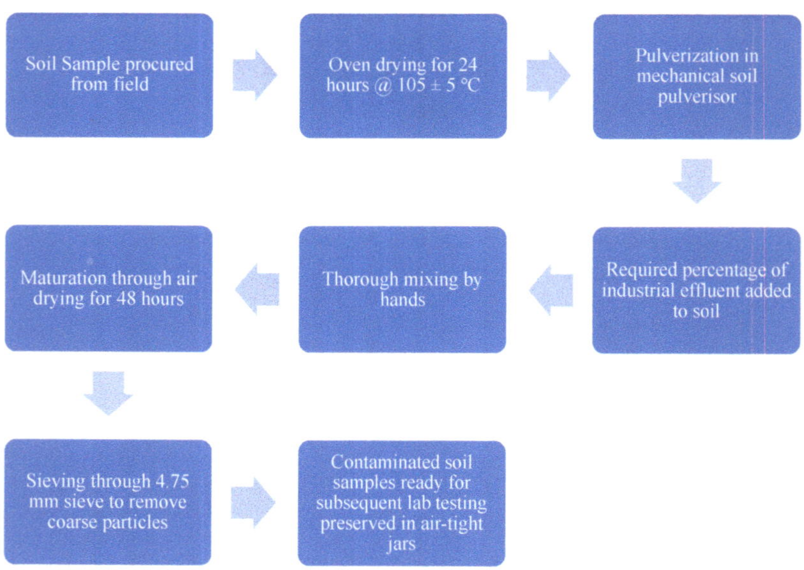

Figure 5. Flow-chart presenting specimen preparation procedure.

A systematic nomenclature was established to represent the contaminated soil samples. The initial two letters indicate soil type, either low plastic (CL) or high plastic (CH) clay. The third letter corresponds to the effluent source, either the dyeing industry (D) or tannery industry (T). The digits depict the percentage of pollutant (e.g., "CH-T5" corresponds to a high plastic clay sample with 5% contamination from the tannery industry). The contaminated soil samples as well as the two virgin soil samples were then subjected to different tests to ascertain the corresponding effects on geotechnical characteristics. Soil properties assessed as a component of this investigation included Atterberg limits, specific gravity, modified compaction, one-dimensional consolidation, unconfined compression, and pH. Chemical analyses on the contaminated soils including total solids, dissolved solids, suspended solids, settleable solids, nitrates, sulfates, chlorides, oxygen demand (OD), biochemical oxygen demand (BOD), and chemical oxygen demand (COD) were also conducted. All tests were performed using the applicable American Society for Testing and Materials (ASTM) standard.

Further, two industrial by-products, marble dust and GGBFS, were added to each soil sample at percentages of 5%, 10%, 15%, and 20% by weight of the dry contaminated soil samples in order to improve soil properties. The effects of different percentages of effluent contamination on soils with the passage of time (i.e., 2 days, 7 days, 15 days, 21 days, 28 days and 2 months) were also investigated.

3. Results and Discussion

3.1. Effects of Contamination on Atterberg Limits

Atterberg limits are considered as index properties of soil and are used to predict their nature, as well as mechanical behavior. Many important characteristics of cohesive soils including shear strength, permeability, consolidation and deformation behavior, etc. are a function of Atterberg limits. A variation in Atterberg limits upon effluent contamination can be used as a precursor to predict change in other soil characteristics.

Increases in the liquid limit and plasticity indices of the soils with the addition of acidic and basic effluents are illustrated in Figure 6. The liquid limit of CH soil was found to increase from 44.58% to 52.77% upon contamination from the acidic effluent. With the basic effluent, CH soil's liquid limit increased to 58.63% at a 20% addition of the contaminant. CL soil showed an increase in liquid limit from 31.18% to 33.82% and 33.86% upon addition of acidic and basic contamination, respectively.

The plasticity index of CH soil also showed a similar trend. That is, the plasticity index increased from 17.97% to 27.03% at 20% contamination of acidic effluent, and 22.28% at 20% contamination of basic effluent.

The increase in Atterberg limits of soil is mainly because of the chemical action between soil particles and effluent. The increase in liquid limit indicates an increase in consolidation potential of contaminated soils. High plasticity of contaminated soils also causes problems related to increased swell potential and high collapsibility. These observations are in-line with the previous findings by [13,14]. Contamination by industrial effluents would therefore deteriorate the quality of soil as an engineering material.

3.2. Effects of Contamination on Specific Gravity

Specific gravity of soil solids indicates how much heavier or lighter the soil particles are compared to water at 25 °C. This characteristic is important, specifically with regards to settlement evaluations in soils. The effects of contaminants on specific gravity are illustrated in Figure 7. The specific gravity of both the soil samples decreased with the addition of contaminants. CH soil showed a decrease in specific gravity by 12.5% and 15.5% for the acidic and basic effluents, respectively. CL-specific gravities decreased by 22.83 and 22.94 for the acidic and basic effluents, respectively.

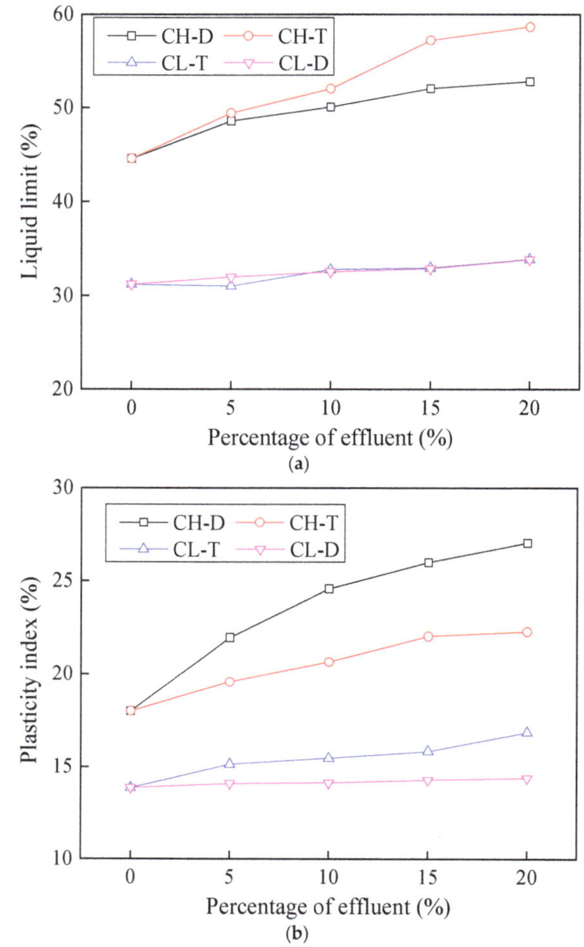

Figure 6. Effects of industrial effluents on (**a**) liquid limit and (**b**) plasticity index of cohesive soils.

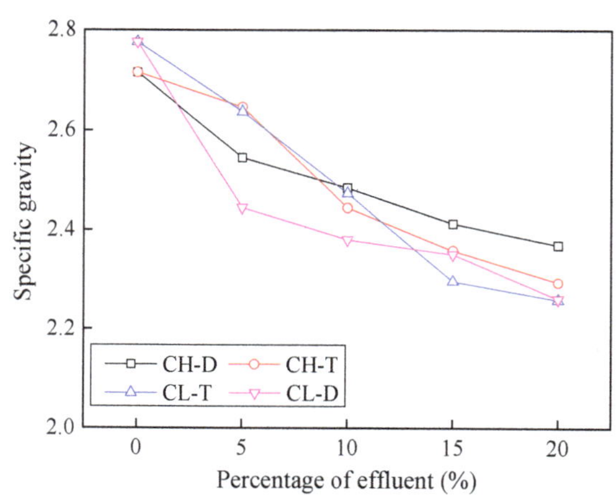

Figure 7. Effects of industrial effluents on specific gravity of cohesive soils.

3.3. pH Value of Contaminated Soils

The two effluents used in this study comprised of samples recovered from dyeing industry and a tannery. Specimen from dyeing industry was acidic (pH = 5.1) in nature whereas the one from tannery was basic (pH = 11.5) in characteristic. Changes in the pH of soil can increase its chemical aggressivity which can cause excessive deterioration of the structural materials placed over it. The effects of contaminants on the pH of cohesive soils appear in Figure 8. At 5% contamination, the pH of all soils showed a sudden increase in pH value compared to unaffected soils. The basic effluent at 20% concentration caused an increase in pH to 8.2 and 8.4 for CL and CH soils, respectively. The acidic effluent at the same concentration caused a decrease to 6.7 and 6.8 for CL and CH soils, respectively.

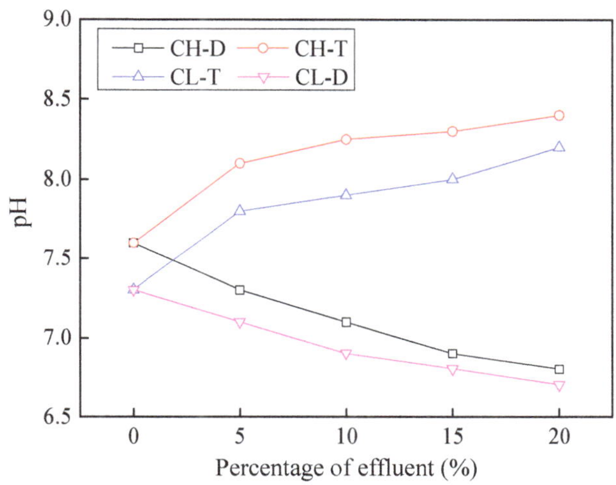

Figure 8. Effects of industrial effluents on pH of cohesive soils.

3.4. Effects on Compaction Characteristics

Compaction characteristics of soil, i.e., optimum moisture content and maximum dry unit weight, are important engineering characteristics with regards to subgrade preparation for roads, highways, and building structures. An increase in optimum moisture content of soil indicates a high water demand to achieve a certain density under same compactive effort. A higher maximum dry unit weight, on the other hand, indicates a relatively stronger material having better engineering utility. The effects of contaminants on the compaction characteristics of cohesive soils are shown in Figure 9. In general, the optimum moisture content was observed to increase by around 21.9% for CL soil and around 13.2% for CH soil with the addition of contaminants. On contrary, the effluents consistently decreased the maximum dry density of cohesive soils with an increase in contaminant percentage for all types of soils.

The variation in the compaction characteristics of cohesive soils with contaminant addition can be explained on the basis of soil plasticity. Optimum moisture content of cohesive soils increases whereas the maximum dry unit weight obtained through compaction tests decreases with plasticity index of soil. Similar phenomenon was previously observed by Pandian et al. [15] and Sridharan et al. [16]. Increase in contamination concentration makes the soil more plastic, thereby leading to an increase in optimum moisture content and a decrease in maximum dry unit weight. From engineering applications perspective, this means a high water demand to attain optimum moisture in the field; which, in general, increases the project cost and is typically undesirable. In other words, soil with high contaminant concentration would be difficult to compact and would yield a lower unit weight compared to uncontaminated soil under the same compactive effort and moisture conditions.

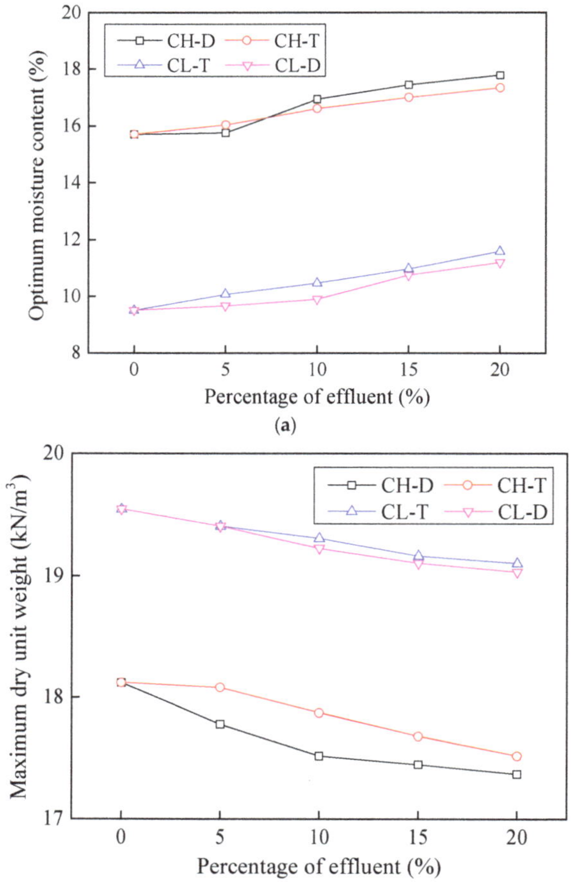

Figure 9. Effects of industrial effluents on the compaction characteristics of the cohesive soils; (**a**) Optimum moisture content, (**b**) Maximum dry density.

3.5. Effects of Contamination on Unconfined Compressive Strength

Unconfined compressive strength is a direct measure of strength for cohesive soils. A higher value indicates a better soil and vice versa. The behavior of cohesive soils upon contamination of acidic and basic effluents in terms of unconfined compressive strength is illustrated in Figure 10. CH soil showed a decrease in unconfined compressive strength from 246.96 kPa to 98.66 kPa and 90.18 kPa for basic and acidic effluents, respectively. Both effluents decreased the unconfined compressive strength of the cohesive soils to almost 60%. The reduction of strength due to contamination can be attributed to the possible breakage of internal bonds as was previously shown by Umesha et al. [17].

The reason for the decrease in unconfined compressive strength may also be due to the decrease in cementing material. Cementing agents in soil help to bond the finer particles together, forming aggregates. Due to leaching, however, these aggregates disintegrate along with the cementing agents. Thus, it stands to reason that both effluents alter the cementing agents in the soil particles and hence decrease unconfined compressive strength.

Industry effluents contain various sulfides in the form of iron pyrites (FeS_2). Oxidation of FeS_2 produces H_2SO_4, which may produce gypsum in the presence of calcium carbonate. Therefore, gypsum may be present in the natural soil. Reactions are shown below.

$$2FeS_2 + 2H_2O + 7O_2 = 2FeSO_4 + 2H_2SO_4 \quad (1)$$

$$CaCO_3 + H_2SO_4 + H_2O = CaSO_4{\cdot}2H_2O + CO_2 \quad (2)$$

In the presence of excess water, hydrated sulfates formed in the reaction are already present in the effluent, and attack the binding material on soil particles. This weakens the bonds thereby decreasing soil strength.

The undrained unconfined compressive strength of soil samples contaminated with the tannery effluent decreased with an increase in tannery percentage. These results correspond with those of Stalin et al. [18], who studied the effect of tannery waste on the behavior of two natural soil samples collected from Madras City, India.

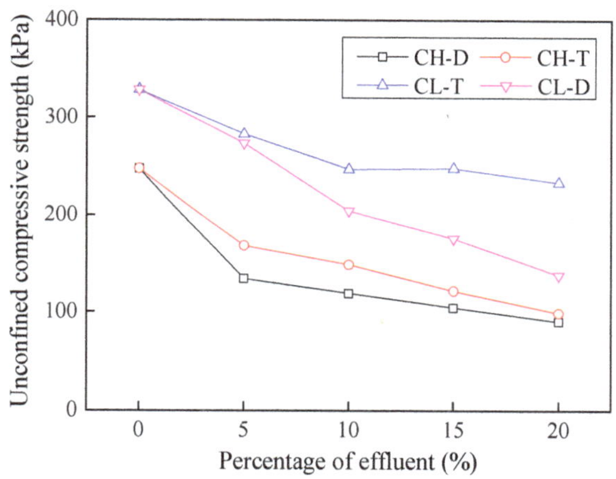

Figure 10. Effects of industrial effluents on unconfined compressive strength of cohesive soils.

3.6. Effects of Contamination on Swelling Characteristics

The swell potential of soil indicates its ability to swell upon wetting. A higher swell potential (and the corresponding swell pressure) is highly undesirable for the structures as it can cause severe cracks and destress in the buildings. The effects of contaminants on soils' swell potential appear in Figure 11. The swell potential of CL and CH soil samples increased by 44.8% and 28.7% with the addition of 20% basic effluent. On the contrary, the swell potential of both the soil samples decreased with the addition of effluent from dyeing industry because of its acidic nature. The observed decrease in swell potential of CL and CH soil samples with 20% contamination of acidic effluent was 52.5% and 82.4% respectively.

The effects of contaminants on soil swell pressure are illustrated in Figure 12. Swell pressure showed a decreasing trend with acidic contamination (i.e., dyeing). In particular, CH swell pressure suddenly decreased with a slight addition of 5% of acidic effluent. Whereas the swell pressure increased with basic effluent.

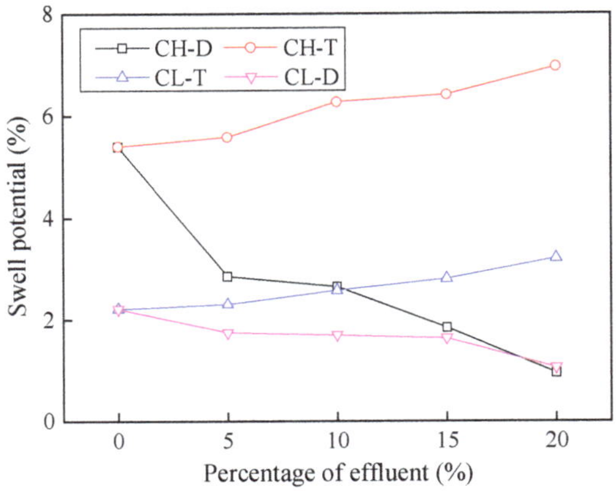

Figure 11. Effects of industrial effluents on swell potential of cohesive soils.

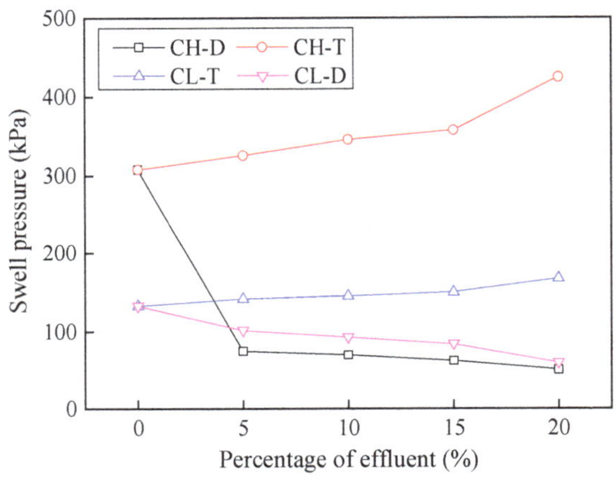

Figure 12. Effects of industrial effluents on swell pressure of cohesive soils.

It can be seen from Figures 11 and 12 that in case of acidic contamination, the swell potential and pressure of both the soils were decreased; whereas the swell potential and pressure of both the soils increased when contaminated with basic effluent. This difference can be attributed to the reactions of the clay soil minerals with effluent contamination. The surfaces of most clay particles carry an unbalanced negative charge, which attracts cations. These positive ions become strongly attracted to the clay surface leading to the formation of diffuse double layer around clay particles. The interaction of the diffuse double layers of neighboring units result in a net repulsive force between them.

An increase in the concentration of H^+ in the pore fluid because of acidic contamination would depress the thickness of the diffuse double layer, thus reducing the repulsion force between clay particles. As a result, the net attractive van der Waal forces would increase [19], leading to the formation of larger clay aggregates in which the clay particles would be closer to each other. Such arrangements are expected to produce "denser" clay microfabrics with a lower void ratio, which leads to the decrease in the swell potential and pressure.

The addition of basic contamination has the opposite effect on the swell potential and pressure of cohesive fine-grained soils. Such changes in the soil behavior can be attributed to the further transformation of clay microfabric. In particular, the charge on the edges of the clay particles becomes negative, leading to the formation of more open, flocculated structures [20]. Sridharan et al. [21]

noted that such flocculated arrangements of the clay particles would enclose large spaces for water entrapment; thus, higher swell potential and pressure would be exhibited.

3.7. Effects of Contamination on Compression Index

Compression index (*Cc*) of cohesive soils is a direct indicative of its tendency to settle under the applied load. Higher compression index indicates a higher tendency to settle which ultimately leads to structural destress. The trends for compression indices in contaminated cohesive soils are shown in Figure 13. In CH soil, the compression index increased from 0.226 to 0.316 and 0.282 for acidic and basic contamination, respectively. In CL soil, the *Cc* value increased from 0.169 to 0.235 and 0.245 for basic and acidic contaminants, respectively. These findings show that the consolidation potential of contaminated soils is greater than that of affected soils. The same behavior can also be seen in soil plasticity with an increase in effluent concentration (see Figure 6), as *Cc* is a direct measure of soil plasticity [13].

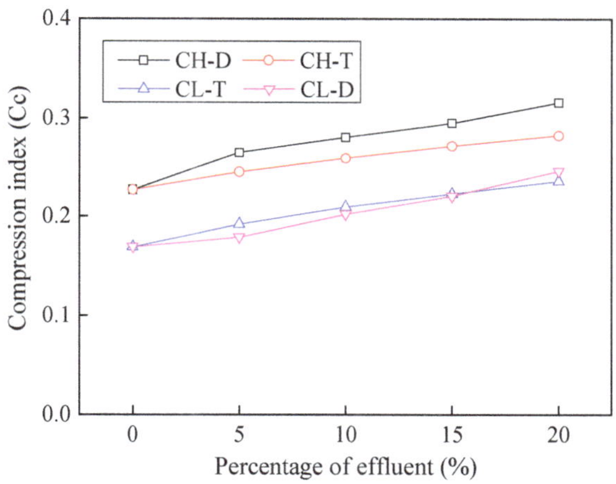

Figure 13. Effects of industrial effluents on compression index of cohesive soils.

3.8. Time-Dependent Variation of Unconfined Compressive

3.8.1. Effects of Acidic Effluent

In order to study the time dependent effects of leachate contamination on cohesive soils, specimens prepared at different contamination percentage were preserved in controlled environment for up to 60 days. The results of unconfined compression tests performed on CL soil samples contaminated with dyeing effluent are shown in Figure 14.

The unconfined compressive strength for 15% CL-dyeing constantly decreased from 2 days to 2 months, while the unconfined compressive strength for the other samples began to increase after 7 days. However, the final strength after 60 days was still less than the unconfined compressive strength of the original (i.e., virgin) soil samples, suggesting that effluent effects diminished after a certain period after which the soil began to regain its strength. As such, dyeing effluent appears to have a temporary effect on CL soil.

The effects of dyeing effluent on unconfined compressive strength of CH soil are shown in Figure 15. The observed behavior is quite similar to the one observed for CL soils, i.e., temporary loss but regaining strength after around 15 days.

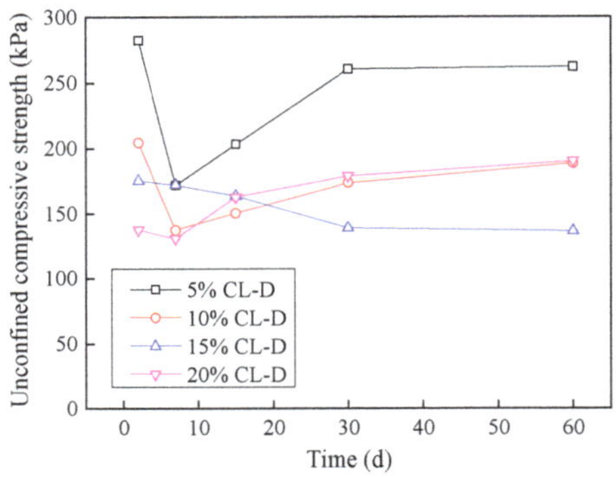

Figure 14. Effects of dyeing effluent on the cohesion of CL soil samples.

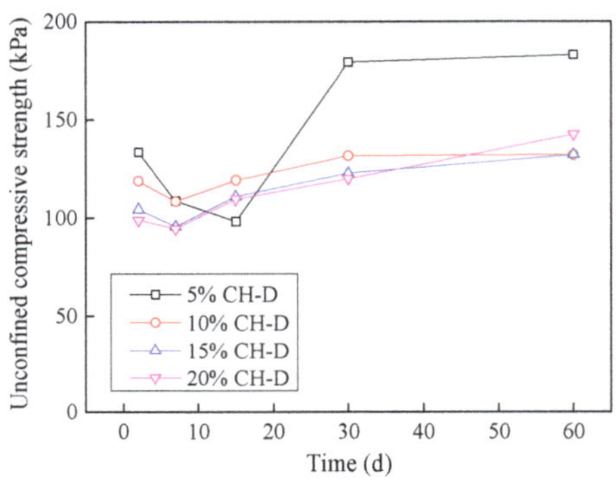

Figure 15. Effects of dyeing effluent on the cohesion of CH soil samples.

3.8.2. Effects of the Basic Effluent

The time-dependent variation of unconfined strength of CL soil samples with addition of tannery effluent is depicted in Figure 16. The unconfined compressive strength for all samples decreased constantly with time. The rate of decrease however, slowed down gradually and strength became nearly constant after 60 days.

The effect of tannery effluent on undrained strength of CH soil samples is similar, but less pronounced compared to CL soils. The summary of time dependent behavior of CH soil contaminated with tannery effluent is shown in Figure 17.

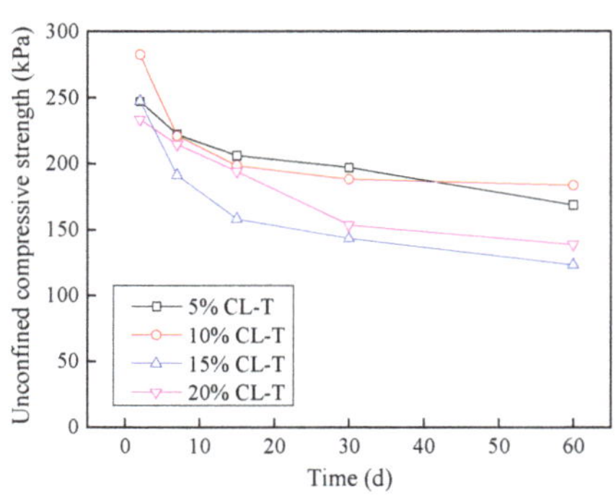

Figure 16. Effects of tannery effluent on the cohesion of CL soil samples.

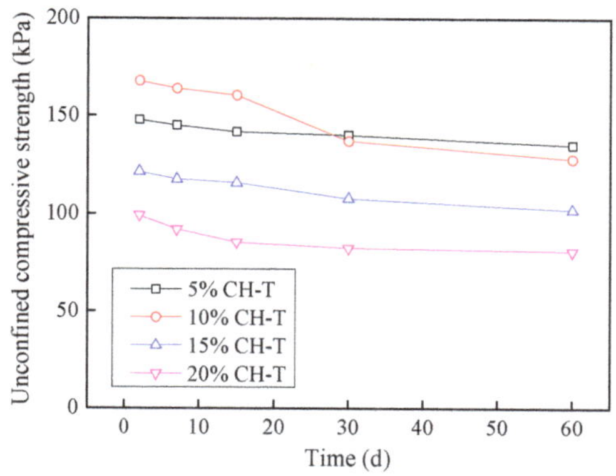

Figure 17. Effects of tannery effluent on the cohesion of CH soil samples.

3.9. *Effect of Industrial by-Products on Leachate Contaminated Soils*

Improvement of weak soil by using various industrial by-products is a well-established practice worldwide. Several studies have successfully shown the significance of using different industrial by-products like marble dust [22,23], glass dust, blast furnace slag [24,25], rice husk ash [26], sugar cane waste, fly ash [27,28], coal ash [29], xanthan gum [30], etc., for soil improvement. As phase-2 of this research, the possibility of stabilization of leachate contaminated cohesive soils (CL-dyeing, CL-tannery, CH-dyeing and CH-tannery) using marble dust and ground granulated blast furnace slag (GGBFS) was explored. Soil specimens (CL and CH) having 5% and 20% effluent contamination were mixed with varying percentages of marble dust as well as GGBFS and their effect on unconfined compressive strength was explored. The main objective was to access possible enhancement in undrained strength of contaminated soils, as is the case with soils having no such contamination.

3.9.1. Stabilization of Acid-Affected CL Soil

Figure 18 illustrates the trends observed when different percentages of marble dust were added to the acid-affected CL soil. An increase in marble dust concentration caused a rapid decline in unconfined compressive strength from 282.42 kPa to 141.55 kPa at 5% acidic contamination. At 20% acidic contamination, unconfined compressive strength decreased from 137.14 kPa to 81.49 kPa when marble concentration reached 20%. A possible reason for this trend is that CaO, a primary and basic component of marble dust, reacted with the acidic part of the effluent present in the soil. This would have produced salts that were responsible for the rapid decrease in unconfined compressive strength. Although it is a renowned soil stabilizer, marble dust was unable to improve soil strength to match that of the original soil "i.e., 328.9 kPa".

Figure 19 shows the trends observed when different percentages of GGBFS were added to the basic-affected CL soil. With an increase in GGBFS, unconfined compressive strength decreased rapidly from 282.42 kPa to 203.33 kPa at 5% acidic contamination. At 20% acidic contamination, it decreased from 137.14 kPa to 45.51 kPa when GGBFS concentration reached 20%.

GGBFS has a pH that varies between 10 and 12. As such, there may have been a possible reaction between acid from the contaminated soil and the basic compounds present in the GGBFS, which could have produced salts that ultimately resulted in a decrease in unconfined compressive strength. It may therefore be concluded that GGBFS is not an effective stabilizer in acid-contaminated soils.

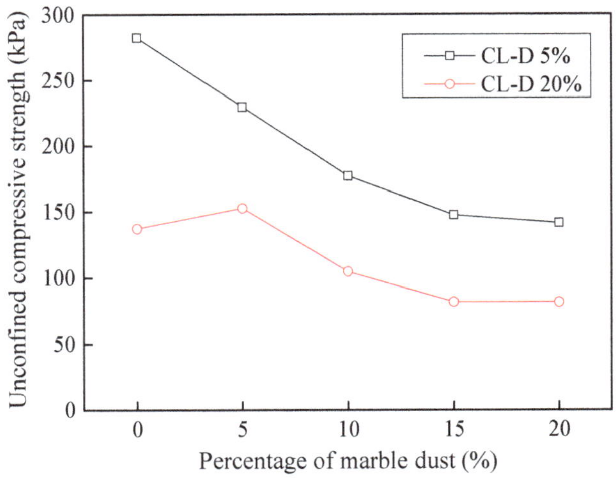

Figure 18. Improvement of dyeing effluent contaminated CL soil using marble dust.

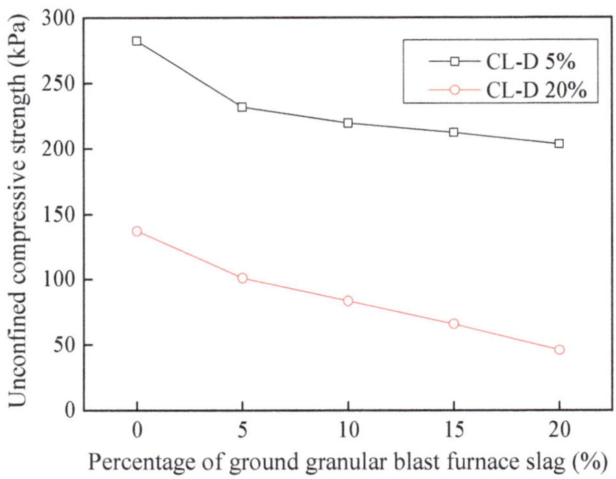

Figure 19. Improvement of dyeing effluent contaminated CL soil using GGBFS.

3.9.2. Stabilization of Tannery-Affected CL Soil

Figure 20 indicates trends observed when different percentages of GGBFS were added to the alkali-affected CL soil. With an increase in the blast furnace slag, unconfined compressive strength decreased rapidly from 246.22 kPa to 209.54 kPa at 5% basic contamination. At 20% basic contamination, it decreased from 232.91 kPa to 109.00 kPa when GGBFS concentration reached 20%. GGBFS was unable to increase soil strength to match that of the original soil. Thus, GGBFS appears not to be an effective stabilizer of base-contaminated soils.

Figure 21 shows the trends observed when different percentages of marble dust were added to alkali-affected CL soil. With an increase in GGBFS, unconfined compressive strength decreased rapidly from 246.22 kPa to 136.45 kPa at 5% basic contamination. At 20% basic contamination, it decreased from 232.91 kPa to 104.80 kPa when GGBFS concentration reached 20%.

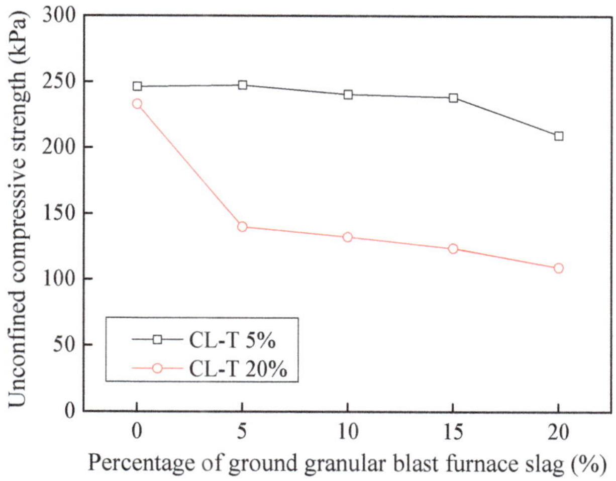

Figure 20. Improvement of tannery effluent contaminated CL soil using GGBFS.

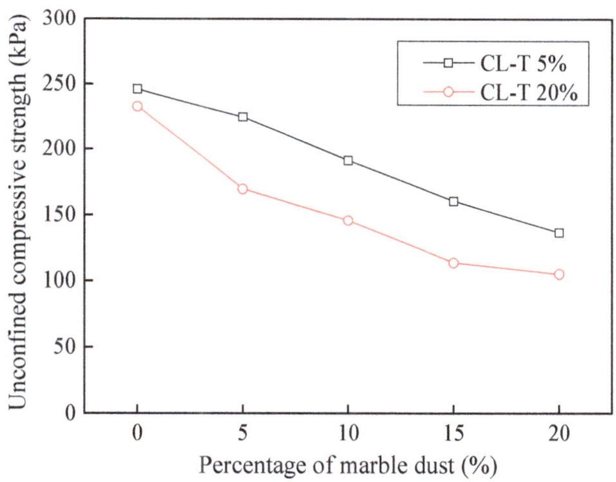

Figure 21. Improvement of tannery effluent contaminated CL soil using marble dust.

3.9.3. Stabilization of Dyeing-Affected CH Soil

Figure 22 displays the trends observed when different percentages of marble dust were added to acid-affected CH soil. With an increase in marble dust concentration, unconfined compressive strength decreased rapidly from 133.69 kPa to 62.88 kPa at 5% acid contamination. At 20% acid contamination, it decreased from 98.81 kPa to 83.29 kPa when marble dust concentration reached 20%.

Figure 23 shows the trends observed when different percentages of GGBFS were added to acid-affected CH soil. With an increase in GGBFS concentration, unconfined compressive strength decreased rapidly from 133.69 kPa to 76.12 kPa at 5% acid contamination. At 20% acid contamination, it decreased from 98.81 kPa to 83.22 kPa when GGBFS concentration reached 20%.

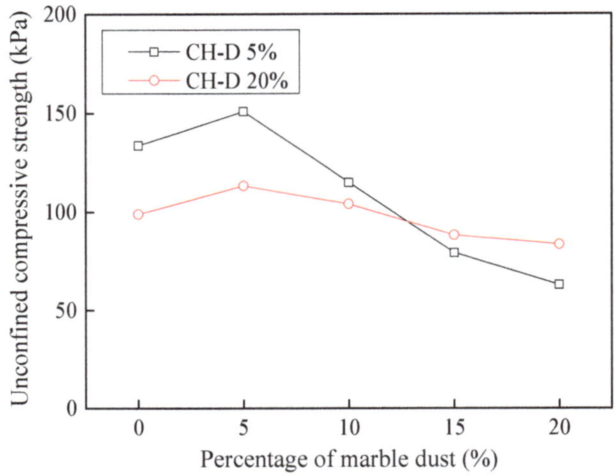

Figure 22. Improvement in dyeing effluent contaminated CH soil using marble dust.

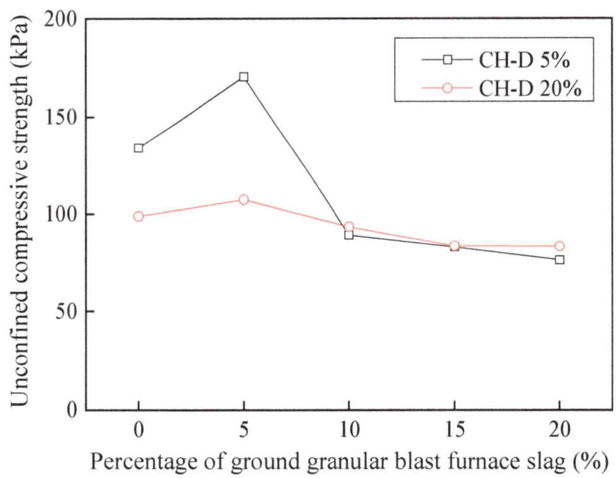

Figure 23. Improvement in dyeing effluent contaminated CH soil using GGBFS.

3.9.4. Stabilization of Tannery-Affected CH Soil

Figure 24 shows the trends observed when different percentages of GGBFS were added to base-affected CH soil. With an increase in GGBFS, unconfined compressive strength decreased rapidly from 148.04 kPa to 141.83 kPa at 5% acid contamination with an optimum value of 220.78 kPa at 10% slag concentration. At 20% acid contamination, unconfined compressive strength decreased from 98.81 kPa to 73.57% kPa when slag concentration reached 20%.

Figure 25 depicts the trends observed when 5% and 20% base-affected CH soil was treated with marble dust. With an increase in marble dust concentration, unconfined compressive strength decreased rapidly from 148.04 kPa to 44.54 kPa at 5% basic contamination. At 20% basic contamination, it decreased from 98.81 kPa to 44.5% kPa when slag concentration reached 20%. Further additions continued to decrease the unconfined compressive strength of soils.

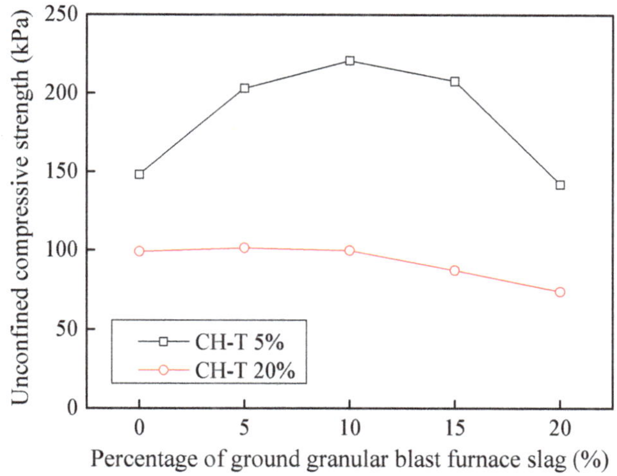

Figure 24. Improvement in tannery effluent contaminated CH soil using GGBFS.

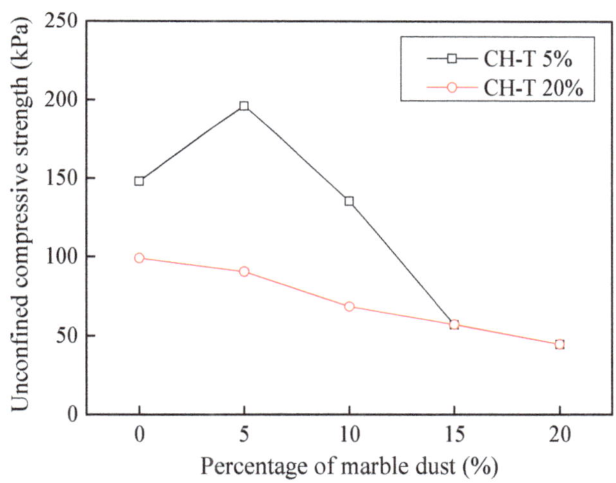

Figure 25. Improvement in tannery effluent contaminated CH soil using marble dust.

4. Conclusions

The study focussed on evaluating the effects of acidic and basic industrial effluents on geotechnical characterisitcs of cohesive soils. In general, the geotechnical characterisitics of cohesive soils were observed to deteriorate upon effluent contamination thereby causing potential risks to future, as well as existing, construction at such sites. A summary of these effects on individual soil characteristics is presented in the following points.

- Effluents drastically reduce mechanical properties of soil, specifically cohesion. The unconfined compressive strength of effluent contaminate soils was observed to decrease by as much as 60% with 20% addition of industrial effluents. Possible disintegration of mineral particles such as Fe_2O_3, causing a decrease in soil density, can be attributed as the primary reason leading to such a reduction in soil strength.

- Both the liquid limit and the plasticity index of contaminated soils showed an increasing trend with contamination. This effect was however more pronounced for CH soils whose liquid limit and plasticity index increased by about 30% compared to around 8% increase for CL soils.
- Owing to the lower specific gravity of industrial effluents, the specific gravity of effluent contaminated soils decreaesed by 12% to 15% with 20% effluent contamination. The contaminated soils were thus found to have lower specific gravity, contained more fine particles, had a lower maximum dry density, and had a lower cohesion and friction angle.
- The behavior of basic effluent (tannery) is unique; it exhibits two distinct behaviors before and after optimum concentration. Prior to optimum concentration, changes in maximum dry density of soil may hypothetically occur because a reaction of the basic effluent's constituents forms oxides of Fe present in the soil. This reaction forms FeO, which is lighter in weight and takes greater volume, thus leading to decreased density. Further increase in the basic effluent concentration post optimum concentration, trigger the remaining constituents to react to certain active masses of soils (likely not FeO), which fill the particle interspaces and increase the soil density.
- Contamination with industrial effluents was observed to cause around 30%–40% increase in compression index *Cc* of soil. An increase in *Cc* value suggests that the tendency of the soil to shrink or expand upon the intrusion of moisture is enhanced, which can lead to distress in structures constructed on such soils.
- The time dependent affect of effluents on cohesive soils was studied. The unconfined compressive strength of both the soils when contaminated with basic (tannery) effluent, kept on decreasing further with time. Both the soils when contaminated with acidic (dyeing) effluent however, show an initial decrease, followed by a gradual increase in compressive strength. Although both the soils start regaining strength with time, yet, in general, they fail to achieve unconfined compressive strength of virgin soil.
- Considering the deterioration in soil characeristics upon effluent contamination, second phase of this study focused on possible stabilization of effluent contaminated soils through addition of marble dust and ground granulated blast furnace slag (GGBFS). The main focus of this phase of research remained enhancement in the mechanical strength of soil, which remains among the most important characteristics for engineering use. Marble dust and GGBFS, which are typically considered as good stabilization agents for cohesive soils were found to possess negligible to low positive effect on the mechanical characteristics of effluent contaminated soils. This key finding of present research can hold significant practical value such that highlighting the need to devise special remediation materials for effluent contaminated soils.

The present investigation examined the effects of textile and dyeing effluents on soil's geotechnical properties. Future studies can be conducted on the influence of these effluents on the California Bearing Ratio values of cohesive soils. The work can also be extended to other contaminants, pollutants, effluents, and/or industrial wastes, namely workshop waste, sugar mill waste, pharmaceutical plant waste, dairy waste, paper and pulp mill waste, fertilizer plant waste, steel mill waste, oil refinery waste, petro-chemical complex waste, soap industry waste, etc.

Author Contributions: M.I. supervised and designed the experiments; Y.C. wrote the paper; M.A. (Muhammad Ali), M.A. (Muhammad Abrar), A.Q. and O.B. performed the experiments.

Funding: This study has been partially funded by China Postdoctoral Science Foundation (Grant Nos. 2017M620048 and 2018T110103).

Conflicts of Interest: The authors declare no conflicts of interest.

References

1. Mitchell, J. *Fundamentals of Soil Behavior*; Wiley: New York, NY, USA, 1993.
2. Gratchev, I.B.; Sassa, K. Cyclic behavior of fine-grained soils at different pH values. *J. Geotech. Geoenviron. Eng.* **2009**, *135*, 271–279. [CrossRef]

3. Anson, R.W.W.; Hawkins, A.B. The effect of calcium ions in pore water on the residual shear strength of kaolinite and sodium montmorillonite. *Geotechnique* **1998**, *48*, 787–800. [CrossRef]
4. Moore, R.; Brunsden, D. Physicochemical effects on the behavior of a coastal mudslide. *Geotechnique* **1998**, *46*, 259–278. [CrossRef]
5. Tiwari, B.; Tuladhar, G.; Marui, H. Variation in residual shear strength of the soil with the salinity of pore fluid. *J. Geotech. Geoenviron. Eng.* **2005**, *131*, 1445–1456. [CrossRef]
6. Ratnaweera, P.; Meegoda, J. Shear strength and stress–strain behavior of contaminated soils. *Geotech. Test. J.* **2006**, *29*, 133–140. [CrossRef]
7. Gratchev, I.; Sassa, K.; Fukuoka, H. How reliable is the plasticity index for estimating the liquefaction potential of clayey sands? *J. Geotech. Geoenviron. Eng.* **2006**, *132*, 124–127. [CrossRef]
8. Gratchev, I.; Sassa, K.; Osipov, V.; Fukuoka, H.; Wang, G. Undrained cyclic behavior of bentonite-sand mixtures and factors affecting it. *Geotech. Geol. Eng.* **2007**, *25*, 349–367. [CrossRef]
9. Gratchev, I.; Towhata, I. Compressibility of soils containing kaolinite in acidic environments. *KSCE J. Civ. Eng.* **2016**, *20*, 623–630. [CrossRef]
10. Sunil, B.; Nayak, S.; Shrihari, S. Effect of pH on the geotechnical properties of laterite. *Eng. Geol.* **2006**, *85*, 197–203. [CrossRef]
11. Sunil, B.; Shrihari, S.; Nayak, S. Shear strength characteristics and chemical characteristics of leachate-contaminated lateritic soil. *Eng. Geol.* **2009**, *106*, 20–25. [CrossRef]
12. Khan, M.I.; Irfan, M.; Aziz, M.; Khan, A.H. Geotechnical characteristics of effluent contaminated cohesive soils. *J. Environ. Eng. Landsc.* **2017**, *25*, 75–82. [CrossRef]
13. Terzaghi, K.; Peck, R.; Mesri, G. *Soil Mechanics in Engineering Practice*; John Wiley & Sons: New York, NY, USA, 1948.
14. Gibbs, H.J.; Bara, J.P. Stability problems of collapsing soil. *J. Soil Mech. Found.* **1967**, *93*, 577–594.
15. Pandian, N.; Nagaraj, T.; Manoj, M. Re-examination of compaction characteristics of fie-grained soils. *Geotechnique* **1997**, *47*, 363–366. [CrossRef]
16. Sridharan, A.; Nagaraj, H. Plastic limit and compaction characteristics of fine grained soils. *Proc. Inst. Civ. Eng. Ground Improv.* **2005**, *9*, 17–22. [CrossRef]
17. Umesha, T.S.; Dinesh, S.V.; Sivapullaiah, P.V. Effects of acids on geotechnical properties of black cotton soil. *Inter. J. Geol.* **2012**, *6*, 69–76.
18. Stalin, V.K.; Muthukumaran, K.; Kartikeyan, A. Effect of Liquid waste on the index and engineering behaviour of soils. In Proceedings of the Indian Geotechnical Conference: IGC-2000: The millennium conference, Mumbai, Indian, 13–15 December 2010; pp. 229–230.
19. Anandarajah, A.; Zhao, D. Triaxial behavior of kaolinite in different pore fluids. *J. Geotech. Geoenviron. Eng.* **2000**, *126*, 148–156. [CrossRef]
20. Van Olphen, H. *An Introduction to Clay Colloid Chemistry*, 2nd ed.; John Wiley & Sons: New York, NY, USA, 1977.
21. Sridharan, A.; Rao, S.M.; Murthy, N.S. Liquid limit of kaolinitic soils. *Geotechnique* **1988**, *38*, 191–198. [CrossRef]
22. Ural, N.; Karakurt, C.; Cömert, A.T. Influence of marble wastes on soil improvement and concrete production. *J. Mater. Cycles Waste Manag.* **2014**, *16*, 500–508. [CrossRef]
23. Tozsin, G.; Arol, A.I.; Oztas, T.; Kalkan, E. Using marble wastes as a soil amendment for acidic soil neutralization. *J. Environ. Manag.* **2014**, *133*, 374–377. [CrossRef] [PubMed]
24. Phummiphan, I.; Horpibulsuk, S.; Rachan, R.; Arulrajah, A.; Shen, S.L.; Chindaprasirt, P. High calcium fly ash geopolymer stabilized lateritic soil and granulated blast furnace slag blends as a pavement base material. *J. Hazard. Mater.* **2017**, *341*, 257. [CrossRef] [PubMed]
25. Hasan, U.; Chegenizadeh, A.; Budihardjo, M.A.; Nikraz, H. Experimental evaluation of construction waste and ground granulated blast furnace slag as alternative soil stabilisers. *Geotech. Geol. Eng.* **2016**, *34*, 1707–1722. [CrossRef]
26. Jaiswal, M.; Lai, B. Stabilization of clayey soil with garlic skin and rice husk ash for flexible pavement construction. *Geotech. Test. J.* **2017**, *40*, 20160227. [CrossRef]
27. Mohanty, S.K.; Pradhan, P.K.; Mohanty, C.R. Consolidation and drainage characteristics of expansive soil stabilized with fly ash and dolochar. *Geotech. Geol. Eng.* **2016**, *34*, 1435–1451. [CrossRef]

28. Phanikumar, B.R.; Shankar, M.U. Heave studies on fly ash-stabilised expansive clay liners. *Geotech. Geol. Eng.* **2017**, *35*, 111–120. [CrossRef]
29. Chiou, J.S.; You, T.R.; Tsai, C.C.; Hwang, J.H. Performance of laterally loaded piles in improved coal ash deposit. *Soils Found.* **2017**, *57*, 872–881. [CrossRef]
30. Rashid, A.S.A.; Latifi, N.; Meehan, C.L.; Manahiloh, K.N. Sustainable improvement of tropical residual soil using an environmentally friendly additive. *Geotech. Geol. Eng.* **2017**, *35*, 2613–2623. [CrossRef]

processes

MDPI

Article

Changes in Particle Size Composition under Seepage Conditions of Reclaimed Soil in Xinjiang, China

Zizhao Zhang [1,2], Wanghua Sui [1,*], Kaikai Wang [2], Guobin Tang [2] and Xiaoping Li [2]

1 School of Resources and Geosciences, China University of Mining and Technology, Xuzhou 221116, China; zhangzizhao@xju.edu.cn

2 School of Geological and Mining Engineering, Xinjiang University, Urumqi 830046, China; wkkdizhi@163.com (K.W.); tangguobin1994@163.com (G.T.); LXPygsz@163.com (X.L.)

* Correspondence: suiwanghua@cumt.edu.cn; Tel.: +86-139-5219-9519; Fax: +86-516-8359-0998

Received: 15 September 2018; Accepted: 17 October 2018; Published: 20 October 2018

Abstract: The distribution of reclaimed soil particle size under seepage conditions after the management period will directly determine the success or failure of reclamation work. The geotechnical experimental method was used in this paper to study the changes in the granulometric composition of soil. The results show that the granulometric composition of the reclaimed soil varied obviously at different depths. The granulometric composition of the soil at a depth of 10 cm was not much different from undisturbed reclaimed soil (URS). At a depth of 30 cm, as the sharp decrease of the content of fine particles resulted in coarser reclaimed soil, the soil became more uniform, with an increase in porosity and water content. At a depth of 50 cm, the fine particle content was generally slightly lower than that of URS. At a depth of 70 cm, the fine particle content of the soil greatly exceeded that of the URS, with the finest soil particles and lowest porosity. The main reason for the above-mentioned changes of granulometric composition in the reclaimed soil was the seepage in soil caused by irrigation during the management period. The research results can provide a reference for management after land reclamation at non-metallic mines in Xinjiang, China.

Keywords: Xinjiang; land reclamation; management period; soil particle size; fluid flow in reclaimed soil

1. Introduction

There are vast non-metallic mineral resources in the northern foothills of the Tianshan Mountains in Xinjiang, and the large-scale exploitation of mineral resources is bound to cause irreversible damage to the fragile geological environment and ecological environment in this region. Non-metallic mines in this region include limestone, dolomite, and granite mines. The open-pit mining method is mainly adopted. Large and deep open pits, and a large number of massive hard waste rocks (Figures 1 and 2), will be formed after mining. For mines whose damaged lands are grassland and woodland, the land will be reclaimed after the pits are closed. To reclaim the land, the waste rock is backfilled into the open pits first, and then the surface of the land is covered with soil and vegetation is planted. In order to ensure the survival rate of the vegetation, a one-year management period is set. The quality of the reclaimed soil after the management period will directly determine the success or failure of the reclamation work [1], and the distribution of soil particle size will directly affect the soil fertility, texture, and water holding capacity [2].

Figure 1. Open-pit mining area.

Figure 2. Waste rock piles.

In recent years, research [3–17] on the physical properties of reclaimed soils has achieved great success. Research on the granulometric composition of reclaimed soil mainly included the following aspects: Using multifractal theory to analyze the granulometric composition of reclaimed soil [18–20], granulometric composition analysis of reclaimed soils under different reclamation methods [21–24], granulometric composition of reclaimed soil before and after reclamation in different years [25–27], and using spectrum analytical methods to analyze the granulometric composition of reclaimed soil [28]. The previous research mainly focused on the granulometric composition of reclaimed soil in coal mining areas, with clay soil or loam soil as the main type of reclaimed soil. Different test methods were used to study the granulometric composition of reclaimed soil under different reclamation modes or in different years. There has been less research directed at sandy loam, taking into account the effect of the irrigation water during the management period, to study the changes in the particle size of the soil at different depths or with different overlaying soil thickness under different compaction circumstance.

In view of the limitations of the past studies and the importance of soil particle size to reclamation work, this study takes the typical sandy loam at the non-metallic mines in the northern foothills of the Tianshan Mountains in Xinjiang as a research object. An in-situ test method was used for the first time to simulate the compaction effect of reclamation machinery on reclaimed soils. Furthermore, field sampling, indoor geotechnical experiments, and mathematical statistical analysis were applied to study the changes in the granulometric composition of the reclaimed soil under seepage conditions after a one-year management period.

2. Materials and Methods

The reclaimed soil in-situ test was conducted at the Changji Groundwater Balance Experiment Site in Xinjiang. The reclaimed soil was taken from a limestone ore mine in Dabancheng, which belongs to the northern foothills of the Tianshan Mountains in Xinjiang. The soil was calcic brown soil and belonged to sandy loam, with a bulk unit weight of 12.646 kN/m^3, a porosity of 57.71%, and a water content of 11.76%. The lower part of the reclaimed soil was backfilled with the waste rock produced by the limestone mining.

A total of 6 test barrels were used in this test (Figures 3–6). The depth of the test barrels was 2 m. The test was divided into 2 groups. In one group, the barrel was backfilled by overlaying soil with a thickness of 30 cm, 50 cm, and 70 cm without compaction, while in the other group, the overlying soil was compacted once by a track-type bulldozer. The lower part of the reclaimed soil was backfilled with limestone waste rocks. The plate load test (Figure 7) was used to simulate the compaction of the bulldozer based on the intensity of pressure of the bulldozer on the ground, which was different from previous conventional drop weight tests. In the management period, watering was conducted in accordance with the empirical value of the irrigation volume and the number of times of irrigation in the land reclamation and management period of the mine area in the foothills of Xinjiang Tianshan Mountains (Figures 8 and 9). The irrigation volume was related to the thickness of the soil layer. The thicker the soil layer was, the greater the irrigation volume would be. Irrigation was carried out a total of three times in the period of management. After one year of management, samples were taken at depths of 10 cm, 30 cm, 50 cm, and 70 cm depending on the thickness of the overlaying soil, and the weight of each sample was 400 g. The samples were then taken into the laboratory for particle size analysis. Sieving analysis was adopted as the soil has a high content of sand and low content of clay (Figure 10). At the same time, the undisturbed soil samples were taken at the corresponding depth for laboratory tests of its density and water content.

Figure 3. Empty cylindrical test barrels.

Figure 4. Backfilled waste rock in lower part of test barrels.

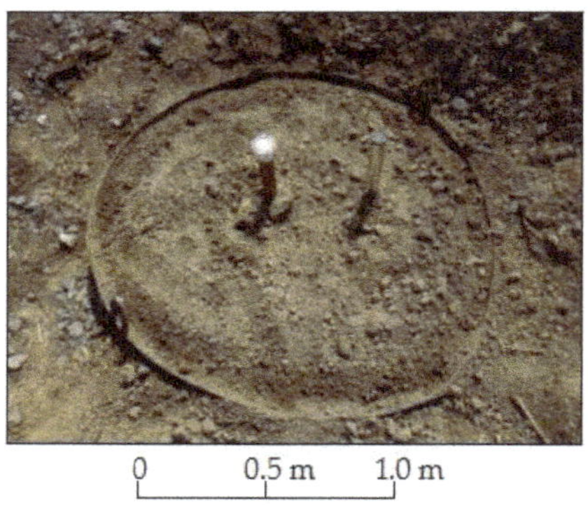

Figure 5. Uncompacted reclaimed soil test barrel.

Figure 6. Reclaimed soil after one-compaction test barrel.

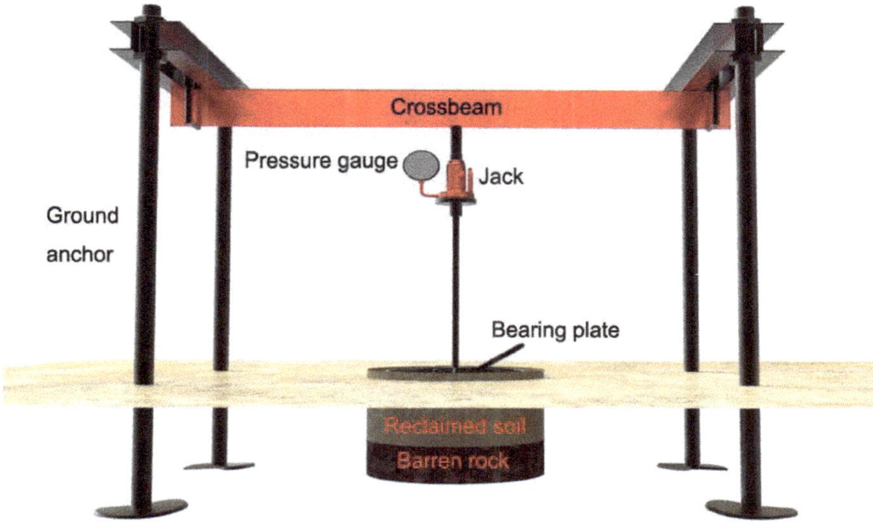

Figure 7. Illustration of the setup for plate load test.

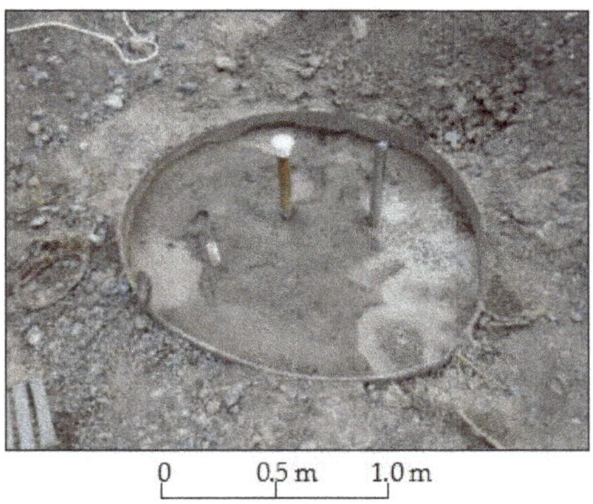

Figure 8. Uncompacted reclaimed soil test barrel after irrigation.

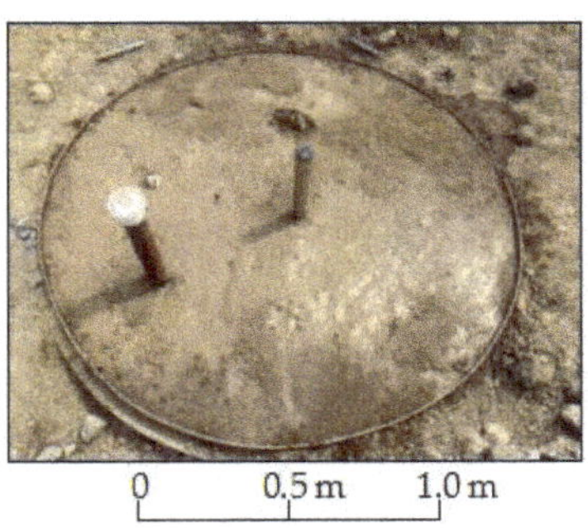

Figure 9. One-compaction reclaimed soil test barrel after irrigation.

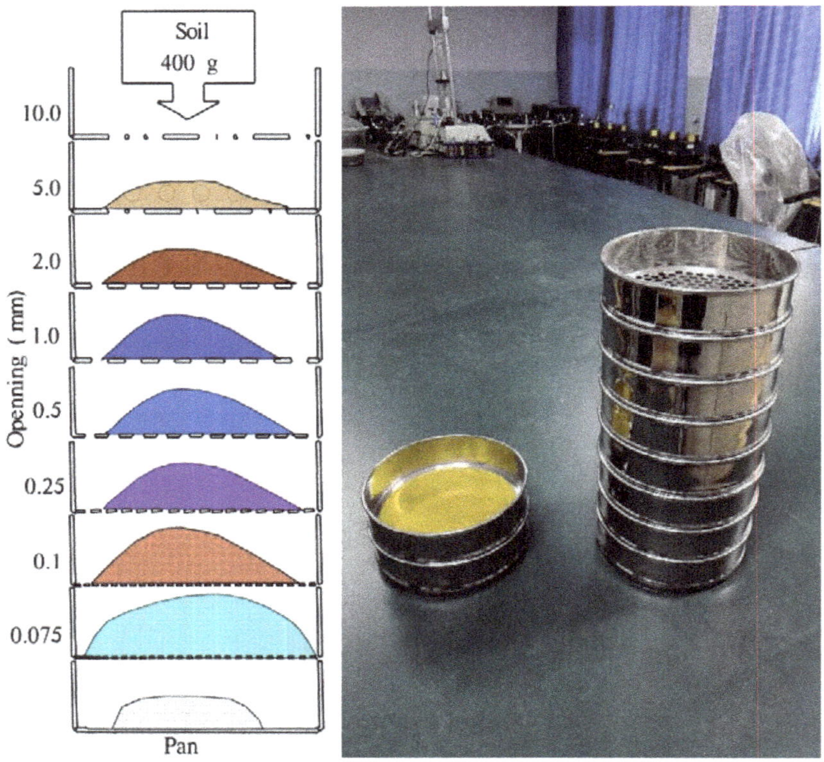

Figure 10. Sieving method.

3. Results and Discussions

3.1. Experimental Data Processing

A total of 18 soil samples were obtained after a one-year management period for particle size analysis. This sampling can be conducted at the depth of the interface between the reclaimed soil and waste rock, mainly because varying degrees of subsidence (Figure 8) occurred in the overlaying soil of the reclaimed soil due to surface irrigation during the one-year period of management, and some of the soil particles could enter the lower part of the waste rock. In order to ensure the accuracy of the physical properties of soil (density, porosity, water content), another three samples were taken from each depth. Table 1 lists the experimental results.

Table 1. Results of the reclaimed soil particle size analysis after a one-year management period.

Test Barrel No.	Sampling Depth (cm)	Percentage (%) of Soil with Particle Size Less than d_x (mm) by Mass								Characteristic Particle Size (mm)				C_u	n	θ (%)
		≤10.0	≤5.0	≤2.0	≤1.0	≤0.5	≤0.25	≤0.1	≤0.075	d_{10}	d_{30}	d_{50}	d_{60}			
A1	10	100	100	93.54	91.11	70.53	40.08	4.43	0.8	0.12	0.20	0.32	0.39	3.20	0.542	13.70
A1	30	100	100	94.19	90.44	71.38	13.96	2.42	0.28	0.23	0.32	0.39	0.43	1.92	0.520	23.28
A2	10	100	100	91.41	83.26	74.58	43.68	6.33	1.03	0.11	0.18	0.29	0.36	3.13	0.489	13.60
A2	30	100	100	94.35	81.15	66.22	26.32	0.85	0.32	0.16	0.28	0.38	0.45	2.85	0.503	23.23
B1	10	100	100	95.76	92.33	73.95	36.60	2.65	0.30	0.13	0.22	0.33	0.39	2.90	0.536	13.93
B1	30	100	100	98.96	93.16	71.91	15.41	4.61	1.13	0.21	0.31	0.39	0.43	2.09	0.523	23.73
B1	50	100	100	98.56	96.11	71.68	37.58	3.83	0.53	0.13	0.22	0.33	0.38	2.88	0.510	21.52
B2	10	100	100	97.01	91.06	69.66	42.71	7.41	0.98	0.11	0.19	0.30	0.39	3.48	0.504	14.00
B2	30	100	100	97.35	93.50	61.90	23.40	3.45	0.87	0.15	0.29	0.41	0.49	3.15	0.520	23.81
B2	50	100	100	89.35	82.90	61.75	35.10	2.15	0.35	0.13	0.22	0.37	0.48	3.58	0.513	16.07
C1	10	100	100	96.10	88.75	57.90	31.95	2.47	0.45	0.14	0.24	0.41	0.53	3.88	0.530	14.09
C1	30	100	100	97.83	92.30	45.20	7.67	0.67	0.32	0.27	0.40	0.54	0.62	2.25	0.525	24.01
C1	50	100	100	98.83	92.98	64.63	26.43	1.68	0.28	0.15	0.28	0.39	0.46	3.04	0.511	23.51
C1	70	100	100	98.70	96.73	80.08	60.18	18.7	8.50	0.08	0.14	0.20	0.26	3.23	0.472	20.93
C2	10	100	100	97.02	95.19	79.14	39.46	6.33	0.95	0.12	0.20	0.31	0.36	3.11	0.515	13.85
C2	30	100	100	96.29	89.74	67.84	25.66	5.73	0.50	0.14	0.28	0.38	0.44	3.24	0.520	24.19
C2	50	100	100	97.90	94.52	68.34	30.24	3.71	0.88	0.14	0.25	0.37	0.43	3.13	0.504	22.28
C2	70	100	100	98.49	96.84	80.16	60.23	21.8	6.67	0.08	0.09	0.19	0.26	3.20	0.492	22.44
URS	-	100	90.97	87.56	81.83	63.70	39.58	5.45	0.70	0.12	0.19	0.34	0.45	3.77	0.577	13.58

Note: Test barrel A represents overlaying soil thickness of 30 cm, B represents overlaying soil thickness of 50 cm, and C represents overlaying soil thickness of 70 cm; 1 represents uncompacted, and 2 represents compacted; A1 represents that the overlaying soil thickness is 10 cm and is uncompacted. d_{10} is the effective particle. d_{50} is the average particle size. $C_u = d_{60}/d_{10}$ is the non-uniform coefficient of the soil; $C_u \geq 5$ indicates non-homogeneous soils, and $C_u < 5$ indicates homogeneous soils. URS represents a sample of undisturbed reclaimed soil. n represents porosity. θ represents volumetric water content.

3.2. Granulometric Composition Analysis of Reclaimed Soil after the Management Period

3.2.1. Analysis of the Factors Affecting the Changes in Granulometric Composition of Reclaimed Soil after the Management Period

According to the Kalkiski soil classification system for soil particle sizes [29], 0.25 mm was used as the classification standard to distinguish between fine sand and medium sand; in the international standard for soil particle size classification, 0.2 mm is used as the classification standard to distinguish between coarse sand and fine sand, while in soil science and soil mechanics, 0.1 mm is taken as the classification standard to distinguish between fine and coarse particles. In order to reflect the change in the granulometric composition of reclaimed soil after the one-year management period, 0.25 mm, 0.1 mm, and d_{50} were studied as the focus of analysis. For the data on the granulometric composition of the reclaimed soil after the one-year management period (Table 1), variance analysis was done with the use of the multivariate in the general linear model in IBM SPSS Statistics (IBM, Chicago, IL, US); the percentage of particles with a size of $\leq$0.25 mm, $\leq$0.1 mm, and d_{50} were used as the dependent variables, and soil thickness and sampling depth as fixed factors. Table 2 lists the results of variance analysis. The results show that soil thickness has an insignificant effect on particle distribution ($p > 0.05$), and sampling depth has a significant effect on particle distribution ($p < 0.05$). The overlying soil of non-metallic mines in the northern foothills of the Tianshan Mountains in Xinjiang is mainly composed of sandy soil with a large amount of sand. In order to reflect the changes in the grain size of reclaimed soils at different depths, especially the changes of the fine particles, the percentage of soil particles with a size of $\leq$0.25 mm was used in the comparison of particle sizes at different depths. Table 3 lists the comparative analysis results. The results show that the fine particle content at the depth of 70 cm is significantly different from those at other depths ($p < 0.05$), and there is also a significant difference between the fine particle content at the depth of 10 cm and that at the depth of 30 cm ($p = 0.039$). The difference between the fine particle content at the depth of 70 cm and the depth at 30 cm is the most significant ($p = 0.002$).

Table 2. Test results of the inter-subject effect after variance analysis.

Source	Dependent Variables	III-Type Sum of Square	*df*	Mean Square	*F*	Sig.
Correction Model	$\leq$0.25 mm	2179.926 [a]	5	435.985	3.987	0.023
	$\leq$0.1 mm	506.317 [b]	5	101.263	25.118	0.000
	d_{50}	0.066 [c]	5	0.013	3.352	0.040
Intercept	$\leq$0.25 mm	17,383.524	1	17,383.524	158.963	0.000
	$\leq$0.1 mm	713.252	1	713.252	176.922	0.000
	d_{50}	1.161	1	1.161	293.865	0.000
Soil Thickness	$\leq$0.25 mm	12.190	2	6.095	0.056	0.946
	$\leq$0.1 mm	2.179	2	1.090	0.270	0.768
	d_{50}	0.005	2	0.003	0.677	0.527
Sampling Depth	$\leq$0.25 mm	2017.396	3	672.465	6.149	0.009
	$\leq$0.1 mm	440.730	3	146.910	36.441	0.000
	d_{50}	0.065	3	0.022	5.492	0.013

Note: [a] R square = 0.624 (adjusted R square = 0.468); [b] R square = 0.913 (adjusted R square = 0.876); [c] R square = 0.583 (adjusted R square = 0.409). Sig. = significance, the values are the statistical p value, if $p < 0.05$, the difference is significant.

Table 3. Results of multiple comparative analyses.

Dependent Variables	(I) Sampling Depth (cm)	(J) Sampling Depth (cm)	Difference in the Means (I−J)	Standard Deviation	Sig.	95% Confidence Interval	
						Lower Limit	Upper Limit
≤0.25 mm	10	30	14.0100 [a]	6.03754	0.039	0.8553	27.1647
		50	12.0758	6.75017	0.099	−2.6315	26.7832
		70	−20.7917 [a]	8.53837	0.031	−39.3952	−2.1882
	30	10	−14.0100 [a]	6.03754	0.039	−27.1647	−0.8553
		50	−1.9342	6.75017	0.779	−16.6415	12.7732
		70	−34.8017 [a]	8.53837	0.002	−53.4052	−16.1982
	50	10	−12.0758	6.75017	0.099	−26.7832	2.6315
		30	1.9342	6.75017	0.779	−12.7732	16.6415
		70	−32.8675 [a]	9.05631	0.003	−52.5995	−13.1355
	70	10	20.7917 [a]	8.53837	0.031	2.1882	39.3952
		30	34.8017 [a]	8.53837	0.002	16.1982	53.4052
		50	32.8675 [a]	9.05631	0.003	13.1355	52.5995

Note: [a] The difference of means is significant at 0.05 level; Sig. = significance.

3.2.2. Analysis of the Granulometric Composition of Reclaimed Soil after the Management Period under Different Overlaying Soil Thickness

The data of Table 1 is plotted into the gradation cumulative curve of the soil particles, where the abscissa is the particle size (mm), and the ordinate is the mass (cumulative percentage) content of the soil which is smaller than a certain particle size. The figure shows the relative content of each particle group in the soil, and it is the basis for calculating d_{10}, d_{30}, and other characteristic values. The uniformity or the gradation of the soil can be roughly judged according to the slope of the curve. A steep curve indicates that the soil particles are relatively uniform, and the quality of corresponding granular group is relatively centralized. The situation is the opposite when the curve is gentle.

The accumulative curve of the soil particle gradation with an overlaying soil thickness of 30 cm is shown in Figure 11. It shows that when the overlaying soil thickness is 30 cm, regardless of if it is compacted or not, the fine-grained soil content at the sampling depth of 30 cm is significantly smaller than that of the undisturbed reclaimed soil (URS). This is because this depth is the interface between soil and waste rock. The 30 cm overlaying soil is thin and the water would seep into the soil during irrigation, and the seepage process can easily carry the fine particles of the soil down to the pores in the large waste rocks. Fine particle content of the uncompacted reclaimed soil is even smaller than the compacted reclaimed soil at the depth of 30 cm. This is because the surface soil is compacted, the water seepage is slow, and less fine particles would be carried down to the waste rock layer compared with the uncompacted reclaimed soil. In the actual irrigation process, it can also be found that the uncompacted reclaimed soil would display a higher settlement in the process of irrigation compared with the compacted one (Figures 8 and 9). At the depth of 10 cm, the content of fine particles of compacted soil are slightly higher than that of the URS and uncompacted soil, which is mainly because the compaction results in an increase in the content of fine particles [20,23].

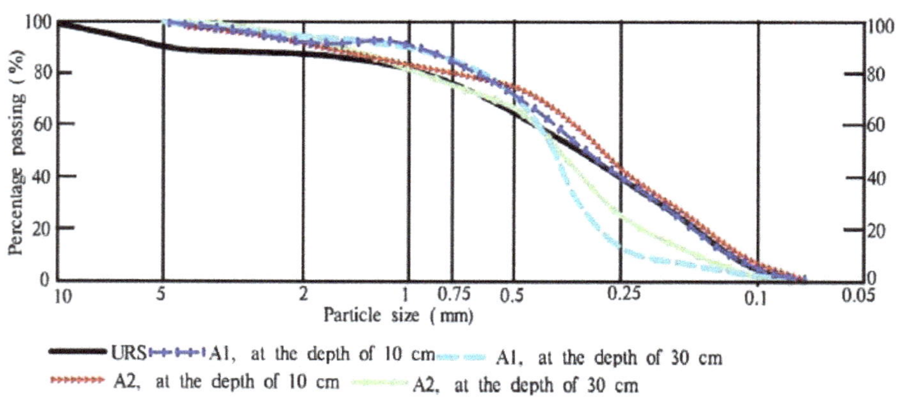

Figure 11. Accumulative curve of particle gradation of soil with overlaying soil thickness of 30 cm (A1 and A2).

The accumulative curve of the soil particle gradation with an overlaying soil thickness of 50 cm is shown in Figure 12. When the overlaying soil is 50 cm thick, the irrigation volume is larger than that of the 30 cm overlaying soil, and when the thickness of the soil overlayer increases, the seepage time in the soil would be longer. The results show that the fine particle content is significantly less than that of the URS at the depth of 30 cm. Furthermore, the fine particle content in the uncompacted reclaimed soil is much less. This is also the result of the seepage in the soil; that is, the water seeps downward after irrigation, and the seepage is fast in uncompacted soil with a large downward seepage force, resulting in a large loss of fine particles at the depth of 30 cm. At the depth of 50 cm, which is the junction of soil and waste rock, the fine particle content is close to that of the URS. This is mainly because the soil layer is thick and the seepage is slow at this depth, with a smaller seepage force, and thus the loss of fine particles is insignificant.

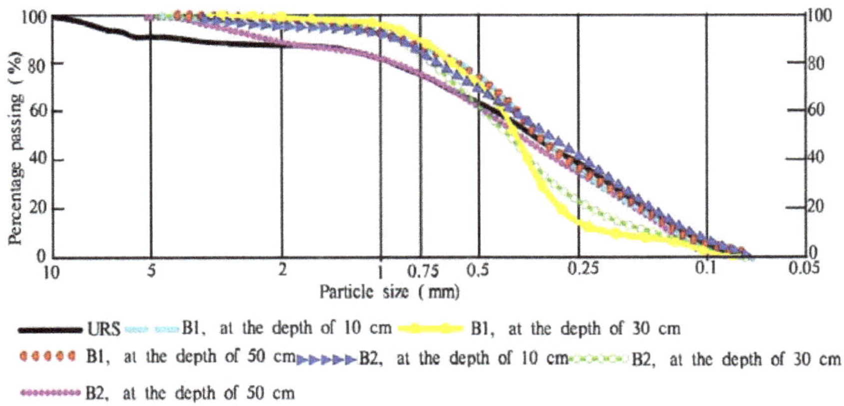

Figure 12. Accumulative curve of particle gradation of soil with overlaying soil thickness of 50 cm (B1 and B2).

The accumulative curve of the soil particle gradation with an overlaying soil thickness of 70 cm is shown in Figure 13. The thickness of the overlaying soil and the irrigation volume are further increased. The greatest change compared with overlaying soil thicknesses of 30 cm and 50 cm is that the fine particles sharply increase at the depth of 70 cm (the junction of soil and waste rock). This is because, as the soil layer becomes thicker, the fine particle content of the soil would increase, and the downward seepage of the soil would get slower. The finer particles are formed into sedimentation at the depth of 70 cm, resulting in a sharp increase in the fine particles.

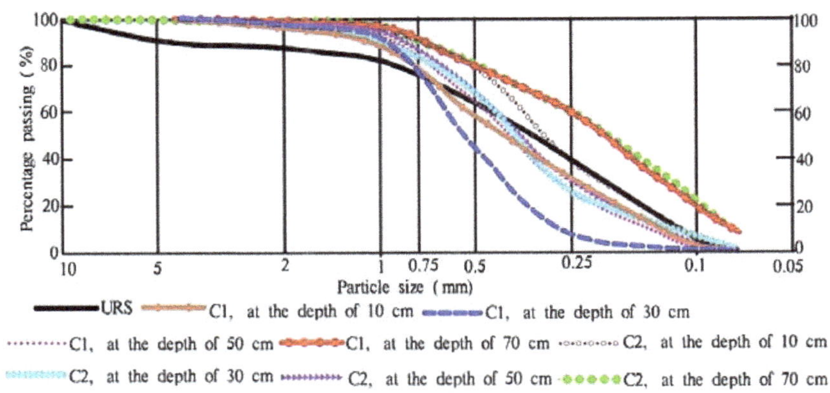

Figure 13. Accumulative curve of particle gradation of soil with overlaying soil thickness of 70 cm (C1 and C2).

3.2.3. Analysis of Change in Granulometric Composition of Reclaimed Soil at Different Depths after the Management Period

Figures 14–16 show that after the one-year management period the content of fine particles at the depth of 10 cm is not significantly different from that of the URS, and the content of fine particles of compacted reclaimed soil is slightly higher than that of the URS. This is because the surface compaction results in an increase in the content of fine particles. At the depth of 30 cm, the content of fine particles is lower than that of the URS, and the lowest value arises in the test barrel of uncompacted overlaying soil with a thickness of 70 cm, followed by the test barrels of the uncompacted overlaying soil with thicknesses of 50 cm and 30 cm. At the depth of 50 cm, the fine particle content is slightly lower than that of the URS, while at the depth of 70 cm, the fine particle content is much higher than that of the URS. The main reason for the above-mentioned phenomenon is the irrigation during the management period. The irrigation volume varies with different overlaying soil thicknesses. The thicker the overlaying soil is, the greater the irrigation volume would be. After irrigation, the water starts to flow downward along the pores of the reclaimed soil from the surface. The seepage speed is fast in uncompacted soil and slow in compacted soil. Therefore, from the surface to the deeper part of the reclaimed soil, the seepage speed gradually slows down, and the seepage volume becomes smaller. The seepage speed on the surface of the reclaimed soil is large, with great seepage volume and large downward seepage force, and it could carry the coarse and fine particles in the soil downward, causing the subsidence of the entire surface of the soil and making it dense. Therefore, the content of the fine-grained soil at the depth of 10 cm does not change much compared with the fine particle content of the URS. At the depth of 30 cm, the seepage speed becomes slower and could only carry fine-grained soil into the pores of deep soil; the uncompacted reclaimed soil was loose and the porosity is large, causing the fine-grained soil to move downward, and resulting in a great loss of fine-grained soil. At the overlaying soil thickness of 70 cm, the fine-grained soil content deceases the most greatly because the irrigation volume reaches the maximum. At a depth of 50 cm, the seepage speed further slows down, thus the fine particle content is only slightly reduced. At the depth of 70 cm, the seepage velocity is the slowest, and most of the fine particles are gathered here.

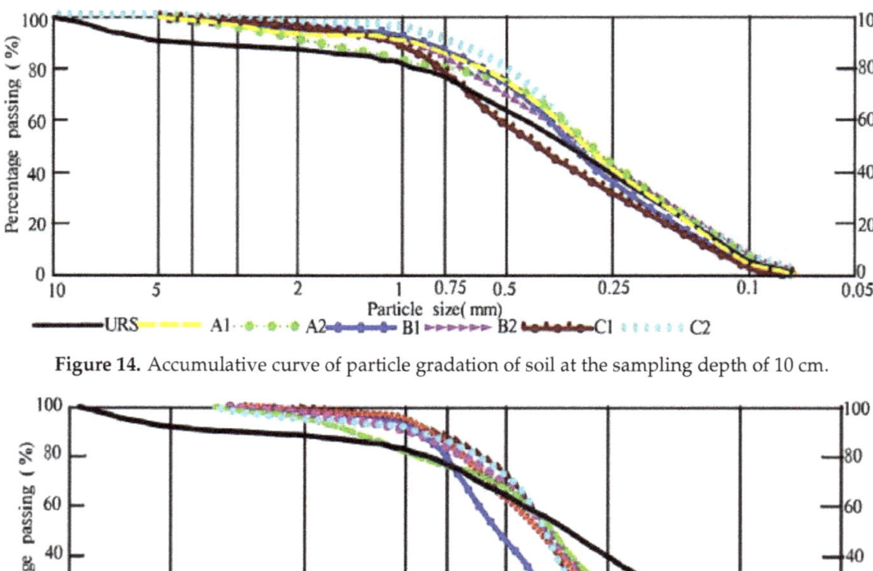

Figure 14. Accumulative curve of particle gradation of soil at the sampling depth of 10 cm.

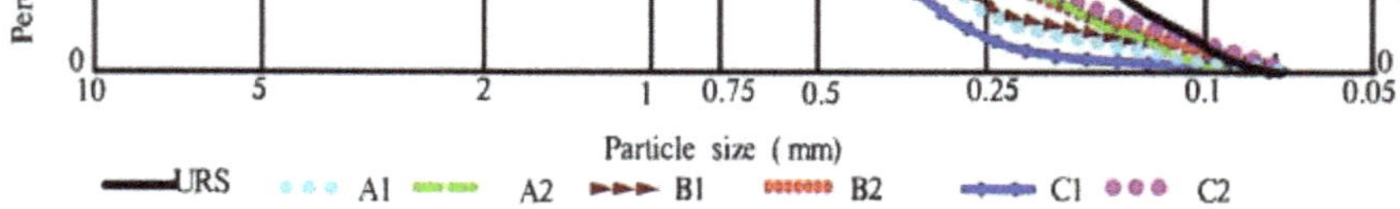

Figure 15. Accumulative curve of particle gradation of soil at the sampling depth of 30 cm.

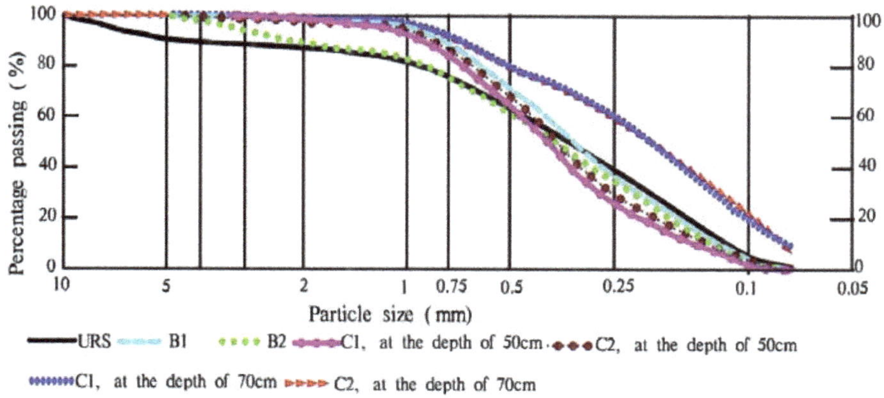

Figure 16. Accumulative curve of particle gradation of soil at the sampling depths of 50 cm and 70 cm.

3.2.4. Effects of Changes in Granulometric Composition of Reclaimed Soil to Physical Properties after the Management Period

The bivariate correlation analysis method in SPSS was used to analyze the correlation between the variation of soil particle size (especially the change of fine particles) and the physical properties, such as porosity, water content, and soil homogeneity of reclaimed soil. Table 4 lists the degree of correlation between the percentage coposition of soil with a particle size less than or equal to 0.25 mm or 0.1 mm, d_{10}, d_{50}, the coefficient of inhomogeneity C_u, the porosity, and the water content of the soil after the management period; the correlation coefficient grading used Spearman. The results show that there is a significant correlation between the percentage coposition of soil with a particle size less than or equal to 0.25 mm or 0.1 mm, d_{10}, d_{50}, C_u, and the water content of the soil after the management period, and they are unrelated only to the porosity of the soil after the management period. The porosity of the soil is related only to d_{50}. The water content of the soil after the one-year management period is only related to the percentage coposition of soil whose particle size is less than or equal to 0.25 mm, d_{10}, and d_{50}.

Table 4. Spearman grading correlation coefficient results.

			≤0.25 mm	≤0.1 mm	d_{10}	d_{50}	C_u	n	θ
Rho of Spearman	≤0.25 mm	Correlation coefficient	1.000	0.727 [a]	−0.967 [a]	−0.904 [a]	0.532 [b]	−0.393	−0.742 [a]
		Sig. (bilateral)	-	0.000	0.000	0.000	0.019	0.096	0.000
		N	19	19	19	19	19	19	19
	≤0.1 mm	Correlation coefficient	0.727 [a]	1.000	−0.812 [a]	−0.806 [a]	0.404	−0.353	−0.365
		Sig. (bilateral)	0.000	-	0.000	0.000	0.087	0.138	0.124
		N	19	19	19	19	19	19	19
	d_{10}	Correlation coefficient	−0.967 [a]	−0.812 [a]	1.000	0.914 [a]	−0.581 [a]	0.368	0.686 [a]
		Sig. (bilateral)	0.000	0.000	-	0.000	0.009	0.121	0.001
		N	19	19	19	19	19	19	19
	d_{50}	Correlation coefficient	−0.904 [a]	−0.806 [a]	0.914 [a]	1.000	−0.279	0.484 [b]	0.611 [a]
		Sig. (bilateral)	0.000	0.000	0.000	-	0.247	0.036	0.005
		N	19	19	19	19	19	19	19
	C_u	Correlation coefficient	0.532 [b]	0.404	−0.581 [a]	−0.279	1.000	0.007	−0.419
		Sig. (bilateral)	0.019	0.087	0.009	0.247	-	0.977	0.074
		N	19	19	19	19	19	19	19
	n	Correlation coefficient	−0.393	−0.353	0.368	0.484 [b]	0.007	1.000	−0.068
		Sig. (bilateral)	0.096	0.138	0.121	0.036	0.977	-	0.781
		N	19	19	19	19	19	19	19
	θ	Correlation coefficient	−0.742 [a]	−0.365	0.686 [a]	0.611 [a]	−0.419	−0.068	1.000
		Sig. (bilateral)	0.000	0.124	0.001	0.005	0.074	0.781	-
		N	19	19	19	19	19	19	19

Note: [a] represents that when the confidence level (double test) is 0.01, the correlation is significant. [b] represents that the correlation is significant when the confidence level (double test) is 0.05.

The data of correlations was formed into a scatter plot with straight lines (Figures 17 and 18). Figure 17 shows that the percentage content of soil with particle size less than or equal to 0.25 mm is basically consistent with the change trend of the C_u values and the percentage content of the soil whose particle size is less than or equal to 0.1 mm; d_{10} and d_{50} are basically consistent with the change trend of the soil water after the management period, and are basically contrary to the change trend of the percentage content of soil with particle size less than or equal to 0.25 and 0.1 mm. Soils with particle size less than or euqal to 0.25 mm and 0.1 mm are all fine-grained soils, and the changes in their contents are mainly affected by the seepage of irrigation water in the reclaimed soil during the period of management, and therefore their changes are positively correlated; the value of the non-uniform coefficient C_u is less than 5, indicating that the reclaimed soil all belongs to homogeneous soil. The non-uniform coefficient of the reclaimed soil after the management period is basically smaller than that of the URS, indicating that the soil becomes homogeneous under the seepage of water; and the reduction of the precentage content of soil whose particle size is less than or equal to 0.25 mm indicates that fine particles have decreased, and the soil is basically composed of coarse particles with a good uniformity. When the fine particle content decreases, the value of the characteristic particle size d_{10} and d_{50} would be greater, reflecting a coarser particle size of the soil on the whole. Table 4 shows that the water content of the reclaimed soil after the one-year management period is corelated to the precentage content of the soil with particle size less than or equivalent to 0.25 mm, as well as to d_{10} and d_{50}. Figure 17 shows that after the one-year management period, the water content of the reclaimed soil is positively corelated to d_{10} and d_{50}, and negatively corelated to precentage content of the soil whose particle size is less than or equivalent to 0.25 mm. This indicates that the coarser the soil is, the higher the soil water content will be. This is because the water content of the reclaimed soil is generally higher at the depth of 30 cm. This position also happens to be the location with the lowest content of fine particles. At the depth of 10 cm, the water content is low due to the impact of evaporation of the ground, while at this position the particles are relatively finer.

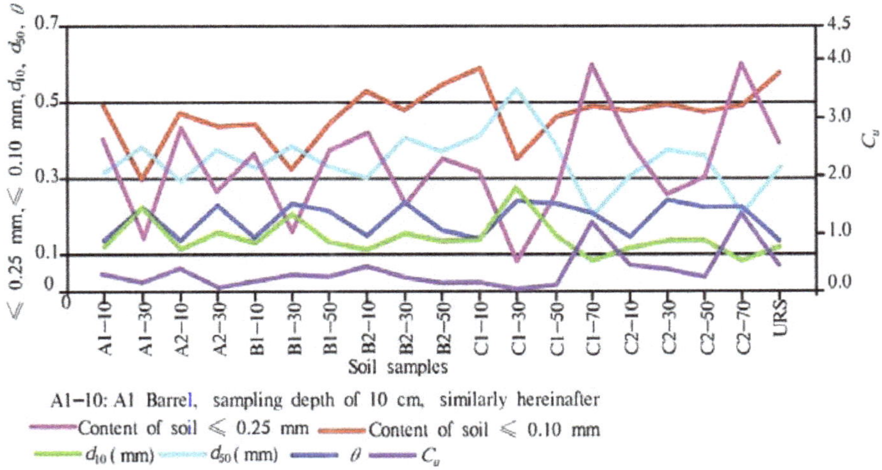

Figure 17. Scatter plot with straight lines of percentage content of soils $\leq$0.25 mm and $\leq$0.1 mm, d_{10}, d_{50}, C_u, and θ.

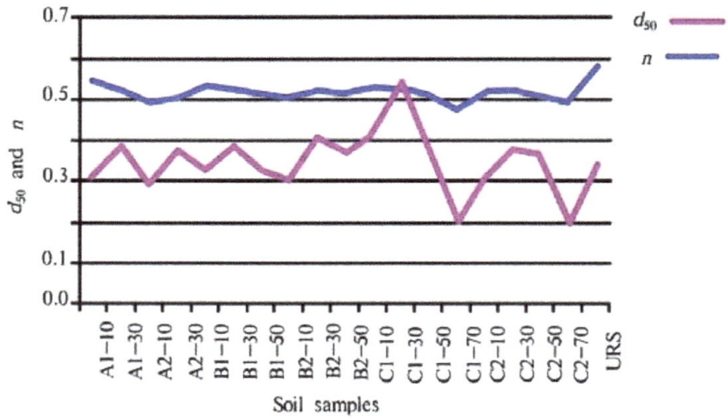

Figure 18. Scatter plot with straight lines of d_{50} and n.

Figure 18 shows that the overall porosity is positively correlated to d_{50}, indicating that the coarser the particles of the reclaimed soil are, the greater the porosity will be. This is quite opposite to the general distribution of porosity in the soil. This is because, if the particle sizes are completely uniform, the porosity of coarse-grained and fine-grained soils will be the same. However, the porosity of coarse-grained soils is usually smaller than that of fine-grained soils, and the pores of coarse-grained soils will be filled with smaller fine particles, while the pores of fine-grained soils will have none or less smaller fine particles to fill in. The porosity of the reclaimed coarse-grained soil is larger than that of the fine-grained soil, because the fine particles in the coarse-grained soil are carried into the deeper part of the reclaimed soil, resulting in a bigger pore volume and porosity.

This paper only includes the uncompacted and one-compaction circumstances, and there were no tests for the situation of multiple compactions. In the future period of this study, the effect of the number of times of compaction on the change in granulometric composition of the reclaimed soil after the management period will be carried out. According to the results of the granulometric composition of the soil after the one-year management period, the most important factor affecting the change in granulometric composition is the transport of the water in the unsaturated zone of the reclaimed soil. The future research could be conducted with a focus on the transport mechanism of the water in reclaimed soil under different overlaying thickness and compaction circumstances.

4. Conclusions

Studying the distribution of fine particles in reclaimed soil is very important for mined land reclamation. This paper uses geotechnical experimental methods to analyze the granulometric composition and the index of physical properties of the reclaimed soil in seepage conditions after a one-year management period. Through experimental results and data analysis, we can conclude that the content of fine particles in reclaimed soil at the depth of the interface between the soil and waste rock varies greatly under seepage after the management period; the thicker the overlaying soil is, the higher the fine particle content at this depth. The sampling depth of reclaimed soil has a significant effect on the granulometric composition of reclaimed soil. The most significant difference is between the depth of 70 cm and 30 cm. The fine particle content varies at different depths of reclaimed soil. The fine particle content at the depth of 10 cm is not much different from that of the URS. The fine particle content of reclaimed soil which has been compacted once at the depth of 10 cm is slightly higher than that of the URS. The fine particle content at the depth of 30 cm is generally lower than that of the URS, and reclaimed soil generally becomes coarser, and the fine particle content of the uncompacted reclaimed soil decreases more significantly. The fine particle content at the depth of 50 cm is slightly lower than that of the URS. Finally, the fine particle content increases sharply at the depth of 70 cm, and is greater than that of the URS.

By comparing the fine grain content of reclaimed soil with the physical properties of soil, we can see that the fine particle content affects the physical properties such as the homogeneity, porosity, and water content; the overall change is that the coarser the particles are, the more uniform the soil particles, the higher the water content, and the larger the porosity will be.

The results of this paper can provide a basis for guaranteeing the survival rate of vegetation in the reclaimed area of Xinjiang during the management period.

Author Contributions: Z.Z. and W.S. conceived the main idea of the paper and designed the test scheme; K.W., G.T., and X.L. performed the test scheme; K.W. and G.T. analyzed the data; Z.Z. wrote the paper; W.S. worked to modify figures and proofread the revised version.

Funding: This research was funded by the Natural Science Foundation of Xinjiang (Grant No. 2018D01C061) and the National Natural Science Foundation of China (Grant No. 41472268).

Conflicts of Interest: The authors declare no conflict of interest.

References

1. Hu, Z.; Wei, Z.; Qin, P. Concept and method of soil reengineering in mine reclamation. *Soils* **2005**, *37*, 8–12. [CrossRef]
2. Lv, S.; Gao, P.; Di, G. Fractal characteristics of soil particles and their relationship with soil organic matter in the the Yellow River Delta. *J. Soil Water Conserv.* **2011**, *25*, 134–138. (In Chinese)
3. Séré, G.; Schwartz, C.; Ouvrard, S.; Sauvage, C.; Renat, J.C.; Morel, J.L. Soil construction: A step for ecological reclamation of derelict lands. *J. Soils Sediments* **2008**, *8*, 130–136. [CrossRef]
4. Tang, Q.; Li, L.; Zhang, S.; Zheng, L.; Miao, C. Characterization of heavy metals in coal gangue-reclaimed soils from a coal mining area. *J. Geochem. Explor.* **2018**, *186*, 1–11. [CrossRef]
5. Yuan, Y.; Zhao, Z.; Li, X.; Wang, Y.; Bai, Z. Characteristics of labile organic carbon fractions in reclaimed mine soils: Evidence from three reclaimed forests in the Pingshuo opencast coal mine, China. *Sci. Total Environ.* **2017**, *613–614*, 1196–1206. [CrossRef]
6. Bao, N.; Wu, L.; Ye, B.; Yang, K.; Zhou, W. Assessing soil organic matter of reclaimed soil from a large surface coal mine using a field spectroradiometer in laboratory. *Geoderma* **2017**, *288*, 47–55. [CrossRef]
7. Kim, K.; Chun, S. Finite element analysis to simulate the effect of impact rollers for estimating the influence depth of soil compaction. *KSCE J. Civ. Eng.* **2016**, *20*, 2692–2701. [CrossRef]
8. Götze, P.; Rücknagel, J.; Jacobs, A.; Märländer, B.; Koch, H.; Christen, O. Environmental impacts of different crop rotations in terms of soil compaction. *J. Environ. Manag.* **2016**, *181*, 54–63. [CrossRef]
9. DeLong, C.; Skousen, J.; Pena-Yewtukhiw, E. Bulk Density of Rocky Mine Soils in Forestry Reclamation. *Soil Sci. Soc. Am. J.* **2012**, *76*, 1810–1815. [CrossRef]
10. Wang, J.; Yang, R.; Feng, Y. Spatial variability of reconstructed soil properties and the optimization of sampling number for reclaimed land monitoring in an opencast coal mine. *ARAB J. Geosci.* **2017**, *10*, 46. [CrossRef]
11. Ciarkowska, K. Organic matter transformation and porosity development in non-reclaimed mining soils of different ages and vegetation covers: A field study of soils of the zinc and lead ore area in SE Poland. *J. Soils Sediments* **2017**, *17*, 2066–2079. [CrossRef]
12. Kołodziej, B.; Bryk, M.; Słowińska-Jurkiewicz, A.; Otremba, K.; Gilewska, M. Soil physical properties of agriculturally reclaimed area after lignite mine: A case study from central Poland. *Soil Till Res.* **2016**, *163*, 54–63. [CrossRef]
13. Papadopoulos, C.; Gekaa, C.; Pavloudakis, F.; Roumpos, C.; Andreadou, S. Evaluation of the Soil Quality on the Reclaimed Lignite Mine Land in West Macedonia, Greece. *Procedia Earth Planet Sci.* **2015**, *15*, 928–932. [CrossRef]
14. Tanushree, D.; Stehouwer, R.C.; Curtis, J.D. Linking Organic Carbon, Moisture content, and Nitrous Oxide Emission in a Reclaimed Coal Mine Soil. *Land Degrad. Dev.* **2015**, *26*, 620–628. [CrossRef]
15. Cheng, W.; Bian, Z.; Dong, J.; Lei, S. Soil properties in reclaimed farmland by filling subsidence basin due to underground coal mining with mineral wastes in China. *Trans. Nonferr. Metal Soc.* **2014**, *24*, 2627–2635. [CrossRef]

16. Cejpek, J.; Kuráž, V.; Frouz, J. Hydrological properties of soils in reclaimed and unreclaimed sites after Brown-Coal mining. *Pol. J. Environ. Stud.* **2013**, *22*, 645–652.
17. Zhao, Z.; Shahrour, I.; Bai, Z.; Fan, W.; Feng, L.; Li, H. Soils development in opencast coal mine spoils reclaimed for 1–13 years in the West-Northern Loess Plateau of China. *Eur. J. Soil Biol.* **2013**, *55*, 40–46. [CrossRef]
18. Min, X.; Li, X.; Li, Q. Influence of mechanical compaction on reclaimed soil particle size distribution multifractal characteristics. *Trans. CSAE* **2017**, *33*, 274–283. [CrossRef]
19. Wang, J.; Zhang, M.; Bai, Z.; Guo, L. Multi-fractal characteristics of the particle distribution of reconstructed soils and the relationship between soil properties and multi-fractal parameters in an opencast coal-mine dump in a loess area. *Environ. Earth Sci.* **2015**, *73*, 4749–4762. [CrossRef]
20. Wang, J.; Zhang, M.; Bai, Z.; Yang, R.; Guo, L. Multifractal characteristics of soil particle composition in reclaimed coal mine dump in Loess Area. *Trans. CSAE* **2014**, *30*, 230–238. [CrossRef]
21. Huang, X.; Li, X.; Liu, N. Effects of different construction machinery on particle composition of reclaimed soil in coal mining area. *J. Soil Water Conserv.* **2014**, *28*, 136–140. [CrossRef]
22. Li, H.; Yu, J.; Fang, F. Effects of reclamation on soil particle composition, fractal dimension and aggregate. *Environ. Sci. Technol.* 2015; 38, 11–16, (In Chinese). [CrossRef]
23. Yu, J.; Fang, L.; Li, H. Particle distribution and fractal characteristics of coal mining subsidence land and its reclaimed soil. *J. China Univ. Min. Technol.* **2014**, *43*, 1095–1101. [CrossRef]
24. Sun, J.; Li, X.; Li, H.; Huang, X. Soil composition mechanism research of different reclamation material in coal mining subsidence area. *J. China Coal Soc.* **2013**, *38*, 2215–2220.
25. Huang, X.; Li, X.; Liu, N. Fractal characteristics of soil particle composition during Different Reclamation Years in coal mining subsidence area. *J. China Coal Soc.* **2014**, *39*, 1140–1146. (In Chinese)
26. Shrestha, R.; Lal, R. Land use impacts on physical properties of 28 years old reclaimed mine soils in Ohio. *Plant Soil* **2008**, *306*, 249–260. [CrossRef]
27. Bartuška, M.; Pawlett, M.; Frouz, J. Particulate organic carbon at reclaimed and unreclaimed post-mining soils and its microbial community composition. *CATENA* **2015**, *131*, 92–98. [CrossRef]
28. Alekseenko, V.L.; Bech, J.; Alekseenko, A.; Shvydkaya, N.; Roca, N. Environmental impact of disposal of coal mining wastes on soils and plants in Rostov Oblast, Russia. *J. Geochem. Explor.* **2018**, *184*, 261–270. [CrossRef]
29. Rousseva, S.S. Data transformations between soil texture schemes. *Eur. J. Soil Sci.* **1987**, *48*, 749–758. [CrossRef]

processes

MDPI

Article

Effect of Dry-Wet Cycling on the Mechanical Properties of Rocks: A Laboratory-Scale Experimental Study

Xiaojie Yang [1,2], Jiamin Wang [1,2,*], Dinggui Hou [3], Chun Zhu [1,*] and Manchao He [1]

1 State Key Laboratory for Geomechanics and Deep Underground Engineering, China University of Mining and Technology, Beijing 100083, China; xyang@cumtb.edu.cn (X.Y.); hemanchao@163.com (M.H.)
2 School of Mechanics and Civil Engineering, China University of Mining and Technology, Beijing 100083, China
3 Construction Engineering Department, North China Institute of Aerospace Engineering, Langfang 065000, China; houdinggui@126.com
* Correspondence: bqt1700620037@student.cumtb.edu.cn (J.W.); zhuchun17@mails.jlu.edu.cn (C.Z.)

Received: 12 September 2018; Accepted: 16 October 2018; Published: 19 October 2018

Abstract: Taking Nanfen open-pit iron mine in Liaoning Province as the engineering background, this study analyzes the effect of water-rock circulation on the mechanical properties of rock through a combination of macro-mechanical experiments and microstructure tests in the laboratory. Uniaxial compression experiments and acoustic wave tests are used to determine the degradation law of the mechanical properties of chlorite under the periodic action of water. The experimental results show that dry-wet cycles have a continuous and gradual effect on the rock sampled: Its uniaxial compressive strength, elastic modulus, and acoustic velocity all decrease gradually with an increase in the number of cycles. After 15 wet-dry cycles, the uniaxial compressive strength and elastic modulus of the rock decreased by 34.21% and 44.63%, respectively. Electron microscope scans of the rock indicate that the particle size, characteristics, and pore distribution at the rock surface had changed significantly after water-rock interaction. Finally, a drainage system and sliding force monitoring devices have been arranged at the mine site that can effectively reduce the impact of water-rock interaction on the stability of the mine. This combination of macro-experiments and micro-analysis allowed the weakening effect of dry-wet cycles on slope rock to be studied quantitatively, providing a theoretical reference for stability evaluation in geotechnical engineering.

Keywords: water-rock interaction; dry-wet cycles; slope stability; laboratory experiment; mechanical properties

1. Introduction

Geotechnical engineering problems have been on the increase in recent years in concert with the continuous expansion of geotechnical projects such as water conservancy projects, deep coal mines, and slope projects [1,2]. Along with environmental impacts, geotechnical engineering problems commonly relate to geotechnical stability, such as foundation stability, slope stability, tunnel stability, etc. [3,4] Therefore, it is essential to evaluate and ensure rock mass stability in order to solve geotechnical engineering problems effectively. In addition to the basic properties of the rock mass such as its hardness and integrity, water, rock structure, stress state, and planes of weakness in the rock also have important effects on the stability of the rock mass. Among these, water is the most common and important factor affecting rock mass stability due to the prevalence of rock-water interaction and the widespread existence of water in nature [5]. According to incomplete statistics, water plays an important role in more than 90% of cases of rock slope instability, 60% of mine accidents, and about 35% of water conservancy accidents [6]. From the perspective of geotechnical engineering, the interaction between water and rock is essentially a continuous physical, chemical, and mechanical process. This process changes the physical and mechanical parameters of the rock, thus causing a loss of stability in geotechnical engineering contexts [7].

Water-rock interaction is an important component of geotechnical engineering stability evaluations, and its study has been at the frontier of this field in recent years [8,9]. Rock is a porous material mainly composing a solid skeleton and pores. The rock is made up of many kinds of minerals, and many cracks and holes are formed in the process of diagenesis [10,11]. Rock is a non-uniform and non-linear material with a discontinuous interior surface inside, and the pore distribution within the rock shows randomness, diversity, and variability. An increase in the microstructural non-uniformity and discontinuity in a rock equates to damage that will inevitably lead to changes in the rock's macro-mechanical properties, thus adversely affecting overall geotechnical stability. Most studies of the influence of water on the mechanical properties of soft rock, both at home and abroad, focus on the changes in the physical and chemical properties of the rock and the law governing strength reduction after water absorption [12–14]. However, the interaction between rock and water occurs not only through soaking and water absorption. Many rocks in practical projects must also face a more complicated and changeable natural environment, such as frequent rainfall and evaporation, rise and fall of underground water level, fluctuation of the water surface in a reservoir area, etc. [15,16]. Periodic water-rock interaction is a kind of "repeated softening" for rocks [17–19]. The consequence of these processes is that the rock repeatedly absorbs and loses water, putting the rock through a frequent alternation of dry and wet cycles. The weakening effect of these cycles on rock is often stronger than that of being soaked in water for a long time and thus a serious influence on the long-term stability of rock masses in engineering. Therefore, further study is required into the influence law of repeated dry and wet cycles on the mechanical properties, mode of crack propagation, and failure mechanism of rock. A systematic and in-depth study of microstructural damage, macro-deformation, and strength deterioration of rock under dry-wet cycles will help to fundamentally establish the mechanism by which rocks are weakened by repeated water-rock interaction.

The Nanfen open-pit iron mine is located in Nanfen District, Benxi City, Liaoning province and is the largest single open-pit mine in Asia [20,21]. Water is an indispensable and important component of a complex geological environment. It can interact with rock through pores, cracks and weak structural planes in the rock mass, thus weakening the mechanical properties and damaging the microstructure of the rock mass. This is particularly an issue for open pit mine slopes, which are very deformable and dynamic. Thus, slope stability problems caused by rainfall and groundwater level rise and fall have become a hot research topic in the field of geotechnical mechanics [22–24]. The mechanical experiment of rock under water-rock interaction can help us to obtain the change law of rock's mechanical properties affected by water at the laboratory scale. For example, uniaxial compression test and acoustic wave test of rock under different dry-wet cycles can obtain the attenuation law of uniaxial compression strength, elastic modulus, and acoustic wave velocity of rock with the number of cycles. In addition, the development of scanning electron microscope makes the quantitative research methods of rock microstructure become more and more mature. Through the macroscopic mechanical experiment and microscopic structure analysis of rock, it is easy to find the deterioration process of slope rock under the different numbers of dry-wet cycles, which can provide a theoretical reference for analyzing the stability of open-pit slope rock in a periodic water environment and the implementation of waterproof engineering.

2. Materials and Methods

2.1. Rock Sample Selection

The Nanfen open-pit iron mine exploits a sedimentary metamorphic iron ore layers [25]. The mining area has an undulating topography, and the footwall slope of the mining area is rocky, with weak development of layering, characteristics that make it easy for geological disasters such as collapse and landsliding to occur. A huge amount of rock stripping and mining is carried out in this mine every year. Due to the characteristics of the terrain and the comprehensive influence of mining, many large-scale landslides have formed on the downdip part of the lower slope, threatening safe production in the mining area. On 3 August 2012, Nanfen open-pit iron mine was hit by heavy rain from typhoon "Dawei", which resulted in large-scale cracks, collapses, and landslides on the slope, as shown in Figure 1a,b. A large area that could potentially slide has been created, which seriously threatens the normal safe production of the mine.

(a)

(b)

Figure 1. Landslides and collapses caused by rainfall. (**a**) local landslide; and (**b**) overall collapse

Rainfall and slope sliding force in the mining area were monitored in real time to study the correlation between deformation and failure of the slope and rainfall in Nanfen open pit mine, as well as to predict the landslide trend. Statistical analysis was carried out on rainfall and significant changes in the slope sliding force at Nanfen open-pit mine, as shown in Table 1 and Figure 2. Changes in sliding force that are greater than or equal to 100 kN within 24 h are considered significant, and heavy rain is defined, as per the rainfall standard set by the Chinese meteorological department, as above 50 mm per day.

Table 1. Statistics on rainfall and sliding force changes in Nanfen open-pit iron mine (From November 2010 to November 2012).

Month	1	2	3	4	5	6	7	8	9	10	11	12
Days of rain	0	0	0	18	11	38	30	37	25	26	3	0
Days of heavy rain	0	0	0	0	2	6	8	9	9	4	0	0
Number of significant changes in sliding force	1	0	0	0	0	4	6	15	2	2	0	0

The statistical data show that rainfall often occurs before a period of significant change in sliding force. Most such changes occur in a month when there is frequent rainfall, especially in months when there is frequent heavy rainfall, which shows that rainfall is an important factor for slope instability. Rainfall will induce landslides and change the physical and mechanical parameters of slope rock and soil. Due to the softening effect of rainwater, the rock and soil will become saturated, the gravity density will increase, and the shear strength and deformation modulus will decrease [26]. The continuous interaction of water and rock will cause cracks in the rock mass to expand and penetrate, eventually leading to rock destruction and slope instability. Sampling was mainly concentrated in chlorite on 478 m steps on the footwall of the slope. A portable knapsack type core drilling rig was used for shallow sampling of complete rock samples. This device is simple and convenient to carry, flexible, and highly efficient. After drilling and taking samples at the site, the samples were sealed in wax and put into a portable sample collection box for storage and transportation. The processed rock samples are cylinders with a diameter of 50 mm and a height of 100 mm (± 1 mm). Rock samples with relatively similar wave velocities were identified non-metallic ultrasonic detectors and screened out as experimental rock samples.

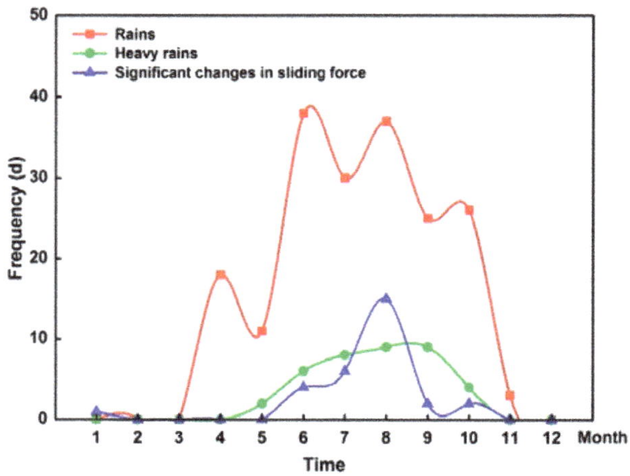

Figure 2. Frequencies of significant changes in sliding force, rain, and heavy rain during two years.

2.2. Experimental Design of Dry and Wet Cycling

Dry-wet cycles were simulated in the laboratory to study their effect on the mechanical properties and microstructure of rock. Taking into account the operability of the dry-wet cycle experiment and the actual situation in the field, the "water saturation-natural air drying" method was implemented for each cycle to avoid the effect that the high temperature used in the oven drying method would have on rock properties and microstructure. The rock samples were divided into 6 groups, 3 in each group, and were subjected to 0 cycles (representing the natural state), 1 cycle, 3 cycles, 6 cycles, 10 cycles, and 15 cycles, respectively. A uniaxial compression test, acoustic wave test, and scanning electron microscope test was carried out on each rock sample. Rock samples were saturated by placing the sample in water, controlling the water level at the 1/4 L, 1/2 L, 3/4 L, and 1 L planes of the sample height L in sequence and soaking them for 2 h at each of these levels. The sample was then completely immersed in water and left to absorb it for 48 h before being taken out. Finally, the surface water was gently absorbed with tissue paper, and the sample was weighed. In order to maintain the uniformity of the water absorption process, the height between the upper surface of the rock sample and the water surface was controlled to be 10 cm in each complete soaking. Rock sample drying was carried out by natural air drying, placing the rock sample in a windless room at 26 degrees Celsius for more than 7 d until the mass became constant.

2.3. Uniaxial Compression Tests

The purpose of uniaxial compression testing of rock is to determine its uniaxial compressive strength, elastic modulus, Poisson's ratio and other parameters of rock [27]. The uniaxial compression experiment was carried out on an XTR01 microcomputer servo-controlled rock compression testing machine (Changchun, China), it is a stiff testing machine specially used for testing materials such as rocks and concrete. This device can automatically record the applied load and the axial and radial deformation values of the rock during the experiment prior to destruction. The stress-strain curve of the rock can be drawn on the basis of the collected data, from which the uniaxial compressive strength of the rock can be obtained.

2.4. Acoustic Velocity Tests

All rock masses in the natural environment contain various joints, fissures, and other structural planes. The deformation of these structural planes in the rock mass under the action of external forces will cause reflection, refraction, diffraction, scattering, and other phenomena during elastic wave penetration, which will affect the propagation path of elastic waves in the rock mass [28–30]. Acoustic wave detection technology analyzes the structural characteristics of a rock mass by using the wave characteristics of elastic waves that pass through it [31]. The acoustic wave velocity test adopted by this study uses an MC-6310 nonmetal ultrasonic detector (Beijing, China). The experimental process is shown in Figure 3a,b. The device is used to excite elastic waves of a certain frequency inside the rock. Elastic waves propagate inside the rock in the form of transverse waves and longitudinal waves, which are transmitted and received by probes at both ends. By analyzing the recorded wave signals, the mechanical properties of the rock and its internal defects are determined. The relationship between the wave velocity value for the rock and the number of dry-wet cycles was analyzed by testing the longitudinal and transverse wave velocities in the rock samples after different numbers of water saturation cycles.

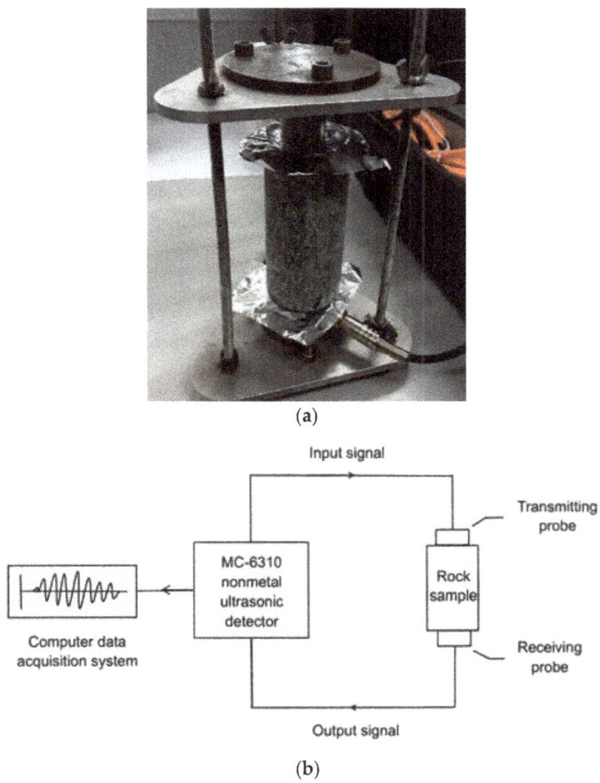

Figure 3. Acoustic wave testing. (**a**) experimental process of wave velocity test in rock; and (**b**) schematic diagram of the acoustic wave test.

2.5. Scanning Electron Microscope Test

The instrument used for scanning electron microscope testing was the Hitachi SU8010 electron microscope made in Tokyo, Japan. The rock sample was prepared as a test block 1 $cm^2 \times 0.5$ cm in size. After pre-treatments such as grinding and drying, the experimental sample was bonded to the sample pile with conductive adhesive, and a layer of gold film was plated to the surface of the sample. It was then put into the sample room for scanning, the scanning result was observed, and a picture was extracted [32,33].

3. Results and Analysis

In order to conveniently illustrate changes in the mass of rock during the dry-wet cycling process, the change in the mass of one rock sample in the first wet-dry cycle is plotted in Figure 4. The four stages a, b, c, and d, in the figure correspond to rock been soaked in water at water levels of 1/4 L, 1/2 L, 3/4 L, and 1 L, respectively. These four stages lasted for 8 h in total, and the rock mass increased from 564.4 g to 565.146 g. After 56 h of soaking, the final saturated rock mass was 565.28 g, the water absorption was 0.755 g, and the ratio of absorbed water to rock was 0.134%. In the process of air-drying, the mass of the rock was measured after 5 d (i.e., after the 176th hour) and 7 d (i.e., after the 224th hour), and the values were found to be equal, indicating that air-drying for 7 d can guarantee the complete drying of the rock. The mass of the dehydrated rock is 564.28 g, which is 0.12 g lower than the initial state; the mass loss ratio is 0.024%. The mass loss ratio of rock is defined as the ratio of the lost mass of rock to the initial mass of rock. The loss curve of rock mass up to 15 cycles is plotted in Figure 5.

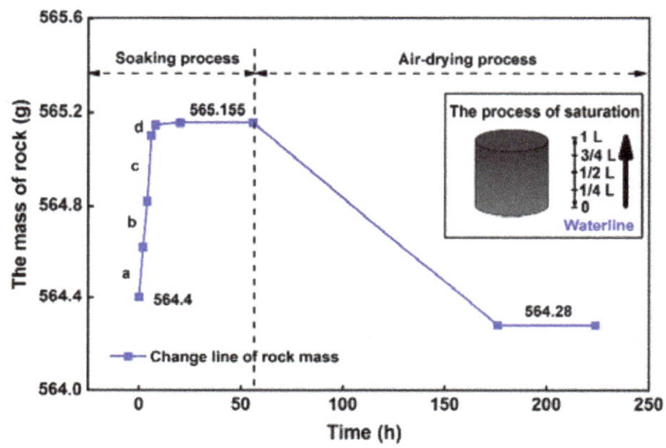

Figure 4. Mass change of one of the rock samples after the first wet-dry cycle.

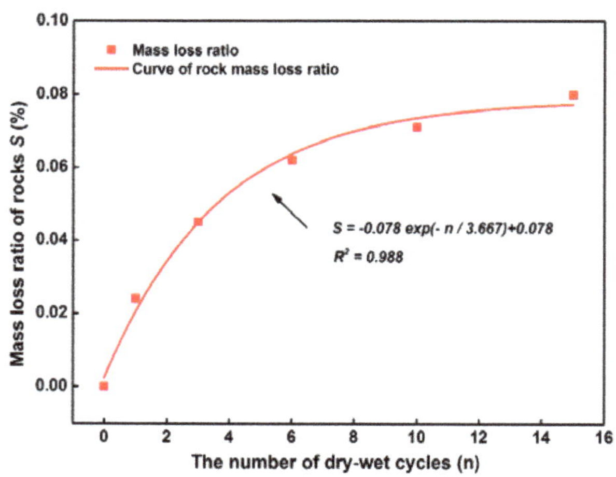

Figure 5. The relationship between rock mass loss ratio and the number of dry-wet cycles.

As can be seen from Figure 5, the mass loss ratio of the rock increases with an increase in the number of dry-wet cycles. After the first three wet-dry cycles, the mass loss ratio of rock is 0.045%, while after 15 cycles, the mass loss ratio increased to 0.08%. This increase is more rapid in the early stages and decelerates in the later stages. According to fitting analysis conducted with Origin software, the two variables are exponentially correlated with the fitting equation:

$$S = -0.078\,exp(-n/3.667) + 0.078,\ R^2 = 0.988 \tag{1}$$

where S represents the mass loss ratio of the rock (%), that is, the ratio of the lost mass of rock to the initial mass, n is the number of dry and wet cycles, and R^2 represents the correlation coefficient between the mass loss ratio of the rock and the number of dry-wet cycles.

Chlorite is an aluminosilicate mineral, and some of the rock debris is physically and chemically unstable, making the rock easily mechanically eroded, transported, and altered in a water-bearing environment. Therefore, from a microscopic point of view, the damage to the rock mass is mainly caused by the migration and diffusion of debris and the dissolution of feldspar and other particles under the influence of repeated wet-dry cycles. This changes the physical structure, creating secondary pores. One of the main macroscopic manifestations of this phenomenon is the mass loss of the rock samples after dry-wet cycling.

3.1. Analysis of Uniaxial Compression Test Results

The uniaxial compression tests allowed stress-strain curves for rock samples subjected to different numbers of dry-wet cycles to be obtained, as well as the uniaxial compressive strength, elastic modulus and other parameters of the rocks. In order to explore the influence of dry-wet cycles on rock strength and deformation, strength loss percentage was defined as the degree of attenuation of uniaxial compressive strength under dry-wet cycling, i.e., $(\sigma_{t0} - \sigma_{ti})/\sigma_{t0} \times 100\%$, and the elastic modulus loss percentage was defined as the degree of attenuation of the elastic modulus under dry-wet cycles, i.e., $(E_0 - E_i)/E_0 \times 100\%$. The statistical results for the average uniaxial compressive strength and elastic modulus of chlorite and their degree of attenuation after different numbers of dry-wet cycles are shown in Table 2.

Table 2. Statistics of experimental results for rock uniaxial compression.

Cycles	Uniaxial Compressive Strength σ_t		Elastic Modulus E	
	Average Value (MPa)	Attenuation Degree D_1 (%)	Average Value (GPa)	Attenuation Degree D_2 (%)
0	102.85	0	31.12	0
1	96.38	6.29	29.83	4.15
3	89.79	12.70	27.13	12.82
6	80.84	21.40	24.86	20.12
10	73.28	28.75	20.94	32.71
15	67.66	34.21	17.23	44.63

It can be seen from Table 2 that the average uniaxial compressive strength and average elastic modulus of the chlorite decrease with an increase in the number of dry-wet cycles. The cumulative degrees of attenuation of the average uniaxial compressive strength and average elastic modulus reach 34.21% and 44.63%, respectively, after 15 dry-wet cycles. After more than 10 cycles, the degree of attenuation of the average elastic modulus of the chlorite gradually exceeds the degree of attenuation of the uniaxial compressive strength, indicating that there is more significant degradation of the average elastic modulus of rock than of its uniaxial compressive strength under the action of multiple dry-wet cycles. Curves for the average uniaxial compressive strength and elastic modulus of the rocks versus the number of dry-wet cycles, together with their degree of attenuation, are given in Figures 6 and 7, respectively.

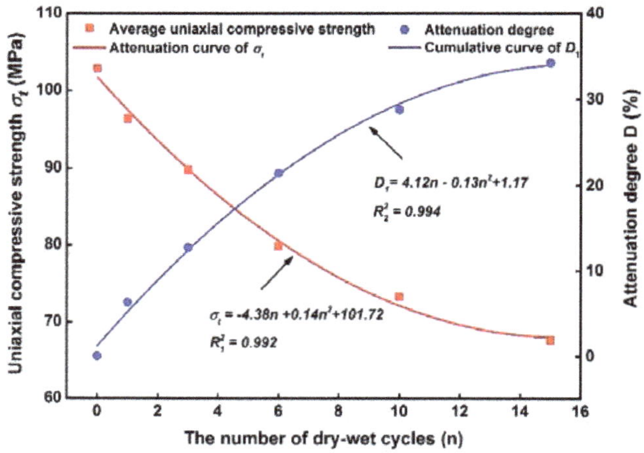

Figure 6. The changes of average uniaxial compressive strength of rock and its degree of attenuation with the number of dry-wet cycles.

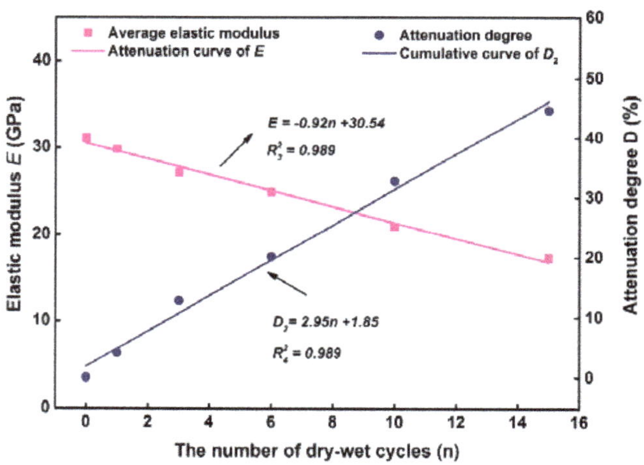

Figure 7. The changes of average elastic modulus of rock and its degree of attenuation with the number of dry-wet cycles.

As can be seen from Figure 6, the average uniaxial compressive strength of the chlorite gradually decreases with an increase in the number of dry-wet cycles, from 102.85 MPa in the initial state to 67.66 MPa after 15 dry-wet cycles. The decrease is most quickly in the early stages and more slowly in the later stages. Correspondingly, the degree of attenuation of average uniaxial compressive strength gradually increases with an increase in the number of dry-wet cycles, from 0% in the dry state to 34.21% after 15 dry-wet cycles. The attenuation trend in the average elastic modulus of the rock shows a similar variation trend, as shown in Figure 7, which is basically uniform and approximately linear. After 15 cycles, the average elastic modulus of the rock is about 45% lower than it was initially, from 31.12 GPa to 17.23 GPa. In order to further quantitatively analyze the softening effect of dry-wet cycling on rock, a fit to the data in the figure was obtained using the function provided by Origin software. The relationship between the average uniaxial compressive strength of rock and the number of dry-wet cycles is found to be well-expressed as a good polynomial function, with the fitting equation

$$\sigma_t = -4.38n + 0.14n^2 + 101.72,\ R_1^2 = 0.992 \tag{2}$$

where σ_t is the uniaxial compressive strength of the rock (MPa), n is the number of dry-wet cycles, and R_1^2 represents the correlation coefficient between the uniaxial compressive strength of the rock and number of dry-wet cycles.

Fitting analysis with Origin software indicates that the average elastic modulus of the rock and the number of dry-wet cycles have a strong linear relationship, with the fitting equation

$$E = -0.92n + 30.54,\ R_3^2 = 0.989 \tag{3}$$

where E is the elastic modulus of the rock (GPa), and R_3^2 represents the correlation coefficient between the uniaxial compressive strength of the rock and the number of dry-wet cycles.

Thus, the uniaxial compressive strength and elastic modulus of the rock decreases with an increase in the number of dry-wet cycles, which shows that dry-wet cycling has a substantial damaging effect on the rock's mechanical strength. The characteristics of the rock itself and the influence of different environments and stress conditions lead to different rock masses showing different failure modes. Figure 8 presents images of chlorite samples from Nanfen destroyed by uniaxial compression tests after different numbers of dry-wet cycles. From left to right, the rock samples in the images were subjected to 0, 1, 3, 6, 10, and 15 dry-wet cycles.

Figure 8. Images of failure in rock samples after different numbers of dry-wet cycles.

Rock failure characteristics are affected by numerous factors such as mineral composition, distribution of structural planes, and the loading and stress conditions in the experimental process. It was clearly seen during the uniaxial compression tests that the post-loading failure surfaces of rocks in dry state and after one dry-wet cycle ran through the entire longitudinal axis of the rock sample. The failures were accompanied by ringing noise characteristics of brittle, tensional failure. With an increase in the number of dry-wet cycles, the characteristics of the brittle failure transitioned from strong to weak. Fissures developed that ran through the weak surfaces of the rock, and friction-induced pulverization appeared at the failure surface, a characteristic of weak surface shear failure. The sound emitted by rock failure became more complex and quieter. The failure characteristics of rocks gradually changed from brittle failure to shear failure with an increase in the number of dry-wet cycles, indicating that dry-wet cycles have a significant impact on the failure characteristics of chlorites.

3.2. Experimental Analysis of Acoustic Wave Velocity Tests

Acoustic velocity tests, including an axial transverse wave test and longitudinal wave test, were conducted the rock samples with an MC-6310 nonmetal ultrasonic detector (Beijing, China) after different numbers of dry-cycles tests to determine the degree of microscopic damage in the rocks. The experimental results are shown in Table 3, and the relationship between the acoustic wave velocities in the rocks and the number of dry-wet cycles is shown in Figure 9.

Table 3. Experimental results of acoustic wave velocity tests.

Cycles	Longitudinal Wave Velocities (m/s)	Absolute Reduction (m/s)	Relative Reduction (%)	Transverse Wave Velocities (m/s)	Absolute Reduction (m/s)	Relative Reduction (%)
0	4358	0	0	2812	0	0
1	4346	12	0.28	2768	44	1.56
3	4298	60	1.34	2706	106	3.77
6	4175	183	4.20	2623	189	6.72
10	3963	395	9.06	2524	288	10.24
15	3752	606	13.91	2446	366	13.02

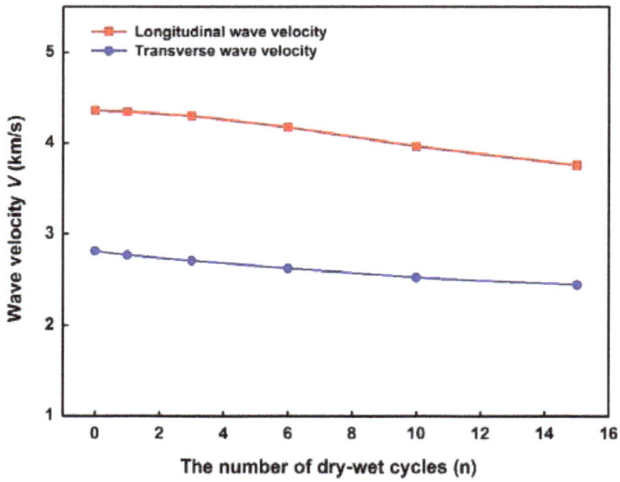

Figure 9. The relationship between the average acoustic velocity in the rocks and the number of dry-wet cycles.

Acoustic wave velocity testing is an effective method for studying the mechanical properties and structural compactness of rock. As can be seen from Table 3, the propagation speed of acoustic waves in solids depends on the direction of particle vibration. The propagation speed of longitudinal waves in rocks is about 1.55 times greater than that of transverse waves. The curves of longitudinal wave velocity and transverse wave velocity versus the number of dry-wet cycles in Figure 9 allow it to be seen intuitively that both longitudinal wave velocity and transverse wave velocity have an attenuation trend with an increase in the number of periodic dry-wet cycles. From the 0 to the 15th cycle, the longitudinal wave velocity decreased by 606 m/s, a decrease of 13.91%, and the transverse wave velocity decreased by 366 m/s, a decrease of 13.02%. The propagation speed of acoustic waves in rock will be greatly affected by the integrity and pore density of the rocks. Chlorite is a hard rock, with high strength and poor water absorption. Acoustic waves propagation through rocks diffract when fissures and pores are encountered, resulting in an increase in the propagation distance and a decrease in the wave velocity. With an increase in the number of dry-wet cycles, the cracks gradually develop and expand inside the rock mass and many new micro-cracks will be formed. This gradual increase in the porosity of the rock explains the decrease in wave velocity. Therefore, the axial wave velocity of rock samples can be used as effective indicators of crack growth inside rock, and the rate of decrease in the acoustic wave velocity can be used to characterize differences in the degree of damage between rocks.

3.3. Analysis of Scanning Electron Microscope (SEM) Images

After obtaining scanning electron microscope images of the rocks after different numbers of dry-wet cycles, an initial observation of the images was carried out with the naked eye. In order to compare the influence of different numbers of dry-wet cycles on the pore structures of the rocks, a consistent magnification was used to generate the scanning electron microscope images. 100×-magnified SEM images of chlorite that have undergone different numbers of cycles are shown in Figure 10. After several dry-wet cycles, there are obvious changes to the characteristics shown in the SEM images and the structure of the rock sample.

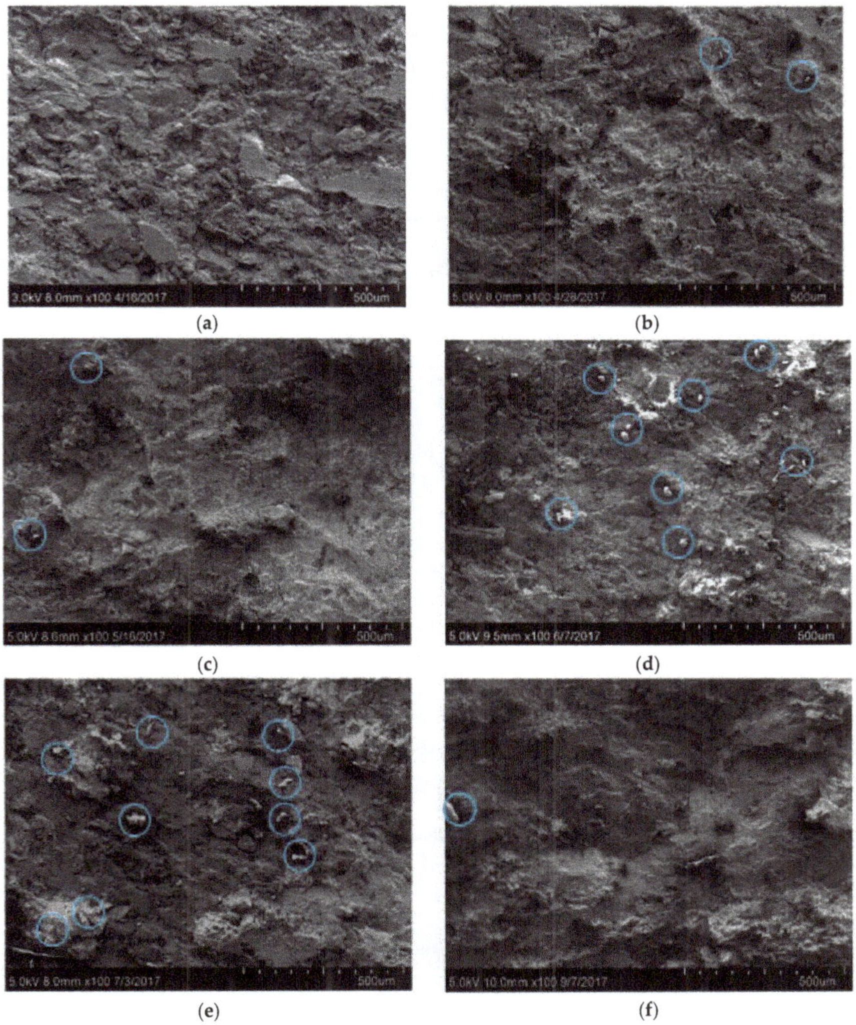

Figure 10. SEM images of rocks after different numbers of dry-wet cycles. (**a**) zero cycles; (**b**) one cycle; (**c**) three cycles; (**d**) six cycles; (**e**) ten cycles; and (**f**) fifteen cycles.

An increase in the number of dry-wet cycles leads to gradual damage accumulation in the rocks. The pore types of chlorites are mainly mineral cleavage crevices and intergranular micropores between clay mineral. Chlorites have a much higher degree of water absorbency than rocks such as sandstone and mudstone. As can be seen from Figure 10, without the action of dry-wet cycles, the rock particles are clear in outline and evenly distributed, and the microstructure is homogeneous and compact. The particles are mainly flat and massive; the macro-mechanical properties of the rocks are also good at this time. After several dry-wet cycles, the microstructures of the rocks are no longer compact and flat, the particles begin to become disordered, and some flaky, curling particles appear. Crystal precipitation also occurs on the surface of the rock (in blue circles in the Figure 10) and gradually increases with an increase in the number of dry-wet-cycle times. After fifteen cycles, fine mud particles have accumulated at the surfaces of the rocks due to water-rock interaction, making them uniform and smooth again. The microscopic structure and morphology of the rock have changed completely, as compared with the chlorites in a dry state, at this time.

Based on the results of the rock mechanics experiments and analyses of microscopic SEM images, we can conclude that water molecules penetrate into the interior of the rock it is soaked in water, water-soluble particles are dissolved by the water, and fine cuttings are stripped off, leaving behind pore fissures. The expansion of pores and fissures creates conditions that are more conducive to the penetration of water molecules into the rock and reduces the cohesion and binding force between rock particles. The microscopic damage to the rock develops continuously, which is exhibited macroscopically in a decrease in mechanical parameters such as the compressive strength, elastic modulus, and acoustic wave velocity. The accumulation of physical, chemical and mechanical damage caused by rock-water interaction is the main reason that slope stability is affected and the deformation and destruction of engineering rock mass occur. Furthermore, rainfall and evaporation and the rise and fall of the groundwater level will subject the rock mass of a slope to dry-wet cycles for a long time, which will accelerate the accumulation of damage.

4. Prevention Measures of Slope Instability

Rainfall is an important cause of hazardous landslides in Nanfen open-pit iron mine slopes. The effect of rainfall can be mitigated by a variety of slope stability control measures including surface drainage engineering, underground drainage engineering, back pressure engineering, weight reduction engineering, and river engineering [34–36]. For landslides with relatively fixed boundaries and a small traction range, permanent interception, and drainage to outside the landslide range can be put in place. The type of surface drainage installed within the landslide range will depend on the specific situation. When the landslide is already in an unstable state, the drainage system should be designed to be temporary, with less investment made and with an emphasis on timely maintenance. Interception ditches and drains are also very effective for intercepting and draining water at the surface. The results of the current study confirm that if effective measures are not taken to drain fissure water on the slope, the strength of the soil and rock mass will be reduced, a landslide will develop more quickly, and the safety and sustainable mining of the open-pit mine will be seriously threatened. One of the ultimate goals of the slope drainage project at Nanfen open-pit iron mine is to discharge surface water and groundwater from the mining site to the region outside the stope. The terrain and rock structure of the stope mean that the drainage pipe can be connected to the drainage interception ditch through a sand basin to form a complete water defense system, as shown in Figure 11.

Figure 11. Pipe draining of rainwater and groundwater away from Nanfen open-pit mine slope.

Deep drainage holes enable the effective and timely discharge of fissure water in the deep rock mass and play a directly beneficial role in maintaining the stability of the slope. When there is no drainage system, rainwater will penetrate into the slope along cracks and form runoff at the surface, which will seriously affect the stability of the local slope. In addition, because the dangerous slope at Nanfen open-pit mine is about 400 m in length, 432 m in height, and 172,800 m^2 in area, joints and fissures are developed, and geological conditions are complex, rainfall and the rise and fall of groundwater level have a major impact on the slope. On the basis of extensive research, our organization has installed sliding force monitoring points and rainfall monitoring points on the footwall slope of Nanfen open-pit iron mine to enable 24-h intelligent monitoring of the hydrogeological environment. This monitoring method allows the comprehensive analysis of slope stability and prediction of slope evolution [37,38]. The installation of drainage system, monitoring, and an early warning system for rainfall and the sliding force of the slope are of great significance for forecasting and analyzing rainfall-induced landslides.

5. Conclusions

Based on the engineering background of Nanfen open-pit iron mine in Liaoning province, this study investigated the mechanical properties of rocks subjected to water-rock circulation through macroscopic mechanical experiments, microstructure tests, and theoretical analysis. Periodic water circulation due to rainfall, evaporation and the fluctuation of groundwater level was simulated in the laboratory, and valuable data were obtained regarding its effect on the mechanical property parameters of these Nanfen chlorite rocks.

Dry-wet cycling is found to accelerate the generation and expansion of cracks in the rock, resulting in a decrease in rock strength. The uniaxial compressive strength of the rock decreases with an increase in the number of dry-wet cycles, showing a large decrease in the early stages and a more gradual decrease with continued cycling. This change trend follows the polynomial function "$\sigma_t = -4.38n + 0.14n^2 + 101.72$". However, there is a linear relationship between the attenuation of the elastic modulus of the rock and the number of dry-wet cycles. Brittle failure in the rock shows a transition from exhibiting the characteristics of strong rock to those of weak rock with an increase in the number of cycles. Acoustic wave testing shows that both longitudinal wave velocity and transverse wave velocity are attenuated to a certain extent under the influence of water-rock interaction. The attenuation rates of longitudinal wave velocity and transverse wave velocity are 13.91% and 13.02%, respectively. The two have a very similar attenuation trend, but the absolute attenuation of longitudinal wave velocity is much greater than that of transverse wave velocity. In the early stages of water absorption, the chlorite itself is hard, strong, and has poor water absorbency. The interaction between water and rock causes cracks in the rock mass to gradually form and expand. This increase in porosity leads to the decrease in wave velocity.

Based on the theory of rock damage mechanics, the microstructure of rock samples was examined by SEM after different numbers of dry-wet cycles. The images show that the rock surface was initially compact and flat, with clear particle outlines. With an increase in the number of dry-wet cycles, crystalline minerals continuously precipitated onto the rock surface and gradually became cohesive. Until after the 15th cycle, the microstructure and morphology of the rock had completely changed, showing a muddy state. The surface was leveled again.

To sum up, there is cumulative physical, chemical, and mechanical damage to the rock mass in Nanfen open-pit iron mine due to rainfall and groundwater. The stability of the rock mass will be gradually weakened over a long period of time, inevitably leading to the deformation and destruction of the mining area slope. Feasible drainage measures should be taken, and early warning and monitoring of slope stability should be strengthened to ensure that mines are mined safely and that lives and property of the workers are protected. It is hoped that further numerical simulation can be conducted and more rock samples can be selected for comparative analysis in the future to further quantify the influence of dry-wet cycling on the mechanical properties of slope rock.

Author Contributions: X.Y. and J.W. designed and performed the experiments, they contributed equally to this work. D.H. and C.Z. analyzed the data; M.H. contributed experimental materials and conditions; X.Y. provided funding; J.W. wrote and edited the paper.

Funding: This research was funded by the National Natural Science Foundation of China, grant number 41672347 and Natural Science Foundation of Beijing Municipality, grant number 8142032.

Acknowledgments: The authors would like to express sincere appreciation to the editors and reviewers for their time and effort to review this paper and provide valuable comments.

Conflicts of Interest: The authors declare no conflict of interest.

References

1. Su, Y.H.; Fang, Y.B.; Li, S.; Su, Y.; Li, X. A one-dimensional integral approach to calculating the failure probability of geotechnical engineering structures. *Comput. Geotech.* **2017**, *90*, 85–95. [CrossRef]
2. Wu, X.Z. Trivariate analysis of soil ranking-correlated characteristics and its application to probabilistic stability assessments in geotechnical engineering problems. *Soils Found.* **2013**, *53*, 540–556. [CrossRef]
3. Yang, J.; Tao, Z.G.; Li, B.L.; Yang, G.; Li, H.F. Stability assessment and feature analysis of slope in Nanfen Open Pit Iron Mine. *Int. J. Min. Sci. Technol.* **2012**, *22*, 329–333. [CrossRef]
4. Ma, C.Q.; Wang, P.; Jiang, L.S.; Wang, C.S. Deformation and control countermeasure of surrounding rocks for water-dripping roadway below a contiguous seam goaf. *Processes* **2018**, *6*, 77. [CrossRef]
5. He, M.C.; Sun, X.M.; Zhao, J. Advances in interaction mechanism of water (gas) on clay minerals in China. *Int. J. Min. Sci. Technol.* **2014**, *24*, 727–735. [CrossRef]
6. Wu, Y.Q.; Zhang, N.Y. *An Introduction to Rock Mass Hydraulics*; Southwest Jiaotong University Press: Chengdu, China, 1995.
7. Zhao, Z.H.; Guo, T.C.; Ning, Z.Y.; Dou, Z.H.; Dai, F.; Yang, Q. Numerical modeling of stability of fractured reservoir bank slopes subjected to water-rock interactions. *Rock Mech. Rock Eng.* **2018**, *51*, 2517–2531. [CrossRef]
8. Phan, T.T.; Vankeuren, A.N.P.; Hakala, J.A. Role of water-rock interaction in the geochemical evolution of Marcellus Shale produced waters. *Int. J. Coal Geol.* **2018**, *191*, 95–111. [CrossRef]
9. Jiang, J.W.; Xiang, W.; Rohn, J.; Zeng, W.; Schleler, M. Research on water–rock (soil) interaction by dynamic tracing method for Huangtupo landslide, Three Gorges Reservoir, PR China. *Environ. Earth Sci.* **2015**, *74*, 557–571. [CrossRef]
10. Zhang, X.M.; Tahmasebi, P. Micromechanical evaluation of rock and fluid interactions. *Int. J. Greenh. Gas Control* **2018**, *76*, 266–277. [CrossRef]
11. Zhu, Y.P. A micromechanics-based damage constitutive model of porous rocks. *Int. J. Rock Mech. Min. Sci.* **2017**, *91*, 1–6. [CrossRef]
12. Hao, R.Q.; Li, J.T.; Cao, P.; Liu, B.; Liao, J. Test of subcritical crack growth and fracture toughness under water-rock interaction in three types of rocks. *J. Cent. South Univ.* **2015**, *22*, 662–668. [CrossRef]
13. Wong, L.N.Y.; Maruvanchery, V.; Liu, G. Water effects on rock strength and stiffness degradation. *Acta Geotech.* **2016**, *11*, 713–737. [CrossRef]
14. Hashiba, K.; Fukui, K. Effect of water on the deformation and failure of rock in uniaxial tension. *Rock Mech. Rock Eng.* **2015**, *48*, 1751–1761. [CrossRef]
15. Doostmohammadi, R.; Moosavi, M.; Mutschler, T.; Osan, C. Influence of cyclic wetting and drying on swelling behavior of mudstone in south west of Iran. *Environ. Geol.* **2009**, *58*, 999. [CrossRef]
16. Zhao, Z.H.; Yang, J.; Zhang, D.F.; Peng, H. Effects of wetting and cyclic wetting–drying on tensile strength of sandstone with a low clay mineral content. *Rock Mech. Rock Eng.* **2017**, *50*, 485–491. [CrossRef]
17. Ma, C.Q.; Li, H.Z.; Niu, Y. Experimental study on damage failure mechanical characteristics and crack evolution of water-bearing surrounding rock. *Environ. Earth Sci.* **2018**, *77*, 23. [CrossRef]
18. Liu, D.Q.; Wang, Z.; Zhang, X.Y.; Wang, Y.; Zhang, X.L.; Li, D. Experimental investigation on the mechanical and acoustic emission characteristics of shale softened by water absorption. *J. Nat. Gas Sci. Eng.* **2018**, *50*, 301–308. [CrossRef]
19. Liu, Y.L.; Wang, L.G.; Sun, X.K.; Wang, J. Experimental study of the influence of water and temperature on the mechanical behavior of mudstone and sandstone. *Bull. Eng. Geol. Environ.* **2017**, *76*, 645–660. [CrossRef]

20. Zhu, M.T.; Dai, Y.P.; Zhang, L.C.; Wang, C.L.; Liu, L. Geochronology and geochemistry of the Nanfen iron deposit in the Anshan-Benxi area, North China Craton: Implications for 2.55 Ga crustal growth and the genesis of high-grade iron ores. *Precambrian Res.* **2015**, *260*, 23–38. [CrossRef]
21. Zhu, C.; Tao, Z.G.; Yang, S.; Zhao, S. V shaped gully method for controlling rockfall on high-steep slopes in China. *Bull. Eng. Geol. Environ.* **2018**. [CrossRef]
22. Yang, X.J.; Hou, D.G.; Hao, Z.L.; Tao, Z.G.; Shi, H.Y.; Jin, K. Research on correlations between slope stability and rainfall of high steep slope on Nanfen open-pit iron ore. *Chin. J. Rock Mech. Eng.* **2016**, *35*, 3232–3240. [CrossRef]
23. Yang, X.J.; Hou, D.G.; Wang, J.M.; Wang, J.X. Study on the stability and remote real-time monitoring for high steep slope in Nanfen open pit iron mine. *J. Min. Saf. Eng.* **2017**, *34*, 1000–1007. [CrossRef]
24. Vinoth, S.; Kumar, L.A.; Kumar, E. Slope stability monitoring by quantification and behavior of microseismic events in an opencast coal mine. *J. Geol. Soc. India* **2015**, *85*, 450–456. [CrossRef]
25. Dai, Y.P.; Zhu, Y.D.; Zhang, L.C.; Zhu, M.T. Meso- and Neoarchean Banded Iron Formations and Genesis of High-Grade Magnetite Ores in the Anshan-Benxi Area, North China Craton. *Econ. Geol.* **2017**, *112*, 1629–1651. [CrossRef]
26. Chueasamat, A.; Hori, T.; Saito, H.; Sato, T.; Kohgo, Y. Experimental tests of slope failure due to rainfalls using 1g physical slope models. *Soils Found.* **2018**, *58*, 290–305. [CrossRef]
27. Li, Y.Y.; Zhang, S.C.; Zhang, X. Classification and fractal characteristics of coal rock fragments under uniaxial cyclic loading conditions. *Arab. J. Geosci.* **2018**, *11*, 201. [CrossRef]
28. Lee, J.S.; Yoon, H.K. Characterization of rock weathering using elastic waves: A Laboratory-scale experimental study. *J. Appl. Geophys.* **2017**, *140*, 24–33. [CrossRef]
29. Ding, P.B.; Di, B.R.; Wang, D.; Wei, J.X.; Zeng, L.B. P- and S-wave velocity and anisotropy in saturated rocks with aligned cracks. *Wave Motion* **2018**, *81*, 1–14. [CrossRef]
30. Amalokwu, K.; Best, A.I.; Sothcott, J.; Chapman, M.; Minshull, T.; Li, X.Y. Water saturation effects on elastic wave attenuation in porous rocks with aligned fractures. *Geophys. J. Int.* **2014**, *197*, 943–947. [CrossRef]
31. Wang, X.; Wen, Z.; Jiang, Y.; Huang, H. Experimental study on mechanical and acoustic emission characteristics of rock-like material under non-uniformly distributed loads. *Rock Mech. Rock Eng.* **2018**, *51*, 729–745. [CrossRef]
32. Lagoeiro, L.; Goncalves, C.C. SEM observation of grain boundary structures in quartz-iron oxide rocks deformed at intermediate metamorphic conditions. *Ann. Braz. Acad. Sci.* **2011**, *83*, 875–889. [CrossRef]
33. Berdnikov, N.V.; Konovalova, N.S.; Zazulina, V.Y. Investigation of precious metal inclusions in highly carbonaceous rocks by the SEM and X-Ray spectrum analysis methods. *Russ. J. Pac. Geol.* **2010**, *4*, 164–170. [CrossRef]
34. Xu, W.Y.; Zhang, Q.; Zhang, J.C.; Wang, R.B.; Wang, R.K. Deformation and control engineering related to huge landslide on left bank of Xiluodu reservoir, South-West China. *Eur. J. Environ. Civ. Eng.* **2013**, *17*, 249–268. [CrossRef]
35. Conte, E.; Troncone, A. A performance-based method for the design of drainage trenches used to stabilize slopes. *Eng. Geol.* **2018**, *239*, 158–166. [CrossRef]
36. Li, J.M.; Huang, Y.L.; Qiao, M.; Chen, Z.W.; Song, T.Q.; Kong, G.Q.; Gao, H.D.; Guo, L. Effects of water soaked height on the deformation and crushing characteristics of loose gangue backfill material in solid backfill coal mining. *Processes* **2018**, *6*, 64. [CrossRef]
37. Yang, X.J.; Hou, D.G.; Hao, Z.L.; Wang, E.Y. Fuzzy comprehensive caused by underground mining subsidence and its monitoring. *Int. J. Environ. Pollut.* **2016**, *59*, 284–301. [CrossRef]
38. Tao, Z.G.; Meng, X.Z.; Ma, C.R.; Zhu, C.; He, M.C.; Wang, C.J.; Zhang, H.J. Analysis of wedge-shaped landslide mechanism and sliding force monitoring warning in Nanfen open pit iron mine. *J. China Coal Soc.* **2017**, *42*, 3149–3158. [CrossRef]

processes

MDPI

Article

Rheological Behavior and Modeling of a Crushed Sandstone-Mudstone Particle Mixture

Zhenfeng Qiu [1,2,*], Ting Cao [1,3,*], Yongsuo Li [3,*], Junjie Wang [1,2,*] and Yulong Chen [4,*]

1 Key Laboratory of Hydraulic and Waterway Engineering of the Ministry of Education, Chongqing Jiaotong University, Chongqing 400074, China
2 National Engineering Research Center for Inland Waterway Regulation, Chongqing Jiaotong University, Chongqing 400074, China
3 School of Civil Engineering, Hunan City University, Yiyang 413000, China
4 State Key Laboratory of Hydroscience and Engineering, Tsinghua University, Beijing 100084, China
* Correspondence: qiuzhenfeng3012@163.com (Z.Q.); forevermelodycao@sina.com (T.C.); liyongsuo@126.com (Y.L.); wangjunjiehhu@163.com (J.W.); chen_yl@tsinghua.edu.cn (Y.C.); Tel.: +86-023-6265-2714 (Z.Q.); +86-023-6290-2665 (T.C.); + 86-073-7462-8311(Y.L.); +86-023-6289-6924 (J.W.); +86-010-6278-1760 (Y.C.)

Received: 15 September 2018; Accepted: 12 October 2018; Published: 16 October 2018

Abstract: A mixture of sandstone and mudstone particles is often used as the main filling material for many agriculture-related and civil engineering projects, including rock-fill dams and foundations. The long-term rheological deformation experienced by rock-fill dams and foundations filled with this mixture is much bigger than that of coarse-grained and cohesive soils, due to the deterioration and softening of the mudstone particles. This study focuses on the rheological deformation of a sandstone-mudstone particle mixture, prepared by mixing sandstone and mudstone particles, based on the content by weight of four mudstone particle types. Confined uniaxial compression tests were performed to test the rheological deformation of 24 samples of the mixture, and a stress-strain curve was obtained for each test. On the basis of compression curves, the rheological process of the mixture was divided into four phases: linear, attenuation rheological, secondary attenuation rheological and stable phases. The three defining features of the curve, namely the rheological attenuation factors, attenuation rheology critical strain and limited rheological strain, were then determined and modeled. A segmented rheological model was then proposed, based on a modified attenuation rheological constitutive model for coarse-grained soil. The modelled results compared well with the experimental data, and the modelled compression-curve prediction was able to describe the two-stage attenuation rheology features (attenuation rheological and secondary attenuation rheological phases) of the sandstone-mudstone particle mixture.

Keywords: sandstone and mudstone particles; rheological deformation; segmented rheological model; rheological limit strain; rheological test

1. Introduction

Interbedded layers of sandstone and mudstone are widespread in Chongqing Province in the People's Republic of China (PRC). The interbedded deposit ranges from 2294–6440 m [1]. The sandstone and mudstone particles become mixed, and a sandstone and mudstone particle mixture (SMPM) is formed, during excavation of the sandstone-mudstone interbeds [2,3]. In southwest China, SMPM is often used as the main filling material in many filling operations such as rock-fill dams, ports and wharfs [4–6].

It is well established that mudstone particles are likely to disintegrate, soften or even become slime when they become saturated [4,5]. Mudstone particles are easier to break than sandstone particles.

The unique physical and mechanical properties of mudstone particles mean that the long-term deformation of a foundation filled with SMPM will be more extensive than that of coarse-grained soil without mudstone particles. Furthermore, this deformation can reduce the strength of SMPM filled in water-front reservoir banks [7,8]. This increased long-term deformation and reduction in strength of the filling material may affect the safety of structures. For example, the horizontal displacement at the top of some piles has exceeded 10 cm in the Huangqi Container Terminal, located in the Three Gorges Reservoir Region. In this case, the deformation seen in the SMPM has been caused by fluctuations in the water level. Another example is Lijiang Airport (Yunnan Province, PRC), where SMPM was used as the runway-foundation filling material. Deformation reached 45.4 cm after a certain period of operation. In addition, the non-uniformity of the foundation settlement resulted in the pavement cracking and serious cavities between the roadbed and pavement, which severely affected the operation of the airport.

Researchers have investigated the rheological properties of coarse-grained soil. With the development of meso-mechanics, granular matter theory and advanced test techniques, Jiang et al. [9] concluded from a triaxial rheological shear test that particle movement, rotation and breakage are the main reasons for the rheological deformation in coarse-grained soil. The particles and structure of the soil tend to be stable when the potential energy of all of the particles and the inter-particle contract force both reach their maximum. In addition, particle breakage during the rheological process of coarse-grained soils may affect deformation, stress distribution, strength, permeability, wetting deformation and creep deformation [10–15]. Some advanced techniques such as computed tomography scanning and magnetic resonance imaging have been used to investigate the rheological mechanisms of coarse-grained soil from a microscopic perspective [9].

Rheological compression tests can be used to investigate long-term deformation. The rheological characteristics of many soil types have been investigated, such as clay [16,17], soft-rock particles [18,19] and coarse-grained soils [15,20,21]. The rheological deformation of an expansive soil can be classified as being instantaneous, linear-elastic and non-linear visco-plastic deformations. According to an investigation by Yang et al. [22], the deformation of cement-improved expansive soil indicated that no plastic deformation occurred before it yielded. They classified the observed yielded plastic creep as being either linear or non-linear visco-plastic. Zhang et al. [21,23] determined that the rheological curve of a coarse-grained material could only be classified into three phases: linear, attenuation rheological and stable. However, the rheological properties of SMPM, coarse-grained soils, clay and other soils differ greatly because of the softening of mudstone particles.

This study focuses on the influence of the mudstone particle content of SMPM on its rheological properties. Rheological compression tests were performed, and on the basis of the compression curves of each sample, the three defining features of the curve (rheological attenuation factors, attenuation rheology critical strain and limited rheological strain) were determined and analyzed. In addition, a segmented rheological model was developed by modifying an attenuation rheological constitutive model for coarse-grained soil.

2. Test Materials and Testing Methods

2.1. Tested Materials

Crushed SMPM was used as the test material. The sandstone and mudstone blocks for the mixture were both collected from the field in the Three Gorges Reservoir Region; specifically, these were lightly weathered rocks from the Jurassic Shaximiao Formation, formed in the Middle Jurassic period [1]. The uniaxial compression strength of the blocks, as determined in the laboratory, were 60.0–72.2 MPa for the sandstone and 8.3–25.8 MPa for the mudstone [3]. The large sandstone and mudstone blocks were artificially crushed into small particles with sizes of less than 5 mm. The small sandstone and mudstone particles were separated by a screen test to obtain sandstone and mudstone particle groups with particle sizes ranges of 5–2, 2–1, 1–0.5, 0.5–0.25, 0.25–0.075 and 0.075–0 mm.

The dry density and moisture content of the samples were 1830 kg/m^3 and 8%, respectively. The particle grading curve is shown in Figure 1; the maximum particle size was 5 mm. The mean particle size of the particle grading curve, D50, was 0.83 mm, and the non-uniformity and curvature coefficients were 25.56 and 1.16, respectively. The particle contents are listed in Table 1, in terms of the various particles size fractions.

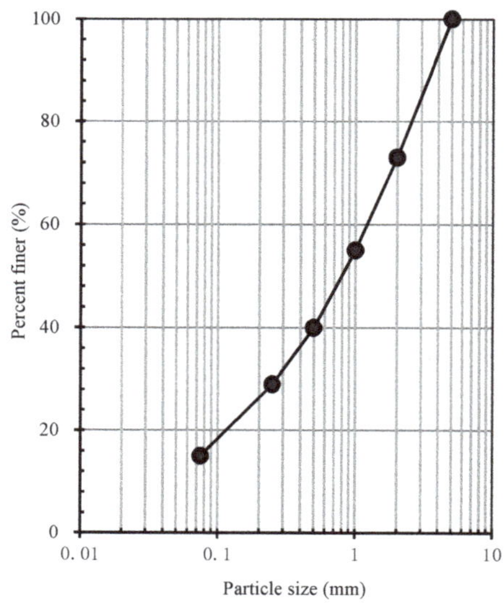

Figure 1. Grain size distribution curves of tested materials.

Table 1. Content of particles of various particle sizes of the sample.

Particle size (mm)	5–2	2–1	1–0.5	0.5–0.25	0.25–0.075	<0.075
Content (%)	27	18	15	11	14	15

2.2. Confined Uniaxial Compression Tests

Confined uniaxial compression tests were performed to investigate the rheological deformation of SMPM. The compression-test instrument consisted of an axial loading system, sample container, temperature sensor, axial-deformation sensor and an automatic test data-acquisition system. The axial deformation of the test instrument ranged between 0 and 15 mm. This test instrument is advantageous in that it can measure deformation to a high precision while acquiring loading and deformation data during the test. The deformation-acquisition precision was 0.001 mm. Axial pressure was introduced using the lever and weights, with a maximum axial pressure of 2.4 MPa. The cylindrical sample was 7 mm high, with a diameter of 100 mm; it was contained in a sample cylinder. Figure 2 shows that the test instrument was equipped with an instrument base, a cover, a loading board and a sample cylinder. At the base, a rigid annular porous stone and a round ceramic plate for drainage were installed. The porous stone and ceramic plate were connected to an outlet valve. O-rubber rings were installed within the base and cover to prevent air and water leakage. The rheological compression test was performed with drainage to investigate the effects of the mudstone particle content and axial pressure on the rheological deformation.

Instrument base   Cover and loading board   Sample cylinder

Figure 2. Test instrument.

2.3. Testing Method

A total number of 24 SMPM samples was prepared, which were then divided into four groups (six samples in each group). The mudstone contents in the four groups were 20%, 40%, 60% and 80%. A compressive rheological test was conducted on each group at six different axial pressure values: 100, 200, 400, 800, 1200 and 1600 kPa. The compressive rheological tests were performed according to the Trade Standard of PRC SL237-015 [24]. Each test was carried out as the follows:

(1) The sandstone and mudstone particles were weighed and poured into a bowl according to their density, the particle-grading curve and the mudstone particle content. These particles were then uniformly mixed. Next, distilled water (the sample moisture content was 8%) was weighed and evenly sprayed onto the sample at three different times, during which the particles were mixed continuously until the moisture was thoroughly integrated into the particles, with a consistent color. Subsequently, the materials were sealed to avoid water evaporation and were left standing for 24 h to ensure thorough mixing of the particles and water.

(2) The prepared materials were divided into three equal parts. Each part was poured into the compaction apparatus for compaction. Each compacted layer was roughened to ensure good contact between the particles at the compaction interfaces. Filter papers and porous stones were orderly placed at the top of every sample.

(3) The sample cylinder was then cleaned. Subsequently, lubricating oil was applied to the base and inner wall of the cylinder using a banister brush. The sample was carefully pushed into the sample cylinder. From the bottom to the top of the sample cylinder, the arrangement was as follows: base (with porous stones), filter paper, sample, porous stone and pressure cover.

(4) The lever position was adjusted to align its axis with that of the counter weight, following which, the lever height was fixed, and the displacement sensor was mounted. A pressure of 1 kPa was introduced for pre-compaction for approximately 10 min, at which point, the axial deformation gauge was reset.

(5) The predetermined axial pressure in the testing scheme was introduced in a stepwise fashion, using six sequential vertical pressure values: 50, 100, 200, 400, 800 and 1200 kPa. Each pressure was applied for 60 min. The sample was allowed to drain freely during the entire testing process.

(6) The sample was considered to be stable when the axial deformation rate was less than 0.005 mm/h. The compressive rheological test was then conducted for 4.5 days after the predetermined axial pressure was applied. The axial stress and deformation during the whole test were recorded per hour.

3. Test Results

3.1. Results of the Rheological Tests

Compressive rheological tests were carried out under different vertical stresses and for different samples to investigate the effects of vertical stress and mudstone particle content on the rheological characteristic of SMPM. The resulting compressive rheological curves under different axial pressure conditions are shown in Figure 3. The test data were recorded per hour. In the figures, the data were selected for every 2 h in the first 10 h and every 6 h in the follows processes. The rheological axial strain of SMPM initially increased before stabilizing. The final rheological limit axial strain ranged from 0.027–0.290% (mudstone particle content of 20%; vertical pressure values ranging from 100–1600 kPa). The rheological limit axial strain increased with increasing vertical pressure and increasing mudstone particle content. When the vertical pressure was 1600 kPa, the rheological limit axial strains were 0.290%, 0.337%, 0.347% and 0.470% for mudstone particle contents of 20%, 40%, 60% and 80%, respectively.

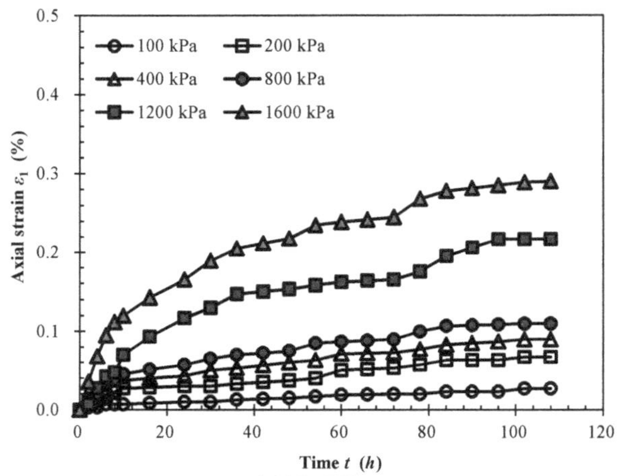

(**a**) M = 20%

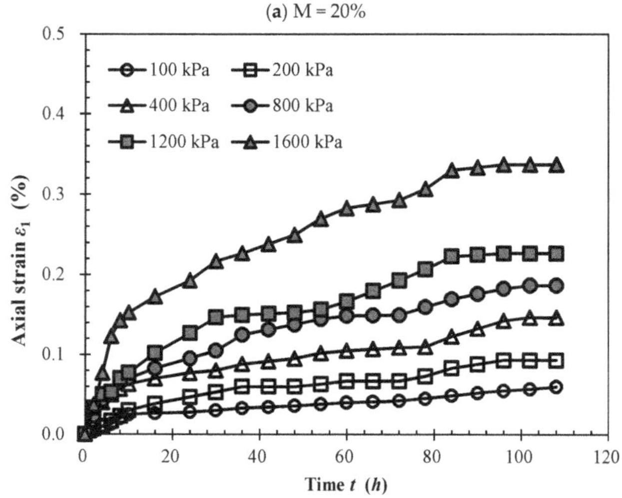

(**b**) M = 40%

Figure 3. *Cont.*

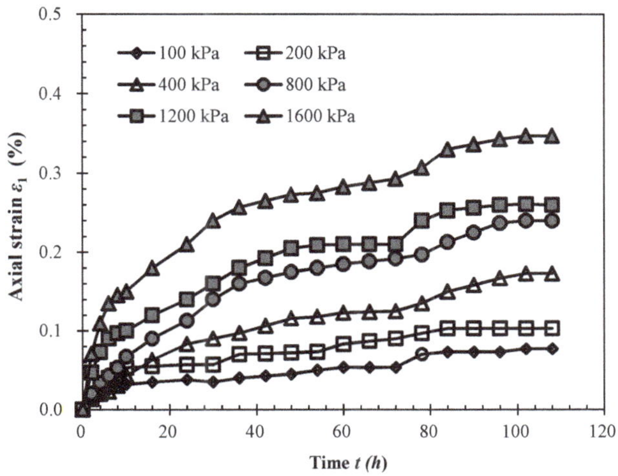

(c) M = 60%

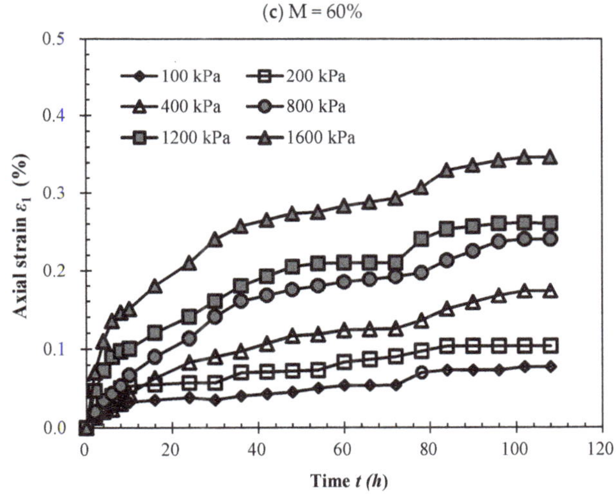

(d) M = 80%

Figure 3. Compressive rheological curves of sandstone and mudstone particle mixture (SMPM) samples; M = mudstone particle content.

Shortly after loading, the axial strain was found to increase in an approximately linear fashion. This period could therefore be considered as the linear rheological phase. The slope of the increasing line was related to both the axial pressure and the mudstone particle content. The final axial strain of the linear rheological phase was 0.057%, 0.065%, 0.080%, 0.120%, 0.157% and 0.250% for vertical pressure values of 100, 200, 400, 800, 1200 and 1600 kPa, respectively (mudstone particle content of 80%). When the vertical pressure was 1600 kPa, the final axial strain values of the linear phase were 0.112%, 0.143%, 0.146% and 0.250% for mudstone particle contents of 20%, 40%, 60% and 80%, respectively.

3.2. Four-Phase Rheological Process

Most of the rheological curve variations of the axial strain and time took the form of the stylized curve shown in Figure 4. The entire process for these tests could be divided into four phases: the linear rheological phase, the attenuation rheological phase, the secondary attenuation rheological phase and the stable phase. The largest difference in the rheological curves between these SMPM phases and those of coarse-grained rock-fill materials was that the attenuation phase for SMPM was followed by a secondary attenuation phase. This was the result of an increase in rheological deformation, which might have been caused by the different crushing behaviors of the sandstone and mudstone particles.

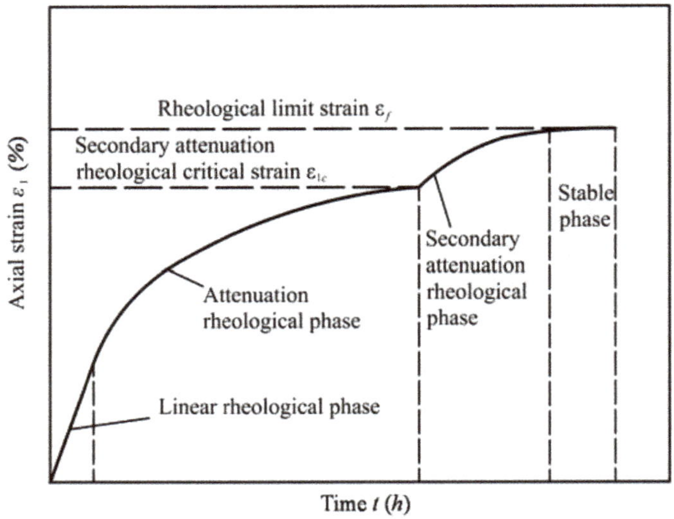

Figure 4. Four-phase compressive rheological curve.

The SMPM samples formed a soil-skeleton-structure during the rheological tests. According to Wang et al. [25], the skeleton mentioned in the soil mechanics is actually a relational grid that characterizes the internal particles of the granular material in contact with each other. The soil skeleton considers whether there is contact extrusion between the particles. The contact extrusion in SMPM can be sandstone-sandstone (SS), sandstone-mudstone (SM) or mudstone-mudstone (MM). In the soil-skeleton-structure, the particles may break with the increasing of the contact stress, while the contact stress avoids the critical crushing stress of the particles. Investigated by Xiao et al. [13], the critical particle crushing stress was directly proportional to the uniaxial compression strength. The mudstone particles in MM and SM may be crushed firstly and no longer bear the role of the skeleton until the complete breakage of sandstone in SS.

In the initial attenuation rheological phase, the contact extrusion include SS, SM and MM. The mudstone particles of the MM and SM were much more heavily crushed than the sandstone particles, leading to a gradual stabilization due to the disintegration of the soft mudstone particles. Then, most of the MM and SS no longer played the role of the skeleton. In the secondary attenuation rheological phase, the sandstone particles of SS were the major bearer of the soil skeleton structure [4]. The stress among the sandstone particles between SS might have caused the particles to yield. Then, the sandstone particles might have been progressively crushed and rearranged in the presence of the skeleton, which increased the deformation in this phase.

Finally, the process reached the stable phase, during which the rheological deformation remained basically unchanged. In this phase, the crushing and rearrangement of the sandstone and mudstone particles had stabilized.

4. Segmented Rheological Model for SMPM

4.1. Segmented Rheological Model

The rheological behavior of most coarse-grained soil could be divided into three phases: the linear rheological phase, the attenuation rheological phase and the stable phase. The whole rheological process is called the attenuation rheological process. The attenuation rheological process of the coarse-grained soil has been described by various rheological models, including power, exponential and hyperbolic functions, as follows:

The exponential model [17,26,27] is

$$\varepsilon_1 = \varepsilon_{1f}\left(1 - e^{bt}\right) \tag{1}$$

The hyperbolic model [21,28] is

$$\varepsilon_1 = t/(a + bt) \tag{2}$$

The power model [26] is

$$\varepsilon_1 = a/\left(1 - t^b\right) \tag{3}$$

where ε_1 is the axial rheological strain; a and b are the model parameters; ε_{1f} is the rheological final strain; t is the time of rheological test.

The rheological limit strain of these three models are $-b\varepsilon_{1f}e^{bt}$, $1/b$, and $-abt^{b-1}$, respectively.

According to the research of Cao [22], the rheological strain value predicted by the exponential function was larger in the early stage than the test data and smaller in the later stage. The slope of the linear rheological phase predicted by the power function was the largest of these three functions.

According to the analysis presented in Section 3, the rheology of the sandstone-mudstone particle mixture was divided into four phases, which differed from the rheological behavior of coarse-grained soils. A segmented rheological model that was appropriate for the sandstone-mudstone particle mixture was therefore developed in this study by modifying an existing rheological model for the coarse-grained soil.

The segmented rheological model was constructed using the following modeling steps. First, rheological strain, ε_{1i}, was calculated using the three coarse-grained soil rheological models (Equations (1) to (3)), if it had not yet reached the initial critical value, ε_{1c}, of the secondary rheology (Figure 4). Otherwise, the rheological time and rheological strain, ε_{2i}, were recalculated, and the rheological parameters were re-determined. The total rheological strain of the segmented rheological model can be obtained as follows:

$$\begin{cases} \varepsilon_1 = \varepsilon_{1i} & \varepsilon_1 \le \varepsilon_{1i} \\ \varepsilon_1 = \varepsilon_{1c} + \varepsilon_{2i} & \varepsilon_1 > \varepsilon_{1c} \end{cases} \tag{4}$$

where ε_1 is the total axial rheological strain and ε_{1i} and ε_{2i} are the attenuation rheology and secondary attenuation rheology, respectively; they can be calculated using the coarse-grained soil rheological models listed in Table 2. ε_{1c} is the critical rheological strain of the secondary attenuation, which can be calculated using the attenuation rheological factor and final rheology, as follows:

$$\varepsilon_{1c} = \alpha\varepsilon_f \tag{5}$$

where ε_f is the rheological limit strain and α is the rheological attenuation factor.

4.2. Rheological Limit Strain ε_f

The rheological limit strain ε_f has been previously investigated by several researchers. Zhang et al. (2010) suggested that ε_f of rock-fill materials was associated with the confining pressure and concluded that the relationship between the vertical stress and the limit strain could be calculated using a hyperbola. Cao [29] argued that ε_f (total strain within a test period) was exponentially related to the vertical stress. In the present study, the relationship between the vertical limit strain (total strain within a test period) and axial pressure is shown in Figure 5; ε_f increased with increasing vertical stress, which could be fitted by the following linear relationship:

$$\varepsilon_f = m\frac{\sigma_1}{p_a} + n \tag{6}$$

where σ_1 and p_a are the axial and atmospheric pressures (kPa), respectively; and m and n are the fitting parameters associated with the material properties. The fitted values of parameters m and n are listed in Table 2.

Table 2. Fitting parameters.

Mudstone Particles Content (%)	m	n	R^2
20	0.0163	0.0164	0.96
40	0.0165	0.0571	0.96
60	0.0168	0.0798	0.96
80	0.0172	0.1711	0.99

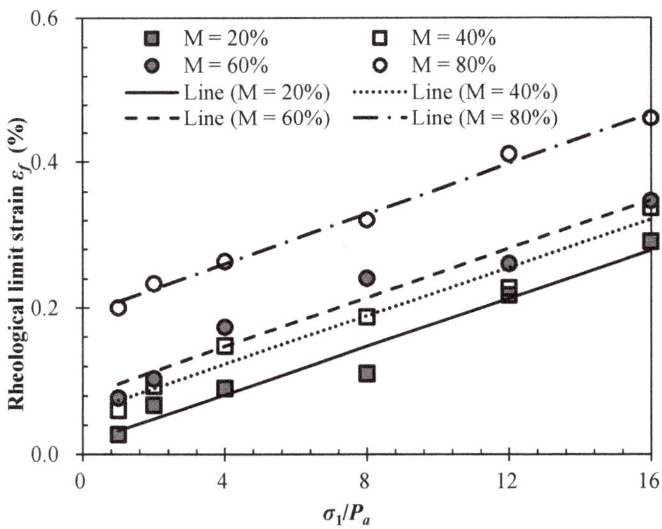

Figure 5. The relationship between rheological limit strain and vertical stress.

The data presented in Table 2 show that no significant relationship existed between the fitting parameter m and the mudstone particle content of the SMPM. Instead, m remained relatively unchanged, showing only a slight increase from 0.0163–0.0172 as the mudstone particle content increased from 20–80% (mean value of 0.0167). Therefore, m did not vary with the mudstone particle content in this study. The second fitting parameter n did increase with increasing mudstone particle content, as shown in Figure 6. The relationship between fitting parameter n and the mudstone particle content is expressed as follows:

$$n = m_0 M + n_0 \tag{7}$$

where n is the fitting parameter of the relationship between the rheological limit strain and vertical stress and parameters m_0 and n_0 are fitting parameters whose values are 0.0024 and 0.0406, respectively.

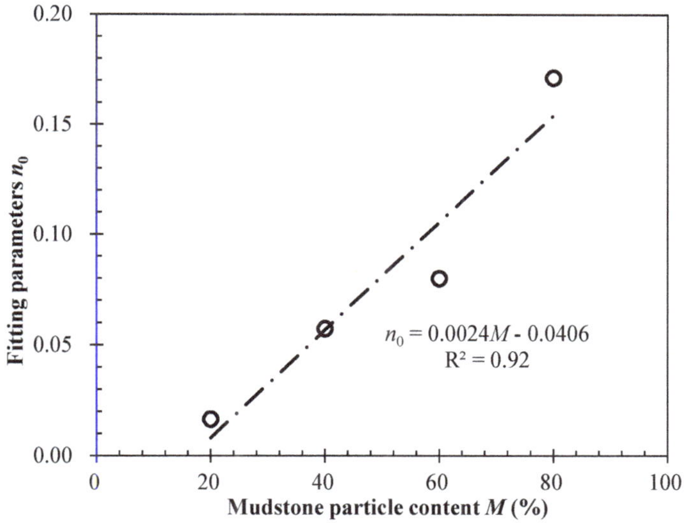

Figure 6. The relationship between parameter n_0 and mudstone particle content M.

In summary, the relationship between the rheological limit strain and mudstone particle content and the vertical stress can be expressed as follows:

$$\varepsilon_f = m\frac{\sigma_1}{p_a} + m_0 M + n_0 \tag{8}$$

where ε_f is the rheological limit strain, M is the mudstone particle content (%) and σ_1 and p_a are the vertical and atmospheric pressures (kPa), respectively. These three fitting parameters for predicting the rheological limit strain, m, m_0 and n_0, are fitting parameters whose values are 0.0163, 0.0024 and 0.0406, respectively. σ_1 is the vertical stress in the present study. The vertical stress is the sum of pore pressure and the horizontal effective stress. The pore pressure in the coarse-soil is a small force as compared with the effective stress. Investigated by Qiu [4], the pore pressure can be ignored due to the permeability coefficient of the SMPM being greater than 0.01 cm/s. In addition to minimizing the pore pressure of the sample, the pore stone with its permeability coefficient greater than 0.05 cm/s was used in the test. The drainage of the sample was free during the entire testing process.

4.3. Rheological Attenuation Factor

The strain values of the attenuation rheology and final rheology were determined from Figures 2 and 3. The ratio between the strain value of the attenuation rheology and that of the final rheology was defined as the attenuation factor α.

$$\alpha = \frac{\varepsilon_{1c}}{\varepsilon_f} \tag{9}$$

where α is the attenuation factor, ε_{1c} is the critical strain value of the attenuation rheology, which can be obtained using the rheological curve, and ε_f is the strain value of the final rheology.

The relationship between the rheological attenuation factor and vertical stress is shown in Figure 7. A certain association appears to exist between the rheological attenuation factor and vertical stress. Furthermore, the rheological attenuation factor increased with increasing vertical stress, before stabilizing towards a constant value. The relationship between the rheological attenuation factor and vertical stress can be described by the following logarithmic function:

$$\alpha = a_\alpha \ln(\sigma_1 / P_a) + b_\alpha \quad (10)$$

where α is the rheological attenuation factor and σ_1 and p_a are the axial and atmospheric pressures (kPa), respectively. a_α and b_α are fitting parameters, and their relationship with the mudstone particle content is shown in Figure 8.

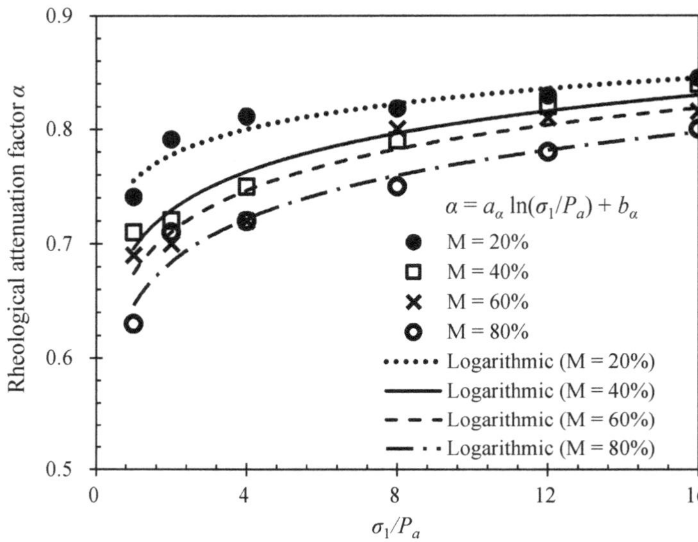

Figure 7. The relationship between rheological attenuation factor and vertical stress.

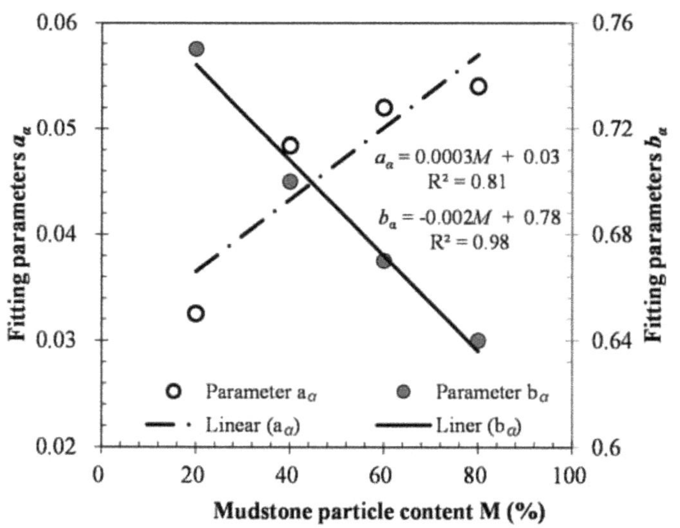

Figure 8. The relationship between fitting parameters and mudstone particle content.

Figure 8 shows that parameter a_α increased with increasing mudstone particle content, whereas b_α decreased. The relationship between a_α, b_α and the mudstone particle content could be fitted with a linearly relationship, as follows:

$$\begin{cases} a_\alpha = c_1 M + d_1 \\ b_\alpha = c_2 M + d_2 \end{cases} \tag{11}$$

where c_1 and d_1 are the fitting parameters of the linear fitting relationship between a_α and the mudstone particle content with values of 0.0003 and 0.03, respectively. c_2 and d_2 are the fitting parameters of the linear fitting relationship between b_α and the mudstone particle content with values of −0.002 and 0.78, respectively.

The relationships between the rheological attenuation factor, mudstone particle content and vertical stress can be obtained by combining Equations (7) and (8) as follows:

$$\alpha = (c_1 M + d_1)\ln(\sigma_1 / P_a) + c_2 M + d_2 \tag{12}$$

where α is the rheological attenuation factor, M is the mudstone particle content and σ_1 and p_a are the axial and atmospheric pressures (kPa), respectively. There are four parameters for predicting the rheological attenuation factor. c_1, d_1, c_2 and d_2 are fitting parameters whose values are 0.0003, 0.03, −0.002 and 0.78, respectively.

5. Discussions

To model the segmented rheological behavior of SMPM, an appropriate function from Table 2 can be chosen by fitting each to the test data. Fitting analysis of four rheological curves (vertical stress was 1600 kPa, M values of 20%, 40%, 60% and 80%) was conducted using the segmented rheological model and the rheological model for the three types of the coarse-grained soil models listed in Table 2. The parameter values are shown in Figure 9.

The predicted values of the exponential and hyperbolic functions (Equations (1) to (3)) are shown in Figure 9a,b, plotted against the experimental data. These show that the secondary attenuation rheological process can be accurately simulated by these two functions after the attenuation rheological phase. The rheological process simulated by the power function is shown in Figure 9c. The characteristics of the segmented rheological process were not favorably reflected by the predicted curves. Therefore, the rheological process simulated by the power function model is not recommended for predicting the rheological process of SMPM.

Figure 9a shows that the predicted strain value calculated by the exponential function was larger than the experimental values in the attenuation rheological phase at a mudstone particle content of 80%. Furthermore, this curve did not conform to the characteristic of the attenuation rheology in the attenuation phase either. In the secondary attenuation rheological phase, however, the predicted value was much smaller than the experimental values. At mudstone particle contents between 20% and 60%, the predicted strain value calculated by the exponential function was much larger than the test value. Engineering designs made using this exponential function would therefore result in excessive rheology, which could lead to additional waste. Therefore, the exponential function does not appear to be suitable to calculate the rheological strain of SMPM.

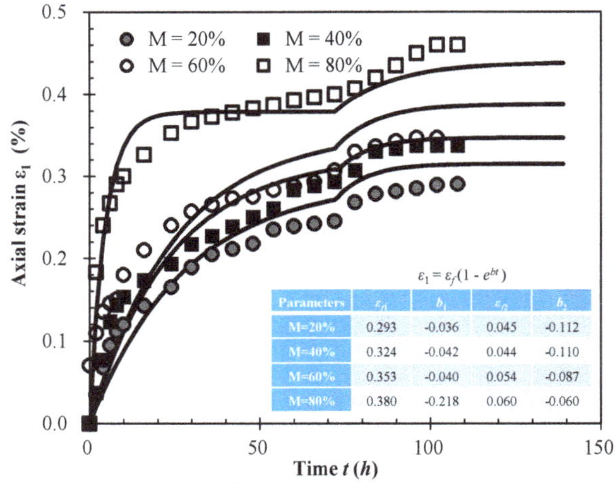

(**a**) Exponential Function.

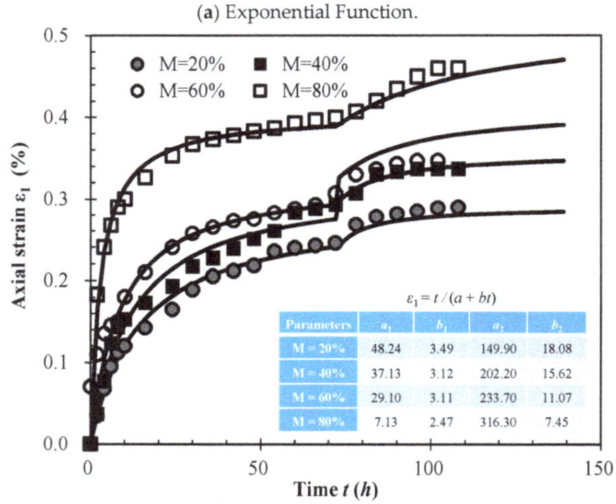

(**b**) Hyperbolic Function.

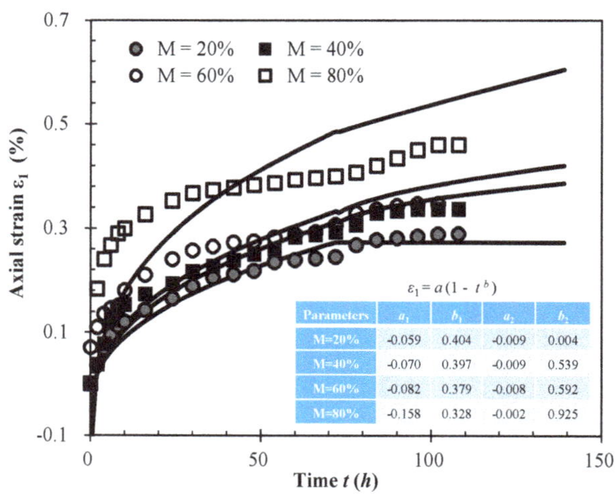

(**c**) Power Function.

Figure 9. Various predicted values of the three rheological models, compared to experimental data.

Figure 9b shows that the hyperbolic rheological model fitted the test values well, and the predicted values were close to the experimental data throughout the rheological process. Therefore, the rheological process of SMPM can be calculated using the hyperbolic function. The segmented rheological model of SMPM can be expressed as:

$$\begin{cases} \varepsilon_1 = \varepsilon_{1i} & \varepsilon_1 \leq \varepsilon_{1i} \\ \varepsilon_1 = \varepsilon_{1c} + \varepsilon_{2i} & \varepsilon_1 > \varepsilon_{1c} \end{cases} \tag{13}$$

$$\varepsilon_f = m\frac{\sigma_1}{p_a} + m_0 M + n_0 \tag{14}$$

$$\varepsilon_{1c} = [(c_1 M + d_1)\ln(\sigma_1/P_a) + c_2 M + d_2]\varepsilon_f \tag{15}$$

$$\varepsilon_{1i} = t/(a_1 + b_1 t) \tag{16}$$

$$\varepsilon_{2i} = (t - T)/[a_2 + b_2(t - T)] \tag{17}$$

where T is the time from the initial moment of rheology to the attenuation rheological phase. a_1, b_1, a_2 and b_2 are model parameters, c_1, d_1, c_2, d_2, m, m_0 and n_0 are seven fitting parameters, and these parameters can be obtained by fitting with the experimental data. There are 11 parameters for the suggested model, which can be used to calculate the rheological deformation filled with SMPM of different mudstone content. There are too many parameters for the suggested model. In further study, research on a model for the rheological deformation of SMPM within fewer parameters is very meaningful. The strength of the mudstone particles is lower than that of the sandstone particles, which suggests that they have different critical crushing stresses. The particle-crushing critical stress of the mudstone particles is lower than that of the sandstone particles. In the rheological compression tests, the deformation initially resulted from the mudstone particles being crushed, with further deformation occurring due to the crushing of sandstone particles at a later time. Mudstone particles are usually disintegrative, and mudstone particle surfaces are often covered with fractures. Therefore, the mudstone particle breakage appears to be more serious than that of the sandstone particles. Compared with other materials such as rock-fill materials and coarse-grained soil, SMPM features a unique rheological mechanism because of the mudstone particles, which are associated with various factors, including the mudstone particle content and the particle strength. The suggested rheological model was proven by the validation exercise. It is not an intuitive method to prove the accuracy of the suggested model. This model can be used to calculating the rheological deformation of a dam or foundations filled with SMPM under different mudstone content. However, it cannot be applied to real engineering works now. In future research, the finite element approach may be used to analyze the rheological behavior of SMPM. When the FEM procedures for the suggested constitutive relationships are worked out, it can be applied to the analysis of a real problem, and it could be generalized while its precision is proven.

6. Conclusions

In this study, rheological tests of a sandstone-mudstone particle mixture with different mudstone particle contents were performed, and a modified rheological deformation model for the sandstone-mudstone particle mixture was proposed. The modified model can be used to calculate the deformation in engineering structures filled with a sandstone-mudstone particle mixture. The conclusions of this study are as follows.

(1) The rheological axial strain of a sandstone-mudstone particle mixture increases with increasing vertical stress and is associated with the mudstone particle content. The higher the mudstone particle content, the larger the rheological limit strain. The rheological limit strain displays linear relationships with both the vertical pressure and the mudstone particle content. The rheological critical strain of the secondary attenuation can be calculated using the rheological attenuation factor and the rheological

limit strain. The rheological attenuation factor increases with increasing vertical pressure, following a logarithmic relationship.

(2) Analysis of the rheological strain reveals that the rheological process of a sandstone-mudstone particle mixture can be divided into four phases: the linear phase, the attenuation rheological phase, the secondary attenuation rheological phase and the stable phase. Modelled rheological compression curves with different mudstone particle contents were fitted at a vertical pressure of 1600 kPa, calculated using different rheological functions; these were compared to the experimental results. The hyperbolic function model used with the proposed segmented rheological model most accurately described the four-phase attenuation rheological characteristics of a sandstone–mudstone particle mixture.

Author Contributions: Conceptualization, Z.Q. and J.W.; Methodology, Z.Q.; Software, Y.L. and Y.C.; Validation, Z.Q., J.W. and Y.C.; Formal Analysis, T.C.; Investigation, Z.Q.; Resources, J.W.; Data Curation, T.C.; Writing-Original Draft Preparation, Z.Q.; Writing-Review & Editing, Z.Q., T.C., and Y.C; Visualization, T.C.; Supervision, J.W.; Project Administration, J.W.; Funding Acquisition, J.W., Y.L., and Z.Q.

Funding: This research was funded by the National Natural Science Foundation of China under Grant No. 51479012 and Grant No. 51479014, the Chongqing Science and Technology Commission of China under Grant No. cstc2017jcyjA1642, the Technology innovation and application demonstration project in Chongqing under Grant No. cstc2018jscx-msybx0328, the Chongqing Municipal Education Commission of China under Grant No. KJ1705123, the National Postdoctoral Science Foundation of China under Grant No. 2017M622978 and the National Postdoctoral Science Foundation of Chongqing under Grant No. Xm2017093.

Conflicts of Interest: The authors declare no conflict of interest.

References

1. Chongqing Geological and Mineral Resource Exploration and Development Corporation (CGMREDC). *Chongqing Geological Map (Scale 1:500,000)*; Chongqing Yangtze River Map Printing Factory: Chongqing, China, 2002.
2. Wang, J.J.; Zhang, H.P.; Deng, D.P. Effects of compaction effort on compaction behavior and particle crushing of a crushed sandstone-mudstone particle mixture. *Soil Mech. Found. Eng.* **2014**, *51*, 67–71. [CrossRef]
3. Wang, J.-J.; Cheng, Y.-Z.; Zhang, H.-P.; Deng, D.-P. Effects of particle size on compaction behavior and particle crushing of crushed sandstone-mudstone particle mixture. *Environ. Earth Sci.* **2015**, *73*, 8053–8059. [CrossRef]
4. Qiu, Z.F. Triaxial Deformation Behavior and Damage Model under Wet-Dry Cycling of a Sandstone-Mudstone Mixture. Ph.D. Thesis, Chongqing Jiaotong University, Chongqing, China, 2016. (In Chinese)
5. Wang, J.J.; Fang, X.S.; Qiu, Z.-F. *Engineering Properties of a Sandstone-Mudstone Particle Mixture*; Science Press: Beijing, China, 2016. (In Chinese)
6. Wang, J.J.; Qiu, Z.F.; Deng, W.J.; Zhang, H.P. Effects of mudstone particle content on shear strength of a crushed sandstone-mudstone particle mixture. *Mar. Georesour. Geotechnol.* **2016**, *34*, 395–402. [CrossRef]
7. Qiu, Z.F.; Wang, J.J.; Huang, S.Y.; Bai, J.P. Wetting-induced axial and volumetric strains of a sandstone mudstone particle mixture. *Mar. Georesour. Geotechnol.* **2018**. [CrossRef]
8. Wang, J.J.; Qiu, Z.F.; Bai, J.P.; Yu, C.; Liu, M.W. Deformation of a sandstone–mudstone particle mixture induced by periodic saturation. *Mar. Georesour. Geotechnol.* **2018**, *36*, 494–503. [CrossRef]
9. Jiang, J.S.; Cheng, Z.L.; Zuo, Y.Z.; Ding, H.S. CT triaxial rheological test on coarse-grained soils. *Rock Soil Mech.* **2014**, *35*, 2507–2514.
10. Indraratna, B.; Thakur, P.K.; Vinod, J.S.; Salim, W. Semi empirical cyclic densification model for ballast incorporating particle breakage. *Int. J. Geomech.* **2012**, *12*, 260–271. [CrossRef]
11. Indraratna, B.; Ngo, N.; Rujikiatkamjorn, C.; Vinod, J.S. Behavior of fresh and fouled railway ballast subjected to direct shear testing: Discrete element simulation. *Int. J. Geomech.* **2014**, *14*, 34–44. [CrossRef]
12. Kong, X.J.; Zhou, Y.; Xu, B.; Zou, D.G. *Analysis on Seismic Failure Mechanism of Zipingpu Dam and Several Reflections of Aseismic Design for High Rock-Fill Dam*; The Workshop on Biennial International Conference on Engineering; ASCE: Reston, VA, USA, 2010; pp. 3177–3189.
13. Xiao, Y.; Sun, Y.F.; Hanif, K.F. A particle-breakage critical state model for rockfill material. *Sci. China Technol. Sci.* **2015**, *58*, 1125–1136. [CrossRef]

14. Zhang, B.Y.; Sun, G.L.; Zhang, Z.L. Degrading deformation of rockfill materials and its constitutive model. *Chin. J. Geotech. Eng.* **2010**, *32*, 98–103.
15. Jia, Y.; Xu, B.; Chi, S.; Xiang, B.; Zhou, Y.; Jia, Y. Research on the particle breakage of rockfill materials during triaxial tests. *Int. J. Geomech.* **2017**, *17*, 04017085. [CrossRef]
16. Carrier, B.; Vandamme, M.; Pellenq RJ, M.; Van Damme, H. Creep of clay: Numerical results at the scale of a layer and experimental results at the scale of thin self-standing films. In Proceedings of the 10th International Conference on Mechanics and Physics of Creep, Shrinkage, and Durability of Concrete and Concrete Structures, Vienna, Austria, 21–23 September 2015; pp. 531–536.
17. Wang, Y.; Yin, Z.Z. Rheology model of rockfill used in the rheology analysis of concrete face rockfill dam. *Rock Soil Mech.* **2000**, *21*, 227–230.
18. Chen, X.B.; Zhang, J.S.; Feng, Z.P. Experimental study on rheological engineering properties of coarsely granular red sandstone soil. *Chin. J. Rock Mech. Eng.* **2007**, *26*, 601–607.
19. Wang, L.Z.; Yin, Z.Y. Stress dilatancy of natural soft clay under an undrained creep condition. *Int. J. Geomech.* **2014**, *15*, A4014002. [CrossRef]
20. Cheng, Z.L.; Ding, H.S. Creep test for rockfill. *Chin. J. Geotech. Eng.* **2004**, *26*, 473–476.
21. Zhang, B.Y.; Chen, T.; Peng, C.; Qian, X.X.; Jie, Y.; Zhang, B. Experimental study on loading-creep coupling effect in rockfill material. *Int. J. Geomech.* **2017**, *17*, 04017059. [CrossRef]
22. Yang, C.; Wen, C.P. Study of nonlinear rheology of improved expansive soil based on model theory and genetic creep theory. *Rock Soil Mech.* **2016**, *S2*, 75–82.
23. Zhang, B.Y.; Zhang, J.H.; Sun, G.L. Deformation and shear strength of rockfill materials composed of soft siltstones subjected to stress, cyclical drying/wetting and temperature variations. *Eng. Geol.* **2015**, *190*, 87–97. [CrossRef]
24. Trade Standard of P. R. China SL237-015. Standard method for one-dimensional consolidation properties of soils. In *Specification of Soil Test*; The Ministry of Water Resources of P.R. China: Beijing, China, 1999. (In Chinese)
25. Wang, Z.J.; Liu, S.H.; Wang, Y.S.; Kong, W.Y. Analysis on microstructure of granular materials under different stress paths. *J. Hydraul. Eng.* **2013**, *44*, 772–778.
26. Shen, Z.J.; Zhao, K.Z. Back analysis of creep deformation of rockfill dams. *J. Hydraul. Eng.* **1988**, *6*, 1–6.
27. Liang, J.; Liu, H.L. Creep test for rockfill of CFRD. *Chin. J. Geotech. Eng.* **2004**, *24*, 257–259.
28. Feda, J. *Creep of Soils and Related Phenomena*; Elsevier: Amsterdam, The Netherlands, 1992.
29. Cao, G.X. Study on Post-Construction Settlement of High Fill Foundation in Mountainous Airport. Ph.D. Thesis, Tsinghua University, Beijing, China, 2011. (In Chinese)

processes

MDPI

Article

Experimental Investigation on the Law of Grout Diffusion in Fractured Porous Rock Mass and Its Application

Donghai Jiang [1,2], Xianzhen Cheng [1,2], Hengjie Luan [1,2,*], Tongxu Wang [1], Mingguang Zhang [1] and Ruiyun Hao [3]

1 College of Mining and Safety Engineering, Shandong University of Science and Technology, Qingdao 266590, China; esriver@163.com (D.J.); lbs_cxz@163.com (X.C.); wtx6383@163.com (T.W.); zmg1981@126.com (M.Z.)

2 State Key Laboratory of Mining Disaster Prevention and Control Co-Founded by Shandong Province and the Ministry of Science and Technology, Shandong University of Science and Technology, Qingdao 266590, China

3 College of Mining Technology, Taiyuan University of Technology, Taiyuan 030024, China; vollxyz989@163.com

* Correspondence: luanjie0330@126.com; Tel.: +86-0532-860-58052

Received: 6 September 2018; Accepted: 12 October 2018; Published: 16 October 2018

Abstract: Because of the limitation of mining techniques and economic conditions, large amounts of residual coal resources have been left in underground coal mines around the world. Currently, with mining technology gradually developing, residual coal can possibly be remined. However, when residual coal is remined, caving areas might form, which can seriously affect the safety of coal mining. Hence, grouting technology is put forward as one of the most effective technologies to solve this problem. To study the grouting diffusion in fractured rock mass, this paper developed a visualization platform of grouting diffusion and a three-dimensional grouting experimental system that can monitor the grout diffusion range, diffusion time and grout pressure; then, a grouting experiment is conducted based on this system. After that, the pattern of the grouting pressure variation, grout flow and grout diffusion surface are analyzed. The relationship among some factors, such as the grouting diffusion radius, compressive strength of the grouted gravel, porosity, water-cement ratio, grouting pressure, grouting time, permeability coefficient and level of grout, is quantitatively analyzed by using MATLAB. The study results show that the flow pattern of the grout in fractured porous rock mass has a parabolic shape from the grouting hole to the bottom. The lower the level is, the larger the diffusion range of the grout is. The grouting pressure has the greatest influence on the grouting diffusion radius, followed by the grouting horizon and water-cement ratio. The grouting permeability coefficient has the least influence on the grouting diffusion radius. The grout water-cement ratio has the greatest influence on the strength of the grouted gravel, followed by the grouting permeability. The grouting pressure coefficient has the least amount of influence on the grouting diffusion radius. According to the results, the grouting parameters are designed, and a layered progressive grouting method is proposed. Finally, borehole observation and a core mechanical property test are conducted to verify the application effect. This grouting technology can contribute to the redevelopment and efficient utilization of wasted underground coal resources.

Keywords: fractured porous rock mass; grouting experiment; visualization system; flow law; layered progressive grouting

1. Introduction

Due to technique limitations and economic conditions, some relatively outdated coal mining methods, such as the room pillar type and lane column type, have been widely used throughout the world for a considerable portion of coal mining history. Hence, large amounts coal resources (residual coal) that were left behind in the gob to maintain stability were wasted [1]. With the development of mining technology, the redevelopment and utilization of residual coal has become possible. Recycling this part of coal resources can fulfill the requirements for the efficient development and utilization of resources. It is also a requirement for coal mining companies to maintain their long-term development and can have a significant impact on the sustainable development of old mining areas. Residual coal repeated mining refers to the use of mechanized coal mining technology to mine residual coal left in old mining areas [2]. However, several collapsed zones, which seriously influence the production and safety of the coal mines, exist in residual coal repeated mining areas. There are many methods that deal with mining faces that repeatedly cross a collapsed zone [3–6]. Grouting in the rock mass of a collapsed zone is one of the most effective technologies for solving this problem. Through grouting, the rock mass of a collapsed zone can be cemented into continuous and stable blocks, and the physical and mechanical parameters of the rock mass can be significantly improved. Therefore, the shearer will be able to pass through the caving area safely. Whether the cemented coal and rock mass can meet the requirements after grouting is mainly affected by the diffusion radius of the grout, the strength of the mass and the layout parameters of the grouting holes. In other words, the key to solving this problem is to determine the law of grouting diffusion in a fractured rock mass.

Over the past few decades, extensive research has been conducted on the relationship between grouting factors, the grout diffusion radius and the strength of the mass. In theoretical research of grout diffusion, Marwan Ahmed et al. established a mathematical model of grout seepage based on the partition of the fractured rock mass and analyzed the diffusion characteristic of grout in different areas [7]. Li Shenju et al. studied the grout penetration law in the surrounding rock containing an orthogonal fracture by the permeability tensor method [8]. Di Nucci systematically studied the diffusion law of grout in a one-dimensional fracture via a model test and numerical simulation [9]. Kelessidis V C et al. proposed a different and optimal methodology to determine the three Herschel–Bulkley rheological parameters of drilling fluids [10]. Li Shucai et al. studied the movement law of grout in porous media based on a theoretical model of the percolation threshold effect. This revealed the influence mechanism of the diafiltration effect on grout diffusion and the effective reinforcement range [11]. In the grout diffusion calculation, Huang Yaoguang et al. deduced the basic equation of the unsteady diffusion of grout under a coupling effect of disturbance stress and grouting pressure, considering the effect of the excavation disturbance and grouting pressure attenuation on the diffusion law of grout [12]. Saada Z et al. established the one-dimensional unidirectional grouting model based on grout seepage and deduced the analytical solution of one-dimensional grout diffusion [13]. Rafi et al. deduced the diffusion radius of the Bingham grout in the fractured rock mass [14]. Shimada Hideki et al. deduced the formula for calculating the effective diffusion radius of the Bingham fluid in the soft and broken seam based on the generalized Darcy law and the spherical diffusion theory model [15]. Funehag J et al. obtained the relationship between grouting pressure, seepage velocity, grouting time and grout diffusion radius by simulating the grout flow process in a single fracture [16,17]. In a grouting test, Minto James M. et al. determined the parameters of the grouting pressure and diffusion radius by the field grouting test [18]. Ge Jialiang et al. analyzed the influence of the grouting material, the grout water-cement ratio and its properties, the grouting pressure and other factors on the grouting volume, the grout diffusion radius and the strength of grouted gravel through a grouting simulation test [19].

A fractured rock consists of large, loose rock masses (Figure 1). The grout flows mainly in a macroscopic gap (void) rather than in mesoscopic pores or fractures. However, the above research is mainly focused on the grouting of a jointed fractured rock mass. Only a few studies have been reported on fractured porous rock masses, especially fractured porous rock masses with large porosities ($n > 0.4$). There is essentially no quantitative formula that can guide grouting construction on-site. Hence, this paper first develops the visualization system of the grouting diffusion range and the three-dimensional grouting experiment system. Based on this system, a grouting changing performance model experiment is performed. This is followed by an analysis of the variation pattern of the grouting pressure, the grout flow pattern, and the diffusion surface pattern of grout. Then, the research results are applied to engineering practice, and the application effect analysis is performed. Finally, discussions and conclusions are presented at the end of this paper. This work can contribute to the redevelopment and efficient utilization of wasted underground coal resources.

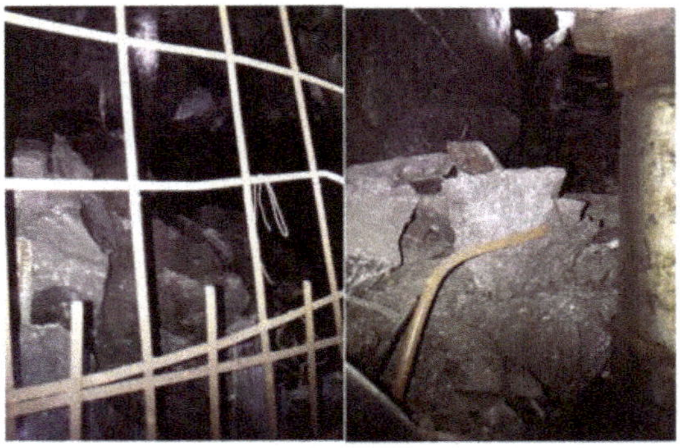

Figure 1. Fractured porous rock mass.

2. Experimental Method

2.1. Experimental System

To study the law of grouting diffusion in a fractured porous rock mass, a series of grouting experiments were performed in the self-developed three-dimensional visualization grouting experimental system, as shown in Figure 2. The system mainly consists of three parts: grouting test-bed, grouting equipment and visualization platform.

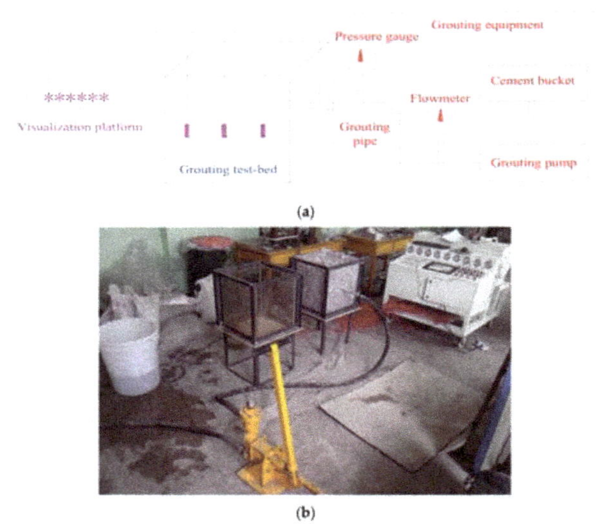

Figure 2. The overall arrangement of the three-dimensional visualization grouting experiment system: (**a**) Sketch Map; (**b**) Practicality picture.

The grouting test-bed, which has dimensions of 0.5 m × 0.5 m × 0.5 m, is a frame welded by angle steel and surrounded by tempered glass. One piece of the tempered glass can be replaced by glass with grouting holes at different positions to simulate different grouting heights. Additionally, the bottom of the test-bed is a piece of wood with five grout overflow holes. The grouting equipment includes a cement bucket, grouting pump, pressure gauge, flowmeter, grouting pipe, stop valve, hole packer, etc. A grouting pump is a kind of manual pressure grouting pump (MG-808, Shangqin High Pressure Grouting Technology Co., Ltd., Shanghai, China). The pressure gauge and flowmeter, which are used to measure the grouting pressure and grouting volume, respectively, are installed on the grouting pipe. The visualization platform is a self-developed platform for monitoring grout diffusion under laboratory conditions. It includes the signal system (Patent No.: ZL 2016200214895), flow sensors, pressure sensors, wire, an industrial personal computer, etc. The visualization system can monitor the grouting pressure, grout diffusion range, the time of grout diffusion to the sensors and the grout diffusion pressure at the sensor during the grouting process. The visualization platform of the grouting diffusion is shown in Figure 3.

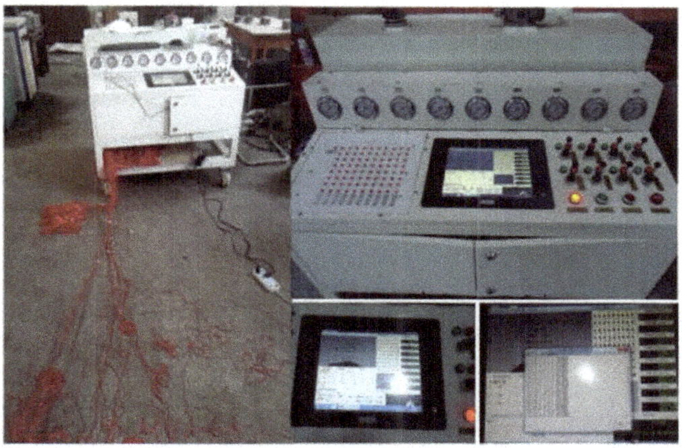

Figure 3. The visualization platform of the grouting diffusion.

The sensors of the visualization platform are embedded in a grouting test-bed during the laying of the similar rock material, after which the spatial position of the sensors can be displayed on the screen of the visualization platform. When conducting the grouting test, once the grout has spread to the position of the sensor, the corresponding signal light in the screen of the visualization platform will light up, and then the distance between the position of the sensor and the grouting port, and the time and pressure at which the grout flows to the sensor can, be recorded.

2.2. Experimental Plan

A similar simulation of fractured porous rock mass was achieved by controlling the proportions of stones with different morphological characteristics and particle sizes. First, the rock mass at the Shenghua coal mine was sampled to measure the morphological characteristics and porosity before the experiment. Then, according to the obtained morphological characteristics and porosity, suitable stones were selected and mixed to simulate fractured porous rock mass. The grouting material used in the experiment was cement grout. Different groups of experiments were tested by adjusting the porosity of the fractured porous rock mass, the water-cement ratio of grout and the grouting pressure.

Five groups of test plans ($U_5(5^3)$) were designed according to the uniform experimental design method [20], as shown in Table 1. The dynamic viscosity coefficients of grouts with different water-cement ratios were measured by the viscometer. The permeability and permeability coefficient of a fractured porous rock mass can be obtained by the following formulae [21].

$$e = \frac{D_p^2}{150} \frac{n^3}{(1-n)^2} \tag{1}$$

$$k = e\rho g/\eta \quad (2)$$

In Formulas (1) and (2), e is the permeability, m^2; D_p is the average particle diameter, m; n is the porosity; k is the permeability coefficient, m/s; ρ is the grout density, kg/m^3; g is the gravity acceleration, m/s^2; and η is the dynamic viscosity coefficient, Pa·s.

Table 1. The parameters of different experiments.

Experiment Number	Porosity n	Water-Cement Ratio$m$	Grouting Pressure p (MPa)	Dynamic Viscosity Coefficient η (Pa·s)	Grout Density ρ (kg/m^3)	Permeability e (m^2)	Permeability Coefficient k (m/s)
1	0.459	0.5:1	0.3	37	1823	2.33×10^{-6}	0.001124
2	0.455	0.6:1	0.5	20	1734	1.41×10^{-6}	0.001195
3	0.447	0.7:1	0.2	13	1662	5.24×10^{-7}	0.000656
4	0.426	0.8:1	0.4	8	1603	1.27×10^{-7}	0.000249
5	0.466	0.9:1	0.6	6	1512	4.27×10^{-6}	0.010553

In this experiment, 26 sensors of the visualization platform were embedded in the grouting test-bed. The sensors were arranged at 50 mm, 200 mm, and 300 mm vertically from the grouting hole. The positions of the sensors in the grouting test-bed are shown in Figure 4, and their coordinates are shown in Table 2.

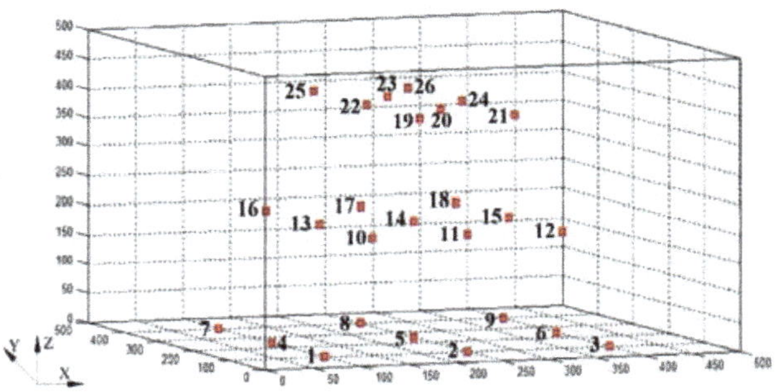

Figure 4. Sensor positions in the grouting test-bed.

Table 2. Visualization system sensor coordinates in the grouting test-bed.

Signal Number		1	2	3	4	5	6	7	8	9	10	11	12	13
Sensor coordinate	X	100	250	400	100	250	400	100	250	400	150	250	350	150
	Y	100	100	100	250	250	250	400	400	400	100	100	100	250
	Z	50	50	50	50	50	50	50	50	50	200	200	200	200
Signal Number		**14**	**15**	**16**	**17**	**18**	**19**	**20**	**21**	**22**	**23**	**24**	**25**	**26**
Sensor coordinate	X	250	350	150	250	350	200	250	300	200	250	300	200	300
	Y	250	250	400	400	400	100	175	100	250	325	250	400	400
	Z	200	200	200	200	200	300	300	300	300	300	300	300	300

2.3. Experimental Procedure

(1) Stones with different particle sizes of 8–10 mm, 15–30 mm and 35–50 mm were selected and mixed evenly to simulate the fractured porous rock mass. Different porosities were obtained by controlling the ratios of stones with different diameters, as shown in Table 3.

(2) The mixed stone material was loaded into the grouting test-bed, and the sensors were arranged at the corresponding position of the grouting test-bed during the loading process.

(3) The coordinates of the sensors were put into visualization platform. After inputting the coordinates, the green dots were displayed at the corresponding positions on the screen. During

the grouting process, the green dots became red dots when the grout reached the corresponding sensor position.

(4) Grouts with different water-cement ratios were prepared according to the experimental requirements. After that, the grout was stirred for 15 min in the grout bucket. The amount of grout in each experiment was 40 kg.

(5) The grouting system was connected, the grouting pipe was inserted into the grouting hole, and then the grouting experiment began. The grouting pressure must be controlled within the required experimental range.

(6) After grouting, the grouting pipe was pulled out from the grouting hole, the hole packer was stuffed into the grouting hole, then the grouting equipment was cleaned.

Table 3. The proportions of stones with different particle sizes.

Experiment Number	The Proportions of Stones with Different Particle Sizes (%)			Porosity
	8–10 mm	15–30 mm	35–50 mm	
1	0	50	50	0.459
2	20	50	30	0.455
3	60	30	10	0.447
4	100	0	0	0.426
5	0	0	100	0.466

3. Result Analysis

3.1. Grouting Pressure Variation Forms

During the grouting process, the data from the pressure gauge installed on the grouting pipe was recorded in the visualization platform. The variation forms of the grouting pressure were drawn according to these data, as shown in Figure 5. It can be seen that the grouting pressure has four kinds of variation forms in the grouting experiment: (a) linear growth; (b) cyclic growth; (c) fluctuation stability and (d) slow growth. On one hand, there is a buffer for each pressure because of the manual grouting pump used. On the other hand, the grout sometimes flows into the macropores of the broken rock mass and sometimes flows into the narrow gap of the contact surface. As a result, the grouting pressure presents unstable fluctuations. Moreover, unstable fluctuations are probably transition-related due to the Herschel-Bulkley material.

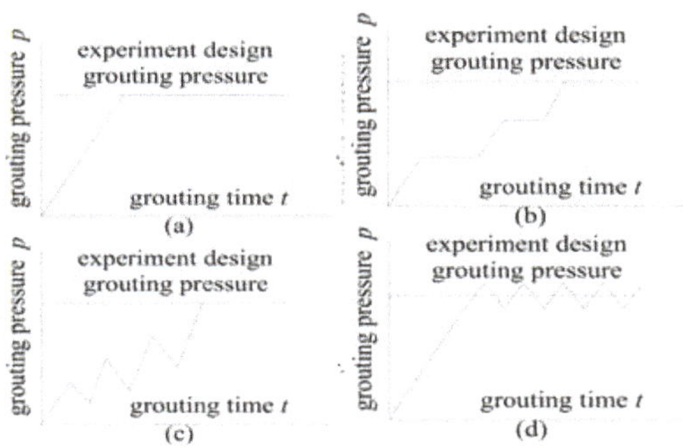

Figure 5. Variation forms of grouting pressure: (**a**) Linear growth; (**b**) Cyclic growth; (**c**) Fluctuation stability; and (**d**) Slow growth.

3.2. Grout Diffusion Pattern

After grouting, the position of the sensors corresponding to the signal lamp light in the visualization platform is shown in Figure 6a. A rough range of grout diffusion can be seen from the visualization system, as shown in Figure 6b. To obtain a more accurate grout diffusion pattern, the three-dimensional coordinate data of the sensors were exported from the visualization platform and put into MATLAB (7.14 version, Branch office of MathWorks in China, Beijing, China, 2012). Through these data, the grout diffusion pattern was synthesized by the interpolation method embedded in MATLAB. There are two kinds of interpolation methods in MATLAB when the two known arrays are used as independent variables, namely, two-dimensional interpolation (Interp2) and two-dimensional scattered point interpolation (Griddata). The Interp2 method is characterized by the gradient distribution of independent variables in the range, which can form a regular matrix. The Griddata method is characterized by the scattered distribution of independent variables in the range, which is not a uniform distribution. In this experiment, the two independent variables are the gradient distribution and the regular matrix, which can be formed. Therefore, the Interp2 method is used. The obtained grout diffusion pattern is as shown in Figure 7. It can be seen that the flow pattern of the grout in fractured porous rock mass has a parabolic shape from the grouting hole to the bottom. The lower the level is, the larger the diffusion range of the grout is. In the horizontal direction, the grout diffusion surface around the grouting hole was an asymmetrical circular surface.

In conclusion, when the height of the grouting pipe, initial velocity of the grout, outlet angle and pore connectivity are determined, the shape of the diffusion surface is certain. Considering that the grout will solidify in a certain period of time, if only through a pipeline grouting, the first injected grout will cement to a solid within a certain range of the outlet, blocking the slurry outlet and reducing grouting efficiency. Therefore, when the spatial shape of the grouting range is determined, the grouting parameters must be designed according to the control range of each grouting pipe.

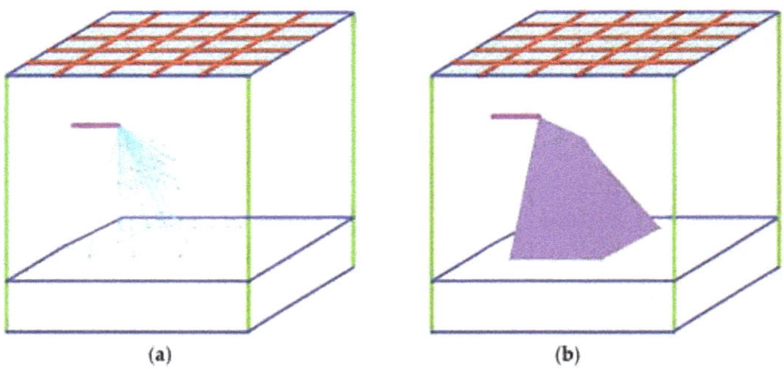

Figure 6. Range of grout diffusion: (**a**) Signal lamp lighted; (**b**) Rough range.

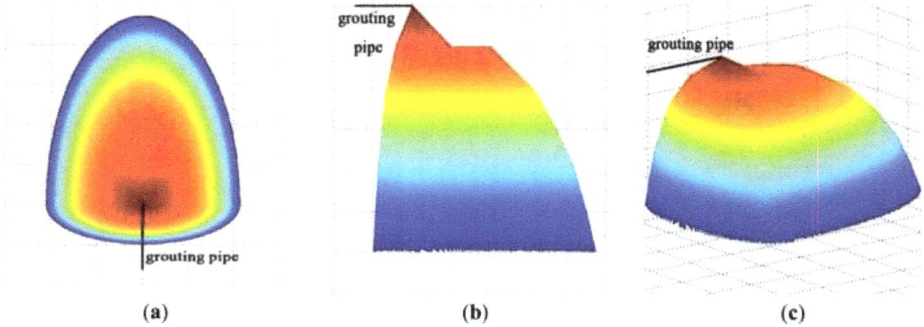

Figure 7. Grout diffusion pattern: (**a**) Top view; (**b**) Side view; and (**c**) Three-dimensional diagram.

To obtain quantitative data of the grouting diffusion radius, plastic nets were arranged on three different layers of grouting. After grouting, the diffusion range of the paste on different plastic layers was measured. Considering the convenience of the measurement and later calculation, the diffusion surface of the different layers was circularly processed. Figure 8 shows the grout diffusion range monitoring diagrams for different levels of experiment 1, and Table 4 shows the range of the grout diffusion for different levels during different experiments.

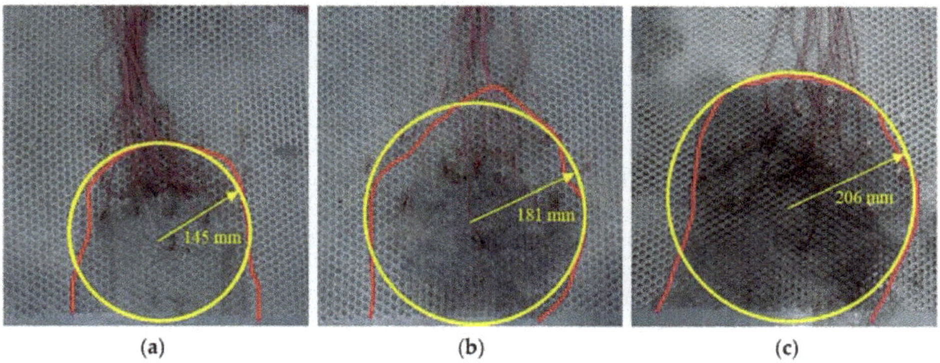

Figure 8. Grout diffusion range diagram for different levels of experiment 1: (**a**) 50 mm; (**b**) 200 mm; and (**c**) 300 mm.

Table 4. Range of grout diffusion for different levels during different experiments.

Experiment Number	Range of Grout Diffusion for Different Level (mm)		
	50 mm	200 mm	300 mm
1	145	181	206
2	157	196	224
3	123	154	175
4	149	186	212
5	173	216	246

The grouting times for experiments 1–5 were 211 s, 178 s, 192 s, 153 s and 121 s, respectively. According to experimental data in Table 1 and the test data listed in Table 4, MATLAB is used to perform a regression analysis between the diffusion radius of the grout and various factors, such as the water-cement ratio, grouting pressure, grouting time, permeability coefficient, and level of grout. From the data analysis of this test, we can see that there is a power function relationship between the test results and the influencing factors. A linear regression is the basis for solving problems. It is necessary to transform the power function relationship into a linear relationship. Therefore, if we suppose its basic model is:

$$Y = A \cdot X_1^B \cdot X_2^C \cdot X_3^D \tag{3}$$

The above model is a nonlinear model, and the two sides of the above formula are logarithmically obtained:

$$\log Y = \log A + B \log X_1 + C \log X_2 + D \log X_3 \tag{4}$$

Assuming $y = \log Y, a = \log A, x_1 = \log X_1, x_2 = \log X_2, x_3 = \log X_3$, the above model can be transformed into a linear regression model:

$$y = a + B \cdot x_1 + C \cdot x_2 + D \cdot x_3 \tag{5}$$

In the formula, a is a constant, and B, C, D are the partial regression coefficients of the dependent variable y on the independent variable x_i $(i = 1, 2, 3)$. The solution is similar to a unitary equation. According to the principle of the least squares method, the square sum of the residuals are determined, and then the solutions of a and B, C, D can be calculated. The resulting regression analysis formula are as follows:

$$R = 0.58m^{0.195}p^{0.24}t^{0.172}k^{0.017}d^{0.22} \quad (6)$$

In Formula (6), the regression fitting exponents for grout water-cement ratio (m), grouting pressure (p), grouting time (t), permeability coefficient (k) and level of grout (d) are 0.195, 0.24, 0.172, 0.017, and 0.22, respectively. The larger the coefficient was, the greater the influence of this factor on the diffusion radius of the grout was. Therefore, the factor with the most substantial effect on the diffusion radius of the grout was the grouting pressure in the fractured porous rock mass. That is, when the other factors were fixed, the larger the grouting pressure was, the larger the grout diffusion radius was. The next most important factors were the level of grout and the water-cement ratio, and the factor with the least impact on the diffusion radius of the grout was the permeability coefficient. Therefore, in the grouting of fractured porous rock mass, the grouting pressure has the greatest influence on the grouting diffusion radius, followed by the grouting horizon and water cement ratio, with the grouting permeability coefficient having the least influence on the grouting diffusion radius.

3.3. Grouting Effect Testing

After one day's worth of grouting, a sampling of the grouting stone bodies was performed, as shown in Figure 9a. It can be determined that in the range of the grout diffusion, the grout diffused evenly, filled densely, and cemented well with the fractured porous rock mass, which can form a skeleton and improve the strength of the rock mass. Without the range of grout diffusion, the fractured porous rock mass was also soaked by water, indicating that the water evolution of the cement slurry is strong.

(a) (b) (c)

Figure 9. Coring of the grouting stone bodies: (**a**) Grouting stone bodies; (**b**) Coring; and (**c**) Cores.

After 28 days' placement in the laboratory, the grouting stone bodies were cored to test the uniaxial compressive strength, as shown in Figure 9b,c. Cores with a diameter of 50 mm and a height of 100 mm were used to obtain the main mechanical properties. The stress-strain curves of the cores of different experiments are presented in Figure 10. The uniaxial compressive strength of a different experiment is listed in Table 5. It can be seen that the stress-strain curves of the five experiments have the same trend, but the uniaxial compressive strength is significantly different.

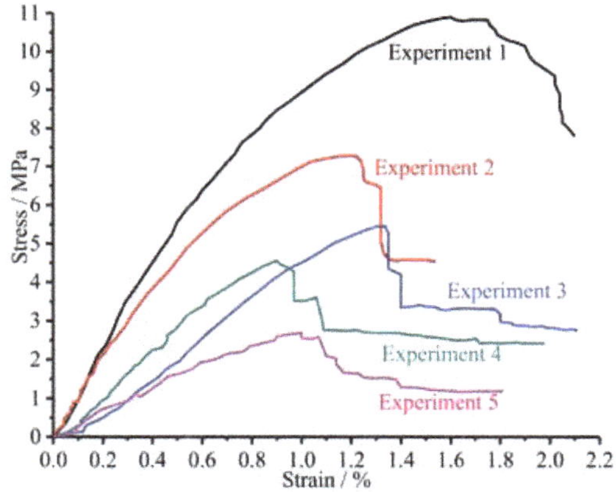

Figure 10. The stress-strain curves of the core for different experiment.

Table 5. Uniaxial compressive strength of cores of different testing programs.

Experiment Number	1	2	3	4	5
Compressive strength (MPa)	10.91	7.28	5.46	4.56	2.69

According to experiment data in Table 1 and the uniaxial compressive strength of cores listed in Table 5, MATLAB was used to perform a regression analysis between the uniaxial compressive strength of specimens and various factors, such as the porosity, water-cement ratio, grouting pressure, grouting time, etc. The regression analysis formula is as follows:

$$P = 0.89n^{0.851}m^{-0.954}p^{0.196}t^{0.705} \quad (7)$$

In Formula (7), the regression fitting exponents for porosity (n), grout water-cement ratio (m), grouting pressure (p), and grouting time (t) are 0.851, −0.954, 0.196, and 0.705, respectively. The negative exponent indicates that the factor and the compressive strength have a negative correlation. Therefore, when grouting the porous fractured coal and rock mass, the grout water-cement ratio has the greatest influence on the strength of the grouted gravel, followed by the grouting permeability. The grouting pressure coefficient has the least influence on the grouting diffusion radius.

4. Field Application Effect and Analysis

In the field, grouting construction often relies only on experience. The result obtained through these experiments can be used to predict the grouting effect and guide the selection of grouting parameters. Therefore, it can overcome the blindness of grouting construction. An engineering practice was performed at the 3101 repeated mining working face of the Shenghua coal mine to verify the experimental results.

4.1. Engineering Geological Conditions

Figure 11 shows the layout of the 3101 repeated mining working face of the Shenghua coal mine. The 3101 repeated mining working face adopts an inclined longwall tilt mining layout, the length of the working face is 300.0 m, and the mining height is 2.5 m. However, there was a collapsed zone in front of the repeated mining working face, which has a tendency length of approximately 25.0 m and a height of approximately 7.0 m.

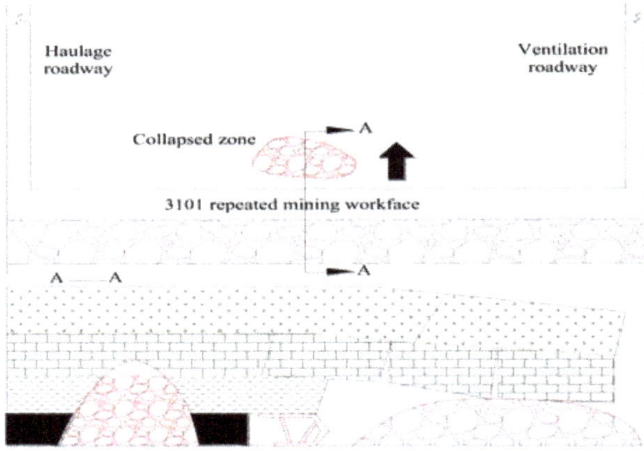

Figure 11. 3101 repeated mining working face.

According to the influence of the grouting consolidation in different layers on safe mining, the consolidation can be divided into a cutting layer and a bearing layer from the bottom to the top. The function of the cut bed is to ensure the stability of the broken surrounding rock after grouting and is easy to cut by a shearer, so the required strength is relatively low. The grouting height of the cutting layer is equal to the mining height, which is 2.5 m. The function of the bearing layer is to form an artificial false roof, which can provide the force surface of the hydraulic support and prevent the influence of gangue caving on the working face. The height of the bearing layer is 4.0 m.

4.2. Grouting Parameters Selection

To ensure that the grout can fully fill the porosity of the fractured surrounding rock and form a stable and continuous cutting layer and bearing layer, the grouting should be injected fully, and its grouting pressure should be significantly increased. The single-hole grouting amount can be estimated according to the following formula [22]:

$$Q = A \cdot L \cdot \pi \cdot R \cdot n \cdot \lambda \tag{8}$$

In Formula (8): Q is the grouting amount of a single hole, m^3; A is the consumption coefficient of the cement grout, taking a value of 1.1; L is the expected filling thickness in the drilling direction, m; R is the diffusion radius of the cement grout, m; n is porosity of collapsed zone; and λ is filling coefficient of the grout, taking a value of 1.2.

Because of the different roles of the upper and lower layers, the grouting parameters are also different. According to the existing data on the mine, the strength of the grouted body of the bearing layer needs to be above 8 MPa, and the strength of the grouted body of the cutting layer needs to above 3 MPa. According to the parameters (R, P, Q, n) that were already known and substituting them into Formulas (3) and (4), the grouting parameters in the collapsed zone were obtained, as shown in Table 6.

Table 6. Single hole grouting parameters of the caving zone.

Action Layer	Diffusion Radius R (m)	Bearing Pressure P (MPa)	Grouting Amount Q (m^3)	Porosity n	Water-Cement Ratio m	Grouting Pressure p (MPa)	Grouting Time t (min)	Permeability k (m/s)
Bearing layer	4	8	34.48	0.52	0.7:1	3	100	0.115
Cutting layer	2.5	3	13.48	0.52	0.9:1	1.5	30	0.226

4.3. Arrangement of Layered Progressive Grouting

To fill the collapsed zone effectively and achieve different effects for different layers, this paper proposed a layered progressive grouting method. The grouting sequence was "from the outside to the inside, from the bottom to the top." After grouting the cutting layer, the same grouting method was used to grout the bearing layer.

The length of the collapsed zone of the 3101 repeated mining face is 25.0 m. According to the grouting diffusion radius mentioned above, the spacing of the grouting holes in the cutting layer was determined to be 5.0 m, the hole depth is 2.5 m, and the distance between the grouting hole and the floor is 1.5 m. The angle between the horizontal plane and grouting hole is 22°, and the height of the final hole is 2.5 m. Three progressive grouting procedures were required to ensure the formation of a continuous and stable cutting layer 2.5 m in height in the repeated mining area. After the cutting layer was constructed, grouting was performed above the cutting layer to form a bearing layer. The spacing of the grouting holes in the cutting layer was determined to be 8.0 m, the hole depth was 4 m, and the distance between the grouting hole and the floor was 4.5 m. The angle between the horizontal plane and grouting hole was 26°, and the height of the final hole was 6.5 m. Two progressive grouting procedures were required to ensure the formation of a continuous and stable bearing layer 4.0 m in height in the repeated mining area. The reinforcement of the collapsed zone can be fully completed by repeating the grouting procedure five times. Figure 12 shows the diagram of the layered progressive grouting method.

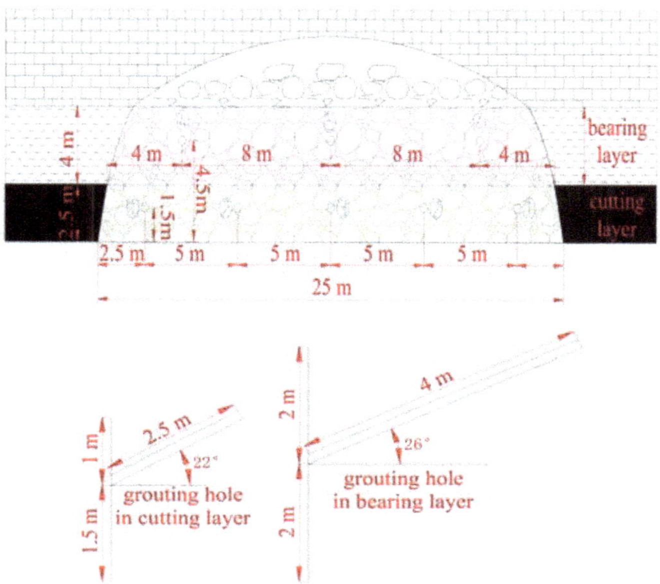

Figure 12. Layered progressive grouting diagram.

The grouting in the collapsed zone used a model ZBSB-148~23 mining grouting pump; the grout mixing equipment adopted a low-speed mechanical mixer; the drilling equipment selected to form the hole was a YT-28 pneumatic drill, with a drilling diameter of ϕ42 mm; and the grouting pipe was composed of a KJR32 high pressure hose and seamless steel tube. Figure 13 shows the partial equipment used for grouting.

(a) (b) (c)

Figure 13. Field grouting equipment: (**a**) Grouting pump; (**b**) Wind coal drill; and (**c**) Grouting pipe.

4.4. Application Effect Testing and Analysis

To ascertain the grout diffusion range and the consolidation effect of the loose fractured rock mass, a borehole observation and core mechanical property test were performed after grouting in the collapsed zone.

The caving area in front of the working face is essentially formed by the caving coal and rock blocks in the early stage of small coal mine mining. After different sizes of coal and rock blocks were stabilized, a large number of voids of different sizes were produced. It can be seen from the borehole images (Figure 14) that the grout is vein-like and discontinuously distributed after filling the voids between the blocks. This shows that the selected cement grout has good fluidity and that the grout diffused into the voids and formed a network skeleton after condensation. The whole structure of loose fractured rock mass was changed by grouting.

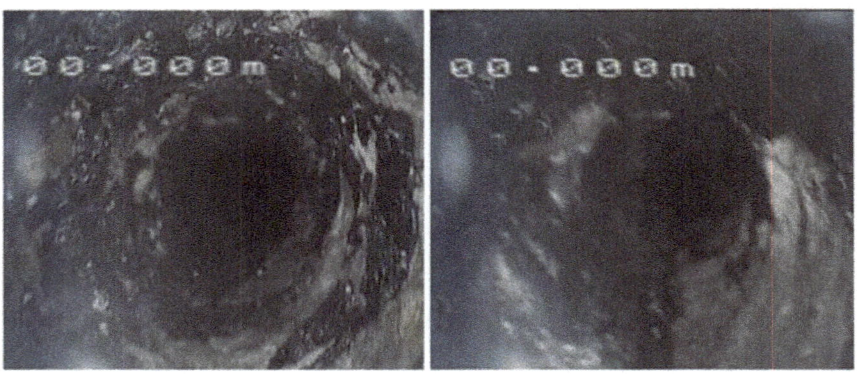

(a) (b)

Figure 14. Borehole images: (**a**) Bearing layer; (**b**) Cutting layer.

To study whether the diffusion radius of grout reaches the required range, a horizontal drilling observation was performed at different layers of the bearing layer and cutting layer to observe the diffusion distances of the grout at different positions, as shown in Figure 15a. It can be seen that the minimum radius of the grout diffusion in the cutting layer is 2.8 m, which is beyond the designed radius of 2.5 m, and the minimum radius of the grout diffusion in the bearing layer is 4.4 m, which is beyond the designed radius of 4.0 m. Therefore, the range of grout diffusion reached the design requirements. Moreover, uniaxial compression experiments on the core specimens taken from drilling were conducted, and the stress-strain curve were obtained, as shown in Figure 15b. It can be seen that the uniaxial compressive strength of the core specimens of the bearing layer and the cutting layer were

8.65 MPa and 3.35 MPa, respectively. The actual strengths are higher than the design strengths, and the strength of the fractured porous rock mass significantly increased.

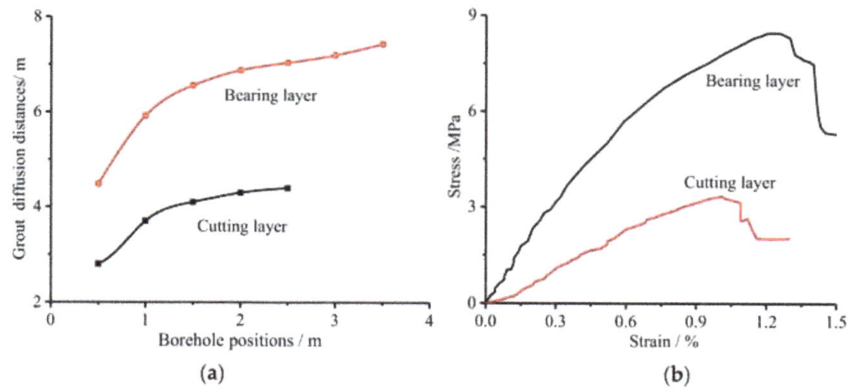

Figure 15. (**a**) Diffusion distances of the grout at different positions; (**b**) Stress-strain curves of the core specimens.

During mining of the grouting area, there was no occurrence of coal wall spalling or tip-to-face roof falling. The working face traversed the old mining caving zone safely and smoothly. The rock mass exposed during the mining is shown in Figure 16. It can be seen that the coal wall and the tip-to-face roof in front of the working face have good integrity. The coal wall is neat after cutting. Thus, a good grouting effect has been achieved.

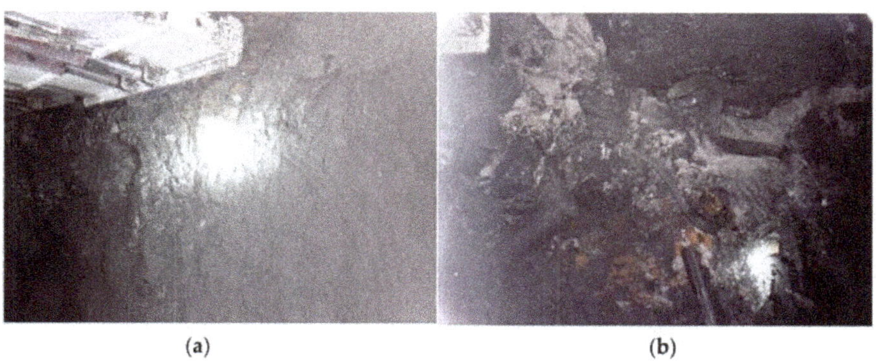

Figure 16. The rock mass exposed in the grouting area: (**a**) Coal wall; (**b**) Tip-to-face roof.

5. Discussion

(1) When grouting in a fractusred porous rock mass, the main diffusion method of grouting is infiltration. The resistance of grout diffusion is mainly caused by friction in the grout and friction between the grout and rock mass medium. The diffusion resistance is small and can be maintained at a low level for an extended period of time. The grouting volume is mainly controlled by the volume and porosity of the fractured rock mass. In engineering practice, the distribution range of the fractured porous rock mass can first be measured, and then the grouting quantity and other related parameters can be calculated. In general, the grouting zone is not a standard sphere, and the equivalent grouting diffusion radius can be used to represent the grouting range.

(2) Grouting in a fractured porous rock mass can improve the strength of the rock mass because grouting can change the properties of the rock mass. On one hand, the grout can improve the

surface mechanical properties of the rock mass, reduce the sliding ability of the fractured rock mass, and enhance the tensile and shear strength parameters. On the other hand, the grout can enter the pores of the rock mass and fill the gaps and cracks in the rock mass. Because of the cohesion of the grout, the cohesion and internal friction angle of the rock mass are improved, and the fractured porous rock mass becomes a strong bearing capacity consolidation body, which is difficult to destroy.

6. Conclusions

(1) To study the law of grouting diffusion in a fractured rock mass, this paper developed a visualization platform of the grouting diffusion range and a three-dimensional grouting experimental system. It can monitor the grout diffusion range, diffusion time, grout pressure, etc.

(2) The grouting pressure has four kinds of variation forms in the grouting experiment: linear growth; cyclic growth; fluctuation stability and slow growth.

(3) The flow pattern of the grout in the fractured porous rock mass has a parabolic shape from the grouting hole to the bottom. The lower the level is, the larger the diffusion range of the grout is. In the horizontal direction, the grout diffusion surface around the grouting hole had an asymmetrical circular surface. When the height of the grouting pipe, initial velocity of the grout, outlet angle and pore connectivity are determined, the shape of diffusion surface is certain.

(4) The grouting pressure has the greatest influence on the grouting diffusion radius, followed by the grouting horizon and water-cement ratio. The grouting permeability coefficient has the least influence on the grouting diffusion radius.

(5) The uniaxial compressive strength is significantly different in different experiments. The grout water-cement ratio has the greatest influence on the strength of the grouted gravel, followed by the grouting permeability. The grouting pressure coefficient has the least amount of influence on the grouting diffusion radius.

(6) The results of the study were applied to old mining caving zone grouting reinforcement engineering. According to the results, the grouting parameters were designed and a layered progressive grouting method was proposed. Borehole observation and a core mechanical property test were performed to analyze the application effect. The working face traversed the old mining caving zone safely and smoothly. A good grouting effect was achieved.

Author Contributions: All the authors contributed to publishing this paper. H.L., X.C. and T.W. contributed to the formulation of overarching research goals and aims; D.J. contributed to the experiment; M.Z. and R.H. provided language and picture support.

Funding: This research was funded by the National Natural Science Foundation of China, grant number 51379116 and the Natural Science Foundation of Shandong Province, grant number ZR2016EEM36).

Acknowledgments: The authors would like to express sincere appreciation to the editor and the anonymous reviewers for their valuable comments and suggestions for improving the presentation of the manuscript.

Conflicts of Interest: The authors declare no conflict of interest.

References

1. Wang, K.; Gong, P.; Zhang, X.; Lian, Q.; Li, J.; Duan, D. Characteristics and control of roof fracture in caving zone for residual coal mining face. *Chin. J. Rock Mech. Eng.* **2016**, *35*, 1–9.
2. Liu, C.; Gong, P.L.; Wang, K.; Zhang, X.Q.; Liu, Y.D. Roof stability for repeated mining workface passing through abandoned parallel gateway. *J. China Coal Soc.* **2015**, *40*, 314–322.
3. Ghosh, G.K.; Sivakumar, C. Application of underground microseismic monitoring for ground failure and secure longwall coal mining operation: A case study in an Indian mine. *J. Appl. Geophys.* **2018**, *150*, 21–39. [CrossRef]

4. Riedel, I.; Guéguen, P. Modeling of damage-related earthquake losses in a moderate seismic-prone country and cost-benefit evaluation of retrofit investments: Application to France. *Nat. Hazards* **2018**, *90*, 639–662. [CrossRef]
5. Xia, X.; Huang, Q. Study on the dynamic height of caved zone based on porosity. *J. Min. Saf. Eng.* **2014**, *31*, 102–107.
6. Feng, G.; Jia, K.; Shang, B. Application and prospect of super-high-water packing material in mining engineering. *Coal Sci. Technol.* **2015**, *43*, 5–9.
7. Marwan, A.; Zhou, M.M.; Abdelrehim, M.Z.; Meschke, G. Optimization of artificial ground freezing in tunneling in the presence of seepage flow. *Comput. Geotech.* **2016**, *75*, 112–125. [CrossRef]
8. Li, S.J.; Wang, L.G.; Lu, Y.L.; Zhang, B. Slurry diffusion within cracked wall rock during the bolt-grouting process. *J. China Univ. Min. Technol.* **2011**, *40*, 874–880.
9. Di Nucci, C. Unsteady free surface flow in porous media: One-dimensional model equations including vertical effects and seepage face. *C. R. Mec.* **2018**, *346*, 366–383. [CrossRef]
10. Kelessidis, V.C.; Maglione, R.; Tsamantaki, C. Optimal determination of rheological parameters for Herschel-Bulkley drilling fluids and impact on pressure drop, velocity profiles and penetration rates during drilling. *J. Pet. Sci. Eng.* **2006**, *5*, 203–224. [CrossRef]
11. Shucai, L.; Zhuo, Z.; Rentai, L. Analysis of diffusion of grout in porous media considering infiltration effects. *Chin. J. Rock Mech. Eng.* **2015**, *34*, 2401–2409.
12. Huang, Y.; Wang, L.; Lu, Y. Study on the law of slurry diffusion within roadway surrounding rock during the whole section bolt-grouting process. *J. Min. Saf. Eng.* **2015**, *32*, 240–246.
13. Saada, Z.; Canou, J.; Dormieux, L.; Dupla, J.C.; Maghous, S. Modeling of cement suspension flow in granular porous media. *Int. J. Numer. Anal. Methods Geomech.* **2005**, *29*, 691–711. [CrossRef]
14. Rafi, J.Y.; Stille, H. Basic mechanism of elastic jacking and impact of fracture aperture change on grout spread, transmissivity and penetrability. *Tunn. Undergr. Space Technol.* **2015**, *49*, 174–187. [CrossRef]
15. Shimada, H.; Hamanaka, A.; Sasaoka, T.; Matsui, K. Behavior of grouting material used for floor reinforcement in underground mines. *Int. J. Mech. Sci.* **2014**, *28*, 133–148.
16. Funehag, J.; Gustafson, G. Design of grouting with silica sol in hard rock: New methods for calculation of penetration length Part I. *Tunn. Undergr. Space Technol.* **2008**, *23*, 1–8. [CrossRef]
17. Funehag, J.; Gustafson, G. Design of grouting with silica sol in hard rock: New methods for calculation of penetration length Part II. *Tunn. Undergr. Space Technol.* **2008**, *23*, 9–17. [CrossRef]
18. Minto, J.M.; MacLachlan, E.; El Mountassir, G.; Lunn, R.J. Rock fracture grouting with microbially induced carbonate precipitation. *Water Resour. Res.* **2016**, *52*, 8810–8827. [CrossRef]
19. Minto, J.M.; MacLachlan, E.; El Mountassir, G.; Lunn, R.J. Study on grouting simulation experiment and its application. *Chin. J. Geotech. Eng.* **1997**, *19*, 28–33.
20. Fang, K.T.; Ma, C.X. *Orthogonal and Uniform Experimental Design*; Science Press: Beijing, China, 2001.
21. Liang, D.; Jiang, Z.; Cao, D. Calculation of hydraulic conductivity in high water-pressure test. *J. Min. Saf. Eng.* **2016**, *33*, 324–328.
22. Yang, Z.Q.; Niu, X.D.; Hou, K.P.; Zhou, Z.H.; Liang, W. Study of grouting diffusion parameters in gravel soil. *Rock Soil Mech.* **2015**, *36*, 397–402.

processes

MDPI

Article

Numerical Investigation of the Failure Mechanism of Transversely Isotropic Rocks with a Particle Flow Modeling Method

Xu-Xu Yang [1,2,*], Hong-Wen Jing [2] and Wei-Guo Qiao [1]

1 Shandong Provincial Key Laboratory of Civil Engineering Disaster Prevention and Mitigation, Shandong University of Science and Technology, Qingdao 266590, China; skd995559@sdust.edu.cn
2 State Key Laboratory for Geomechanics and Deep Underground Engineering, China University of Mining and Technology, Xuzhou 221116, China; yangxucumt@126.com
* Correspondence: yangxu@sdust.edu.cn; Tel.: +86-0532-8605-7646

Received: 30 August 2018; Accepted: 14 September 2018; Published: 17 September 2018

Abstract: Transversely isotropic rocks are commonly encountered in rock engineering practices, and their strength and failure behavior is often governed by the property of anisotropy. The particle flow modeling method was utilized to investigate the failure mechanism of transversely isotropic rocks subject to uniaxial compressive loading. The details for establishing transversely isotropic rock models were first presented, and then a parametric study was carried out to look into the effect of interface properties on the failure mode and strength of transversely isotropic rock models by varying the interface dip angle. The smooth joint model was incorporated to create interfaces for the completeness of establishing transversely isotropic rock models with the particle flow modeling method. Accordingly, three failure modes observed in transversely isotropic rock models with varying dip angles were tensile failure across interfaces, shear failure along interfaces, and tensile failure along interfaces. Furthermore, the interface mechanical parameters were found to differently influence the failure behavior of transversely isotropic rock models. The bonded joint cohesion and bonded joint friction angle that contribute to the shear strength of interfaces have considerable influence on the uniaxial compressive strength (UCS) values, while the joint coefficient of friction and joint tensile strength have a slight influence on the UCS values. The findings in this paper indicated the importance of interfaces in estimating failure behavior of transversely isotropic rocks.

Keywords: transversely isotropic rocks; failure mechanism; particle flow modeling; interface

---

1. Introduction

The dominant anisotropy or transverse isotropy of geological materials, especially of foliated metamorphic rocks, such as slates, gneisses, schist, and sedimentary rocks with bedding planes, leads to complicated failure behaviors [1]. Rock anisotropy is one of the most significant characteristics that should be taken into consideration for underground engineering. The design and stability analysis of underground structures excavated in anisotropic rock masses, for instance, require a complete understanding of the failure behavior of the rock materials. Thus, the failure mechanism of anisotropic or transversely isotropic rocks has been a significant topic in rock mechanics.

Engineering practice has suggested that the rock anisotropy is of importance in the stabilization of underground excavations in bedded rock masses [2,3]. Previous studies indicate that drilling through bedding planes was quite dangerous. The boreholes might become unstable if the deviation angle of drilled well applied to sub-horizontal bedding planes is very high due to rock strength anisotropy [4–8]. Particularly, Okland et al. [4] carried out hollow cylinder tests to study the critical bedding inclination angle of shale that can induce severe borehole damage during extended reach drilling. Meier et al. [6] also claimed that the lowest stress is required to induce borehole breakout when the borehole is sub-parallel to the bedding planes. Moreover, Zhou et al. [9] reported that in the Wudongde hydropower station on the Jinsha River in China, instability of high sidewalls, particularly separation of bedding planes and the accompanying remarkable deformation, is likely to happen in those sections where strata are almost parallel to the cavern axis with a thickness less than 10 cm. For better engineering design, the mechanical properties of aforementioned anisotropic rock materials are suggested to be determined through the field tests or through rock mass classification [10].

However, due to the inherent difficulty and inaccuracy in conducting field tests or rock mass classification, many investigators tend to investigate the mechanical properties of anisotropic rock materials through laboratory tests. To be specific, Gatelier et al. [11], according to laboratory tests, stated a decrease in uniaxial compressive strength with increasing inclination angle in Adamswiller sandstone. Kim et al. [12] through X-ray CT (Computed Tomography) study and uniaxial compression tests found that Berea sandstone in northern Ohio is composed of cross-bedded loose layers and relatively thin tightly packed layers, and its uniaxial compressive strength (UCS) value decreases with increasing porosity as well as with increasing inclination of the bedding plane. Heng et al. [13] investigated the directional shale samples obtained from the Longmaxi Formation in Shizhu County, China, and three types of shear failure modes were identified that were dependent on the shearing angle, which were sliding failure across the bedding plane, sliding failure along the bedding planes, and sliding failure across the bedding planes combined with tensile splitting along the bedding planes. Besides these real rock materials prepared from engineering sites, some rock-like materials have also been utilized to investigate the anisotropic or transversely isotropic mechanism. Kulatilake et al. [14] carried out uniaxial compressive tests on the jointed rock-like material blocks having existing joint sets. Their findings show that for rock samples with dip angles of 0° to 15°, the failure mechanism was the tensile failure through the intact model material, while for rock samples with joint dip angles of 40° to 60°, the main failure mechanism was the combined shear and tensile failure through the joints. A mixed mechanism of the above two modes accounted for the failure of jointed blocks having dip angles of 20° to 35°. Yang et al. [15] performed a series of physical model tests for jointed rock masses with persistent discontinuities and indicated that the failure modes of these models with different dip angles can be divided into split mode, sliding mode, and mixed mode. Recently, an experimental investigation was implemented by Yang et al. [16] on jointed rock models made of rock-like materials. In accordance with the experimental results, the failure of simulated rock models with varied orientations is classified into one of four modes: (a) tensile failure across the joint plane, (b) shear failure along the joint plane, (c) tensile failure along the joint plane, and (d) intact material failure. These above-mentioned researchers explained the anisotropic or transversely isotropic mechanism of rock materials with weak planes/layers presence. These studies broaden our understanding of weak planes/layers geometry on the mechanical property of rock materials.

As one important research methodology in rock mechanics as well as in geotechnical engineering, the numerical modeling method has significantly enriched the data bank of anisotropic rock behavior besides those obtained from laboratory experiments. Particularly, the particle flow modeling method with the basis of the discrete element theory has obtained a pronounced development from being applied to rock and soil mechanics to greatly broader applications. This approach is capable of deriving the rock mass response based on the relatively simple particle contact laws at joints and in rock instead of more complex constitutive models [17,18]. Therefore, a particle flow modeling approach could be utilized directly to extend the experimental results obtained in the laboratory. For example, Chiu et al. [19] put forward a modified smooth-joint model utilizing the particle flow modeling approach to mimic the anisotropic behavior of a rock mass, and the failure modes of the rock mass obtained by Yang et al. [15] in the laboratory were perfectly reproduced. The load-deformation curves in different joint orientations were also well mimicked. Bahaaddini et al. [18], by using particle flow modeling, thoroughly investigated the effect of geometric parameters of joints on the rock mass failure mechanism, deformation modulus, and uniaxial compression strength, and critically compared these with the physical experiments [20]. Park and Min [21] also conducted particle flow simulations with embedded smooth joints for mimicking the strength and failure behavior of transversely isotropic rocks with systematic verification and extensions to laboratory and field problems. They succeeded in capturing the failure modes observed in anisotropic rock in which weak planes significantly matter. Furthermore, the particle flow modeling has been utilized as well to study the failure behavior of transversely isotropic rock mass through Brazilian tests [1,22]. Although the particle flow modeling promotes the understanding of the anisotropic or transversely isotropic rock mass mechanical behavior, the concentration is still mainly put on the geometrically structural effect. On the other hand, the influence of bedded planes/layers' mechanical properties on the failure response of anisotropic or transversely isotropic rock masses is not well understood.

This study aims to better understand the influence of mechanical properties of bedded layers (interfaces) on the failure behavior of transversely isotropic rocks through a particle flow modeling method. Thus, this paper is organized in the following matter. First, the physical experiment on simulated transversely isotropic rocks is briefly introduced in Section 2.1 as a benchmark for particle flow modeling. After that, the procedure to establish particle flow models, which consists of intact materials and interfaces, and the calibration of their mechanical parameters, are illustrated in Section 2.2. With these numerically simulated models, the failure mechanism of transversely isotropic rocks is analyzed at the mesoscale under uniaxial compressive loading in Section 3.1. In Section 3.2, a sensitivity study is carried out systematically to estimate the influence of mechanical parameters of interfaces on the failure strength of transversely isotropic rocks.

2. Materials and Methods

2.1. Physical Experiment on Simulated Transversely Isotropic Rocks

To investigate the failure mechanism of the transversely isotropic rocks through particle flow modeling, associated experimental results are required. The present modeling study refers to artificial transversely isotropic rocks, as implemented by Tien et al. [23], and utilized their test data for the base of the particle flow simulations. The physical experiment program and results are described as follows.

Two model materials, referred to as material A and material B, respectively, were chosen to combine into the transversely isotropic rock samples. Material A was composed of cement, kaolinite, and water in a weight ratio of 4:1:1.2, while material B was composed of cement, kaolinite, and water in a weight ratio of 1:1:0.6. According to the test results of the materials' mechanical properties, the uniaxial compressive strength (UCS), Young's modulus, and Poisson's ratio of material A are 104.2 MPa, 21.7 GPa, and 0.23, respectively, representing rocks with high strength and stiffness. On the other hand, the uniaxial compressive strength, Young's modulus, and Poisson's ratio of material B are 43.3 MPa, 11.9 GPa, and 0.21, respectively, representing rocks with relatively low strength and stiffness.

Tien et al. [23] incorporated eight steps to develop these transversely isotropic rock samples. After the mold assembly and model material preparation steps (steps 1 and 2), the preliminary compaction step (step 3) was conducted by using the MTS (Mechanical Testing and Simulation) servo-controlled loading frame to apply an axial load of 20 kN on the model materials to make them into a preliminary square rock mass soft enough for cutting into slices. In the cutting into slices step (step 4), the preliminarily compacted model materials (material A and material B) were put on a cutting platform and cut with a slice cutter into slices with a thickness of 5 mm in each layer. After that, a suction lifter was utilized to pile material A slices and material B slices up in sequence, which is referred to as the stacking slices step (step 5). Following the repairing step (step 6), the final compaction step (step 7) was implemented to apply a final load of 114 kN on those sliced materials, and then it was held for one hour until a steady settlement was achieved. Finally, upon the curing and drilling step (step 8), the transversely isotropic rock samples having different dip angles ($\alpha = 0°, 15°, 30°, 45°, 60°, 75°$, and $90°$) were created for the subsequent compression tests. These artificial transversely isotropic rock samples were developed to represent the anisotropic behavior of sandstone-shale interlayered rock masses. Figure 1 shows the physical dimension of a typical transversely isotropic rock sample.

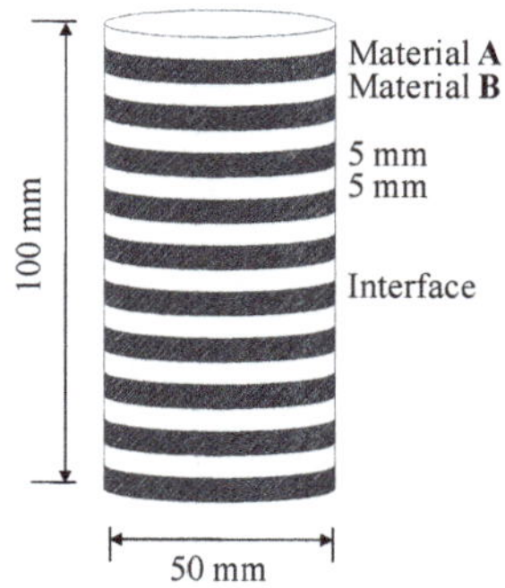

Figure 1. A typical transversely isotropic rock sample (α = 0°).

These artificial transversely isotropic rock samples were then subjected to compression tests. In accordance with the test results [23], with zero confining pressure, three failure modes were identified. When the dip angles were relatively low (α = 0°, 15°, and 30°), the artificial transversely isotropic rock samples failed in a mode of tensile fracture across discontinuities; whereas, when the dip angles were medium (α = 45°, 60°, and 75°), the artificial transversely isotropic rock samples failed in a mode of sliding failure along discontinuities. Moreover, for artificial transversely isotropic rock sample having a vertical dip angle, its failure mode was defined as the tensile-split along discontinuities.

2.2. Particle Flow Modeling of Material A and Material B

The particle flow code in three dimensions (PFC^{3D}) is a commercial software package based on the discrete element method [24,25]. In this study, PFC^{3D} was utilized to develop the particle flow models of transversely isotropic rocks. In PFC^{3D}, the intact materials (material A and material B) were mimicked through a composite of particles that interact with each other at contacts (Figure 2a). A linear contact model provides an elastic relationship between the relative displacements and forces of particles at the point contact (Figure 2b). This model consists of the contact normal force component, F_n, contact overlap, U_n, shear increment, ΔF_s, and shear displacement increment, ΔU_s, as follows:

$$F_n = k_n U_n \tag{1}$$

$$\Delta F_s = -k_s \Delta U_s \tag{2}$$

with k_n and k_s being the contact normal and shear stiffness.

The frictional resistance of the contact is as follows:

$$F_s \le \mu F_n \tag{3}$$

with μ being the friction coefficient between particles.

To numerically mimic a relatively rock-like material, it is required to stick these granular particles through a bonded model [26]. The bonded model, herein, is a parallel bond model that resists not only the contact forces but also the moments between the particles at a cemented contact (Figure 2c). The function mechanism of the parallel bond model is described as:

$$\Delta \overline{F}_n = \overline{k}_n A \Delta U_n \tag{4}$$

$$\Delta \overline{F}_s = -\overline{k}_s A \Delta U_s \tag{5}$$

and

$$\Delta\overline{M}_n = -\overline{k}_s J \Delta\theta_n \tag{6}$$

$$\Delta\overline{M}_s = -\overline{k}_n I \Delta\theta_s \tag{7}$$

with $\overline{F}_n$, $\overline{F}_s$, $\overline{M}_n$, and $\overline{M}_s$ being the force components and moments about the center of the cemented-contact zone, respectively; $\overline{k}_n$ and $\overline{k}_s$ being the normal and shear bond stiffness per unit area, respectively; θ_n and θ_s being the rotation angle components, respectively; and A, J, and I being the area, polar moment of inertia, and moment of inertia of the bond contact cross-section, respectively. The strength value of the bonded contact is given by:

$$\overline{\sigma}_{\max} = \frac{-\overline{F}_n}{A} + \frac{|\overline{M}_s|\overline{R}}{I} < \overline{\sigma}_c \tag{8}$$

$$\overline{\tau}_{\max} = \frac{-\overline{F}_s}{A} + \frac{|\overline{M}_n|\overline{R}}{J} < \overline{\tau}_c \tag{9}$$

with $\overline{R}$ being the radii of the cemented contact plane between particles (Figure 2c); and $\overline{\sigma}_c$ and $\overline{\tau}_c$ being the tensile and shear strength of the bond contact, respectively.

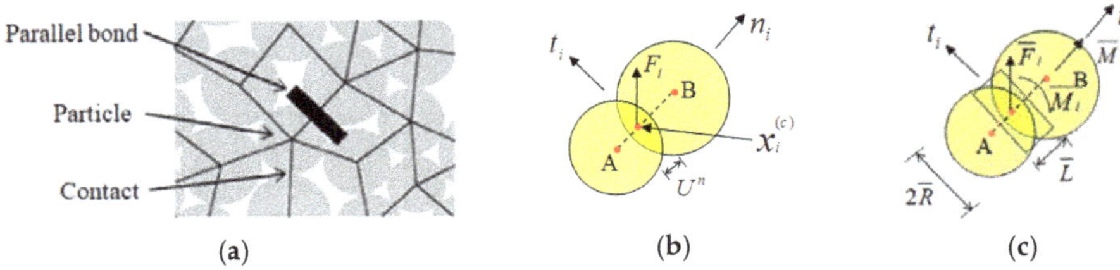

Figure 2. The contact model and parallel bond model: (**a**) intact material, (**b**) particle contact, and (**c**) parallel bond.

Tensile cracks occur when the applied normal stress exceeds the specified tensile strength of the parallel bond, $\overline{\sigma}_c$, while shear cracks happen as the applied shear stress surplus the shear strength, $\overline{\tau}_c$, either through rotation or through sliding of particles. The tensile strength at the contact soon falls to zero once the cracks occur, while the shear strength decreases as the residual friction strength (Equation (3)).

For presenting the aforesaid mechanical properties of material A and material B, this numerical modeling method requires a calibration of the microparameters to describe contact and bond deforming and strength behavior, and the selected package of microparameters are expected to derive the macro mechanical properties of simulated material used in physical experiments [23]. Therefore, a calibration procedure was carried out to determine these microparameters for intact rock materials (material A and material B). Several steps [25,27–29] are followed to reproduce certain properties of the artificial materials. First, the particle and parallel bond moduli and the ratios of normal to shear stiffness are set to be equal between the particles and parallel bonds, which aims to decrease the number of independent parameters. Second, the Young's modulus is determined through setting the material strengths to a large value and varying E_c (particle Young's modulus) and $\overline{E}_c$ (parallel bond Young's modulus) to match the Young's modulus between the particle flow modeling and laboratory samples (with the size of φ 50 mm × 100 mm). Then, through varying k_n/k_s (particle normal stiffness/shear stiffness) and $\overline{k}_n/\overline{k}_s$ (parallel bond normal stiffness/shear stiffness), the Poisson's ratio of the intact synthetic cylindrical sample was matched to the laboratory samples. Afterwards, the strength between the numerical and laboratory samples were matched by decreasing the normal and shear bond strengths of the parallel bonds. During this procedure, it is of importance to fix the ratio of normal to shear bond strength ($\overline{\sigma}_c/\overline{\tau}_c$) for the reason that it influences the failure pattern of the sample. The determined microparameters of the intact rock material were listed in Table 1. A comparison between particle flow modeling and experiment results in Table 2 claims a good capability of this numerical modeling method to reproduce these simulated mechanical properties under uniaxial compression loading. The deviations for uniaxial compressive strength (UCS), Young's modulus (E), and Poisson's ratio (ν) of material A and material B were all less than 5.0%.

Table 1. Microparameter values of intact rock materials (material A and material B).

Property	Parameter	Value1 (Material A)	Value2 (Material B)
Particle	ρ (kg/m^3)	2150	1760
	k_n/k_s	1.65	1.55
	E_c (GPa)	18.8	10.3
	μ	0.554	0.466
	$R_{rat} = R_{\max}/R_{\min}$	1.66	1.66
	$R_{\min}$ (mm)	0.65	0.65
Parallel bond	λ	1.0	1.0
	$\overline{k}_n/\overline{k}_s$	1.65	1.55
	$\overline{E}_c$ (GPa)	18.8	10.3
	$\overline{\sigma}_c$ (mean $\pm$ std.dev., MPa)	76.0 $\pm$ 19.0	31.7 $\pm$ 7.9
	$\overline{\tau}_c$ (mean $\pm$ std.dev., MPa)	152.0 $\pm$ 38.0	63.4 $\pm$ 15.9

Note that ρ is the density of the synthetic rock material; λ is the radius multiplier used to set the parallel bond radii; R_{rat} is the radius of the particle; $R_{\max}$ and $R_{\min}$ are the maximum radius and minimum radius of the particle, respectively.

Table 2. Comparison of mechanical properties of intact material between the physical experiment and the particle flow modeling results.

Material	Macro Properties	Experimental Results	Numerical Results	Abs. Deviation
Material A	UCS (MPa)	104.2	105.8	1.53%
	E (GPa)	21.7	20.8	4.15%
	ν	0.230	0.224	2.61%
Material B	UCS (MPa)	43.3	44.1	1.85%
	E (GPa)	11.9	11.4	4.20%
	ν	0.210	0.202	3.81%

2.3. *Particle Flow Modeling of Interface between Material A and Material B*

2.3.1. Smooth Joint Model

As aforementioned, the simulated transversely isotropic rock samples were prepared through piling up the material A slices and material B slices in sequence [23]. The interfaces between material A and material B (as shown in Figure 1) were discontinuities that behave quite differently from the adjacent intact materials. Hence, to better describe the mechanical behavior of transversely isotropic rocks the interfaces were intersected between material A and material B slices and represented with the smooth joint model. The numerically developed transversely isotropic rock samples having different interface dip angles are shown in Figure 3.

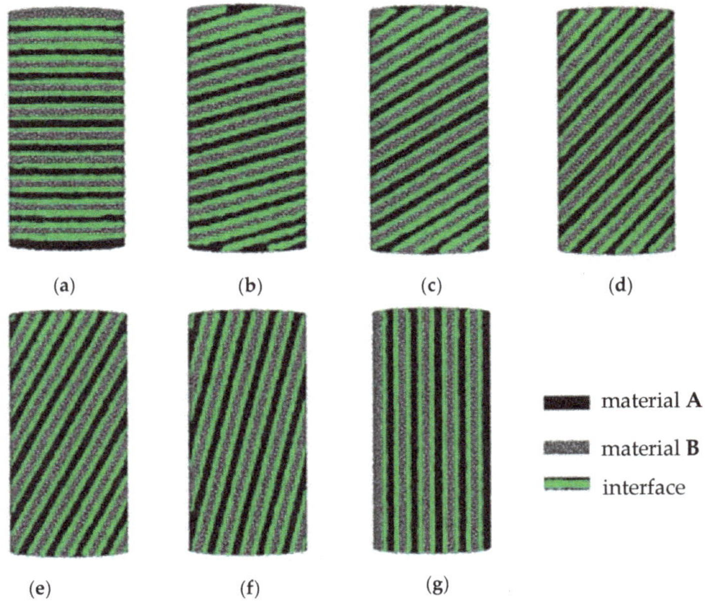

Figure 3. Numerically developed transversely isotropic rock samples having different interface dip angles: (**a**) $\alpha = 0°$, (**b**) $\alpha = 15°$, (**c**) $\alpha = 30°$, (**d**) $\alpha = 45°$, (**e**) $\alpha = 60°$, (**f**) $\alpha = 75°$, and (**g**) $\alpha = 90°$.

The smooth joint model is shown in Figure 4. Once a joint plane is defined, a smooth joint is defined at contacts between the particles whose centers are lying on the opposite sides of the joint plane. At the contacts, the existing parallel bonds will be removed first, and the smooth joints are assigned in the direction parallel to the joint plane. These contacts will act in accordance with the rules defined through the smooth joint model with particular parameter values assigned by the user. The particles having such contacts may overlap or overpass through each other rather than to roll around one another (Figure 4).

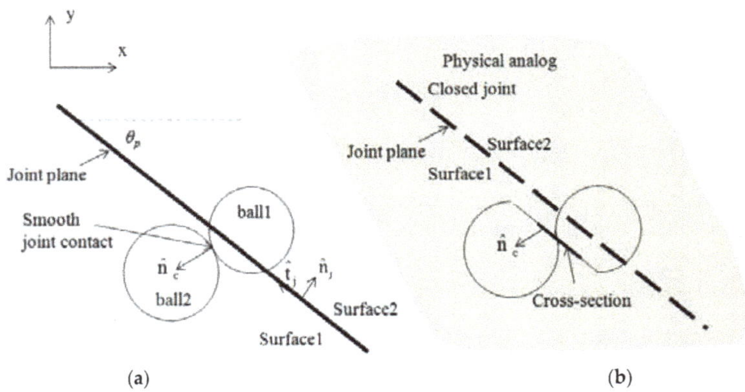

Figure 4. Smooth joint model: (**a**) numerical analog, and (**b**) physical analog.

The newly defined contacts act mechanically like a set of elastic springs uniformly distributed on a circular cross-section disc with the center at the contact point. The area A of the smooth joint disc is defined as:

$$A = \pi \overline{R}^2 \tag{10}$$

where $\overline{R}$ is the disc radius:

$$\overline{R} = \lambda \min(R_1, R_2) \tag{11}$$

where λ is a radius multiplier (usually set as 1.0), and R_1 and R_2 are particle radii, as shown in Figure 4.

The smooth joint model can behave with respect to the following basic modes: (i) not bonded and never fail mode, (ii) not bonded and fail in tension mode, (iii) not bonded and fail in shear mode, and (iv) bonded mode. Through these modes, the smooth joint model could be used to simulate various discontinuity conditions. In the present paper, we use the smooth joint model with the bonded mode (mode (iv)) to mimic the interfaces between material A slices and material B slices. Besides the stiffness provided by the aforementioned springs, the bonded smooth joint model also possesses both the normal and shear strengths. The shear strength is given by:

$$\overline{\tau}_{cj} = \overline{\sigma}_{nj} \tan \varphi_j + c_j \tag{12}$$

In Equation (12), $\overline{\sigma}_{nj}$ implies the normal stress acted on the joint surface; φ_j implies the bonded joint friction angle; while c_j implies the bonded joint cohesion, given in MPa. If the normal or shear stress exceeds the corresponding bond strength, the joint bond between the joint surfaces fails and the bond stiffness will be removed. As long as the joint bond fails either in shear or tensile pattern, the shear strength reduces as its residual value and the tensile strength will be set to be zero. The residual shear strength is a function of the normal stress, $\overline{\sigma}_{nj}$, as well as the joint friction coefficient, μ_j. Figure 5 shows the constitutive law for the bonded smooth joint model.

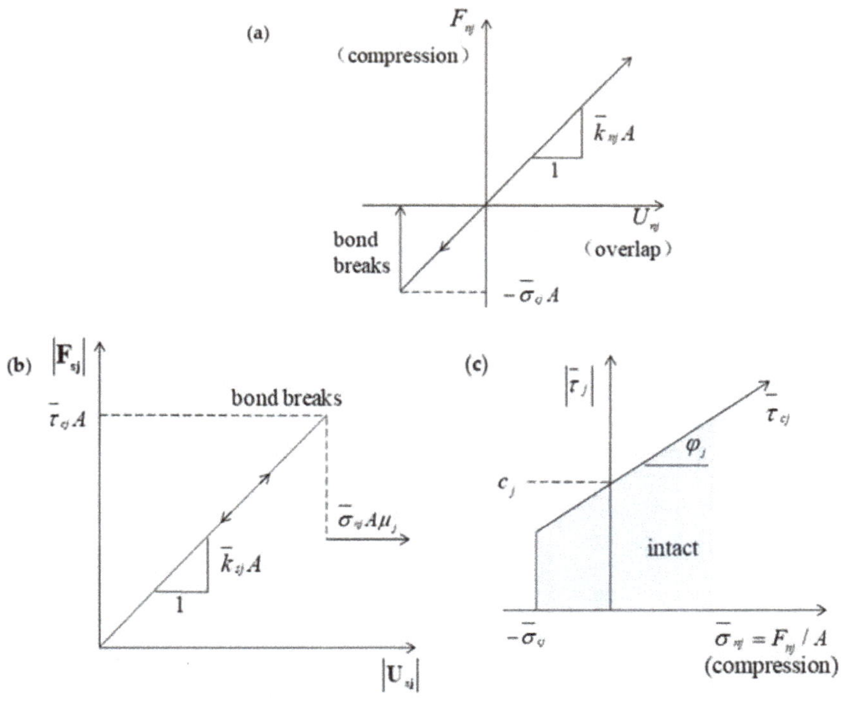

Figure 5. Constitutive law for a bonded smooth joint model: **(a)** normal force, F_{nj}, versus normal displacement, U_{nj}; **(b)** shear force, $\mathbf{F_{sj}}$, versus shear displacement, $\mathbf{U_{sj}}$; and **(c)** strength envelope.

2.3.2. Calibration and Validation of Smooth Joint Mechanical Parameters

The smooth joint parameters consist of the joint normal stiffness, $\overline{k}_{nj}$; joint shear stiffness, $\overline{k}_{sj}$; joint tensile strength, $\overline{\sigma}_{cj}$; joint shear strength, $\overline{\tau}_{cj}$; joint coefficient of friction, μ_j; and joint dilation angle,

ψ_j. These mechanical parameters have been determined through a calibration procedure utilizing the data of experiments carried out by Tien et al. [23] on simulated transversely isotropic rock samples. The calibration process is illustrated in Figure 6. During the calibration procedure, sets of trial-and-error tests were implemented on numerically developed transversely isotropic rock samples (Figure 3) to reproduce the uniaxial compressive strength data with those obtained by Tien et al. [23].

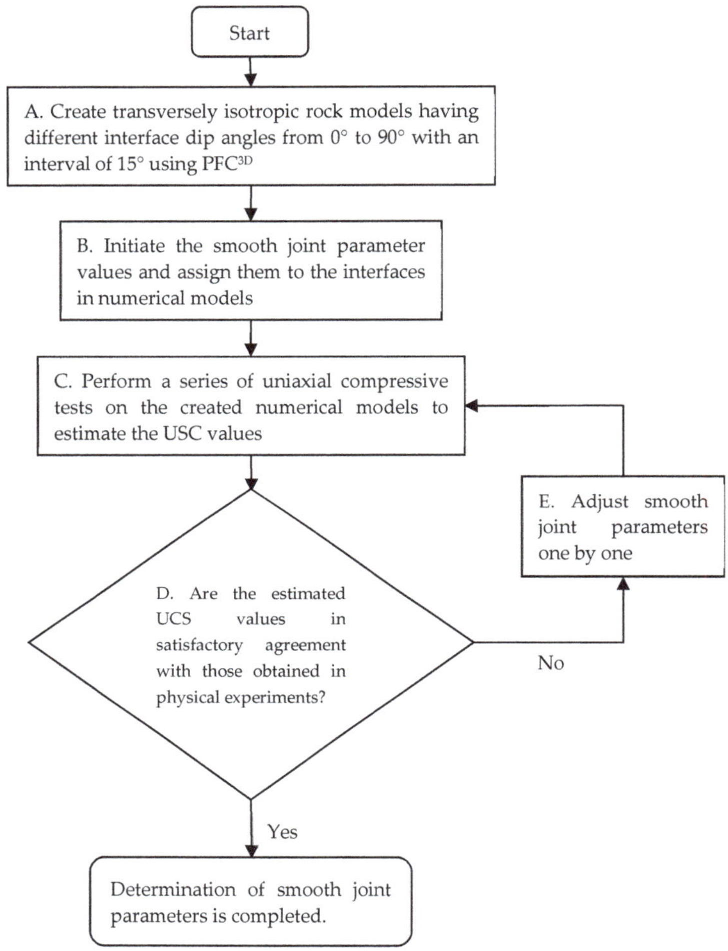

Figure 6. Calibration process for determination of smooth joint parameters.

Through the above calibration procedure, the mechanical parameters of the smooth joint model for simulating interfaces are determined as listed in Table 3. Note that the value for each smooth joint parameter is not unique. That is to say, each joint parameter in Table 3 is changeable along with re-adjusting other joint parameters. Strong coupling effects exist between various joint parameters such that the macro mechanical properties of the simulated transversely isotropic rock models is a result of the complicated interaction among these joint parameters [30]. Based on the calibrated smooth joint mechanical parameters, the UCS of transversely isotropic rock samples are matched well with those obtained by Tien et al. [23]. Table 4 shows a comparison of unconfined strength values (UCS) between the experiments and the particle flow simulations. The comparison given claims the capability of the particle flow modeling method to derive the strength property of transversely isotropic rock models under uniaxial compressive loading. The uniaxial compressive strength values obtained through particle flow simulations agree well with those obtained through physical experiments. The particle flow models with the determined intact material and smooth joint mechanical parameters are subsequently utilized to investigate the failure mechanism of transversely isotropic rocks.

Table 3. Mechanical parameter values used for the smooth joint model.

Mechanical Parameter	Determined Value	Mechanical Parameter	Determined Value
$\bar{k}_{nj}$ (N/m^3)	1.81×10^{12}	$\bar{\sigma}_{cj}$ (MPa)	23.8
$\bar{k}_{sj}$ (N/m^3)	0.79×10^{12}	c_j (MPa)	13.0
μ_j	0.78	φ_j	22.5°
ψ	0	-	-

Table 4. Comparison of UCS values between physical experiment and numerical simulation results.

Comparison Condition	UCS Value for Different Interface Dip Angle						
Dip angle, α (deg.)	0	15	30	45	60	75	90
Physical experiment (MPa)	53.0	47.0	50.0	38.0	30.0	38.0	66.0
Numerical simulation (MPa)	51.6	56.8	51.5	36.7	32.0	36.2	52.4

3. Results and Discussion

3.1. Failure Mechanism of Transversely Isotropic Rocks

For better understanding of the failure mechanism of transversely isotropic rocks under uniaxial compressive loading, the created particle flow models, of which the mechanical parameters are determined through the above calibration procedure, were investigated under differing interface dip angles. The interface dip angles were varied from 0° to 90° with an interval of 15°. In this section, failure modes of transversely isotropic rock models are analyzed in terms of newly generated cracks which present the breakage of parallel bonds as well as the smooth joint bonds. Figure 7 plots the failure process of transversely isotropic rock models having different interface dip angles. Note that the breakage of parallel bonds is shown as red (tensile) or blue (shear) crack items, whereas the breakage of smooth joint bonds is marked as magenta (tensile) or black (shear) crack items. As shown in Figure 7, when the interface dip angles, α, are 0° and 15°, under uniaxial compressive loading the cracks initiate in the soft material (material B) rather than in the hard material (material A). With subsequent compression, the cracks (tensile and shear) increasingly occur both in material A and material B, and eventually coalesce into a macro failure plane penetrating the transversely isotropic rock models. This macro failure plane comes across the simulated interfaces, the rock models fail, and resistance capability was lost. At α = 30°, the cracks also initiated in material B under uniaxial compressive loading, as shown in Figure 7. With increasing compression stress, the cracks develop not only in intact materials (material A and material B), but also in interfaces. As a result, the final failure planes were generated along the soft intact material as well as the interfaces. At α = 45°, although considerable cracks developed in the soft intact material and interfaces, the failure planes that penetrated and failed the whole rock samples only occurred along the interfaces. When the interface dip angle increased to 60° and 75°, similar failure processes were observed, as shown in Figure 7, and the final failure planes were generated along the interfaces. When the beddings were vertical (α = 90°), numerous cracks occurred both in material A and material B, and were mainly concentrated in the top and bottom ends of the simulated rock sample at the final stage.

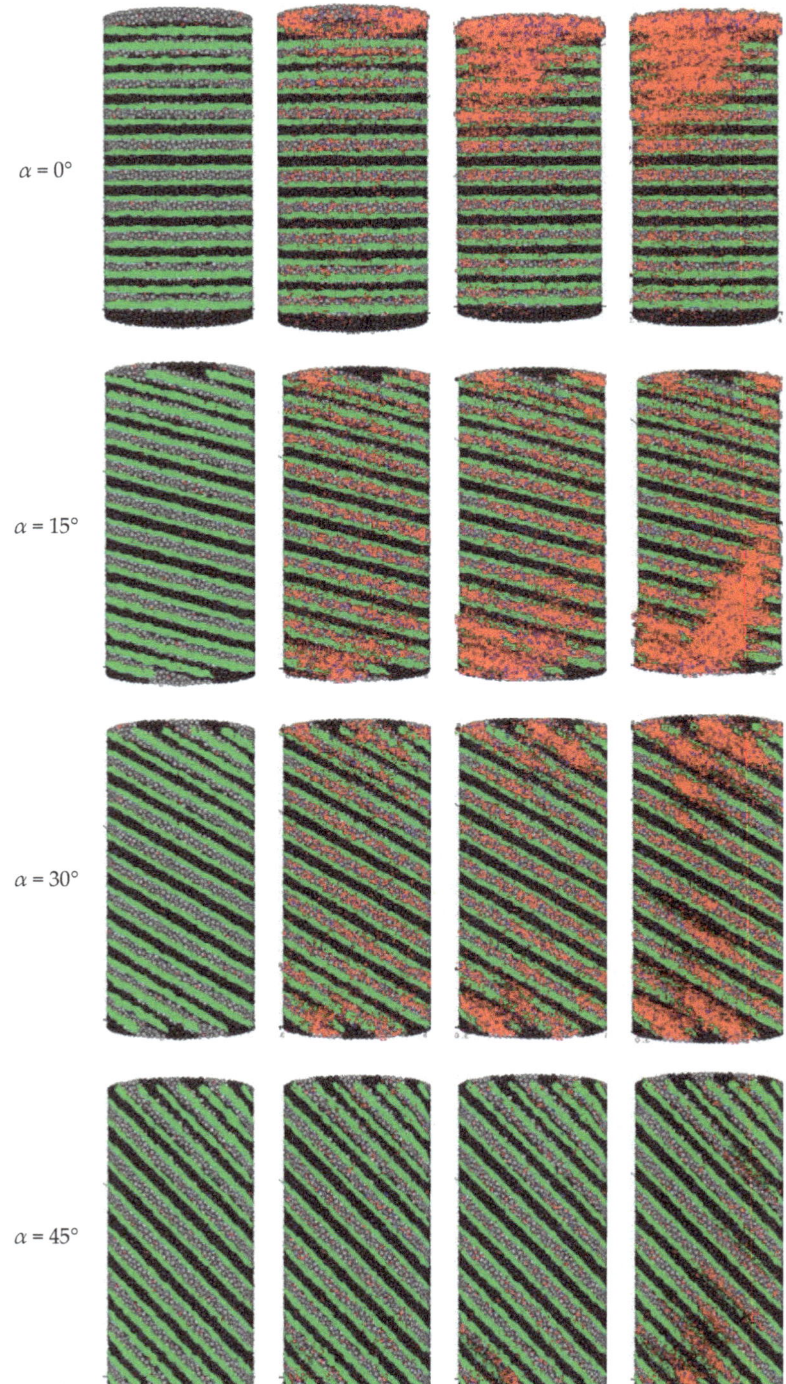

Figure 7. *Cont.*

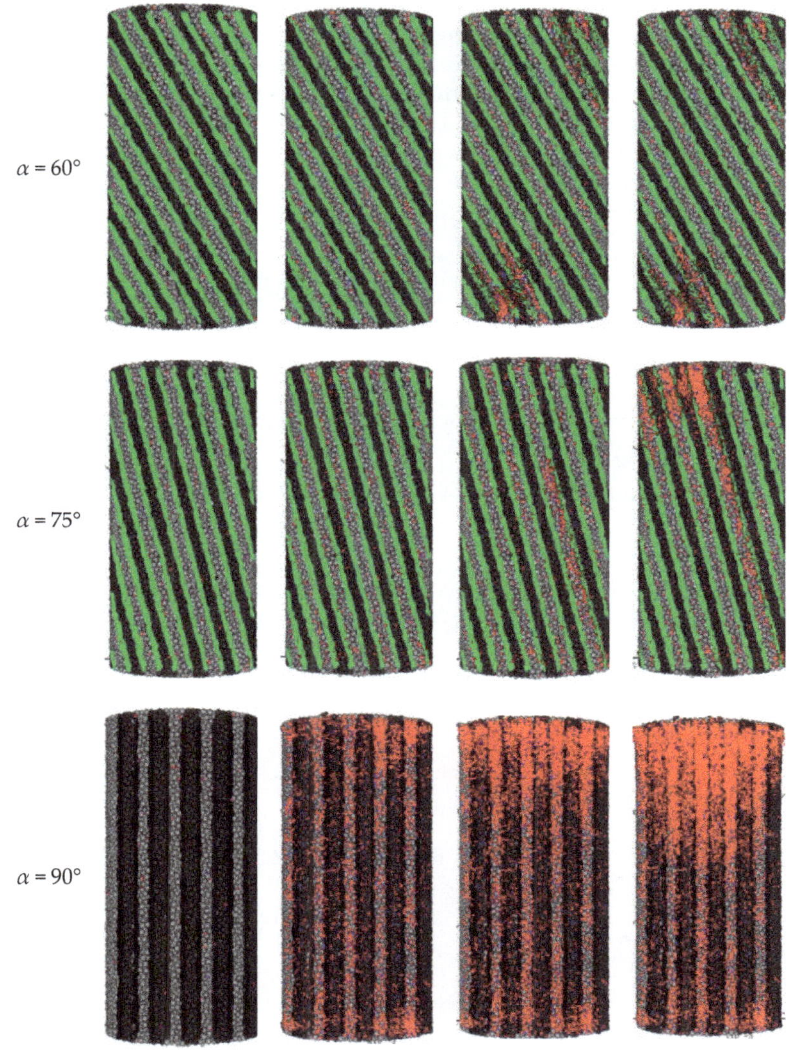

Figure 7. Failure process of transversely isotropic jointed rock models.

Figure 8 shows a comparison of the failure modes of artificial and simulated transversely isotropic rock samples with dip angles of 0°, 15°, and 30°. This figure shows that the numerically simulated failure modes matched well with those obtained by Tien et al. in the physical experiments [23]. Therefore, the numerically simulated failure modes could be referred to as tensile fractures across discontinuities as defined in physical experiments. To be more specific, in this study the failure mechanism was further analyzed at the mesoscale. Figure 9 illustrates the microcrack number evolution of transversely isotropic rock models having interface dip angles of 0°, 15°, and 30°. As shown in this figure, under uniaxial compressive loading, the crack number increased slowly at the beginning, and increased sharply when the axial stress reached the peak. Additionally, the crack number in intact materials was consistently larger than that in interfaces. That is to say, the intact materials rather than the interfaces dominated the resistance capability of the whole transversely isotropic rock models. Moreover, as shown in Figure 10, in intact materials, the tensile crack number was dramatically larger than that of shear cracks. Therefore, tensile failure of an intact material is thought to be the main cause of the transversely isotropic rock models failing when the interface dip angles are relatively low, e.g., $\alpha = 0°$ and 15°. Taking the effect of interfaces into consideration, this type of failure mode, herein, was redefined as the tensile failure across interfaces. As the interface dip angle increased from 0° to 30° through 15°, the difference of crack number in intact materials and in interfaces decreased, as shown in Figure 9. Furthermore, at $\alpha = 30°$, the crack number in interfaces almost equaled that in interfaces. Therefore, when the interface dip angle was 30° the transversely isotropic rock sample partially fails in the mode of tensile failure across interfaces.

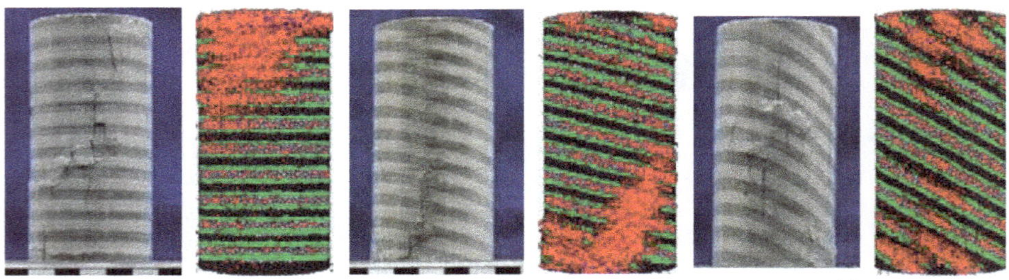

(a) (b) (c)

Figure 8. Failure modes of transversely isotropic jointed rock models having dip angles of (**a**) $\alpha = 0°$, (**b**) $\alpha = 15°$, and (**c**) $\alpha = 30°$. (Note that the failure images of physical samples were adopted from Ref. [23]).

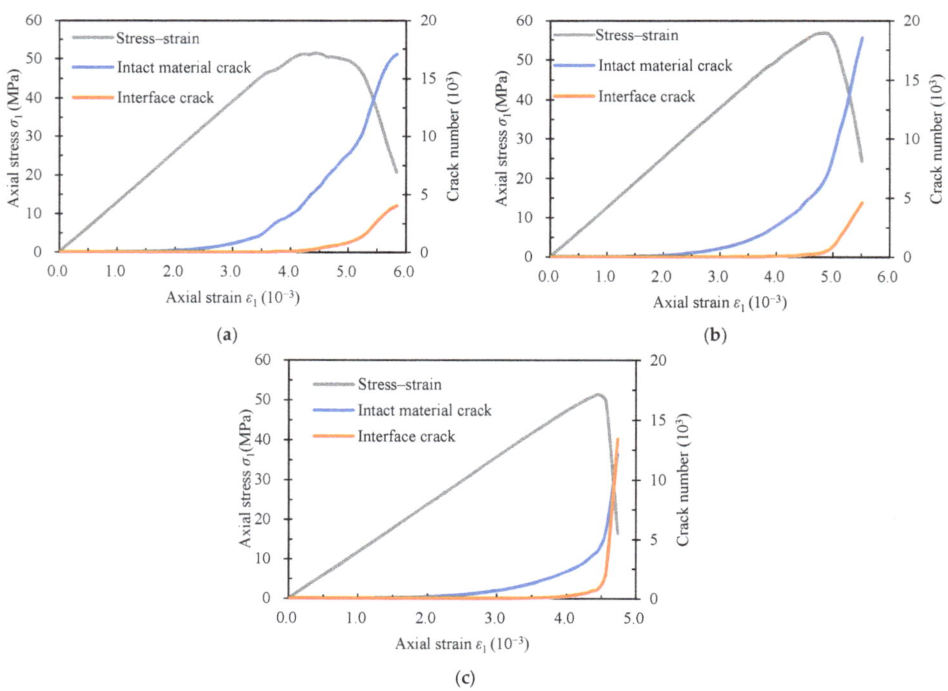

Figure 9. Crack number evolution of transversely isotropic rock models with dip angles of (**a**) $\alpha = 0°$, (**b**) $\alpha = 15°$, and (**c**) $\alpha = 30°$.

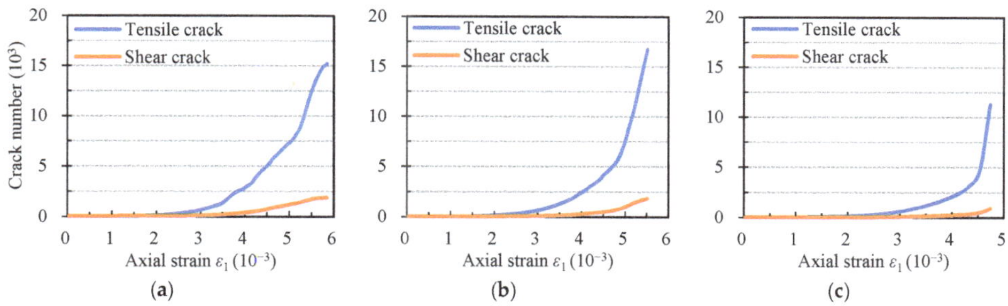

Figure 10. Intact material crack number evolution of transversely isotropic jointed rock models having dip angles of (**a**) $\alpha = 0°$, (**b**) $\alpha = 15°$ and (**c**) $\alpha = 30°$.

Figure 11 displays the comparison of failure modes of artificial and simulated transversely isotropic rock models having dip angles of 45°, 60°, and 75°. It shows that the numerically simulated failure modes were also in good agreement with those obtained through physical experiments [23]. The numerically simulated failure modes could be defined as a sliding failure along discontinuities in accordance with the physical experiments. Furthermore, Figure 12 displays the crack number evolution of transversely isotropic rock models having interface dip angles of 45°, 60°, and 75°. As shown in this figure, under compressive loading, the crack number increases slowly for a relatively long period, and increases sharply when the axial stress arrives at peak stress. Contrary to transversely isotropic rock samples having low dip angles (α = 0°, 15°, and 30°), the transversely isotropic rock samples having medium dip angles (α = 45°, 60°, and 75°) generated more cracks in interfaces than in intact materials. That is to say, the interfaces rather than the intact materials dominated the resistance capability of the whole rock models. Moreover, as shown in Figure 13, in interfaces, the shear crack number was much larger than that of the tensile crack. Therefore, the shear failure of interfaces was the main mechanism of the transversely isotropic rock models failing when the interface dip angles were medium, e.g., α = 45°, 60°, and 75°. To emphasize the failure mechanism of the transversely isotropic rock models, this type of failure mode, herein, is redefined as the shear failure along interfaces.

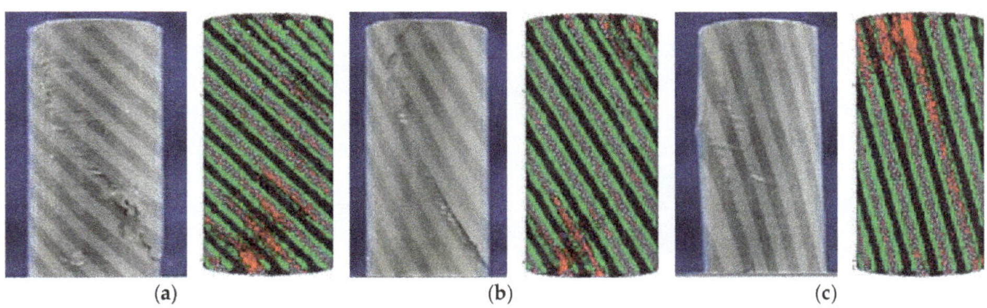

Figure 11. Failure modes of transversely isotropic jointed rock models having dip angles of (**a**) $\alpha = 45°$, (**b**) $\alpha = 60°$, and (**c**) $\alpha = 75°$. (Note that the failure images of physical samples were adopted from Ref. [23]).

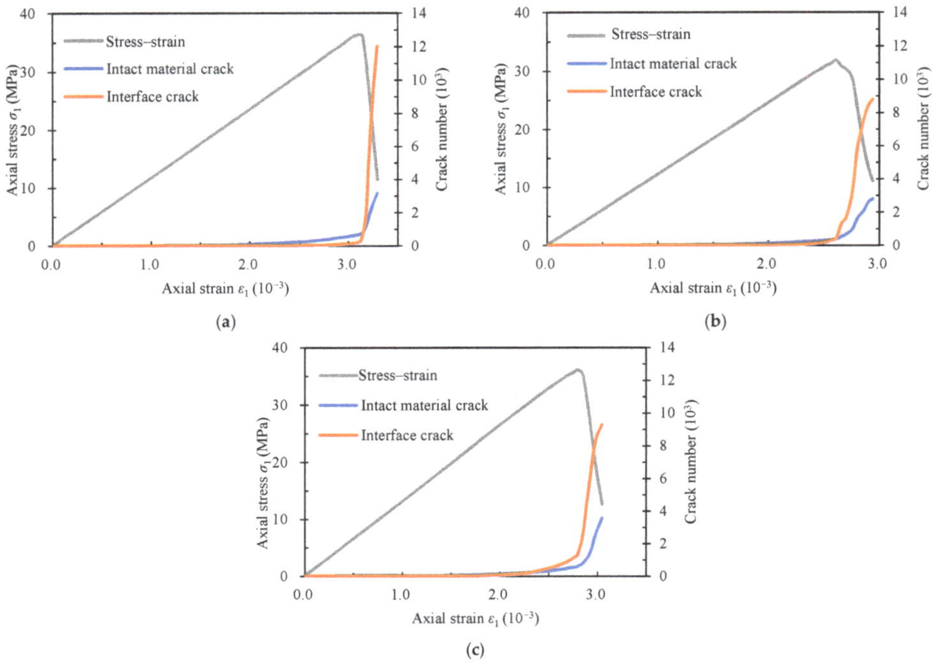

Figure 12. Crack number evolution of transversely isotropic rock models with dip angles of (**a**) $\alpha = 45°$, (**b**) $\alpha = 60°$ and (**c**) $\alpha = 75°$.

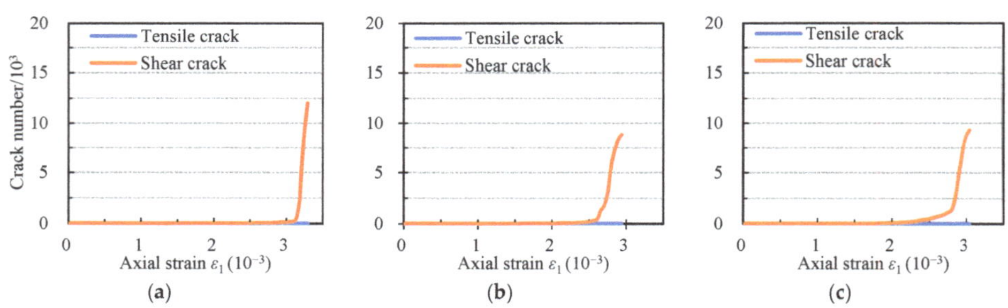

Figure 13. Intact material crack number evolution of transversely isotropic jointed rock models having dip angles of (**a**) $\alpha = 45°$, (**b**) $\alpha = 60°$ and (**c**) $\alpha = 75°$.

Figure 14 plots the comparison of failure modes of transversely isotropic rock models with vertical interfaces in numerical simulations and in physical experiments [23]. In the numerically simulated rock model, there were many more cracks observed than those in the physical rock model. This might be because the particle flow modeling method could recognize more microcracks than the eyes [18]. On the other hand, Figure 15 displays that before the peak stress, the number of interface cracks was larger than that of intact material crack, which accounts for the failure planes mainly observed in vertical interfaces of artificial rock models having a dip angle of 90°. This failure mode, herein, is referred to as tensile failure along interfaces.

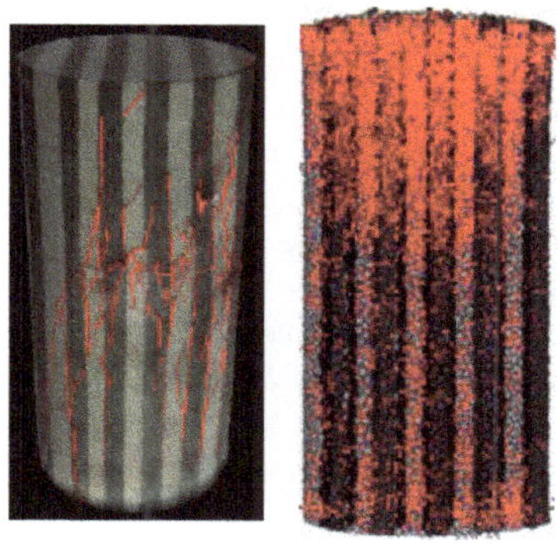

(a) (b)

Figure 14. Failure modes of transversely isotropic jointed rock model having dip angle of 90°: (**a**) experiment sample, and (**b**) numerical sample. (Note that the failure images of physical samples were adopted from Ref. [23]).

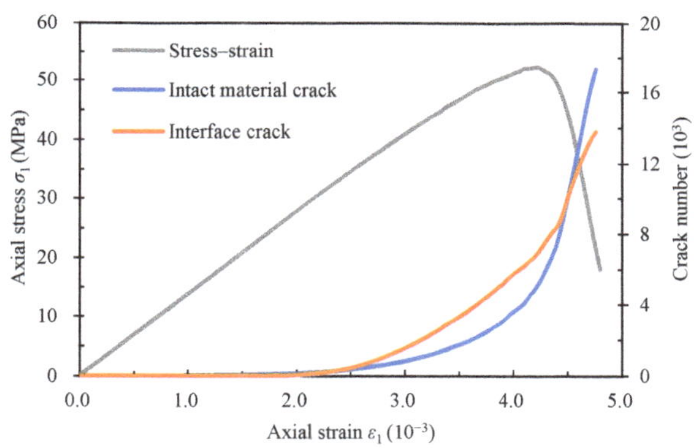

Figure 15. Evolution of crack number in transversely isotropic jointed rock models having dip angle of 90°.

3.2. Effect of Interfaces on Strength of Transversely Isotropic Rocks

As aforesaid, the role of interfaces played an influence on the failure behavior of transversely isotropic rocks. The failure mode of transversely isotropic rocks were classified into three types, which are tensile failure across interfaces, shear failure along interfaces, and tensile failure along interfaces. The failure mode are significantly dependent on the interface dip angle. However, in laboratory experiments, the interfaces between material A and material B were somehow ignored without taking their mechanical properties into account [23]. This section thus focuses on the effect of interfaces on the strength behavior of transversely isotropic rock models.

For comparison, Figure 16 displays the failure modes of transversely isotropic rock models without interfaces. This figure shows that the rock models having only inter-beddings of material A and material B still behaved anisotropically. The microcracks mainly concentrated in the soft material, i.e., material B, and most of them were tensile cracks (red), which was also observed by Kim et al. [12] through experimental tests on Berea sandstone. However, as shown in Figure 17, which illustrates the comparison of UCS of transversely isotropic rocks with and without weak interfaces, the UCS values of rock models without weak interfaces were highly beyond those obtained in physical experiments. On the contrary, the UCS values of rock models with weak interfaces agreed very well with the physical experiment results. In addition, the shape of the compressive strength interface dip angle curve obtained in physical experiments could be classified as a shoulder type [31], which means that the compressive strength was relatively high with low and high dip angles, while the compressive strength was relatively low with medium dip angles. Some typical examples of such curves are verified by Yasar [32], Autio et al. [33], Nasseri et al. [34], Cho et al. [35], and Fjaer and Nes [36]. In these studies, the research objects included various rocks such as sandstone, siltstone, shale, gneiss, and schist. In most of these works, it was noted that bedding planes acted as planes of weakness. The particle flow modeling results of transversely isotropic rocks with interfaces fit this type of the compressive strength interface dip angle curve. That is to say, the interfaces were very necessary elements to build transversely isotropic rock models with particle flow modeling method. To further understand the effect of interfaces on the failure behavior of transversely isotropic rocks, the mechanical properties of interfaces were systematically investigated as follows.

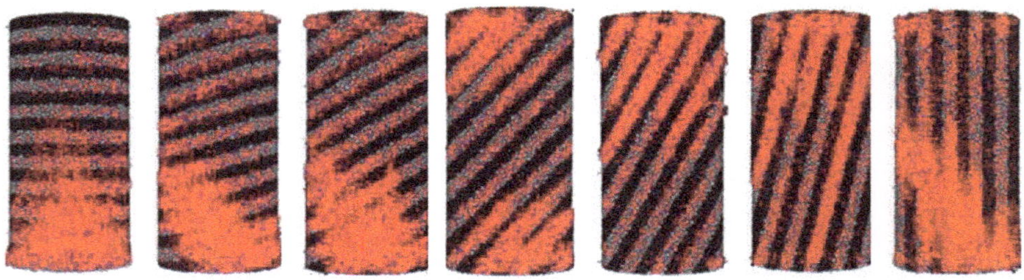

Figure 16. Failure modes of transversely isotropic rocks without weak interfaces: (**a**) $\alpha = 0°$, (**b**) $\alpha = 15°$, (**c**) $\alpha = 30°$, (**d**) $\alpha = 45°$, (**e**) $\alpha = 60°$, (**f**) $\alpha = 75°$, and (**g**) $\alpha = 90°$.

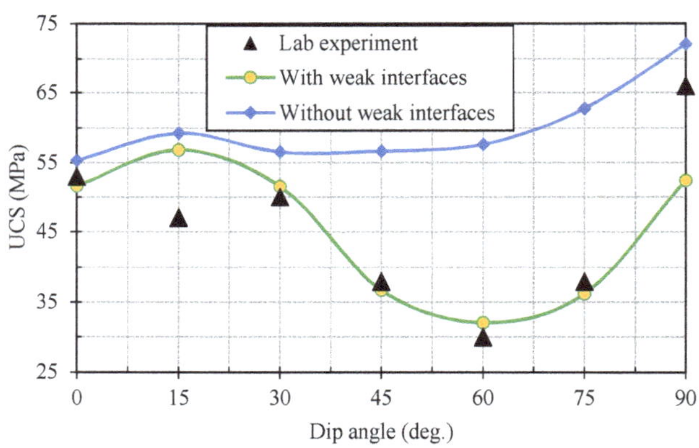

Figure 17. Comparison of UCS of transversely isotropic rocks with and without weak interfaces.

3.2.1. Joint Normal Stiffness

To look into the effect of the interface property on the failure behavior of transversely isotropic rock models, the joint normal stiffness of interfaces, $\overline{k}_{nj}$, was first investigated. A set of joint normal stiffness values of 0.15×10^{13} N/m^3, 0.50×10^{13} N/m^3, 1.50×10^{13} N/m^3, and 5.00×10^{13} N/m^3 were numerically tested, respectively, for the transversely isotropic rock samples having various interface dip angles from 0° to 90° with an interval of 15° (Figure 3). Meanwhile, the joint shear stiffness, $\overline{k}_{sj}$, was set to be $1/3\overline{k}_{nj}$ for all tests. Furthermore, the other smooth joint parameters were kept constant, i.e., $\overline{\sigma}_{cj} = 23.8$ MPa, $\varphi_j = 22.5°$, $c_j = 13.0$ MPa, and $\mu_j = 0.40$. Due to the planar surface of the interfaces in the physical experiments, the joint dilation angle, ψ_j, was always set to be 0. Figure 18 shows the uniaxial compressive strength variation of transversely isotropic rock samples with different joint normal stiffness. It is shown that the joint normal stiffness had a dramatic effect on the uniaxial compressive strength (UCS) of transversely isotropic rocks. Particularly, when the interface dip angle was less than 30° (i.e., α = 0° and 15°), the UCS basically increased with the increment of joint normal stiffness. Whereas, when the interface dip angle was equal or more than 30° ($\alpha \geq 30°$), the UCS decreased with the increment of joint normal stiffness. Moreover, as the joint normal stiffness, $\overline{k}_{nj}$, was set to be 0.50×10^{13} N/m^3, the numerically obtained set of UCS values of transversely isotropic rock samples were close to those obtained in physical experiments [23].

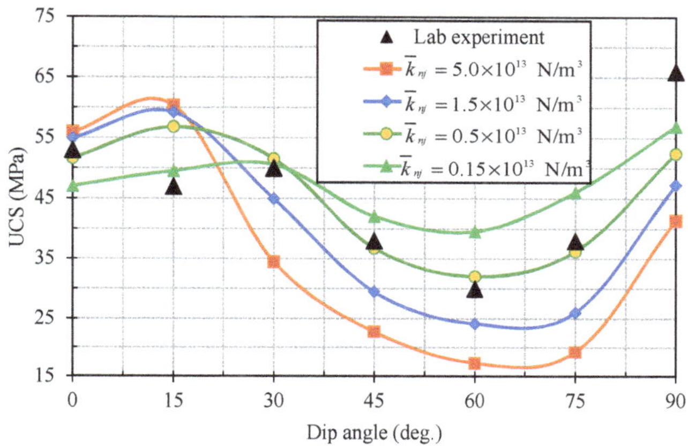

Figure 18. UCS variation of transversely isotropic rock samples with different joint normal stiffness.

3.2.2. Joint Shear Stiffness

Keeping the joint normal stiffness constant (i.e., $\overline{k}_{nj} = 0.5 \times 10^{13}$ N/m^3), the joint shear stiffness, $\overline{k}_{sj}$, is subsequently adjusted from $1/10\overline{k}_{nj}$ to $1.0\overline{k}_{nj}$ through $1/3\overline{k}_{nj}$. Furthermore, the other smooth joint parameters were set to be constant, that is to say, $\overline{\sigma}_{cj} = 23.8$ MPa, $\varphi_j = 22.5°$, $c_j = 13.0$ MPa, and $\mu_j = 0.40$. Figure 19 shows the UCS variation of transversely isotropic rock samples with different joint shear stiffness. As shown in this figure, the joint shear stiffness possessed a similar influence to the joint normal stiffness with respect to the UCS values of transversely isotropic rock samples. When the interface dip angle was less than 30°, the UCS basically increased with the increment of joint shear stiffness. Meanwhile, when the interface dip angle was equal to or more than 30° ($\alpha \geq 30°$), the UCS decreased with the increment of joint shear stiffness. Specially, as the joint shear stiffness, $\overline{k}_{sj}$, was set to be $1/3\overline{k}_{nj}$, the numerically obtained set of UCS values of transversely isotropic rock samples agreed well with those obtained by Tien et al. in the physical experiments [23].

The joint normal stiffness and joint shear stiffness had a similar influence on the UCS variation of transversely isotropic rock models. When the interface dip angle was low (e.g., α = 0° and 15°), the uniaxial loading compressed the inter-layered materials (material A, material B, and interfaces) to deform normally, which was dominated by the normal stiffness, and further to deform tangentially, which was dominated by the shear stiffness because of the Poisson's ratio effect, which means that the normal deformation would induce the tangential deformation. If the joint stiffness was low, the interfaces would have a large stiffness difference compared with the adjacent intact materials (material A and material B), which resulted in a significant deforming difference between them. Figure 20 shows the force chains in the transversely isotropic rock models having horizontal interfaces (α = 0°) with different joint normal stiffness values at the peak stress. It shows that the deforming difference caused tensile force chains in the interfaces, and the lower joint stiffness (e.g., $\overline{k}_{nj} = 0.15 \times 10^{13}$ N/m^3) introduced more tensile force chains. These tensile force chains further cause tensile cracks and eventually lower the UCS values of transversely isotropic rock samples. Therefore, the UCS decreased with the reduction of joint stiffness at low dip angles (i.e., α = 0° and 15°), as shown in Figures 18 and 19. On the other hand, when the interface dip angle was higher ($\alpha \geq 30°$), both the high joint stiffness and the low joint stiffness induced substantial tensile force chains in the interfaces. However, the lower joint stiffness (e.g., $\overline{k}_{nj} = 0.15 \times 10^{13}$ N/m^3) increased the deforming capability of these interfaces and postponed their failure. This process motivated more intact material to contribute to the bearing capability of the whole rock samples. Therefore, the UCS increased with the reduction of joint stiffness at higher dip angles (i.e., $\alpha \geq 30°$).

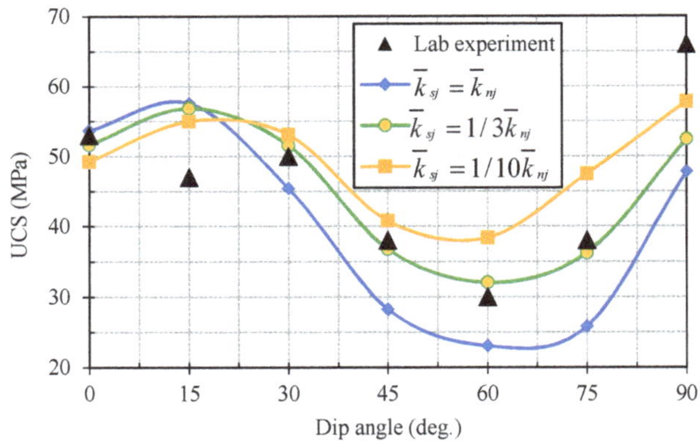

Figure 19. UCS variation of transversely isotropic rock samples with different joint shear stiffness.

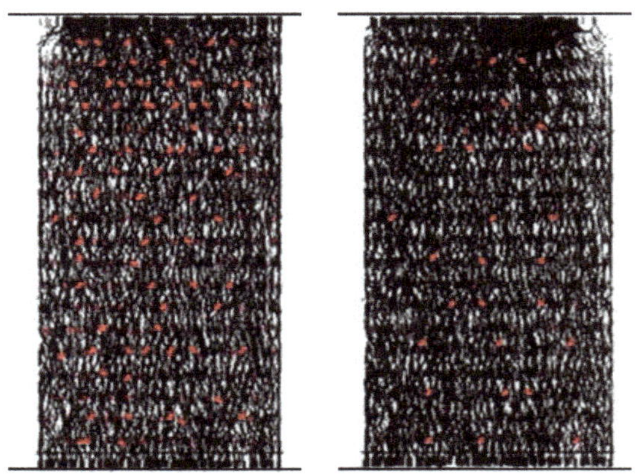

(a) (b)

Figure 20. Force chains in the transversely isotropic rock models having horizontal interfaces with different joint normal stiffness at peak stress: (**a**) $\overline{k}_{nj} = 0.15 \times 10^{13}$ N/m^3, and (**b**) $\overline{k}_{nj} = 5.0 \times 10^{13}$ N/m^3. Note that the black chain and red chain imply compression force and tension force, respectively, and the force quantity is proportional to the chain thickness.

3.2.3. Bonded Joint Cohesion

Further investigation was implemented to look into the effect of interface strength parameters, including joint shear strength, $\overline{\tau}_{cj}$, joint coefficient of friction, μ_j, and joint tensile strength, $\overline{\sigma}_{cj}$, on the strength behavior of transversely isotropic rock models. In accordance with Equation (9), the joint shear strength consisted of bonded joint cohesion, c_j, and bonded joint friction angle, φ_j. Herein, the bonded joint cohesion was first adjusted to be 4.0 MPa, 8.0 MPa, 13.0 MPa, and 16.0 MPa, while the other joint parameters were kept constant, i.e., $\overline{k}_{nj} = 0.5 \times 10^{13}$ N/m^3, $\overline{k}_{sj} = 1/3\overline{k}_{nj}$, $\overline{\sigma}_{cj} = 23.8$ MPa, $\varphi_j = 22.5°$, and $\mu_j = 0.40$.

Figure 21 shows the UCS variation of transversely isotropic rock samples with different bonded joint cohesion. As shown in this figure, the bonded joint cohesion, c_j, had a varying influence on the UCS values depending on the interface dip angle. Specifically, when the interface dip angles, α, were 0° and 15°, the UCS values decreased slightly with the decrement of bonded joint cohesion. Even though the transversely isotropic rock samples with dip angles of 0° and 15° failed in the mode of tensile failure across interfaces (Figure 8), the shear cracks occurred in interfaces, as shown in Figure 22. This figure displays the interface shear crack proportion at peak stress varying with bonded joint cohesion of 13.0 MPa and 4.0 MPa. For a better understanding, the shear crack number in interfaces was normalized by the total crack number in the whole rock model. With the decrement of bonded joint cohesion, the shear crack number in the interfaces increased significantly. At α = 0°, when the bonded joint cohesion decreased from 13.0 MPa to 4.0 MPa, the interface shear crack proportion increased from 5.1% to 31.8%. The occurrence of more interface shear cracks accelerated the failure of the whole rock model. Moreover, at α = 15°, when the bonded joint cohesion decreased from 13.0 MPa to 4.0 MPa, the interface shear crack proportion increased from 5.9% to 63.0%. Therefore, the shear failure, in this case, became dominated when the bonded joint cohesion was small, and the failure mode changed from tensile failure across interfaces to shear failure along interfaces.

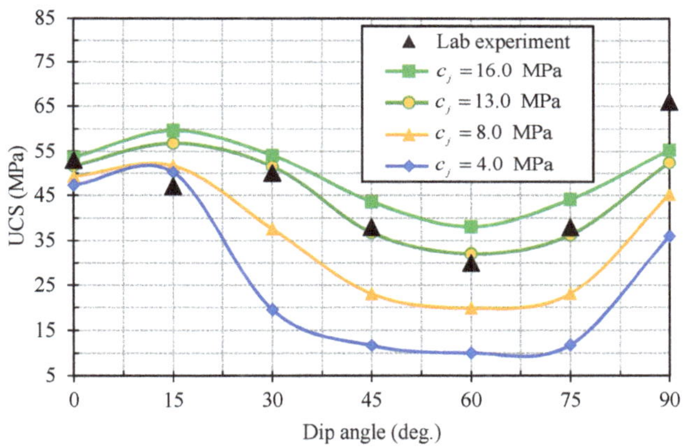

Figure 21. Strength variation of transversely isotropic rock samples with different bonded joint cohesion.

When the interface dip angles were medium (α = 30°, 45°, 60°, and 75°), the UCS values decreased heavily with the decrement of the bonded joint cohesion (Figure 21). The decrement of bonded joint cohesion directly reduced the shear resistance capability of interfaces, and made them much easier to fail. As shown in Figure 22, with the bonded joint cohesion decreasing from 13.0 MPa to 4.0 MPa, the shear cracks in the interfaces having medium dip angles were the main cracks occurring in the whole rock models. The intact materials (material A and material B) could not play their role effectively to contribute to the resistance capability of the whole models, which accounted for the decrement of UCS values in Figure 21. When the interfaces were vertical (α = 90°), even though the interface shear crack proportion increased from 57.8% to 93.8% with the bonded joint cohesion decreasing from 13.0 MPa to 4.0 MPa, the UCS value decreased less significantly, as shown in Figure 21. This was because the intact materials (material A and material B) could still resist the top and bottom loading platens moving towards each other, and even their connections (interfaces) were very weak.

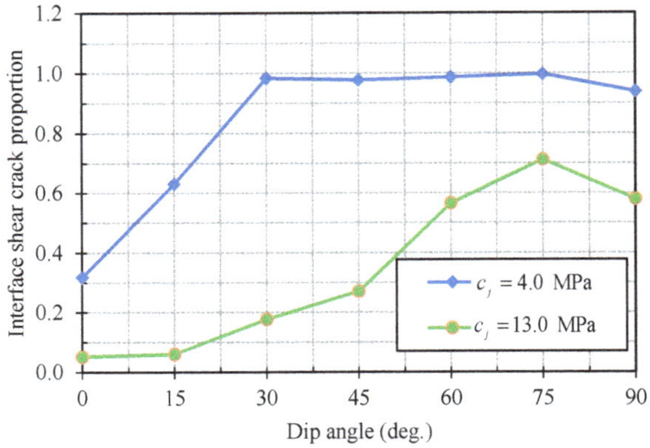

Figure 22. Interface shear crack proportion variation with different bonded joint cohesion.

3.2.4. Bonded Joint Friction Angle

The bonded joint friction angle, φ_j, was then adjusted to 15°, 22.5°, and 30°, respectively, while the other joint parameters were kept constant, i.e., $\overline{k}_{nj} = 0.5 \times 10^{13}$ N/m^3, $\overline{k}_{sj} = 1/3\overline{k}_{nj}$, $\overline{\sigma}_{cj} = 23.8$ MPa, $c_j = 13.0$ MPa, and $\mu_j = 0.40$. Figure 23 shows the UCS variation of transversely isotropic rock samples with different bonded joint friction angles. As shown in this figure, the modification of the bonded joint fiction angle had certain influences on the UCS values of transversely isotropic rock samples when the interface joint angles were medium. However, this influence was negligible when the interface was horizontal (α = 0°) or vertical (α = 90°). Like bonded joint cohesion, the bonded joint friction angle contributed to the shear strength of interfaces. However, the contribution of the bonded joint friction angle depended on the normal stress component in accordance with Equation (9). Thus, the bonded joint friction had a negligible influence on the shear resistance of the interfaces when the dip angles were high, i.e., α = 75° and 90°, due to the small normal stress component. With respect to the case of α = 0°, although the normal stress component was considerable, which enhanced the interface shear strength, the shear behavior of the interfaces was not the main mechanism for the failure of the whole rock model. This accounted for the negligible variation of the UCS values with the bonded joint friction angle at α = 0° in Figure 23. Figure 24 further indicates that the change of bonded joint friction angle mainly influenced the failure behavior of transversely isotropic rock samples when the dip angles were medium.

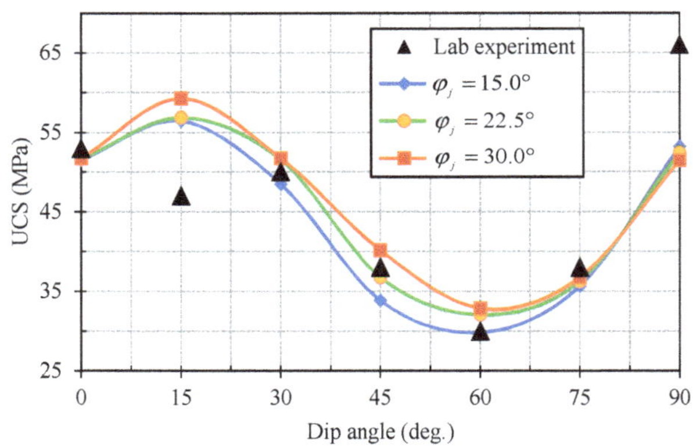

Figure 23. Strength variation of transversely isotropic rock samples with bonded joint friction angle.

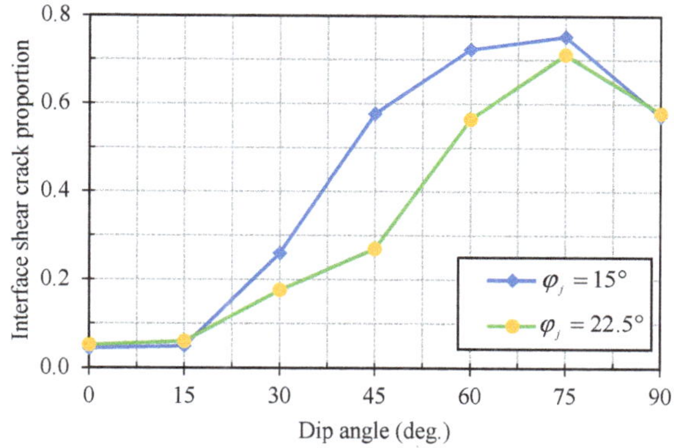

Figure 24. Interface shear crack proportion variation with different bonded joint friction angle.

3.2.5. Joint Coefficient of Friction

The joint coefficient of friction began to work only after the joint bond broke to provide the residual shear strength. To learn about the effect of the interface residual shear strength on the failure behavior of transversely isotropic rock models, the joint coefficient of friction, μ_j, was set to be 0.10, 0.25, and 0.40, whereas, the other joint parameters were kept constant, i.e., $\overline{k}_{nj} = 0.5 \times 10^{13}$ N/m^3, $\overline{k}_{sj} = 1/3\overline{k}_{nj}$, $\overline{\sigma}_{cj} = 23.8$ MPa, $c_j = 13.0$ MPa, and $\varphi_j = 22.5°$. Figure 25 displays the UCS variation of transversely isotropic rock samples with different joint coefficients of friction. As shown in this figure, the joint coefficient of friction had a negligible influence on the resistance capability of transversely isotropic rock samples. For rock samples having low dip angles, e.g., $\alpha = 0°$ and 15°, the increment of the interface shear strength was not effective enough to affect UCS values because the rock samples tended to fail in the mode of tensile failure across interfaces. Furthermore, when the dip angle increased to be medium, such as $\alpha = 30°$, 45°, 60°, and 75°, even though the shear strength of interfaces mattered, the joint coefficient of friction could not contribute to the peak stress of the whole rock model. This was due to the fact that the interface shear cracks began to occur quite close to the peak stress stage, as shown in Figure 12. Therefore, the joint coefficient of friction mainly contributed to the residual strength rather than to the UCS of the whole rock models, which agreed very well with Ref. [25]. For $\alpha = 90°$, the transversely isotropic rock model failed according to tensile failure along interfaces, which accounted for the negligible influence of the joint coefficient of friction on UCS values.

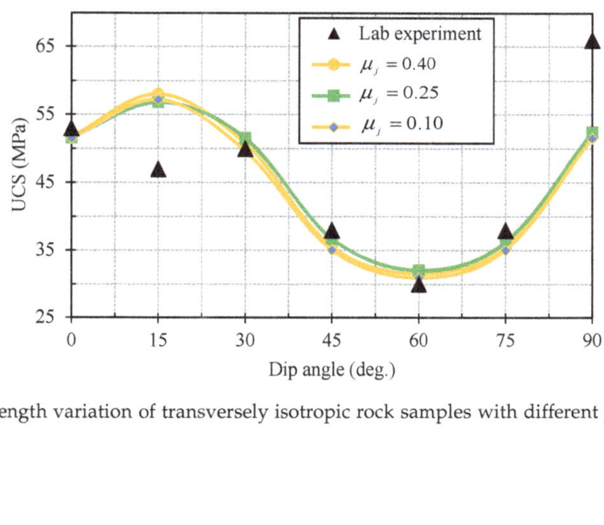

Figure 25. Strength variation of transversely isotropic rock samples with different joint coefficient of friction.

3.2.6. Joint Tensile Strength

A parametric study on the influence of the joint tensile strength, $\overline{\sigma}_{cj}$, on the strength behavior of transversely isotropic rock models was carried out as well through varying this joint micro-mechanical parameter from 15.90 MPa to 31.80 MPa through 23.80 MPa while setting the other parameters constant. In accordance with Figure 26, the joint tensile strength had a slight influence on the UCS values of transversely isotropic rock models. For rock samples having low dip angles (α = 0°, 15°, and 30°), although the increment of the joint tensile strength somehow affected the strength behavior of transversely isotropic rock models, the influence was not significant. When the dip angle was medium, because the rock models failed in the mode of shear failure along interfaces, the increment of the interface tensile strength had a negligible influence on the strength of the whole rock models. Furthermore, although the rock model failed in the mode of tensile failure along interfaces, the UCS of the transversely isotropic rock sample was mainly contributed by the vertical intact materials. Thus, the change of joint tensile strength also had a negligible influence on the UCS value.

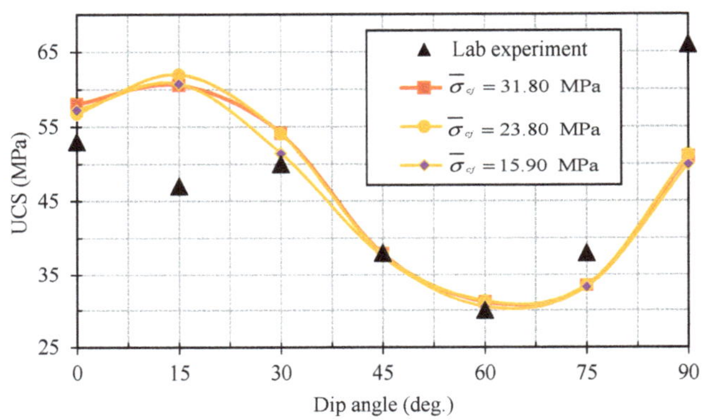

Figure 26. Strength variation of transversely isotropic rock samples with different joint tensile strength.

4. Conclusions

Particle flow modeling was undertaken to investigate the failure mechanism of transversely isotropic rocks under uniaxial compressive loading. After setting up a particle flow model with the PFC^{3D} software package according to the conceptual model proposed by Tien et al. [23], a parametric study was then carried out to investigate the effect of the interface dip angle and interface mechanical parameters on the failure mode and uniaxial compressive strength of the transversely isotropic rock models. The following conclusions can be drawn from the obtained particle flow modeling results:

(1) The interfaces interspaced in intact materials were pivotal elements to successfully build transversely isotropic rock models with a particle flow modeling method. With careful calibrations of the interface mechanical parameters, these simulated transversely isotropic rock models derived quite similar failure modes and UCS values to those obtained in physical experiments.

(2) To highlight the effect of interfaces, the failure mode of transversely isotropic rock models was redefined according to the observed crack revolution at the meso level. Three basic failure modes were identified in the transversely isotropic rock models under uniaxial compressive loading: (a) tensile failure across interfaces, (b) shear failure along interfaces, and (c) tensile failure along interfaces.

(3) The joint normal stiffness and joint shear stiffness had a dramatic influence on the failure strength of transversely isotropic rock models. The difference of mechanical response to uniaxial compressive loading for each layered material accounted for the UCS variation with varying stiffness values.

(4) The mechanical parameters for the bonded joint shear strength property had quite a different influence on the failure behavior of transversely isotropic rock models. The bonded joint cohesion and bonded joint friction angle, which contributed to the shear strength of interfaces, had a considerable influence on the UCS values, while the joint coefficient of friction, which contributed to the residual strength of interfaces, had a negligible influence on the UCS values.
(5) The shear failure of interfaces was the dominant mechanism for anisotropic behavior of layered rock models, the change of joint tensile strength had a negligible influence on the UCS values of transversely isotropic rock models.

The present study put the emphasis on the failure mechanism of transversely isotropic rocks under the unconfined compression condition, whilst a further study on the anisotropic behavior of transversely isotropic rocks under the triaxial stress condition is undergoing. Moreover, the hydromechanical behavior of the transversely isotropic rocks is also a challenge topic that is of our interest. The following studies would provide useful guidelines for the stabilization control of engineering structures constructed in or on the transversely isotropic rocks.

Author Contributions: Funding acquisition, X.-X.Y. and W.-G.Q.; Investigation, X.-X.Y.; Methodology, X.-X.Y.; Supervision, H.-W.J. and W.-G.Q.

Funding: This research was funded by the National Natural Science Foundation of China, grant numbers 51704183, 51774192, and 51474135. This research was also partially funded by the Natural Science Foundation of Shandong Province, grant number ZR2017BEE020.

Acknowledgments: The authors would like to express sincere appreciation to the editor and the anonymous reviewers for their valuable comments and suggestions for improving the presentation of the manuscript.

Conflicts of Interest: The authors declare no conflict of interest.

References

1. Tan, X.; Konietzky, H.; Frühwirt, T.; Quoc Dan, D. Brazilian tests on transversely isotropic rocks: Laboratory testing and numerical simulations. *Rock Mech. Rock Eng.* **2015**, *48*, 1341–1351. [CrossRef]
2. Feng, X.T.; An, H. Hybrid intelligent method optimization of a soft rock replacement scheme for a large cavern excavated in alternate hard and soft rock strata. *Int. J. Rock Mech. Min. Sci.* **2004**, *41*, 655–667. [CrossRef]
3. Hudson, J.A.; Feng, X.T. Updated flowcharts for rock mechanics modelling and rock engineering design. *Int. J. Rock Mech. Min. Sci.* **2007**, *44*, 174–195. [CrossRef]
4. Okland, D.; Hydro, N.; Cook, J.M. Bedding-related borehole instability in high angle wells. In Proceedings of the EUROCK '98-Rock Mechanics in Petroleum Engineering, Trondheim, Norway, 8–10 July 1998.
5. Meier, T.; Rybacki, E.; Backers, T.; Dresen, G. Influence of bedding angle on borehole stability: A laboratory investigation of transversely isotropic oil shale. *Rock Mech. Rock Eng.* **2015**, *48*, 1535–1546. [CrossRef]
6. Zoback, M.D.; Barton, C.A.; Brudy, M.; Castillo, D.A.; Finkbeiner, T.; Grollimund, B.R.; Moos, D.B.; Peska, P.; Ward, C.D.; Wiprut, D.J. Determination of stress orientation and magnitude in deep wells. *Int. J. Rock Mech. Min. Sci.* **2003**, *40*, 1049–1076. [CrossRef]
7. Zhang, J. Borehole stability analysis accounting for anisotropies in drilling to weak bedding planes. *Int. J. Rock Mech. Min. Sci.* **2013**, *60*, 160–170. [CrossRef]
8. Liu, R.C.; Li, B.; Jiang, Y.J. Critical hydraulic gradient for nonlinear flow through rock fracture networks: The roles of aperture, surface roughness, and number of intersections. *Adv. Water Res.* **2016**, *88*, 53–65. [CrossRef]
9. Zhou, Y.Y.; Feng, X.T.; Xu, D.P.; Fan, Q.X. Experimental investigation of the mechanical behavior of bedded rocks and its implication for high sidewall caverns. *Rock Mech. Rock Eng.* **2016**, *49*, 3643–3669. [CrossRef]
10. Cai, M.; Kaiser, P.K.; Tasaka, Y.; Minami, M. Determination of residual strength parameters of jointed rock masses using the GSI system. *Int. J. Rock Mech. Min. Sci.* **2007**, *44*, 247–265. [CrossRef]
11. Gatelier, N.; Pellet, F.; Loret, B. Mechanical damage of an anisotropic porous rock in cyclic triaxial tests. *Int. J. Rock Mech. Min. Sci.* **2002**, *39*, 335–354. [CrossRef]

12. Kim, K.Y.; Zhuang, L.; Yang, H.; Kim, H.; Min, K.B. Strength anisotropy of Berea sandstone: Results of X-ray computed tomography, compression tests, and discrete modeling. *Rock Mech. Rock Eng.* **2016**, *49*, 1201–1220. [CrossRef]
13. Heng, S.; Guo, Y.T.; Yang, C.H.; Daemen, J.K.; Li, Z. Experimental and theoretical study of the anisotropic properties of shale. *Int. J. Rock Mech. Min. Sci.* **2015**, *74*, 58–68. [CrossRef]
14. Kulatilake, P.H.S.W.; He, W.; Um, J.; Wang, H. A physical model study of jointed rock mass strength under uniaxial compressive loading. *Int. J. Rock Mech. Min. Sci.* **1997**, *34*, 165.e1–165.e15. [CrossRef]
15. Yang, Z.Y.; Chen, J.M.; Huang, T.H. Effect of joint sets on the strength and deformation of rock mass models. *Int. J. Rock Mech. Min. Sci.* **1998**, *35*, 75–84. [CrossRef]
16. Yang, X.X.; Jing, H.W.; Tang, C.A.; Yang, S.Q. Effect of parallel joint interaction on mechanical behavior of jointed rock mass models. *Int. J. Rock Mech. Min. Sci.* **2017**, *92*, 40–53. [CrossRef]
17. Bahaaddini, M.; Sharrock, G.; Hebblewhite, B.K.; Mitra, R. Direct shear tests to model the shear behavior of rock joints by PFC2D. In Proceedings of the 46th US Rock Mechanics/Geomechanics Symposium, American Rock Mechanics Association, Chicago, IL, USA, 24–27 June 2012.
18. Bahaaddini, M.; Sharrock, G.; Hebblewhite, B.K. Numerical investigation of the effect of joint geometrical parameters on the mechanical properties of a non-persistent jointed rock mass under uniaxial compression. *Comput. Geotech.* **2013**, *49*, 206–225. [CrossRef]
19. Chiu, C.C.; Wang, T.T.; Weng, M.C.; Huang, T.H. Modeling the anisotropic behavior of jointed rock mass using a modified smooth-joint model. *Int. J. Rock Mech. Min. Sci.* **2013**, *62*, 14–22. [CrossRef]
20. Prudencio, M.; Van Sint Jan, M. Strength and failure modes of rock mass models with non-persistent joints. *Int. J. Rock Mech. Min. Sci.* **2007**, *44*, 890–902. [CrossRef]
21. Park, B.; Min, K.B. Bonded-particle discrete element modeling of mechanical behavior of transversely isotropic rock. *Int. J. Rock Mech. Min. Sci.* **2015**, *76*, 243–255. [CrossRef]
22. Yang, S.Q.; Huang, Y.H. Particle flow study on strength and meso-mechanism of Brazilian splitting test for jointed rock mass. *Acta Mech. Sin.* **2014**, *30*, 547–558. [CrossRef]
23. Tien, Y.M.; Kuo, M.C.; Juang, C.H. An experimental investigation of the failure mechanism of simulated transversely isotropic rocks. *Int. J. Rock Mech. Min. Sci.* **2006**, *43*, 1163–1181. [CrossRef]
24. Potyondy, D.O.; Cundall, P.A. A bonded-particle model for rock. *Int. J. Rock Mech. Min. Sci.* **2004**, *41*, 1329–1364. [CrossRef]
25. Itasca Consulting Group Inc. *PFC3D-Particle Flow Code in Three Dimensions*, version 4.0; Itasca Consulting Group Inc.: Minneapolis, MN, USA, 2008.
26. Mas Ivars, D.M.; Pierce, M.E.; Darcel, C.; Juan, R.M.; Potyondy, D.O.; Young, R.P.; Cundall, P.A. The synthetic rock mass approach for jointed rock mass modelling. *Int. J. Rock Mech. Min. Sci.* **2011**, *48*, 219–244. [CrossRef]
27. Kulatilake, P.H.S.W.; Malama, B.; Wang, J.L. Physical and particle flow modeling of jointed rock block behavior under uniaxial loading. *Int. J. Rock Mech. Min. Sci.* **2001**, *38*, 641–657. [CrossRef]
28. Yang, X.X.; Kulatilake, P.H.S.W.; Jing, H.W.; Yang, S.Q. Numerical simulation of a jointed rock block mechanical behavior adjacent to an underground excavation and comparison with physical model test results. *Tunn. Undergr. Space Technol.* **2015**, *50*, 129–142. [CrossRef]
29. Yang, X.X.; Kulatilake, P.H.S.W.; Chen, X.; Jing, H.W.; Yang, S.Q. Particle flow modeling of rock blocks with nonpersistent open joints under uniaxial compression. *Int. J. Geomech.* **2016**, *16*, 04016020-1-17. [CrossRef]
30. Yang, X.X.; Qiao, W.G. Numerical investigation of the shear behavior of granite materials containing discontinuous joints by utilizing the flat-joint model. *Comput. Geotech.* **2018**, *104*, 69–80. [CrossRef]
31. Tien, Y.M.; Kuo, M.C. A failure criterion for transversely isotropic rocks. *Int. J. Rock Mech. Min. Sci.* **2001**, *38*, 399–412. [CrossRef]
32. Yasar. Failure and failure theories for anisotropic rocks. In Proceedings of the 17th International Mining Congress and Exhibition of Turkey (IMCET), Ankara, Turkey, 19–22 June 2001; pp. 417–424.
33. Autio, J.; Wanne, T.; Potyondy, D. Particle mechanical simulation of the effect of schistosity on strength and deformation of hard rock. In Proceedings of the 5th North American Rock Mechanics Symposium & 17th Tunneling Association of Canada Conference: NARMS-TAC, Toronto, ON, Canada, 7–10 July 2002; Volume 1, pp. 275–282.
34. Nasseri, M.H.B.; Rao, K.S.; Ramamurthy, T. Anisotropic strength and deformation behavior of Himalayan schists. *Int. J. Rock Mech. Min. Sci.* **2002**, *40*, 3–23. [CrossRef]

35. Cho, J.W.; Kim, H.; Jeon, S.; Min, K.B. Deformation and strength anisotropy of Asan gneiss, Boryeong shale, and Yeoncheon schist. *Int. J. Rock Mech. Min. Sci.* **2012**, *50*, 158–169. [CrossRef]
36. Fjaer, E.; Nes, O.M. Strength anisotropy of Mancos shale. In Proceedings of the 47th US Rock Mechanics/Geomechanics Symposium, San Francisco, CA, USA, 23–26 June 2013.

MDPI
St. Alban-Anlage 66
4052 Basel
Switzerland
Tel. +41 61 683 77 34
Fax +41 61 302 89 18
www.mdpi.com

Processes Editorial Office
E-mail: processes@mdpi.com
www.mdpi.com/journal/processes

www.ingramcontent.com/pod-product-compliance
Lightning Source LLC
Chambersburg PA
CBHW060758240926
51847CB00020B/154
* 9 7 8 3 0 3 9 2 1 4 2 3 5 *